Essentials of
Genetics

Essentials of
Genetics

Third Edition

William S. Klug
College of New Jersey

Michael R. Cummings
University of Illinois at Chicago

With essays contributed by

Mark Shotwell, Slippery Rock University

Charlotte Spencer, University of Alberta

Prentice Hall
Upper Saddle River, New Jersey 07458

Library of Congress Cataloging-in-Publication Data

Klug, William S.
   Essentials of genetics / William S. Klug, Michael R. Cummings. — 3rd ed.
     p. cm.
   Includes bibliographical references and index.
   ISBN 0-13-080017-1 (pbk.)
   1. Genetics.   I. Cummings, Michael R.   II. Title.
QH430.K576   1999
576.5—dc21   98-25676
                                                   CIP

Executive Editor: Sheri L. Snavely
Editor in Chief: Paul F. Corey
Editorial Director: Tim Bozik
Assistant Vice President of Production and Manufacturing: David W. Riccardi
Executive Managing Editor: Kathleen Schiaparelli
Assistant Managing Editor: Lisa Kinne
Marketing Manager: Jennifer Welchans
Manufacturing Manager: Trudy Pisciotti
Buyer: Benjamin Smith
Art Manager: Gus Vibal
Art Editor: Karen Branson
Director of Creative Services: Paula Maylahn
Associate Creative Director: Amy Rosen
Art Director: Heather Scott
Assistant to Art Director: John Christiana
Assistant Editor: Mary Hornby
Cover Designer: Ann France
Photo Editor: Lorinda Morris-Nantz
Photo Researcher: Mary Teresa Giancoli
Editorial Assistants: Lisa Tarabokjia and Nancy Gross
Art Studio: Boston Graphics
Copyediting and Text Composition: Electronic Publishing Services Inc., NYC
Cover Art: © Gregor Zlokarnik, Ph.D. / Aurora Biosciences Corporation. Reprinted (abstracted/excerpted) with permission from *Science,* January 2 edition Vol. 279. Copyright © 1998 The American Association for the Advancement of Science.

Printed in the United States of America

10 9 8 7 6 5 4 3 2 1

ISBN 0-13-080017-1

Prentice-Hall International (UK) Limited, *London*
Prentice-Hall of Australia Pty. Limited, *Sydney*
Prentice-Hall of Canada, Inc., *Toronto*
Prentice-Hall Hispanoamericana, S. A., *Mexico*
Prentice-Hall of India Private Limited, *New Delhi*
Prentice-Hall of Japan, Inc., *Tokyo*
Simon & Schuster Asia Pte. Ltd., *Singapore*
Editora Prentice-Hall do Brasil, Ltda., *Rio de Janeiro*

Dedication

To Dr. Joseph A. Vena, previously of The College of New Jersey, who always provided interest, enthusiasm, support and encouragement during the preparation of the many editions of both Essentials of Genetics *and* Concepts of Genetics. *The world is a lesser place without you.*

About the Authors

William S. Klug is currently a Professor of Biology at The College of New Jersey at Hillwood Lakes in Ewing, New Jersey. He served as Chairman of the Biology Department for 17 years, a position to which he was first elected in 1974. He received his B.A. at Wabash College in Crawfordsville, Indiana, and his Ph.D. at Northwestern University in Evanston, Illinois. Before coming to The College of New Jersey, he returned to Wabash College as an Assistant Professor, where he first taught genetics as well as electron microscopy. His research interests have centered around the genetic control of development in *Drosophila*. He has taught the general genetics course to undergraduates during each of the last 30 years.

Michael R. Cummings is currently Associate Professor of Biological Sciences and Associate Professor of Genetics at the University of Illinois at Chicago. He has also served on the faculty at Northwestern University and Florida State University. He received his B.A. at St Mary's College in Winona, Minnesota, and his M.S. and Ph.D. degrees from Northwestern University in Evanston, Illinois. He has also written textbooks in human genetics and general biology for non-majors. His research interests center on the molecular organization and physical mapping of human acrocentric chromosomes. At the undergraduate level, he teaches courses in Mendelian genetics, human genetics, and general biology for non-majors.

Brief Contents

Contents

GENETICS, TECHNOLOGY, AND SOCIETY   104

The Green Revolution Revisited

6   Linkage and Chromosome Mapping   110

7   Extranuclear Inheritance   138

GENETICS, TECHNOLOGY, AND SOCIETY   149

Mitochondrial DNA and the Mystery of the
Romanovs

8   Chromosome Variation and Sex Determination   154

9 DNA—The Physical Basis of Life 184

10 DNA—Replication and Synthesis 210

GENETICS, TECHNOLOGY, AND SOCIETY 228

Telomerase: The Key to Immortality?

11 Organization of DNA in Chromosomes and Genes 232

GENETICS, TECHNOLOGY, AND SOCIETY 316

Mad Cows and Heresies: The Prion Story

15 DNA—Mutation, Repair, and Transposable Elements 320

GENETICS, TECHNOLOGY, AND SOCIETY 343

Chernobyl's Legacy

16 Genetics of Bacteria and Bacteriophages 348

17 DNA Biotechnology: Techniques and Analysis 372

GENETICS, TECHNOLOGY, AND SOCIETY 391

DNA Fingerprints in Forensics: The Case of the Telltale Palo Verde

18 DNA Biotechnology: Applications and Ethics 396

GENETICS, TECHNOLOGY, AND SOCIETY 417

Beyond Dolly: The Cloning of Humans

19 Genetics of Cancer and Immunology 422

GENETICS, TECHNOLOGY, AND SOCIETY 441

The Double-Edged Sword of Genetic Testing: The Case of Breast Cancer.

GENETICS, TECHNOLOGY, AND SOCIETY **468**
Coming to Terms with the Heritability of IQ

GENETICS, TECHNOLOGY, AND SOCIETY **489**
What Can We Learn from the Failure of the Eugenics Movement?

Preface

Essentials of Genetics is designed for courses requiring a text that is shorter and more basic than its more comprehensive companion, *Concepts of Genetics*. While coverage is still thorough and of high quality, *Essentials* is written to be more accessible to students with less background in biology and chemistry. While it will serve well courses offered to biology majors early in their undergraduate careers, it will also be useful in courses populated by a mixture of majors who are pursuing careers in agriculture, forestry, wildlife management, chemistry, psychology, etc. Because the text is shorter than many other available books, *Essentials of Genetics* will be more manageable in courses of both one-quarter and one-semester length.

Goals

Although *Essentials of Genetics* is almost 300 pages shorter than its companion volume, our goals are the same for both books. Specifically, we have sought to

- Emphasize concepts rather than excess detail
- Write clearly and directly to the students of genetics in order to provide straightforward explanations of complex, analytical topics
- Establish a careful organization within and between chapters
- Maintain constant emphasis on scientific analysis as the means to illustrate how we know what we know
- Propagate the rich history of genetics that so beautifully illustrates how information is acquired and extended within a discipline as it develops and grows
- Design inviting, pedagogically sound full-color figures supplemented by equally useful photographs to support the textual material

These goals serve as the cornerstones of this text. Providing such a pedagogic foundation allows *Essentials of Genetics* to accommodate courses with many different organizational approaches and lecture formats. Chapters are written to be as independent of one another as possible, allowing instructors to utilize them in various sequences. We believe that by combining these various approaches, we have created a text that provides optimal support for students as they study genetics.

Features of the Third Edition

- Length—a more manageable, streamlined text of 515 pages apportioned into 22 chapters.

- Revised Organization—an improved chapter sequence designed to provide a smooth flow of topics from start to finish, including a cohesive sequence of chapters centering on the genetic role of DNA, its structure, replication, expression, and regulation.
- New Chapters—the Recombinant DNA chapter of the second edition has been replaced by two new chapters: *DNA Biotechnology-Techniques and Analysis* and *DNA Biotechnology-Applications and Ethics* in order to provide thorough, modern coverage of this contemporary topic
- Redesign of the Entire Art Program—to further improve its pedagogic value.
- New Photographs—many new images that supplement the beautiful art program, often linked directly to a topic illustrated in a figure.
- Inclusion of Emerging Topics—extensive coverage of cutting edge subjects, including the genetic involvement in cancer, the genetic basis of immunology, and the genetic control of development and behavior.
- Modernization of Topics—to remain current in areas that are at the forefront of genetic studies, coverage of human genetics has been updated throughout the text, particularly where the Human Genome Project has impacted on our knowledge of our own species. Additionally, more extensive information has been provided on the more specific topics including cell cycle regulation, genomic imprinting, genetic anticipation, gene therapy, diagnosis of human genetic disorders, trinucleotide repeat mutations, and RNA editing.
- New "Genetics, Technology, and Society" Essays—to introduce relevant and sometimes controversial topics where advances in the technology of genetics interfaces with society. Among others, topics such as the creation of edible vaccines, the use of DNA analysis to establish the fate of Tsar Nicholas II and his family, the link between prions and Mad Cow disease, the impact of radiation produced at Chernobyl, the use of antisense RNAs as therapeutic agents, and telomeres and aging are presented.
- Emphasis on Problem Solving—by including both the "Insights and Solutions" feature as an aid to obtaining solutions to problems and the extensive "Problems and Discussion Questions" section in each chapter.

Emphasis on Concepts

As in its companion volume, *Essentials of Genetics* continues to emphasize the conceptual framework of this discipline. The concepts derived from each major topic represent those ideas and information that students take with them once the course is completed. To aid the student in identifying the conceptual aspects of a major topic, each chapter begins with a section called "Chapter Concepts," which in a few sentences captures the essence of the most important ideas about to be presented. Then, each chapter ends with a Chapter Summary, which enumerates the five to ten key points that have been established. These two sections help to ensure that the text emphasizes concepts rather than detail. Reinforcing examples and carefully designed figures have been selected to support this approach.

Insights and Solutions

Genetics, more than any other discipline within biology, lends itself to problem solving and analytical thinking. At the end of each chapter, we have included the popular and successful section "Insights and Solutions," first developed in *Concepts of Genetics*. In this section, we stress:

Problem solving
Quantitative analysis
Analytical thinking
Experimental rationale

Problems or questions are posed, and detailed solutions or answers are provided. This section is designed to prime the students as they move on to the "Problems and Discussion Questions" section concluding each chapter.

Problems and Discussion Questions

Following the "Insight and Solutions" section in each chapter, an extensive collection of problems and discussion questions are presented. These represent various levels of difficulty, with the most challenging appearing at the end of each section. In this section, we have attempted to optimize the opportunities for student growth in the important area of problem solving and analytical thinking skills. Brief answers to half the problems are in Appendix A. For those faculty wishing to expose their students to detailed answers to all problems and questions, the Student's Handbook is available (see supplements section).

Acknowledgments

All comprehensive texts are dependent on the valuable input provided by many reviewers and colleagues. While we take full responsibility for any errors herein, we gratefully acknowledge the help provided by those individuals involved in this and the previous edition. We wish to specifically recognize:

Jim Bricker, *The College of New Jersey*
Nancy H. Ferguson, *Clemson University*
Mark L. Hammond, *Campbell University*
John A. Hunt, *University of Hawaii, Honolulu*
Beth A. Krueger, *Monroe Community College*
Paul F. Lurquin, *Washington State University*
Sally Mackenzie, *Purdue University*
Cynthia J. Moore, *Washington University*
Michelle A. Murphy, *University of Notre Dame*
Marcia L. O'Connell, *The College of New Jersey*
Ralph Seelke, *University of Wisconsin, Superior*
Gurel S. Sidhu, *California State University*
Gerald Schlink, *Missouri Southern State College*
Randy Scholl, *Ohio State University*
Christine Tachibana, *Pennsylvania State University*
Sarah Ward, *Colorado State University*
Marie W. Wooten, *Auburn University*

Many thanks are also extended to Monica Zrada at The College of New Jersey for her administrative assistance throughout the preparation of the text and the production process.

At Prentice Hall, we wish to acknowledge the skillful editorial management provided throughout the project by Sheri Snavely. We would also like to acknowledge Ray Mullaney, whose developmental input was very helpful. At Electronic Publishing Services Inc., Anthony Calcara provided critical guidance during the production process. We are grateful also to Mary Teresa Giancoli for procuring many new photos that enhance the text, as well as to Arlene Larson of the University of Colorado at Denver and Anthony Maffia of Dupont Pharmaceuticals, who provided the accuracy checks.

Superb design and rendering of the various components of the art program was performed by Paul Foti and his staff at Boston Graphics, Inc. One need only page briefly through the text to appreciate the value of the contribution.

Special recognition has been saved for last to personally thank Charlotte Spencer and Mark Shotwell for their valuable contributions. They provided the initial drafts of the essays for the "Genetics, Technology, and Society" sections. Their efforts and these essays greatly enhance the pedagogic value of the text. We are particularly grateful to both of these colleagues.

It has been very pleasurable working with all of the above individuals, who deserve to share in any success enjoyed by this text. Our gratitude is equaled only by the extreme dedication evident during their many efforts.

William S. Klug
Michael R. Cummings

Supplements and New Media Tools

For the Student

Student Handbook and Solutions Manual

Harry Nickla, Creighton University

This valuable handbook provides a detailed step-by-step solution or lengthy discussion for every problem in the text in a chapter-by-chapter format. The handbook also contains extra study problems and a thorough review of concepts and vocabulary.

New York Times Themes of the Times: Genetics and Molecular Biology

Coordinated by Harry Nickla, Creighton University

This exciting newspaper-format supplement brings together recent genetics and molecular biology articles from the pages of the world-renowned *New York Times*. This free supplement, available through your local representative, encourages students to make the connections between the genetic concepts and the latest research and breakthroughs in science. This resource is updated regularly.

Life on the Internet: Biology

Andrew Stull, California State University at Fullerton

The perfect guide to help your students take advantage of our *Essentials of Genetics* home page on the World Wide Web. This unique resource gives clear steps to access our regularly updated genetics resource area as well as an overview of general navigation and research strategies.

For the Instructor

Instructor's Resource Manual with Testbank

Harry Nickla, Creighton University

This manual/testbank contains over 800 questions and problems an instructor can use to prepare exams. The manual also provides optional course sequences, a guide to audiovisual supplements, and several "starter references" for term papers and special research projects. The testbank portion of the manual is also available in IBM Windows and Macintosh formats (see below).

Prentice Hall Custom Test—WIN
Prentice Hall Custom Test—MAC

Harry Nickla, Creighton University

Available for Windows and Macintosh, *Prentice Hall Custom Test* allows instructors to create and tailor exams to their own needs. With the Online Testing option, exams can also be administered online and data can then be automatically transferred for evaluation. A comprehensive desk reference guide is included, along with online assistance.

Transparencies

110 four-color large type transparencies from the text are available for adopters.

Prentice Hall CD-ROM Image Bank for Genetics

This unique image bank contains illustrations from the third edition of *Essentials of Genetics* as well as animations and video in a digitized format for use in the classroom. The CD-ROM includes a navigational tool to allow instructors to customize lecture presentations. Additional features include keyword searches and the ability to incorporate lecture notes based on custom presentations.

Genetics in Action Web Site

www.prenhall.com/klug

Our companion Web site is tightly integrated with the textbook, forming a dynamic, technology-rich teaching and learning system. Throughout the site, textbook references will help students maximize their use of both the textbook and the technology. The tools on the Web site include:

- PROBLEMS—a non-threatening, self-paced tool to help you reinforce your understanding of key concepts and topics in genetics

- DESTINATIONS—accurate, dependable Web sites which enhance the topics and concepts presented in your textbook, by your instructor

- GENETICS & SOCIETY—a collection of thought-provoking issues to challenge your view and understanding of genetics in our society

- POSTINGS—an open, timely message board where you can share your knowledge with fellow students, seek help for a problem from your instructor, or discuss issues of genetics in our society with students from around the world

- SEARCH TOOLS—expeditious tools to help you actively search the Web for information that is most important to you

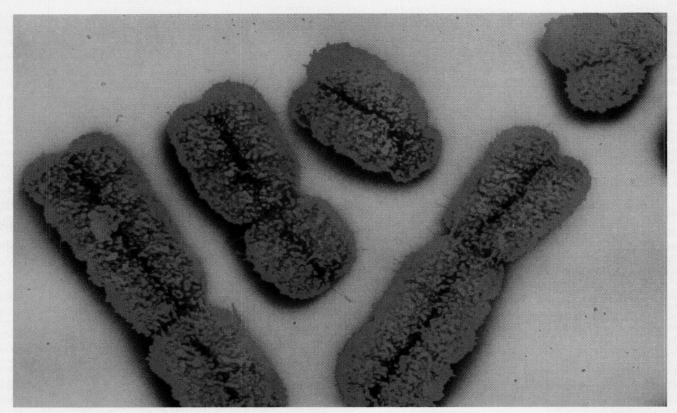

Scanning electron micrograph of human mitotic chromosomes.

CHAPTER OUTLINE

- **The Historical Context of Genetics**
  Prehistoric Times: Domesticated Animals and
    Cultivated Plants
  The Greek Influence: Hippocrates and Aristotle
  1600–1850: The Dawn of Modern Biology
  Charles Darwin and Evolution

- **Basic Concepts of Genetics**
- **Investigative Approaches in Genetics**
- **Genetics and Society**
  Eugenics and Euphenics
  Genetic Advances in Agriculture and Medicine

CHAPTER

1

An Introduction to Genetics

Chapter Concepts

Genetics is the science of heredity. The discipline has a rich history and involves investigations of molecules, cells, organisms, and populations, using many different experimental approaches. Not only does genetic information play a significant role during evolution, its expression influences the functioning of individuals at all levels. Genetics thus unifies the study of biology and has had a profound impact on human affairs.

Welcome to the study of genetics. You are about to explore a subject that many students before you have found to be the most interesting in the field of biology. This is not surprising, because an understanding of genetic processes is fundamental to the comprehension of life itself. Genetic information directs cellular function, determines an organism's external appearance, and serves as the link between generations in every species. Knowing how these roles are accomplished is important to understanding the living world. Knowledge of genetic concepts also helps us to understand other disciplines of biology. The topics studied in genetics overlap directly with molecular biology, cell biology, physiology, evolution, ecology, systematics, and behavior. The study of each of these disciplines is incomplete without knowledge of the genetic component underlying each of them. Genetics thus unifies biology and serves as its "core."

Fascination with this discipline stems further from the fact that, in genetics, many initially vague and abstract concepts have been so thoroughly investigated that subsequently they have become clearly and definitively understood. During these many investigations, genetics has established a rich history that exemplifies the nature of scientific inquiry and the analytical approach used to acquire information. Scientific analysis, moving from the unknown to the known, is one of the major forces that attracts students to biology.

The study of genetics is also extremely dynamic. Every year large numbers of significant findings are made. Over the past five decades, major research frontiers in genetics have been continually pushed forward, and new areas of investigation have resulted. In each case, the ensuing advances become part of an ever-expanding body of knowledge upon which further progress is based. It is exciting to be in the midst of these developments, whether you are studying or teaching genetics.

The Historical Context of Genetics

In the chapters that follow, we will discuss the nature of chromosomes, the way in which genetic information is transmitted from one generation to the next, and the way in which this information is stored, altered, expressed, and regulated. The most significant scientific findings, which serve as the foundation for our discussions, were obtained in the nineteenth century. As the twentieth century dawned, discoveries were made that began to clarify the understanding of the physical basis of living organisms and their relationship to one another. Several related ideas were gaining acceptance and were particularly significant: (1) Matter is composed of atoms; (2) cells are the fundamental units of living organisms; (3) nuclei somehow serve as the "life force" of cells; and (4) chromosomes housed within nuclei somehow play an important role in heredity. When these ideas were correlated with the newly rediscovered genetic findings of Gregor Mendel and integrated

3

with Darwin's theory of natural selection and the origin of species, a more complete picture of life at the level of the individual and of the population emerged. The era of modern-day biology was initiated on this foundation.

But what of the many important ideas and hypotheses that served as forerunners of nineteenth-century thought? In the following short section, we consider some of these, several of which can be traced back well over 1000 years!

Prehistoric Times: Domesticated Animals and Cultivated Plants

We may never know when people first recognized the existence of heredity. However, a variety of archeological evidence (e.g., primitive art, preserved bones and skulls, and dried seeds) have provided many insights. Such evidence documents the successful domestication of animals and cultivation of plants thousands of years ago. These efforts represent artificial selection of genetic variants within populations.

For example, between 8000 and 1000 B.C., horses, camels, oxen, and various breeds of dogs (derived from the wolf family) were domesticated to serve various roles. Cultivation of many plants, including maize, wheat, rice, and the date palm, is thought to have been initiated around 5000 B.C. Remains of maize dating to this period have been recovered in caves in the Tehucan Valley of Mexico. Assyrian art depicts artificial pollination of the date palm, thought to have originated in Babylonia (Figure 1–1). Such cultivation very likely led to the types of date palms found in that region today, where over 400 varieties exist in just four oases in the Sahara Desert. They differ from one another in various traits such as fruit taste.

Prehistoric evidence of cultivated plants and domesticated animals documents our ancient ancestors' suc-

■ Figure 1–1   Relief carving depicting artificial pollination of date palms during the reign of Assyrian King Assurnasirpal II (883–859 B.C.).

cessful attempts to manipulate the genetic composition of useful species. There is little doubt that people soon learned that desirable and undesirable traits are passed to successive generations and more desirable varieties of animals and plants could be selected. Human awareness of heredity was thus apparent during prehistoric times.

The Greek Influence: Hippocrates and Aristotle

Although few, if any, significant ideas were put forward to explain heredity during prehistoric times, philosophers directed much more attention to this subject during the Golden Age of Greek culture. They paid considerable attention to the subjects of reproduction and heredity, particularly as related to the origin of humans. This is particularly evident in the writings of the Hippocratic school of medicine (500–400 B.C.) and subsequently of the philosopher and naturalist Aristotle (384–322 B.C.).

Central to their explanation of the hereditary basis of reproduction of animals were hypotheses concerning (1) the source of the *physical substance* of the offspring and (2) the nature of the *generative force* that directs the physical substance as it materializes (develops) into an adult organism.

For example, the Hippocratic treatise *On the Seed* argues that male semen is formed in numerous parts of the body and is transported through blood vessels to the testicles. Active "humors" act as the bearers of hereditary traits and are drawn from various parts of the body to the semen. These humors could be healthy or diseased, the latter condition accounting for the appearance of newborns exhibiting congenital disorders or deformities. Furthermore, it was believed that these humors could be altered in individuals and, in their new form, could be passed on to offspring. In this way, newborns could "inherit" traits that their parents had "acquired" because of their environment.

Aristotle, who had studied under Plato for some 20 years, was more critical and more expansive than Hippocrates in his analysis of human origins and heredity. Aristotle proposed that male semen was formed from blood rather than from each organ and that its generative power resided in a "vital heat" that it contained. This vital heat had the capacity to produce offspring of the same "form" (i.e., basic structure and capacities) as the parent. Aristotle believed that it generated offspring by cooking and shaping the menstrual blood produced by the female, which was the "physical substance" giving rise to an offspring. The embryo developed from the initial "setting" of the menstrual blood by the semen into a mature offspring, not because it already contained the parts in miniature (as some Hippocratics had thought), but because of the shaping power of the vital heat. These ideas constituted only one part of Aristotelian philosophy of order in the living world.

Although the ideas of Hippocrates and Aristotle may sound primitive and naive today, we should recall that, prior to the 1800s, neither sperm nor eggs had yet been observed in mammals. Thus, the Greek philosophers' ideas were worthy ones in their time and for centuries to come. As we will see, their thinking was not so different from that of Charles Darwin in his formal proposal of the theory of pangenesis put forward during the nineteenth century.

1600–1850: The Dawn of Modern Biology

During the ensuing 1900 years (300 B.C.–1600 A.D.), the theoretical understanding of genetics was not extended by significant, new ideas. However, interest in applied genetics remained strong. As early as Roman times, plant grafting and animal breeding were much emphasized. By the Middle Ages, naturalists, well aware of the impact of heredity on organisms they studied, were faced with reconciling their findings with current religious beliefs. The theories of Hippocrates and Aristotle still prevailed and, when applied to humans, they no doubt conflicted with the prevailing religious doctrines.

Between 1600 and 1850, major strides were made that provided much greater insights into the biological basis of life, setting the scene for the revolutionary work and principles presented by Charles Darwin and Gregor Mendel. In the 1600s the English anatomist William Harvey (1578–1657), better known for his experiments demonstrating that the blood is pumped by the heart through a circulatory system made up of the arteries and veins, also wrote a treatise on reproduction and development, patterned after Aristotle's work. In it he is credited with the earliest statement of the theory of **epigenesis**—that an organism is derived from substances present in the egg, which differentiate into adult structures during embryonic development. Patterned after Aristotle's ideas, epigenesis holds that new structures, such as body organs, are not present initially but instead are formed *de novo* in the embryo. Indeed, Harvey had studied Aristotle and was well aware of his ideas.

The theory of epigenesis conflicts directly with the **theory of preformation**, first put forward in the seventeenth century. Stating that sex cells contain a complete miniature adult called the **homunculus** (Figure 1–2), perfect in every form, preformation was popular well into the eighteenth century. However, work by the embryologist Casper Wolff (1733–1794) and others clearly disproved this theory, thus favoring epigenesis. Wolff was convinced that several structures, such as the alimentary canal, were not initially present in the earliest embryos he studied, but instead were formed later during development.

During this same period, other significant chemical and biological findings affected future scientific thinking. In 1808, John Dalton expounded his **atomic theory**, which stated that all matter is composed of small invisible units called atoms. Improved microscopes became available, and around 1830 Matthias Schleiden and

■ Figure 1–2 Depiction of the "homunculus," a sperm containing a miniature adult, perfect in proportion and fully formed.

Theodor Schwann proposed the **cell theory**: All organisms are composed of basic visible units called cells, which are derived from similar preexisting structures. By this time, the idea of **spontaneous generation**, the creation of living organisms from nonliving components, had clearly been disproved by the experiments of Francesco Redi (1621–1697), Lazzaro Spallanzani (1729–1799), and Louis Pasteur (1822–1895), among others. Thus, living organisms were considered to be derived from preexisting organisms and to consist of cells made up of atoms.

Another prevailing notion had a major influence on nineteenth century thinking: the **fixity of species**. According to this doctrine, animal and plant groups remain unchanged in form from the moment of their appearance on earth. Embraced particularly by those who also adhered to a belief in special creation, this doctrine was popularized by several people, including the Swedish physician and plant taxonomist Carolus Linnaeus (1707–1778), who is better known for devising the binomial system of classification.

The influence of this tenet is illustrated by considering the work of the German plant breeder Joseph Gottlieb Kolreuter (1733–1806), who worked with tobacco. He cross-bred two groups and derived a new hybrid form, which he then converted back to one of the parental types by repeated backcrosses. In other breeding experiments using carnations, he clearly observed segregation of traits, which was to become one of Mendel's principles of genetics. These results seemed to contradict the idea of "fixed species" that do not change with time. Because of Kolreuter's belief in both special creation and the fixity of species, he was puzzled about these outcomes and failed to recognize the real significance of his findings.

Charles Darwin and Evolution

With the above information as background, we conclude our coverage of the historical context of genetics with a brief discussion of the work of Charles Darwin, who in 1859 published the book-length statement of his evolutionary theory, *The Origin of Species*. Darwin's many geological, geographical, and biological observations convinced him that existing species arose by descent with modification from other ancestral species. Greatly influenced by his now famous voyage on the H.M.S. *Beagle* (1831–1836), Darwin's thinking culminated in his formulation of the **theory of natural selection**, a theory that attempted to explain the causes of evolutionary change. Formulated and proposed at the same time, but independently, by Alfred Russel Wallace, natural selection is based on the observation that populations tend to consist of more offspring than the environment can support, leading to a struggle for survival among them. Those organisms with heritable traits that allow them to adapt to their environment are better able to survive and reproduce than those with less adaptive traits. Over a long period of time, slight but advantageous variations will accumulate. If a population bearing these inherited variations becomes reproductively isolated, a new species may result.

The primary gap in Darwin's theory was a lack of understanding of the genetic basis of variation and inheritance, a gap that left it open to reasonable criticism well into the twentieth century. Aware of this weakness in his theory of evolution, in 1868 Darwin published a second book, *Variations in Animals and Plants under Domestication*, in which he attempted to provide a more definitive explanation of how heritable variation arises gradually over time. Two of his major ideas, pangenesis and the inheritance of acquired characteristics, have their roots in the theories involving "humors," as put forward by Hippocrates and Aristotle.

In his provisional hypothesis of **pangenesis**, Darwin coined the term **gemmules** (rather than humors) to describe the physical units representing each body part that were gathered by the blood into the semen. Darwin felt that these gemmules determined the nature or form of each body part. He further believed that gemmules could respond in an adaptive way to an individual's external environment. Once altered, such changes would be passed on to offspring, allowing for the inheritance of acquired characteristics. Lamarck had previously formalized this idea in his 1809 treatise, *Philosophie Zoologique*. Lamarck's theory, which became known as the **doctrine of use and disuse**, proposed that organisms acquire or lose characteristics that then become heritable.

Even though Darwin never understood the basis for inherited variation, his ideas concerning evolution may be the most influential theory ever put forward in the history of biology. He was able to distill his extensive observations and synthesize his ideas into a cohesive hypothesis describing the origin of diversity of organisms populating the earth.

As Darwin's work ensued, the experiments of Gregor Johann Mendel (Figure 1–3) were performed between 1856 and 1863, forming the basis for his classic 1866 paper. In it, Mendel demonstrated a number of statistical patterns underlying inheritance and developed a theory involving hereditary factors in the germ cells to explain these patterns. His research was virtually ignored until it was partially duplicated and then cited by Carl Correns, Hugo de Vries, and Eric Von Tschermak around 1900 and subsequently championed by William Bateson.

By the early part of the twentieth century, chromosomes were discovered and support for the epigenetic interpretation of development had grown considerably.

■ Figure 1–3 Gregor Johann Mendel, who in 1866 put forward the major postulates of transmission genetics as a result of experiments with the garden pea.

It gradually became clear that heredity and development were dependent on "information" contained in chromosomes, which were contributed by gametes to each individual. The "gap" in Darwin's theory had narrowed considerably.

In Chapter 3, we will return to a thorough analysis of Mendel's findings, which have served to this day as the foundation of genetics. His work was but one important part of the body of knowledge that would initiate the era of modern biological thought in the twentieth century.

Basic Concepts of Genetics

We now turn to a review of some of the simple but basic concepts in genetics, which you have undoubtedly already studied. By reviewing them at the outset, we can establish an initial vocabulary and proceed through the text with a common foundation of knowledge. We shall approach these basic concepts by asking and answering a series of questions. You may wish to write or think through an answer before reading the explanation of each question. A number of issues that are addressed below are summarized in Figure 1–4, so you may wish to refer to this figure as you read through this section. Throughout the text, the answers to these questions will be expanded as more detailed information is presented.

What does "genetics" mean? **Genetics** is the branch of biology concerned with heredity and variation. This discipline involves the study of cells, individuals, their offspring, and the populations within which organisms live. Geneticists investigate all forms of inherited variation as well as the molecular basis underlying such characteristics.

■ Figure 1–4 Depiction of (a) the storage of genetic information in homologous chromosomes, which contain genes made up of DNA; and (b) genetic expression involving transcription of DNA into mRNA, which can be translated on a ribosome into a protein.

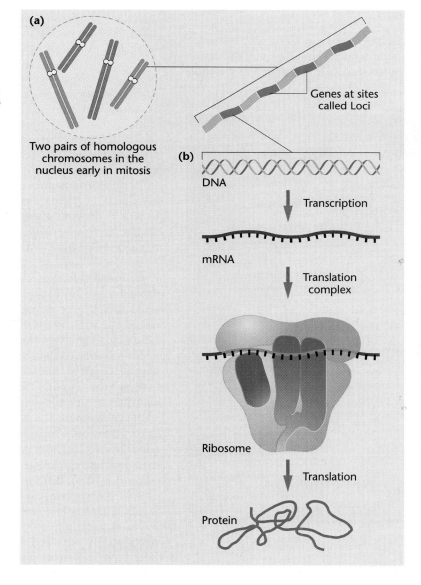

What is the center of heredity in a cell? In eukaryotic organisms, the **nucleus** contains the genetic material. In prokaryotes, such as bacteria, the genetic material exists in an unenclosed but recognizable area of the cell called the **nucleoid region** (Figure 1–5). In viruses, which are not true cells, the genetic material is ensheathed in the protein coat, together constituting the viral head or capsid.

What is the genetic material? In eukaryotes and prokaryotes, **DNA** serves as the molecule storing genetic information. In viruses, either DNA or **RNA** serves this function.

What do DNA and RNA stand for? DNA and RNA are abbreviations for **deoxyribonucleic acid** and **ribonucleic acid**, respectively. These are the two types of nucleic acids found in organisms. Nucleic acids, along with carbohydrates, lipids, and proteins, compose the four major classes of organic biomolecules found in living things.

How is DNA organized to serve as the genetic material? DNA, although single-stranded in a few viruses, is usually a double-stranded molecule organized as a **double**

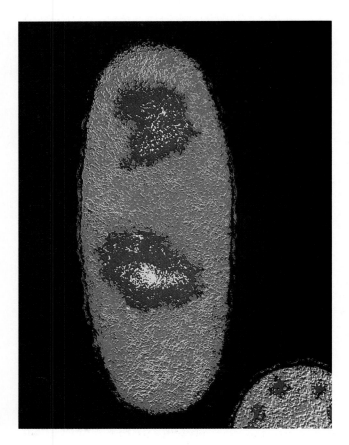

■ Figure 1–5 Enhanced electron micrograph of *Escherichia coli*, demonstrating the nucleoid regions (shown in blue). The bacterium has replicated its DNA and is about to begin cell division.

helix. Contained within each DNA molecule are hereditary units called **genes**, which are part of larger elements, the **chromosomes**.

What is a gene? In simplest terms, the gene is the functional unit of heredity. In chemical terms, it is a linear array of nucleotides—the chemical building blocks of DNA and RNA. A more conceptual approach is to consider it as an **informational storage unit** capable of undergoing **replication**, **mutation**, and **expression**. As investigations have progressed, the gene has been found to be a very complex element.

What is a chromosome? In viruses and bacteria, which have only a single chromosome, a chromosome is most simply thought of as a long, usually circular DNA molecule organized into genes. Most eukaryotes have many chromosomes that are composed of linear DNA molecules intimately associated with proteins. In addition, eukaryotic chromosomes contain many nongenic regions. It is not yet clear what role, if any, is played by many of these regions. Our knowledge of the chromosome, like that of the gene, is continually expanding.

When and how can chromosomes be visualized? If the chromosomes are released from the viral head or the bacterial cell, they can be visualized under the electron microscope (Figure 1–6). In eukaryotes, chromosomes are most easily visualized under the light microscope when they are undergoing **mitosis** or **meiosis**. In these division processes, the material constituting chromosomes is tightly coiled and condensed, giving rise to the characteristic image of chromosomes. Following division, this material, called **chromatin**, uncoils during interphase, where it can be studied under the electron microscope.

How many chromosomes does an organism have? Although there are exceptions, members of most eukaryotic species have a specific number of chromosomes, called the **diploid number (2n)**, in each somatic cell. Upon close analysis, these chromosomes are found to occur in pairs, each member of which shares a nearly identical appearance when visible during cell division. Called **homologous chromosomes**, the members of each pair are identical in their length and in the location of the **centromere**, the point of spindle fiber attachment during division. They also contain the same sequence of gene sites, or **loci**, and pair with one another during gamete formation (the process of meiosis).

The number of different *types* of chromosomes in any diploid species is equal to half the diploid number and is called the **haploid (n)** number. Some organisms, such as yeasts, are haploid during most of their life cycle and contain only one "set" of chromosomes. Other organisms, especially many plant species, are sometimes characterized by more than two sets of chromosomes and are said to be **polyploid**.

■ Figure 1–6 DNA constituting the chromosome of a bacterial virus (a bacteriophage) viewed under the electron microscope.

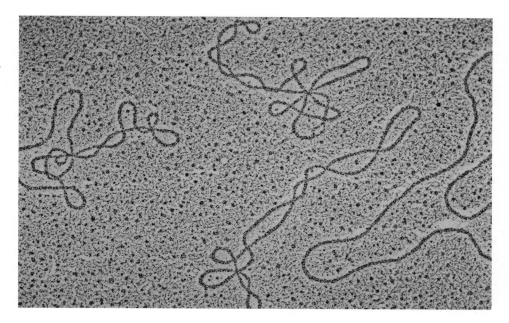

What is accomplished during the processes of mitosis and meiosis? Mitosis is the process by which the genetic material of eukaryotic cells is duplicated and distributed during cell division. Meiosis is the process whereby cell division produces gametes in animals and spores in most plants. While mitosis occurs in somatic tissue and yields two progeny cells with an amount of genetic material identical to that of the progenitor cell, meiosis creates cells with precisely one-half of the genetic material. Each gamete receives one member of each homologous pair of chromosomes and is haploid. This reduction in chromosome number is essential if offspring arising from two gametes are to maintain a constant number of chromosomes characteristic of their parents and other members of the species.

What are the sources of genetic variation? Classically, there are two sources of genetic variation: **chromosomal mutations** and **gene mutations**. The former, also called chromosomal **aberrations**, include duplication, deletion, or rearrangement of chromosome segments. Gene mutations result from a change in the stored chemical information in DNA, collectively referred to as an organism's **genotype**. Such a change may include substitution, duplication, or deletion of nucleotides, which compose this chemical information. Alternative forms of the gene, which result from mutation, are called **alleles**. Genetic variation frequently, but not always, results in a change in some characteristic of an organism, referred to as its **phenotype**.

How does DNA store genetic information? There are four different forms of chemical building blocks called **nucleotides** in a segment of DNA constituting a gene. The sequence of nucleotides making up a gene encodes the chemical nature (the amino acid composition) of a protein, the end product of genetic expression. **Mutations** are produced when the nucleotide sequence (the genetic code) is altered, creating alternate forms of the genes, called **alleles**.

How is the genetic code organized? There are four different nucleotides in DNA, each varying in one of its components, the **nitrogenous base**. The genetic code is a triplet; therefore, each combination of three nucleotides constitutes a code word. Almost all possible codes specify one of 20 **amino acids**, the chemical building blocks of proteins.

How is the genetic code expressed? The coded information in DNA is first transferred during a process called **transcription** into a **messenger RNA (mRNA)** molecule. The mRNA subsequently associates with the cellular organelle, the **ribosome**, where it is **translated** into a protein molecule.

Are there exceptions where proteins are not the end product of a gene? Yes. For example, genes coding for **ribosomal RNA (rRNA)**, which is part of the ribosome, and for **transfer RNA (tRNA)**, which is involved in the translation process, are transcribed but not translated. Therefore, RNA is sometimes the end product of stored genetic information.

Why are proteins so important to living organisms that they serve as the end product of the vast majority of genes? Many proteins serve as highly specific biological catalysts, or enzymes. In this role, these proteins control cellular metabolism, determining which carbohydrates, lipids, nucleic acids, and other proteins are present in the cell. Many other proteins perform nonenzymatic roles. For example, hemoglobin, collagen, immunoglobulins, and some hormones are proteins that play diverse roles in living organisms.

Why are enzymes necessary in living organisms? As biological catalysts, enzymes lower the **activation energy** required for most biochemical reactions and speed the attainment of equilibrium. Otherwise, these reactions would proceed so slowly as to be ineffectual in organisms living under the physical conditions on earth. Some genes control the variety of enzymes present in any cell type, dictating its overall biochemical composition.

Investigative Approaches in Genetics

The scope of topics encompassed in the field of genetics is enormous. Studies have involved viruses, bacteria, and a wide variety of plants and animals and have spanned all levels of biological organization, from molecules to populations. It is helpful, before we embark on a detailed study of genetics, to categorize the types of investigations that have been used most often in this field. Although some overlap exists, most have used one of four basic approaches.

The most classical investigative approach is the study of **transmission genetics**, in which the patterns of inheritance of traits are examined. Experiments are designed so that the transmission of traits from parents to offspring can be analyzed through several generations. Patterns of inheritance are sought that will provide insights into genetic principles. The first significant experimentation of this kind to have a major impact on understanding heredity was performed by Gregor Mendel in the middle of the nineteenth century. Information derived from his work serves today as the foundation of transmission genetics. In human studies, where designed matings are neither possible nor desirable, **pedigree analysis** is often useful. In pedigree analysis, patterns of inheritance are traced through as many generations as possible, leading to inferences concerning the mode of inheritance of the trait under investigation.

The second approach involves **cytogenetics**—the study of chromosomes. The earliest such studies used light microscopy. The initial discovery of chromosome behavior during mitosis and meiosis, early in the twentieth century, was a critical event in the history of genetics. In addition to playing an important role in the rediscovery and acceptance of Mendelian principles, these observations served as the basis of the **chromosomal theory of inheritance**. This theory, which viewed the chromosome as the carrier of genes and the functional unit of transmission of genetic information, was the cornerstone for further studies in genetics throughout the first half of this century.

The light microscope continues to be useful in the investigation of chromosome structure and abnormalities and is instrumental in preparing **karyotypes**, which illustrate the chromosomes characteristic of any species arranged in a standard sequence.

With the advent of electron microscopy, the repertoire of investigative approaches in genetics has grown. In high-resolution microscopy, genetic molecules and their behavior during gene expression can be visualized directly.

The third general approach involves **molecular genetic analysis**, which has had the greatest impact on the recent growth of genetic knowledge. Molecular studies, initiated in the early 1940s, have consistently expanded our knowledge of the role of genetics in life processes. Although experiments initially relied on bacteria and the viruses that infect them, extensive information is now available concerning the nature, expression, replication, and regulation of the genetic information in eukaryotes as well. The precise nucleotide sequence has been determined for many genes cloned in the laboratory. **Recombinant DNA** studies (Figure 1–7), in which genes from another organism are literally spliced into bacterial or viral DNA and cloned *en masse*, serve as the basis of a far-reaching research technology used in molecular genetic investigations. Building on this approach, the new field of **DNA biotechnology** now exists whereby genes are identified, sequenced, cloned, and manipulated. Furthermore, using the most recent technology, it is now possible to probe gene function in extreme detail. Such molecular and biochemical analysis has had profound implications in medicine, agriculture, and bioethics.

The final approach involves the study of **population genetics**. In these investigations, scientists attempt to define how and why certain genetic variation is maintained in populations, while other variation diminishes or is lost with time (Figure 1–8). Such information is critical to the understanding of evolutionary processes. Population genetics also allows us to predict gene frequencies in future generations.

Together these varied approaches used in investigative genetics have transformed a subject that was only poorly understood in 1900 into one of the most advanced scientific disciplines today. As a result, the impact of genetics on society has been immense. We shall discuss many examples of the applications of genetics in the following section and throughout the text.

■ Figure 1–7 Visualization of DNA fragments under ultraviolet light. The bands were produced using recombinant DNA technology.

■ Figure 1–8 Lady-bird beetles atop a daisy, illustrating genetic variation in populations.

Genetics and Society

In addition to acquiring information for the sake of extending knowledge in any discipline of science—an experimental approach called **basic research**—scientists conduct investigations to solve problems facing society or simply to improve the well-being of members of our society—an approach called **applied research**. Together, both types of genetic research have combined to: (1) enhance the quality of our existence on this planet; and (2) provide a more thorough understanding of life processes. As we shall see throughout this text, there is very little in our lives that genetics fails to touch.

Eugenics and Euphenics

There is always the danger that scientific findings will be used to formulate policies and/or actions that are unjust or even tragic. This section reviews such a case, which began near the end of the nineteenth century. At that time, Darwin's theory of natural selection provided a major influence on some people's thinking concerning the human condition. Our story recounts the initial attempt to apply genetic knowledge directly for the improvement of human existence. Championed in England by Francis Galton, the general approach is called **eugenics**, a term Galton coined in 1883.

Galton, a cousin of Charles Darwin, believed that many human characteristics were inherited and subject to artificial selection if human matings could be controlled. **Positive eugenics** encouraged parents displaying favorable characteristics to have large families. Superior intelligence, intellectual achievement, and artistic talent are examples. **Negative eugenics**, on the other hand, at-tempted to restrict reproduction of parents displaying unfavorable characteristics. Low intelligence, mental retardation, and criminal behavior are examples.

In the United States, the eugenics movement was a significant social force and led to state and federal laws that required sterilization of those considered "genetically inferior." Over half of the states passed such laws, commencing in 1907 with Indiana. Sterilization was mandated for "imbeciles, idiots, convicted rapists, and habitual criminals." By 1931, involuntary sterilization also applied to "sexual perverts, drug fiends, drunkards, and epileptics." Immigration to the United States from certain areas of Europe and from Asia was also restricted, to prevent the influx of what were regarded as genetically inferior people.

In addition to the violation of individual human rights, such policies were seriously flawed by an inadequate understanding of the genetic basis of various characteristics. Formulation of eugenic policies was premised on the mistaken notions that "superior" and "inferior" traits are totally under genetic control and that genes deemed unfavorable could be removed from a population by selecting against (sterilizing) individuals expressing those traits. The potential impact of the environment as well as genetic theory underlying population genetics were largely ignored as eugenic policies were developed.

In Nazi Germany in the 1930s, the concept of achieving a superior, racially pure group was an extension of the eugenics movement. Initially applied to those considered socially and physically defective, the underlying rationale of negative eugenics was soon applied to entire ethnic groups, including Jews and Gypsies. Fueled by various forms of racial prejudice, Adolf Hitler and the Nazi regime took eugenics to its extreme by instituting policies aimed at the extinction of these "impure" human populations. This deplorable disregard for human life was preceded by incremental policies involving forced sterilization and mercy killings. This movement, based on scientifically invalid premises, soon led to mass murder.

Even before the Nazi Party came to power in 1933, English and American geneticists began separating themselves from the eugenics movement. They were concerned about the validity of the premises underlying the movement and the evidence in support of these premises. Thus, many geneticists chose not to study human genetics for fear of being grouped with those who supported eugenics.

However, since the end of World War II, tremendous strides have been made in human genetics research. Today, a new term, **euphenics**, has replaced eugenics. Euphenics refers to medical and/or genetic intervention designed to reduce the impact of defective genotypes on individuals. The use of insulin by diabetics and the dietary control of newborn phenylketonurics are long-standing examples. Today, "genetic surgery" to replace defective genes rests clearly on the horizon. Furthermore, social policies now have a solid genetic foundation on which they may be based. Nevertheless, caution is still required to ensure that our expanded knowledge of human genetics does not obscure the role played by the environment in determining an individual's phenotype.

Genetic Advances in Agriculture and Medicine

As a result of research in genetics, major benefits have accrued to society in the fields of agriculture and medicine. Although cultivation of plants and domestication of animals had begun long before, the rediscovery of Mendel's work in the early twentieth century spurred scientists to apply genetic principles to these human endeavors. The use of selective breeding and hybridization techniques has had the most significant impact in agriculture.

Plants have been improved in four major ways: (1) enhanced potential for more vigorous growth and increased yields, including the unique genetic phenomenon of **hybrid vigor (heterosis)**; (2) increased resistance to natural predators and pests, including insects and disease-causing microorganisms; (3) production of hybrids exhibiting a combination of superior traits derived from two different strains or even two different species (Figure 1-9); and (4) selection of genetic variants with desirable qualities such as increased protein value, increased content of limiting amino acids, which are essential in the human diet, or smaller plant size, reducing vulnerability to adverse weather conditions.

Over the past five decades, these improvements have resulted in a tremendous increase in yield and nutrient value in such crops as barley, beans, corn, oats, rice, rye, and wheat. It is estimated that in the United States the use of improved genetic strains has led to a threefold increase in crop yield per acre. In Mexico, where corn is the staple crop, a significant increase in protein content and yield has occurred. A substantial effort has also been made to improve the growth of Mexican wheat. Led by Norman Borlaug, a team of researchers developed varieties of wheat that incorporated favorable genes from other strains found in various parts of the world, revolutionizing wheat production in Mexico and other underdeveloped countries. Because of this effort, which led to the well-publicized "Green Revolution," Borlaug received the Nobel Peace Prize in 1970. There is little question that this application of genetics has contributed to the well-being of our own species by improving the quality of nutrition worldwide.

Applied research in genetics has also resulted in the development of superior breeds of livestock (Figure 1-10). Selective breeding has produced chickens that grow faster, produce more high-quality meat per chicken, and lay greater numbers of larger eggs. In larger animals, including pigs and cows, the use of artificial insemination has been particularly important. Sperm samples derived from a single male with superior genetic traits may now be used to fertilize thousands of females located in all parts of the world.

Equivalent strides have been made in medicine as a result of advances in genetics, particularly since 1950. Numerous disorders in humans have been discovered

■ Figure 1–9 *Triticale*, a hybrid grain derived from wheat and rye, produced as a result of applied genetic research.

to result from either a single mutation or a specific chromosomal abnormality. For example, the genetic basis of disorders such as sickle-cell anemia, erythroblastosis fetalis, cystic fibrosis, hemophilia, muscular dystrophy, Tay–Sachs disease, Down syndrome, and many metabolic disorders is now well documented and often understood at the molecular level. The importance of acquiring knowledge of inherited disorders is underscored by the estimate that more than 10 million children or adults in the United States suffer from some form of genetic affliction and that every child-bearing

■ Figure 1–10 The effects of breeding and selection, as illustrated by the production of this Vietnamese pot-bellied pig.

GENETICS, TECHNOLOGY, AND SOCIETY

The Fruits of Plant Biotechnology: Edible Vaccines

Almost lost amid the furor over the cloned sheep Dolly was the original purpose of the undertaking: to genetically engineer an animal that would produce a valuable pharmaceutical product, eventually leading to a herd of genetically identical drug-producing animals. The goal, in other words, was to be able to turn sheep, cattle, and other animals into living drug factories.

Meanwhile, almost unnoticed by the general public, the genetic engineering of plants to serve a similar role is much closer to reality. When foreign genes are inserted into the plant genome, transgenic plants are created that produce the foreign gene product. One of the most intriguing and potentially life-saving objectives is the genetic engineering of plants to produce vaccines against human diseases. Immunization would then involve eating foods altered to contain a protein that acts as an antigen and stimulates the production of antibodies to protect against bacterial or viral infection.

Plants have many advantages as vaccine-producers. Since plants can be grown in large numbers, plant-produced vaccines should be less expensive than conventional vaccines. Further, extensive purification and refrigerated transport and storage of vaccines will not be required. This is important in parts of the world where the supply of electricity is unreliable. Finally, since people would simply eat the vaccine-containing food, there would be no need for syringes and needles, or medical staff to give injections.

Leading the effort to develop edible vaccines in plants is Charles J. Arntzen, formerly at Texas A&M University and now president of the Boyce Thompson Institute for Plant Research at Cornell University. Arntzen's research team is focusing on intestinal diseases, especially cholera. Cholera may at first seem an odd target, since it is a disease that has not been a major public health problem in this country for over a century. But cholera remains a leading cause of death of infants and children throughout the Third World, where basic sanitation is lacking and water supplies are often contaminated. For example, in July of 1994, 70,000 cases of cholera were reported among the Rwandans crowded into refugee camps in Goma, Zaire, leading to 12,000 fatalities. A cholera epidemic struck East Africa in the fall of 1997 killing nearly 3000. And after an absence of over 100 years, cholera has reappeared in Latin America in 1991, spreading from Peru to Mexico and claiming more than 10,000 lives.

The causative agent of cholera is *Vibrio cholerae,* a curved, rod-shaped bacterium found mostly in rivers and oceans. Most strains of *V. cholerae* are harmless; only one strain, called O1, is pathogenic. Infection occurs when a person drinks water or eats food contaminated with this strain. Once in the digestive system, the bacteria colonize the small intestine and begin producing proteins called enterotoxins. The cholera enterotoxin binds to the surface of the mucosal cells lining the intestine, triggering the massive secretion of water and dissolved salts from these cells. This results in violent diarrhea, which, if untreated, is followed by severe dehydration, muscle cramps, lethargy, and often death.

The cholera enterotoxin consists of two polypeptides, called the A and B subunits, which individually have no effect. For the toxin to be active, one A subunit must be linked to five B subunits. Since it is the B subunit of this complex that binds to intestinal cells, Arntzen's group decided to use this polypeptide as their antigen, reasoning that antibodies against it would potentially prevent toxin binding and render the bacteria harmless.

Their test of this idea involved the B subunit of an *E. coli* enterotoxin, which is similar in structure and immunological properties to the cholera protein. The DNA coding sequence of the B-subunit gene was obtained, to which they attached a promoter that would prompt transcription in all tissues of the plant. They then introduced this hybrid gene into potato plants by means of *Agrobacterium*-mediated transformation. They chose the potato not only because methods for transformation and regeneration of this plant are fairly routine, but also so they could assay the effectiveness of the antigen in the edible part of the plant, the tuber. Analysis showed that the engineered plants expressed their new gene and produced the enterotoxin.

After feeding mice a few grams of these genetically engineered tubers that had produced the B subunit, they found that the mice began to produce specific antibodies against it and to secrete them into the small intestinal. And, most critically, mice later fed purified enterotoxin were protected from its effects. They did not develop the symptoms of cholera. Clinical trials are now planned to test the efficacy of the potato-produced vaccine in humans.

In the meantime, the Arntzen group is also working towards producing edible vaccines in bananas, which have several advantages over potatoes. For one thing, bananas can be grown almost anywhere throughout the tropical or sub-tropical developing countries of the world, exactly where they are needed the most. And unlike potatoes, bananas are usually eaten raw, avoiding the potential inactivation of the antigenic proteins by cooking. Finally, bananas are well liked by infants and children, making this approach to immunization a more feasible one.

Procedures are now being perfected for the transformation of banana cells with genes encoding the cholera enterotoxin and the regeneration of transgenic plants. It will be some time before the engineered bananas are ready to test, however, since it takes three years to grow a banana crop. If all goes as planned, it may someday be possible to immunize all Third World children against cholera and other intestinal diseases, saving untold thousands of young lives.

References

Arntzen, C.J. 1997. Edible vaccines. *Public Health Rep.* 112: 190–197.

Haq, T.A., Mason, H.S., Clements, J.D., and Arntzen, C.J. 1995. Oral immunization with a recombinant bacterial antigen produced in transgenic plants. *Science* 268: 714–716.

Sanchez, J.L., and Taylor, D.N. 1997. Cholera. *Lancet* 349: 1825–1830.

couple stands an approximately 3 percent risk of having a child with some form of genetic anomaly.

Additionally, it has gradually become clear during the current decade that most, if not all, forms of **cancer** have a genetic basis. Although cancer is not usually an inherited disorder, it is now very clear that *cancer is a genetic disorder at the somatic cell level*. That is, most cancers are derived from somatic cells that have undergone some type of genetic change; malignant tumors are then derived from the genetically altered cell. In some cases, such changes are inherited, conferring a genetic predisposition to cancer.

Recognition of the molecular basis of human genetic disorders and cancer has provided the impetus for the development of methods for detection and treatment. **Genetic counseling** provides couples with objective information on which they can base rational decisions about child-bearing. In the case of cancer, recent genetic discoveries have already led to more effective early detection and more efficient approaches to treatment.

Applied research in genetics has also provided other medical benefits. Advances in **immunogenetics** have made possible compatible blood transfusions as well as organ transplants. In conjunction with immunosuppressive drugs, transplant operations involving human organs, including the heart, liver, pancreas, and kidney, are increasing annually and in many cases are now considered routine surgery.

The most recent advances in human genetics have been dependent on the application of **DNA biotechnology**. First developed in the 1970s, **recombinant DNA techniques** paved the way for manipulating and cloning a variety of genes, including those that encode many medically important molecules, such as insulin, blood clotting factors, growth hormone, and interferon. Human genes were isolated and spliced into vectors and transferred to host cells that serve as "production centers" for the synthesis of these proteins.

Recombinant DNA techniques have now been extended considerably. DNA of any organism of interest is routinely manipulated in the laboratory. Human genes responsible for inherited disorders such as **cystic fibrosis** and **Huntington disease** have been identified, isolated, cloned, and studied. It is hoped that such research will pave the way for **gene therapy**, whereby genetic disorders are treated by inserting normal copies of genes into the cells of afflicted individuals.

Perhaps the most far-reaching utilization of DNA biotechnology involves the **Human Genome Project**, in which the entire genetic complement (the **genome**) of several species, including our own, is being sequenced. The genomic sequencing of several bacterial species as well as that of yeast is now completed. The sequencing of the entire 3.2 billion nucleotides constituting the human genome is scheduled for completion in the year 2003.

In later chapters, the applications of DNA biotechnology to agriculture and medicine are discussed in greater detail. Although other scientific disciplines are also expanding in knowledge, none has paralleled the growth of information that is occurring in genetics. By the end of this course, we are confident you will agree that the present truly represents the "Age of Genetics."

Chapter Summary

1. The history of genetics, which emerged as a fundamental discipline of biology early in the twentieth century, dates back to prehistoric times.

2. Numerous concepts and a basic vocabulary essential to the study of genetics have been presented.

3. Four investigative approaches are most often used in the study of genetics, including transmission genetic studies, cytogenetic analyses, molecular experimentation, and inquiries into the genetic structure of populations.

4. Genetic research can be either basic or applied. Basic genetic research extends our knowledge of the discipline; the objective of applied genetics research is to solve specific problems affecting the quality of our lives and society in general.

5. Eugenics, the application of knowledge of genetics for the improvement of human existence, has a long and controversial history. Euphenics, genetic intervention designed to ameliorate the impact of genotypes on individuals, represents the modern eugenic approach.

6. Genetic research has had a highly positive impact on many facets of agriculture.

7. DNA biotechnology has greatly expanded our research capability. It also has had a profound impact in elucidating the basis of inherited diseases, has made possible the mass production of medically important gene products, and will serve as the foundation on which gene therapy is developed.

Key Terms

activation energy, 10
allele, 9

amino acid, 9
applied research, 11

basic research, 11
cell theory, 5

Problems and Discussion Questions

1. Describe and contrast the ideas of Hippocrates and Aristotle relating to the genetic basis of life.
2. Define and contrast pangenesis, epigenesis, and preformationism.
3. Which ideas and doctrines that preceded Darwin were central to his thinking?
4. Describe Darwin's and Wallace's theory of natural selection. What information was lacking from it; that is, what gap remained in it?
5. Contrast chromosomes and genes and describe their role in heredity.
6. Describe the four major investigative approaches used in studying genetics.
7. Contrast basic and applied research.
8. Norman Borlaug received the Nobel Peace Prize for his work in genetics. Why do you think he was awarded this prize?
9. Contrast positive and negative eugenics. Which of these categories includes the approach called euphenics? Define this approach.
10. How has genetic research been applied to agriculture and to medicine?

Selected Readings

Allen, G. E. 1996. Science misapplied: The eugenics age revisited. *Technol. Rev.* 99:23–31.

Anderson, W. F., and Dircumakos, E. G. 1981. Genetic engineering in mammalian cells. *Sci. Am.* (July) 245:106–21.

Borlaug, N. E. 1983. Contributions of conventional plant breeding to food production. *Science* 219:689–93.

Bowler, P. J. 1989. *The Mendelian revolution: The emergence of hereditarian concepts in modern science and society.* London: Athione.

Cocking, E. C., Davey, M. R., Pental, D., and Power, J. B. 1981. Aspects of plant genetic manipulation. *Nature* 293:265–70.

Day, P. R. 1977. Plant genetics: Increasing crop yield. *Science* 197:1334–39.

Dunn, L. C. 1965. *A short history of genetics.* New York: McGraw-Hill.

Gardner, E. J. 1972. *History of biology,* 3rd ed. New York: Macmillan.

Garver, K. L., and Garver, B. 1991. Eugenics: Past, present, and future. *Am. J. Hum. Genet.* 49:1109–18.

Gasser, C. S., and Fraley, R. T. 1989. Genetically engineering plants for crop improvement. *Science* 244:1293–99.

—————. 1992. Transgenic crops. *Sci. Am.* (June) 266:62–69.

Horgan, J. 1993. Eugenics revisited. *Sci. Am.* (June) 268:123–31.

King, R. C., and Stansfield, W. D. 1997. *A dictionary of genetics,* 5th ed. New York: Oxford University Press.

Olby, R. C. 1985. *Origins of Mendelism,* 2nd ed. London: Constable.

Stubbe, H. 1972. *History of genetics: From prehistoric times to the rediscovery of Mendel* (transl. by T. R. W. Waters). Cambridge, MA: MIT Press.

Torrey, J. G. 1985. The development of plant biotechnology. *Am. Sci.* 73:354–63.

Vasil, I. K. 1990. The realities and challenges of plant biotechnology. *Bio/Technology* 8:296–301.

Weinberg, R. A. 1985. The molecules of life. *Sci. Am.* (Oct.) 253:48–57.

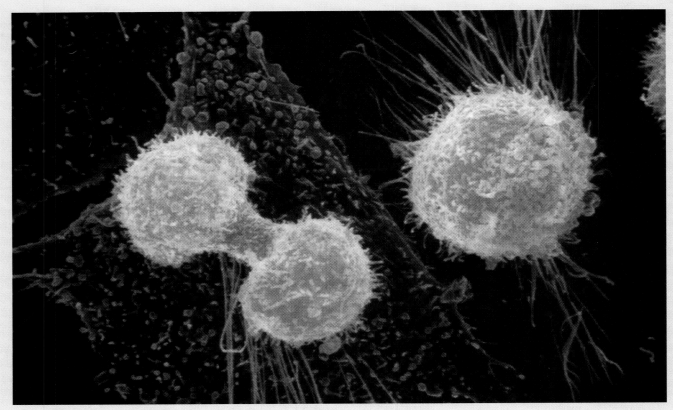

Scanning electron micrograph of human ovarian cells undergoing cell division.

CHAPTER OUTLINE

2

Cell Division and Chromosomes

Chapter Concepts

Genetic continuity between cells and organisms of any sexually reproducing species is maintained by the processes of mitosis and meiosis. The processes are orderly and efficient, serving to produce diploid somatic cells and haploid gametes, respectively. It is during these division stages that the genetic material is condensed into discrete, visible structures called chromosomes.

In every living thing there exists a substance referred to as the **genetic material**. Except in certain viruses, this material is composed of the nucleic acid DNA. A molecule of DNA is organized into units called genes, the products of which direct the metabolic activities of cells. DNA, with its array of genes is organized into structures called **chromosomes**, which serve as vehicles for transmission of genetic information. The manner in which chromosomes are transmitted from one generation of cells to the next, and from organisms to their descendants, must be exceedingly precise. In this chapter, we consider exactly how genetic continuity is maintained between cells and organisms.

Two major processes are involved in eukaryotes: **mitosis** and **meiosis**. Although the mechanisms of the two processes are similar in many ways, the outcomes are quite different. Mitosis leads to the production of two cells, each with the same number of chromosomes as the parent cell. Meiosis, on the other hand, reduces the genetic content and the number of chromosomes by precisely half. This reduction is essential if sexual reproduction is to occur without doubling the amount of genetic material at each generation. Strictly speaking, mitosis is that portion of the cell cycle during which the hereditary components are precisely and equally divided into daughter cells. Meiosis is part of a special type of cell division leading to the production of sex cells: **gametes** or **spores**. This process is an essential step in the transmission of genetic information from an organism to its offspring.

Normally, chromosomes are visible during a cell's life only during mitosis and meiosis. When cells are not undergoing division, the genetic material making up chromosomes unfolds and uncoils into a diffuse network within the nucleus, generally referred to as **chromatin**. In this chapter, we will examine this transition and also consider two specialized cases where chromosomes can be seen in nondividing cells: **polytene** and **lampbrush chromosomes**. These specialized structures extend our knowledge of genetic organization and its relation to genetic function.

Cell Structure

Before describing mitosis and meiosis, we will briefy review the structure of cells. As we shall see, many components, such as the nucleolus, ribosome, and centriole, are involved directly or indirectly with genetic processes. Other components, the mitochondria and chloroplasts, contain their own unique genetic information. It is also useful for us to compare the structural differences between the prokaryotic bacterial cell and the eukaryotic cell. Variation in the structure and function of cells is dependent on specifc genetic expression by each cell type.

Before 1940, knowledge of cell structure was based on information obtained with the light microscope. Around 1940 the transmission electron microscope was in its early stages of development, and by 1950 many details of cell ultrastructure were unveiled. Under the electron microscope, cells were seen as highly organized, precise structures. A new world of whorling membranes, organelles, microtubules, granules, and filaments was revealed. These discoveries revolutionized thinking in the entire field of biology. We will be concerned with those aspects of cell structure relating to genetic study. Figure 2–1 depicts a typical animal cell, illustrating most of the structures discussed below.

All cells are surrounded by a **plasma membrane**, an outer covering that defines the cell boundary and delimits the cell from its immediate external environment. This membrane is not passive, but instead actively controls the movement of materials into and out of the cell. In addition to this membrane, plant cells have an outer covering called the **cell wall**. One of the major components of this rigid structure is a polysaccharide called **cellulose**.

Many, if not most, animal cells have a covering over the plasma membrane, referred to as the **cell coat**. Consisting of glycoproteins and polysaccharides, its chemical composition differs from comparable structures in either plants or bacteria. One function served by the cell coat is to provide biochemical identity at the surface of cells. These forms of cellular identity at the cell surface are under genetic control. For example, various antigenic determinants such as the **AB** and **MN antigens** are found on the surface of red blood cells. In other cells, **histocompatibility antigens**, which elicit an immune response during tissue and organ transplants, are present. A variety of **receptor molecules** are also important components at the surface of cells. These constitute recognition sites that transfer specific chemical signals across the cell membrane into the cell.

The presence of a nucleus and other membranous organelles characterizes eukaryotic cells. The nucleus houses the genetic material, DNA, which is found in association with large numbers of acidic and basic proteins. During nondivisional phases of the cell cycle, this DNA/protein complex exists in an uncoiled, dispersed state called **chromatin**. As we will soon discuss, during mitosis and meiosis this material coils up and condenses into structures called **chromosomes**. Also present in the nucleus is the **nucleolus**, an amorphous component where ribosomal RNA is synthesized and where the initial stages of ribosomal assembly occur. The areas of DNA encoding rRNA are collectively referred to as the **nucleolus organizer** region or the **NOR**.

The lack of a nuclear envelope and membraneous organelles is characteristic of **prokaryotes**. In bacteria such as *Escherichia coli*, the genetic material is present as a long, looped DNA molecule that is compacted into an area referred to as the **nucleoid region**. Part of the DNA may be attached to the cell membrane, but in general the nucleoid constitutes a large area throughout the cell. Although the DNA is compacted, it does not undergo the extensive coiling characteristic of the stages of mitosis where, in eukaryotes, chromosomes become visible. Nor is the DNA in these organisms associated as extensively with proteins as is eukaryotic DNA. Figure 2–2 shows the formation of two bacteria during cell division and illustrates the bacterial chromosomes in the nucleoid regions. Prokaryotic cells do not have a distinct nucleolus, but do contain genes that specify rRNA molecules.

The remainder of the eukaryotic cell enclosed by the plasma membrane, excluding the nucleus, is composed of **cytoplasm** and all associated **cellular organelles**. Cytoplasm consists of a nonparticulate, colloidal material referred to as the cytosol, which surrounds and encompasses the cellular organelles. Beyond these components, an extensive system of tubules and filaments comprising the cytoskeleton provides a lattice of support structures within the cytoplasm. Consisting primarily of tubulin-derived microtubules and actin-derived microfilaments, this structural framework maintains cell shape, facilitates cell mobility, and anchors the various organelles.

One such organelle, the membranous **endoplasmic reticulum (ER)**, compartmentalizes the cytoplasm, greatly increasing the surface area available for biochemical synthesis. The ER may appear smooth, in which case it serves as the site for synthesis of fatty acids and phospholipids, or it may appear rough because it is studded with ribosomes. Ribosomes serve as sites for the translation of genetic information contained in messenger RNA (mRNA) into proteins.

Three other cytoplasmic structures are very important in the eukaryotic cell's activities: **mitochondria**, **chloroplasts**, and **centrioles**. Mitochondria are found in both animal and plant cells and are the sites of the oxidative phases of **cell respiration**. These chemical reactions generate large amounts of **adenosine triphosphate (ATP)**, an energy-rich molecule. Chloroplasts are found in plants, algae, and some protozoans. This organelle is associated with **photosynthesis**, the major energy-trapping process on earth. Both mitochondria and chloroplasts contain a type of DNA that is distinct from that found in the nucleus. Furthermore, these organelles can duplicate themselves and transcribe and translate their genetic information. It is interesting to note that the genetic machinery of mitochondria and chloroplasts closely resembles that of prokaryotic cells. This and other observations have led to the proposal that these organelles were once primitive free-living organisms that established a symbiotic relationship with a primitive eukaryotic cell. This theory, which describes the evolutionary origin of these organelles, is called the **endosymbiont hypotheis**.

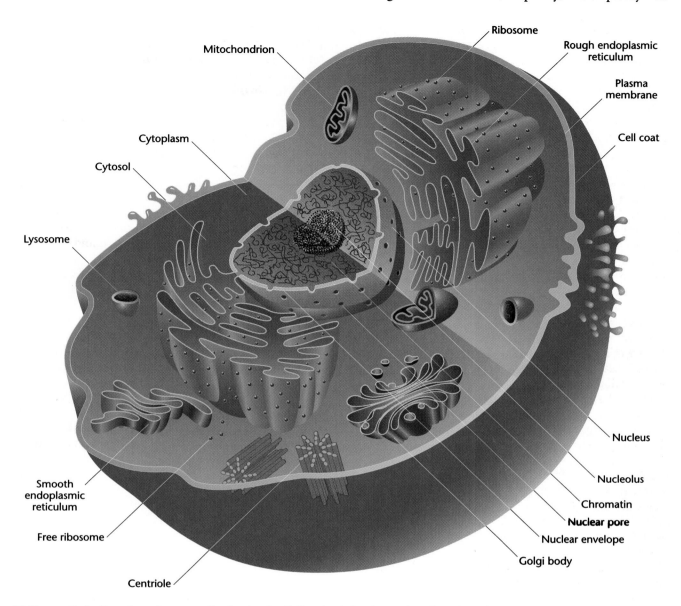

Mitochondrion

Cytoplasm

Cytosol

Lysosome

Smooth
endoplasmic
reticulum

Free ribosome

Centriole

Ribosome

Rough endoplasmic
reticulum

Plasma
membrane

Cell coat

Nucleus

Nucleolus

Chromatin

Nuclear pore

Nuclear envelope

Golgi body

■ Figure 2–1   Drawing of a generalized animal cell. Emphasis has been placed on the cellular components discussed in the text.

Animal cells and some plant cells also contain a pair of complex structures called the **centrioles**. These cytoplasmic bodies, contained within a specialized region called the **centrosome**, are associated with the organization of spindle fibers that function in mitosis and meiosis. In some organisms, the centriole is derived from another structure, the **basal body**, which is associated with the formation of cilia and flagella. Over the years, there have been many reports suggesting that centrioles and basal bodies contain DNA, which could be involved in the replication of these structures. This is currently thought not to be the case.

The organization of spindle fibers by the centrioles occurs during the early phases of mitosis and meiosis. Composed of arrays of microtubules, these fibers play

an important role in the movement of chromosomes as they separate during cell division. The microtubules consist of polymers of polypeptide subunits of the protein tubulin. The interaction of the chromosomes and spindle fibers will be considered later in this chapter.

Homologous Chromosomes, Haploidy, and Diploidy

To discuss mitosis and meiosis, we must employ the concept of **homologous chromosomes**. Understanding this concept will also be of critical importance in our future discussions of Mendelian genetics. Chromosomes are most easily visualized during mitosis. When they are

■ Figure 2–2 Color-enhanced electron micrograph of *E. coli* undergoing cell division. Particularly prominent are the two chromosomal areas (shown in red), called nucleoids, that have been partitioned into the daughter cells.

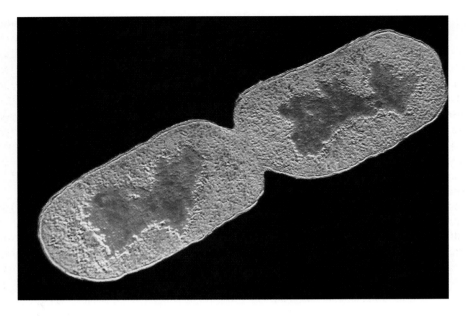

examined carefully, they are seen to take on distinctive lengths and shapes. Each contains a condensed or constricted region called the **centromere**, which establishes the general appearance of each chromosome. Figure 2–3 illustrates chromosomes with centromere placements at different points along their lengths. Extending from either side of the centromere are the arms of the

chromosome. Depending on the position of the centromere, different arm ratios are produced. As Figure 2–3 illustrates, chromosomes are classified as **metacentric**, **submetacentric**, **acrocentric**, or **telocentric** on the basis of the centromere location. The shorter arm, by convention, is shown above the centromere and is called the **p arm** (p stands for "petite"). The longer arm

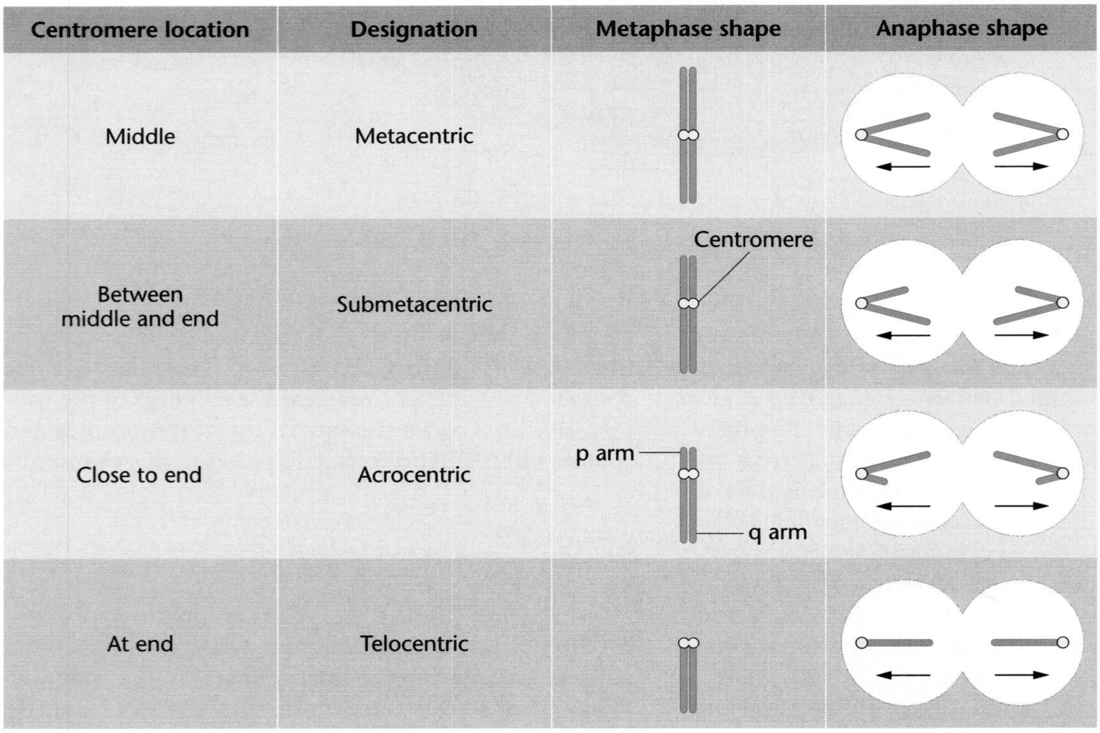

Centromere location	Designation	Metaphase shape	Anaphase shape
Middle	Metacentric		
Between middle and end	Submetacentric	Centromere	
Close to end	Acrocentric	p arm / q arm	
At end	Telocentric		

■ Figure 2–3 Centromere locations and designations of chromosomes based on their location. Note that the shape of the chromosome during anaphase is determined by the position of the centromere.

is shown below the centromere and is called the **q arm** (the next letter in the alphabet).

When studying mitosis, several other important observations are relevant. First, all somatic cells derived from members of the same species contain an identical number of chromosomes. In most cases this represents the **diploid number (2n)**. When the lengths and centromere placements of all such chromosomes are examined, a second general feature is apparent. Nearly all of the chromosomes exist in pairs with regard to these two criteria. The members of each pair are called **homologous chromosomes**. For each chromosome exhibiting a specific length and centromere placement, another exists with identical features. There are, of course, exceptions to the above descriptions, illustrated by organisms such as yeasts and molds, where the predominant phase of the life cycle is spent in the haploid stage.

Figure 2–4 illustrates the nearly identical physical appearance of members of homologous chromosome pairs. There, the human mitotic chromosomes have been photographed, cut out of the print, and matched up, creating a **karyotype**. As you can see, humans have a 2n number of 46 and exhibit a diversity of sizes and centromere placements. Note also that each of the 46 chromosomes is clearly a double structure consisting of two parallel **sister chromatids** connected by a common centromere. Had these chromosomes been allowed to continue dividing, the sister chromatids, which are replicas of one another, would have separated into the two new cells as division continued.

Collectively, the total set of genes contained on one member of each homologous pair of chromosomes constitutes the **haploid genome** of the species. The haploid number (*n*) of chromosomes is equal to one-half the diploid number. Table 2.1 illustrates the wide range of *n* values found in a variety of plants and animals.

Homologous pairs of chromosomes have important genetic similarities. They contain identical gene sites along their lengths called **loci**. Thus, they have identical genetic potential. In sexually reproducing organisms, one member of each pair is derived from the maternal parent (through the ovum) and one is derived from the paternal parent (through the sperm). Therefore, each diploid organism contains two copies of each gene as a consequence of **biparental inheritance**. As we will see in the following chapters on transmission genetics, the members of each pair of genes, while influencing the same characteristic or trait, need not be identical. Alternative forms of the same gene are called **alleles**. In a

■ Figure 2–4 A metaphase preparation of chromosomes derived from a human male and the karyotype derived from it. All but the X and Y chromosomes are present in homologous pairs. Each chromosome is clearly a double structure, constituting a pair of sister chromatids joined by a common centromere.

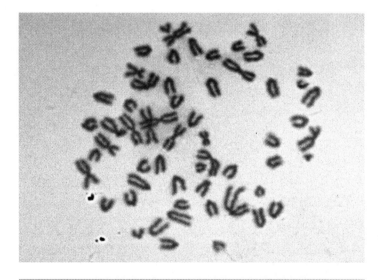

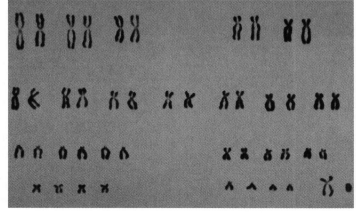

TABLE 2.1	The haploid number of chromosomes for a variety of organisms	
Common Name	**Scientific Name**	**Haploid No.**
Black Bread mold	*Aspergillus nidulans*	8
Broad bean	*Vicia faba*	6
Cat	*Felis domesticus*	19
Cattle	*Bos taurus*	30
Chicken	*Gallus domesticus*	39
Chimpanzee	*Pan troglodytes*	24
Corn	*Zea mays*	10
Cotton	*Gossypium hirsutum*	26
Dog	*Canis familiaris*	39
Evening primrose	*Oenothera biennis*	7
Frog	*Rana pipiens*	13
Fruit fly	*Drosophila melanogaster*	4
Garden onion	*Allium cepa*	8
Garden pea	*Pisum sativum*	7
Grasshopper	*Melanoplus differentialis*	12
Green alga	*Chlamydomonas reinhardi*	18
Horse	*Equus caballus*	32
House fly	*Musca domestica*	6
House mouse	*Mus musculus*	20
Human	*Homo sapiens*	23
Jimson weed	*Datura stramonium*	12
Mosquito	*Culex pipiens*	3
Pink bread mold	*Neurospora crassa*	7
Potato	*Solanum tuberosum*	24
Rhesus monkey	*Macaca mulatta*	21
Roundworm	*Caenorhabditis elegans*	6
Silkworm	*Bombyx mori*	28
Slime mold	*Dictyostelium discoidium*	7
Snapdragon	*Antirrhinum majus*	8
Tobacco	*Nicotiana tabacum*	24
Tomato	*Lycopersicon esculentum*	12
Water fly	*Nymphaea alba*	80
Wheat	*Triticum aestivum*	21
Yeast	*Saccharamyces cerevisiae*	16

population of members of the same species, many different alleles of the same gene may exist.

The concepts of haploid number, diploid number, and homologous chromosomes may be related to the process of meiosis. During the formation of gametes or spores, meiosis converts the diploid number of chromosomes to the haploid number. As a result, haploid gametes or spores contain precisely one member of each homologous pair of chromosomes—that is, one complete set. Following fusion of two gametes in fertilization, the diploid number is reestablished. The zygote contains two complete sets of chromosomes. The constancy of genetic material is thus maintained from generation to generation.

There is one important exception to the concept that chromosomes exist in physically identical homologous pairs. In many species, one pair, the **sex-determining chromosomes**, is often not homologous in size, centromere placement, arm ratio, or genetic content. For example, in humans, males contain one Y chromosome in addition to one X chromosome (Figure 2–4), whereas females carry two homologous X chromosomes. The X and Y chromosomes are not strictly homologous. The Y is considerably smaller and lacks most of the gene sites contained on the X. Nevertheless, in meiosis they behave as homologs so that gametes produced by males receive either one X or one Y chromosome.

Mitosis and Cell Division

The process of mitosis is critical to all eukaryotic organisms. In some that are single-celled, such as protozoans and some fungi and algae, mitosis, as a part of cell division, provides the basis for **asexual reproduction**. Multicellular diploid organisms begin life as single-celled fertilized eggs called **zygotes**. The mitotic activity of the zygote and the subsequent daughter cells is the foundation for development and growth of the organism. In adult organisms, mitotic activity is prominent in wound healing and other forms of cell replacement in certain tissues. For example, the epidermal skin cells of humans are continuously being sloughed off and replaced. Cell division is also critical to the continuous production of reticulocytes, which shed their nuclei and replenish the supply of red blood cells in vertebrates. In abnormal situations, somatic cells may exhibit uncontrolled cell divisions, resulting in cancer.

Usually, following cell division, the initial size of new daughter cells is approximately one-half the size of their parent cell. However, the nucleus of each new cell is not appreciably smaller than the nucleus of the original cell. Quantitative measurements of DNA confirm that there are equivalent amounts of genetic material in the parental and daughter nuclei.

The process of cytoplasmic division is called **cytokinesis**. The division of cytoplasm requires a mechanism that results in a partitioning of the volume into two parts, followed by the enclosure of both new cells within a distinct plasma membrane. Cytoplasmic organelles either replicate themselves, arise from existing membrane structures, or are synthesized *de novo* (anew) in each cell. The subsequent proliferation of these structures aids in reconstituting the cytoplasm of daughter cells.

Nuclear division **(karyokinesis)**, where the genetic material is partitioned into daughter cells, is more complex than cytokinesis and requires greater precision. The chromosomes must first be exactly replicated and then accurately partitioned into daughter cells. The end result is the production of two daughter cells, each with a chromosome composition identical to the parent cell.

Interphase and the Cell Cycle

Many cells undergo a continuous alternation between division and nondivision. The interval between each mitotic division is called **interphase**. It was once thought that the biochemical activity during interphase was devoted solely to the cell's growth and its normal function. However, we now know that another biochemical step critical to the ensuing mitosis occurs during interphase: *the replication of the DNA of each chromosome*. Occurring well before the cell enters mitosis, this period during which DNA is synthesized is called the **S phase**. The initiation and completion of synthesis can be detected by monitoring the incorporation of radioactive precursors into DNA.

Investigations of this nature have demonstrated two periods during interphase when no DNA synthesis occurs, one before and one after S phase. These are designated **G1 (gap I)** and **G2 (gap II)**, respectively. During both of these periods, as well as during S, intensive metabolic activity, cell growth, and cell differentiation occur. By the end of G2, the volume of the cell has roughly doubled, DNA has been replicated, and mitosis **(M)** is initiated. Following mitosis, continuously dividing cells then repeat this cycle (G1, S, G2, M) over and over. The stages of the **cell cycle** are illustrated in Figure 2–5.

Much is known about the cell cycle based on *in vitro* (test tube) studies. When grown in culture, many cell types in different organisms traverse the complete cycle in about 16 hours. The actual process of mitosis occupies only a small part of the overall cycle, usually about an hour. The length of the S and G2 stages of interphase are fairly consistent among different cell types. Most variation is seen in the length of time spent in the G1 stage. Figure 2–6 illustrates the relative length of these periods in a typical cell.

G1 is of great interest in the study of cell proliferation and its control. At a point late in G1, all cells follow one of two paths. They either withdraw from the cycle, become quiescent, and are said to enter the **G0 stage**, or they become committed to initiate DNA synthesis and complete the cycle. The time when this decision is made has been referred to as the **G1 checkpoint**. Cells that

■ Figure 2–5

Diagrammatic representation of the stages comprising an arbitrary cell cycle. Following mitosis (M), cells initiate a new cycle (G1). Cells may become nondividing (G0) or continue through G1, where they become committed to begin DNA synthesis (S) and complete the cycle (G2 and M). Following mitosis, two daughter cells are produced.

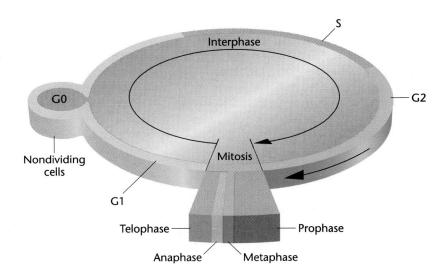

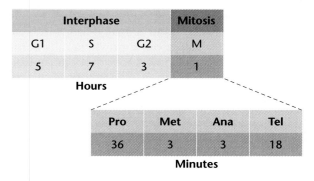

Interphase			Mitosis
G1	S	G2	M
5	7	3	1

Hours

Pro	Met	Ana	Tel
36	3	3	18

Minutes

■ Figure 2–6 The time spent in each phase of one complete cell cycle of a human cell in culture. Times vary according to cell types and conditions.

enter G0 remain viable and metabolically active but are nonproliferative. Cancer cells apparently avoid entering G0 or pass through it quickly. Other cells enter G0 and never reenter the cell cycle. Still others can remain quiescent in G0, but they may be stimulated to return to G1, reentering the cycle.

Cytologically, interphase is characterized by the absence of visible chromosomes. Instead, the nucleus is filled with chromatin that has formed as the chromosomes have unfolded and uncoiled following the previous mitosis. Such a nucleus is depicted diagramatically in Figure 2–7(a).

summary

Prophase

Once the G1, S, and G2 stages of interphase are completed, mitosis is initiated. Mitosis is a dynamic period of vigorous and continual activity. For discussion purposes, the entire process is subdivided into discrete phases, and specific events are assigned to each stage. These stages, in order of occurrence, are **prophase**, **prometaphase**, **metaphase**, **anaphase**, and **telophase**. Each of these stages is depicted diagramatically in Figure 2–7. A photograph of each stage is also shown in Figure 2–8.

Often, over half of mitosis is spent in prophase, a stage characterized by several significant activities. One of the early events in prophase of all animal cells involves the migration of two pairs of centrioles to opposite ends of the cell. These structures are found just outside the nuclear envelope in an area of differentiated cytoplasm called the **centrosome**. It is thought that each pair of centrioles consists of one mature unit and a smaller, newly formed centriole.

The direction of migration of the centrioles is such that two poles are established at opposite ends of the cell. This creates an axis along which chromosomal separation occurs. Following their migration, the centrioles are responsible for the organization of cytoplasmic microtubules into a series of **spindle fibers** that run between the poles of the cell.

Interestingly, cells of most plants (there are a few exceptions), fungi, and certain algae seem to lack centrioles. Spindle fibers are nevertheless apparent during mitosis. Therefore, centrioles are not universally responsible for the organization of spindle fibers.

As the centrioles migrate, the nuclear envelope begins to break down and gradually disappears. In a similar fashion, the nucleolus disintegrates within the nucleus. While these events are taking place, the diffuse chromatin begins to condense, continuing until distinct threadlike structures, or chromosomes, become visible. It becomes apparent near the end of prophase that each chromosome is actually a double structure split longitudinally except at a single point of constriction, the **centromere** (see Figure 2–4). The two parts of each chromosome are called **chromatids**. Because the DNA contained in each pair of chromatids represents the duplication of a single chromosome, these chromatids are genetically identical. Therefore, they are called **sister chromatids**. In humans, with a diploid number of 46, a cytological preparation of late prophase will reveal 46 chromosomes randomly distributed in the area formerly occupied by the nucleus.

In addition, spindle fibers form between the centrioles, which are now at opposite ends (poles) of the cell. The nucleolus and nuclear envelope are no longer visible, and sister chromatids are apparent. Part (b) of Figures 2–7 and 2–8 illustrate prophase.

Prometaphase and Metaphase

The distinguishing event of the ensuing stages is the migration of each chromosome, led by the centromeric region, to the equatorial plane. In some descriptions the term **metaphase** is applied strictly to the chromosome configuration following migration. In such descriptions, **prometaphase** refers to the period of chromosome movement, as depicted in part (c) of Figures 2–7 and 2–8. The equatorial plane, also referred to as the **metaphase plate**, is the midline region of the cell, a plane that lies perpendicular to the axis established by the spindle fibers.

Migration is made possible by the binding of spindle fibers to a structure associated with the centromere of each chromosome called the **kinetechore**. This structure, consisting of multilayered plates, forms on opposite sides of each centromere, intimately associating with the two sister chromatids of each chromosome. Once attached to microtubules making up the spindle fibers, the sister chromatids are now ready to be pulled to opposite poles during the ensuing anaphase stage.

At the completion of metaphase, each centromere is aligned at the plate with the chromosome arms extending outward in a random array. This configuration is shown in part (d) of Figures 2–7 and 2–8.

Mitosis

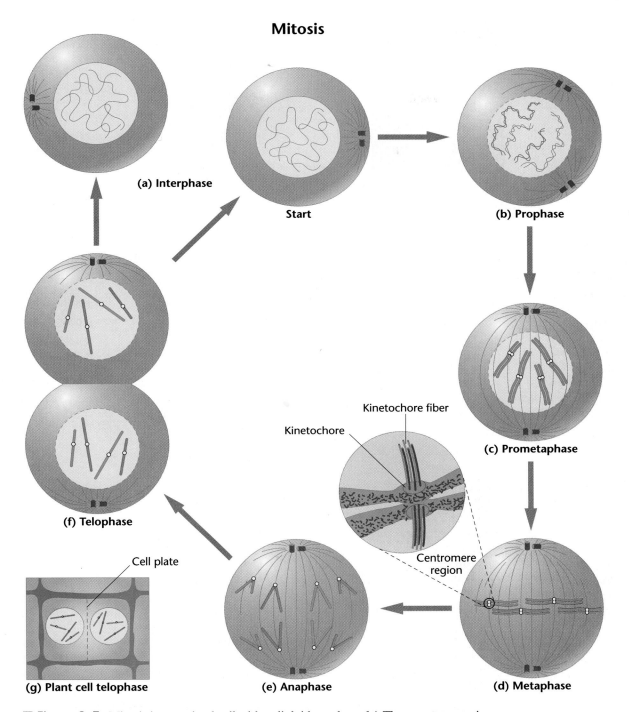

(a) Interphase

Start

(b) Prophase

(c) Prometaphase

Kinetochore fiber

Kinetochore

(f) Telophase

Centromere region

Cell plate

(g) Plant cell telophase

(e) Anaphase

(d) Metaphase

■ Figure 2–7 Mitosis in an animal cell with a diploid number of 4. The events occurring in each stage are described in the text. Of the two homologous pairs of chromosomes, one contains longer, metacentric members and the other shorter, submetacentric members. The maternal chromosomes and the paternal chromosomes are shown in different colors. The insert (g), showing the telophase stage in a plant cell, illustrates the formation of the cell plate and lack of centrioles.

(a) Heamanthus flower

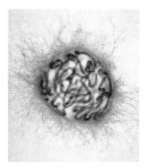

(b) Prophase

(c) Prometaphase

(g) Late telophase

(f) Early telophase

(e) Anaphase

(d) Metaphase

■ Figure 2–8   Light micrographs illustrating the stages of mitosis depicted in Figure 2–7. These stages are derived from the flower of *Haemanthus*, shown in the part (a).

Anaphase

Events critical to chromosome distribution during mitosis occur during the relatively brief stage, **anaphase**. During this phase, sister chromatids of each chromosome *disjoin* (separate) from each other and migrate to opposite ends of the cell. For complete disjunction to occur, each centromeric region must be divided into two. This event signals the initiation of anaphase. Once it has occurred, each chromatid is now referred to as a **daughter chromosome**. Depending on the location of the centromere along the chromosome, different shapes are assumed during migration as the centromeres lead the way to opposite poles.

As the chromosomes begin their migration, the spindle elongates, extending the distance between the two centrosome regions. At the completion of anaphase, the chromosomes have migrated to the opposite poles of the cell. The steps occurring during anaphase are criti-

cal in providing each subsequent daughter cell with an identical set of chromosomes. In human cells there would now be 46 chromosomes at each pole, one from each pair of sister chromatids. Part (e) of Figures 2–7 and 2–8 illustrates anaphase prior to its completion.

Telophase

Telophase is the final stage of mitosis. At its beginning, there are two complete sets of chromosomes, one at each pole. The most significant event is **cytokinesis**, the division or partitioning of the cytoplasm. Cytokinesis is essential if two new cells are to be produced from one. The mechanism differs greatly in plant and animal cells. in plant cells, a **cell plate** is synthesized and laid down across the region of the metaphase plate. Animal cells, however, undergo a constriction of the cytoplasm in much the same way a loop of string might be tightened around the middle of a balloon. The end result is the same: Two distinct cells are formed.

It is not surprising that the process of cytokinesis varies among cells of different organisms. Plant cells, which are more regularly shaped and structurally rigid, require a mechanism for the deposition of new cell wall material around the plasma membrane. The cell plate, laid down during telophase, becomes the **middle lamella**. Subsequently, the primary and secondary layers of the cell wall are deposited between the cell membrane and middle lamella on both sides of the boundary between the two daughter cells. In animals, complete constriction of the cell membrane produces the **cell furrow** characteristic of newly divided cells.

Other events necessary for the transition from mitosis to interphase are initiated during late telophase. They represent a general reversal of those that occurred during prophase. In each new cell, the chromosomes begin to uncoil and become diffuse chromatin once again, while the nuclear envelope re-forms around them. The nucleolus gradually reforms and becomes visible in the nucleus during early interphase. The spindle fibers also disappear. Telophase in animal and plant cells is illustrated in part (f) and part (g), respectively, of Figures 2–7 and 2–8.

Cell Cycle Control

Having provided a detailed account of the events characterizing each step of the cell cycle, it is important to provide a brief overview of cell cycle control. Some cells continually traverse the cell cycle and remain mitotically active. Other cells permanently arrest in the G0 stage and do not divide. Still others are arrested at a point in the cell cycle, but may be stimulated to reenter the cycle and divide. These observations suggest that cellular mechanisms exist that regulate the cell cycle.

That these mechanisms are under genetic control has been demonstrated by the discovery of mutations that disrupt that control. Many mutations are now known that exert their effect at various stages of the cell cycle. First discovered in yeast but now evident in all organisms, including humans, such mutations were originally designated as *cdc* **mutations (cell division cycle muta-**

tions). The study of these mutations has established that during the cell cycle, at least three major **checkpoints** exist, where the cell is monitored or "checked" before it can proceed to the next stage of the cycle.

The products of many of these genes are enzymes called *cdc* **kinases** that can add phosphates to other proteins. They serve as "master control" molecules that work in conjunction with proteins called **cyclins.** These kinases phosphorylate cyclins and influence their activity at the cell cycle checkpoints. These activities thus regulate the cell cycle. When a *cdc* kinase works in conjunction with a cyclin, it is called a **Cdk protein**, for **Cy-clin-*d*ependent *k*inase protein**.

Figure 2–9 identifies the location of three "checkpoints" within the cell cycle. The first is the **G1/S checkpoint**, which monitors the size that the cell has achieved following the previous mitosis, and whether the DNA has been damaged. If the cell has not achieved an adequate size, or if the DNA has been damaged, further progress through the cycle is arrested until these conditions are "corrected," so to speak. If both conditions are initially "normal," then the checkpoint is traversed and the cell proceeds to the S phase of the cycle.

The second important checkpoint is the **G2/M checkpoint**, where physiological conditions in the cell are monitored prior to entering mitosis. If DNA replication or repair to any DNA damage has not been completed, the cell cycle is arrested until these processes are completed. The final checkpoint occurs during mitosis and is called the **M checkpoint**. Here, both the successful formation of the spindle fiber system and the attachment of spindle fibers to the kinetechores associated with the centromeres are monitored. If spindle fibers are not properly formed or attachment is inadequate, mitosis is arrested.

The importance of cell cycle control and these checkpoints can be illustrated by considering what happens when this regulatory system is impaired. If, for example, a cell has incurred damage to its DNA and is allowed to proceed through the cell cycle, it may begin a series of uncontrolled cell divisions—precisely the definition of a cancerous cell. As we saw above, such a damaged cell would normally be arrested at either the G1/S or the G2/M checkpoint.

An interesting related finding involves the protein product of the *p53* gene in humans and its involvement

■ Figure 2–9 Illustration of the three major checkpoints within the cell cycle that regulate its progress.

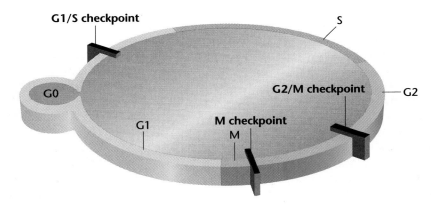

during scrutiny at the G1/M checkpoint. This protein functions during the regulation of **apoptosis**, the genetic process whereby programmed cell death occurs. When the normal *p53* gene product is present, a proliferative cell that has incurred severe damage to its DNA will be targeted for programmed cell death at the G1/M checkpoint and, thus, effectively removed from the cell population. However, if the *p53* gene has mutated, resulting in abnormal function of the *p53* gene product, the damaged cell may proceed through the checkpoint and continue to proliferate in an uncontrolled manner.

In fact, a high percentage of human cancers has been found to contain mutations in the *p53* gene. These include a wide variety of cancers, including colon, breast, lung, and bladder malignancies. In the language of cancer genetics, *p53* is referred to as a **tumor-supressor gene**.

The role of systems that control the cell cycle in cancer has stirred great interest in genetic research. We shall return to this topic, which will be discussed in much greater detail in Chapter 19.

Meiosis and Sexual Reproduction

The process of meiosis, unlike mitosis, reduces the amount of genetic information. Whereas in diploid cells, mitosis produces daughter cells with a full diploid complement, meiosis produces haploid gametes or spores with only one set of chromosomes. During sexual reproduction, gametes then combine in fertilization to reconstitute the diploid complement found in parental cells. The process itself must be very specific because, in order to ensure genetic continuity from generation to generation, each gamete must receive precisely one member of each homologous pair of chromosomes.

Sexual reproduction also ensures genetic variety among members of a species. As you study meiosis, it will become apparent that this process results in gametes with many unique combinations of maternally and paternally derived chromosomes among the haploid complement. With a tremendous genetic variation among the gametes derived from two genetically unique individuals, a large number of chromosome combinations are possible at fertilization. Furthermore, we will see that the meiotic event referred to as **crossing over** results in genetic exchange between members of each homologous pair of chromosomes. This creates intact chromosomes that are mosaics of the maternal and paternal homologs from which they are derived. This has the effect of further enhancing the potential genetic variation in gametes and the offspring derived from them. Sexual reproduction, therefore, reshuffles the genetic material, producing offspring that often differ greatly from either parent. This process constitutes the major form of genetic recombination within eukaryotic species.

An Overview of Meiosis

We have already established what must be accomplished during meiosis. Before systematically considering the stages of this process, we will briefly describe the genetic events that occur as diploid cells give rise to haploid gametes or spores. This overview is depicted in Figure 2–10.

Unlike mitosis, in which each paternally and maternally derived member of any given homologous pair of chromosomes behaves autonomously during division, in meiosis homologous chromosomes pair together; that is, they **synapse**. Each synapsed structure, called a **bivalent**, gives rise to a unit, the **tetrad**, that consists of four chromatids. The presence of four chromatids demonstrates that both chromosomes have duplicated. In order to achieve haploidy, two divisions are necessary. In the first, described as a **reductional division** (because the number of chromosomes, each representing one chromosome, is "reduced" by one-half), components of each tetrad representing the two homologs separate, yielding two **dyads**. Each dyad is composed of two sister chromatids joined at a common centromere. During the second division, described as **equational** (because the number of centromeres remains "equal"), each dyad splits into two **monads**, each composed of one chromosome from each tetrad. Thus, the two divisions may potentially produce four haploid cells. As in mitosis, meiosis is a continuous process. We give names to each stage of division only for convenience in describing and learning about the process.

The First Meiotic Division: Prophase I

From a genetic standpoint, there are two critical events during prophase I. First, members of each homologous pair of chromosomes somehow find one another and undergo synapsis. Second, the exchange process referred to above as crossing over occurs between synapsed homologs. Because of the importance of these genetic events, this stage of meiosis has been subjected to extensive investigation. Most recently, the study of yeasts has provided a model for our understanding of these critical events. As we discuss them, you should be aware that replication of DNA has preceded meiosis, even though chromosomes do not immediately reveal their duplication. Recall that this also occurs in mitosis. The first meiotic prophase is complex and has been further subdivided into five substages: leptonema, zygonema, pachynema, diplonema,* and diakinesis [Figure 2–11(a)].

Leptonema   During the **leptotene stage**, the interphase chromatin material begins to condense, and the

---

*These are the noun forms of these stages. The adjective forms (leptotene, zygotene, pachytene, and diplotene) are also used in the text.

■ Figure 2–10 Summary of the events that each pair of homologous chromosomes undergoes during meiosis, converting the diploid (2*n*) number to the haploid (*n*) number.

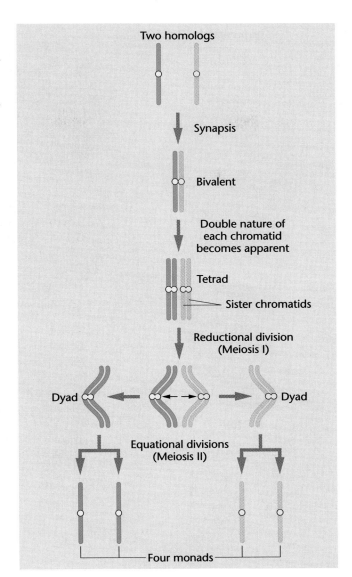

chromosomes, although still extended, become visible. Along each chromosome are **chromomeres**, localized condensations that resemble beads on a string. Recent evidence suggests that it is during leptonema that a process called **homology search** begins, one that precedes initial pairing of homologs.

Zygonema The chromosomes continue to shorten and thicken during the **zygotene stage**. During the process of homology search, homologous chromosomes undergo initial alignment with one another. This alignment is considered to be a **rough pairing**, and by the end of zygonema it is complete. In yeast, homologs are separated by about 300 nm, and near the end of zygonema, structures referred to as **lateral elements** are visible between paired homologs. As meiosis proceeds, the overall length of the lateral elements increases and a more extensive ultrastructural component, the **synaptonemal complex**, begins to form between the homologs.

At the completion of zygonema, the paired homologs represent structures referred to as **bivalents**. Although both members of each bivalent have already replicated their DNA, it is not yet visually apparent that each member is a double structure. The number of bivalents in each species is equal to the haploid (*n*) number.

Pachynema In the transition from the zygotene to the **pachytene stage**, coiling and shortening of chromosomes continues, and further development of the synaptonemal complex occurs between the two members of each bivalent. This leads to a more intimate pairing referred to as **synapsis**. Compared to the rough pairing characteristic of yeast zygonema, homologs are now separated by only 100 nm.

During pachynema, it first becomes evident that each homolog is a double structure, providing visual evidence of the earlier replication of the DNA of each chromosome. Thus, each bivalent contains four

Meiosis I

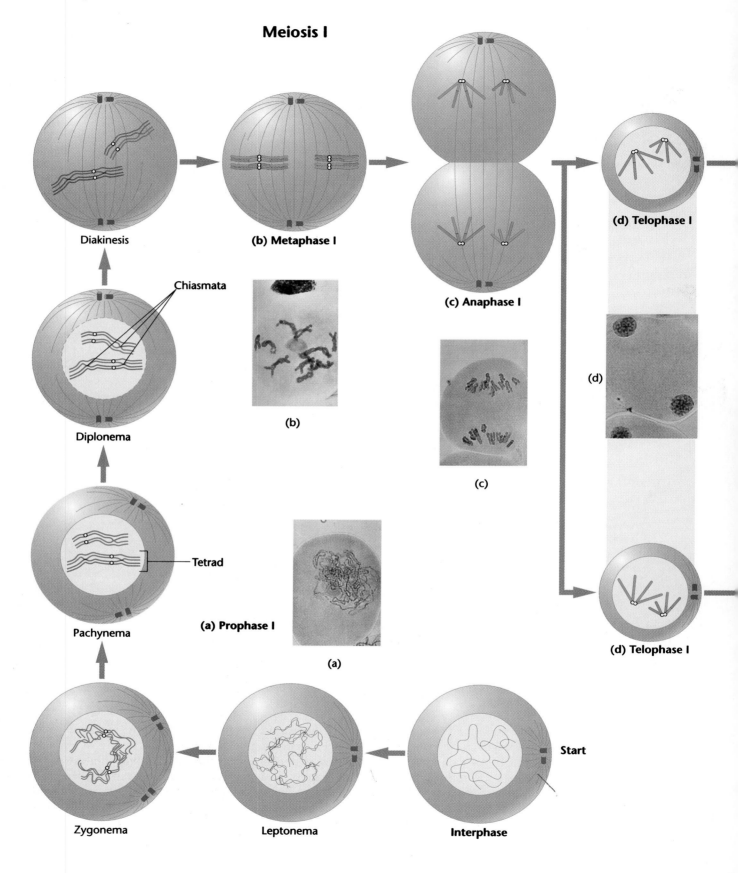

Diakinesis

(b) Metaphase I

(c) Anaphase I

(d) Telophase I

Chiasmata

Diplonema

(b)

(d)

(c)

Tetrad

Pachynema

(a) Prophase I

(a)

(d) Telophase I

Zygonema

Leptonema

Interphase

Start

Meiosis II

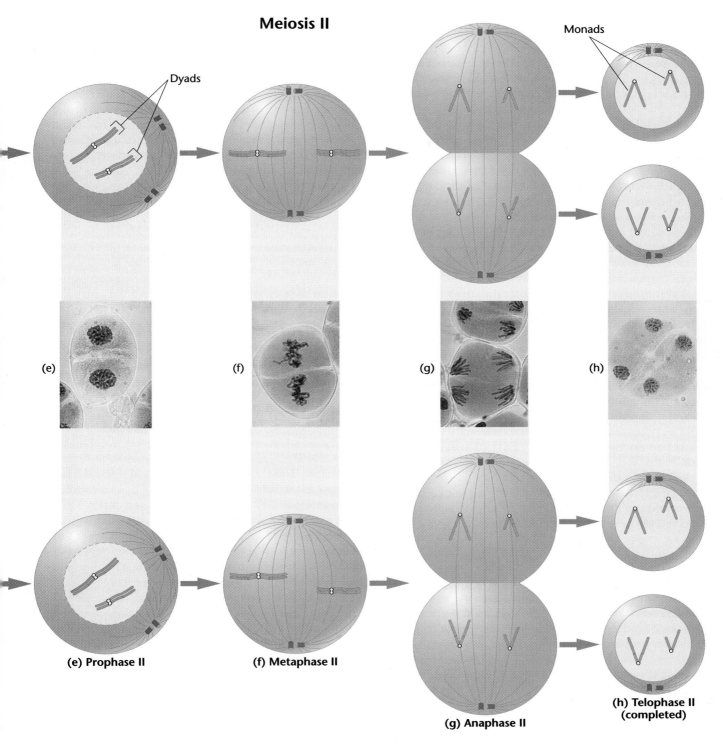

■ Figure 2–11 (left and above) A diagrammatic representation of the major events occurring during meiosis in a male animal with a diploid number of 4. The same chromosomes depicted in Figure 2–7 are followed, as described in the legend there. Note that the combination of chromosomes contained in the cells produced following telophase II is dependent on the random alignment of each tetrad and dyad on the equatorial plate during metaphase I and metaphase II. Several other combinations, which are not shown, can be formed. The events depicted in this figure should be correlated with the description in the text. Light micrographs illustrate many of the stages of meiosis that are diagrammed.

members, called **chromatids**. As in mitosis, replicates are called **sister chromatids**, while chromatids from maternal vs. paternal members of a homologous pair are called **nonsister chromatids**. The four-membered structure is also referred to as a **tetrad**, and each tetrad contains two pairs of sister chromatids.

Diplonema During observation of the ensuing **diplotene stage**, it is even more apparent that each tetrad consists of two pairs of sister chromatids. Within each tetrad, each pair of sister chromatids begins to separate. However, one or more areas remain in contact where chromatids are intertwined. Each such area, called a **chiasma** (pl. **chiasmata**), is thought to represent a point where nonsister chromatids have undergone genetic exchange through the process referred to above as **crossing over**. Although the physical exchange between chromosome areas occurred during the previous pachytene stage, the result of crossing over is visible only when the duplicated chromosomes begin to separate. Crossing over is an important source of genetic variability. New combinations of genetic material are formed during this process.

Diakinesis The final stage of prophase I is **diakinesis**. The chromosomes pull farther apart, but nonsister chromatids remain loosely associated via the chiasmata. As separation proceeds, the chiasmata move toward the ends of the tetrad. This process, called **terminalization**, begins in late diplonema, and is completed during diakinesis. During this final period of prophase I, the nucleolus and nuclear envelope break down, and the two centromeres of each tetrad become attached to the recently formed spindle fibers.

Metaphase I, Anaphase I, and Telophase I

Following the first meiotic prophase stage, steps similar to those of mitosis occur. In the **metaphase stage of the first division**, the chromosomes have maximally shortened and thickened. The terminal chiasmata of each tetrad are visible and appear to be the only factor holding the nonsister chromatids together. Each tetrad interacts with spindle fibers, facilitating movement to the metaphase plate.

During the first division, a single centromere holds each pair of sister chromatids together. It does *not* divide. At the **anaphase stage of the first division**, one-half of each tetrad (one pair of sister chromatids—called a dyad) is pulled toward each pole of the dividing cell. This separation process is the physical basis of what we refer to as **disjunction**, meaning the separation of chromosomes from one another. Occasionally, errors in meiosis occur and separation is not achieved. The term **nondisjunction** describes such an error, the consequences of which will soon be discussed. At the completion of the normal anaphase I, a series of dyads equal to the haploid number is present at each pole.

If no crossing over had occurred in the first meiotic prophase, each dyad at each pole would consist solely of either paternal or maternal chromatids. However, the exchanges produced by crossing over create mosaic chromatids of paternal and maternal origin. The alignment of each tetrad prior to this first anaphase stage is random. One-half of each tetrad is pulled to one or the other pole at random, and the other half moves to the opposite pole. This random **segregation of dyads** is the basis for the Mendelian principle of **independent assortment**, which we will discuss in Chapter 3. You may wish to return to this discussion when you study this principle.

In many organisms, **telophase of the first meiotic division** reveals a nuclear membrane forming around the dyads. Next, the nucleus enters into a short interphase period—sometimes called **interkinesis**. In other cases, the cells go directly from the first anaphase into the second meiotic division. If an interphase period occurs, the chromosomes do not replicate because they already consist of two chromatids. In general, meiotic telophase is much shorter than the corresponding stage in mitosis.

The Second Meiotic Division

A second division of the sister chromatids making up each dyad is essential if each gamete or spore is to receive only one chromatid from each original tetrad. During **prophase II**, each dyad is composed of one pair of sister chromatids attached by a common centromere. During **metaphase II**, the centromeres are directed to the equatorial plate. Then, the centromeres divide, and during **anaphase II** the sister chromatids of each dyad are pulled to opposite poles. Because the number of dyads is equal to the haploid number, **telophase II** reveals one member of each pair of homologous chromosomes present at each pole. Each chromosome is referred to as a **monad**. At the conclusion of meiosis, not only has the haploid state been achieved, but if crossing over has occurred, each monad is a combination of maternal and paternal genetic information. As a result, the offspring produced by any gamete will receive from it a mixture of genetic information originally present in his or her grandparents. Potentially, following cytokinesis in telophase II, four haploid gametes may result from a single meiotic event.

Spermatogenesis and Oogenesis

Although events that occur during the meiotic divisions are similar in all cells that participate in gametogenesis, there are certain differences between the production of a male gamete (spermatogenesis) and a female gamete (oogenesis) in most animal species.

Spermatogenesis takes place in the testes, the male reproductive organs. The process begins with the expanded growth of an undifferentiated diploid germ cell called a spermatogonium. This cell enlarges to become a **primary spermatocyte**, which undergoes the first meiotic division.

The products of this division are called **secondary spermatocytes**. Each secondary spermatocyte contains the haploid number of dyads. The secondary spermatocytes then undergo the second meiotic division, and each of these cells produces two haploid **spermatids**. Spermatids go through a series of developmental changes, **spermiogenesis**, and become highly specialized, motile **spermatozoa** or **sperm**. All sperm cells produced during spermatogenesis receive equal amounts of genetic material and cytoplasm. Figure 2–12 summarizes these steps.

In animal **oogenesis**, the formation of **ova** (singular: **ovum**), or eggs, occurs in the ovaries, the female reproductive organs. The daughter cells resulting from the two meiotic divisions receive equal amounts of genetic material, but they do *not* receive equal amounts of cytoplasm. Instead, during each division, almost all the cytoplasm of the **primary oocyte**, itself derived from the **oogonium**, is concentrated in one of the two daughter cells. The concentration of cytoplasm is necessary because the function of the mature ovum is to nourish the developing embryo following fertilization.

During the first meiotic anaphase in oogenesis, the tetrads of the primary oocyte separate, and the dyads move toward opposite poles. During the first telophase, the dyads present at one pole are pinched off with very little surrounding cytoplasm to form **the first polar body**. The other daughter cell produced by this first meiotic division contains most of the cytoplasm and is called the **secondary oocyte**. The first polar body may or may not divide again to produce two small haploid cells. The mature ovum is produced from the secondary oocyte during the second meiotic division. During this division, the cytoplasm of the secondary oocyte again divides unequally, producing an **ootid** and a **second polar body**. The ootid then differentiates into the **mature ovum**. Figure 2–12 illustrates the steps leading to formation of the mature ovum and polar bodies.

Unlike the continuous divisions of spermatogenesis, the two meiotic divisions of oogenesis may not be continuous. In some animal species they may follow each other directly. In others, including the human species, the first division (meiosis I) of all oocytes is initiated in

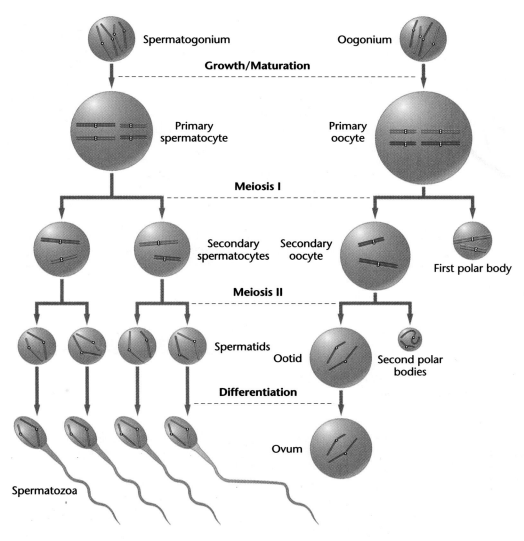

■ Figure 2–12 Spermatogenesis and oogenesis in animal cells.

the embryonic ovary, but arrests in prophase I. Many years later, the process resumes in each oocyte just prior to ovulation. The second division (meiosis II) is completed only after fertilization, if it occurs.

The Significance of Meiosis

The process of meiosis is critical to the successful sexual reproduction of all diploid organisms. It is the mechanism by which the diploid amount of genetic information is reduced to the haploid amount. In animals, meiosis leads to the formation of gametes, while in plants haploid spores are produced, which in turn lead to the formation of haploid gametes.

Each diploid organism contains its genetic information in the form of homologous pairs of chromosomes. Each pair consists of one member derived from the maternal parent and one from the paternal parent. Following meiosis, haploid cells potentially contain either the paternal or maternal representative of each homolo-gous pair of chromosomes. However, the process of crossing over, which occurs in the first meiotic prophase stage, reshuffles the genetic information. Crossing over occurs between the maternal and paternal members of each homologous pair, which then assort independently into gametes. This results in the great amounts of genetic variation in gametes.

It is important to elaborate briefly on the significant role that meiosis plays in the life cycles of fungi and plants. In many fungi, which we shall use as an illustration, the predominant stage of the life cycle consists of haploid vegetative cells. They arise through meiosis and proliferate by mitotic cell division. In multicellular plants, the life cycle alternates between the diploid **sporophyte stage** and the haploid **gametophyte stage** (Figure 2–13). While one or the other predominates in different plant groups during this "alternation of generations," the processes of meiosis and fertilization constitute the "bridge" between the sporophyte and gametophyte generations. Therefore, meiosis is an essential component of the life cycle of plants.

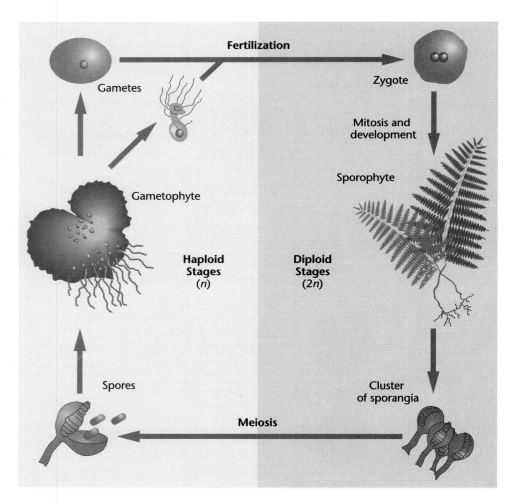

■ Figure 2–13   Alternation of generations between the diploid sporophyte (2*n*) and the haploid gametophyte (*n*) in a multicellular plant. The process of meiosis bridges the two phases of the life cycle.

The Cytological Origin of the Mitotic Chromosome

Thus far in this chapter, we have focused on mitotic and meiotic chromosomes, emphasizing their behavior during cell division and gamete formation. Initially, biologists knew about these chromosomes only from observations made with the light microscope. Although chromosomes are invisible during interphase, they appear during the prophase stage of mitosis and meiosis. How this happens has long intrigued geneticists. Based primarily on studies using the electron microscope, we are now quite clear as to why these chromosomes become visible only during mitosis and meiosis.

During interphase, only dispersed chromatin fibers are present in the nucleus. **Chromatin** consists of DNA and associated proteins, particularly the proteins called **histones**. It is now believed that during interphase, starting in G1, mitotic chromosomes unwind to form these long chromatin fibers. Figure 2–14(a) illustrates chromatin fibers viewed under the electron microscope. This physical arrangement facilitates transcription of RNA and DNA replication.

Once mitosis begins, the fibers coil and fold up, condensing into visible chromosomes as seen in Figure 2–14(b). Electron microscopic observations of such chromosomes reveal the chromatin fibers looping in and out of the mitotic chromosome, as seen in Figure 2–14(c). Coiling is extensive. It is estimated that a 5000-fold contraction occurs in the length of DNA as the chromatin fiber condenses into a mitotic chromosome! This process must indeed be extremely precise, given the highly ordered nature and consistent appearance of mitotic chromosomes in all eukaryotes.

Figure 2–14(d) summarizes the current interpretation of the mitotic chromosome and its various components. In a later chapter, after we have provided a more thorough description of DNA structure, we will return to this topic and explore more thoroughly the molecular basis of the chromatin fiber.

Specialized Chromosomes

We conclude this chapter by discussing two unique types of chromosomes visible under the light microscope. These examples demonstrate the specialized forms that chromosomes may achieve in addition to the mitotic-like structures discussed above. Both specialized types were studied extensively long before we understood how chromosomes "appear" during mitosis.

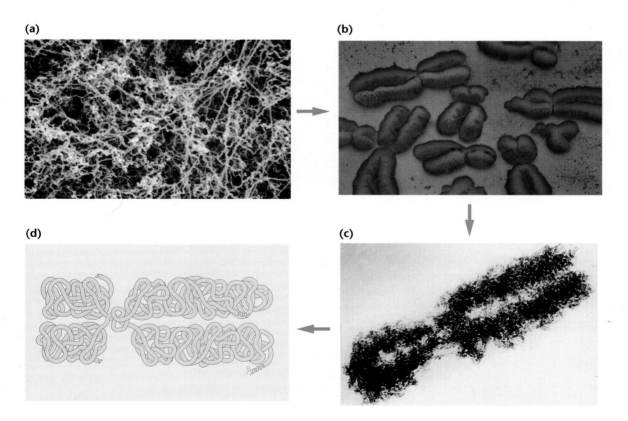

(a) **(b)** **(d)** **(c)**

■ Figure 2–14 A comparison of: (a) the chromatin fibers characteristic of the interphase nucleus with (b) and (c) metaphase chromosomes that are derived from chromatin during mitosis. Parts (a) and (c) are transmission electron micrographs, while part (b) is a scanning electron micrograph. Part (d) depicts diagrammatically the mitotic chromosome and its various components, illustrating how chromatin is condensed into it.

Polytene Chromosomes

Cells from a variety of organisms contain giant **polytene chromosomes** that are 200–600 μm in length. First observed by E. G. Balbiani in 1881, they are found in various tissues (e.g., salivary glands, midgut, and malpighian excretory tubules) in the larvae of some flies and in several protozoans and plants. What is particularly intriguing about these chromosomes is that they can be seen in the nuclei of interphase cells!

As revealed in Figure 2–15, each polytene chromosome consists of a linear series of alternating **bands** and **interbands**. The banding pattern is distinctive for each chromosome in any given species. Following extensive study using electron microscopy and radioactive tracers, it is now clear how such giant chromosomes are formed. Their large size and distinctiveness result from the many DNA strands that comprise them. The DNA of paired homologs undergoes many rounds of replication, but newly formed replicas do not separate as they normally do in mitosis. Thus, as replication proceeds, chromosomes are created that have 1000–5000 DNA strands, which remain in precise parallel register with one another. It is apparently this arrangement of so many DNA strands that gives rise to the distinctive band pattern along the axis of the chromosome.

The relationship between the structure of these chromosomes and the genes contained within them is intriguing. The presence of bands was initially interpreted as the visible manifestation of individual genes. When it became clear that the strands present in bands undergo localized uncoiling during genetic activity, this view was further strengthened. Each uncoiling event results in what is called a **puff** because of its appearance (see Figure 2–16). Puffs are visible manifestations of gene activity (transcription) as evidenced by their high rate of incorporation of radioactively labeled RNA precursors, as assayed by autoradiography. Bands that are not extended into puffs incorporate much less radioactive precursor or none at all.

The study of polytene chromosomes in insects such as the fruit fly *Drosophila* and the midge fly *Chironomus* reveals a characteristic pattern of "puffing" during development. Depending on the time during development, specific puffs form, some regress, other puffs form, etc. This pattern of puffing is equated with genes turning on and off as development proceeds. This topic is pursued in more detail in Chapter 20.

Lampbrush Chromosomes

Another type of specialized chromosome that has provided insights into chromosomal structure is the **lampbrush chromosome**. It was given this name because its appearance is similar to the brushes used to clean lamps in the nineteenth century. Lampbrush chromosomes were first discovered in 1892 in the oocytes of sharks and are now known to be characteristic of most vertebrate oocytes as well as spermatocytes of some insects. Therefore, they are meiotic chromosomes. Most experimental work has been done with material taken from oocytes of amphibians such as *Xenopus*.

These unique chromosomes are easily isolated from primary oocytes in the diplotene stage of the first

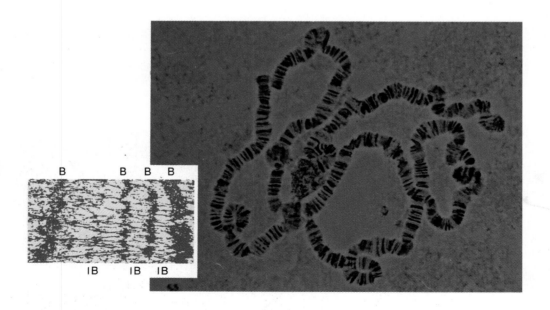

■ Figure 2–15 Polytene chromosomes derived from larval salivary gland cells of *Drosophila*. The insert depicts alternating band (B) and interband (IB) regions along the axis of these giant chromosomes.

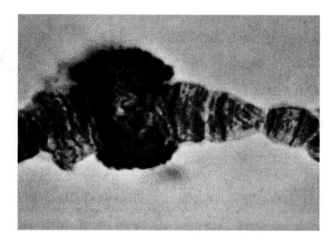
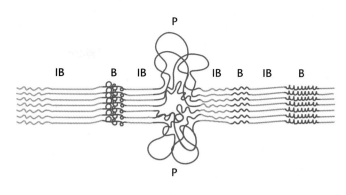

■ Figure 2–16 Photomicrograph of a puff within a polytene chromosome. The schematic representation depicts the uncoiling of strands within a band region to produce a puff (P) in polytene chromosomes. Band (B) and interband (IB) regions are also labeled.

prophase of meiosis, where they are active in directing the metabolic activities of the developing cell. As is evident in Figure 2–17(a), homologous chromosomes are held together by chiasmata. Instead of condensing, as do most meiotic chromosomes, lampbrush chromosomes remain extended, often to lengths of 500–800 μm. Later in meiosis they revert to their normal length of 15–20 μm. Thus, lampbrush chromosomes are interpreted as uncoiled and unfolded versions of the normal meiotic chromosomes.

Figure 2–17 shows two views of these structures, providing significant insights into their morphology. In Part (a), the meiotic configuration, as described above, is seen under the light microscope. In Part (b), a scanning electron microscope (SEM) reveals that the linear axis of each chromosome contains large numbers of condensed areas referred to as **chromomeres**. Emanating

from each chromomere is a pair of **lateral loops**, which give the chromosome its distinctive appearance.

Each loop is thought to be composed of one DNA double helix, while the central axis of the chromosome is made up of two DNA helices. This hypothesis is consistent with the belief that each chromosome is composed of a pair of sister chromatids. Studies using radioactive RNA precursors reveal that the loops are active in transcription, where the synthesis of RNA occurs. The lampbrush loops, in a way similar to puffs in polytene chromosomes, represent the DNA constituting genes that has been reeled out from the central chromomere axis during transcription. As with polytene chromosomes, the study of lampbrush chromosomes has provided many insights into the arrangement and function of the genetic information.

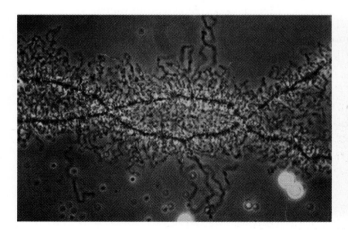
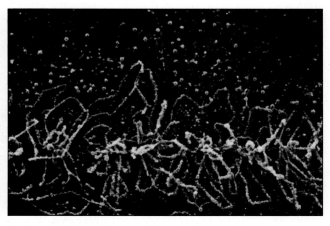

■ Figure 2–17 Lampbrush chromosomes derived from amphibian oocytes. Part (a) is a photomicrograph, while part (b) is a scanning electron micrograph.

Chapter Summary

1. The structure of cells is elaborate and complex. Many components of cells are involved directly or indirectly with genetic processes.

2. In diploid organisms, chromosomes exist in homologous pairs. Each pair shares the same size, centromere placement, and gene sites. One member of each pair is derived from the maternal parent and one is derived from the paternal parent.

3. Mitosis and meiosis are mechanisms by which cells distribute genetic information contained in their chromosomes to progeny cells in a precise, orderly fashion.

4. Mitosis, or nuclear division, is part of the cell cycle and is the basis of cellular reproduction. Daughter cells are produced that are genetically identical to their progenitor cell.

5. Mitosis is subdivided into discrete phases: prophase, prometaphase, metaphase, anaphase, and telophase. Condensation of chromatin into chromosome structures occurs during prophase. During prometaphase, chromosomes take on the appearance of double structures, each represented by a pair of sister chromatids. In metaphase, chromosomes line up on the equatorial plane of the cell. During anaphase, sister chromatids of each chromosome are pulled apart and directed toward opposite poles. Telophase completes daughter cell formation and is characterized by cytokinesis, the division of the cytoplasm.

6. The cell cycle is characteristic of all eukaryotes and is tightly regulated at three checkpoints: G1/S, S/G2, and M.

7. Meiosis, the underlying basis of sexual reproduction, results in the conversion of a diploid cell to a haploid gamete or spore. As a result of chromosome duplication and two subsequent divisions, each haploid cell receives one member of each homologous pair of chromosomes.

8. In animals, a major difference exists during meiosis in males and females. Spermatogenesis partitions cytoplasmic volume equally and produces four haploid sperm cells. Oogenesis, on the other hand, accumulates the cytoplasm around one egg cell and reduces the other haploid sets of genetic material to polar bodies. The extra cytoplasm contributes to zygote development following fertilization.

9. Meiosis results in extensive genetic variation by virtue of the exchange during crossing over between maternal and paternal chromatids and their random segregation into gametes. In addition, meiosis plays an important role in the life cycles of fungi and plants, serving as the bridge between alternating generations.

10. Mitotic chromosomes are produced as a result of the coiling and condensation of chromatin fibers characteristic of interphase.

11. Polytene and lampbrush chromosomes are examples of specialized structures that have extended our knowledge of genetic organization and function.

Key Terms

acrocentric, 20
anaphase, 24
antigen, 18
bivalent, 28
cdc gene, 27
cell cycle, 27
centriole, 18
centromere, 20
centrosome, 19
chiasma (chiasmata), 32
chromatid, 24
chromatin, 17
chromomere, 29
crossing over, 28
cyclin, 27
cytokinesis, 23
daughter chromosome, 26
diakinesis, 32
diploid, 21
diplotene, 32
dyad, 28
endoplasmic reticulum (ER), 18
endosymbiont hypothesis, 18

G0 stage, 23
G1 stage, 23
G1 checkpoint, 23
G2 stage, 23
G2/M checkpoint, 27
gamete, 17
haploid genome, 21
homologous chromosome, 19
independent assortment, 32
interphase, 23
karyokinesis, 23
karyotype, 21
kinetochore, 24
lampbrush chromosome, 17
lateral elements, 29
leptotene, 28
M checkpoint, 27
metacentric, 20
metaphase, 24
metaphase plate, 24
monad, 28
nonsister chromatids, 32
oocyte, 33

oogenesis, 33
p arm, 20
p53 gene, 27
pachytene, 29
polar body, 33
polytene chromosome, 17
prokaryote, 18
prometaphase, 24
q arm, 21
S phase, 23
segregation, 32
sex-determining chromosome, 22
sister chromatids, 21
spermatogenesis, 32
spindle fiber, 24
submetacentric, 20
synapsis, 29
synaptonemal complex, 29
telocentric, 20
telophase, 24
tetrad, 28
zygote, 22
zygotene, 29

INSIGHTS
and
SOLUTIONS

With this initial appearance of "Insights and Solutions," it is appropriate to provide a brief description of its value to you as a student. This section will precede the "Problems and Discussion Questions" in each chapter. One or more examples will be provided. Solutions to problems and answers to questions will be provided in order to illustrate approaches that are useful in genetic analysis. Our initial emphasis will be on insights that will help you arrive at correct solutions to ensuing problems.

1. In an organism with a haploid number of 3, how many individual chromosomal structures will align on the metaphase plate during (a) mitosis; (b) meiosis I; and (c) meiosis II? Describe each configuration.

Solution: (a) In mitosis, where homologous chromosomes do not synapse, there will be 6 double structures, each consisting of a pair of sister chromatids. The number of structures is equivalent to the diploid number. (b) In meiosis I, the homologs have synapsed, reducing the number of structures to 3. Each is called a tetrad and consists of two pairs of sister chromatids. (c) In meiosis II, the same number of structures exist (3), but in this case they are called dyads. Each consists of a pair of sister chromatids. When crossing over has occurred, each

chromatid may contain parts of one of its nonsister chromatids, obtained during exchange in prophase I.

2. For the chromosomes illustrated in Figure 2–11, draw all possible alignment configurations that may occur during metaphase of meiosis I.

Solution: As shown in the illustration below, there are four configurations possible when $n = 2$.

3. Describe the composition of a meiotic tetrad, as it exists during prophase I, assuming no crossing over event has occurred. What impact would a single crossover event have on this structure?

Solution: Such a tetrad contains four chromatids, existing as two pairs. Members of each pair are replicas of one another and are called sister chromatids. They are held together by a common centromere. Members of one pair are maternally derived, whereas members of the other are paternally derived. Maternal and paternal members are called nonsister chromatids. A single crossover event has the effect of exchanging a portion of a maternal and a paternal chromatid, leading to a chiasma, where the two involved chromatids overlap physically in the tetrad. The process of exchange is referred to as crossing over.

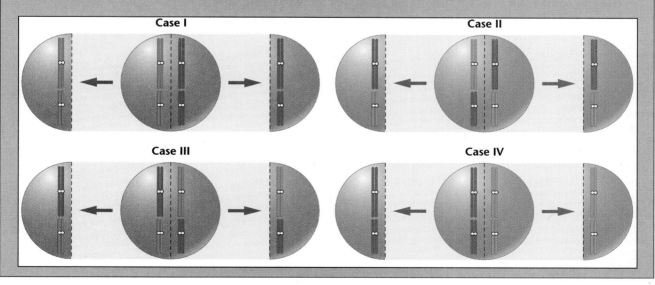

Problems and Discussion Questions

1. What role do the following cellular components play in the storage, expression, or transmission of genetic information: (a) chromatin, (b) nucleolus, (c) ribosome, (d) mitochondrion, (e) centriole, (f) centromere?

2. Discuss the concepts of homologous chromosomes, diploidy, and haploidy. What characteristics are shared between two chromosomes considered to be homologous?

3. If two chromosomes of a species are the same length and have similar centromere placements yet are not homologous, what is different about them?

4. Describe the events that characterize each stage of mitosis.

5. If an organism has a diploid number of 16, how many chromatids are visible at the end of mitotic prophase? How many chromosomes are moving to each pole during anaphase of mitosis?

6. How are chromosomes generally named on the basis of centromere placement?

7. Contrast telophase in plant and animal mitosis.

8. Describe the events of the cell cycle and the way in which it is regulated.

9. Examine Figure 2–12 showing oogenesis in animal cells. Will the genotype of the second polar body (derived from meiosis II) always be identical to that of the ootid? Why or why not?

10. Contrast the end results of meiosis with those of mitosis.

11. Define and discuss the following terms: (a) synapsis, (b) bivalents, (c) chiasmata, (d) crossing over, (e) chromomeres, (f) sister chromatids, (g) tetrads, (h) dyads, (i) monads, and (j) synaptonemal complex.

12. If an organism has a diploid number of 16 in an oocyte: (a) How many tetrads are present in the first meiotic prophase? (b) How many dyads are present in the second meiotic prophase? (c) How many monads migrate to each pole during the second meiotic anaphase?

13. Contrast spermatogenesis and oogenesis. What is the significance of the formation of polar bodies?

14. Explain why meiosis leads to significant genetic variation while mitosis does not.

15. Assume that a diploid cell contains three pairs of homologous chromosomes designated C1 C2, M1 M2, and S1 S2 and that no crossing over occurs. What possible configurations of chromosomes will be present in: (a) two daughter cells following mitosis? (b) the first meiotic metaphase? (c) haploid cells following meiosis?

16. Predict the number of different haploid cells that will occur if a fourth chromosome pair (W1 W2) is considered in addition to C, M, and S chromosomes considered in the preceding problem.

17. Describe the role of meiosis in the life cycle of a vascular plant.

18. Contrast the chromatin fiber with the mitotic chromosome. How are the two structures related?

19. Describe and compare polytene and lampbrush chromosomes.

20. Describe the role of the *p53* gene product in the regulation of the cell cycle.

Selected Readings

Alberts, B., et al. 1994. *Molecular biology of the cell,* 3rd ed. New York: Garland.

Angelier, N., et al. 1984. Scanning electron microscopy of amphibian lampbrush chromosomes. *Chromosoma* 89:243–53.

Beerman, W., and Clever, U. 1964. Chromosome puffs. *Sci. Am.* (April) 210:50–58.

Brachet, J., and Mirsky, A. E. 1961. *The cell: Meiosis and mitosis,* Vol. 3. Orlando, FL: Academic Press.

Callan, H. G. 1986. *Lampbrush chromosomes.* New York: Springer-Verlag.

DuPraw, E. J. 1970. *DNA and chromosomes.* New York: Holt, Rinehart & Winston.

Golomb, H. M., and Bahr, G. F. 1971. Scanning electron microscopic observations of surface structures of isolated human chromosomes. *Science* 171:1024–26.

Glover, D. M., Gonzalez, C., and Raff, J. W. 1993. The centrosome. *Sci. Am.* (June) 268:62–68.

Hartwell, L. H., and Karstan, M. B. 1994. Cell cycle control and cancer. *Science* 266:1821–28.

Hartwell, L .H., and Weinert, T.A. 1989. Checkpoint controls that ensure the order of cell cycle events. *Science* 246:629–34.

Hawley, R. S., and Arbel, T. 1993. Yeast genetics and the fall of the classical view of meiosis. *Cell* 72:301–303.

Mazia, D. 1961. How cell divide. *Sci. Am.* (Jan.) 205:101–20.

—————————. 1974. The cell cycle. *Sci. Am.* (Jan.) 235:54–64.

McIntosh, J.R., and McDonald, K. L. 1989. The mitotic spindle. *Sci. Am.* (Oct.) 261:48–56.

Murray, A. W., and Kirschner, M. 1993. *The cell cycle: An introduction.* New York: Oxford Univ. Press.

Prescott, D. M., and Flexer, A. S. 1986. *Cancer, the misguided cell,* 2nd ed. Sunderland, MA:Sinauer.

Swanson, C. P., Merz, T., and Young, W. J. 1981. *Cytogenetics, the chromosome in division, inheritance, and evolution,* 2nd ed. Englewood Cliffs, NJ: Prentice-Hall.

Westergard, M., and von Wettstein, D. 1972. The synaptinemal complex. *Annu. Rev. Genet.* 6:71–110.

Wheatley, D. N. 1982. *The centriole: A central enigma of cell biology.* New York: Elsevier/North-Holland Biomedical.

Mendel's garden, as seen in the 1980s.

CHAPTER OUTLINE

CHAPTER
3

Mendelian Genetics

Chapter Concepts

Inherited characteristics are under the control of particulate factors called genes that are transmitted from generation to generation on vehicles called chromosomes, according to rules first described by Gregor Mendel. The outcomes of crosses subject to these rules are affected by chance deviation and may be evaluated using statistical analysis. Human traits are initially studied using the pedigree method.

Although inheritance of biological traits has been recognized for thousands of years, the first significant insights into the mechanisms involved occurred about 130 years ago. In 1866, Gregor Johann Mendel published the results of a series of experiments that would lay the foundation for the formal discipline of genetics. Although Mendel's work went largely unnoticed until the turn of the century, in the ensuing years the concept of the gene as a distinct hereditary unit was established. Ways in which genes, as members of chromosomes, are transmitted to offspring and control traits were clarified. Research has continued unabated throughout the twentieth century. It is safe to say that studies in genetics, most recently at the molecular level, have remained continually at the forefront of biological research since the early 1900s.

In this chapter we focus on the development of the principles established by Mendel, now referred to as **Mendelian** or **transmission genetics**. These principles describe how genes are transmitted from parents to offspring and were derived directly from his experimentation.

When Mendel began his studies of inheritance using *Pisum sativum*, the garden pea, there was no knowledge of chromosomes nor of the role and mechanism of meiosis. Nevertheless, Mendel was able to determine that distinct **units of inheritance** exist and to predict their behavior during the formation of gametes. Subsequent investigators, with access to cytological data, were able to relate their observations of chromosome behavior during meiosis to Mendel's principles of inheritance. Once this correlation was made, Mendel's postulates were accepted as the underlying basis for the study of transmission genetics. Even today, they serve as the cornerstone of the study of inheritance.

Gregor Johann Mendel

Johann Mendel was born in 1822 to a peasant family in the European village of Heinzendorf, now part of the Czech Republic. An excellent student in high school, Mendel studied philosophy for several years afterward and was admitted to the Augustinian Monastery of St. Thomas in Brno in 1843. There, he took the name of Gregor and received support for his studies and research throughout the rest of his life. In 1849, he was relieved of pastoral duties and received a teaching appointment that lasted several years. From 1851 to 1853 he attended the University of Vienna, where he studied physics and botany. In late 1853 he returned to Brno, where, for the next 15 years, he taught physics and natural science.

In 1856 Mendel performed his first set of hybridization experiments with the garden pea. The research phase of his career lasted until 1868, when he was elected abbot of his religious chapter. Although his interest in genetics remained, his new administrative and public service responsibilities demanded most of his time. His health declined during the ensuing years, and

43

in 1884 Mendel died of chronic kidney disease and heart failure. The local newspaper paid him the following tribute: "His death deprives the poor of a benefactor, and mankind at large of a man of the noblest character, one who was a warm friend, a promoter of the natural sciences, and an exemplary priest."

Mendel's Experimental Approach

In 1865, Mendel first reported the results of some simple genetic crosses between certain strains of the garden pea. Although, as we saw in Chapter 1, his was not the first attempt to provide experimental evidence pertaining to inheritance, Mendel's work is an elegant model of experimental design and analysis.

Mendel showed remarkable insight into the methodology necessary for good experimental biology. He chose an organism that was easy to grow and to hybridize artificially. The pea plant is self-fertilizing in nature, but is easy to cross-breed during experiments. It reproduces well and grows to maturity in a single season. Mendel worked with seven unit characters, visible features that were each represented by two contrasting forms or traits. For the character stem height, for example, he experimented with the traits *tall* and *dwarf*. He selected six other pairs of contrasting traits involving seed shape and color, pod shape and color, and pod and flower arrangement. True-breeding strains were available from local seed merchants. Each trait appeared unchanged generation after generation in self-fertilizing plants; that is, the strains exhibiting them "bred true."

Mendel's success in an area where others had failed may be attributed to several factors in addition to the choice of a suitable organism. He restricted his examination to one or very few pairs of contrasting traits in each experiment. He also kept accurate quantitative records, a necessity in genetic experiments. From the analysis of his data, certain postulates can be derived that have become the principles of transmission genetics.

While the results of Mendel's experiments were cited repeatedly for decades after their publication, they nevertheless went unappreciated until the turn of the century. The significance of his work is far-reaching. He had discovered the basis for the transmission of hereditary traits!

The Monohybrid Cross

The simplest crosses performed by Mendel involved only one pair of contrasting traits. Each such breeding experiment is called a **monohybrid cross**. A monohybrid cross is made by mating individuals from two parent strains, each of which exhibits one of the two contrasting forms of the character under study. Initially we will examine the first generation of offspring of such a cross, and then we will consider the offspring of **selfing** or **self-fertilizing** individuals from this first genera-

tion. The original parents are called the P_1 or **parental generation**, their offspring are the F_1 or **first filial generation**, and the individuals resulting from the selfing of the F_1 generation are called the F_2 or **second filial generation**. We can, of course, continue to follow subsequent generations, if desired.

The cross between true-breeding pea plants with tall stems and dwarf stems is representative of Mendel's monohybrid crosses. *Tall* and *dwarf* represent contrasting forms or traits of the character of stem height. Unless tall or dwarf plants are crossed together or with another strain, they will undergo self-fertilization and breed true, producing their respective trait generation after generation. However, when Mendel crossed tall plants with dwarf plants, the resulting F_1 generation consisted only of tall plants. When members of the F_1 generation were selfed, Mendel observed that 787 of 1064 F_2 plants were tall, while 277 of 1064 were dwarf. Note that in this cross (Figure 3–1) the dwarf trait disappears in the F_1, only to reappear in the F_2 generation.

Genetic data are usually expressed and analyzed as ratios. In this particular example, many identical P_1 crosses were made and many F_1 plants—all tall—were produced. Of the 1064 F_2 offspring, 787 were tall and 277 were dwarf—a ratio of approximately 2.8:1.0, or about 3:1.

Mendel made similar crosses between pea plants exhibiting other pairs of contrasting traits. The results of these crosses are also shown in Figure 3–1. In every case, the outcome was similar to the tall/dwarf cross just described. All F_1 offspring were identical to one of the parents. In the F_2, an approximate ratio of 3:1 was obtained. Three-fourths appeared like the F_1 plants, while one-fourth exhibited the contrasting trait, which had disappeared in the F_1 generation.

It is appropriate to point out one further aspect of the monohybrid crosses. In each, the F_1 and F_2 patterns of inheritance were similar regardless of which P_1 plant served as the source of pollen (sperm), and which served as the source of the ovum (egg). The crosses could be made either way—that is, pollination of dwarf plants by tall plants, or vice versa. These are called **reciprocal crosses**. Therefore, the results of Mendel's monohybrid crosses were not sex-dependent.

To explain these results, Mendel proposed the existence of particulate **unit factors** for each trait. He suggested that these factors serve as the basic units of heredity and are passed unchanged from generation to generation, determining various traits expressed by each individual plant. Using these general ideas, Mendel proceeded to hypothesize precisely how such factors could account for the results of the monohybrid crosses.

Mendel's First Three Postulates

Using the consistent pattern of results in the monohybrid crosses, Mendel derived the following three postulates or principles of inheritance.

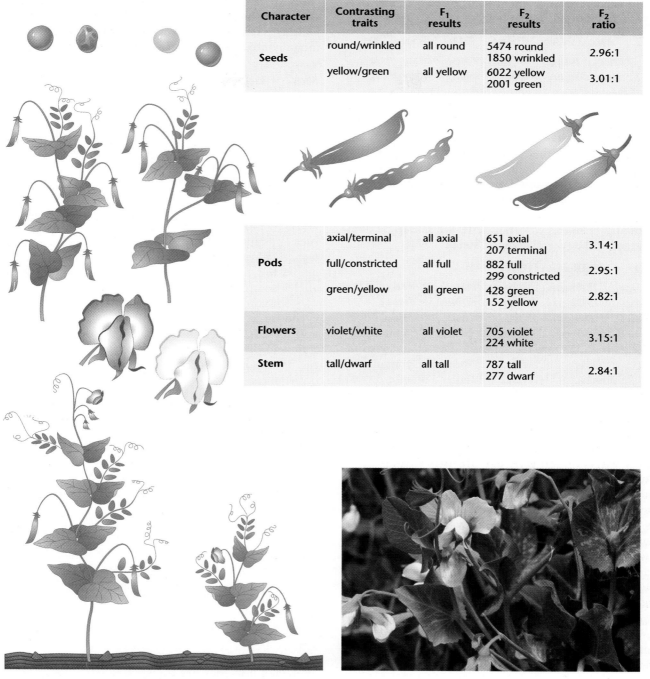

Character	Contrasting traits	F₁ results	F₂ results	F₂ ratio
Seeds	round/wrinkled	all round	5474 round 1850 wrinkled	2.96:1
	yellow/green	all yellow	6022 yellow 2001 green	3.01:1
Pods	axial/terminal	all axial	651 axial 207 terminal	3.14:1
	full/constricted	all full	882 full 299 constricted	2.95:1
	green/yellow	all green	428 green 152 yellow	2.82:1
Flowers	violet/white	all violet	705 violet 224 white	3.15:1
Stem	tall/dwarf	all tall	787 tall 277 dwarf	2.84:1

■ Figure 3–1 A summary of the seven pairs of contrasting traits and the results of Mendel's seven monohybrid crosses. In each case, pollen derived from plants exhibiting one contrasting trait was used to fertilize the ova of plants exhibiting the other contrasting trait. In the F₁ generation one of the two traits, referred to as dominant, was exhibited by all plants. The contrasting trait, referred to as recessive, then reappeared in approximately one-fourth of the F₂ plants. The garden pea (*Pisum sativum*) is shown in the photograph.

1.   UNIT FACTORS IN PAIRS

Genetic characters are controlled by unit factors that exist in pairs in individual organisms.

In the monohybrid cross involving tall and dwarf stems, a specifc unit factor exists for each trait. Because the factors occur in pairs, three combinations are possible: two factors for tallness, two factors for dwarfness, or one for each factor. Every indivdual contains one of these three combinations, which determines stem height.

2. DOMINANCE/RECESSIVENESS

When two unlike unit factors responsible for a single character are present in a single individual, one unit factor is dominant to the other, which is said to be recessive.

In each monohybrid cross, the trait expressed in the F_1 generation is controlled by the dominant unit factor. The trait that is not expressed is controlled by the recessive unit factor. Note that this dominance/recessiveness relationship pertains only when unlike unit factors are present in pairs. The terms dominant and recessive are also used to designate the traits. In this case, tall stems are said to be dominant to the recessive dwarf stems.

3. SEGREGATION

During the formation of gametes, the paired unit factors separate or segregate randomly so that each gamete receives one or the other with equal likelihood.

If an individual contains a pair of like unit factors (e.g., both are specific for tall), then all gametes receive one tall unit factor. If an individual contains unlike unit factors (e.g., one for tall and one for dwarf), then each gamete has a 50 percent probability of receiving either the tall or the dwarf unit factor.

These postulates provide a suitable explanation for the results of the monohybrid crosses. The tall/dwarf cross will be used to illustrate this explanation. Mendel reasoned that P_1 tall plants contained identical paired unit factors, as did the P_1 dwarf plants. The gametes of tall plants all received one tall unit factor as a result of segregation. Likewise, the gametes of dwarf plants all received one dwarf unit factor. Following fertilization, all F_1 plants received one unit factor from each parent, a tall factor from one and a dwarf factor from the other, reestablishing the paired relationship. Because tall is dominant to dwarf, all F_1 plants were tall.

When F_1 plants form gametes, the postulate of segregation demands that each gamete randomly receive either the tall or the dwarf unit factor. Following random fertilization events during F_1 selfing, four F_2 combinations will result in equal frequency:

(1) tall/tall
(2) tall/dwarf
(3) dwarf/tall
(4) dwarf/dwarf

Combinations (1) and (4) will result in tall and dwarf plants, respectively. According to the postulate of dominance/recessiveness, combinations (2) and (3) will both yield tall plants. Therefore, the F_2 is predicted to consist of three-fourths tall and one-fourth dwarf, or a

ratio of 3:1. This is approximately what Mendel observed in the cross between tall and dwarf plants. A similar pattern was observed in each of the other monohybrid crosses (Figure 3–1).

Modern Genetic Terminology

In order to illustrate the monohybrid cross and Mendel's first three postulates, we must introduce several new terms as well as a set of symbols for the unit factors. Traits such as tall or dwarf are visible expressions of the information contained in unit factors. The physical appearance of a trait is called the **phenotype** of the individual.

All unit factors represent units of inheritance called genes by modern geneticists. For any given character, such as plant height, the phenotype is determined by alternative forms of a single gene called **alleles**. For example, the unit factors representing tall and dwarf are alleles determining the height of the pea plant.

By one convention, the first letter of the recessive trait is chosen to symbolize the character in question. The lowercase letter designates that allele for the recessive trait, and the uppercase letter designates the allele for the dominant trait. Therefore, we let *d* stand for the dwarf allele and *D* represent the tall allele. When alleles are written in pairs to represent the two unit factors present in any individual (*DD, Dd,* or *dd*), these symbols are referred to as the **genotype**. This term reflects the genetic makeup of an individual whether it is haploid or diploid. By reading the genotype, it is possible to know the phenotype of the individual: *DD* and *Dd* are tall, and *dd* is dwarf. When both alleles are the same (*DD* or *dd*), the individual is said to be **homozygous** or a **homozygote**; when the alleles are different (*Dd*), we use the term **heterozygous** or **heterozygote**. These symbols and terms are used in Figure 3–2 to illustrate the complete monohybrid cross.

Because he operated without the hindsight that modern geneticists enjoy, Mendel's analytical reasoning must be considered a truly outstanding scientific achievement. On the basis of rather simple but precisely executed breeding experiments, he proposed that discrete particulate units of heredity exist, and he explained how they are transmitted from one generation to the next!

Punnett Squares

The genotypes and phenotypes resulting from the recombination of gametes during fertilization can be easily visualized by constructing a **Punnett square**, so named after the person who first devised this approach, Reginald C. Punnett. Figure 3–3 illustrates this method of analysis for the $F_1 \times F_1$ monohybrid cross. Each of the possible gametes is assigned to a column or a row, with the vertical column representing those of the female parent and the horizontal row those of the male

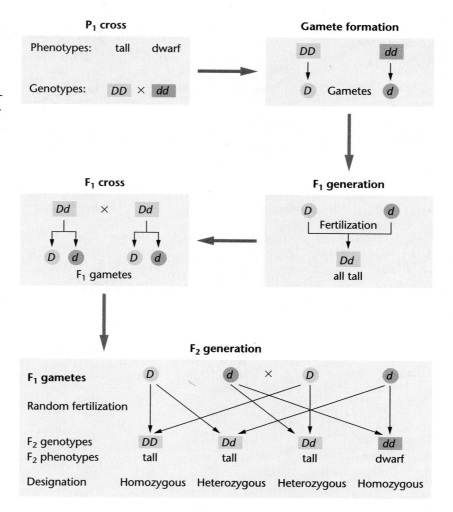

■ Figure 3–2 An explanation of the monohybrid cross between tall and dwarf pea plants. The symbols *D* and *d* are used to designate the tall and dwarf unit factors, respectively, in the genotypes of mature plants and gametes. All individuals are shown in rectangles. All gametes are shown in circles.

parent. After entering the gametes in rows and columns, the new generation is predicted by combining the male and female gametic information for each combination and entering the resulting genotypes in the boxes. This process represents all possible random fertilization events. The genotypes and phenotypes of all potential offspring are ascertained by reading the entries in the boxes.

The Punnett square method is particularly useful when one is first learning about genetics and how to solve problems. In Figure 3–3, note the ease with which the 3:1 phenotypic ratio and the 1:2:1 genotypic ratio may be derived in the F$_2$ generation.

The Test Cross: One Character

Tall plants produced in the F$_2$ generation are predicted to be of either the *DD* or *Dd* genotype. We might ask if there is a way to distinguish the genotype. Mendel devised a rather simple method that is still used today in breeding plants and animals: the **test cross**. The organism expressing the dominant phenotype, but of unknown genotype, is crossed to a homozygous recessive individual. For example, as shown in Figure 3–4(a), if a

tall plant of genotype *DD* is test-crossed to a dwarf plant, which must have the *dd* genotype, all offspring will be tall phenotypically and *Dd* genotypically. However, as shown in Figure 3–4(b), if a tall plant is *Dd* and it is crossed to a dwarf plant (*dd*), then one-half of the offspring will be tall (*Dd*) and the other half will be dwarf (*dd*). Therefore, a 1:1 tall/dwarf ratio demonstrates the heterozygous nature of the tall plant of unknown genotype. The test cross reinforced Mendel's conclusion that separate unit factors control the tall and dwarf traits.

The Dihybrid Cross

A natural extension of performing monohybrid crosses was for Mendel to design experiments in which two characters were examined simultaneously. Such a cross, involving two pairs of contrasting traits, is a **dihybrid cross**. It is also called a **two-factor cross**. For example, if pea plants having yellow seeds that are also round are bred with those having green seeds that are also wrinkled, the results shown in Figure 3–5 will occur. The F$_1$ offspring will be all yellow and round. It is therefore

■ Figure 3–3 The use of a Punnett square in generating the F_2 ratio of the $F_1 \times F_1$ cross shown in Figure 3–2.

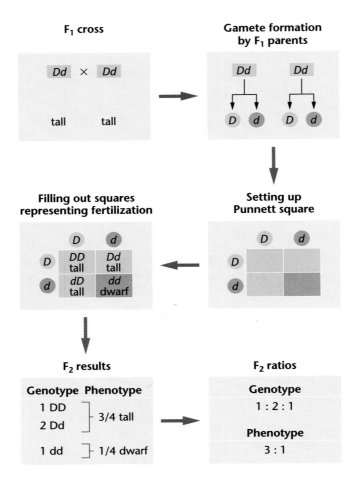

F_1 cross

Dd × Dd

tall   tall

Gamete formation by F_1 parents

Dd   Dd

D d   D d

Setting up Punnett square

	D	d
D		
d		

Filling out squares representing fertilization

	D	d
D	DD tall	Dd tall
d	dD tall	dd dwarf

F_2 results

Genotype	Phenotype
1 DD	
2 Dd	3/4 tall
1 dd	1/4 dwarf

F_2 ratios

Genotype

1 : 2 : 1

Phenotype

3 : 1

■ Figure 3–4 The test cross illustrated with a single character. In (a), the tall parent is homozygous. In (b), the tall parent is heterozygous. The genotype of each tall parent may be determined by examining the offspring when each is crossed to the homozygous recessive dwarf plant.

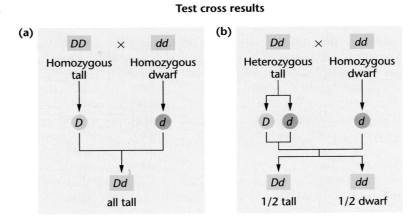

Test cross results

(a)

DD × dd

Homozygous tall   Homozygous dwarf

D   d

Dd

all tall

(b)

Dd × dd

Heterozygous tall   Homozygous dwarf

D d   d

Dd   dd

1/2 tall   1/2 dwarf

apparent that yellow is dominant to green and that round is dominant to wrinkled. When the F_1 individuals are selfed, approximately 9/16 of the F_2 plants express yellow and round, 3/16 express yellow and wrinkled, 3/16 express green and round, and 1/16 express green and wrinkled.

A variation of this cross is also shown in Figure 3–5. Instead of crossing one P_1 parent with both dominant traits (yellow, round) and one with both recessive traits (green, wrinkled), plants with yellow, wrinkled seeds are crossed to those with green, round seeds. In spite of

the change in the P_1 phenotypes, both the F_1 and F_2 results remain unchanged. It will become clear in the next section why this is so.

Mendel's Fourth Postulate: Independent Assortment

We can most easily understand the results of a dihybrid cross if we consider it theoretically as consisting of two monohybrid crosses conducted separately. Think of the two sets of traits as being inherited independently of

■ Figure 3–5 The F_1 and F_2 results of Mendel's dihybrid crosses between yellow, round and green, wrinkled pea plants and between yellow, wrinkled and green, round pea plants.

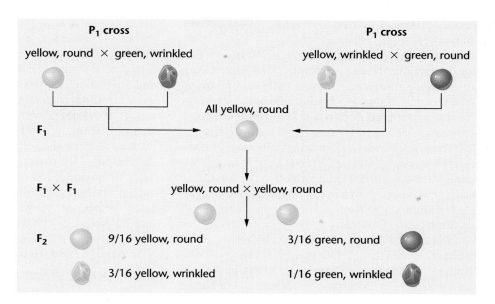

each other; that is, the chance of any plant becoming tall or dwarf is not at all influenced by the chance that this plant will have round or wrinkled seeds. Thus, because yellow is dominant to green, all F_1 plants in the first theoretical cross would have yellow seeds. In the second theoretical cross, all F_1 plants would have round seeds because round is dominant to wrinkled. When Mendel examined the F_1 plants of the dihybrid cross, all were yellow and round, as we just predicted.

The predicted F_2 results of the first cross are 3/4 yellow and 1/4 green. Similarly, the second cross would yield 3/4 round and 1/4 wrinkled. Figure 3–5 shows that in the dihybrid cross, 12/16 F_2 plants are yellow while 4/16 are green, exhibiting the expected 3:1 (3/4:1/4) ratio. Similarly, 12/16 F_2 plants have round seeds while 4/16 have wrinkled seeds, again revealing the 3:1 (3/4:1/4) ratio.

Because it is evident that the two pairs of contrasting traits are inherited independently, we can predict the frequencies of all possible F_2 phenotypes by applying the **product law** of probabilities: **When two independent events occur simultaneously, the combined probability**

of the two outcomes is equal to the product of their individual probabilities of occurrence. For example, the probability of an F_2 plant having yellow *and* round seeds is (3/4)(3/4), or 9/16, because 3/4 of all F_2 plants should be yellow and 3/4 of all F_2 plants should be round.

In a like way, the probabilities of the other three F_2 phenotypes may be calculated: yellow (3/4) and wrinkled (1/4) are predicted to be present together 3/16 of the time; green (1/4) and round (3/4) are predicted 3/16 of the time; and green (1/4) and wrinkled (1/4) are predicted 1/16 of the time. These calculations are illustrated in Figure 3–6. It is now apparent why the F_1 and F_2 results are identical whether the initial cross is yellow, round bred with green, wrinkled or if yellow, wrinkled are bred with green, round. In both crosses, the F_1 genotype of all plants is identical. Each plant is heterozygous for both gene pairs. As a result, the F_2 generation is also identical in both crosses.

On the basis of similar results in numerous dihybrid crosses, Mendel proposed a fourth postulate called **independent assortment: During gamete formation,**

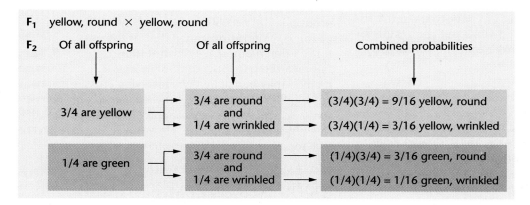

■ Figure 3–6 The determination of the combined probabilities of each F_2 phenotype for two independently inherited characters. The probability of each plant being yellow or green is independent of the probability of it being round or wrinkled.

segregating pairs of unit factors assort independently of each other.

This postulate stipulates that segregation of any pair of unit factors occurs independently of all others. As a result of segregation, each gamete receives one member of every pair of unit factors. For one pair, whichever unit factor is received does not influence the outcome of segregation of any other pair. Thus, according to the postulate of independent assortment, all possible combinations of gametes will be formed in equal frequency.

Independent assortment is illustrated in the formation of the F_2 generation, shown in the Punnett square in Figure 3–7. Examine the formation of gametes by the F_1 plants. Segregation prescribes that every gamete receives either a G or g allele and a W or w allele. Independent assortment stipulates that all four combinations (GW, Gw, gW, and gw) will be formed with equal probabilities.

In every $F_1 \times F_1$ fertilization event, each zygote has an equal probability of receiving one of the four combinations from each parent. If many offspring are produced, 9/16 are yellow and round, 3/16 are yellow and wrinkled, 3/16 are green and round, and 1/16 are green and wrinkled, yielding what is designated as **Mendel's 9:3:3:1 dihybrid ratio**. This ratio is based on probability events involving segregation, independent assortment, and random fertilization. Therefore, it is an ideal ratio. Because of deviation due strictly to chance, particularly if small numbers of offspring are produced, the ideal ratio will seldom be approached.

The Test Cross: Two Characters

The test cross may also be applied to individuals that express two dominant traits but whose genotypes are unknown. For example, the expression of the yellow, round phenotype in the F_2 generation just described may result from the $GGWW$, $GGWw$, $GgWW$, and $GgWw$ genotypes. If an F_2 yellow, round plant is crossed with a homozygous recessive green, wrinkled plant ($ggww$), analysis of the offspring will indicate the actual genotype of that yellow, round plant. Each of the above genotypes will result in a different set of gametes, and in a test cross, a different set of phenotypes in the resulting offspring. You should work out the results of each of the four crosses in order to demonstrate this.

The Trihybrid Cross

We have thus far considered inheritance by individuals of up to two pairs of contrasting traits. Mendel demonstrated that the identical processes of segregation and independent assortment apply to three pairs of contrasting traits in what is called a **trihybrid cross**, also referred to as a **three-factor cross**.

Although a trihybrid cross is somewhat more complex than a dihybrid cross, its results are easily calculated if the principles of segregation and independent assortment are followed. For example, consider the cross shown in Figure 3–8, where the gene pairs representing theoretical contrasting traits are represented by the symbols A/a, B/b, and C/c. In the cross between $AABBCC$ and $aabbcc$ individuals, all F_1 individuals are heterozygous for all three gene pairs. Their genotype, $AaBbCc$, results in the phenotypic expression of the dominant A, B, and C traits. When F_1 individuals serve as parents, each produces 8 different gametes in equal frequencies. At this point, we could construct a Punnett square with 64 separate boxes and read out the phenotypes. Because such a method is cumbersome in a cross involving so many factors, another method has been devised that may also be used to calculate the predicted ratio.

The Forked-Line Method, or Branch Diagram

It is much less difficult to consider each contrasting pair of traits separately and then to combine these results using the **forked-line method**, which was first illustrated in Figure 3–6. This method, also called a **branch diagram**, relies on the simple application of the laws of probability established for the dihybrid cross. Each gene pair is assumed to behave independently during gamete formation.

When the monohybrid cross $AA \times aa$ is made, we know that:

1. All F_1 individuals have the genotype Aa and express the phenotype represented by the A allele, which is called the A phenotype in the following discussion.

2. The F_2 generation consists of individuals with either the A phenotype or the a phenotype in the ratio of 3:1.

The same generalizations may be made for the $BB \times bb$ and $CC \times cc$ crosses. Thus, in the F_2 generation, 3/4 of all organisms will express phenotype A, 3/4 will express B, and 3/4 will express C. Similarly, 1/4 of all organisms will express phenotype a, 1/4 will express b, and 1/4 will express c. The proportions of organisms that express each phenotypic combination may be predicted by assuming that fertilization, following the independent assortment of these three gene pairs during gamete formation, is a random process. We must simply apply once again the product law of probabilities. The phenotypic proportions of the F_2 generation calculated in this way using the forked-line method are illustrated in Figure 3–9. They fall into the trihybrid ratio 27:9:9:9:3:3:3:1. The same method may be applied when solving crosses involving any number of gene pairs, *provided that all gene pairs assort independently from each other*. We will see later that this is not always the case. However, it appeared to be true for all of Mendel's characters.

The Rediscovery of Mendel's Work

Mendel's work, initiated in 1856, was presented to the Brünn Society of Natural Science in 1865 and published the following year. However, while his findings

■ Figure 3–7 Diagram of the dihybrid crosses shown in Figure 3–5. The F$_1$ heterozygous plants are self-fertilized to produce an F$_2$ generation, which is computed using a Punnett square. Both the phenotypic and genotypic F$_2$ ratios are shown.

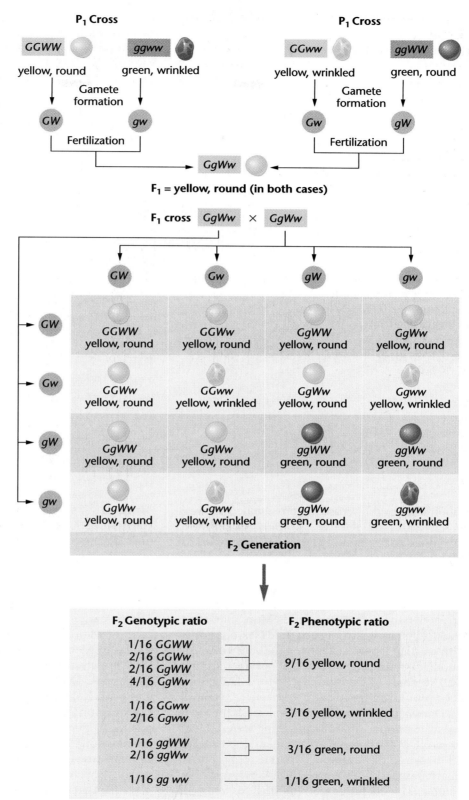

■ Figure 3–8 The formation of P₁ and F₁ gametes in a trihybrid cross.

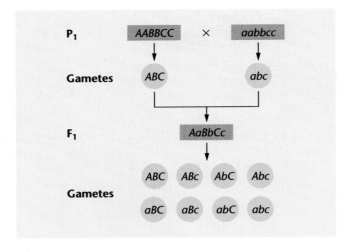

■ Figure 3–9 The generation of the F₂ trihybrid phenotypic ratio using the forked-line, or branch diagram, method.

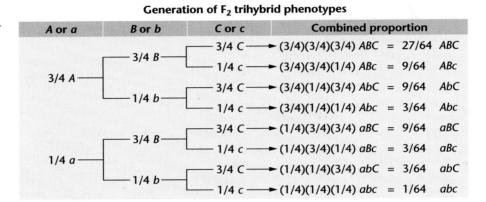

were often cited and discussed, their significance went unappreciated for about 35 years! Many reasons have been suggested to explain why the signifcance of his research was not immediately recognized.

First of all, Mendel's adherence to mathematical analysis of probability events was quite an unusual approach in biological studies. Perhaps his approach seemed foreign to his contemporaries. More important, his conclusions drawn from such analyses did not fit well with existing theories involving the cause of variation among organisms. The source of natural variation intrigued students of evolutionary theory. These individuals, stimulated by the proposal developed by Charles Darwin and Alfred Russel Wallace, believed in **continuous variation**, whereby offspring were a blend of their parents' phenotypes. As we mentioned earlier, Mendel theorized that variation was due to discrete or particulate units, resulting in **discontinuous variation**. For example, Mendel proposed that the F₂ offspring of a dihybrid cross were merely expressing traits produced by new combinations of previously existing unit factors. As a result, Mendel's theories did not fit well with the evolutionists' preconceptions about causes of variation.

Near the end of the nineteenth century, a remarkable observation set the scene for the rebirth of Mendel's work: Walter Flemming's discovery of chromosomes in the nuclei of salamander cells. In 1879, Flemming was able to describe the behavior of these threadlike structures during cell division. As a result of the findings of Flemming and many other cytologists, the presence of discrete units within the nucleus soon became an integral part of ideas about inheritance. It was in this setting that scientists were able to reexamine Mendel's findings.

In the early twentieth century, research led to the rebirth of Mendel's work. Hybridization experiments similar to Mendel's were performed independently by three botanists: Hugo de Vries, Karl Correns, and Erich Tschermak. De Vries's work, for example, focused on unit characters, and he demonstrated the principle of segregation in his experiments with several plant species. Apparently, he had searched the existing literature and found that Mendel's work anticipated his own conclusions! Correns and Tschermak independently reached conclusions similar to those of Mendel.

In 1902 two cytologists, Walter Sutton and Theodor Boveri, independently published papers linking their discoveries of the behavior of chromosomes during meiosis to the Mendelian principles of segregation and independent assortment. They pointed out that the separation of chromosomes during meiosis could serve as the cytological basis of these two postulates. Although they thought that Mendel's unit factors were probably chromosomes rather than genes on chromosomes, their findings reestablished the importance of Mendel's work, which served as the foundation of ensuing genetic investigations. Sutton and Boveri are credited with the initiation of the **chromosomal theory of inheritance**, which was developed during the next two decades.

Unit Factors, Genes, and Homologous Chromosomes

Because the correlation between Sutton's and Boveri's observations and Mendelian principles is the foundation for the modern interpretation of transmission genetics, we will examine this correlation before moving on to the topics of the next several chapters.

As pointed out in Chapter 2, each species possesses a specific number of chromosomes in each somatic cell nucleus (all but gametes). For diploid organisms, this number is called the **diploid number ($2n$)** and is characteristic of that species. During the formation of gametes, this number is precisely halved (n), and when two gametes combine during fertilization, the diploid number is reestablished. During meiosis, however, the chromosome number is not reduced in a random manner. It was apparent to early cytologists that the diploid number of chromosomes is composed of homologous pairs identifiable by their morphological appearance and behavior. The gametes contain one member of each pair. The chromosome complement of a gamete is thus quite specific, and the number of chromosomes in each gamete is equal to the haploid number.

With this basic information, we can see the correlation between the behavior of unit factors and chromosomes and genes. Figure 3–10 shows three of Mendel's postulates and the accepted explanation of each. Unit factors are really genes located on homologous pairs of chromosomes [Figure 3–10(a)]. Members of each pair of homologs separate, or segregate, during gamete formation [Figure 3–10(b)]. Two different alignments are possible, both of which are shown.

To illustrate the principle of independent assortment, it is important to distinguish between members of any given homologous pair of chromosomes. One member of each pair is derived from the **maternal parent**, while the other comes from the **paternal parent**. We represent the different parental origins by different colors. As shown in Figure 3–10(c), following independent segregation of each pair of homologs, each gamete receives one member from each pair of chromosomes. All possible combinations are formed. If we add the symbols used in Mendel's dihybrid cross (G, g and W, w) to the diagram, we see why equal numbers of the four types of gametes are formed. The independent behavior of Mendel's pairs of unit factors (G and W in this example) was due to the fact that they were on separate pairs of homologous chromosomes.

From observations of the phenotypic diversity of living organisms, we see that it is logical to assume that there are many more genes than chromosomes. Therefore, each homolog must carry genetic information for more than one trait. The currently accepted concept is that a chromosome is composed of a large number of linearly ordered, information-containing units called **genes**. Mendel's unit factors (which determine tall or dwarf stems, for example) actually constitute a pair of genes located on one pair of homologous chromosomes. The location on a given chromosome where any particular gene occurs is called its **locus** (pl. **loci**). The different forms taken by a given gene, called **alleles** (G or g), contain slightly different genetic information that determines the same character (stem length). Alleles are alternative forms of the same gene. Although we have only discussed genes with two alternative alleles, most genes have more than two allelic forms. We will discuss this concept of **multiple alleles** in Chapter 4.

We conclude this section by reviewing the criteria necessary to classify two chromosomes as a homologous pair:

1. During mitosis and meiosis, when chromosomes are visible as distinct figures, both members of a homologous pair are the same size and exhibit identical centromere locations.

2. During early stages of meiosis, homologous chromosomes pair together, or synapse.

3. Although not generally microscopically visible, homologs contain identical, linearly ordered gene loci.

Independent Assortment and Genetic Variation

One of the major consequences of independent assortment is the production by an individual of genetically dissimilar gametes. Genetic variation results because the two members of any homologous pair of chromosomes are rarely, if ever, genetically identical. Therefore, because independent assortment leads to the production of all possible chromosome combinations, extensive genetic diversity results.

The number of possible gametes, each with different chromosome compositions, is 2^n, where n equals the haploid number. Thus, if a species has a haploid number of 4, then 2^4 or 16 different gamete combinations can be formed as a result of independent assortment. Although this number is not high, consider the human species, where $n = 23$. If 2^{23} is calculated, we find that in excess of 8×10^6, or over 8 million, different types of gametes are represented. Because fertilization represents an event involving only one of approximately 8×10^6 possible gametes from each of two parents, each offspring represents only one of $(8 \times 10^6)^2$, or 64×10^{12}, potential genetic combinations! It is no wonder that, except for identical twins, each member of the human species demonstrates a distinctive appearance and individuality. This number of combinations is far greater than the number of humans

(a)

**Unit factors
In
pairs**

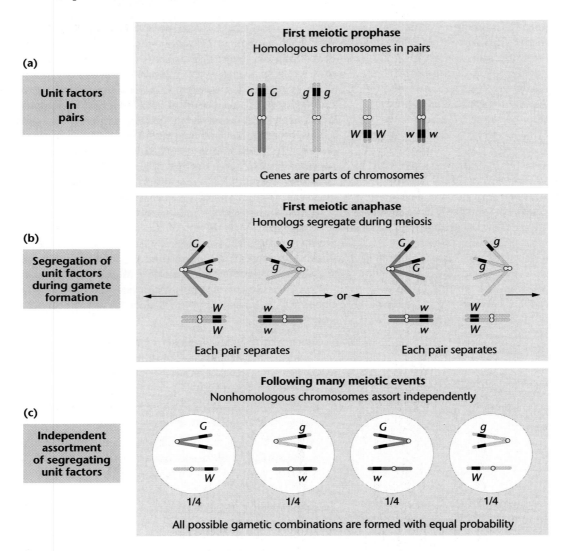

First meiotic prophase
Homologous chromosomes in pairs

G G g g

W W w w

Genes are parts of chromosomes

(b)

**Segregation of
unit factors
during gamete
formation**

First meiotic anaphase
Homologs segregate during meiosis

G g G g

G g G g

or

W w w W

W w w W

Each pair separates   Each pair separates

(c)

**Independent
assortment
of segregating
unit factors**

Following many meiotic events
Nonhomologous chromosomes assort independently

G g G g

W w w W

1/4   1/4   1/4   1/4

All possible gametic combinations are formed with equal probability

■ Figure 3–10 The correlation between the Mendelian postulates of (a) unit factors in pairs, (b) segregation, and (c) independent assortment, and the presence of genes located on homologous chromosomes and their behavior during meiosis.

who have ever lived on earth! Genetic variation resulting from independent assortment has been extremely important to the process of evolution in all organisms.

Probability and Genetic Events

Genetic ratios are most properly expressed as probabilities—for example, 3/4 tall:1/4 dwarf. These values predict the outcome of each fertilization event, such that the probability of each zygote having the genetic potential for becoming tall is 3/4, while the potential for becoming dwarf is 1/4. Probabilities range from 0, when an event is *certain not to occur*, to 1.0, when an event is *certain to occur*.

When two or more events occur independently but at the same time, we can calculate the probability of possible outcomes when they occur together. This is accomplished by applying the **product law**. As mentioned in our

earlier discussion of independent assortment, the law states that the probability of two or more events occurring simultaneously is equal to the product of their individual probabilities. Two or more events are independent of one another if the outcome of each one does not affect the outcome of any of the others under consideration.

To illustrate the use of the product law, consider the possible results if you were to toss a penny (P) and a nickel (N) at the same time and examined all combinations of heads (H) and tails (T) that can occur. There are four possible outcomes:

$$(P_H:N_H) = (1/2)(1/2) = 1/4$$
$$(P_T:N_H) = (1/2)(1/2) = 1/4$$
$$(P_H:N_T) = (1/2)(1/2) = 1/4$$
$$(P_T:N_T) = (1/2)(1/2) = 1/4$$

The probability of obtaining a head or a tail in the toss of either coin is 1/2 and is unrelated to the outcome of the toss of the other coin. All four possible combinations are predicted to occur with equal probability.

If we are interested in calculating the probability where the possible outcomes of two events are independent of one another but can be accomplished in more than one way, we apply the **sum law**. For example, we can ask: What is the probability of tossing our penny and nickel and obtaining one head and one tail? In such a case, we do not care whether it is the penny or the nickel that comes up heads, provided that the other coin has the alternate outcome. As we have seen, there are two ways in which the desired outcome can be accomplished, each with a probability of 1/4. Thus, according to the sum law, the overall probability is

$$(1/4) + (1/4) = 1/2$$

One-half of all such tosses are predicted to yield the desired outcome.

These simple probability laws will be useful throughout our discussions of transmission genetics, and as you solve genetics problems. In fact, we have already applied the product law, when we used the forked-line method to calculate the phenotypic results of Mendel's dihybrid and trihybrid crosses. When we wish to know the results of a cross, we need only to calculate the probability of each possible outcome. The results of this calculation then allow us to predict the proportion of offspring expressing each phenotype or each genotype.

There is a very important point to remember when dealing with probability. Predictions of possible outcomes are based on large sample sizes. If we predict that 9/16 of the offspring of a dihybrid cross will express both dominant traits, it is very unlikely that, in a small sample, exactly 9 of every 16 offspring will express this phenotype. Instead, our prediction is that, of a large number of offspring, approximately 9/16 of them will do so. The deviation from the predicted ratio in smaller sample sizes is attributed to chance, a subject we deal with in our discussion of statistics in the next section. As we will see, the impact of deviation due strictly to chance is diminished as the sample size increases.

Evaluating Genetic Data: Chi-Square Analysis

Mendel's 3:1 monohybrid and 9:3:3:1 dihybrid ratios are hypothetical predictions based on the following assumptions: (1) Each allele is dominant or recessive; (2) segregation is operative; (3) independent assortment occurs; and (4) fertilization is random. The last three assumptions are influenced by chance events and therefore are subject to random fluctuation. This concept, called **chance deviation**, is most easily illustrated by tossing a single coin numerous times and recording the number of heads and tails observed. In each toss, there is a probability of 1/2 that a head will occur and a probability of 1/2 that a tail will occur. Therefore, the expected ratio of many tosses is 1:1. If a coin is tossed 1000 times, usually *about* 500 heads and 500 tails will be observed. Any reasonable fluctuation from this hypothetical ratio (e.g., 486 heads and 514 tails) would be attributed to chance.

As the total number of tosses is reduced, the impact of chance deviation increases. For example, if a coin were tossed only 4 times, you wouldn't be too surprised if all 4 tosses resulted in only heads or only tails. For 1000 tosses, however, 1000 heads or 1000 tails would be most unexpected. In fact, you might believe that such a result would be impossible. Actually, all heads or all tails in 1000 tosses can be predicted to occur with a probability of only $(1/2)^{1000}$. Because $(1/2)^{20}$ is equivalent to less than one in a million times, an event occurring with a probability of only $(1/2)^{1000}$ would be virtually impossible to occur.

Two major points are significant here:

1. The outcomes of segregation, independent assortment, and fertilization, like coin tossing, are subject to random fluctuations from their predicted occurrences as a result of chance deviation.

2. As the sample size increases, the average deviation from the expected results decreases. Therefore, a larger sample size diminishes the impact of chance deviation on the final outcome.

It is important in genetics to be able to evaluate observed deviation. When we assume that data will fit a given ratio such as 1:1, 3:1, or 9:3:3:1, we establish what is called the **null hypothesis**. It is so named because the hypothesis assumes that there is *no real difference* between the *measured values* (or ratio) and the *predicted values* (or ratio). The apparent difference can be attributed purely to chance. Evaluation of the null hypothesis is accomplished by statistical analysis. On this basis, the null hypothesis may either (1) be rejected or (2) fail to be rejected. If it is rejected, the observed deviation from the expected is not attributed to chance alone. The null hypothesis and the underlying assumptions leading to it must be reexamined. If the null hypothesis fails to be rejected, any observed deviations are attributed to chance.

One of the simplest statistical tests devised to assess the null hypothesis is **chi-square (χ^2) analysis**. This test takes into account the observed deviation in each component of an expected ratio as well as the sample size and reduces them to a single numerical value. This value (χ^2) is then used to estimate how frequently the observed deviation can be expected to occur strictly as a result of chance. The formula used in chi-square analysis is

$$\chi^2 = \Sigma \frac{(o - e)^2}{e}$$

In this equation, *o* is the observed value for a given category and *e* is the expected value for that category. Σ represents the sum of the calculated values for each category of the ratio. Because $(o - e)$ is the deviation (d) in each case, the equation can be reduced to

$$\chi^2 = \Sigma \frac{d^2}{e}$$

Table 3.1(a) illustrates the step-by-step procedure necessary to make the χ^2 calculation for the F_2 results of a hypothetical monohybrid cross. If you were analyzing these data, you would work from left to right, calculating and entering the appropriate numbers in each column. Regardless of whether the calculated deviation $(o - e)$ is positive or negative, it becomes positive after the number is squared. Table 3.1(b) illustrates the analysis of the F_2 results of a hypothetical dihybrid cross. Based on your study of the calculations involved in the monohybrid cross, check to make certain that you understand how each number was calculated in the dihybrid example.

The final step in chi-square analysis is to interpret the χ^2 value. To do so, you must initially determine the value of the **degrees of freedom (df)**, which is equal to $n - 1$, where n is the number of different categories into which each datum point may fall. For the 3:1 ratio, $n = 2$, so $df = 2 - 1 = 1$. For the 9:3:3:1 ratio, $df = 3$. Degrees of freedom must be taken into account because the greater the number of categories, the more deviation is expected as a result of chance.

With this accomplished, the χ^2 value must now be interpreted in terms of a corresponding **probability** value (p). Because this calculation is complex, the p value is usually located on a table or graph. Figure 3–11 shows the wide range of χ^2 and p values for numerous degrees of freedom in both forms. We will use the graph to determine the p value. The caption for Figure 3–11(b) explains how to use the table.

These simple steps must be followed to determine p:

1. Locate the χ^2 value on the abscissa (the horizontal or X axis).
2. Draw a vertical line from this point up to the line on the graph representing the appropriate df.
3. Extend a horizontal line from this point to the left until it intersects the ordinate (the vertical or Y axis).
4. Estimate, by interpolation, the corresponding p value.

For our first example (the monohybrid cross) in Table 3.1, the p value of 0.48 may be estimated in this way and is illustrated in Figure 3–11(a). For the dihybrid cross, use this method to see if you can determine the value. χ^2 is 4.16 and df equals 3. A p value of 0.26 is the approximate value. Use of the table rather than the graph confirms that both p values are between 0.20 and 0.50. Examine the table to confirm this.

Thus far, we have been concerned only with the determination of p. The most important aspect of χ^2 analysis is understanding what the p value actually means. We will use the example of the dihybrid cross ($p = 0.26$) to illustrate. In these discussions, it is simplest to think of the p value as a percentage (e.g., 0.26 = 26 percent).

In our example, one interpretation is that the p value indicates the probability that the observed deviation occurred by chance alone. The probability is 0.26 (26 percent) that the observed deviation (or more) can be attributed to chance. The nearer the p value is to 1.0, the closer the data are to the predicted or ideal ratio.

Another description provides more specific statistical information. In our example, the p value indicates that, were the same experiment repeated many times, 26 percent of the trials would be expected to exhibit chance deviation as great or greater than that seen in

TABLE 3.1 ·	Chi-square analysis				

(a) Monohybrid Cross

Expected Ratio	Observed (o)	Expected (e)	Deviation (o − e)	Deviation² (d²)	Deviation²/ Expected (d²/e)
3/4	740	3/4 (1000) = 750	740 − 750 = −10	$(-10)^2 = 100$	100/750 = 0.13
1/4	260	1/4 (1000) = 250	260 − 250 = +10	$(+10)^2 = 100$	100/250 = 0.40
TOTAL = 1000					$\chi^2 = 0.53$
					$p = 0.48$

(b) Dihybrid Cross

Expected Ratio	o	e	o − e	d²	d²/e
9/16	587	567	+20	400	0.71
3/16	197	189	+ 8	64	0.34
3/16	168	189	−21	441	2.33
1/16	56	63	− 7	49	0.78
TOTAL = 1008					$\chi^2 = 4.16$
					$p = 0.26$

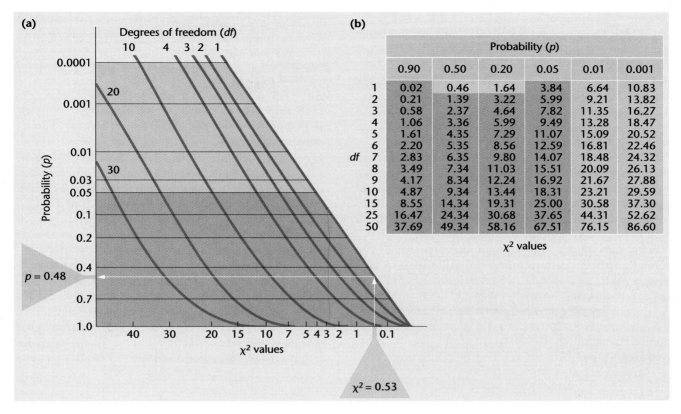

(a)

Degrees of freedom (*df*)
10   4   3   2   1

(b)

		Probability (*p*)					
		0.90	0.50	0.20	0.05	0.01	0.001
	1	0.02	0.46	1.64	3.84	6.64	10.83
	2	0.21	1.39	3.22	5.99	9.21	13.82
	3	0.58	2.37	4.64	7.82	11.35	16.27
	4	1.06	3.36	5.99	9.49	13.28	18.47
	5	1.61	4.35	7.29	11.07	15.09	20.52
	6	2.20	5.35	8.56	12.59	16.81	22.46
df	7	2.83	6.35	9.80	14.07	18.48	24.32
	8	3.49	7.34	11.03	15.51	20.09	26.13
	9	4.17	8.34	12.24	16.92	21.67	27.88
	10	4.87	9.34	13.44	18.31	23.21	29.59
	15	8.55	14.34	19.31	25.00	30.58	37.30
	25	16.47	24.34	30.68	37.65	44.31	52.62
	50	37.69	49.34	58.16	67.51	76.15	86.60

χ^2 values

$p = 0.48$

$\chi^2 = 0.53$

■ Figure 3–11   (a) A graph used to convert χ^2 values to *p* values. The interpolation of a χ^2 value of 0.53 with 1 degree of freedom to an estimated probability value of 0.48 is illustrated. (b) A table showing χ^2 values for a variety of combinations of *df* and *p*. Any χ^2 value greater than that shown at the $p = 0.05$ level for a particular *df* serves as the basis to reject the null hypothesis in question. In our example, a χ^2 value of 0.53 for a *df* of 1 is converted to a probability value between 0.20 and 0.50. From our graph in (a), the more precise value ($p = 0.48$) was estimated by interpolation. All values that serve to fail to reject the null hypothesis are shaded in both the graph and the chart.

the initial trial. Conversely, 74 percent of the repeats would be expected to show less deviation as a result of chance than initially observed.

These interpretations of the *p* value reveal that a hypothesis (a 9:3:3:1 ratio in this case) is never proved or disproved absolutely. Instead, a relative standard is set that serves as the basis for either rejecting or failing to reject the hypothesis. This standard is most often a probability value of 0.05. If used in chi-square analysis, a *p* value less than 0.05 means the probability is less than 5 percent that the observed deviation could be obtained by chance alone. Such a *p* value indicates that the difference between the observed and predicted results is substantial and thus serves as the basis for rejecting the null hypothesis.

On the other hand, *p* values of 0.05 or greater (1.0–0.05) indicate that the probability of the observed deviation being due to chance is 5 percent or more (5–100 percent). The conclusion is not to reject the null hypothesis. In our example, where $p = 0.26$, the hypothesis that independent assortment accounts for the results fails to be rejected based on the analysis of the experimental data. Therefore, the observed deviation can be reasonably attributed to chance.

Human Pedigrees

In all crosses discussed so far, one of the two traits for each character has been dominant to the other. Based on this observation, two significant questions may be asked:

1. Does the expression of all genes occur in this fashion?

2. Is it possible to ascertain the mode of inheritance of genes in organisms where designed crosses and the production of large numbers of offspring are not practical?

The answer to the first question is no. As we will see in Chapter 4, many modes of inheritance exist that modify the monohybrid and dihybrid ratios observed by Mendel.

The answer to the second question is yes. The pattern of inheritance of a specific phenotype can be studied even in humans.

The simplest way to study this pattern is to construct a family tree indicating the phenotype of the trait in question for each member. Such a family tree is called a **pedigree**. By analyzing the pedigree, we may be able to predict how the gene controlling the trait is inherited. If many similar pedigrees for the same trait are found, the prediction is strengthened.

Figure 3–12 shows the conventions used in constructing a pedigree. Circles represent females, and squares designate males. If the sex is unknown, a diamond may be used (II-2 in Figure 3–12). If a pedigree traces only a single trait, as Figure 3–12 does, the circles, squares, and diamonds are shaded if the phenotype being considered is expressed. Those who fail to express a recessive trait, when known with certainty to be heterozygous, have only the left half of their square or circle shaded (see II-3 and II-4).

The parents are connected by a horizontal line, and vertical lines lead to their offspring. All such offspring are called **sibs** and are connected by a horizontal **sibship line**. Sibs are placed from left to right according to birth order and are labeled with Arabic numerals. Each generation is indicated by a Roman numeral.

Twins are indicated by diagonal lines stemming from a vertical line connected to the sibship line. For **monozygotic** or **identical twins**, the diagonal lines are linked by a horizontal line [see III-5,6 in Figure 3–12(a)]. **Dizygotic** or **fraternal twins** lack this connecting line [see III-8,9 in Figure 3–12(a)]. A number within one of the symbols [see II-10–13) in Figure 3–12(b)] represents numerous sibs of the same or unknown phe-

notypes. A male whose phenotype drew the attention of a physician or geneticist is called the **propositus**, while a female in the same circumstance is called a **proposita**. Such an individual is indicated by an arrow [see III-4 in Figure 3–12(a)].

The pedigree shown in Figure 3–12 traces the pattern of inheritance of the human trait **albinism**. By analyzing the pedigree, we will see that albinism is inherited as a recessive trait.

One of the parents of the first generation, I-1, is affected. Because none of his offspring show the disorder, we might conclude that the unaffected female parent (I-2) was a homozygous normal individual. Had she been heterozygous, one-half of the offspring would be expected to exhibit albinism. However, such a small sample (three offspring) prevents any certainty in the matter.

An unaffected second generation is characteristic of a rare recessive trait. If albinism were inherited as a dominant trait, individual II-3 would have to express the disorder in order to pass it to his offspring (III-3 and III-4). He does not. Inspection of the offspring constituting the third generation [row III in Figure 3–12(a)] provides further support for the hypothesis that albinism is a

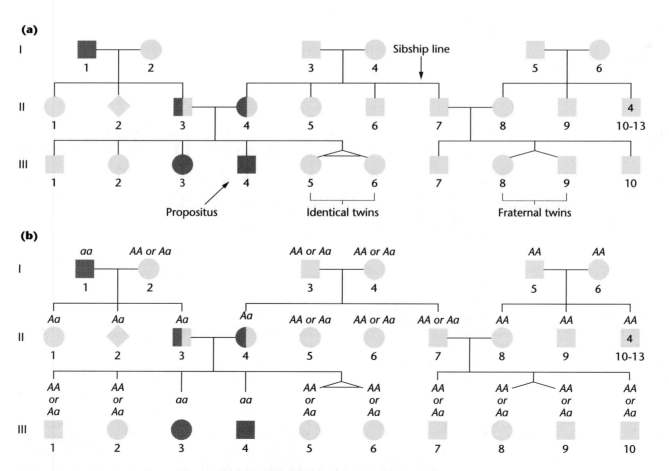

■ Figure 3–12 (a) A representative pedigree for a single character through three generations. (b) The most probable genotypes of each individual in the pedigree.

TABLE 3.2	Representative recessive and dominant human traits
Recessive Traits	**Dominant Traits**
Albinism	Achondroplasia
Alkaptonuria	Brachydactyly
Ataxia telangiectasia	Congenital stationary night blindness
Color blindness	Ehler-Danlos syndrome
Cystic fibrosis	Fascio-scapulo-humeral muscular dystrophy
Duchenne muscular dystrophy	Huntington disease
Galactosemia	Hypercholestrolemia
Hemophilia	Marfan syndrome
Lesch-Nyhan syndrome	Neurofibromatosis
Phenylketonuria	Phenylthiocarbamide tasting (PTC)
Sickle-cell anemia	Porphyria
Tay-Sachs disease	Widow's peak

recessive trait. If so, parents II-3 and II-4 are both heterozygous, and approximately one-fourth of their offspring should be affected. Two of the six offspring do show albinism. This deviation from the expected ratio is not unexpected in crosses with few offspring.

Based on this pedigree analysis and the conclusion that albinism is a recessive trait, the most probable genotypes of all individuals can be predicted. For both the first and second generations, this can be done with some certainty in only a few cases. In the third generation, for most normal individuals, we cannot determine whether they are homozygous or heterozygous.

Pedigree analysis of many traits has been an extremely valuable research technique in human genetic studies. However, this approach does not usually provide the certainty in drawing conclusions that is afforded by designed crosses yielding large numbers of offspring. Nevertheless, when many independent pedigrees of the same trait or disorder are analyzed, consistent conclusions can often be drawn. Table 3.2 lists numerous human traits and classifies them according to their recessive or dominant expression. As we will see in Chapter 4, the genes controlling some of these traits are located on the sex-determining chromosomes.

Chapter Summary

1. Over a century ago, Mendel studied inheritance patterns in the garden pea, establishing the principles of transmission genetics.

2. Mendel's postulates described the basis for the inheritance of phenotypic expression. He showed that unit factors, later called alleles, exist in pairs and exhibit a dominant/recessive relationship in determining the expression of traits.

3. Mendel postulated that unit factors (now called alleles) must segregate during gamete formation, such that each gamete receives only one of the two factors with equal probability.

4. Mendel's final postulate, of independent assortment, states that each pair of unit factors segregates independently of other such pairs. As a result, all possible combinations of gametes will be formed with equal probability.

5. The discovery of chromosomes in the late 1800s and subsequent studies of their behavior during meiosis led to the rebirth of Mendel's work, linking the behavior of his unit factors with that of chromosomes during meiosis.

6. The Punnett square and the forked-line methods are used to predict the probabilities of phenotypes (and genotypes) from crosses involving two or more gene pairs.

7. Genetic ratios are expressed as probabilities. Thus, deriving outcomes of genetic crosses relies on an understanding of the laws of probability.

8. Statistical analysis is used to test the validity of experimental outcomes. In genetics, variations from the expected ratios are anticipated owing to chance deviation.

9. Chi-square analysis provides the basis for assessing the null hypothesis, namely, that there is no real difference between the expected and observed values. As such, it tests the probability of whether observed variations may be attributed to chance deviation.

10. Pedigree analysis provides a method for studying the inheritance pattern of human traits over several generations. This often provides the basis for determining the mode of inheritance of human characteristics and disorders.

GENETICS, TECHNOLOGY, AND SOCIETY

Improving the Genetic Fate of Purebred Dogs

For a dog-lover, there is nothing quite so heart-breaking as watching a dog slowly go blind, standing by helplessly as he struggles to adapt to a life of perpetual darkness. That's what happens in progressive retinal atrophy (PRA), an inherited disorder first described in Gordon setters in 1911. Since then, PRA has been found in many other breeds of dogs, including Irish setters, border collies, Norwegian elkhounds, toy poodles, miniature schnauzers, cocker spaniels, and Siberian huskies.

The products of many genes are required for the development and maintenance of a healthy retina, and a defect in any one of them would have the potential to cause retinal dysfunction. Decades of research have led to the identification of four such genes (*rcd1*, *rcd2*, *erd*, and *prcd*), and more are likely to be discovered. Different genes are mutated in different breeds,e.g., *rcd1* in Irish setters and *prcd* in Labrador retrievers. In all dogs examined so far, PRA shows a recessive pattern of inheritance.

Whichever mutation is responsible, PRA is more common in certain purebred dogs than in mixed breeds. The development of distinct breeds of dogs has involved the intensive selection for desirable attributes, whether a particular size, shape, color, or behavior. Since most desired characteristics are determined by recessive alleles, the fastest way to increase the homozygosity of these alleles and establish the characteristics in the population is to mate close relatives, which are likely to carry the same alleles. In the practice known as line breeding, for example, dogs may be mated to a cousin or a grandparent.

Unfortunately, the generations of inbreeding that have established favorable characteristics in pure breeds have also increased the homozygosity of certain harmful recessive alleles such as in PRE, resulting in other inherited diseases. Many breeds are plagued with inherited hip dysplasia, which is particularly prevalent in German shepherds. Deafness and kidney disorders are common genetic maladies in Dalmatians. Over 300 such diseases have been characterized in purebred dogs, and most breeds of dogs are affected by one or more of them. By some estimates, fully 25 percent of purebred dogs are affected with one genetic ailment or another. Inbreeding is not the cause of these genetic diseases, but there is general agreement that breeding practices, if not executed properly, can increase the frequency of disease.

Inbreeding is the likely explanation for the elevated frequency of PRA in certain breeds. Fortunately, advances in canine genetics are beginning to provide new tools for the breeding of healthy dogs. Since 1995, a genetic test has been available to identify mutations in the *rcd1* gene, which is responsible for the form of PRA that occurs in Irish setters. This test is now being used to identify heterozygous carriers of *rcd1* mutations, dogs that show no symptoms of PRA but, if mated with other carriers, pass the trait on to about 25 percent of their offspring. Eliminating PRA carriers from breeding programs could, in theory, eradicate this condition from Irish setters in just a few generations.

It took many years to identify *rcd1* and other genes responsible for PRA. In the future, the isolation of genes underlying canine inherited disease should be faster, thanks to the Dog Genome Project, a collaborative effort involving scientists at the University of California, the University of Oregon, the Fred Hutchinson Cancer Research Center in Seattle, and other research centers. Its goal is the creation of a complete genetic map of the 39 chromosomes in the dog. In late 1997, preliminary stages were completed, paving the way for the development of a more comprehensive map. The identification of genes that cause inherited disease will allow genetic tests for the diagnosis of disease before symptoms develop, as well as for ensuring that breeding animals are free of harmful recessive alleles.

The Dog Genome Project may have benefits to humans beyond the reduction in disease in their canine companions. That's because about 85 percent of the genes in the dog genome have equivalents in humans, so identification of a disease-causing dog gene may be a short-cut to the isolation of the corresponding gene in humans. For example, some forms of PRA in dogs appear to be equivalent to retinitis pigmentosa (RP) in humans, which afflicts about 1.5 million people worldwide. Despite much research, RP remains poorly understood, and current treatments only slow its progress. Understanding the genetic basis of PRA could lead to breakthroughs in the diagnosis and treatment of RP, potentially saving the sight of thousands of people every year. By contributing to the cure of human diseases, dogs may prove to be "man's best friend" in an entirely new way.

References
Berson, E. L. 1996. Retinitis pigmentosum: Unfolding its mystery. *Proc. Natl. Acad. Sci. USA.* 93: 4526–28.
Ray, K., Baldwin, V., Acland, G., and Aquirre, G. 1995. Molecular diagnostic tests for ascertainment of genotype at the rod cone dysplasia (*rcd1* locus) in Irish setters. *Curr. Eye Res.* 14: 243.
Smith, C.A. 1994. New hope for overcoming canine inherited disease. *Am. J. Vet. Med. Assoc.* 204: 41–46.

INSIGHTS
and
SOLUTIONS

Students demonstrate their knowledge of transmission genetics by solving genetics problems. Success at this task represents not only comprehension of theory but its application to more practical genetic situations. Most students find problem solving in genetics to be challenging but rewarding. This section is designed to provide basic insights into the reasoning essential to this process. Genetics problems are in many ways similar to algebraic word problems. The approach taken should be identical: (1) Analyze the problem carefully; (2) translate words into symbols, defining each one first; and (3) choose and apply a specific technique to solve the problem. The first two steps are critical. The third step is largely mechanical.

The simplest problems are those that state all necessary information about the P_1 generation and ask you to find the expected ratios of the F_1 and F_2 genotypes and/or phenotypes. The following steps should always be followed when you encounter this type of problem:

1. Determine insofar as possible the genotypes of the individuals in the P_1 generation.

2. Determine what gametes may be formed by the P_1 parents.

3. Recombine gametes either by the Punnett square method, by the forked-line method, or, if the situation is very simple, by inspection. Read the F_1 phenotypes directly.

4. Repeat the process to obtain information about the F_2 generation.

Determining the genotypes from the given information requires an understanding of the basic theory of transmission genetics. For example, consider the following problem.

A recessive mutant allele, *black,* causes a very dark body in *Drosophila* when homozygous. The wild-type (normal) color is described as gray. What F_1 phenotypic ratio is predicted when a black female is crossed to a gray male whose father was black?

To work this problem, you must understand dominance and recessiveness as well as the principle of segregation. Furthermore, you must use the information about the male parent's father. You can work out the problem as follows.

1. Because the female parent is black, she must be homozygous for the mutant allele (*bb*).

2. The male parent is gray; therefore, he must have at least one dominant allele (*B*). Because his father was black (*bb*) and he received one of the chromosomes

bearing these alleles, the male parent must be heterozygous (*Bb*).

From here, the problem is simple:

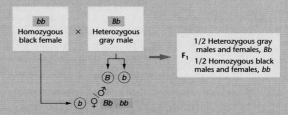

Apply this approach to the following problems.

1. In Mendel's work, he found that full pods are dominant to constricted pods while round seeds are dominant to wrinkled seeds. One of his crosses was between full, round plants and constricted, wrinkled plants. From this cross, he obtained an F_1 that was all full and round. In the F_2 Mendel obtained his classic 9:3:3:1 ratio. Using the above information, determine the expected F_1 and F_2 results of a cross between homozygous constricted, round and full, wrinkled plants.

Solution:

First of all, define gene symbols for each pair of contrasting traits. Select the lowercase forms of the first letter of the recessive traits to designate those phenotypes, and use the uppercase forms to designate the dominant traits. Thus, use *C* and *c* to indicate full and constricted, and use *W* and *w* to indicate the round and wrinkled phenotypes, respectively.

Now, determine the genotypes of the P_1 generation, form gametes, reconstitute the F_1 generation, and read off the phenotype(s):

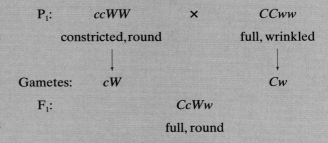

You can see immediately that the F_1 generation expresses both dominant phenotypes and is heterozygous for both gene pairs. Thus, we can expect that the F_2 generation will yield the classic Mendelian ratio of 9:3:3:1. Let's work it out

(continues)

(continued)

anyway, just to confirm this, using the forked-line method. Because both gene pairs are heterozygous and can be expected to assort independently, we can predict the F_2 outcomes from each gene pair separately and then proceed with the forked-line method.

Every F_2 offspring is subject to the following probabilities:

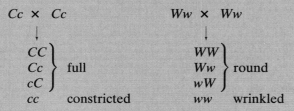

The forked-line method then allows us to confirm the 9:3:3:1 phenotypic ratio. Remember that this represents proportions of 9/16:3/16:3/16:1/16. Note that we are applying the product law as we compute the final probabilities:

3/4 full ⟨ — 3/4 round $\xrightarrow{(3/4)(3/4)}$ 9/16 full, round
— 1/4 wrinkled $\xrightarrow{(3/4)(1/4)}$ 3/16 full, wrinkled

1/4 constricted ⟨ — 3/4 round $\xrightarrow{(1/4)(3/4)}$ 3/16 constricted, round
— 1/4 wrinkled $\xrightarrow{(1/4)(1/4)}$ 1/16 constricted, wrinkled

2. In another cross, involving parent plants of unknown genotype and phenotype, the offspring shown below were obtained. Determine the genotypes and phenotypes of the parents.

> Offspring: 3/8 full, round
> 3/8 full, wrinkled
> 1/8 constricted, round
> 1/8 constricted, wrinkled

Solution:

This problem is more diffcult and requires keener insight because you must work backward. The best approach is to consider the outcomes of pod shape separately from those of seed texture.

Of all plants, 6/8 (3/4) are full and 2/8 (1/4) are constricted. Of the various genotypic combinations that can serve as parents, which will give rise to a ratio of 3/4:1/4? Because this ratio is identical to Mendel's monohybrid F_2 results, we can propose that both unknown parents share the same genetic characteristic as the monohybrid F_1 parents; they must both be heterozygous for the genes controlling pod shape and thus are

$$Cc$$

Before accepting this hypothesis, let's consider the possible genotypic combinations that control seed texture. If we consider this characteristic alone, we can see that the traits are expressed in a ratio of 4/8 (1/2) round:4/8 (1/2) wrinkled. In order to generate such a ratio, the parents cannot both be heterozygous or their offspring would yield a 3/4:1/4 phenotypic ratio. They cannot both be homozygous or all offspring would express a single phenotype. Thus, we are left with testing the hypothesis that one parent is homozygous and one is heterozygous for the alleles controlling texture. The potential case of $WW \times Ww$ will not work, since it also would yield only a single phenotype. This leaves us with the potential case of $Ww \times ww$. Offspring in such a mating will yield 1/2 Ww (round):1/2 ww (wrinkled), exactly the outcome we are seeking.

Now, let's combine our hypotheses and predict the outcome of crossing. In our solution, we will use a dash (−) to indicate that the second allele may be either dominant or recessive, since we are only predicting phenotypes.

3/4 $C-$ ⟨ — 1/2 Ww ⟶ 3/8 $C-Ww$ full, round
— 1/2 ww ⟶ 3/8 $C-ww$ full, wrinkled

1/4 cc ⟨ — 1/2 Ww ⟶ 1/8 $ccWw$ constricted, round
— 1/2 ww ⟶ 1/8 $ccww$ constricted, wrinklec

As we can see, this cross produces offspring according to our initial information. Thus, we have solved the problem. Note that in this solution, we have used genotypes in the forked-line method, in contrast to the use of phenotypes in the earlier solution.

3. Determine the probability that a plant of genotype $CcWw$ will be produced from parental plants of the genotypes $CcWw$ and $Ccww$.

Solution:

Because the two gene pairs demonstrate straightforward dominance and recessiveness and assort independently during gamete formation, we need only calculate the individual probabilities of obtaining the two separate outcomes (Cc and Ww) and apply the product law to calculate the final probability:

$Cc \times Cc \longrightarrow$ 1/4 CC:1/2 Cc:1/4 cc
$Ww \times ww \longrightarrow$ 1/2 Ww:1/2 ww

$$p = (1/2\ Cc)\ (1/2\ Ww) = 1/4\ CcWw$$

4. In the laboratory, a genetics student crossed flies with normal, long wings to flies expressing the *dumpy* mutation (truncated wings), which she believed was a recessive trait. In the F_1, all flies had long wings. In the F_2, the following results were obtained:

792 long-winged flies
208 dumpy-winged flies

The student tested the hypothesis that the dumpy wing is inherited as a recessive trait by performing χ^2 analysis of the F_2 data.

(a) What ratio was hypothesized?

(b) Did the χ^2 analysis support the hypothesis?

(c) What do the data suggest about the *dumpy* mutation?

Solution:

(a) The student hypothesized that the F_2 data (792:208) fit Mendel's 3:1 monohybrid ratio for recessive genes.

(b) The initial step in χ^2 analysis is to calculate the expected results (e) if the ratio is 3:1 and the deviations (d) between the expected and observed data are:

Ratio	o	e	d	d^2	d^2/e
3/4	792	750	+42	1764	2.35
1/4	208	250	−42	1764	7.06

Total = 1000

$$\chi^2 = \sum \frac{d^2}{e}$$
$$= 2.35 + 7.06$$
$$= 9.41$$

Consulting Figure 3–11 allows us to determine the probability (p). This value will allow us to determine whether the deviations can be attributed to chance. There are two possible outcomes (n), so the degrees of freedom (df) = $n - 1$ or 1. The table in Figure 3–11 shows that p = 0.01 to 0.001. The graph gives an estimate of about 0.001. That p is less than 0.05 causes us to reject the null hypothesis. The data do not fit a 3:1 ratio statistically.

(c) When we accept Mendel's 3:1 ratio as a valid expression of the monohybrid cross, numerous assumptions are made. One of these may explain why the null hypothesis was rejected. We must assume that all genotypes are equally viable. That is, genotypes yielding long wings are equally likely to survive from fertilization through adulthood, when the data were collected, as the genotype yielding dumpy wings. Further study would reveal that dumpy flies are somewhat less viable than normal flies. As a result, we would expect less than 1/4 of the total offspring to express dumpy. This observation is borne out in the data, although we have not proven that this is the reason.

Key Terms

albinism, 58
allele, 46
branch diagram, 50
chance deviation, 55
chi-square (χ_2) analysis, 55
chromosomal theory of inheritance, 53
continuous variation, 52
degrees of freedom (df), 56
dihybrid cross, 47
discontinuous variation, 52
dizygotic twins, 58
dominant, 46
F_1 (first filial) generation, 44

F_2 (second filial) generation, 44
forked-line method, 50
genotype, 46
heterozygote, 46
homozygote, 46
independent assortment, 49
maternal parent, 53
monohybrid cross, 44
monozygotic twins, 58
multiple alleles, 53
null hypothesis, 55
P_1 (parental) generation, 44
paternal parent, 53
pedigree, 57

phenotype, 46
probability, 56
product law, 49
propositus (proposita), 58
recessive, 46
reciprocal cross, 44
segregation, 46
sum law, 55
test cross, 47
transmission genetics, 43
trihybrid cross, 50
unit factors, 44
unit of inheritance, 43

Problems and Discussion Questions

When working genetics problems in this and succeeding chapters, always assume that members of the P_1 generation are homozygous, unless the information given indicates or requires otherwise.

1. In a cross between a black and a white guinea pig, all members of the F_1 generation are black. The F_2 generation is made up of approximately 3/4 black and 1/4 white guinea pigs. Diagram this cross, showing the genotypes and phenotypes.

2. Albinism in humans is inherited as a simple recessive trait. For the following families, determine the genotypes of the parents and offspring. When two alternative genotypes are possible, list both. (a) Two nonalbino (normal) parents have five children, four normal and one albino. (b) A normal male and an albino female have six children, all normal.

3. In a problem involving albinism (such as Problem 2), which of Mendel's postulates are illustrated?

4. Why was the garden pea a good choice as an experimental organism in Mendel's work?

5. Pigeons may exhibit a checkered or plain pattern. In a series of controlled matings, the following data were obtained:

		F_1 Progeny	
	P_1 cross	Checkered	Plain
(a)	checkered × checkered	36	0
(b)	checkered × plain	38	0
(c)	plain × plain	0	35

How are the checkered and plain patterns inherited? Predict the results of the $F_1 \times F_1$ mating from cross (b).

6. Mendel crossed peas having round seeds and yellow cotyledons with peas having wrinkled seeds and green cotyledons. All the F_1 plants had round seeds with yellow cotyledons. Diagram this cross through the F_2 generation using both the Punnett square and forked-line, or branch diagram, methods.

7. Determine the genotypes of the parental plants shown below by analyzing the phenotypes of the offspring from these crosses.

	Parental Plants	*Offspring*
(a)	round, yellow × round, yellow	3/4 round, yellow 1/4 wrinkled, yellow
(b)	round, yellow × wrinkled, yellow	6/16 wrinkled, yellow 2/16 wrinkled, green 6/16 round, yellow 2/16 round, green
(c)	round, yellow × wrinkled, green	1/4 round, yellow 1/4 round, green 1/4 wrinkled, yellow 1/4 wrinkled, green

8. Are any of the crosses in Problem 7 a test cross? If so, which?

9. Which of Mendel's postulates can be demonstrated in the crosses of Problem 7, but not in those in Problems 1 and 5? State this postulate.

10. Correlate Mendel's four postulates with what is now known about homologous chromosomes, genes, alleles, and the process of meiosis.

11. What is the basis for homology among chromosomes?

12. Distinguish between homozygosity and heterozygosity.

13. In *Drosophila*, gray body color is dominant to ebony body color, while long wings are dominant to vestigial wings. Work the following crosses through the F_2 gen-

eration and determine the genotypic and phenotypic ratios for each generation. Assume that the P_1 individuals are homozygous: (a) gray, long × ebony, vestigial; (b) gray, vestigial × ebony, long; (c) gray, long × gray, vestigial.

14. How many different types of gametes can be formed by individuals of the following genotypes: (a) *AaBb*, (b) *AaBB*, (c) *AaBbCc*, (d) *AaBBcc*, (e) *AaBbcc*, and (f) *AaBbCcDdEe*? What are they in each case?

15. Using the forked-line, or branch diagram, method, determine the genotypic and phenotypic ratios of the trihybrid crosses (a) *AaBbCc × AaBBCC*, (b) *AaBBCc × aaBBCc*, and (c) *AaBbCc × AaBbCc*.

16. Mendel crossed peas with round, green seeds to ones with wrinkled, yellow seeds. All F_1 plants had seeds that were round and yellow. Predict the results of test-crossing these F_1 plants.

17. Below are shown F_2 results of two of Mendel's monohybrid crosses. State a null hypothesis to be tested using χ^2 analysis. Calculate the χ^2 value and determine the p value for both. Interpret the p values. Which of the two crosses shows a greater amount of deviation?

(a) Full pods 882
Constricted pods 299

(b) Violet flowers 705
White flowers 224

18. In one of Mendel's dihybrid crosses, he observed 315 round, yellow, 108 round, green, 101 wrinkled, yellow, and 32 wrinkled, green F_2 plants. Analyze these data using the chi-square test to see if (a) they fit a 9:3:3:1 ratio; (b) the round:wrinkled traits fit a 3:1 ratio; (c) the yellow:green traits fit a 3:1 ratio.

19. A geneticist, in assessing data that fell into two phenotypic classes, observed values of 250:150. She decided to perform chi-square analysis using two different null hypotheses: (a) the data fit a 3:1 ratio; and (b) the data fit a 1:1 ratio. Calculate the χ^2 values for each hypothesis. What can be concluded about each hypothesis?

20. The basis for rejection of any null hypothesis is arbitrary. The researcher can set more or less stringent standards by deciding to raise or lower the critical p value used to reject or fail to reject the hypothesis. Would the use of a standard of $p = 0.10$ be more or less stringent in failing to reject the null hypothesis? Explain.

21. For the following pedigree, predict the mode of inheritance and the resulting genotypes of each individual. Assume that the alleles A and a control the expression of the trait.

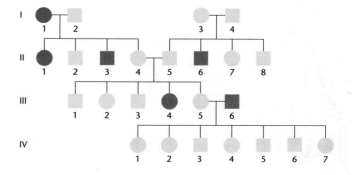

22. Which of Mendel's postulates are illustrated by the pedigree in Problem 21? List and define these postulates.

23. The following pedigree follows the inheritance of myopia (near-sightedness) in humans. Predict whether the disorder is inherited as a dominant or a recessive trait. Based on your prediction, indicate the most probable genotype for each individual.

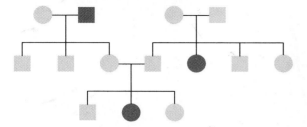

24. Two true-breeding pea plants were crossed. One parent is round, terminal, violet, constricted, while the other expresses the contrasting phenotypes of wrinkled, axial, white, full, respectively. The four pairs of contrasting traits are controlled by four genes, each located on a separate chromosome. In the F_1, only round, axial, violet, and full were expressed. In the F_2 all possible combinations of these traits were expressed in ratios consistent with Mendelian inheritance.

(a) What conclusion about the inheritance of these traits can be drawn based on the F_1 results?

(b) In the F_2 results, which phenotype appeared most frequently? Write a mathematical expression that predicts the frequency of occurrence of this phenotype.

(c) Which F_2 phenotype is expected to occur least frequently? Write a mathematical expression that predicts this frequency.

(d) In the F_2 generation, how often is either of the P_1 phenotypes likely to occur?

(e) If the F_1 plant were testcrossed, how many different phenotypes would be produced? How does this number compare to the number of different phenotypes in the F_2 generation discussed above?

25. Tay–Sachs disease (TSD) is an inborn error of metabolism that results in death by the age of 2. You are a genetic counselor and one day you interview a phenotypically normal couple who consult you because the man had a female first cousin (on his father's side) who died from TSD and the woman had a maternal uncle with TSD. There are no other known cases in either family, and none of the matings were/are between related individuals. Assume that this trait is rare in this population.

(a) Using standard pedigree symbols, draw a pedigree of these individuals' families showing the relevant individuals.

(b) This couple wants to know what the probability is that they both are heterozygous for the TSD allele. Calculate these probabilities.

(c) They also want to know what the probability is that neither of them is heterozygous.

(d) Finally, they ask the probability that one of them is heterozygous but the other is not. (Hint: The answers to b, c, and d should add up to 1.)

Selected Readings

Cummings, M. R. 1997. *Human heredity: principles and issues,* 4th ed. St. Paul, MN: West.

Dunn, L. C. 1965. *A short history of genetics.* New York: McGraw-Hill.

Olby, R. C. 1985. *Origins of Mendelism,* 2nd ed. London: Constable.

Orel, V. 1996. *Gregor Mendel. The first geneticist.* New York: Oxford Univ. Press.

Peters, J., ed. 1959. *Classic papers in genetics.* Englewood Cliffs, NJ: Prentice-Hall.

Sokal, R. R., and Rohlf, F. J. 1987. *Introduction to biostatistics,* 2nd ed. New York: W. H. Freeman.

Soudek, D. 1984. Gregor Mendel and the people around him. *Am. J. Hum. Genet.* 36:495–98.

Stern, C. 1950. *The birth of genetics.* (Supplement to *Genetics* 35.)

Stern, C., and Sherwood, E. 1966. *The origin of genetics: a Mendel source book.* San Francisco: W. H. Freeman.

Stubbe, H. 1972. *History of genetics: from prehistoric times to rediscovery of Mendel's laws.* Cambridge, MA: MIT Press.

Sturtevant, A. H. 1965. *A history of genetics.* New York: Harper & Row.

Voeller, B. R., ed. 1968. *The chromosome theory of inheritance: Classical papers in development and heredity.* New York: Appleton-Century-Crofts.

Welling, F. 1991. Historical study: Johann Gregor Mendel 1822–1884. *Am. J. Med. Genet.* 40:1–25.

Mice expressing the agouti, yellow, and black phenotypes.

CHAPTER OUTLINE

CHAPTER
4

Modification of Mendelian Ratios

Chapter Concepts

Specific phenotypes are often controlled by one or more gene pairs whose alleles exhibit modes of expression other than dominance and recessiveness. In all such cases, however, the Mendelian principles of segregation and independent assortment are operative during the distribution of the alleles into gametes.

In Chapter 3, we discussed the simplest principles of transmission genetics. We saw that genes are present on homologous chromosomes and that these chromosomes **segregate** from each other and **assort independently** with other segregating chromosomes during gamete formation. These two postulates are the fundamental principles of gene transmission from parent to offspring. However, when gene expression does not adhere to a simple dominant/recessive mode or when more than one pair of genes influences the expression of a single character, the classic 3:1 and 9:3:3:1 ratios are usually modified. Although more complex modes of inheritance result, the fundamental principles set down by Mendel still hold true in these situations.

In this chapter, our discussion will initially be restricted to the inheritance of traits that are under the control of only one set of genes. In diploid organisms, which have homologous pairs of chromosomes, two copies of each gene influence such traits. The copies need not be identical because alternative forms of genes, or **alleles**, occur within populations. How alleles influence a given phenotype will be our major consideration. Then we will proceed to consider how a single phenotype may be controlled by more than one set of genes. This general phenomenon is referred to as **gene interaction**, indicating that phenotypes are frequently under the influence of the expression of more than one gene pair.

We will conclude by examining cases where genes are present on the X chromosome, illustrating **X-linkage**. Prior to that point, in this and preceding chapters, we have restricted our discussion to chromosomes other than the X and Y pair, referred to as **autosomes**. As we will see, X-linkage provides yet another modification of Mendelian ratios.

Potential Function of an Allele

Following the rediscovery of Mendel's work in the early 1900s, research focused on the many ways in which genes can influence an individual's phenotype. This course of investigation, stemming from Mendel's findings, is called **neo-Mendelian genetics** (*neo* from the Greek word meaning since or new).

Each type of inheritance described in this chapter was investigated when observations of genetic data did not conform precisely to the expected Mendelian ratios. Hypotheses that modified and extended the Mendelian principles were proposed and tested with specifically designed crosses. Explanations for these observations were in accordance with the principle that a phenotype is under the control of one or more genes located at specific loci on one or more pairs of homologous chromosomes. If we adhere to the principles of segregation and independent assortment, we can predict accurately the transmission of any number of allele pairs.

To understand the various modes of inheritance, we must first examine the potential function of an **allele**. Alleles are alternative forms of the same gene, containing modified genetic information and often specifying

67

an altered gene product. For example, in human populations there are well over 100 known alleles of the genes that specify the protein portions of hemoglobin. All such alleles store information necessary for the synthesis of a polypeptide chain of hemoglobin, but each allele specifies a modification of the chemical composition. Once manufactured, however, the product of an allele may or may not have its function altered.

The allele that occurs most frequently in a population, or the one that is arbitrarily designated as normal, is often called the **wild-type allele**. In diploids, it is usually dominant, such as the allele for tall plants in the garden pea, and its product is functional in the cell. Wild-type alleles are, of course, responsible for the corresponding wild-type phenotype and thus serve as standards for comparison against all mutations occurring at a particular locus.

The process of **mutation** is the source of new alleles. Each new allele *may* be recognized by a change in the phenotype. A new phenotype results from a change in functional activity of the cellular product controlled by that gene. Usually, the alteration or mutation is expressed as a loss of the specific wild-type function. For example, if a gene is responsible for the synthesis of a specific enzyme, a mutation in the gene *may* change the conformation of this enzyme and eliminate its affinity for the substrate. Such a mutation will result in a total loss of function. Conversely, another organism may have a different mutation in the same gene that results in an enzyme with either a reduced or an increased affinity for binding the substrate. This mutation, representing a separate allele, may reduce or enhance rather than eliminate the functional capacity of the gene product. In either case, the overt phenotype of the organism *may or may not* be altered in a discernible way. It is important to realize that mutation and the resultant allele does not always produce a clearly altered phenotype. Of course, those that we use as examples do express altered phenotypes, so that we may follow them in our discussion!

Symbols for Alleles

In Chapter 3 we learned to symbolize alleles for very simple Mendelian traits. We used the lowercase form of the initial letter of the name of a recessive trait to denote the recessive allele and used the same letter in uppercase form to refer to the corresponding dominant allele (an approach used by Mendel). Thus, for *tall* and *dwarf*, where dwarf is recessive, D and d represent the alleles responsible for these respective traits. Mendel used upper- and lowercase letters to symbolize alleles.

Another useful system was developed in *Drosophila* genetics to discriminate between wild-type and mutant traits. In this system, the initial letter, or a combination of several letters, of the name of the mutant trait is selected. If the trait is recessive, the lowercase form of the letters is used; if it is dominant, the uppercase form of the initial letter is used. The contrasting wild-type trait is denoted by the same letter, but with a + as a superscript. This system works nicely for other organisms as well, provided that there is a distinct wild-type phenotype for the character under consideration.

For example, *ebony* is a recessive body color mutation in the fruit fly, *Drosophila melanogaster*. The normal wild-type body color is gray. Using the above system, *ebony* is denoted by the symbol e, while gray is denoted by e^+. If we focus on the ebony mutation, the responsible locus may be occupied by either the wild-type allele (e^+) or the mutant allele (e). A diploid fly may thus exhibit three possible genotypes:

e^+/e^+	gray homozygote (wild type)
e^+/e	gray heterozygote (wild type)
e/e	ebony homozygote (mutant)

The slash is used to indicate that the two allele designations represent the same locus on two homologous chromosomes. If we were instead considering a dominant wing mutation such as *Wrinkled* (*Wr*) wing in *Drosophila*, the three possible designations would be Wr^+/Wr^+, Wr^+/Wr, and Wr/Wr. The latter two genotypes express the wrinkled-wing phenotype.

One advantage of this system is that further abbreviation may be used when convenient: the wild-type allele may simply be denoted by the + symbol. Using *ebony* as an example under consideration in a cross, the designations of the three possible genotypes become:

+/+	gray homozygote (wild type)
+/e	gray heterozygote (wild type)
e/e	ebony homozygote (mutant)

As we will see in Chapter 5, this abbreviation is particularly useful when two or three genes are linked together on the same chromosome and considered simultaneously.

Still other allele designations are sometimes useful, particularly when no dominance exists. In such a case, we may simply use uppercase letters and superscripts to denote alleles—for example, R^1 and R^2, L^M and L^N, I^A and I^B. These and other conventions will become apparent in the ensuing sections of this chapter.

Finally, note that in each of the many crosses discussed in the next few chapters, only one or a few gene pairs are involved. It may be useful for you to remember that in each cross, all other genes, which are not under consideration, are assumed to have no effect on the inheritance patterns described.

Incomplete (Partial) Dominance

Incomplete, or **partial**, **dominance** in the offspring is based on the observation of intermediate phenotypes generated by a cross between parents with contrasting traits. For example, if plants such as four-o'clocks or snapdragons with red flowers are crossed with plants with white flowers, offspring have pink flowers. It appears that neither red nor white flower color is dominant. Because some red pigment is produced in the F_1 intermediate-colored pink flowers, dominance appears to be incomplete or partial.

If this phenotype is under the control of a single pair of alleles and neither is dominant, the results of the F_1 (pink) $\times$ F_1 (pink) cross can be predicted. The resulting F_2 generation is shown in Figure 4–1, confirming the hypothesis that only one pair of alleles determines these phenotypes. The genotypic ratio (1:2:1) of the F^2 generation is identical to that of Mendel's monohybrid cross. **Because there is no dominance, however, the phenotypic ratio is identical to the genotypic ratio.** Note here that because neither of the alleles is recessive, we have chosen not to use upper- and lowercase letters. Instead, we have chosen R^1 and R^2 to denote the red and white alleles. We could have chosen W^1 and W^2.

Codominance

If two alleles are responsible for the production of two distinct, detectable gene products, a situation different from incomplete dominance or dominant/recessive arises. In such a case, the joint expression of both alleles in a heterozygote is called **codominance**. The **MN blood group** in humans illustrates this phenomenon and is characterized by a molecule called a **glycoprotein**, found on the surface of red blood cells. In the human population, two forms of this glycoprotein exist, designated M and N. An individual may exhibit either one or both of them.

The MN system is under the control of an autosomal locus found on chromosome 4 and two alleles designated L^M and L^N. Because humans are diploid, three combinations are possible, each resulting in a distinct blood type:

Genotype	Phenotype
$L^M L^M$	M
$L^M L^N$	MN
$L^N L^N$	N

As predicted, a mating between two MN parents may produce children of all three blood types:

$$L^M L^N \quad \times \quad L^M L^N$$

$$\downarrow$$

$$1/4 \quad L^M L^M$$
$$1/2 \quad L^M L^N$$
$$1/4 \quad L^N L^N$$

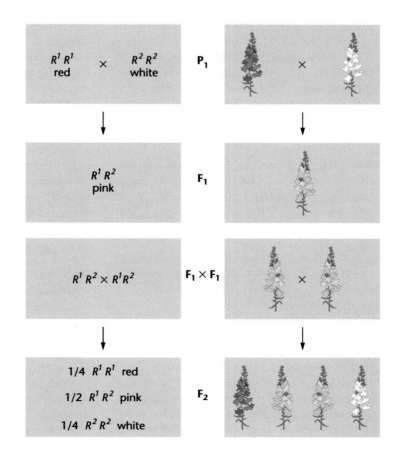

■ Figure 4–1 Incomplete dominance illustrated by flower color. The photograph illustrates red, white, and pink snapdragons.

Codominant inheritance is characterized by distinct expression of the gene products of both alleles. This characteristic distinguishes it incomplete dominance, where heterozygotes express an intermediate, blended phenotype.

Multiple Alleles

Because the information stored in any gene is extensive, mutations may modify this information in many ways. Each change has the potential for producing a different allele. Therefore, for any gene, the number of alleles of a specific gene within members of a population need not be restricted to only two. When three or more alleles of the same gene are found, the mode of inheritance is called **multiple allelism**.

The concept of multiple alleles can be studied only in populations. Any individual diploid organism has, at most, two homologous gene loci that may be occupied by different alleles of the same gene. However, among members of a species, many alternative forms of the same gene may exist.

The ABO Blood Group

The simplest case of multiple alleles is that in which there are three alternative alleles of one gene. This situation is illustrated by the **ABO blood group** in humans, discovered by Karl Landsteiner in the early 1900s. The ABO system, like the MN blood types, is characterized by the presence of **antigens** on the surface of red blood cells. The A and B antigens are distinct from the MN antigens and are under the control of a different gene, located on chromosome 9. As in the MN system, one combination of alleles in the ABO system exhibits a codominant mode of inheritance.

When individuals are tested using antisera containing antibodies against the A or B antigen, four phenotypes are revealed. Each individual has either the A antigen (A phenotype), the B antigen (B phenotype), the A and B antigens (AB phenotype), or neither antigen (O phenotype). In 1924, it was hypothesized that these phenotypes were inherited as the result of three alleles of a single gene. This hypothesis was based on studies of the blood types of many different families.

Although different designations may be used, we will use the symbols I^A, I^B, and I^O for the three alleles. The I designation stands for **isoagglutinogen**, another term for antigen. If we assume that the I^A and I^B alleles are responsible for the production of their respective A and B antigens and that I^O is an allele that does not produce any detectable A or B antigens, the various genotypic possibilities can be listed and the appropriate phenotype assigned to each:

Genotype	Antigen	Phenotype
$I^A I^A$	A	
$I^A I^O$	A	A
$I^B I^B$	B	
$I^B I^O$	B	B
$I^A I^B$	A and B	AB
$I^O I^O$	Neither	O

Note that in these assignments the I^A and I^B alleles behave dominantly to the I^O allele, but **codominantly** to each other. Our knowledge of human blood types has several practical applications, the most important of which is compatible blood transfusions.

The Bombay Phenotype

The biochemical basis of the ABO blood type system has now been carefully worked out. The A and B antigens are actually carbohydrate groups (sugars) that are bound to lipid molecules (fatty acids) protruding from the membrane of the red blood cell. The specificity of the A and B antigens is based on the terminal sugar of the carbohydrate group. Both the A and B antigens are derived from a precursor molecule called the **H substance**, to which one or two terminal sugars is added.

In extremely rare instances, first recognized in a woman in Bombay, the H substance is incompletely formed. As a result, it is an inadequate substrate for the enzyme that normally adds either terminal sugar. This condition results in blood type O and is called the **Bombay phenotype**. It has been shown to be due to a rare recessive mutation, h, at a locus separate from that controlling the A and B antigens. Thus, even though an individual may contain the I^A and/or I^B alleles, if he or she has the hh genotype, neither the A or B antigen can be added. This information helped explain why the woman in Bombay was typed as O even though one of her parents was type AB (and thus she should not have been type O), and why she was able to donate the I^B allele to her children (Figure 4–2).

Multiple Alleles in Drosophila: *The* white *Locus*

Many other phenotypes in plants and animals are known to be controlled by multiple allelic inheritance. In *Drosophila*, for example, many alleles are known at practically every locus. The recessive mutation that causes white eyes, discovered by Thomas H. Morgan and Calvin Bridges in 1912, is one of over 100 alleles that may occupy this locus. In this allelic series, eye colors range from complete absence of pigment in the *white* allele, to deep ruby in the *white-satsuma* allele, to orange in the *white-apricot* allele, to a buff color in the *white-buff* allele. These alleles are designated w, w^{sat}, w^a, and w^{bf}, respectively (Table 4.1). In each of these cases,

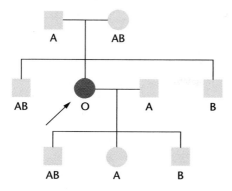

■ Figure 4–2 A partial pedigree of a woman displaying the Bombay phenotype. Functionally, her ABO blood group behaves as type O. Genetically, she is type B.

TABLE 4.1	Some of the alleles present at the *white* locus of *Drosophila melanogaster* and their eye color phenotype

Allele	Name	Eye Color
w	white	pure white
w^a	white-apricot	yellowish orange
w^{bf}	white-buff	light buff
w^{bl}	white-blood	yellowish ruby
w^{cf}	white-coffee	deep ruby
w^e	white-eosin	yellowish pink
w^{mo}	white-mottled orange	light mottled orange
w^{sat}	white-satsuma	deep ruby
w^{sp}	white-spotted	fine grain, yellow mottling
w^t	white-tinged	light pink

the total amount of pigment in these mutant eyes is reduced to less than 20 percent of that found in the brick-red, wild-type eye.

Lethal Alleles

Many gene products are essential to an organism's survival. Mutations resulting in the synthesis of a gene product that is nonfunctional can sometimes be tolerated in the heterozygous state; that is, one wild-type allele may be sufficient to produce enough of the essential product to allow survival. However, such a mutation behaves as a **recessive lethal allele**, and homozygous recessive individuals will not survive. The time of death will depend on when the product is essential, in development or in adulthood.

In some cases, the allele responsible for a lethal effect when homozygous may also result in a distinctive mutant phenotype when present heterozygously. Such an allele is behaving as a recessive lethal but is dominant with respect to the phenotype. For example, a mutation

that causes a yellow coat in mice was discovered in the early part of this century. The yellow coat varies from the normal agouti (wild-type) coat phenotype, as illustrated in Figure 4–3. Crosses between the various combinations of the two strains yield unusual results:

Crosses			
A: agouti	× agouti	$\longrightarrow$	all agouti
B: yellow	× yellow	$\longrightarrow$	2/3 yellow: 1/3 agouti
C: agouti	× yellow	$\longrightarrow$	1/2 agouti: 1/2 yellow

These results are explained on the basis of a single pair of alleles. With regard to coat color, the mutant *yellow* allele (A^Y) is dominant to the wild-type *agouti* allele (A), so heterozygous mice will have yellow coats. However, the *yellow* allele also behaves as a homozygous lethal. When present in two copies, the mice die before birth. Thus, no homozygous yellow mice are ever recovered. The genetic basis for these three crosses is provided in Figure 4–3.

Alleles in other organisms are known to behave as **dominant lethals**, where the presence of one copy of the allele will result in the death of the individual. In humans, a disorder called **Huntington disease** (previously referred to as Huntington's chorea) is due to a dominant autosomal allele H, where the onset of the disease in heterozygotes (Hh) is delayed, usually well into adulthood. Affected individuals then undergo gradual nervous and motor degeneration until they die. This lethal disorder is particularly tragic because it has such a late onset, typically at about age 40. By that time, the affected individual may have produced a family. Each child has a 50 percent probability of inheriting the lethal allele, transmitting the allele to his or her offspring, and eventually developing the disorder. The American folk singer and composer Woody Guthrie died from this disease at age 39.

Dominant lethal alleles are rarely observed. For them to exist in a population, the affected individual must reproduce before lethality caused by the allele is expressed, as can occur in Huntington disease. If all affected individuals die before reaching reproductive age, the mutant gene will not be passed to future generations and the mutation will disappear from the population unless it recurs again as a result of a new mutation.

Combinations of Two Gene Pairs

Each example discussed so far modifies Mendel's 3:1 F_2 monohybrid ratio. Therefore, combining any two of these modes of inheritance in a dihybrid cross will likewise modify the classical 9:3:3:1 ratio. Having established the foundation for the modes of inheritance of incomplete dominance, codominance, multiple alleles, and lethal genes, we can now deal with the situation of two modes of inheritance occurring simultaneously.

■ Figure 4–3 Inheritance patterns in three crosses involving the wild-type agouti allele (*A*) and the mutant yellow allele (*A^Y*) in the mouse. Note that the mutant allele behaves dominantly to the normal allele (*A*) in controlling coat color, but it also behaves as a homozygous lethal allele. The genotype *A^YA^Y* does not survive.

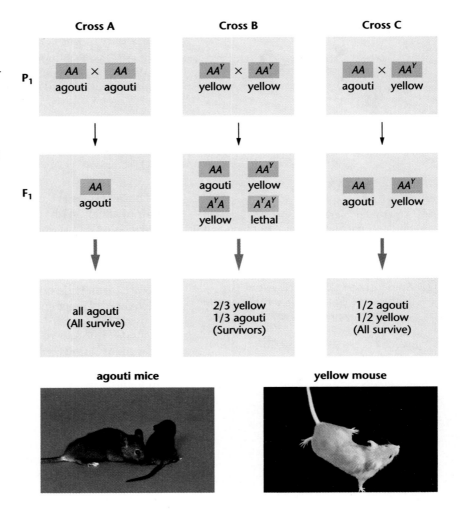

Mendel's principle of independent assortment applies to these situations, provided that the genes controlling each character are not linked on the same chromosome.

Suppose, for example, that a mating occurs between two humans who are both heterozygous for the autosomal recessive gene that causes albinism and who are both of blood type AB. What is the probability of any particular phenotypic combination occurring in each of their children? Albinism is inherited in the simple Mendelian fashion, and the blood types are determined by the series of three multiple alleles, I^A, I^B, and I^O. The solution to this problem is diagrammed in Figure 4–4, using the forked-line method.

Instead of this dihybrid cross yielding the classical four phenotypes in a 9:3:3:1 ratio, six phenotypes occur in a 3:6:3:1:2:1 ratio, establishing the expected probability for each phenotype. This is just one of many variants of modified ratios possible when different modes of inheritance are combined.

Gene Interaction: Discontinuous Variation

Soon after the rediscovery of Mendel's work, experimentation revealed that individual characteristics displaying **discrete**, or **discontinuous**, **phenotypes** are often under the control of more than one gene. This was

a significant discovery because it revealed for the first time that genetic influence on the phenotype is much more complex and sophisticated than envisioned by Mendel. Instead of single genes controlling the development of individual parts of the plant and animal body, it soon became clear that phenotypic characters may be influenced by the interactions of many different genes and their products.

The concept of **gene interaction** does not mean that two or more genes, or their products, necessarily interact *directly* to influence a particular phenotype. Instead, this concept implies that the cellular function of numerous gene products is related to the development of a common phenotype. For example, the development of an organ such as the eye of an insect is exceedingly complex and leads to a structure with multiple phenotypic manifestations. The compound eye of an adult insect, most simply described, exhibits a particular size, shape, texture, and color. The development of the eye can be envisioned as occurring as the result of a **complex cascade of developmental events leading to its formation**. This process illustrates the developmental concept of **epigenesis** (first introduced in Chapter 1), whereby each ensuing step of development increases the complexity of this sensory organ and is under the control and influence of one or more genes.

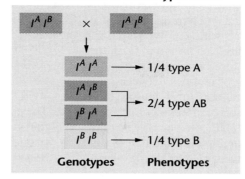

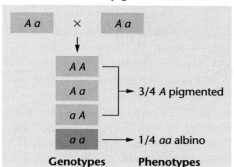

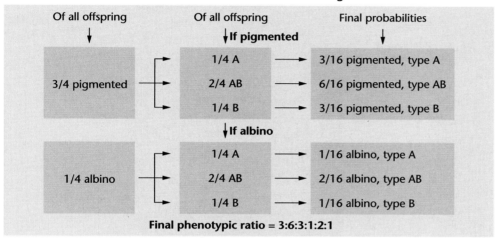

■ Figure 4–4 The calculation of the probabilities in a mating involving the ABO blood type and albinism in humans using the forked-line method.

When more than one such gene is studied simultaneously, "gene interaction" is being examined. When discrete phenotypic categories are produced that vary from one another in a qualitative way, **discontinuous variation** is said to occur. In the current chapter, we will focus our attention on examples that illustrate such variation. In the ensuing chapter, we will continue our coverage of gene interaction by considering **continuous variation**, in which phenotypic categories vary in a quantitative way. In this case, alleles provide an additive effect on the phenotype. As we illustrate discontinuous variation with numerous examples, the difference between these types of inheritance and phenotypic expression will become clearer.

Epistasis

Perhaps the best examples of gene interaction leading to discontinuous variation are those illustrating the phenomenon of **epistasis**. Derived from the Greek word meaning "stoppage," epistasis occurs when the expression of one gene or gene pair *masks* or *modifies* the expression of another gene or gene pair. The genes involved control the expression of the same general phenotypic characteristic, sometimes in an antagonistic manner, as when masking occurs. In other cases, however, the genes involved may exert their influence on one another in a complementary, or cooperative, fashion.

The phenomenon of epistasis may occur under different conditions. For example, homozygous presence of a recessive allele may prevent or override the expression of other alleles at a second locus (or several other loci). In this case, the alleles at the first locus are said to be **epistatic** to those at the second locus. The alleles at the second locus are described as being **hypostatic** to those at the first locus. In another example, a single dominant allele at the first locus may influence the expression of the alleles at a second gene locus. In a third case, two gene pairs may complement one another such that at least one dominant allele at each locus is required to express a particular phenotype. This is an example of a cooperative effect on the development of a phenotype.

Before moving into an extensive discussion of epistasis, we shall introduce an example to illustrate the above discussion. A case in which the homozygous recessive condition at one locus masks the expression of a second locus was examined earlier in this chapter when we discussed human blood types and the Bombay phenotype. There, the homozygous condition (*hh*) masked the expression of the I^A and I^B alleles. Only individuals with the *H–* genotype can form and express the A or B antigen. As a result, individuals with genotypes that include the I^A or I^B allele who are also *hh* fail to form the A or B antigen and are therefore blood type O.

An example of the outcome of matings between individuals who are heterozygous at both loci is illustrated in Figure 4–5. If many children are examined whose parents are both genotypically $I^A I^B Hh$, the phenotypic ratio of 3 A:6 AB:3 B:4 O can be predicted. Note that in this genetic cross, only one characteristic—blood type—is examined. Even so, the final phenotypic ratio is expressed in sixteenths, characteristic of crosses involving two gene pairs. In our case, the two genes are "interacting" during the expression of the phenotype under consideration.

As we discuss other examples of gene interaction, we make several assumptions and adopt certain conventions.

1. In each case, distinct phenotypic classes are produced, each clearly discernible from all others. Such traits illustrate **discontinuous variation**, in which phenotypic categories are *qualitatively* different from one another.

2. The genes considered in each cross are not linked on the same chromosome and therefore assort independently of one another during gamete formation. So that you may easily compare results, the alleles at each gene pair of all crosses will be designated *A, a* and *B, b* in each case.

$$I^A I^B \; Hh \; \times \; I^A I^B \; Hh$$

Consideration of blood types

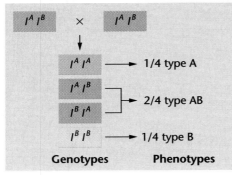

Consideration of H substance

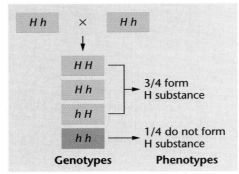

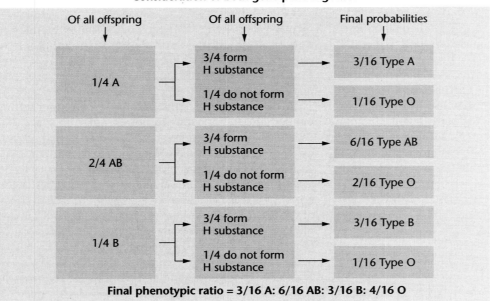

Final phenotypic ratio = 3/16 A: 6/16 AB: 3/16 B: 4/16 O

■ Figure 4–5 The outcome of a mating between individuals who are heterozygous at two genes determining their ABO blood type. Final phenotypes are calculated by considering both genes separately and then combining the results using the forked-line method.

3. When we assume that dominance exists between the alleles of any gene pair, such that *AA* and *Aa* or *BB* and *Bb* are equivalent in their genetic effects, the designations A– or B– will be used for both combinations. Therefore, the dash (–) indicates that either the dominant or recessive allele may be present, without consequence to the phenotype.

4. All P_1 crosses involve homozygous individuals, and in all cases we will use the same allele designations (e.g., *AABB × aabb*, *AAbb × aaBB*, or *aaBB × AAbb*). Therefore, each F_1 generation will consist of only heterozygotes of genotype *AaBb*.

5. In each example, the F_2 generation produced from these heterozygous parents will be the main focus of analysis. When two genes are involved (Figure 4–6), the F_2 genotypes fall into four main groupings: 9/16 *A–B–*, 3/16 *A–bb*, 3/16 *aaB–*, and 1/16 *aabb*. Provided that dominance is operable, all genotypes in each group are equivalent in their effect on the phenotype.

The study of gene interaction has revealed inheritance patterns in which these four groups are recombined in various ways, modifying the 9:3:3:1 phenotypic ratio, characteristic of Mendel's dihybrid cross (Figure 4–7). In the next several sections, we shall discuss a number of these examples.

Our first specific example is seen in the inheritance of coat color in mice (case 1 of Figure 4–7). Normal wild-type coat color is agouti, a grayish pattern formed by alternating bands of pigment on each hair. Agouti is dominant to black (nonagouti) hair, which is caused by a recessive mutation, *a*. Thus, *A-* results in agouti, while *aa* yields black coat color. When it is homozygous, a recessive mutation, *b*, at a separate locus, eliminates pigmentation altogether, yielding albino mice (*bb*), regardless of the genotype at the other locus. The presence of at least one *B* allele allows pigmentation to occur in much the same way that the *H* allele in humans allows the expression of the ABO blood types. In a cross between agouti (*AABB*) and albino (*aabb*), members of the F_1 are all *AaBb* and have agouti coat color. In the F_2 progeny of a cross between two F_1 heterozygotes, the following genotypes and phenotypes are observed:

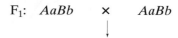

F_1:   *AaBb*   ×   *AaBb*

F_2 Ratio	Genotype	Phenotype	Final Phenotypic Ratio
9/16	*A−B−*	agouti	9/16 agouti
3/16	*A−bb*	albino	4/16 albino
3/16	*aaB−*	black	
1/16	*aabb*	albino	3/16 black

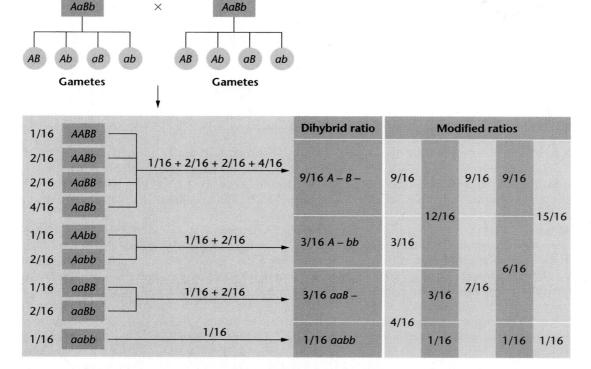

■ Figure 4–6 Generation of the various modified dihybrid ratios from the nine unique genotypes produced in a cross between individuals who are heterozygous at two genes.

			AABB 1/16	AABb 2/16	AaBB 2/16	AaBb 4/16	AAbb 1/16	Aabb 2/16	aaBB 1/16	aaBb 2/16	aabb 1/16	Final phenotypic ratio
F₁ AaBb × AaBb →						F₂ genotypes						
	Organism	Character										
Case	Pea	Mendel's dihybrid	9/16				3/16		3/16		1/16	9:3:3:1
1	Mouse	Coat color	agouti				albino		black		albino	9:3:4
2	Squash	Color	white						yellow		green	12:3:1
3	Pea	Flower color	purple						white			9:7
4	Squash	Fruit shape	disc				sphere				long	9:6:1
5	Chicken	Color	white						colored		white	13:3
6	Mouse	Color	white-spotted				white		colored		white-spotted	10:3:3
7	Shepherd's purse	Seed capsule	triangular								ovoid	15:1
8	Flour beetle	Color	red	sooty	red	sooty	black		jet		black	6:3:3:4

■ Figure 4–7   The basis of modified dihybrid F₂ phenotypic ratios, resulting from crosses between doubly heterozygous F₁ individuals. The four groupings of the F₂ genotypes shown in Figure 4–6 and across the top of this figure are combined in various ways to produce these ratios.

Gene interaction yielding the observed 9:3:4 F₂ ratio can be envisioned as a two-step process:

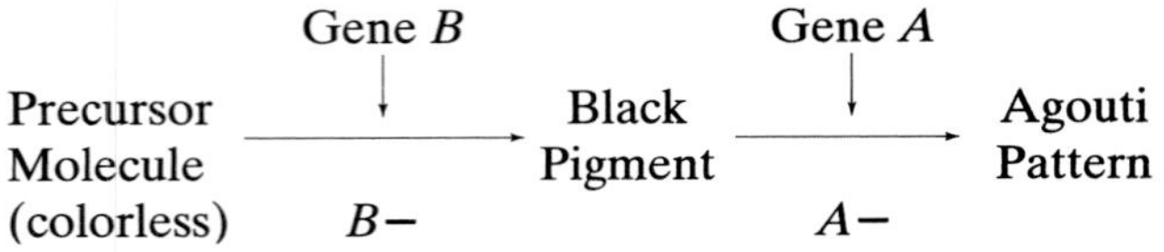

```
                Gene B              Gene A
                  ↓                   ↓
Precursor  ───────────→  Black  ───────────→  Agouti
Molecule                 Pigment              Pattern
(colorless)      B–                    A–
```

In the presence of a B allele, black pigment can be made from a colorless substance. In the presence of an A allele, the black pigment is deposited during the development of hair in a pattern that produces the agouti phenotype. If the aa genotype occurs, all of the hair remains black. If the bb genotype occurs, no black pigment is produced, regardless of the presence of the A or a alleles, and the mouse is albino. Therefore, the bb genotype masks or suppresses the expression of the A gene, illustrating epistasis.

A second type of epistasis occurs when a dominant allele at one genetic locus masks the expression of the alleles at a second locus. For instance, case 2 of Figure 4–7 deals with the inheritance of fruit color in summer squash. Here, the dominant allele A results in white fruit color regardless of the genotype at a second locus, B. In the absence of the dominant A allele (the aa genotype), BB or Bb results in yellow color, while bb results in green color. Therefore, if two white-colored double heterozygotes (AaBb) are crossed, an interesting phenotypic ratio occurs because of this type of epistasis:

F₁: AaBb × AaBb
           ↓

F₂ Ratio	Genotype	Phenotype	Final Phenotypic Ratio
9/16	A–B–	white ⎫	12/16 white
3/16	A–bb	white ⎭	
3/16	aaB–	yellow	3/16 yellow
1/16	aabb	green	1/16 green

Of the offspring, 9/16 are A–B– and are thus white. The 3/16 bearing the genotypes A–bb are also white. Of the remaining squash, 3/16 are yellow (aaB-), while 1/16 are green (aabb). When combined, the modified ratio of 12:3:1 occurs.

Our third example (case 3 of Figure 4–7), first discovered by William Bateson and Reginald Punnett (of Punnett square fame), is demonstrated in a cross between two strains of white-flowered sweet peas. Unexpectedly, the F₁ plants were all purple, and the F₂ occurred in a ratio of 9/16 purple to 7/16 white. The proposed explanation for these results suggests that the presence of at least one dominant allele of each of two gene pairs is essential in order for flowers to be purple. All other genotype combinations yield white flowers because the homozygous condition of either recessive allele masks the expression of the dominant allele at the other locus.

The cross is shown as follows:

P$_1$:   *AAbb*   ×   *aaBB*
    (white) | (white)

F$_1$:   All *AaBb* (purple)

F$_2$ Ratio	Genotype	Phenotype	Final Phenotypic Ratio
9/16	*A−B−*	purple	9/16 purple
3/16	*A−bb*	white	
3/16	*aaB−*	white	7/16 white
1/16	*aabb*	white	

We can envision the way in which two gene pairs might yield such results:

Gene *A*   Gene *B*
Precursor | Intermediate | Final
Substance ——————→ Product ——————→ Product
(colorless)   *A−*   (colorless)   *B−*   (purple)

At least one dominant allele from each pair of genes is necessary to ensure both biochemical conversions to the final product, yielding purple flowers. In the cross above, this will occur in 9/16 of the F$_2$ offspring. All other plants (7/16) have flowers that remain white.

Novel Phenotypes

Other cases of gene interaction yield novel, or new, phenotypes in the F$_2$ generation, in addition to producing modified dihybrid ratios. Case 4 in Figure 4–7 depicts the inheritance of fruit shape in the summer squash *Cucurbita pepo*. When plants with disc-shaped fruit (*AABB*) are crossed to plants with long fruit (*aabb*), the F$_1$ generation all have disc fruit. However, in the F$_2$ progeny, fruit with a novel shape—*sphere*—appear, as well as fruit exhibiting the parental phenotypes. The F$_2$ generation, with a modified 9:6:1 ratio, is generated as follows:

F$_1$:   *AaBb*   ×   *AaBb*
    (disc) | (disc)

F$_2$ Ratio	Genotype	Phenotype	Final Phenotypic Ratio
9/16	*A−B−*	disc	9/16 disc
3/16	*A−bb*	sphere	
3/16	*aaB−*	sphere	6/16 sphere
1/16	*aabb*	long	1/16 long

In this example of gene interaction, both gene pairs influence fruit shape equally. A dominant allele at either locus ensures a sphere-shaped fruit. In the absence of dominant alleles, the fruit is long. However, if both dominant alleles (*A* and *B*) are present, the fruit displays a flattened, disc shape.

Other Modified Dihybrid Ratios

The remaining cases (5–8) in Figure 4–7 illustrate additional modifications of the dihybrid ratio and provide still other examples of gene interactions. All the cases (1–8) have two things in common. First, in arriving at a suitable explanation of the inheritance pattern of each one, we have not violated the principles of segregation and independent assortment. Therefore, the added complexity of inheritance in these examples does not detract from the validity of Mendel's conclusions. Second, the F$_2$ phenotypic ratio in each example has been expressed in sixteenths. When similar observations are made in crosses where the inheritance pattern is unknown, it suggests to geneticists that two gene pairs are controlling the observed phenotypes. You should make the same inference in the analysis of genetics problems. Other insights into solving genetics problems are provided in the "Insights and Solutions" section at the conclusion of this chapter.

Complementation Analysis and Alleles

An interesting situation arises when two mutations are independently isolated in an organism, both of which produce a similar phenotype. For example, in *Drosophila*, numerous recessive mutations have been recovered that eliminate or severely reduce the size of the wings. Suppose that a genetics laboratory in the United States and a genetics laboratory in Canada independently isolated and established a true-breeding strain of wingless flies and demonstrated that each was due to a recessive mutation.

One might assume that both strains contained mutations in the same gene. This might be true, particularly if only one gene in the fly was responsible for wing formation. However, we know this is not the case. Many genes are involved in the formation of wings. Mutations in any one of them may inhibit wing formation during development.

Therefore, we need to devise an experimental approach that allows us to determine whether two such mutations are in the same gene—that is, to determine if they are alleles—or if they represent mutations in separate genes. Fortunately, this approach, called **complementation analysis**, has been worked out and is rather easy to execute.

The analysis seeks to answer this simple question: *Are two mutations that yield similar phenotypes present in the same gene or in two different genes?* As we will see, the answer is forthcoming if we cross the two mutant strains together and analyze the F_1 generation.

There are two alternative outcomes and interpretations of such a cross, as illustrated in Figure 4–8. The mutation isolated in the United States is designated m^{usa}, while the mutation isolated in Canada is designated m^{can}. These alternatives are discussed below.

Case 1. *All offspring develop normal wings.*

INTERPRETATION: The two recessive mutations are in separate genes and are not alleles of one another. Following the cross, all F_1 flies become heterozygous for both genes. Since a normal wild-type allele is present at each locus, both genes function normally and wings are formed.

Case 2. *All offspring fail to develop wings.*

INTERPRETATION: The two mutations affect the same gene and are alleles of one another. The diploid flies have two copies of this gene. In F_1 flies, one copy is the m^{usa} allele and the other copy is the m^{can} allele. Thus, all offspring are homozygous for mutations of this gene and fail to develop wings.

In case 1, **complementation** occurs. Because each mutation is in a separate gene and each F_1 fly is heterozygous at both loci, the normal products of both genes are produced (by the one normal copy of each gene). Wings develop.

In case 2, **no complementation** occurs because the two mutations affect the same gene. The F_1 flies are homozygous for the mutant alleles. No normal product of the gene is produced. Wings do not form.

Complementation analysis, as originally devised by Edward B. Lewis, is often called the ***cis–trans* test.** "*Cis*" refers to the case when the two mutations are on the same homolog (they are alongside one another). "*Trans*" refers to the case when the mutations are on separate homologs (they are opposite one another). These designations were borrowed from nomenclature used in organic chemistry. Figure 4–8 illustrates the *trans* configuration, critical to the determination of whether the two mutations are alleles of the same gene or not. In our example, if we assume that the two mutations are indeed alleles, and if they are present as heterozygotes in the *cis* configuration, flies will develop wings. This observation serves as an important control in analyses such as the one described above.

Complementation analysis may be utilized to screen any number of individual mutations that result in the same phenotype. Such an analysis may reveal that only

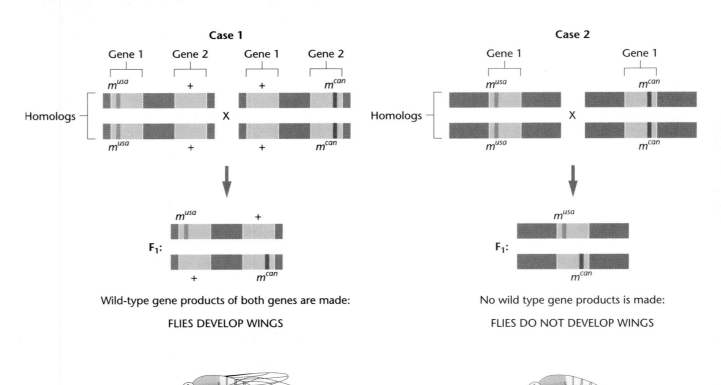

■ Figure 4–8   Complementation analysis by which alternative outcomes of testing two wingless mutations (m^{usa} and m^{can}) are investigated. In case 1, the mutations are not alleles of the same gene, while in case 2, the mutations are alleles of the same gene.

a single gene is involved, or that two or more genes are involved. All mutations determined to be present in any single gene are said to fall into the same **complementation group**. All mutations in the *same* complementation group *complement* mutations in all other groups.

If large numbers of mutations affecting the same trait are available and studied using complementation analysis, it is possible to predict the total number of genes involved in the determination of that trait. We will return to the use of complementation analysis in Chapter 15, where the inherited human disorder xeroderma pigmentosum is discussed. There, complementation analysis reveals that mutations in any of seven complementation groups (genes) lead to the disorder.

Genes on the X Chromosome

We near the conclusion of this chapter with a discussion of still one other mode of neo-Mendelian inheritance: **X-linkage**. This phenomenon results from the fact that one of the sexes in many animal and a few plant species contains a pair of *unlike* chromosomes, the X and Y, which are involved in sex determination. For example, in both fruit flies (*Drosophila*) and humans, males contain an X and a Y chromosome, while females contain two X chromosomes. As we will see, the unique pattern of inheritance stems from the fact that the Y, while behaving as a homolog to the X in meiosis, contains only a few genes. X-linkage, therefore, involves the transmission and expression of the genes located on the X chromosome. To distinguish this pair of sex chromosomes, all other pairs of chromosomes are referred to as **autosomal chromosomes** or **autosomes**.

X-Linkage in Drosophila

One of the first cases of X-linkage was documented by Thomas H. Morgan around 1920 during his studies of the *white* mutation in the eyes of *Drosophila* (Figure 4–9). We will use this case to illustrate X-linkage. The normal wild-type red eye color is dominant to white.

■ Figure 4–9 The F_1 and F_2 results of T. H. Morgan's reciprocal crosses involving the X-linked *white* mutation in *Drosophila melanogaster*. The actual F_2 results are shown in parentheses. The photograph contrasts white eyes with the brick-red wild-type eye color.

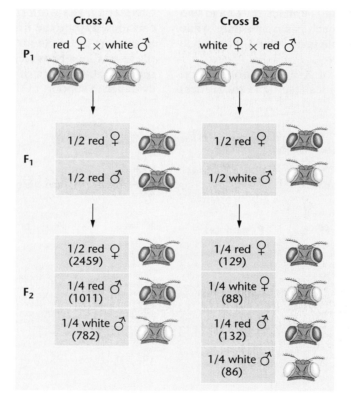

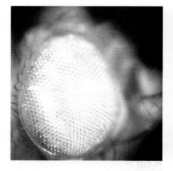
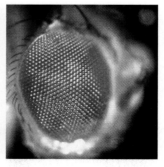

Morgan's work established that the inheritance pattern of the white-eye trait was clearly related to the sex of the parent carrying the mutant allele. Unlike the outcome of the typical monohybrid cross, reciprocal crosses between white- and red-eyed flies did not yield identical results. In contrast, in all of Mendel's monohybrid crosses, F$_1$ and F$_2$ data were very similar regardless of which P$_1$ parent exhibited the recessive mutant trait. Morgan's analysis led to the conclusion that the *white* locus is present on the X chromosome rather than on one of the autosomes. As such, both the gene and the trait are said to be **X-linked**.

Results of reciprocal crosses between white-eyed and red-eyed flies are shown in Figure 4–9. The obvious differences in phenotypic ratios in both the F$_1$ and F$_2$ generations are dependent on whether or not the P$_1$ white-eyed parent was male or female.

Morgan was able to correlate these observations with the difference found in the sex chromosome composition between male and female *Drosophila*. He hypothesized that the recessive allele for white eye is found on the X chromosome, but its corresponding locus is absent from the Y chromosome. Females thus have two available gene sites, one on each X chromosome, while males have only one available gene site on their single X chromosome.

Morgan's interpretation of X-linked inheritance, shown in Figure 4–10, provides a suitable theoretical explanation for his results. Since the Y chromosome lacks homology with most genes on the X chromosome, whatever alleles are present on the X chromosome of the males will be expressed directly in their phenotype. Because males cannot be either homozygous or heterozygous for X-linked genes, this condition is referred to as being **hemizygous**.

One result of X-linkage is the **crisscross pattern of inheritance**, whereby phenotypic traits controlled by recessive X-linked genes are passed from homozygous mothers to all sons. This pattern occurs because females exhibiting a recessive trait must carry the mutant allele on both X chromosomes. Because male offspring receive one of their mother's two X chromosomes and are hemizygous for all alleles present on that X, all sons will express the same recessive X-linked traits as their mother.

Morgan's work has taken on great historical significance. By 1910, the correlation between Mendel's work and the behavior of chromosomes during meiosis had provided the basis for the **chromosome theory of inheritance**, as postulated by Sutton and Boveri (see Chapter 2). The work involving the X chromosome is considered to be the first solid experimental evidence in support of this theory. In the ensuing two decades, these findings provided the impetus for further research, the findings of which provided indisputable evidence in support of this theory.

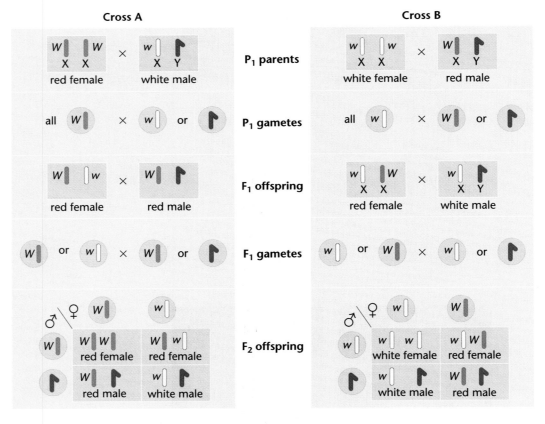

■ Figure 4–10 The chromosomal explanation of the results of the X-linked crosses shown in Figure 4–9.

X-Linked Inheritance in Humans

In humans, many genes and the respective traits controlled by them are recognized as being linked to the X chromosome. These X-linked traits may be easily identified in a **pedigree** because of the crisscross pattern of inheritance. A pedigree for one form of human color blindness is shown in Figure 4–11. The mother in generation I passes the trait to all her sons but to none of her daughters. If the offspring in generation II marry normal individuals, the color-blind sons will produce all normal male and female offspring (III-1, 2, and 3); the normal-visioned daughters will produce normal-visioned female offspring (III-4, 6, and 7), as well as color-blind (III-8) and normal-visioned (III-5) male offspring.

Many X-linked human genes have now been identified, as shown in Table 4.2. For example, the genes controlling two forms of hemophilia and one form of muscular dystrophy are located on the X chromosome. Other genes, whose expression yields enzymes, are X-linked. **Glucose-6-phosphate dehydrogenase (G6PD)** and **hypoxanthine-guanine-phosphoribosyl transferase**

(HGPRT) are two examples. In the former case, during exposure to certain chemicals, an anemic reaction may occur if the G6PD gene is in mutant form. In the latter case, the **Lesch–Nyhan syndrome** results from the mutant form of the X-linked HGPRT gene product.

Because of the way in which X-linked genes are transmitted, unusual circumstances may be associated with recessive X-linked disorders, in comparison to recessive autosomal disorders. For example, if an X-linked disorder debilitates or is lethal to the affected individual prior to reproductive maturation, the disorder occurs exclusively in males. This is the case because the only sources of the lethal allele in the population are females who carry it heterozygously but do not express the disorder. They pass the allele to one-half of their sons, who develop the disorder because they are hemizygous but who rarely, if ever, reproduce. Heterozygous females also pass the allele to one-half of their daughters, who become carriers but do not develop the disorder. An example of such an X-linked disorder is **Duchenne muscular dystrophy**. The disease has an onset prior to age 6 and is often lethal prior to age 20. It normally occurs only in males.

■ Figure 4–11 (a) A human pedigree of the X-linked color blindness trait. (b) The most probable genotypes of each individual in the pedigree. The photograph is of an Ishihara color blindness chart. Red-green color blind individuals see a 3 rather than an 8 visualized by those with normal color vision.

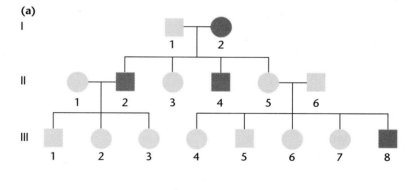

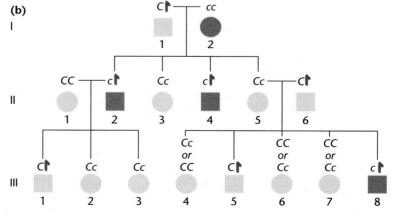

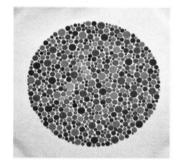

Symbols
c = color blindness
C = normal vision
⸸ = Y chromosome

TABLE 4.2	Human X-linked traits
Condition	**Characteristics**
Color blindness, deutan type	Insensitivity to green light.
Color blindness, protan type	Insensitivity to red light.
Fabry's disease	Deficiency of galactosidase A; heart and kidney defects, early death.
G-6-PD deficiency	Deficiency of glucose-6-phosphate dehydrogenase, severe anemic reaction following intake of primaquines in drugs and certain foods, including fava beans.
Hemophilia A	Classical form of clotting deficiency; lack of clotting factor VIII.
Hemophilia B	Christmas disease; deficiency of clotting factor IX.
Hunter syndrome	Mucopolysaccharide storage disease resulting from iduronate sulfatase enzyme deficiency; short stature, clawlike fingers, coarse facial features, slow mental deterioration, and deafness.
Ichthyosis	Deficiency of steroid sulfatase enzyme; scaly dry skin, particularly on extremities.
Lesch–Nyhan syndrome (HGPRT)	Deficiency of hypoxanthine-guanine phosphoribosyl transferase enzyme leading to motor and mental retardation, self-mutilation, and early death.
Muscular dystrophy (Duchenne type)	Progressive, life-shortening disorder characterized by muscle degeneration and weakness; sometimes associated with mental retardation; deficiency of the protein dystrophin.

Sex-Limited and Sex-Influenced Inheritance

Our final topics involve inheritance affected by the sex of the organism, but not necessarily by genes on the X chromosome. There are numerous examples in different organisms where the sex of the individual plays a determining role in the expression of certain phenotypes. In some cases, the expression of a specific phenotype is absolutely limited to one sex; in others, the sex of an individual influences the expression of a phenotype that is not limited to one sex or the other. This distinction differentiates **sex-limited inheritance** from **sex-influenced inheritance**.

In domestic fowl, tail and neck plumage is often distinctly different in males and females (Figure 4–12),

demonstrating sex-limited inheritance. Cock feathering is longer, more curved, and pointed, while hen feathering is shorter and more rounded. Inheritance of these feather phenotypes is controlled by a single pair of autosomal alleles whose expression is modified by the individual's sex hormones.

As shown in the following chart, hen feathering is due to a dominant allele, H; but regardless of the homozygous presence of the recessive h allele, all females remain hen-feathered. Only in males does the hh genotype result in cock feathering.

■ Figure 4–12   Hen feathering (left) vs. cock feathering (right) in domestic fowl. The feathers in the hen are shorter and less curved.

■ Figure 4–13   Pattern baldness, a sex-influenced autosomal trait in humans.

Genotype	Phenotype	
	♀	♂
HH	Hen-feathered	Hen-feathered
Hh	Hen-feathered	Hen-feathered
hh	Hen-feathered	Cock-feathered

In the development of certain breeds of fowl, one allele or the other has become fixed in the population. In the Leghorn breed, all individuals are of the *hh* genotype; as a result, males always differ from females in their plumage. Sebright bantams are all *HH*, resulting in no sexual distinction in feathering phenotypes.

Cases of sex-influenced inheritance include **pattern baldness** in humans, horn formation in sheep, and certain coat patterns in cattle. In such cases, autosomal genes are responsible for the contrasting phenotypes displayed by both males and females, but the expression of these genes is dependent on the hormone constitution of the individual. Thus, the heterozygous genotype may exhibit one phenotype in one sex and the contrasting one in the other. For example, pattern baldness in humans, where the hair is very thin on the top of the head (Figure 4–13), is inherited in the following way:

Genotype	Phenotype	
	♀	♂
BB	Bald	Bald
Bb	Not bald	Bald
bb	Not bald	Not bald

Even though females may display pattern baldness, this phenotype is much more prevalent in males. When females do inherit the *BB* genotype, the phenotype is much less pronounced than in males and is expressed later in life.

Chapter Summary

1. Since Mendel's work was rediscovered, the study of transmission genetics has been expanded to include many alternative modes of inheritance. In many cases, phenotypes may be influenced by two or more genes.

2. Incomplete or partial dominance is exhibited when an intermediate phenotypic expression of a trait occurs in an organism that is heterozygous for two alleles.

3. Codominance is exhibited when distinctive expression of two alleles occurs in a heterozygous organism.

4. The concept of multiple allelism applies to populations, since a diploid organism may host only two alleles at any given locus. Within a population, however, many alternative alleles of the same gene may occur.

5. Lethal mutations usually result in the inactivation or the lack of synthesis of gene products that are essential during development. Such mutations may be recessive or dominant. Some lethal genes, such as the one that causes Huntington disease, are not expressed until adulthood.

6. Mendel's classic F_2 ratio is often modified in instances where gene interaction results in discontinuous variation.

7. Epistasis may occur when two or more genes influence a single characteristic. Usually, the expression of one of the genes masks the expression of the other gene or genes.

8. Complementation analysis allows the determination as to whether independently isolated mutations producing similar phenotypes are alleles of one another or whether they represent separate genes.

9. Genes located on the X chromosome display a unique mode of inheritance referred to as X-linkage.

10. Sex-limited and sex-influenced inheritance occur when the sex of the organism affects the phenotype controlled by a gene located on an autosome.

Key Terms

ABO blood group, 70
allele, 67
antigen, 70
autosome, 67
Bombay phenotype, 70
chromosome theory of inheritance, 80
cis-trans test, 78
codominance, 69
complementation analysis, 77
complementation group, 79

continuous variation, 73
discontinuous variation, 73
Duchenne muscular dystrophy, 81
epigenesis, 72
epistasis, 73
gene interaction, 67
hemizygous, 80
Huntington disease, 71
incomplete dominance, 68
Lesch–Nyhan syndrome, 81
lethal allele, 71

MN blood group, 69
multiple alleles, 70
mutation, 68
neo-Mendelian genetics, 67
pattern baldness, 83
pedigree, 81
sex-influenced inheritance, 82
sex-limited inheritance, 82
wild-type allele, 68
X-linkage, 79

INSIGHTS
and
SOLUTIONS

Genetic problems take on added complexity if they involve two independent characters and multiple alleles, incomplete dominance, or epistasis. The most difficult types of problems are those that were faced by pioneering geneticists during laboratory or field studies. They had to determine the mode of inheritance by working backwards from the observations of offspring to parents of unknown genotype.

1. For example, consider the problem of comb shape inheritance in chickens, where walnut, rose, pea, and single are the observed distinct phenotypes. *How is comb shape inherited, and what are the genotypes of the P_1 generation of each cross?* Use the following data to answer these questions.

Cross 1: single	× single	⟶	all single
Cross 2: walnut	× walnut	⟶	all walnut
Cross 3: rose	× pea	⟶	all walnut
Cross 4: F_1	× F_1 (of Cross 3)		
(walnut)	(walnut)	⟶	93 walnut
			28 rose
			32 pea
			10 single

Solution:

At first glance, this problem may appear quite difficult. The approach used in solving it must involve two steps. First, analyze the data carefully for any useful information. Once you determine something concrete, follow an empirical approach; that is, formulate a hypothesis and, in a sense, test it against the given data. Look for a pattern of inheritance that is consistent with all cases.

(a) In this problem there are two immediately useful facts. First, in cross 1, P_1 singles breed true. Second, while P_1 walnut breeds true (cross 2), a walnut phenotype is also produced in crosses between rose and pea (cross 3). When these F_1 walnuts are crossed together (cross 4), all four comb shapes are produced in a ratio that approximates 9:3:3:1. This observation should immediately suggest a cross involving two gene pairs, because the resulting data display the same ratio as in Mendel's dihybrid crosses. Because only one trait is involved (comb shape), epistasis may be occurring. This may serve as a working hypothesis, and we must now propose how the two gene pairs "interact" to produce each phenotype.

(b) If we call the allele pairs A, a and B, b, we might predict that because walnut represents 9/16 in cross 4, A–B– will produce walnut. We might also hypothesize that in the case of cross 2, the genotypes were $AABB × AABB$, where walnut was seen to breed true. (Recall that A– and B– mean AA or Aa and BB or Bb, respectively.)

(c) Because single is the phenotype representing 1/16 of the offspring of cross 4, we could predict that this phenotype is the result of the $aabb$ genotype. This is consistent with cross 1.

(d) Now we have only to determine the genotypes for rose and pea. The most logical prediction would be that at least one dominant A or B allele combined with the double recessive condition of the other allele pair can account for these phenotypes. For example,

$$A–bb \longrightarrow \text{rose}$$
$$aa\ B– \longrightarrow \text{pea}$$

If, in cross 3, $AAbb$ (rose) were crossed with $aaBB$ (pea), all offspring would be $AaBb$ (walnut). This is consistent with the data, and we must now only

Walnut

Pea

Rose

Single

look at cross 4. We predict these walnut genotypes to be AaBb (as above), and from the cross

$$AaBb \text{ (walnut)} \times AaBb \text{ (walnut)}$$

we expect:

$9/16\ A-B-$ (walnut)

$3/16\ A-bb$ (rose)

$3/16\ aaB-$ (pea)

$1/16\ aabb$ (single)

Our prediction is consistent with the information we were given. The initial hypothesis of the epistatic interaction of two gene pairs proves consistent throughout, and the problem has been solved.

This example illustrates the need to have a basic theoretical knowledge of transmission genetics. Then, you must search for the appropriate clues so that you can proceed in a stepwise fashion toward a solution. Mastering problem solving requires practice, but provides a great deal of satisfaction. Apply this general approach to the following problems.

2. In radishes, flower color may be red, purple, or white. The edible portion of the radish may be long or oval. When only flower color is studied, red × white yields all purple. If these F_1 purples are interbred, no dominance is evident, and the F_2 generation consists of 1/4 red:1/2 purple:1/4 white. Regarding radish shape, long is dominant to oval in a normal Mendelian fashion.

(a) Determine the F_1 and F_2 phenotypes from a cross between a true-breeding red long radish and one that is white oval. Be sure to define all gene symbols initially.

Solution:

This is a modified dihybrid cross in which the gene pair controlling color exhibits incomplete dominance. Shape is controlled conventionally. First, establish gene symbols:

RR = red; Rr = purple; rr = white

$O-$ = long; oo = oval

$P_1: RROO \times rroo$

(red long) (white oval)

F_1: all $RrOo$ (purple long)

$F_1 \times F_1: RrOo \times RrOo$

$$F_2 \begin{cases} 1/4\ RR \begin{cases} 3/4\ O- \longrightarrow 3/16\ RRO- & \text{red long} \\ 1/4\ oo \longrightarrow 1/16\ RRoo & \text{red oval} \end{cases} \\ 2/4\ Rr \begin{cases} 3/4\ O- \longrightarrow 6/16\ RrO- & \text{purple long} \\ 1/4\ oo \longrightarrow 2/16\ Rroo & \text{purple oval} \end{cases} \\ 1/4\ rr \begin{cases} 3/4\ O- \longrightarrow 3/16\ rrO- & \text{white long} \\ 1/4\ oo \longrightarrow 1/16\ rroo & \text{white oval} \end{cases} \end{cases}$$

Note that to generate the F_2 results, we have used the forked-line method. First, the outcome of crossing F_1 parents for the color genes is considered ($Rr \times Rr$). Then the outcome of shape is considered ($Oo \times Oo$).

(b) A red oval plant was crossed with a plant of unknown genotype and phenotype, yielding the data shown below. Determine the genotype and phenotype of the unknown plant.

Offspring: 103 red long:101 red oval:98 purple long:100 purple oval

Solution:

Since the two characters are inherited independently, consider them separately. The data indicate a 1/4:1/4:1/4:1/4 proportion. First, consider color:

P_1: red × ??? (unknown)

F_1: 204 red (1/2)   198 purple (1/2)

Because the red parent must be RR, the unknown must have a genotype of Rr to produce these results. It is thus purple. Now, consider shape:

P_1: oval × ??? (unknown)

F_1: 201 long (1/2)   201 oval (1/2)

Because the oval plant must be oo, the unknown plant must have a genotype of Oo to produce these results. It is thus long. The unknown plant is thus

$RrOo$ purple long

3. In humans, red-green color blindness is inherited as an X-linked recessive trait. A woman with normal vision, but whose father is color-blind, marries a male who has normal vision. Predict the color vision of their male and female offspring.

Solution:

The female is heterozygous because she inherited an X chromosome with the mutant allele from her father. Her husband is normal. Therefore, the parental genotypes are

$Cc \times C\uparrow$ ($\uparrow$ is the Y chromosome)

All female offspring are normal (CC or Cc). One-half of the male children will be color-blind ($c\uparrow$) and the other half will have normal vision ($C\uparrow$).

Problems and Discussion Questions

1. In shorthorn cattle, coat color may be red, white, or roan. Roan is an intermediate phenotype expressed as a mixture of red and white hairs. The following data were obtained from various crosses:

red	×	red	→ all red
white	×	white	→ all white
red	×	white	→ all roan
roan	×	roan	→ 1/4 red: 1/2 roan: 1/4 white

How is coat color inherited? What are the genotypes of parents and offspring for each cross?

2. Contrast incomplete dominance and codominance.

3. With regard to the ABO blood types in humans, determine the genotypes of the male parent and female parent:

Male Parent: Blood type B whose mother was type O

Female Parent: Blood type A whose father was type B

Predict the blood types of the offspring that this couple may have and the expected proportion of each.

4. Distinguish between discontinuous and continuous variation. To which category does epistasis belong?

5. Three gene pairs located on separate autosomes determine flower color and shape as well as plant height. The first pair exhibits incomplete dominance, where color can be red, pink (the heterozygote), or white. The second pair leads to personate (dominant) or peloric (recessive) flower shape, while the third gene pair produces either the dominant tall trait or the recessive dwarf trait. Homozygous plants that are red, personate, and tall are crossed to those that are white, peloric, and dwarf. Determine the F_1 genotype(s) and phenotype(s). If the F_1 plants are interbred, what proportion of the offspring will exhibit the same phenotype as the F_1 plants?

6. As in Problem 5, color may be red, white, or pink, and flower shape may be personate or peloric. For the following crosses, determine the P_1 and F_1 genotypes.

(a) red peloric × white personate	→ F_1: all pink personate
(b) red personate × white peloric	→ F_1: all pink personate
(c) pink personate × red peloric	→ F_1: 1/4 red personate
	1/4 red peloric
	1/4 pink personate
	1/4 pink peloric
(d) pink personate × white peloric	→ F_1: 1/4 white personate
	1/4 white peloric
	1/4 pink personate
	1/4 pink peloric

What phenotype ratios would result from crossing the F_1 of (a) to the F_1 of (b)?

7. In some plants a red pigment, cyanidin, is synthesized from a colorless precursor. The addition of a hydroxyl group (—OH) to the cyanidin molecule causes it to become purple. In a cross between two randomly selected purple plants, the following results were obtained:

94 purple:31 red:43 colorless

How many genes are involved in the determination of these flower colors? Which genotypic combinations produce which phenotypes? Diagram the purple × purple cross.

8. In rats, the following genotypes of two independently assorting autosomal genes determine coat color:

A–B– (gray); *A–bb* (yellow); *aaB–* (black); *aabb* (cream)

A third gene pair on a separate autosome determines whether or not any color will be produced. The *CC* and *Cc* genotypes allow color according to the expression of the *A* and *B* alleles. However, the *cc* genotype results in albino rats regardless of the *A* and *B* alleles present. Determine the F_1 phenotypic ratio of the following crosses: (a) *AAbbCC* × *aaBBcc*; (b) *AaBBCC* × *AABbcc*; (c) *AaBbCc* × *AaBbcc*.

9. Given the inheritance pattern of coat color in rats as described in Problem 8, predict the genotype and phenotype of the parents that produced the following F_1 offspring: (a) 9/16 gray:3/16 yellow:3/16 black:1/16 cream; (b) 9/16 gray:3/16 yellow:4/16 albino; (c) 27/64 gray:16/64 albino:9/64 yellow:9/64 black:3/64 cream.

10. A husband and wife have normal vision, although both of their fathers are red-green color-blind, which is inherited as an X-linked recessive condition. What is the probability that their first child will be: (a) a normal son? (b) a normal daughter? (c) a color-blind son? (d) a color-blind daughter?

11. In humans, the ABO blood type is under the control of autosomal multiple alleles. Red-green color blindness is a recessive X-linked trait. If two parents who are both type A and have normal vision produce a son who is color-blind and is type O, what is the probability that their next child will be a female who has normal vision and is type O?

12. In spotted cattle, the colored regions may be mahogany or red. If a red female and a mahogany male, both derived from separate true-breeding lines, are mated and the cross is carried to an F_2 generation, the following results are obtained:

F_1: 1/2 mahogany males:1/2 red females

F_2: 3/8 mahogany males:1/8 red males:1/8 mahogany females:3/8 red females

When the reciprocal of the initial cross is performed (mahogany female and red male), identical results are obtained. Explain these results by postulating how the color is genetically determined. Diagram the crosses.

13. In cats, yellow coat color is determined by the *b* allele and black coat color is determined by the *B* allele. The heterozygous condition results in a color known as

tortoise shell. These genes are X-linked. What kinds of offspring would be expected from a cross of a black male and a tortoise-shell female? What are the chances of getting a tortoise-shell male?

14. In *Drosophila*, an X-linked recessive mutation, *scalloped (sd)* causes irregular wing margins. Diagram the F_1 and F_2 results if:

    (a) A scalloped female is crossed with a normal male.

    (b) A scalloped male is crossed with a normal female.

    Compare these results to those that would be obtained if the *scalloped* gene were autosomal.

15. Another recessive mutation in *Drosophila, ebony (e)*, is on an autosome (chromosome 3) and causes darkening of the body compared with wild-type flies. What phenotypic F_1 and F_2 male and female ratios will result if a scalloped-winged female with normal body color is crossed with a normal-winged ebony male? Work this problem by both the Punnett square method and the forked-line method.

16. While *vermilion* is X-linked and causes the eye color to be bright red, *brown* is an autosomal recessive mutation that causes the eye to be brown. Flies carrying both mutations lose all pigmentation and are white-eyed. Predict the F_1 and F_2 results of the following crosses:

    (a) vermilion females × brown males

    (b) brown females × vermilion males

    (c) white females × wild males

17. In pigs, coat color may be sandy, red, or white. A geneticist had spent several years mating true-breeding pigs of all different color combinations, even going so far as to obtain true-breeding lines from different parts of the country. For crosses 1 and 4 below, a major problem was encountered. She experienced a computer crash and lost the F_2 data. She nevertheless persevered, and using the limited data shown here, she was able to predict the mode of inheritance, the number of genes involved, and assign genotypes to each coat color. Based on the available data generated from the crosses shown, attempt to duplicate her analysis.

P_1	F_1	F_2
1 sandy × sandy	all red	data lost
2 red × sandy	all red	3/4 red: 1/4 sandy
3 sandy × white	all sandy	3/4 sandy: 1/4 white
4 white × red	all red	data lost

    Once you have formulated a hypothesis to explain the mode of inheritance and assigned genotypes to the respective coat colors, predict the outcomes of the F_2 generations where the data were lost.

18. An alien geneticist from a planet that prohibits genetic research brought with him two true-breeding lines of frogs. One line croaked by *uttering* "rib-it rib-it" and had purple eyes. The other line croaked by *muttering* "knee-deep knee-deep" and had green eyes. He mated the two types, producing F_1 frogs, all of whom uttered "rib-it" and had blue eyes. A large F_2 generation yielded the following ratio:

27/64 blue-eyed utterers
12/64 green-eyed utterers
9/64 blue-eyed mutterers
9/64 purple-eyed utterers
4/64 green-eyed mutterers
3/64 purple-eyed mutterers

    (a) How many total gene pairs are involved in the inheritance of both eye color and croaking?

    (b) Of these, how many are controlling eye color? croaking?

    (c) Assign gene symbols for all phenotypes and indicate the genotypes of the P_1, F_1, and F_2 frogs.

19. After many years, the frog geneticist (see Problem 18) isolated true-breeding lines of all six F_2 phenotypes. Indicate the F_1 and F_2 phenotypic ratios of a cross between a blue-eyed mutterer and a purple-eyed utterer.

20. In another cross, the frog geneticist mated two purple-eyed utterers, with the results shown below. What were the genotypes of the parents?

9/16 purple-eyed utterers
3/16 purple-eyed mutterers
3/16 green-eyed utterers
1/16 green-eyed mutterers

21. In the guinea pig, a locus controlling coat color may be occupied by any of four alleles: C (*black*), c^k (*sepia*), c^d (*cream*), or c^a (*albino*). A progressive order of dominance exists among these alleles when they are present heterozygously: $C > c^k > c^d > c^a$. In the following crosses, determine the genotype of each individual and predict the phenotypic ratios of the offspring.

    (a) sepia × cream, where both had an albino parent.

    (b) sepia × cream, where the sepia individual had an albino parent and the cream individual had two sepia parents.

    (c) sepia × cream, where the sepia individual had two full-color parents and the cream individual had two sepia parents.

    (d) sepia × cream, where the sepia individual had two full color parents and the cream individual had two full color parents.

22. In the parakeet, two autosomal genes (located on different chromosomes) control the production of feather pigment. Gene B controls the production of a blue pigment, and gene Y controls the production of a yellow pigment. Recessive mutations in each gene are known that result in the loss of synthesis of the respective pigment. Two green parakeets are mated and produce green, blue, yellow, and albino progeny.

    (a) Based on the information given above, explain this pattern of inheritance. Be sure to include in your answer:

    (i) The genotypes of the green parents

    (ii) The genotypes of all four phenotypic classes

(iii) The fraction of total progeny that each phenotypic class represents

(b) The parental (green) parakeets were the progeny of a cross between two true-breeding strains. What two types of crosses between true-breeding strains could have produced the green parents? Indicate the genotypes and phenotypes for each cross.

23. Consider the three pedigrees below, all involving a single human trait.

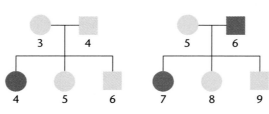

(a) Which sets of conditions (see below), if any, can be excluded?

> *Conditions:* dominant *and* X-linked
> dominant *and* autosomal
> recessive *and* X-linked
> recessive *and* autosomal

(b) For any set of conditions that you excluded, indicate the *single individual* in generation II (e.g., II-1, II-2, . . .) that was most instrumental in your decision to exclude that condition. If none were excluded, answer "none apply."

(c) Given your conclusions above, indicate the *genotype* of the individuals listed below. If more than one possibility applies, list all possibilities. Use the symbols *A* and *a* for the genotypes.

II-1; II-6; II-9

24. Three autosomal recessive mutations in *Drosophila*, all with tan eye color (*r1*, *r2*, and *r3*) were independently isolated and subjected to complementation analysis. The results are shown below. Which, if any, are alleles of one another? Predict the results of the cross that is not shown, i.e., *r2 × r3*.

Cross 1: *r1 × r2* → F₁: all wild-type eyes
Cross 2: *r1 × r3* → F₁: all tan eyes

Selected Readings

Brink, R. A., ed. 1967. *Heritage from Mendel.* Madison: University of Wisconsin Press.

Bultman, S. J., Michaud, E. J., and Woychik, R. P. 1992. Molecular characterization of the mouse *agouti* locus. *Cell* 71:1195–1204.

Carlson, E. A. 1987. *The gene: A critical history*, 2nd ed. Philadelphia: Saunders.

Drayna, D., and White, R. 1985. The genetic linkage map of the human X chromosome. *Science* 230:753–58.

Dunn, L. C. 1966. *A short history of genetics.* New York: McGraw-Hill.

Foster, M. 1965. Mammalian pigment genetics. *Adv. Genet.* 13:311–39.

Grant, V. 1975. *Genetics of flowering plants.* New York: Columbia University Press.

McKusick, V. A. 1962 On the X chromosome of man. *Quart. Rev. Biol.* 37:69–175.

Morgan, T. H. 1910. Sex-limited inheritance in *Drosophila. Science* 32:120–22.

Peters, J. A., ed. 1959. *Classic papers in genetics.* Englewood Cliffs, NJ: Prentice-Hall.

Race, R. R., and Sanger, R. 1975. *Blood groups in man,* 6th ed. Oxford, England: Blackwell.

Vogel, F., and Motulsky, A. G. 1997. *Human genetics: Problems and approaches,* 3rd ed. New York: Springer-Verlag.

Watkins, M. W. 1966. Blood group substances. *Science* 152:172–81.

Yoshida, A. 1982. Biochemical genetics of the human blood group ABO system. *Am. J. Hum. Genet.* 34:1–14.

A field of tobacco plants.

CHAPTER OUTLINE

Quantitative Genetics

Chapter Concepts

Traits that exhibit quantitative phenotypic variation are often under the genetic control of additive alleles and are classified as polygenic traits. Such traits may be analyzed and characterized using statistical methods, which also allow the assessment of the relative importance of genetic factors during phenotypic expression. Such calculations establish the heritability of traits in a population. In humans, twin studies provide a similar, but less precise, estimate. Recently developed techniques allow chromosomal mapping of polygenes.

In the preceding chapter, numerous examples were discussed that illustrated gene interaction leading to phenotypic variation that was classified into distinct traits. Pea plants were tall or dwarf; squash shape was spherical, disc-shaped, or elongated; and fruit-fly eye color was red or white. These phenotypes are examples of **discontinuous variation,** in which discrete phenotypic categories exist. Many other traits in a population demonstrate considerably more variation and are not easily categorized into a small number of distinct classes. Such phenotypes demonstrate **continuous variation**.

It is now known that traits that exhibit continuous variation are often controlled by two or more genes that provide an additive component to the phenotype. Such traits are examples of quantitative or polygenic inheritance. In this chapter, we examine such patterns of inheritance and outline some statistical techniques used to study traits that exhibit continuous variation. In addition, we consider how geneticists assess the relative importance of genetic versus environmental factors as they contribute to phenotypic variation, and we discuss an approach used to map these genes.

with large peas were crossed to those with small peas, the F_1 plants contained peas that were all of an intermediate diameter. When the F_2 generation was examined, peas were of many sizes, as large or as small as the original parents, and many sizes in between!

This example illustrates a pattern of inheritance encountered by other investigators, including at least one cross made by Mendel. The F_1 generations were an intermediate blend of the parental phenotypes, and the F_2 generation exhibited a more or less continuous variation of phenotypic expression. In each case, the traits under investigation behaved in a quantitative fashion expressed as size, height, weight, color, and so on.

Not surprisingly, these traits were difficult to study, and their mode of inheritance was not clarified until early in the twentieth century. Because these results were exceptions to the patterns observed by Mendel in most of his crosses, they failed to support his hypotheses and no doubt delayed the acceptance of his work. Nevertheless, the genetic explanation of continuous variation serves as the foundation for our current understanding of the field of genetics called **quantitative inheritance**.

Quantitative Inheritance

Throughout the eighteenth and nineteenth centuries, traits were studied that exhibited a continuous gradation of phenotypes. For example, Sir Francis Galton investigated the diameter of sweet peas. When plants

Polygenes and the Multiple-Factor Hypothesis

One of the first cases of continuous phenotypic variation was encountered by Josef Gottlieb Kolreuter when he crossed tall and dwarf tobacco plants. The plants of the F_1 generation were all intermediate in height. When the F$_2$ generation was examined, individuals showed continuous variation in height, ranging from tall to dwarf, like the original parents, including many heights in between. A critical observation involved the distribution of phenotypes in the second generation: The majority of the F$_2$ plants were intermediate like the F$_1$, but only a few were as tall or dwarf as the P$_1$ parents. These distributions are depicted in histograms in Figure 5–1. Note that the F$_2$ data demonstrate a normal distribution, as evidenced by the bell-shaped curve in the histogram.

At the beginning of the twentieth century, geneticists noted that many characters in different species had similar patterns of inheritance, such as height and stature in humans, seed size in the broad bean, grain color in wheat, and kernel number and ear length in corn. In each case, offspring in the succeeding generation seemed to be a blend of their parents' characteristics.

The issue of whether continuous variation could be accounted for in Mendelian terms caused considerable controversy in the early 1900s. William Bateson and Gudny Yule, who adhered to the Mendelian explanation of inheritance, suggested that a large number of factors or genes could account for the observed patterns. This proposal, called the **multiple-factor** or **multiple-gene hypothesis**, implied that many factors or genes contribute to the phenotype in a *cumulative* or *quantitative* way. However, other geneticists argued that Mendel's unit factors could not account for the blending of parental phenotypes characteristic of these patterns of inheritance and were thus skeptical of these ideas.

By 1920, the conclusions of several critical sets of experiments largely resolved the controversy and demonstrated that Mendelian factors could account for continuous variation. In one experiment, Edward M. East performed crosses between two strains of the tobacco plant *Nicotiana longiflora*. The fused inner petals of the flower, or corollas, of strain A were decidedly shorter than the corollas of strain B. With only minor variation, each strain was true-breeding. Thus, the differences between them were clearly under genetic control.

When plants from the two strains were crossed, the F$_1$, F$_2$, and selected F$_3$ data (Figure 5–2) demonstrated a very distinct pattern. The F$_1$ generation displayed corollas that were intermediate in length compared with the P$_1$ varieties, and showed only minor variability among individuals. While corolla lengths of the P$_1$ plants were about 40 mm and 94 mm, the F$_1$ generation contained plants with corollas that were all about 64 mm. In the F$_2$ generation, lengths varied much more, ranging from 52 mm to 82 mm. The majority of individuals were similar to their F$_1$ parents, and as the deviation from this average increased, fewer and fewer plants were observed. When the data are plotted graphically (frequency vs. length), a bell-shaped curve results.

East further experimented with this population by selecting F$_2$ plants of various corolla lengths and allowing

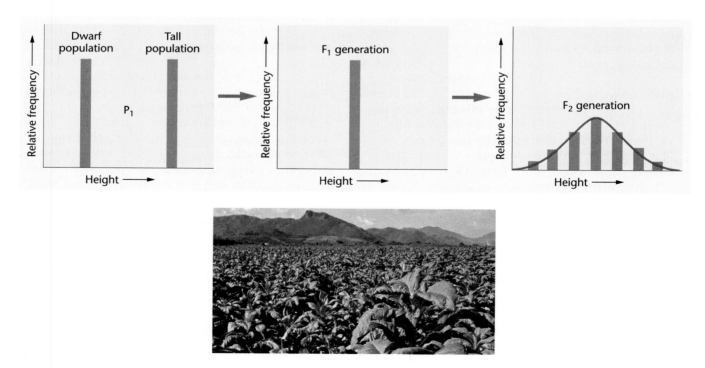

■ Figure 5–1 Histograms showing the relative frequency of individuals expressing various height phenotypes derived from Kolreuter's cross between dwarf and tall tobacco plants carried to the F$_2$ generation. The photograph shows a field of tobacco plants.

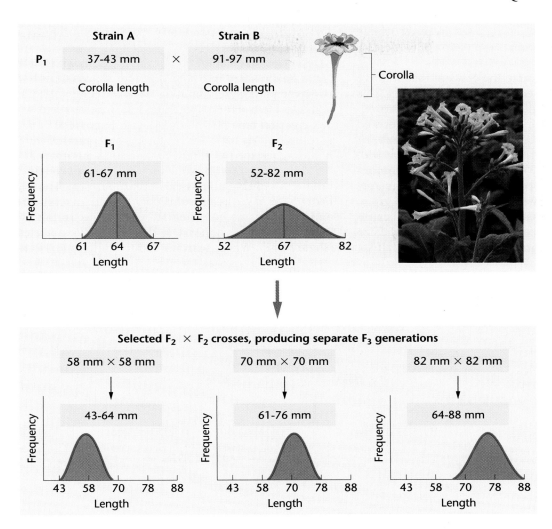

■ Figure 5–2 The F_1, F_2, and selected F_3 results of East's cross between two strains of *Nicotiana* with different corolla lengths. Plants of strain A vary from 37 to 43 mm, while plants of strain B vary from 91 to 97 mm. The photograph and drawing illustrate the flower and corolla of a tobacco plant.

them to produce separate F_3 generations. Several are illustrated in Figure 5–2. In each case, a bell-shaped distribution was observed, with most individuals similar in height to the selected F_2 parents, but with considerable variation around this value.

East's experiments demonstrated that although the variation in corolla length seemed continuous, experimental crosses resulted in the segregation of distinct phenotypic classes as observed in the three independent F_3 categories. This finding strongly suggested that the multiple-factor hypothesis could account for traits that deviate considerably in their expression.

Additive Alleles: The Basis of Continuous Variation

The multiple-factor hypothesis, suggested by the observations of East and others, embodies the following major points:

1. Characters that exhibit continuous variation can usually be quantified by measuring, weighing, counting, and so on.

2. Two or more pairs of genes, located throughout the genome, account for the hereditary influence on the phenotype in an *additive way*. Because many genes may be involved, inheritance of this type is often called *polygenic*.

3. Each gene locus may be occupied by either an **additive allele**, which contributes a set amount to the phenotype, or by a **nonadditive allele**, which does not contribute quantitatively to the phenotype.

4. The total effect of each additive allele at each locus, while small, is approximately equivalent to all other additive alleles at other gene sites.

5. Together, the genes controlling a single character produce substantial phenotypic variation.

6. Analysis of polygenic traits requires the study of large numbers of progeny from a population of organisms.

These points center around the concept that additive alleles at numerous loci control quantitative traits. This is illustrated by examining Herman Nilsson-Ehle's experiments involving grain color in wheat performed early in the twentieth century. In one set of experiments, wheat with red grain was crossed to wheat with white grain (Figure 5–3). The F_1 generation demonstrated an intermediate color. In the F_2, approximately 15/16 of the plants showed some degree of red grain, while 1/16 of the plants showed white grain. Because the ratio occurred in sixteenths, we can hypothesize that two gene pairs control the phenotype and, if so,

they segregate independently from one another in a Mendelian fashion.

Upon careful examination of the F_2, grain with color could be classified into four different shades of red. If two gene pairs were operating, each with one potential additive allele and one potential nonadditive allele, we can envision how the multiple-factor hypothesis could account for this variation. In the P_1, both parents were homozygous; the red parent contains only additive alleles (uppercase letters), while the white parent contains only nonadditive alleles (lowercase letters). The F_1, being heterozygous, contains only two additive alleles and expresses an intermediate phenotype. In the F_2, each offspring has either 4, 3, 2, 1, or 0 additive alleles (Figure 5–3). Wheat with no additive alleles (1/16) is

■ Figure 5–3 An illustration of how the multiple-factor hypothesis can account for the 1:4:6:4:1 phenotypic ratio of grain color when all alleles designated by an uppercase letter are additive and contribute an equal amount of pigment to the phenotype.

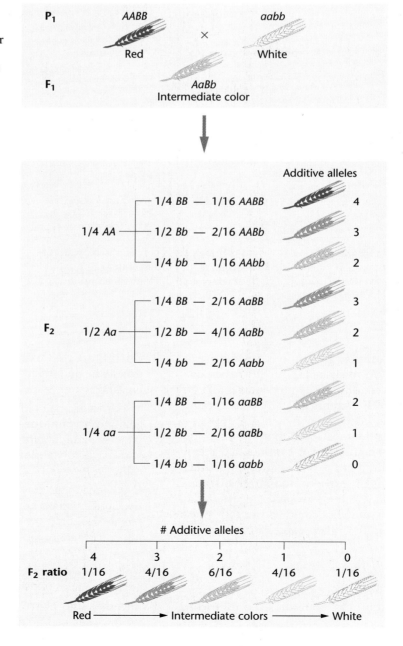

white like one of the P_1 parents, while wheat with 4 additive alleles is red like the other P_1 parent. Plants with 3, 2, or 1 additive alleles constitute the other three categories of red color observed in the F_2, with most (6/16) having 2 additive alleles like the F_1 plants.

Therefore, continuous variation can be explained in a Mendelian fashion. Multiple-factor inheritance, where additive alleles influence the phenotype in a quantitative manner, results in continuous variation. Although the Nilsson-Ehle experiment involved two gene pairs, there is no reason why three, four, or more gene pairs cannot function in controlling various phenotypes. As we saw in Nilsson-Ehle's initial cross, if two gene pairs were involved, only five F_2 phenotypic categories, in a 1:4:6:4:1

ratio, would be expected. On the other hand, as three, four, five, or more gene pairs become involved, greater and greater numbers of classes would be expected to appear in more complex ratios. The number of phenotypes and the expected F_2 ratios of crosses involving up to five gene pairs are illustrated in Figure 5–4.

Calculating the Number of Polygenes

When additive effects control polygenic traits, it is of interest to determine the number of genes that are involved. If the ratio (proportion) of F_2 individuals resembling *either* of the two most extreme phenotypes (the parental phenotypes) can be determined, then the

Figure 5–4 The results of crossing two heterozygotes when polygenic inheritance is in operation with one to five gene pairs. Each histogram bar indicates a distinct phenotypic class from one extreme (left end) to the other extreme (right end). Each phenotype results from a different number of additive alleles.

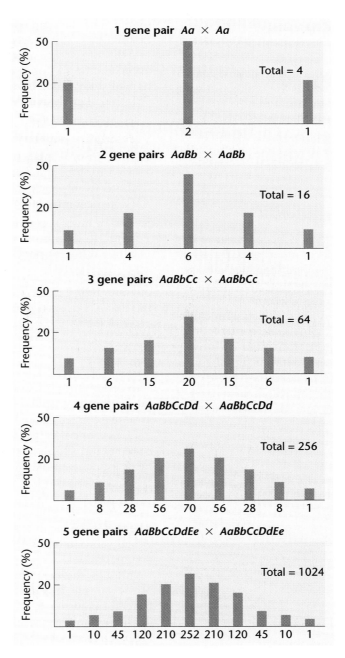

number of gene pairs involved (n) may be calculated using the following simple formula:

$$\frac{1}{4^n} = \text{ratio of F}_2 \text{ individuals expressing either extreme phenotype}$$

In our previous example, the P_1 phenotypes represent these two extremes. In Figure 5–3, for example, 1/16 of the F_2 are either red *or* white like the P_1 classes; this ratio can be substituted on the right side of the equation before solving for n:

$$\frac{1}{4^n} = \frac{1}{16}$$
$$\frac{1}{4^2} = \frac{1}{16}$$
$$n = 2$$

Table 5.1 lists the ratio and the number of F_2 phenotypic classes produced in crosses involving up to five gene pairs.

For low numbers of gene pairs, it is sometimes easier to use the ($2n + 1$) rule. If n equals the number of gene pairs, $2n + 1$ will determine the total number of categories of possible phenotypes. When $n = 2$, $2n + 1 = 5$. Each phenotypic category can have 4, 3, 2, 1, or 0 additive alleles. When $n = 3$, $2n + 1 = 7$, since each phenotypic category could have 6, 5, 4, 3, 2, 1, or 0 additive alleles, and so on.

The Significance of Polygenic Control

Polygenic control is a significant concept because it is believed to serve as the mode of inheritance for a vast number of traits involved in animal breeding and agriculture. For example, height, weight, and physical stature in animals, size and grain yield in crops, beef and milk production in cattle, and egg production in chickens are thought to be under polygenic control. Polygenic traits are also an important part of human genetics; traits such as skin pigmentation, intelligence, obesity, and predisposition to certain diseases are thought to be under polygenic control. In most cases, it is important to note that the genotype, which is fixed at fertilization, establishes the potential range in which a particular phenotype may fall. However, environmen-

tal factors determine how much of the potential will be realized. In the crosses described thus far in this chapter, we have assumed an optimal environment, which minimizes variation from external sources.

Analysis of Polygenic Traits

Analysis of any given polygenic trait involves quantitative measurements, usually from many offspring generated from many crosses. The outcome can be expressed as a frequency diagram that often demonstrates a normal (bell-shaped) distribution (Figure 5–5) . While it is hoped that each series of crosses is representative of the population at large, variation in samples due strictly to chance may influence the data gathered. To assess the experimental validity of the data, statistical techniques must be employed. Such techniques were first devised by Galton early in this century in order to assess the inheritance of traits exhibiting continuous variation. Galton's efforts served as the initial basis of the field of study called **biometry**.

Statistical analysis serves three purposes:

1. Data can be mathematically reduced to provide a **descriptive summary** of the sample.

2. Data from a small but random sample can be utilized to infer information about groups larger than those from which the original data were obtained (**statistical inference**).

3. Two or more sets of experimental data may be compared to determine whether they represent significantly different populations of measurements.

Several statistical methods are useful in the analysis of traits that exhibit a normal distribution, including the mean, variance, standard deviation, and standard error of the mean.

The Mean

The distribution of two sets of phenotypic measurements that is graphed in Figure 5–6 tends to cluster around a central value. This clustering is called a **central tendency**, one measurement of which is the **mean ($\overline{X}$)**. The mean is simply the arithmetic average of a set of measurements or data and is calculated as

$$\overline{X} = \frac{\sum X_i}{n}$$

where $\overline{X}$ is the mean, $\sum X_i$ represents the sum of all individual values in the sample, and n is the number of individual values.

Although the mean provides a descriptive summary of the sample, it is of itself of limited value. As illustrated in Figure 5–6, a symmetrical distribution of values in the sample may, in one case, be clustered near the mean. Or, a set of values may have the same mean, but be distributed widely around it. These contrasting

TABLE 5.1	Determination of the number of gene pairs (n) involved in polygenic crosses	
n	**Ratio of Individuals Expressing an Extreme Phenotype**	**Number of Distinct F_2 Phenotypic Classes**
1	1/4	3
2	1/16	5
3	1/64	7
4	1/256	9
5	1/1024	11

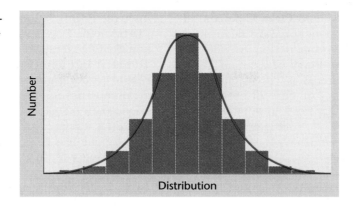

■ Figure 5–5 A normal frequency distribution characterized by a bell-shaped curve.

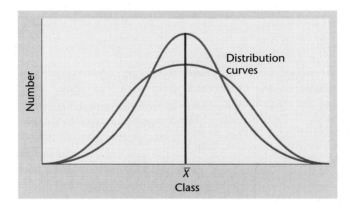

■ Figure 5–6 Two normal frequency distributions with the same mean but different amounts of variation.

conditions represent different types of variation within each sample called the **frequency distribution**. Whether due to chance or to one or more experimental variables, such variation creates the need for methods to describe sample measurements statistically.

Variance

As seen in Figure 5–6, the range and distribution of values on either side of the mean determines the shape of the distribution curve. The degree to which values within this distribution diverge from the mean is called the sample **variance** (s^2) and is used as an estimate of the variation present in an infinitely large population. The variance for a sample is calculated as

$$s^2 = \frac{\Sigma(X_i - \bar{X})^2}{n - 1}$$

where the sum (Σ) of the squared differences between each measured value (X_i) and the mean ($\bar{X}$) is divided by one less than the total sample size ($n - 1$). To avoid the numerous subtraction functions necessary in calculating s^2 for a large sample, we can convert the equation to its algebraic equivalent:

$$s^2 = \frac{\Sigma X_i^2 - n\bar{X}^2}{n - 1}$$

The variance is a valuable measure of sample variability. Two distributions may have identical means ($\bar{X}$), yet vary considerably in their frequency distribu-

tion around the mean. The variance represents the average squared deviation of the measurements from the mean. The estimation of variance has been particularly valuable in determining the degree of genetic control of traits when the immediate environment also influences the phenotype.

Standard Deviation

Because the variance is a squared value, its unit of measurement is also squared (cm^2, mg^2, etc.). To express variation around the mean in the original units of measurement, it is necessary to calculate the square root of the variance, a term called the **standard deviation** (s):

$$s = \sqrt{s^2}$$

Table 5.2 shows what percentage of the individual values within a normal distribution is included with different multiples of the standard deviation. The mean plus or minus one standard deviation ($\bar{X} \pm 1s$) includes 68

TABLE 5.2	Sample inclusion for various *s* values
Multiples of s	*Percent of Sample Included*
$\bar{X} \pm 1s$	68.3%
$\bar{X} \pm 1.96s$	95.0
$\bar{X} \pm 2s$	95.5
$\bar{X} \pm 3s$	99.7

percent of all values in the sample. Over 95 percent of all values are found within two standard deviations ($\overline{X} \pm 2s$). As such, the standard deviation provides an important descriptive summary of a set of data. Furthermore, s can be interpreted as a probability. The $\overline{X} \pm 1s$ indicates that there is a 68 percent probability that a measured value picked at random will fall within that range.

Standard Error of the Mean

To estimate how much the means of other similar samples drawn from the same population might vary, we can calculate the **standard error of the mean ($S_{\overline{X}}$):**

$$S_{\overline{X}} = \frac{s}{\sqrt{n}}$$

where s is the standard deviation, and $\sqrt{n}$ is the square root of the sample size. The standard error of the mean is a measure of the accuracy of the sample mean—that is, the variation of sample means in replications of the experiment. Because the standard error of the mean is computed by dividing s by $\sqrt{n}$, it is always a smaller value than the standard deviation.

Analysis of a Quantitative Character

Many characteristics of importance in livestock and crop plants are controlled by polygenic systems. To illustrate how biometric methods are used to analyze such quantitative characters statistically, we will consider a simplified example involving fruit weight in tomatoes. Let us assume that fruit weight is a quantitative character, and that one highly inbred strain (one that is highly homozygous) produces tomatoes averaging 18 oz. in weight, and another highly inbred strain produces fruit averaging 6 oz. in weight. These two varieties are crossed and produce an F_1 generation with weights ranging from 10 oz. to 14 oz. The F_2 population contains individuals that produce fruit ranging from 6 oz. to 18 oz. The results characterizing both generations are shown in Table 5.3.

The mean value for the fruit weight in the F_1 generation can be calculated as

$$\overline{X} = \frac{\sum X_i}{n} = \frac{626}{52} = 12.04$$

Similarly, the mean value for fruit weight in the F_2 generation is calculated as

$$\overline{X} = \frac{\sum X_i}{n} = \frac{872}{72} = 12.11$$

Average fruit weight is 12.04 oz. in the F_1 generation and 12.11 oz. in the F_2 generation. Although these mean values are similar, it is apparent from the frequency distributions (Table 5.3) that there is more variation pre-

sent in the F_2 generation. Fruit weight ranges from 6 oz. to 18 oz. in the F_2 generation, but only from 10 oz. to 14 oz. in the F_1 generation.

To assist in quantifying the amount of variation present in each generation, we can calculate the variance (Table 5.4). As noted above, the sample variance can be calculated as the sum of the squared differences between each value and the mean, divided by one less than the total number of observations. However, in the case where a number of observations (f) have been grouped into representative classes (x), the variance can be calculated according to the formula

$$s^2 = \frac{n \sum f(x^2) - (\sum fx)^2}{n(n-1)}$$

As shown in Table 5.4, the value for the F_1 generation is 1.29, and for the F_2 generation it is 4.27. When converted to the standard deviation ($\sqrt{s^2}$) the values become 1.13 and 2.06, respectively. Therefore, the distribution of tomato weight in the F_1 generation can be described as 12.04 ± 1.13, and that in the F_2 generation can be described as 12.11 ± 2.06. This analysis indicates that the mean fruit weight of the F_1 is identical to that of the F_2, but that the F_2 generation shows greater variability in the distribution of weights than does the F_1.

Observations about the inheritance of fruit weight in crosses between these two strains of tomatoes meet the expectations for polygenic traits. For the sake of this example, if we assume that each parental strain is homozygous for the additive or nonadditive alleles that control fruit weight, we can estimate the number of gene pairs involved in controlling fruit weight in these two strains of tomatoes. Since 1/72 of the F_2 offspring have a phenotype that overlaps one of the parental strains (72 total F_2 offspring, one weighs 6 oz., one weighs 18 oz.; see Table 5.3), the use of the formula $1/4^n = 1/72$ indicates that n is between 3 and 4, indicative of the number of genes that control fruit weight in these tomato strains. If this experiment were repeated many times, with similar results, our confidence in this conclusion would be bolstered.

Heritability

Having just introduced several ways in which quantitative or continuous variation can be measured and characterized in populations, we now consider how we assess the extent to which genetic factors contribute to such phenotypic variation. Often, much of the variation can be attributed to genetic factors, with the total environment having less impact. In other cases, the environment may have a greater impact on phenotypic variation within a population. The following discussion considers

| TABLE 5.3 | *Distribution of F₁ and F₂ progeny* | | | | | | | | | | | | |

							Weight in oz.						
		6	**7**	**8**	**9**	**10**	**11**	**12**	**13**	**14**	**15**	**16**	**17**	**18**
Number of	F_1:					4	14	16	12	6				
Individuals	F_2:	1	1	2	0	9	13	17	14	7	4	3	0	1

TABLE 5.4	Calculation of variance

	F_1					F_2		
x	f	$f(x)$	$f(x)^2$	x	f	$f(x)$	$f(x)^2$
6				6	1	6	36
7				7	1	7	48
8				8	2	16	128
9				9	0	0	0
10	4	40	400	10	9	90	900
11	14	154	1694	11	13	143	1573
12	16	192	2304	12	17	204	2448
13	12	156	2028	13	14	182	2366
14	6	84	1176	14	7	98	1372
15				15	4	60	900
16				16	3	48	768
17				17	0	0	0
18				18	1	18	324
	$n = 52$	$\Sigma fx = 626$	$\Sigma fx^2 = 7602$		$n = 72$	$\Sigma fx = 872$	$\Sigma fx^2 = 10{,}864$

For F_1:

$$s^2 = \frac{52 \times 7602 - (626)^2}{52(52 - 1)}$$

$$= \frac{395{,}304 - 391{,}876}{2652}$$

$$= 1.29$$

For F_2:

$$s^2 = \frac{72 \times 10{,}864 - (872)^2}{72(72 - 1)}$$

$$= \frac{782{,}208 - 760{,}384}{5112}$$

$$= 4.27$$

how geneticists attempt to define the impact of heredity versus environment on phenotypic variation.

Broad-Sense Heritability

Provided that a trait can be measured quantitatively, experiments on many plants and animals can test the causes of variation. One approach is to use inbred strains containing individuals of a relatively homogeneous (highly homozygous) **genetic background**. Experiments are then designed to test the effects of the range of prevailing environmental conditions on phenotypic variability. Variation observed *between* different inbred strains reared in a constant environment is due predominantly to genetic factors. Variation observed *among* members of the same inbred strain reared under different environmental conditions is due to nongenetic factors, which are generally categorized as "environmental."

The relative importance of genetic versus environmental factors may be formally assessed by examining the **heritability index (H^2)**, which can be calculated using an analysis of variance among individuals of a known genetic relationship. (In the ensuing discussion, we will assign the term V to designate variance, also designated as s^2.) This is an important approach when investigating organisms with long generation times. Also called **broad-sense heritability**, H^2 measures the degree to which **phenotypic variance (V_P)** is due to variation in genetic factors for a single population under the limits of environmental variation during the study. An important distinction can be made regarding H^2. Its calculation *does not* determine the proportion of the total phenotype attributed to genetic factors. It *does* estimate the proportion of observed variation in the phenotype attributed to genetic factors, in comparison to environmental factors.

Phenotypic variance is due to the sum of three components: **environmental variance (V_E), genetic variance (V_G)**, and variance resulting from the interaction of genetics and environment (V_{GE}). Therefore, phenotypic variance (V_P) is theoretically expressed as

$$V_P = V_E + V_G + V_{GE}$$

Because V_{GE} is often negligible, it is usually omitted. Therefore, the simpler equation is generally used:

$$V_P = V_E + V_G$$

Broad heritability expresses that proportion of variance due to the genetic component:

$$H_2 = \frac{V_G}{V_P}$$

An H^2 value that approaches 1.0 indicates that the environmental conditions have had little impact on phenotypic variance in the population studied. An H^2 value close to 0.0 indicates that variation in the environment has been almost solely responsible for the observed phenotypic variation within the population studied.

It is not possible to obtain an absolute H^2 value for any given character. If measured in a different population under a greater or lesser degree of environmental variability, H^2 might well change for that character. Additionally, broad-sense heritability estimates are not very accurate in estimating the selection potential of a quantitative trait. Therefore, another type of calculation, narrow-sense heritability, has been devised that is of more practical use.

Narrow-Sense Heritability

Information regarding heritability is most useful in animal and plant breeding as a measure of potential response to selection. In this case, a different estimate of heritability must be used, based on a subcomponent of V_G referred to as **additive variance (V_A)**:

$$V_G = V_A + V_D + V_I$$

where V_A represents additive variance that results from the average effect of additive components of genes. V_D represents **dominance variance**, deviation from the additive components that results when phenotypic expression in heterozygotes is not precisely intermediate between the two homozygotes. V_I reflects **interactive variance**, deviation from the additive components that occurs when two or more loci behave epistatically. V_I reflects the variance that is not associated with the average effect of V_A and is often negligible. Thus, it is often excluded from calculations.

When V_G is partitioned into V_A and V_D, a new assessment of heritability, h^2, or **narrow-sense heritability**,

may be calculated. It is h^2 that is useful in assessing selection potential in randomly breeding animal and plant populations:

$$h^2 = \frac{V_A}{V_P}$$

Because $V_P = V_E + V_G$ and $V_G = V_A + V_D$, we obtain:

$$h^2 = \frac{V_A}{V_E + V_A + V_D}$$

Artificial Selection

A relatively high h^2 value is a prediction of the impact that selection will have in altering a population. As you can see from our preceding discussion, measuring the components necessary to calculate h^2 is a complex task. A more simplified approach involves measurement of the central tendencies (the means) of a trait from (1) a parental population exhibiting a bell-shaped distribution (M), (2) a "selected" segment of the parental population that expresses the most desirable quantitative phenotypes (M1), and (3) the offspring resulting from interbreeding the selected M1 group (M2). When this is accomplished, the following relationship of the three means and h^2 exists:

$$M2 = M + h^2(M1 - M)$$

Solving this equation for h^2 leads to the formula:

$$h^2 = \frac{M2 - M}{M1 - M}$$

Once the above relationships are established, the equation can be further simplified by defining $M2 - M$ as the **response (R)** and $M1 - M$ as the **selection differential (S)**, where h^2 reflects the ratio of the response observed to the total response possible. Thus,

$$h^2 = \frac{R}{S}$$

To illustrate the assessment of selection, assume that we measured a population of corn kernels, whose mean diameter (M) was much larger than desirable (30 mm), and from that population we selected a group with the smallest diameters, for which the mean (M1) equaled 15 mm. If plants that yielded this selected population are interbred and the progeny kernels yield a mean (M2) of 20 mm, then we can calculate h^2 in order to estimate the potential for artificial selection on kernel size:

$$h^2 = \frac{20 - 30}{15 - 30}$$

$$h^2 = \frac{-10}{-15} = 0.67$$

On the basis of this calculation, we may conclude that the selection potential for kernel size is relatively high.

The longest-running artificial selection experiment known is still being conducted at the State Agricultural Laboratory in Illinois. Since 1896, corn has been selected for both high and low oil content. After 76 generations, selection continues to increase oil content (Figure 5–7). As selection has progressed, heritability of increased oil content has declined (see parenthetical values at generations 9, 25, 52, and 76 in Figure 5–7). The process will continue to proceed until all individuals in the population contain a uniform genotype for the additive alleles responsible for oil content. At that point, heritability will be reduced to zero and artificial selection will cease. Examination of the selection for low oil content shows that heritability is approaching this point.

Similar calculations involving artificial selection for other traits in a variety of organisms have established whether selection will be an effective approach in obtaining populations exhibiting desirable phenotypes. Table 5.5 provides estimates of narrow heritability for a variety of traits in various organisms. These proportions are expressed as percentage values. As you can see, heritability varies considerably between traits.

In general, heritability is low for traits that are essential to an organism's survival, primarily because the genetic component has to be largely optimized during evolution. Egg production, litter size, and conception rate are examples where such physiological limitations on selection have already been established. Nevertheless, traits that are less critical to survival, such as body weight, tail length, and wing length, show higher heritability values. Narrow-sense heritability estimates are

TABLE 5.5	Estimates of heritability for traits in different organisms
Trait	**Heritability (h^2)**
Mice	
Tail length	60%
Body weight	37
Litter size	15
Drosophila	
Abdominal bristle number	52
Wing length	45
Egg production	18
Chickens	
Body weight	50
Egg production	20
Egg hatchability	15
Cattle	
Birth weight	51
Milk yield	44
Conception rate	3

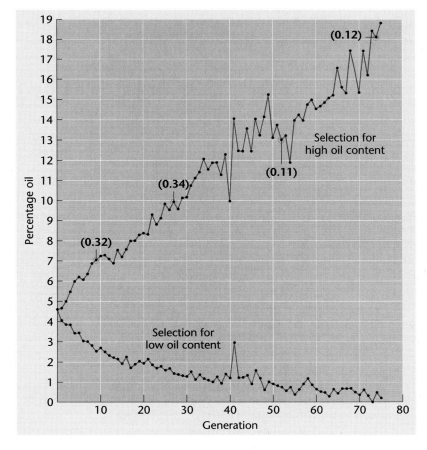

■ Figure 5–7 Response of corn selected for high and low oil content over 76 generations. The numbers in parentheses at generations 9, 25, 52, and 76 for the "high oil" line indicate the calculation of heritability at these points in the continuing experiment.

more valuable when they have been based on data collected in many populations and environments and where a clear trend is established. Based on such estimates, selection techniques have led to vast improvements in the quality of animal and plant products.

Twin Studies in Humans

Traditional heritability studies are not possible in humans, for obvious reasons. However, human twins are very useful subjects for studying the heredity-versus-environment question. **Monozygotic** or **identical twins**, derived from the division and splitting of a single egg following fertilization, are identical in their genetic compositions. Although most identical twins are reared together and are exposed to very similar environments, some pairs are separated and raised in different settings. For any particular trait, average similarities or differences can be investigated. Such an analysis is particularly useful because characteristics that remain similar in different environments are believed to have a strong genetic component. These data can then be compared with a similar analysis of **dizygotic** or **fraternal twins**, who originate from two separate fertilization events. Dizygotic twins are thus no more genetically similar than any two siblings, sharing (on average) one-half of their genes.

A form of quantitative analysis of characteristics of twins reared together may also be pursued. Twins are said to be **concordant** for a given trait if both express it or neither expresses it, and **discordant** if one shows the trait and the other does not. Table 5.6 lists concordance values for various traits in both types of twins.

These data must be examined very carefully before any conclusions are drawn. If the concordance value approaches 90 to 100 percent with monozygotic twins, we might be inclined to interpret this value as indicating a large genetic contribution to the expression of the trait. In some cases—for example, blood types and eye color—we know that this is indeed true. In the case of measles, however, a high concordance value merely indicates that the trait is almost always induced by a factor in the environment—in this case, a virus.

It is more meaningful to compare the difference between the concordance values of monozygotic and dizygotic twins. If these values are significantly higher for monozygotic twins than for dizygotic twins, we suspect that there is a strong genetic component involved in the determination of the trait. We reach this conclusion because monozygotic twins, with identical genotypes, would be expected to show a greater concordance than genetically related, but not genetically identical, dizygotic twins. In the case of measles, where concordance is high in both types of twins, the environment is assumed to contribute significantly.

Even though a particular trait may demonstrate considerable genetically based variation, it is often difficult

TABLE 5.6	A comparison of concordance of various traits between monozygotic (MZ) and dizygotic (DZ) twins	
	Concordance	
Trait	*MZ*	*DZ*
Blood types	100%	66%
Eye color	99	28
Mental retardation	97	37
Measles	95	87
Idiopathic epilepsy	72	15
Schizophrenia	69	10
Diabetes	65	18
Identical allergy	59	5
Tuberculosis	57	23
Cleft lip	42	5
Club foot	32	3
Mammary cancer	6	3

to formulate a precise mode of inheritance based on available data. In many cases, the trait is considered to be controlled by multiple-factor inheritance. However, when the environment is also exerting a partial influence, such a conclusion is particularly difficult to prove.

Mapping Quantitative Trait Loci

As we will see in the next two chapters, the identification of the location of genes within the genome represents an important step in establishing genetic identity of organisms. Referred to as **chromosome mapping**, this experimental exercise initially relies on the localization of genes of interest on a particular chromosome. Chromosome mapping has been of great interest regarding the genes controlling the expression of quantitative traits, which in the context of this discussion are called **quantitative trait loci (QTLs)**. Because each such trait is influenced by numerous genes, geneticists would like to know where these genes are located in the genome. Are they linked on a single chromosome, or scattered throughout the genome among many chromosomes? As we will see, some rather ingenious methods have been devised to localize these genes.

As an example of this type of analysis, we will consider the phenotypic trait in *Drosophila* of resistance to the insecticide DDT, which has been shown to be under polygenic control. To find the loci responsible, strains selected for resistance to DDT and strains selected for sensitivity to DDT were crossed to flies carrying dominant genes that serve as markers on each of *Drosophila's* four chromosomes. Following a variety of crosses, offspring were produced that contained many different combinations of marker chromosomes and chromosomes from either resistant or sensitive strains, as shown in Figure 5–8, flies with various chromosome combinations were then tested for resistance (their survival) when exposed to DDT. Results indicate that each of the chromosomes in *Drosophila* contains genes that contribute to resistance. In other words, the loci bearing the genes that control DDT resistance are scattered throughout the genome.

It is possible to map the position of these genes more specifically along each chromosome in *Drosophila* because of the presence of molecular markers on each chromosome. The locations of QTLs are determined relative to the known positions of these markers. These markers, called **restriction fragment length polymorphisms (RFLPs)**, represent specific nuclease cleavage sites (see Chapter 17 for a detailed description of RFLPs). They provide a new approach to enumerating and mapping the loci responsible for quantitative traits. RFLP markers are now available for many organisms of agricultural importance, making possible systematic mapping of QTLs.

For example, in the tomato, the locations of hundreds of DNA markers are known, which are spaced along all 12 chromosomes contained in the genome of this organism. Analysis is performed by crossing plants with extreme but opposite polygenically controlled phenotypes and through several generations, looking for cosegregation of specific RFLPs along with the phenotypic expression of the trait of interest. Whenever both an RFLP marker and a QTL responsible for the trait under investigation are closely linked along a single chromosome, they are more likely to demonstrate an association together throughout a pedigree than if they are not closely linked. When both the marker and the phenotypic trait are expressed together, they are said to cosegregate. When this occurs consistently, it establishes the presence of a QTL at or near the RFLP marker along the chromosome. When numerous QTLs are located, a genetic map is created for the involved genes.

Using RFLPs in conjunction with some newly derived analytical methods, QTLs for fruit weight, soluble solids, and acidity have been mapped in the tomato. Six loci responsible for fruit weight were found on six different chromosomes, four loci for soluble solids were identified on five chromosomes, and five loci for acidity were found on four different chromosomes. Note that each locus may indeed house only a single gene, although the RFLP method simply allows the identification of chromosomal regions, not individual genes. Several chromosomes contained loci for all three traits. Further investigation has revealed that the genes present at these loci account for approximately 50 percent of the phenotypic variation expressed in these traits.

The determination of the locations of QTLs for agriculturally important characteristics has permitted increased efficiency in selection and will open the door to their future genetic manipulation and transfer between organisms. As these mapping techniques became more routine in the 1990s, they have become an important tool in the repertoire of genetic engineering. Eventually, such methods can be used to isolate and characterize the genes that control quantitative traits.

■ Figure 5–8 The differential survival of *Drosophila* carrying combinations of chromosomes from DDT-resistant and DDT-sensitive (control) strains when exposed to DDT. The results indicate that DDT resistance is polygenic, with genes on each of the major chromosomes making a major contribution. Chromosome 4 carries only a few *Drosophila* genes and was omitted from this analysis.

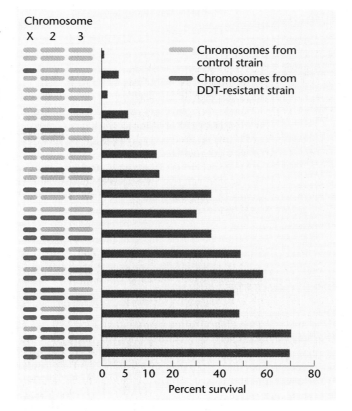

GENETICS, TECHNOLOGY, AND SOCIETY

The Green Revolution Revisited

Of the 5.9 billion people now living on earth, about 750 million don't have enough to eat. And despite efforts to limit population growth, an additional million people are expected to go hungry each year for the next several decades.

Will we be able to solve this problem? The past gives us some reason to be optimistic. In the 1950s and 1960s, in the face of looming population increases, plant scientists around the world set about to increase the production of crop plants, including the three most important grains, wheat, rice, and maize. These efforts became known as the Green Revolution. The approach was three-pronged: (1) to increase the use of fertilizers, pesticides, and irrigation water; (2) to bring more land under cultivation; and (3) to develop improved varieties of crop plants by intensive plant breeding. While highly successful, in recent years the rate of increase in grain yields has slowed. If food production is to keep pace with the projected increase in the world's population, plant breeders will have to depend more and more on the genetic improvement of crop plants to make them higher yielding. But is this possible? Are we approaching the theoretical limits of yield in important crop plants? Recent work with rice suggests that the answer to this question is a resounding no.

Rice ranks third in worldwide production, just behind wheat and maize. About 2 billion people, fully one-third of the earth's population, depend on rice for their basic nourishment. The majority of the world's rice crop is grown and consumed in Asia, but it is also a dietary staple in Africa and Central America. The Green Revolution for rice began in 1960 with the establishment of the International Rice Research Institute (IRRI), headquartered at Los Baños, the Philippines. The goal was to breed rice with improved disease resistance and higher yield. Breeders were almost too successful: The first high-yielding varieties were so top-heavy with grain that they tended to fall over. (Plant breeders call this "lodging.") To reduce lodging, IRRI breeders crossed a high-yielding line with a dwarf native variety to create semi-dwarf lines, which were introduced to farmers in 1966. Due in large part to the adoption of the semi-dwarf lines, the world production of rice doubled in the next 25 years.

Rice breeders cannot afford to rest on their laurels, however, as the yield of modern rice varieties has not improved much in recent years. Predictions suggest that a 70 percent increase in the annual rice harvest may be necessary to keep pace with anticipated population growth during the next 30 years. Breeders are now looking to wild rice varieties for further crop improvement. Leading the way are Susan McCouch and Steven Tanksley and their co-workers at Cornell University. To test the hypothesis that wild rice species carry genes that will improve the yield of cultivated varieties, they crossed cultivated rice (*Oryza sativa*) with a low-yielding wild ancestor species (*Oryza rufipogon*) and then successively backcrossed the interspecific hybrid to cultivated rice for three generations. In theory, this would create lines whose genomes were about 95 percent from *O. sativa* and 5 percent from *O. rufipogon*. When testing these backcross lines for grain yield, they found that several of them outproduced cultivated rice by as much as 30 percent. These results demonstrated strikingly that even though wild rice relatives have low yields and appear to be inferior to cultivated rice, they still carry genes that will increase the yield of elite rice varieties. It will now be up to breeders to exploit the wild rice relatives.

But introducing favorable genes from a wild relative into a cultivated variety by conventional breeding is a long and involved process, often requiring a decade or more of crossing, selection, backcrossing, and more selection. Future improvements in cultivated species must be quicker if crop yields are to keep up with population growth. Fortunately, modern gene mapping techniques are now leading to the identification of *quantitative trait loci* (QTL) that control complex traits such as yield and disease resistance. This makes possible a more direct approach to crop improvement, which has been termed the *advanced backcross QTL method*.

First, a cultivated variety is crossed with a wild relative, just as *O. sativa* was interbred with *O. rufipogon*. The hybrid is then backcrossed to the cultivated variety to generate lines that contain only a small fraction of the "wild" genome. The backcross lines with the best qualities (highest yield, most disease resistance, etc.) are selected, and the "wild QTL" responsible for the superior performance are identified using a detailed molecular linkage map. Once beneficial QTL are discovered by this method, they may be introduced into other cultivated varieties.

In order for this strategy to succeed, it is essential that wild crop relatives be preserved as a storehouse of potentially useful genes. Efforts were begun in the 1970s to protect existing crop relatives of many plants in their natural habitats and to preserve them in seed banks. As the work of McCough and Tanksley and others has shown, it is not possible to predict which wild varieties may be needed decades or even centuries from now to contribute their beneficial alleles to cultivated varieties. To prevent the loss of superior genes, the widest possible spectrum of wild species must be preserved, even those that have no obvious favorable characteristics.

Almost 60 years ago, the great Russian plant geneticist N.I. Vavilov suggested that wild relatives of crop plants could be the source of genes to improve agriculture. In the coming century Vavilov's vision may finally be realized, as genes from long-neglected wild crop relatives, identified by new molecular methods, spark a revitalized Green Revolution.

References

Mann, C. 1997. Reseeding the Green Revolution. *Science* 277: 1038-1043.

Ronald, P.C. 1997. Making rice disease-resistant. *Sci. Am.* (Nov.) 277: 98-105.

Tanksley, S.D., and McCouch, S.R. 1997. Seed banks and molecular maps: Unlocking genetic potential from the wild. *Science* 277: 1063-1066.

Xiao, J., et. al. 1996. Genes from wild rice improve yield. *Nature* 384: 223-224.

Chapter Summary

1. Continuous variation is exhibited in crosses involving traits under polygenic control. Such traits are quantitative in nature and are inherited as a result of the cumulative impact of additive alleles.

2. Polygenic characteristics can be analyzed using statistical methods, which include the mean, the variance, the standard deviation, and the standard error of the mean. Such statistical analysis can be descriptive, can be used to make inferences about a population, or can be used to compare and contrast sets of data.

3. Variation due to genetic and environmental factors in the establishment of a given phenotype is difficult to ascertain. Heritability, the estimate of the relative importance of genetic versus nongenetic factors in determining genetic variation in populations, can be calculated for many characters and is especially useful in selective breeding of commercially valuable plants and animals.

4. Studies involving twins are aimed at resolving the question of heredity versus environment in human traits. The degree of concordance of a trait may be compared in monozygotic (identical) and dizygotic (fraternal) twins raised together or apart.

5. Loci bearing genes that control a quantitative trait are called quantitative trait loci (QTLs). Using either genetic or molecular markers, QTLs may be mapped in order to ascertain their location and distribution within the genome.

INSIGHTS and SOLUTIONS

1. In a plant, height varies from 6 to 36 cm. When 6-cm and 36-cm plants were crossed, all F_1 plants were 21 cm. In the F_2 generation, a continuous range of heights was observed. Most were around 21 cm, and 3 of 200 were as short as the 6-cm P_1 parent.

(a)   What mode of inheritance is illustrated, and how many gene pairs are involved?

Solution:

Polygenic inheritance is illustrated, where a continuous trait is involved and where alleles contribute additively to the phenotype. The 3/200 ratio of F_2 plants is the key to determining the number of gene pairs. This reduces to a ratio of 1/66.7, very close to 1/64. Using the formula $1/4^n = 1/64$ (where 1/64 is equal to the proportion of F_2 phenotypes as extreme as either P_1 parent), $n = 3$. Therefore, three gene pairs are involved.

(b)   How much does each additive allele contribute to height?

Solution:

The variation between the two extreme phenotypes is

$$36 - 6 = 30 \text{ cm}$$

Because there are six potential additive alleles (*AABBCC*), each contributes

$$30/6 = 5 \text{ cm}$$

to the base height of 6 cm, which results when no additive alleles (*aabbcc*) are part of the genotype.

(c)   List all genotypes that give rise to plants that are 31 cm.

Solution:

All genotypes that include 5 additive alleles:

(5 alleles × 5 cm + 6 cm base height) = 31 cm

Therefore, *AABBCc*, *AABbCC*, and *AaBBCC* will be 31 cm.

| *Length of Ear in cm* | | | | | | | | | | | | | | | | |
	5	6	7	8	9	10	11	12	13	14	15	16	17	18	19	20	21
Parent A	4	21	24	8													
Parent B									3	11	12	15	26	15	10	7	2
F_1					1	12	12	14	17	9	4						

(continues)

(continued)

2. The results in the table on page 105 were recorded for ear length in corn:

For each of the parental strains and the F_1, calculate the mean values for ear length.

Solution:

The mean values can be calculated as follows:

$$\bar{X} = \frac{\sum X_i}{n} \quad P_A : \bar{X} = \frac{\sum X_i}{n} = \frac{378}{57} = 6.63$$

$$P_B : \bar{X} = \frac{\sum X_i}{n} = \frac{1697}{101} = 16.80$$

$$F_1 : \bar{X} = \frac{\sum X_i}{n} = \frac{836}{69} = 12.11$$

3. Compare the mean of the F_1 with that of each parental strain. What does this tell you about the type of gene action involved?

Solution:

The F_1 mean (12.11) is almost midway between the parental means of 6.63 and 16.80. This indicates that the genes in question may be additive in effect.

4. The mean and variance of corolla length in two highly inbred strains of *Nicotiana* and their progeny are shown below. One parent (P_1) has a short corolla and the other parent (P_2) has a long corolla.

Strain	Mean (mm)	Variance
P_1 short	40.47	3.12
P_2 long	93.75	3.87
F_1 ($P_1 \times P_2$)	63.90	4.74
F_2 ($F_1 \times F_1$)	68.72	47.70

Calculate the heritability (H^2) of corolla length in this plant.

Solution:

The formula for estimating heritability is $H^2 = V_G/V_P$, where V_G and V_P are the genetic and phenotypic components of variation, respectively. The main issue in this problem is obtaining some estimate of two components of phenotypic variation: genetic and environmental factors. V_P is the combination of genetic and environmental variance. Because the two parental strains are true-breeding, they are assumed to be homozygous and the variance of 3.12 and 3.87 is considered to be the result of environmental influences. The average of these two values is 3.50. The F_1 is also genetically homogeneous and gives us an additional estimate of the environmental factors. By averaging with the parents,

$$\frac{3.50 + 4.74}{2} = 4.12$$

we obtain a relatively good idea of environmental impact on the phenotype. The phenotypic variance in the F_2 is the sum of the genetic (V_G) and environmental (V_E) components. We have estimated the environmental input as 4.12, so 47.70 minus 4.12 gives us an estimate of V_G, which is 43.58. Heritability then becomes 43.58/47.70 or 0.91. This value, when viewed in percentage form, indicates that about 91 percent of the variation in corolla length is due to genetic influences.

Key Terms

Problems and Discussion Questions

1. Distinguish between discontinuous and continuous variation. Which type relates to inheritance of a quantitative nature?

2. Define and discuss the following: (a) polygenes, (b) additive alleles, (c) multiple-factor hypothesis, (d) monozygotic versus dizygotic twins, (e) concordance versus discordance, and (f) heritability.

3. A dark red strain and a white strain of wheat are crossed and produce an intermediate, medium red F_1. When the F_1 plants are interbred, an F_2 generation is produced in a ratio of 1 dark red:4 medium-dark red:6 medium red:4 light red:1 white. Further crosses reveal that the dark red and white F_2 plants are true-breeding.

   (a) Based on the ratio of offspring in the F_2, how many genes are involved in the production of color?

   (b) How many additive alleles are needed to produce each possible phenotype?

   (c) Assign symbols to these alleles and list possible genotypes that give rise to the medium red and the light red phenotypes.

   (d) Predict the outcome of the F_1 and F_2 generations in a cross between a true-breeding medium red plant and a white plant.

4. Height in humans depends on the additive action of genes. Assume that this trait is controlled by the four loci R, S, T, and U and that environmental effects are negligible. Instead of additive versus nonadditive alleles, assume that additive and partially additive alleles exist. Additive alleles contribute two units and partially additive alleles contribute one unit to height.

   (a) Can two individuals of moderate height produce offspring that are much taller or shorter than either parent? If so, how?

   (b) If an individual with the minimum height specified by these genes marries an individual of intermediate or moderate height, will any of their children be taller than the tall parent? Why or why not?

5. An inbred strain of plants has a mean height of 24 cm. A second strain of the same species from a different geographical region also has a mean height of 24 cm. When plants from the two strains are crossed together, the F_1 plants are the same height as the parent plants. However, the F_2 generation shows a wide range of heights; the majority are like the P_1 and F_1 plants, but approximately 4 of 1000 are only 12 cm high, and about 4 of 1000 are 36 cm high. (a) What mode of inheritance is occurring here? (b) How many gene pairs are involved? (c) How much does each gene contribute to plant height? (d) Indicate one possible set of genotypes for the original P1 parents and the F_1 plants that could account for these results. (e) Indicate three possible genotypes that could account for F_2 plants that are 18 cm high and three that account for F_2 plants that are 33 cm high.

6. Erma and Harvey were a compatible barnyard pair, but a curious sight. Harvey's tail was only 6 cm, while Erma's was 30 cm. Their F_1 piglet offspring all grew tails that were 18 cm. When inbred, an F_2 generation resulted in many piglets (Erma and Harvey's grandpigs), whose tails ranged in 4-cm intervals from 6 to 30 cm (6, 10, 14, 18, 22, 26, 30). Most had 18-cm tails, while 1/64 had 6-cm tails and 1/64 had 30-cm tails. (a) Explain how tail length is inherited by describing the mode of inheritance, indicating how many gene pairs are at work, and designating the genotypes of Harvey, Erma, and their 18-cm offspring. (b) If one of the 18-cm F_1 pigs is mated with the 6-cm F_2 pigs, what phenotypic ratio would be predicted if many offspring resulted? Diagram the cross.

7. In the following table, average differences of height and weight between monozygotic twins (reared together and apart), dizygotic twins, and siblings are compared. Draw as many conclusions as you can concerning the effects of genetics and the environment in influencing these human traits.

Trait	MZ Reared Together	MZ Reared Apart	DZ Reared Together	Sibs Reared Together
Height (cm)	1.7	1.8	4.4	4.5
Weight (kg)	1.9	4.5	4.5	4.7

8. Discuss the use of monozygotic and dizygotic twins reared together and apart in assessing the genetic component responsible for phenotypic variation in humans.

9. List as many human traits as you can that are likely to be under the control of a polygenic mode of inheritance.

10. Corn plants from a test plot are measured, and the distribution of heights at 10-cm intervals is recorded in the following table:

Height (cm)	Plants (no.)
100	20
110	60
120	90
130	130
140	180
150	120
160	70
170	50
180	40

Calculate (a) the mean height, (b) the variance, (c) the standard deviation, and (d) the standard error of the mean. Plot a rough graph of plant height versus frequency. Do the values represent a normal distribution? Based on your calculations, how would you assess the variation within this population?

11. Contrast broad-sense heritability (H^2) and narrow-sense heritability (h^2). To what type of population is each calculation applicable? Which is useful in artificial selection procedures? Why?

12. The mean and variance of plant height of two highly inbred strains (P_1 and P_2) and their progeny (F_1 and F_2) are shown below. Calculate the broad-sense heritability (H^2) of plant height in this species.

Strain	Mean (cm)	Variance
P_1	34.2	4.2
P_2	55.3	3.8
F_1	44.2	5.6
F_2	46.3	10.3

13. In a hypothetical situation, the study of vitamin A content and cholesterol content of eggs from a large population of chickens was investigated. The variances (V) were calculated, as shown below:

	Trait	
Variance	Vitamin A	Cholesterol
V_P	123.5	862.0
V_E	96.2	484.6
V_A	12.0	192.1
V_D	15.3	185.3

(a) Calculate the narrow heritability (h^2) for both traits.

(b) Which trait, if either, is likely to respond to selection?

14. In an assessment of learning in *Drosophila*, flies may be trained to avoid certain olfactory cues. In one population, a mean of 8.5 trials was required. A subgroup of this parental population that was trained most quickly (mean = 6.0) was interbred and their progeny examined. These flies demonstrated a mean training value of 7.5. Calculate and interpret narrow-sense heritability for olfactory learning in *Drosophila*.

15. A population of tomato plants is studied whose mean fruit weight is 60 g and where h^2 is 0.3. Predict the results of artificial selection (the mean weight of the progeny) if tomato plants whose fruit averaged 80 g were selected from the original population and interbred.

Selected Readings

Brink, R., ed. 1967. *Heritage from Mendel.* Madison: Univ. of Wisconsin Press.

Dudley, J. W. 1977. 76 generations of selection for oil and protein percentage in maize. In: Pollack, E., Kempthorne, O., and Bailey, T. (eds.), *Proceedings of the International Conference on Quantitative Genetics,* pp. 459–73. Ames: Iowa State Univ. Press.

Falconer, D. 1989. *Introduction to quantitative genetics,* 3rd ed. Harlow, England: Longman.

Farber, S. 1980. *Identical twins reared apart.* New York: Basic Books.

Feldman, M. W., and Lewontin, R. C. 1975. The heritability hangup. *Science* 190:1163–66.

Fowler, C., and Mooney, P. 1990. *Shattering: Food, politics, and the loss of genetic diversity.* Tucson: Univ. of Arizona Press.

Haley, C. 1991. Use of DNA fingerprints for the detection of major genes for quantitative traits in domestic species. *Anim. Genet.* 22:259–77.

Lander, E., and Botstein, D. 1989. Mapping mendelian factors underlying quantitative traits using RFLP linkage maps. *Genetics* 121:185–99.

Lewontin, R. C. 1974. The analysis of variance and the analysis of causes. *Am. J. Hum. Genet.* 26:400–11.

Newman, H. H., Freeman, F. N., and Holzinger, K. T. 1937. *Twins: A study of heredity and environment.* Chicago: Univ. of Chicago Press.

Paterson, A., Deverna, J., Lanini, B., and Tanksley, S. 1990. Fine mapping of quantitative traits loci using selected overlapping recombinant chromosomes in an interspecific cross of tomato. *Genetics* 124:735–42.

Plomin, R., McClearn, G., Gora-Maslak, G., and Neiderhiser, J. 1991. Use of recombinant inbred strains to detect quantitative trait loci associated with behavior. *Behav. Genet.* 21:99–116.

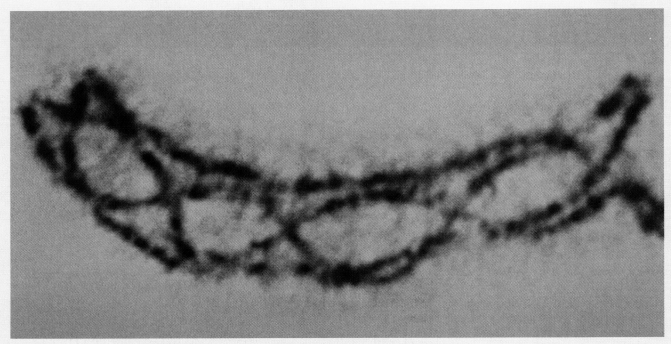

Chiasmata present between synapsed homologues during the first meiotic prophase.

CHAPTER OUTLINE

CHAPTER

6

Linkage and
Chromosome Mapping

Chapter Concepts

Many genes reside on each chromosome. Unless they are separated by crossing over, alleles at the loci on each homolog segregate as a unit during gamete formation. Recombinant gametes resulting from crossing over enhance genetic variability within a species and serve as the basis for constructing chromosome maps.

s early as 1903, Walter Sutton, who along with Theodor Boveri united the fields of cytology and genetics, pointed out the likelihood that there must be many more "unit factors" than chromosomes in most organisms. Soon thereafter, genetic investigations revealed that certain genes segregate as if they were somehow joined or linked together. Further investigations showed that such genes are part of the same chromosome and are indeed transmitted as a single unit. We now know that most chromosomes contain a very large number of genes. Those that are part of the same chromosome are said to be **linked** and to demonstrate **linkage** in genetic crosses.

Because the chromosome, not the gene, is the unit of transmission during meiosis, linked genes are not free to undergo independent assortment. Instead, the alleles at all loci of one chromosome should, in theory, be transmitted as a unit during gamete formation. However, in many instances this does not occur. During the first meiotic prophase, when homologs are paired or synapsed, a reciprocal exchange of chromosome segments may take place. This event, called **crossing over**, results in the reshuffling or **recombination** of the alleles between homologs.

The degree of crossing over between any two loci on a single chromosome is proportional to the distance between them. Thus, depending on which loci are being studied, the percentage of recombinant gametes varies. This correlation serves as the basis for the construction

of **chromosome maps**, which provide the relative locations of genes on chromosomes.

In this chapter, we will discuss linkage, crossing over, and chromosome mapping; the latter topic was introduced at the conclusion of the previous chapter. We will conclude this chapter by entertaining the rather intriguing question of why Mendel, who studied seven genes in an organism with seven chromosomes, did not encounter linkage. Or did he?

Linkage Versus Independent Assortment

To provide an overview of the major theme of this chapter, Figure 6–1 illustrates and contrasts the meiotic consequences of independent assortment, linkage without crossing over, and linkage with crossing over. Two homologous pairs of chromosomes are considered in the case of independent assortment and one homologous pair in the two cases of linkage.

Figure 6–1(a) illustrates the results of independent assortment of two pairs of nonhomologous chromosomes, each containing one heterozygous gene pair. In this case, no linkage is exhibited by the two genes. When a large number of meiotic events is observed, four genetically different gametes are formed in equal proportions.

We can compare these results with those that occur if the same genes are instead linked on the same chromosome. If no crossing over occurs between the two genes

■ Figure 6–1 A comparison of the results of gamete formation where two heterozygous genes are (a) on two different pairs of chromosomes, (b) on the same pair of homologs, but with no exchange occurring between them, and (c) on the same pair of homologs, with an exchange between two nonsister chromatids.

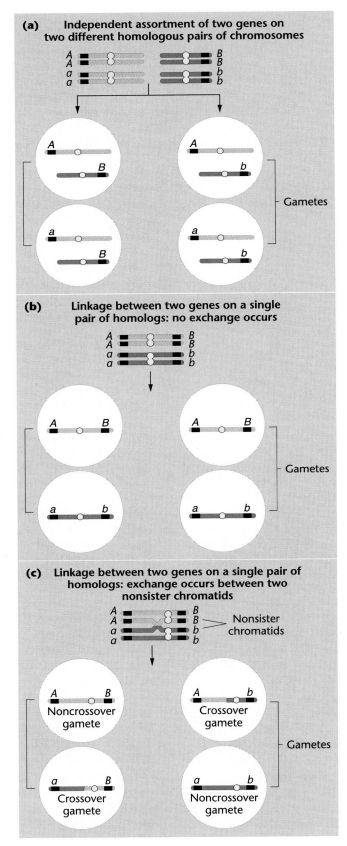

(a) Independent assortment of two genes on two different homologous pairs of chromosomes

Gametes

(b) Linkage between two genes on a single pair of homologs: no exchange occurs

Gametes

(c) Linkage between two genes on a single pair of homologs: exchange occurs between two nonsister chromatids

Nonsister chromatids

Noncrossover gamete

Crossover gamete

Crossover gamete

Noncrossover gamete

Gametes

[Figure 6–1(b)], only two genetically different gametes are formed. Each gamete receives the alleles present on one homolog *or* the other, which has been transmitted intact as the result of segregation. This case illustrates **complete linkage**, which results in the production of only **parental** or **noncrossover** gametes. The two parental gametes are formed in equal proportions. While complete linkage between two genes rarely occurs, consideration of the theoretical consequences of this concept is useful when studying crossing over.

Figure 6–1(c) illustrates the results when crossing over occurs between two linked genes. As you will note, this crossover involves only two nonsister chromatids of the four chromatids present in the tetrad. This exchange generates two new allele combinations, called **recombinant** or **crossover gametes**. The two chromatids that are not involved in the exchange result in noncrossover gametes [like those in part (b) of this figure].

The frequency with which crossing over occurs between any two linked genes is generally proportional to the distance separating the respective loci along the chromosome. In theory, two randomly selected genes could be so close to each other that crossover events would be too infrequent to be detected easily. This circumstance, **complete linkage,** would result in the production of only parental gametes, as shown in case (b). On the other hand, if a distinct but small distance separates two genes, few recombinant and many parental gametes will be formed. As the distance between two genes increases, the proportion of recombinant gametes increases and that of the parental gametes decreases. As a result, in case (c), the proportion of crossover gametes will vary depending on the distance between the genes studied.

As we will explore again later in this chapter, when two linked genes whose loci are far apart are considered, the number of recombinant gametes approaches, but does not exceed, 50 percent. If 50 percent recombinants occurred, a 1:1:1:1 ratio of the four types (two parental and two recombinant gametes) would result. In such a case, transmission of two linked genes would be indistinguishable from that of two unlinked, independently assorting genes. That is, the proportion of the four possible genotypes would be identical in cases (a) and (c) of Figure 6–1.

The Linkage Ratio

If complete linkage exists between two genes because of their close proximity, as in case (b) of Figure 6–1, and organisms with mutant alleles representing these genes are mated, a unique F_2 phenotypic ratio results that is characteristic of linkage. To illustrate this ratio, we will consider a cross involving the closely linked recessive mutant genes *brown (bw)* eye and *heavy (hv)* wing vein in *Drosophila melanogaster* (Figure 6–2). The normal, wild-type alleles bw^+ and hv^+ are both dominant and result in red eyes and thin wing veins, respectively.

In this cross, flies with mutant brown eyes and normal thin veins are mated to flies with normal red eyes and mutant heavy veins. In more concise terms, brown-eyed

flies are crossed with heavy-veined flies. If we extend the system of using genetic symbols established in Chapter 4, linked genes may be represented by placing their allele designations above and below a single or double horizontal line. Those placed above the line are located at loci on one homolog and those placed below at the homologous loci on the other homolog. Thus, we may represent the P_1 generation as follows:

$$P_1: \quad \frac{bw \; hv^+}{bw \; hv^+} \quad \times \quad \frac{bw^+ \; hv}{bw^+ \; hv}$$

brown, thin   red, heavy

Because the genes are located on an autosome, no distinction for male and female is necessary.

In the F_1 generation, each fly receives one chromosome of each pair from each parent; all flies are heterozygous for both gene pairs and exhibit the dominant traits of red eyes and thin veins:

$$F_1: \quad \frac{bw \;\; hv^+}{bw^+ \; hv}$$

red, thin

As shown in Figure 6–2, when the F_1 generation is interbred, because of complete linkage, each F_1 individual forms only parental gametes. Following fertilization, the F_2 generation is produced in a 1:2:1 phenotypic and genotypic ratio. One-fourth of this generation will show brown eyes and thin veins; one-half will show both wild-type traits, namely, red eyes and thin veins; and one-fourth will show red eyes and thick veins. In more concise terms, the ratio is 1 brown:2 wild:1 thick. Such a ratio is characteristic of complete linkage. Complete linkage is typically observed only when genes are very close together and the number of progeny is relatively small.

Figure 6–2 also demonstrates the results of a test cross with the F_1 flies. Such a cross produces a 1:1 ratio of brown, thin and red, thick flies. Had the genes controlling these traits been incompletely linked or located on separate autosomes, four phenotypes rather than two would have been produced.

When large numbers of mutant genes present in any given species are investigated, genes located on the same chromosome will show evidence of linkage to one another. As a result, **linkage groups** can be established, one for each chromosome. In theory, the number of linkage groups should correspond to the haploid number of chromosomes. In diploid organisms in which large numbers of mutant genes are available for genetic study, this correlation has been upheld.

Incomplete Linkage, Crossing Over, and Chromosome Mapping

If two genes linked on the same chromosome are selected at random, it is highly improbable that they will demonstrate complete linkage. Instead, crosses involv-

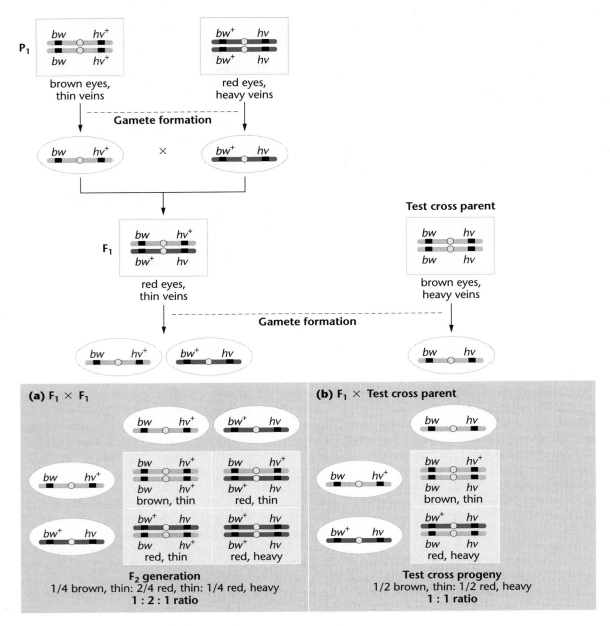

■ Figure 6–2   The results of a cross involving two genes located on the same chromosome, demonstrating complete linkage. Part (a) generates the F_2 results of the cross, while part (b) generates the results of a test cross involving the F_1 progeny.

ing two randomly selected linked genes will almost always produce a percentage of offspring resulting from recombinant gametes. This percentage is variable, depending upon which two genes are involved in the cross and the distance between them along the chromosome. This phenomenon was first explained by two *Drosophila* geneticists, Thomas H. Morgan and his undergraduate student Alfred H. Sturtevant.

Morgan and Crossing Over

As you may recall from our discussion in Chapter 4, Morgan first discovered the phenomenon of X-linkage. In his studies, he investigated numerous *Drosophila*

mutations located on the X chromosome. When he analyzed crosses involving only one trait, he was able to deduce the mode of X-linked inheritance. However, when he made crosses involving two X-linked genes, his results were at first puzzling. For example, as shown in Cross A of Figure 6–3, he crossed female flies expressing mutant *yellow* body (*y*) and *white* eyes (*w*) with wild-type males (gray bodies and red eyes). The F_1 females were wild type, while the F_1 males expressed both mutant traits. In the F_2, 98.7 percent of the offspring showed the parental phenotypes—either yellow-bodied, white-eyed flies or wild-type (gray-bodied, red-eyed). The remaining 1.3 percent of the flies were either yellow-bodied with red eyes or gray-bodied with white

■ Figure 6–3 The F_1 and F_2 results of crosses involving the *yellow* body, *white* eye mutations and the *white* eye, *miniature* wing mutations. In the F_2 generation of cross A, 1.3 percent of the flies demonstrate recombinant phenotypes, which express either *white* or *yellow*. In the F_2 generation of cross B, 37.2 percent of the flies demonstrate recombinant phenotypes, which express either *miniature* or *white*.

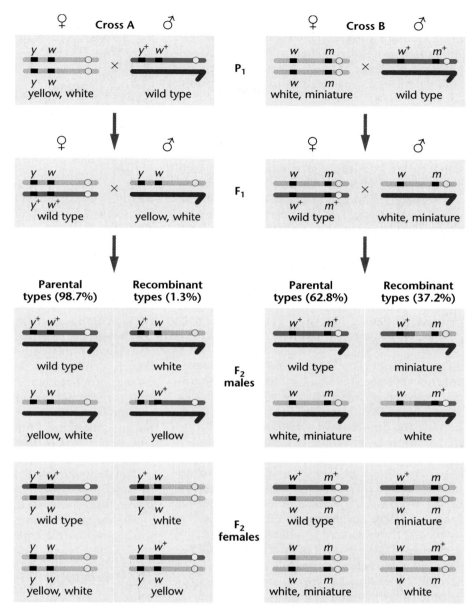

eyes. It was as if the genes had somehow separated from each other during gamete formation in the F_1 flies.

When Morgan made crosses involving other X-linked genes, the results were even more puzzling (Cross B of Figure 6–3). The same basic pattern was observed, but the proportion of F_2 phenotypes differed; for example, in a cross involving white-eye, miniature-wing mutants, only 62.8 percent of the F_2 showed the parental phenotypes, while 37.2 percent of the offspring appeared as if the mutant genes had been separated during gamete formation.

In 1911, Morgan was faced with two questions: (1) What was the source of gene separation? and (2) Why did the frequency of the apparent separation vary depending on the genes being studied? The proposed answer to the first question was based on his knowledge of earlier cytological observations made by F. A. Janssens and others. Janssens had observed that synapsed homologous chromosomes in meiosis wrap

around each other, creating **chiasmata** (singular: chiasma), where points of overlap are evident. Morgan proposed that these chiasmata could represent points of genetic exchange.

In the crosses shown in Figure 6–3, Morgan postulated that if an exchange occurred between the mutant genes on the two X chromosomes of the F_1 females, it would lead to the observed results. He suggested that such exchanges led to 1.3 percent recombinant gametes in the *yellow-white* cross and 37.2 percent in the *white-miniature* cross. On the basis of this and other experiments, Morgan concluded that linked genes exist in a linear order along the chromosome and that a variable amount of exchange occurs between any two genes.

As an answer to the second question involving the frequency of gene separation, Morgan proposed that two genes located relatively close to each other along a chromosome are less likely to have a chiasma form between them than if the two genes are farther apart on the chromosome. Therefore, the closer two genes are, the less likely it is that a genetic exchange will occur between them. Morgan proposed the term **crossing over** to describe the physical exchange leading to recombination.

Sturtevant and Mapping

Morgan's student, Alfred H. Sturtevant, was the first to realize that his mentor's proposal could be used to map the sequence of, as well as the distance between, linked genes. For example, Sturtevant compiled further data on recombination between the genes represented by the *yellow, white,* and *miniature* mutants initially studied by Morgan. Frequencies of crossing over between each pair of these three genes were observed in separate crosses to be

(1) *yellow, white*   0.5%

(2) *white, miniature*   34.5%

(3) *yellow, miniature*   35.4%

Because the sum of (1) and (2) is approximately equal to (3), Sturtevant suggested that the recombination frequencies between linked genes are additive. On this basis, he predicted that the order of the genes on the X chromosome was *yellow–white–miniature*. In arriving at this conclusion, he reasoned as follows: The *yellow* and *white* genes are apparently close to each other because the recombination frequency is low. However, both of these genes are quite far apart from *miniature* because the *white, miniature* and *yellow, miniature* combinations show large recombination frequencies. Because *miniature* shows more recombination with *yellow* than with *white* (35.4 vs. 34.5), it follows that *white* is between the other two genes, not outside of them.

Sturtevant knew from Morgan's work that the frequency of exchange could be taken as an estimate of the distance between two genes or loci along the chromosome. He constructed a map of the three genes on the X chromosome, with one map unit being equated with 1 percent recombination between two genes.*

In the preceding example, the distance between *yellow* and *white* would be 0.5 map unit, and that between *yellow* and *miniature* would be 35.4 map units. It follows that the distance between *white* and *miniature* should be (35.4 – 0.5) or 34.9. This estimate is close to the actual frequency of recombination between *white* and *miniature* (34.5). The simple map for these three genes is shown in Figure 6–4.

In addition to these three genes, Sturtevant considered two other genes on the X chromosome and produced a more extensive map including all five genes. He and a colleague, Calvin Bridges, soon began a search for autosomal linkage in *Drosophila*. By 1923, they had

---

■ Figure 6–4 A simple map of the *yellow* (*y*), *white* (*w*), and *miniature* (*m*) genes on the X chromosome of *Drosophila melanogaster*. Each number represents the percentage of recombinant offspring produced in the three crosses, each involving two different genes.

clearly shown that linkage and crossing over are not restricted to X-linked genes, but also could be demonstrated with autosomes.

During this work, they made another interesting observation. In *Drosophila*, crossing over was shown to occur only in females. The fact that no crossing over occurs in males made genetic mapping much less complex to analyze in *Drosophila*. However, crossing over does occur in both sexes in most other organisms.

Although many refinements in chromosome mapping have been developed since Sturtevant's initial work, his basic principles are accepted as correct. They have been used to produce detailed chromosome maps of organisms for which large numbers of linked mutant genes are known. In addition to providing the basis for chromosome mapping, Sturtevant's findings were historically significant to the field of genetics. In 1910, the **chromosomal theory of inheritance** was still being widely disputed. Even Morgan was skeptical of this theory prior to the time he conducted the bulk of his experimentation. Research has now firmly established that chromosomes contain genes in a linear order and that these genes are the equivalent of Mendel's unit factors.

Single Crossovers

Why should the relative distance between two loci influence the amount of recombination and crossing over observed between them? The basis for this variation is explained in the following analysis.

During meiosis, a limited number of crossover events occurs in each tetrad. These recombinant events occur randomly along the length of the tetrad. Therefore, the closer two loci reside along the axis of the chromosome, the less likely it is that any **single crossover** event will occur between them. The same reasoning suggests that the farther apart two linked loci are, the more likely it is that a random crossover event will occur between them.

In Figure 6–5(a), a single crossover occurs between two nonsister chromatids, but not between the two loci; therefore, the crossover goes undetected because no recombinant gametes are produced. In Figure 6–5(b), where two loci are quite far apart, the crossover occurs between them, yielding recombinant gametes.

When a single crossover occurs between two nonsister chromatids, the other two strands of the tetrad will not be involved in this exchange and will enter the gamete

---

*In honor of Morgan's work, one map unit is often referred to as a centimorgan (cM).

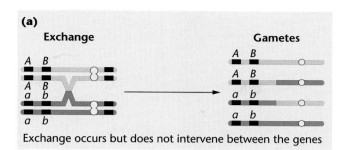

(a)

Exchange occurs but does not intervene between the genes

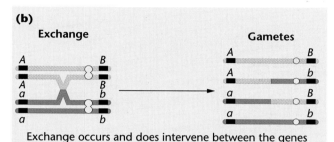

(b)

Exchange occurs and does intervene between the genes

■ **Figure 6–5** Two cases of exchange between two nonsister chromatids and the gametes subsequently produced. In part (a) the exchange does not separate the alleles of the two genes, only parental gametes are formed, and the exchange goes undetected. In part (b) the exchange separates the alleles, resulting in recombinant gametes, which are detectable.

unchanged. Even if a single crossover occurs 100 percent of the time between two linked genes, recombination will subsequently be observed in only 50 percent of the potential gametes formed. This concept is diagrammed in Figure 6–6. Theoretically, when only single exchanges are considered and when 20 percent recombinant gametes are observed, crossing over actually occurs between these two loci in 40 percent of the tetrads. The general rule is that under these conditions, the percentage of tetrads involved in an exchange between two genes is twice as great as the percentage of recombinant gametes produced. Therefore, the theoretical limit of recombination due to crossing over is 50 percent.

When two linked genes are more than 50 map units apart, a crossover can theoretically be expected to occur between them in 100 percent of the tetrads. If this prediction were to be achieved, each tetrad would yield equal proportions of the four gametes shown in Figure

6–6, just as if the genes were on different chromosomes and assorting independently. For a variety of reasons, this theoretical limit is seldom achieved.

Multiple Crossovers

It is possible that in a single tetrad, two, three, or more result of several crossing-over events. Double exchanges of genetic material result from **double crossovers**, as shown in Figure 6–7. For a double exchange to be studied, three gene pairs must be investigated, each heterozygous for two alleles. Before we can determine the frequency of recombination among all three loci, we must review some simple probability calculations.

The probabilities of a single exchange occurring between the A and B or the B and C genes are related directly to the distance between the loci. The closer A is to B and B is to C, the less likely it is that a single exchange will occur between either of the two sets of loci. In the case of a double crossover, two separate and independent events or exchanges must occur simultaneously. The mathematical probability of two independent events occurring simultaneously is equal to the product of the individual probabilities. This is the product law previously introduced in Chapter 3.

Suppose that crossover gametes resulting from single exchanges between A and B are recovered 20 percent of the time ($p = 0.20$) and between B and C 30 percent of the time ($p = 0.30$). The probability of recovering a double-crossover gamete arising from two exchanges, between A and B and between B and C, is predicted to be $(0.20)(0.30) = 0.06$, or 6 percent. It is apparent from this calculation that the frequency of double-crossover gametes is always expected to be much lower than that of either single-crossover class of gametes.

If three genes are relatively close together along one chromosome, the expected frequency of double-crossover gametes is extremely low. For example, consider the A–B distance in Figure 6–7 to be 3 map units and the B–C distance in that figure to be 2 map units. The expected double-crossover frequency is $(0.03)(0.02) = 0.0006$, or 0.06 percent. This translates to only 6 events in 10,000. Thus, in a mapping experiment such as this, where closely linked genes are involved, very large numbers of offspring are required in order to detect double-crossover events. In this example, it would

■ **Figure 6–6** The consequences of a single exchange between two nonsister chromatids occurring in the tetrad stage. Two noncrossover (parental) and two crossover (recombinant) gametes are produced.

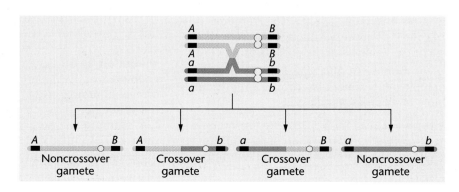

■ Figure 6–7 The results of a double exchange occurring between nonsister chromatids. Because the exchanges involve only two strands, two noncrossover gametes and two double-crossover gametes are produced. The photograph illustrates chiasmata found in a tetrad isolated during the first meiotic prophase stage.

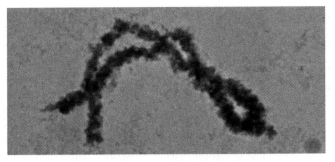

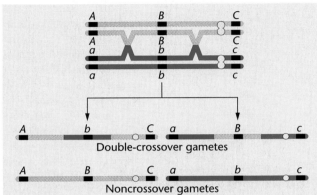

be unlikely that a double crossover would be observed even if 1000 offspring were examined. If these probability considerations are extended, it is evident that if four or five genes are being mapped, even fewer **triple** and **quadruple crossovers** can be expected to occur.

Three-Point Mapping in Drosophila

The information presented in the previous section serves as the basis for mapping three or more linked genes in a single cross. To illustrate this, we will examine a situation involving three linked genes.

Three criteria must be met for a successful mapping cross:

1. The genotype of the organism producing the crossover gametes must be heterozygous at all loci under consideration.

2. The cross must be constructed so that genotypes of all gametes can be determined accurately by observing the phenotypes of the resulting offspring. This is necessary because the gametes and their genotypes can never be observed directly. To overcome this problem, each phenotypic class must reflect the genotype of the gametes of the parents producing it.

3. A sufficient number of offspring must be produced in the mapping experiment to recover a representative sample of all crossover classes.

These criteria are met in the three-point mapping cross from *Drosophila melanogaster* shown in Figure 6–8. In this cross, three sex-linked recessive mutant genes—*yellow* body color, *white* eye color, and *echinus* eye shape—are considered. In order to diagram the cross, we must assume some theoretical sequence, even though we do not yet know if it is correct. In Figure 6–8, we will initially assume the sequence of the three genes to be *y–w–ec*. If this is incorrect, our analysis will reveal the correct sequence.

In the P_1 generation, males hemizygous for all three wild-type alleles are crossed to females that are homozygous for all three recessive mutant alleles. Therefore, the P_1 males are wild type with respect to body color, eye color, and eye shape. They are said to have a *wild-type phenotype*. The females, on the other hand, exhibit the three mutant traits—yellow body color, white eyes, and echinus eye shape.

This cross produces an F_1 generation consisting of females that are heterozygous at all three loci and males that, because of the Y chromosome, are hemizygous for the three mutant alleles. Phenotypically, all F_1 females are wild type, while all F_1 males are *yellow, white,* and *echinus*. The genotype of the F_1 females fulfills the first criterion for constructing a map of the three linked genes; that is, it is heterozygous at the three loci and may serve as the source of recombinant gametes generated by crossing over. Note that because of the genotypes of the P_1 parents, all three mutant alleles are on one homolog, and all three wild-type alleles are on the other homolog. *Other arrangements are possible.* For example, the heterozygous F_1 female might have the *y* and *ec* mutant alleles on one homolog and the *w* allele on the other. This would occur if, in the P_1 cross,

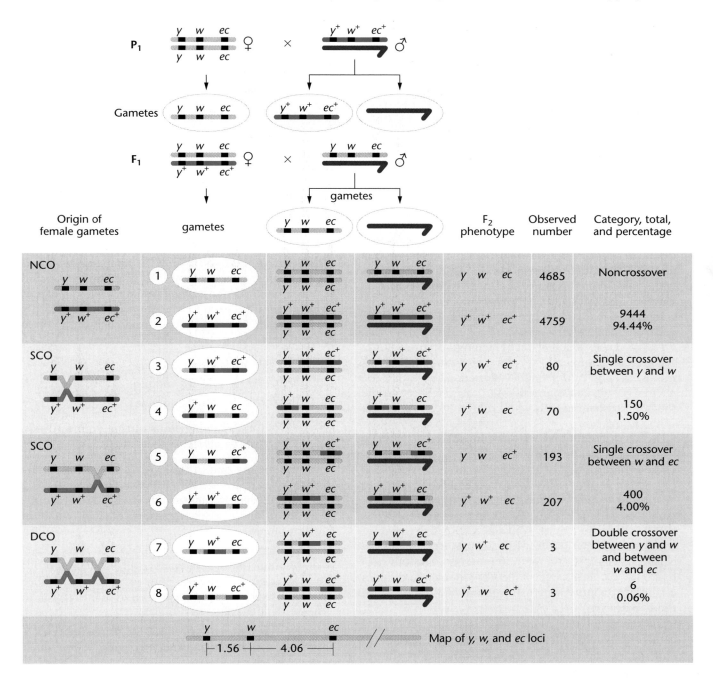

■ Figure 6–8 A three-point mapping cross involving the *yellow* (*y* or *y⁺*), white (*w* or *w⁺*), and *echinus* (*ec* or *ec⁺*) genes in *Drosophila melanogaster*. NCO, SCO, and DCO refer to non-crossover, single-crossover, and double-crossover groups, respectively. Because of the complexity of this and several of the ensuing figures, centromeres have not been included on the chromosomes, and only two nonsister chromatids are shown initially. Crossing over, of course, always occurs in the four-strand tetrad stage.

one parent was *yellow* and *echinus* and the other parent was *white*.

In our cross, the second criterion is met by virtue of the gametes formed by the F_1 males. Every gamete contains either an X chromosome bearing the three mutant alleles or a Y chromosome, which is genetically inert for the three loci being considered. Whichever

type participates in fertilization, the genotype of the gamete produced by the F_1 female will be expressed phenotypically in the F_2 male and female offspring derived from it. Thus, all F_1 noncrossover and crossover gametes can be detected by observing the F_2 phenotypes.

With these two criteria met, we can construct a chromosome map from the crosses illustrated in Figure 6–8.

First, we must determine which F_2 phenotypes correspond to the various noncrossover and crossover categories. Two of these can be determined immediately.

The **noncrossover phenotypes** are determined by the combination of alleles present in the parental gametes formed by the F_1 female. Each such gamete contains one or the other of the X chromosomes *unaffected by crossing over*. As a result of segregation, approximately equal proportions of the two types of gametes, and subsequently the F_2 phenotypes, are produced. Because they are derived from a heterozygote, the genotypes of the two parental gametes and the phenotypes of the two F_2 phenotypes complement one another. For example, if one is wild type, the other is completely mutant. This is the case in the cross under consideration. In other situations, if one chromosome shows one mutant allele or trait, the other shows the other two mutant traits, and so on. They are therefore called **reciprocal classes** of gametes and phenotypes.

The two noncrossover phenotypes are most easily recognized because *they exist in the greatest proportion*. Figure 6–8 shows that classes (1) and (2) are present in the greatest numbers. Therefore, flies that express *yellow, white*, and *echinus* phenotypes and those that are normal (or wild type) for all three characters constitute the noncrossover category and represent 94.44 percent of the F_2 offspring.

The second category that can be easily detected is represented by the **double-crossover phenotypes**. Because of their low probability of occurrence, *they must be present in the least numbers*. Remember that this group represents two independent but simultaneous single-crossover events. Two reciprocal phenotypes can be identified: class (7), which shows the mutant traits *yellow* and *echinus* but normal eye color; and class (8), which shows the mutant trait white but normal body color and eye shape. Together these double-crossover phenotypes constitute only 0.06 percent of the F_2 offspring.

The remaining four phenotypic classes represent two categories resulting from single crossovers. Classes (3) and (4), reciprocal phenotypes produced by single-crossover events occurring between the *yellow* and *white* loci, are equal to 1.50 percent of the F_2 offspring. Classes (5) and (6), constituting 4.00 percent of the F_2 offspring, represent the reciprocal phenotypes resulting from single-crossover events occurring between the *white* and *echinus* loci.

The map distances separating the three loci can now be calculated. The distance between *y* and *w*, or between *w* and *ec*, is equal to the percentage of all detectable exchanges occurring between them. For any two genes under consideration, this will include all appropriate single crossovers as well as all double crossovers. The latter are included because they represent two simultaneous single crossovers. For the *y* and *w* genes this includes classes (3), (4), (7), and (8), totaling 1.50 percent

+ 0.06 percent, or 1.56 map units. Similarly, the distance between *w* and *ec* is equal to the percentage of offspring resulting from an exchange between these two loci: classes (5), (6), (7), and (8), totaling 4.00 percent + 0.06 percent, or 4.06 map units. The map of these three loci on the X chromosome, based on these data, is shown at the bottom of Figure 6–8.

Determining the Gene Sequence

In the preceding example, the sequence (or order) of the three genes along the chromosome was assumed to be *y–w–ec*. Our analysis shows this sequence to be consistent with the data. However, in most mapping experiments the gene sequence is not known, and this constitutes another variable in the analysis. In our example, had the gene sequence been unknown, it could have been determined using a straightforward method.

This method is based on the fact that there are only three possible orders, each containing one of the three genes in between the other two:

(I) *w–y–ec* (*y* is in the middle)

(II) *y–ec–w* (*ec* is in the middle)

(III) *y–w–ec* (*w* is in the middle)

If you use the following steps during your analysis, you will be able to determine the gene order.

1. Assuming any one of the three orders, first determine the **arrangement** of alleles along each homolog of the heterozygous parent giving rise to noncrossover and crossover gametes (the F_1 female in our example).

2. Determine whether a double-crossover event occurring within that arrangement will produce the **observed** double-crossover phenotypes. Remember that these phenotypes occur least frequently and can be easily identified.

3. If this order does not produce the predicted phenotypes, try each of the other two orders. One must work!

These steps are illustrated in Figure 6–9, using the cross discussed above. The three possible orders are labeled **I**, **II**, and **III**, as shown previously. Either *y, ec*, or *w* must be in the middle.

1. Assuming that *y* is between *w* and *ec*, the arrangement of alleles along the homologs of the F_1 heterozygote is

$$\textbf{I} \quad \frac{w \quad\quad y \quad\quad ec}{w^+ \quad y^+ \quad ec^+}$$

We know this because of the way in which the P_1 generation was crossed. The P_1 female contributed an X chromosome bearing the *w, y*, and *ec* alleles,

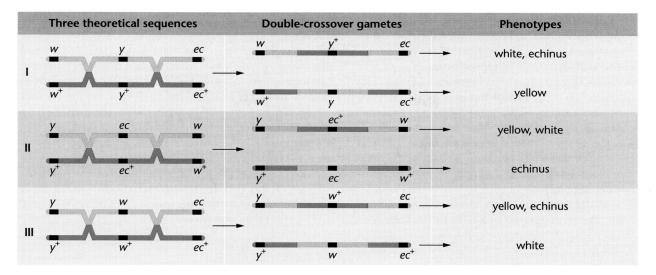

Figure 6–9 A summary of the three possible orders of the *white, yellow,* and *echinus* genes, the results of a double crossover in each case, and the resulting phenotypes produced in a test cross. The two noncrossover chromatids of each tetrad have been omitted to simplify the illustrations.

while the P_1 male contributed an X chromosome bearing the w^+, y^+, and ec^+ alleles.

2. A double crossover within the above arrangement would yield the following gametes:

$$\underline{w \quad y^+ \quad ec} \text{ and } \underline{w^+ \quad y \quad ec^+}$$

Following fertilization, if y is in the middle, the F_2 double-crossover phenotypes will correspond to the above gametic genotypes, yielding offspring that express *white, echinus* and offspring that express *yellow*. Instead, however, determination of the actual double crossovers reveals them to be *yellow, echinus* flies and *white* flies. Therefore, our assumed order is incorrect.

3. If we consider the other orders, one with the ec/ec^+ alleles in the middle (**II**) or one with the w/w^+ alleles in the middle (**III**),

$$\textbf{II} \quad \frac{y \quad ec \quad w}{y^+ \quad ec^+ \quad w^+} \quad \text{or} \quad \textbf{III} \quad \frac{y \quad w \quad ec}{y^+ \quad w^+ \quad ec^+}$$

we see that arrangement **II** again provides **predicted** double-crossover phenotypes that *do not* correspond to the **actual** (observed) double-crossover phenotypes. The predicted phenotypes are *yellow, white* flies and *echinus* flies in the F_2 generation. Therefore, this order is also incorrect. However, arrangement **III** will produce the observed phenotypes—*yellow, echinus* flies and *white* flies. Therefore, this order, where the w gene is in the middle, is correct.

To summarize, utilizing this method is rather straightforward. Determine the arrangement of alleles on the homologs of the heterozygote yielding the crossover gametes. This is done by locating the reciprocal noncrossover phenotypes. Then, test each of three possible orders to determine which one yields the observed double-crossover phenotypes. Whichever of the three does so represents the correct order.

A Mapping Problem in Maize

Having established the basic principles of chromosome mapping, we will now consider a related problem in maize (corn), in which the gene sequence and interlocus distances are unknown.

This analysis differs from the preceding discussion in several ways and therefore will expand your knowledge of mapping procedures. First, the previous mapping cross involved sex-linked genes. Here, autosomal genes are considered. Second, in the discussion of this cross, we will make a change in the use of symbols, as first suggested in Chapter 4. Instead of using the gene symbols and superscripts (e.g., bm^+, v^+, and pr^+), we will simply use + to denote each wild-type allele. This symbol is less complex to manipulate, but requires a better understanding of mapping procedures.

When we consider three autosomally linked genes in maize, the experimental cross must still meet the same three criteria established for the X-linked genes in *Drosophila:* (1) One parent must be heterozygous for all traits under consideration; (2) the gametic genotypes produced by the heterozygote must be apparent from observing the phenotypes of the offspring; and (3) a sufficient sample size must be available for complete analysis.

In maize, the recessive mutant genes *bm (brown midrib), v (virescent seedling),* and *pr (purple aleurone)* are linked on chromosome 5. Assume that a female plant is known to be heterozygous for all three traits. We do not know (1) the arrangement of the mutant alleles on the maternal and paternal homologs of this heterozygote, (2) the sequence of genes, or (3) the map distances between the genes. What genotype must the male plant have to allow successful mapping? In order to meet the second criterion, the male must be homozygous for all three recessive mutant alleles. Otherwise, offspring of this cross showing a given phenotype might represent more than one genotype, making accurate mapping impossible.

Figure 6–10 diagrams this cross. As shown, we do not know either the arrangement of alleles or the sequence

of loci in the heterozygous female. Some of the possibilities are shown, but we have yet to determine which is correct. In the test-cross male parent, *we do not know the sequence*, and so we must write it down initially as a random selection. Note that we have chosen initially to place *v* in the middle. This may or may not be correct.

The offspring have been arranged in groups of two for each pair of reciprocal phenotypic classes. The two members of each reciprocal class are derived from either no crossing over **(NCO)**, one of two possible single-crossover events **(SCO)**, or a double crossover **(DCO)**.

To solve this problem, consider the following three questions. It will be helpful to refer to Figures 6–10 and 6–11 as you consider them.

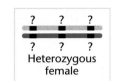

(a) Possible allele arrangements and gene sequences in a heterozygous female

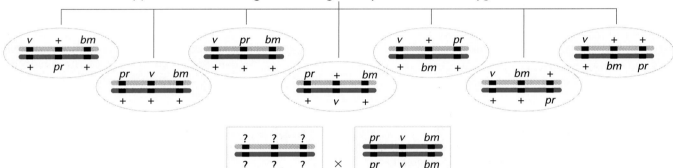

(b) Actual results of mapping cross

Phenotypes of offspring			Number	Total and percentage	Exchange classification
+	v	bm	230	467	Noncrossover
pr	+	+	237	42.1%	(NCO)
+	+	bm	82	161	Single crossover
pr	v	+	79	14.5%	(SCO)
+	v	+	200	395	Single crossover
pr	+	bm	195	35.6%	(SCO)
pr	v	bm	44	86	Double crossover
+	+	+	42	7.8%	(DCO)

■ Figure 6–10 In part (a) are shown some of the possible combinations of allele arrangements and gene sequences in a heterozygous female. The data from a three-point mapping cross, depicted in part (b), where the female is test-crossed, provide the basis for determining which combination of arrangement and sequence is correct.

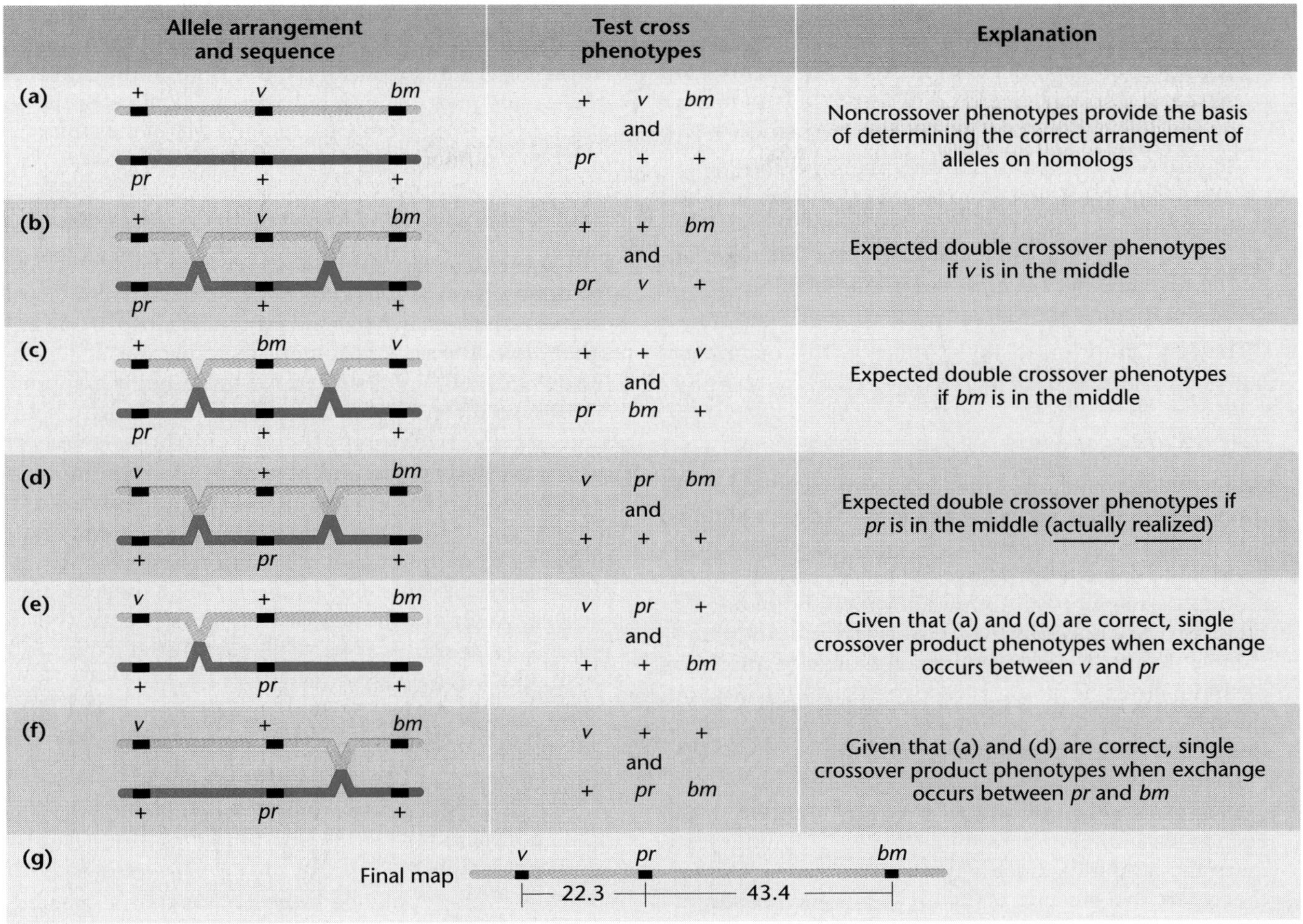

Allele arrangement and sequence	Test cross phenotypes	Explanation
(a)	+ v bm and pr + +	Noncrossover phenotypes provide the basis of determining the correct arrangement of alleles on homologs
(b)	+ + bm and pr v +	Expected double crossover phenotypes if v is in the middle
(c)	+ + v and pr bm +	Expected double crossover phenotypes if bm is in the middle
(d)	v pr bm and + + +	Expected double crossover phenotypes if pr is in the middle (actually realized)
(e)	v pr + and + + bm	Given that (a) and (d) are correct, single crossover product phenotypes when exchange occurs between v and pr
(f)	v + + and + pr bm	Given that (a) and (d) are correct, single crossover product phenotypes when exchange occurs between pr and bm
(g) Final map	v pr bm — 22.3 — 43.4 —	

■ Figure 6–11   The steps used in producing a map of the three genes involved in the cross shown in Figure 6–10, where neither the arrangement of alleles nor the sequence of genes is initially known in the heterozygous female parent.

1. *What is the correct heterozygous arrangement of alleles in the female parent?*

   Determine the two noncrossover classes, those that occur with the highest frequency. In this case, they are $\underline{+\ v\ bm}$ and $\underline{pr\ +\ +}$. Therefore, the arrangement of alleles on the homologs of the female parent must be as shown in Figure 6–11(a). These homologs segregate into gametes, unaffected by any recombination event. Any other arrangement of alleles could not yield the observed noncrossover classes. (Remember that $\underline{+\ v\ bm}$ is equivalent to $\underline{pr^+\ v\ bm}$ and that $\underline{pr\ +\ +}$ is equivalent to $\underline{pr\ v^+\ bm^+}$.)

2. *What is the correct sequence of genes?*

   We know that the arrangement of alleles is

   $$\frac{+\quad v\quad bm}{pr\quad +\quad +}$$

   But is the assumed sequence correct? That is, will a double-crossover event yield the observed double-crossover phenotypes following fertilization? Simple observation shows that it will not [Figure

6–11(b)]. Try the other two orders [Figures 6–11(c) and 6–11(d)], *keeping the same arrangement:*

$$\frac{+\quad bm\quad v}{pr\quad +\quad +}\qquad\qquad \frac{v\quad +\quad bm}{+\quad pr\quad +}$$

Only the latter case will yield the observed double-crossover classes [Figure 6–11(d)]. Therefore, the *pr* gene is in the middle. From this point on, work the problem using the above arrangement and sequence, with the *pr* locus in the middle.

3. *What is the distance between each pair of genes?*

   Having established the sequence of loci as *v pr bm*, we can now determine the distance between *v* and *pr* and between *pr* and *bm*. Remember that the map distance between two genes is calculated on the basis of all detectable recombinational events occurring between them. This includes both the single- and double-crossover events involving the two genes being considered.

   Figure 6–11(e) shows that the phenotypes *v pr +* and *+ + bm* result from single crossovers between the *v* and *pr* loci, accounting for 14.5 percent of the offspring. By adding the percentage of double crossovers (7.8 percent) to the number obtained for single crossovers, the total distance between the *v* and *pr* loci is calculated to be 22.3 map units.

Figure 6–11(f) shows that the phenotypes $v + +$ and $+ pr bm$ result from single crossovers between the *pr* and *bm* loci, totaling 35.6 percent. With the addition of the double-crossover classes (7.8 percent), the distance between *pr* and *bm* is calculated to be 43.4 map units.

The final map for all three genes in this example is shown in Figure 6–11(g).

The Accuracy of Mapping Experiments

So far, we have assumed that crossover frequencies are directly proportional to the distance between any two loci along the chromosome. However, it is not always possible to detect all crossover events. A case in point is a double exchange that occurs between the two loci in question. As shown in Figure 6–12(a), if a double exchange occurs, the original arrangement of alleles on each nonsister homolog is recovered. Therefore, even though crossing over occurs, it is impossible to detect in these cases. This factor holds for all even-numbered exchanges between two loci.

As a result of complications posed by multiple strand exchanges, mapping determinations usually underestimate the actual distance between two genes. The farther apart two genes are, the greater is the probability that undetected crossovers will occur. While the discrepancy is minimal for two genes that are relatively

close together, the degree of inaccuracy increases as the distance becomes greater, as noted in Figure 6–12(b), where the relationship between recombination frequency and map distance is graphed. The most accurate maps are derived from experiments with genes that are relatively close together.

Interference and the Coefficient of Coincidence

Still another factor tends to affect mapping data. This factor involves the reduction of the expected number of double crossovers when genes are quite close to one another along the chromosome. This reduction, called **interference**, will be illustrated for three-point mapping.

We have already considered the probability relationships between single- and double-crossover events. In theory, the percentage of **expected double crossovers** (DCOs) is predicted by multiplying the percentage of the total crossovers between each pair of genes. Remember that a DCO represents two single-crossover events. For example, the expected double-crossover frequency of the cross illustrated in Figures 6–10 and 6–11 may be calculated in the following manner:

$$DCO_{exp} = (0.223) \times (0.434) = 0.097 = 9.7\%$$

Frequently, this predicted figure does not correspond precisely with the observed DCO frequency. Generally, there are fewer DCOs observed than predicted. In the maize cross, only 7.8 percent DCOs were observed. In some cases there may be more DCOs than expected. These disparities are explained by the concept of **interference**, which is quantified by calculating the ratio of

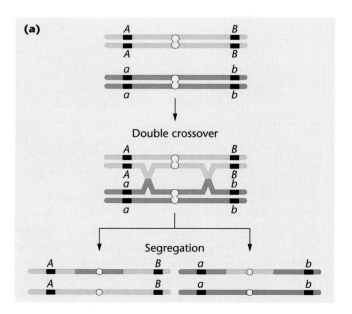
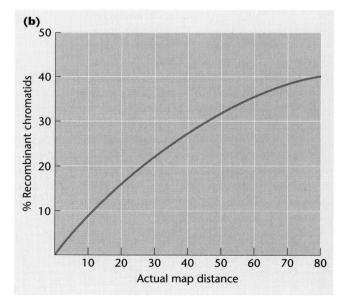

■ Figure 6–12 (a) An illustration of a double crossover that goes undetected because no rearrangement of alleles occurs. (b) The relationship between the frequency of recombination and map distance, as studied in *Drosophila*, *Neurospora*, and *Zea mays* (maize).

observed DCOs to expected DCOs. This ratio is called the **coefficient of coincidence** *(C):*

$$C = \frac{\text{observed DCO}}{\text{expected DCO}}$$

In the maize cross, we have

$$C = \frac{0.078}{0.097} = 0.804$$

Once *C* is calculated, interference *(I)* may be quantified using the simple equation

$$I = 1.0 - C$$

In the maize cross, we have

$$I = 1.000 - 0.804 = 0.196$$

If interference is complete and no double crossovers occur, then *I* = 1.0. If fewer observed DCOs than expected DCOs occur, *I* is a positive number (as above) and **positive interference** has occurred. If more observed DCOs than expected DCOs occur, *I* is a negative number and **negative interference** has occurred.

In eukaryotic systems, positive interference is most often observed. It appears that a crossover event in one region of a chromosome inhibits a second crossover in neighboring regions of the chromosome. In general, the closer genes are to one another along the chromosome, the more positive interference is observed (and the lower is the *C* value). One explanation is that a mechanical stress is imposed on chromatids during crossing over such that one chiasma inhibits the formation of a second chiasma in the neighboring region.

The Genetic Map of Drosophila

In organisms such as *Drosophila*, maize, and the mouse, where large numbers of mutations have been discovered and where mapping crosses are possible, extensive chromosome maps have been constructed. Illustrated in Figure 6–13 are partial maps of each of the four chromosomes of *Drosophila*. Virtually every morphological feature of the fruit fly has been subjected to mutations. The locus of each mutant gene is first localized to one of the four chromosomes (or linkage groups) and then mapped in relation to all other genes present on that chromosome. Based on cytological evidence, the relative lengths of these genetic maps correlate roughly with the relative physical lengths of these chromosomes.

Somatic Cell Hybridization and Human Gene Mapping

In humans, where neither designed matings nor large numbers of offspring are available, the earliest linkage studies were based on pedigree analysis. Attempts were made to establish whether a trait was X-linked or autosomal. For autosomal traits, geneticists tried to distinguish clearly whether pairs of traits demonstrated linkage or independent assortment. In this way, it was hoped that a human gene map could be created.

The difficulty arises, however, when two genes of interest are separated on a chromosome such that recombinant gametes are formed, obscuring linkage in a pedigree. In such cases, the demonstration of linkage is enhanced by the use of an approach relying on probability calculations, called the **lod score method**. First devised by J. B. S. Haldane and C. A. Smith in 1947 and refined by Newton Morton in 1955, the lod score (standing for log of the odds favoring linkage) assesses the probability that a particular pedigree involving two traits reflects linkage or not. First, the probability is calculated that family data (pedigrees) concerning two or more traits conform to the transmission of traits without linkage. Then the probability is calculated that the identical family data following these same traits result from linkage with a specified recombination frequency. The ratio of these probability values expresses the "odds" for and against linkage. Accuracy using the lod score method is limited by the extent of the family data, but nevertheless represented an important advance in assigning human genes to specific chromosomes and constructing preliminary human chromosome maps.

The initial results were discouraging because of the limitations of this approach and because of the relatively high haploid number of human chromosomes (23). By 1960, almost no linkage or mapping information had become available.

However, in the 1960s, a new technique, **somatic cell hybridization**, was developed that aided immensely in assigning human genes to their respective chromosomes. This technique, first discovered by Georges Barsky, relies on the fact that two cells in culture can be induced to fuse into a single hybrid cell. While Barsky used two mouse cell lines, it soon became evident that cells from different organisms will also fuse together. When this event occurs, an initial cell type called a **heterokaryon** is produced. The hybrid cell contains two nuclei in a common cytoplasm. Using the proper techniques, it is possible to fuse human and mouse cells, for example, and isolate the hybrids from the parental cells.

As the heterokaryons are cultured *in vitro*, two interesting changes occur. Eventually, the nuclei fuse together, creating what is termed a **synkaryon**. Then, as culturing is continued for many generations, chromosomes from one of the two parental species are gradually lost. In the case of the human–mouse hybrid, human chromosomes are lost randomly until eventually the synkaryon has a full complement of mouse chromosomes and only a few human chromosomes. As we will see, it is the preferential loss of human chromosomes (rather than mouse chromosomes) that makes

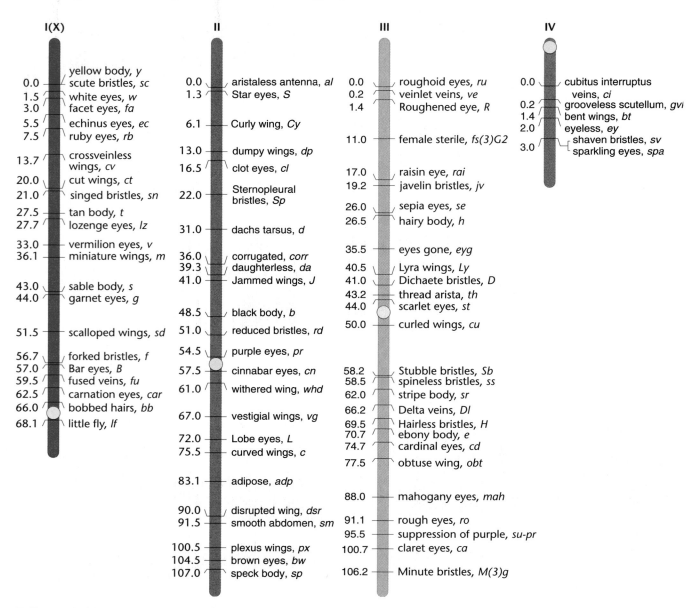

■ Figure 6–13   A partial genetic map of the four chromosomes of *Drosophila melanogaster*. The circle on each chromosome represents the position of the centromere.

possible the assignment of human genes to the chromosomes on which they reside.

The experimental rationale is straightforward. If a specific human gene product is synthesized in a synkaryon containing one to three human chromosomes, then the gene responsible for that product must reside on one of the human chromosomes remaining in the hybrid cell. On the other hand, if the human gene product is not synthesized in a synkaryon, the gene responsible is not present on those human chromosomes that remain in this hybrid cell. Ideally, a panel of 23 hybrid cell lines, each with but one unique human chromosome, would allow the immediate assignment of any human gene for which the product could be characterized.

In practice, a panel of cell lines, each with several remaining human chromosomes, is most often utilized. The correlation of the presence or absence of each chromosome with the presence or absence of each gene product is called **synteny testing**. Consider, for example, the hypothetical data provided in Figure 6–14, where four gene products (A, B, C, and D) are tested in relationship to eight human chromosomes. Let us carefully analyze the gene that produces product A.

1. Product A is not produced by cell line 23, but chromosomes 1, 2, 3, and 4 are present in cell line 23. Therefore, we can rule out the presence of gene *A* on those four chromosomes and conclude that it must be on chromosome 5, 6, 7, or 8.

2. Product A is produced by cell line 34, which contains chromosomes 5 and 6, but not 7 and 8. Therefore, gene *A* is on chromosome 5 or 6, but cannot be on 7 or 8 because they are absent, even though product A is produced.

Hybrid cell lines	Human chromosomes present								Gene products expressed			
	1	2	3	4	5	6	7	8	*A*	*B*	*C*	*D*
23	●—	●—	●—	●—					−	+	−	+
34	●—	●—			●—	●—			+	−	−	+
41	●—		●—		●—		●—		+	+	−	+

■ **Figure 6–14** A hypothetical grid of data used in synteny testing to assign genes to their appropriate human chromosomes. Three somatic hybrid cell lines, designated 23, 34, and 41, have each been scored for the presence or absence of human chromosomes 1 through 8, as well as for their ability to produce the hypothetical human gene products A through D.

3. Product A is also produced by cell line 41, which contains chromosome 5 but not chromosome 6. Using similar reasoning to that described above, gene *A* must be on chromosome 5.

Using a similar approach, gene *B* can be assigned to chromosome 3. You should perform this analysis to demonstrate for yourself that this is correct. Gene *C* presents a unique situation. The data indicate that it is not present on any of the first seven chromosomes (1–7). While it might be on chromosome 8, no direct evidence supports this conclusion. Other panels are needed. We shall leave gene *D* for you to analyze. On what chromosome does it reside?

Using the approach described above, literally hundreds of human genes have been assigned to one chromosome or another. Some of the assignments shown in Figure 6–15 were either derived or confirmed with the use of somatic cell hybridization techniques. To map some other genes, for which the products have yet to be discovered, researchers have had to rely on other approaches. For example, by combining a rather sophisticated technique using recombinant DNA technology

with pedigree analysis, it has been possible to assign the genes responsible for **Huntington disease**, **cystic fibrosis**, and **neurofibromatosis** to their respective chromosomes, 4, 7, and 17. This approach will be discussed in Chapter 18.

The Use of Haploid Organisms in Linkage and Mapping Studies

Many of the single-celled eukaryotes are haploid during the vegetative stages of their life cycle. The alga *Chlamydomonas* and the mold *Neurospora* illustrate this genetic condition. These organisms do form reproductive cells that fuse during fertilization, producing a diploid zygote. However, this structure soon undergoes meiosis, resulting in haploid vegetative cells that are then propagated by mitotic divisions.

Small, haploid organisms have several important advantages in genetic studies compared with diploid eukaryotes. They can be cultured and manipulated in genetic crosses much more easily. In addition, a hap-

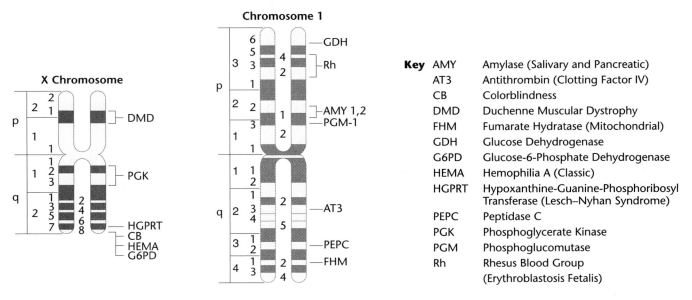

■ **Figure 6–15** Representative regional gene assignments for human chromosome 1 and the X. Many assignments were initially derived using somatic cell hybridization techniques.

loid organism contains only a single allele of each gene, which is expressed directly in the phenotype. This fact greatly simplifies genetic analysis. As a result, organisms such as *Chlamydomonas* and *Neurospora* have served as subjects of research investigations in many areas of genetics, including linkage and mapping studies.

In order to perform genetic experiments with such organisms, crosses are made, and following fertilization the meiotic structures are isolated. Because in each of these structures all four meiotic products give rise to spores, each structure is called a **tetrad**. *The term "tetrad" has a different meaning here than when it was used earlier to describe a precise chromatid configuration in meiosis.* In any case, individual tetrads are isolated, and the resultant cells are grown and analyzed separately from those of other tetrads. In the results to be described, the data reflect the proportion of tetrads that showed one combination of genotypes, the proportion that showed another combination, and so on. Such experimentation is called **tetrad analysis**.

Gene-to-Centromere Mapping

When a single gene in *Neurospora* is analyzed, as diagrammed in Figure 6–16, the data can be used to calculate the map distance between the gene and the centromere. This process is sometimes referred to as **mapping the centromere**. It may be accomplished by experimentally determining the frequency of recombination using tetrad data. Note that once the four meiotic products of the tetrad are formed, a mitotic division occurs, resulting in eight ordered products (ascospores).

If no crossover event occurs between the gene under study and the centromere, the pattern of ascospores (contained within an ascus) appears as shown in Figure 6–16(a) (*aaaa++++*).[*]

This pattern represents **first division segregation** because the two alleles are separated during the first meiotic division. However, a crossover event will alter this pattern, as shown in Figures 6–16(b) (*aa++aa++*) and 6–16(c) (*++aaaa++*). Actually, two other recombinant patterns may occur, depending on the chromatid orientation during the second meiotic division: *++aa++aa* and *aa++++aa*. All four of the latter patterns reflect **second division segregation** because the two alleles are not separated until the second meiotic division. Usually, the ordered tetrad data are condensed to reflect the genotypes of the identical ascospore pairs. Six combinations are possible:

First Division Segregation

a	a	+	+
+	+	a	a

In order to calculate the distance between the gene and the centromere, a large number of asci resulting from a controlled cross must be counted. Using these data, the distance (*d*) is calculated:

Second Division Segregation

a	+	a	+
+	a	+	a
+	a	a	+
a	+	+	a

$$d = 1/2 \left(\frac{\text{second division segregant asci}}{\text{total asci}} \right)$$

The distance (*d*) reflects the percentage of recombination and is only one-half the number of second division segregant asci. This is because crossing over in each of them has occurred in only two of the four strands during meiosis.

To illustrate, assume that *a* represents albino and + represents wild type in *Neurospora*. In crosses between the two genetic types, suppose the following data were observed:

65 first division segregants
70 second division segregants

The distance between *a* and the centromere is

$$d = 1/2 \left(\frac{70}{135} \right) = 0.259$$

or about 26 map units.

As the distance increases up to 50 units, in theory all asci should reflect second division segregation. However, numerous factors prevent this. As in diploid organisms, mapping accuracy based on crossover events is greatest when the gene and centromere are relatively close together.

Analysis in haploid organisms can also be performed in order to distinguish between linkage and independent assortment of two genes. Mapping distances can be calculated between gene loci once linkage is established. As a result, detailed maps of organisms such as *Neurospora* and *Chlamydomonas* are now available.

Other Aspects of Genetic Exchange

We have established that careful analysis of crossing over during gamete formation can serve as the basis for the construction of chromosome maps in both diploid and haploid organisms. We should not, however, lose sight of the real biological significance of the process, which is to generate genetic variation in gametes and, subsequently, in the offspring derived from the resultant eggs and sperm. Because of the critical role of crossing over in generating variation, the study of genetic exchange has remained an important topic for

---

[*]The pattern (*++++aaaa*) can also be formed. However, it is indistinguishable from (*aaaa++++*).

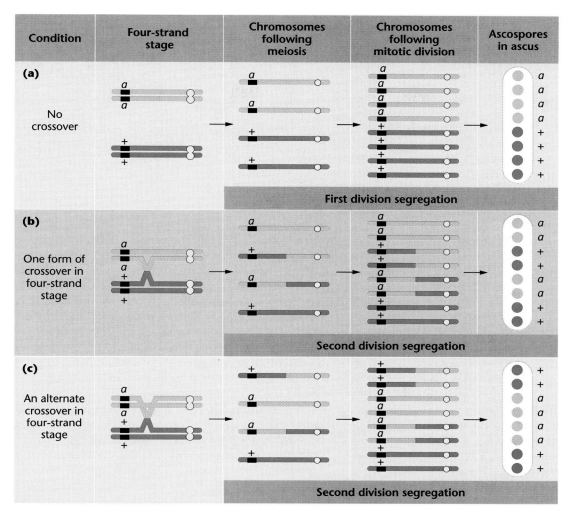

Condition	Four-strand stage	Chromosomes following meiosis	Chromosomes following mitotic division	Ascospores in ascus

(a)

No crossover

First division segregation

(b)

One form of crossover in four-strand stage

Second division segregation

(c)

An alternate crossover in four-strand stage

Second division segregation

■ Figure 6–16   Three ways in which ascospore patterns can be generated in the asci of *Neurospora*. Analyses of these patterns serve as the basis for "gene-to-centromere" mapping. The photograph shows various ascopore patterns in asci of *Neurospora*.

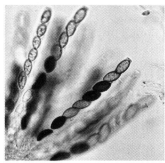

study in genetics. Many questions need to be addressed. For example, does crossing over involve an actual exchange of chromosome arms? Does exchange occur between paired sister chromatids during mitosis? We will briefly consider observations that attempt to answer these questions.

Cytological Evidence for Crossing Over

Visual proof that genetic crossing over is accompanied by an actual physical exchange between homologous chromosomes has been demonstrated independently by Curt Stern in *Drosophila* and by Harriet Creighton and Barbara McClintock in *Zea mays* (corn). Because

the experiments are similar, we will consider only the work with corn. In Creighton and McClintock's work, two linked genes on chromosome 9 were studied. At one locus, the alleles *colorless (c)* and *colored (C)* control endosperm coloration (endosperm is stored material in the kernel). At the other locus, the alleles *starchy (Wx)* and *waxy (wx)* control the carbohydrate characteristics of the endosperm.

A corn plant was obtained that was heterozygous at both loci and contained two unique cytological markers on one of the homologs. The markers consisted of a densely stained knob at one end of the chromosome and a translocated piece of another chromosome (8) at the other end. The arrangement of alleles and cytologi-

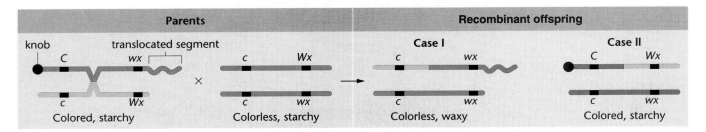

■ Figure 6–17   The phenotypes and chromosome compositions observed in Creighton and McClintock's experiment in maize demonstrating that crossing over involves a breakage and rejoining process. The knob and translocated segment served as cytological markers to establish that crossing over involves an actual exchange of chromosome arms.

cal markers in this plant is shown in Figure 6–17. Creighton and McClintock crossed this plant to one that was homozygous for the color alleles and heterozygous for the endosperm alleles. They obtained several different phenotypes in the offspring, but they were most interested in one that occurred as a result of crossing over between the unique cytological markers. The chromosomes of this plant, with the colorless, waxy phenotype (Case I in Figure 6–17), were examined for the presence of the cytological markers. As expected, if genetic crossing over is accompanied by a physical exchange between homologs, the translocated chromosome will still be present, but the knob will not. This was the case! In a second plant (Case II), the phenotype colored, starchy can result from either nonrecombinant gametes or from crossover gametes. When the latter is the case, chromosomes from it should contain the dense knob but not the translocated piece of chromosome. This was the case, and again the findings supported the conclusion that a physical exchange had taken place. Similar findings by Stern using *Drosophila* leave no doubt about this conclusion involving the cytological basis of crossing over.

Sister Chromatid Exchanges

Knowing that crossing over occurs between synapsed homologs in meiosis, we might ask whether such a physical exchange occurs during mitosis. While homologous chromosomes do not usually pair up or synapse in somatic cells (*Drosophila* is an exception), each individual chromosome in prophase and metaphase of mitosis consists of two identical sister chromatids, joined at a common centromere. Surprisingly, several experimental approaches have demonstrated that reciprocal exchanges similar to crossing over occur between sister chromatids. While these **sister chromatid exchanges** (SCEs) do not produce new allelic combinations, evidence is accumulating that attaches significance to these events.

Identification and study of SCEs are facilitated by several modern staining techniques. In one approach, cells are allowed to replicate for two generations in the presence of the thymidine analog **bromodeoxyuridine** (BUdR). Following two rounds of replication, each pair

of sister chromatids has one member with one strand of DNA "labeled" with BUdR and one with both strands labeled. Using a differential staining technique, chromatids with BUdR in both strands stain less brightly than chromatids with only one strand of the DNA helix containing the analog. As a result, SCEs are readily detectable, if they occur. In Figure 6–18, numerous instances of SCE events are clearly evident. Because of the patchlike appearance, these sister chromatids are sometimes referred to as **harlequin chromosomes.**

While the significance of SCEs is still uncertain, several observations have led to great interest in this phenomenon. It is known, for example, that agents that induce chromosome damage (such as viruses, X-rays, ultraviolet light, and certain chemical mutagens) also increase the frequency of SCEs. The frequency of SCEs

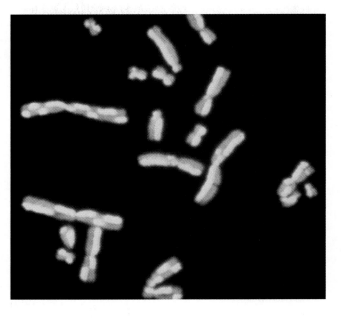

■ Figure 6–18   Demonstration of sister chromatid exchanges (SCEs) in mitotic chromosomes. Chromatids containing the thymidine analog BUdR in both DNA strands fluoresce less brightly than those with the analog in only one strand. These chromosomes were stained with 33258-Hoechst reagent and acridine orange and then viewed under fluorescence microscopy.

is also elevated in **Bloom syndrome**, an autosomal recessive disorder in humans. This rare genetic disease is characterized by retardation of growth, a great sensitivity of skin to the sun, immunodeficiency, and a predisposition to the formation of both benign and malignant tumors. Increased breaks and rearrangements between nonhomologous chromosomes are observed in addition to excessive amounts of sister chromatid exchanges.

Did Mendel Encounter Linkage?

We conclude this chapter by examining a modern-day interpretation of the experiments that serve as the cornerstone of transmission genetics—the crosses with garden peas performed by Mendel.

It has been said often that Mendel had extremely good fortune in his classical experiments with the garden pea. In none of his crosses did he encounter apparent linkage relationships between any of the seven mutant characters. Had Mendel obtained highly variable data characteristic of linkage and crossing over, these unorthodox ratios might have hindered his successful analysis and interpretation.

The article by Stig Blixt, reprinted in its entirety in the box below, demonstrates the inadequacy of this hypothesis that Mendel dealt with seven genes on seven chromosomes. As we shall see, some of Mendel's genes were indeed linked. We shall leave it to Stig Blixt to enlighten you as to why Mendel did not detect linkage.

Why Didn't Gregor Mendel Find Linkage?

It is quite often said that Mendel was very fortunate not to run into the complication of linkage during his experiments. He used seven genes, and the pea has only seven chromosomes. Some have said that had he taken just one more, he would have had problems. This, however, is a gross oversimplification. The actual situation, most probably, is shown in Table 1. This shows that Mendel worked with three genes in chromosome 4, two genes in chromosome 1, and one gene in each of chromosomes 5 and 7. It seems at first glance that, out of the 21 dihybrid combinations Mendel theoretically

could have studied, no less than four (that is, *a–i, v–fa, v–le, fa–le*) ought to have resulted in linkages. As found, however, in hundreds of crosses and shown by the genetic map of the pea, *a* and *i* in chromosome 1 are so distantly located on the chromosome that no linkage is normally detected. The same is true for *v* and *le* on the one hand, and *fa* on the other, in chromosome 4. This leaves *v–le*, which ought to have shown linkage.

Mendel, however, seems not to have published this particular combination and thus, presumably, never made the appropriate cross to obtain both genes

segregating simultaneously. It is therefore not so astonishing that Mendel did not run into the complication of linkage, although he did not avoid it by choosing one gene from each chromosome.

STIG BLIXT

Weibullsholm Plant Breeding Institute, Landskrona, Sweden, and Centro Energia Nucleate na Agricultura, Piracicaba, SP, Brazil.

Source: Reprinted by permission from *Nature,* Vol. 256, p. 206. Copyright © 1975 Macmillan Journals Limited.

Table 1	Relationship between modern genetic terminology and character pairs used by Mendel	
Character Pair Used by Mendel	*Alleles in Modern Terminology*	*Located in Chromosome*
Seed color, yellow–green	*I–i*	1
Seed coat and flowers, colored–white	*A–a*	1
Mature pods, smooth expanded–wrinkled indented	*V–v*	4
Inflorescences, from leaf axis–umbellate in top of plant	*Fa–fa*	4
Plant height, 1m–around 0.5 m	*Le–le*	4
Unripe pods, green–yellow	*Gp–gp*	5
Mature seeds, smooth–wrinkled	*R–r*	7

Received March 5; accepted June 4, 1975.

[1] Blixt, S. 1974. In *Handbook of genetics,* ed. R.C. King. New York: Plenum Press.

INSIGHTS and SOLUTIONS

1. In rabbits, *black (B)* is dominant to *brown (b)*, while *full color (C)* is dominant to *chinchilla (c^{ch})*. The genes controlling these traits are linked. Rabbits that are heterozygous for both traits and express *black, full color*, were crossed to rabbits that express *brown, chinchilla*, with the following results:

31 *brown, chinchilla*

34 *black, full*

16 *brown, full*

19 *black, chinchilla*

Determine the arrangement of alleles in the heterozygous parents and the map distance between the two genes.

Solution:

This is a two-point map problem, where the two reciprocal phenotypes most prevalent are the noncrossovers. The less frequent reciprocal phenotypes arise from a single crossover. The arrangement of alleles is derived from the non-crossover phenotypes because they enter gametes intact.

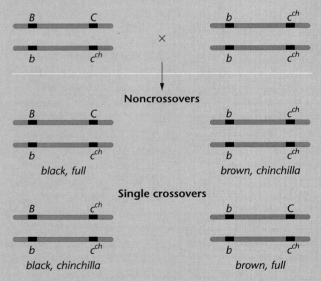

The single crossovers give rise to 35/100 offspring (35 percent). Therefore, the distance between the two genes is 35 map units (mu).

2. In *Drosophila, Lyra (Ly)* and *Stubble (Sb)* are dominant mutations located at locus 40 and 58, respectively, on chromosome 3. A recessive mutation with bright red eyes was discovered and shown also to be on chromosome 3. A map was obtained by crossing a female who

was heterozygous for all three mutations to a male that was homozygous for the bright red mutation (which we will call *br*). The following data were obtained:

	Phenotype			Number
(1)	Ly	Sb	br	404
(2)	+	+	+	422
(3)	Ly	+	+	18
(4)	+	Sb	br	16
(5)	Ly	+	br	75
(6)	+	Sb	+	59
(7)	Ly	Sb	+	4
(8)	+	+	br	2
				1000

Determine the location of the bright red mutation on chromosome 3.

Solution:

First, determine the arrangement of the alleles on the homologs of the heterozygous crossover parent (the female in this case). This is done by locating the most frequent reciprocal phenotypes, which arise from the noncrossover gametes. These are phenotypes (1) and (2). Each one represents the arrangement of alleles on one of the homologs. Therefore, the arrangement is

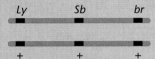

Second, determine the correct sequence of the three loci along the chromosome. This is done by determining which sequence will yield the observed double-crossover phenotypes, which are the least frequent reciprocal phenotypes (7 and 8). If the sequence is correct as written, then a double crossover as depicted below,

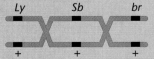

will yield *Ly + br* and *+ Sb +* as phenotypes. Inspection shows that these categories (5 and 6) are actually single crossovers, not double crossovers. Therefore, the sequence, as written, is incorrect.

There are only two other possible sequences. The *br* gene is either to the left of *Ly*, or it is between *Ly* and *Sb*:

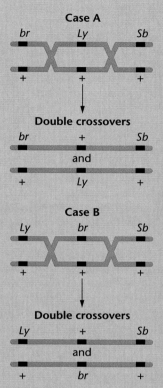

Comparison with the actual data shows that case B is correct. The double-crossover gametes yield flies that express *Ly* and *Sb* but not *br*; or express *br* but not *Ly* and *Sb*. Therefore, the correct arrangement and sequence is

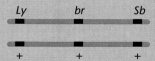

Once this has been determined, it is possible to determine the location of *br* relative to *Ly* and *Sb*. A single crossover between *Ly* and *br*, as shown below,

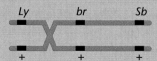

yields flies that are *Ly* + + and + *br Sb* (categories 3 and 4). Therefore, the distance between the *Ly* and *br* loci is equal to

$$\frac{18 + 16 + 4 + 2}{1000} = \frac{40}{1000} = 0.04 = 4 \text{ map units}$$

Remember that we must add in the double crossovers because they represent two single crossovers occurring simultaneously. Because we need to know the frequency of all crossovers between *Ly* and *br*, they must be included. Similarly, the distance between the *br* and *Sb* loci is derived mainly from single crossovers between them:

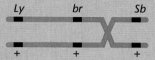

This event yields *Ly br* + and + + *Sb* phenotypes (categories 5 and 6). Therefore, the distance equals

$$\frac{75 + 59 + 4 + 2}{1000} = \frac{140}{1000} = 0.14 = 14 \text{ map units}$$

The final map shows that *br* is located at locus 44, since *Lyra* and *Stubble* are known:

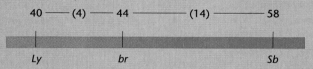

3. By making reference to Figure 6–13, predict what gene was discovered in Problem 2 (which we referred to as *br*). Suggest an experimental cross that could confirm your prediction.

Solution:

Inspection of Figure 6–13 reveals that the mutation *scarlet (st)* is present at locus 44.0, so it is reasonable to hypothesize that the bright-red eye mutation is an allele at the *scarlet* locus.

To test this hypothesis, you could perform complementation analysis (see Chapter 4) by crossing females expressing the bright red mutation with known *scarlet* males. If the two mutations are alleles, no complementation will occur, and all progeny will reveal a bright-red mutant eye phenotype. If complementations occurs, all progeny will express normal brick red (wild-type) eyes, since the bright red mutation and *scarlet* are at different loci (they are probably very close together). In such a case, all progeny will be heterozygous at both the bright eye and the *scarlet* loci and will not express either of the mutations, since they are both recessive. This type of complementation analysis is called an **allelism test**.

Chapter Summary

1. Genes located on the same chromosome are said to be linked. Alleles located on the same homolog, therefore, can be transmitted together during gamete formation. However, the mechanism of crossing over between homologs during meiosis results in the reshuffling of alleles, thereby contributing to genetic variability within gametes.

2. Early in this century, geneticists realized that crossing over could provide an experimental basis for mapping the location of linked genes relative to one another along the chromosome.

3. Somatic cell hybridization techniques have made possible linkage and mapping analysis of human genes.

4. Mapping studies are also possible in haploid organisms such as *Chlamydomonas* and *Neurospora*.

5. Cytological investigations of both corn and *Drosophila* have revealed that crossing over involves a physical exchange of segments between nonsister chromatids.

6. An exchange of genetic material between sister chromatids may occur during mitosis as well. These events are referred to as sister chromatid exchanges (SCEs). An elevated frequency of such events is seen in the human disorder Bloom syndrome.

7. Evidence now suggests that several of the genes studied by Mendel are linked. However, in such cases, the genes are sufficiently far apart along the chromosme to prevent the detection of linkage.

Key Terms

allelism test, 133
Bloom syndrome, 131
chiasma (chiasmata), 115
chromosomal theory of inheritance, 116
chromosome map, 111
coefficient of coincidence (*C*), 125
complete linkage, 113
crossing over, 111

cystic fibrosis, 127
double crossover (DCO), 117
harlequin chromosomes, 130
heterokaryon, 125
Huntington's disease, 127
interference (I), 124
linkage, 111
linkage group, 113
lod score method, 125

neurofibromatosis, 127
recombination, 111
sister chromatid exchange (SCE), 130
somatic cell hybridization, 125
synkaryon, 125
synteny testing, 126
tetrad analysis, 128

Problems and Discussion Questions

1. Why does more crossing over occur between two distantly linked genes than between two genes that are very close together on the same chromosome?

2. Why are double-crossover events expected in lower frequency than single-crossover events?

3. What essential criteria must be met in order to execute a successful mapping cross?

4. The genes *dumpy* wing *(dp)*, *clot eye (cl)*, and *apterous* wing *(ap)* are linked on chromosome 2 of *Drosophila*. In a series of two-point mapping crosses, the following genetic distances were determined:

    dp–ap   42
    dp–cl   3
    ap–cl   39

    What is the sequence of the three genes?

5. In corn, colored aleurone (in the kernels) is due to the dominant allele *R*. The recessive allele *r*, when homozygous, produces colorless aleurone. The plant color (not the kernel color) is controlled by the gene pair *Y* and *y*. The dominant *Y* gene results in green color, while the homozygous presence of the recessive *y* gene causes the plant to appear yellow. In a test cross between a plant of unknown genotype and phenotype and a plant that is homozygous recessive for both traits, the following progeny were obtained:

    Colored green 88

    Colored yellow 12

    Colorless green 8

    Colorless yellow 92

    Explain how these results were obtained by determining the exact genotype and phenotype of the unknown plant, including the precise association of the two genes on the homologs and the distance between them.

6. With two pairs of genes involved (*P/p* and *Z/z*), a testcross parent (*ppzz*) was crossed to an organism of unknown genotype. Analysis of the data indicated that the gametes were produced in the following proportions:

    PZ 42.4% *Pz* 6.9% *pZ* 7.1% *pz* 43.6%

Draw all possible conclusions about the location of these genes.

7. Two different female *Drosophila* were isolated, each heterozygous for the autosomally linked genes *b* (*black* body), *d* (*dachs* tarsus), and *c* (*curved* wings). These genes are in the order *d–b–c*, with *b* being closer to *d* than to *c*. Shown below is the genotypic arrangement for each female along with the various gametes formed by both. Identify which categories are noncrossovers (NCOs), single crossovers (SCOs), and double crossovers (DCOs) in each case. Which category is likely to have the fewest gametes? Which will have the most gametes?

Female A	Female B
$\dfrac{d \quad b \quad +}{+ \quad + \quad c}$	$\dfrac{d \quad + \quad +}{+ \quad b \quad c}$

⇓ Gametes ⇓

Female A		Female B	
(1) *d b c*	(5) *d + +*	(1) *d b +*	(5) *d b c*
(2) *+ + +*	(6) *+ b c*	(2) *+ + c*	(6) *+ + +*
(3) *+ + c*	(7) *d + c*	(3) *d + c*	(7) *d + +*
(4) *d b +*	(8) *+ b +*	(4) *+ b +*	(8) *+ b c*

8. In *Drosophila*, a cross was made between females expressing the three X-linked recessive traits, *scute (sc)* bristles, *sable* body *(s)*, and *vermilion* eyes *(v)*, and wild-type males. In the F_1, all females were wild type, while all males expressed all three mutant traits. The cross was carried to the F_2 generation, and 1000 offspring were counted, with the results shown below. No determination of sex was made in the F_2 data.

Phenotype			Offspring
sc	*s*	*v*	314
+	+	+	280
+	*s*	*v*	150
sc	+	+	156
sc	+	*v*	46
+	*s*	+	30
sc	*s*	+	10
+	+	*v*	14

(a) Determine the genotypes of the P_1 and F_1 parents, using proper nomenclature.

(b) Determine the sequence of the three genes and the map distance between them.

(c) Are there more or fewer double crossovers than expected? Calculate the coefficient of coincidence. Does this represent positive or negative interference?

9. In *Drosophila, Dichaete (D)* is a chromosome 3 mutation with a dominant effect on wing shape. It is lethal when homozygous. The genes *ebony (e)* and *pink (p)* are chromosome 3 recessive mutations affecting body and eye color, respectively. Flies from a *Dichaete* stock were crossed to homozygous *ebony, pink* flies, and the F_1 progeny, with a *Dichaete* phenotype, were backcrossed to the *ebony, pink* homozygotes. The results of this back cross were

Phenotype	Number
Dichaete	401
ebony, pink	389
Dichaete, ebony	84
pink	96
Dichaete, pink	2
ebony	3
Dichaete, ebony, pink	12
wild type	13

(a) Diagram this cross, showing the genotypes of the parents and offspring of both crosses.

(b) What is the sequence of and interlocus distance between these three genes?

10. In a cross in *Neurospora*, involving two alleles *B* and *b*, the following tetrad patterns were observed. Calculate the distance between the locus and the centromere.

Tetrad Pattern	Number
BBbb	36
bbBB	44
BbBb	4
bBbB	6
BbbB	3
bBBb	7

11. A female of genotype

$$\frac{a \quad b \quad c}{+ \quad + \quad +}$$

produces 100 meiotic tetrads. Of these, 68 show no crossover events. Of the remaining 32, 20 show a crossover between *a* and *b*, 10 show a crossover between *b* and *c*, and 2 show a double crossover between *a* and *b* and between *b* and *c*. Of the 400 meiotic products, how many of each of the 8 different genotypes will be produced? Assuming the order *a–b–c* and the allele arrangement shown above, what is the map distance between these loci?

12. In *Drosophila*, two mutations, *Stubble (Sb)* and *curled (cu)*, are linked on chromosome 3. *Stubble* is a dominant gene that is lethal in the homozygous state, and *curled* is a recessive gene. If a female of genotype

$$\frac{sb \quad cu}{+ \quad +}$$

is to be mated to detect recombinants among her offspring, what male genotype would you choose as a mate?

13. Another cross in *Drosophila* involved the recessive, X-linked genes *yellow (y), white (w)*, and *cut (ct)*. A female that was yellow-bodied and white-eyed with normal wings was crossed to a male whose eyes and body were normal but whose wings were cut. The F_1 females were wild type for all three traits, while the F_1 males expressed the yellow-body, white-eye traits. The

cross was carried to an F_2, and only male offspring were tallied. On the basis of the data shown below, a genetic map was constructed.

Phenotype			Male Offspring
y	+	ct	9
+	w	+	6
y	w	ct	90
+	+	+	95
+	+	ct	424
y	w	+	376
y	+	+	0
+	w	ct	0

(a) Diagram the genotypes of the F_1 parents.

(b) Construct a map, assuming that *white* is at locus 1.5 on the X chromosome.

(c) Were any double-crossover offspring expected?

(d) Could the F_2 female offspring be used to construct the map? Why or why not?

14. In a series of two-point map crosses involving five genes located on chromosome II in *Drosophila*, the following recombinant (single-crossover) frequencies were observed:

pr–adp	29		*adp–c*	8
pr–vg	13		*adp–vg*	16
pr–c	21		*vg–b*	19
pr–b	6		*vg–c*	8
adp–b	35		*c–b*	27

(a) If the *adp* gene is present near the end of the chromosome II (locus 83), construct a map of these genes.

(b) In another set of experiments, a sixth gene, *d*, was tested against *b* and *pr*:

$$d–b \ 17\%$$

$$d–pr \ 23\%$$

Predict the results of two-point maps between *d* and *c*, *d* and *vg,* and *d* and *adp*.

15. An organism of genotype *AaBbCc* was test crossed. The genotypes of the progeny were as follows:

Genotype	Number of Progeny
AaBbCc	90
AaBb cc	10
aa bb cc	90
aa bb Cc	10

(a) If these three genes were all assorting independently of each other, how many genotypic classes would you expect in the progeny of the testcross?

(b) If these three genes were so tightly linked that crossing over never occurred between them, how many genotypic classes would you expect in the progeny of the test cross?

(c) What can you conclude from the actual data?

16. *Drosophila melanogaster* has one pair of sex chromosomes (XX or XY) and three autosomes, referred to as chromosomes 2, 3, and 4. A male fly with very short legs was discovered by a genetics student. Using this male, the student was able to establish a pure breeding stock of this mutant and found that it was recessive. This mutant was then incorporated into a stock containing the recessive gene *black* (body color, located on chromosome 2) and the recessive gene *pink* (eye color, located on chromosome 3). A female from the homozygous black, pink, short stock was then mated to a wild-type male. The F_1 males of this cross were all wild type and were then backcrossed to the homozygous *b p sh* females. The F_2 results appeared as shown in the following table. No other phenotypes were observed.

	Wild	Pink*	Black, Short	Black, Pink, Short
Females	63	58	55	69
Males	59	65	51	60

*Pink indicates that the other two traits are wild type, and so on.

(a) Based on these results, the student was able to assign *short* to a linkage group (a chromosome). Which one was it? Include step-by-step reasoning.

(b) The experiment was subsequently repeated making the reciprocal cross, F_1 females backcrossed to homozygous *b p sh* males. It was observed that 85 percent of the offspring fell into the above classes, but 15 percent of the offspring were equally divided among *b + p*, *b + +*, *+ sh p*, and *+ sh +* phenotypic males and females. How can these results be explained, and what information can be derived from the data?

Selected Readings

Allen, G. E. 1978. *Thomas Hunt Morgan: The man and his science.* Princeton, NJ: Princeton University Press.

Chaganti, R., Schonberg, S., and German, J. 1974. A manyfold increase in sister chromatid exchange in Bloom syndrome lymphocytes. *Proc. Natl. Acad. Sci.* 71:4508–12.

Creighton, H. S., and McClintock, B. 1931. A correlation of cytological and genetical crossing over in *Zea mays. Proc. Natl. Acad. Sci.* 17:492–97.

Douglas, L., and Novitski, E. 1977. What chance did Mendel's experiments give him of noticing linkage? *Heredity* 38:253–57.

Ephrussi, B., and Weiss, M. C. 1969. Hybrid somatic cells. *Sci. Am.* (April) 220:26–35.

Morgan, T. H. 1911. An attempt to analyze the constitution of the chromosomes on the basis of sex-linked inheritance in *Drosophila. J. Exp. Zool.* 11:365–414.

Morton, N.E. 1995. *LODs*—past and present. *Genetics* 140:7–12.

Neuffer, M. G., Jones, L., and Zober, M. 1968. *The mutants of maize.* Madison, WI: Crop Science Society of America.

Perkins, D. 1962. Crossing-over and interference in a multiply marked chromosome arm of *Neurospora. Genetics* 47:1253–74.

Ruddle, F. H., and Kucherlapati, R. S. 1974. Hybrid cells and human genes. *Sci. Am.* (July) 231:36–49.

Stahl, F. W. 1979. *Genetic recombination.* New York: W. H. Freeman.

Sturtevant, A. H. 1913. The linear arrangement of six sex-linked factors in *Drosophila*, as shown by their mode of association. *J. Exp. Zool.* 14:43–59.

———. 1965. *A history of genetics.* New York: Harper & Row.

Voeller, B. R., ed. 1968. *The chromosome theory of inheritance: Classical papers in development and heredity.* New York: Appleton-Century-Croft.

Wolff, S., ed. 1982. *Sister chromatid exchange.* New York: Wiley-Interscience.

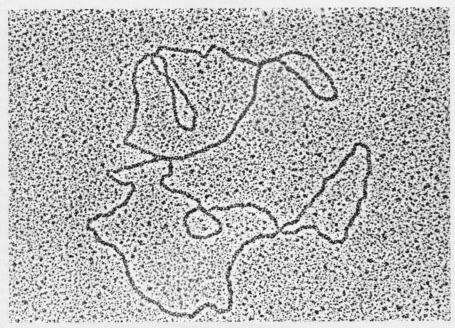

Electron micrograph of mitochondrial DNA (mtDNA).

CHAPTER OUTLINE

CHAPTER
7
Extranuclear Inheritance

Chapter Concepts

Some traits in eukaryotes do not adhere to expected patterns associated with the biparental inheritance characteristic of Mendelian genetics. Transmission is often through the female parent, whose gametes either contain maternal gene products that influence development or are the sole source of chloroplasts and mitochondria that affect the offspring's phenotype. Other cases involve infectious particles in eukaryotes as well as the phenomenon of genomic imprinting. Collectively, all of the above cases exhibit what is called extranuclear inheritance.

Throughout the history of genetics, occasional reports have challenged the basic tenets of Mendelian transmission genetics—the production of the phenotype through the transmission and expression of nuclear genes located on chromosomes of both parents. Instead, genetic results have been observed that do not reflect either Mendelian or neo-Mendelian principles, which we have introduced in the previous chapters. Some of these reports have indicated an apparent extranuclear (or extrachromosomal) influence on the phenotype. Such observations were often regarded with skepticism. However, with the increasing knowledge of molecular genetics and the discovery of DNA in mitochondria and chloroplasts, extranuclear inheritance is now recognized as an important aspect of genetics.

There are many diverse examples of these unusual modes of inheritance. Initially in this chapter, we will focus on three general types of extrachromosomal genetic phenomena: (1) maternal influence resulting from the effect of stored products of nuclear genes of the female parent during early development; (2) organelle heredity resulting from the expression of the genetic information stored in DNA contained in mitochondria and chloroplasts; and (3) infectious heredity resulting from the symbiotic or parasitic association of microorganisms with eukaryotic cells. Each has the effect of producing inheritance patterns that vary from those predicted by the concepts of Mendelian genetics.

We will conclude this chapter by examining several examples of a phenomenon called **genomic imprinting**. In such cases, biparental transmission of genes occurs, but their expression depends on which parent transmitted them, leading to still another example of non-Mendelian genetics.

Maternal Effect

Maternal effect, also referred to as *maternal influence*, implies that an offspring's phenotype for a particular trait is strongly influenced by the nuclear genotype of the gamete of the maternal parent. This is in contrast to most cases, where expression of traits is the result of both the maternal and paternal contributions to the nuclear genotype. In cases of maternal effect, the genetic information of the female gamete is transcribed, and these genetic products (either proteins or yet untranslated mRNAs) are present in the egg cytoplasm. Following fertilization, these products influence patterns or traits established during early development. Two examples will illustrate the influence of the maternal genome on particular traits.

Ephestia *Pigmentation*

In the Mediterranean meal moth, *Ephestia kuehniella*, the wild-type larva has a pigmented skin and brown eyes as a result of the dominant gene *A*. The pigment is derived from a precursor molecule, kynurenine, which is in turn a derivative of the amino acid tryptophan. A mutation, *a*, results in red eyes and little pigmentation in larvae when homozygous. As illustrated in Figure 7–1, different results are obtained in the cross *Aa* × *aa*, depending on which parent carries the dominant gene. When the male is the heterozygous parent, a 1:1 brown/red-eyed ratio is observed in larvae, as predicted by Mendelian segregation. When the female is heterozygous (*Aa*), however, all larvae are pigmented and have brown eyes. As these larvae develop into adults, one-half of them gradually develop red eyes, reestablishing the 1:1 ratio.

One explanation of these results is that the *Aa* oocytes synthesize kynurenine or an enzyme necessary for its synthesis and accumulate it in the ooplasm prior to the completion of meiosis. Even in *aa* progeny (whose mothers were *Aa*), this pigment is distributed in the cytoplasm of the cells of the developing larvae—thus, they develop pigmentation and brown eyes. Eventually, the brown pigment is diluted in the cells of the eye as new pigment is synthesized, resulting in the conversion to red eyes as adults. The *Ephestia* example demonstrates the maternal effect in which a cytoplasmically stored nuclear gene product influences the larval phenotype, and at least temporarily, overrides the biparental genotype of the progeny.

Limnaea *Coiling*

Shell coiling in the snail *Limnaea peregra* represents a permanent rather than a transitory maternal effect on the individual. Some strains of this snail have left-handed or sinistrally coiled shells (*dd*), while others have right-handed or dextrally coiled shells (*DD* or *Dd*). These snails are hermaphroditic and may undergo either cross- or self-fertilization, providing a variety of types of matings.

As shown in Figure 7–2, the coiling pattern of the progeny snails is determined by *the genotype of the parent producing the egg, regardless of the phenotype of*

that parent. Maternal parents that are *DD* or *Dd* produce only dextral-coiling progeny. Maternal parents that are *dd* produce only sinistral-coiled progeny.

Investigation of the developmental events in these snails reveals that the orientation of the mitotic spindle in the first cleavage division after fertilization apparently determines the direction of coiling. Spindle orientation appears to be controlled by maternal genes acting on the developing eggs in the ovary. The orientation of the spindle, in turn, influences cell divisions following fertilization and quickly establishes the permanent adult coiling pattern.

The dextral allele (*D*) produces an active gene product that causes right-handed coiling. If ooplasm from dextral eggs is injected into uncleaved sinistral eggs, they cleave in a dextral pattern. However, in the converse experiment, sinistral ooplasm has no effect when injected into dextral eggs. Apparently, the sinistral allele is the result of a classical recessive mutation that encodes a nonfunctional gene product.

Embryonic Development in Drosophila

A more recently documented example of maternal effect involves various genes that function to control embryonic development in *Drosophila melanogaster*. As a result of extensive work by Edward B. Lewis, Christiane Nüsslein-Volhard, and Eric Wieschaus (who jointly shared the 1995 Nobel Prize for Physiology or Medicine for their findings), the role of these and other genes has become clear. Those that illustrate maternal effect are genes whose products are synthesized by the developing egg and stored in the oocyte prior to fertilization. Following fertilization, these products specify molecular gradients in the early embryo that determine spatial organization as development proceeds.

For example, the gene *bicoid* (*bcd*) plays an important role in specifying the development of the anterior portion of the fly. This conclusion is based on the observation that embryos derived from mothers who are homozygous for this mutation (*bcd⁻/bcd⁻*) fail to develop anterior areas that normally give rise to the head

■ Figure 7–1 Illustration of maternal influence in the inheritance of eye pigment in the meal moth *Ephestia kuehniella*. Multiple light receptors (eyes) are present on the anterior portion of larvae.

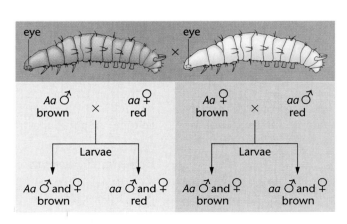

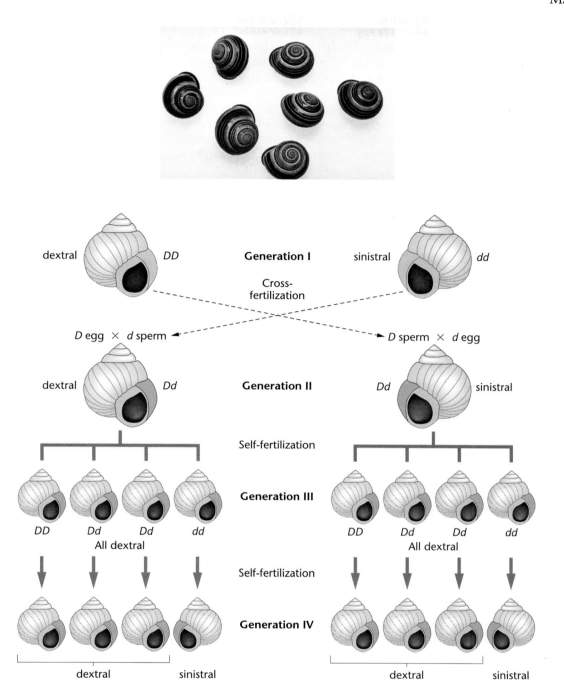

■ Figure *7–2*   Inheritance of coiling in the snail *Limnaea peregra*. Coiling is either dextral (right-handed) or sinistral (left-handed). A maternal effect is evident in generations II and III, where the genotype of the maternal parent controls the phenotype of the offspring. The photograph illustrates coiling in another species, *Ceprea nemoralis*.

and thorax of the adult fly. Embryos whose mothers contain at least one wild-type allele (*bcd*⁺) develop normally, even if the genotype of the embryo is homozygous for the mutation. Consistent with the concept of maternal effect, *the genotype of the female parent, not the genotype of the embryo, determines the phenotype of the offspring.*

The genetic control of embryonic development in *Drosophila* is a fascinating story. The protein products of the maternal effect genes function to activate other genes, which may in turn activate still other genes. This cascade of gene activity leads to a normal embryo whose subsequent development yields a normal adult fly. We shall return for a more detailed discussion of this general topic in Chapter 20. There we will see examples of other genes that illustrate maternal effect, as well as many so-called zygotic genes, whose expression occurs during early development and which behave genetically in a conventional Mendelian fashion.

Organelle Heredity

In this section we will examine examples of inheritance patterns of phenotypes related to chloroplast and mitochondrial function. Our interpretation of these patterns is based on the discovery over three decades ago of the presence of DNA in these organelles and the subsequent characterization of genetic processes occurring within them. Prior to those findings, certain mutant phenotypes appeared to be inherited through the transmission of information from the cytoplasm rather than through the genetic information of chromosomes contained within the nucleus. Transmission was most often from the maternal parent through the ooplasm; in such cases, these phenotypes are now considered examples of **organelle heredity**, which is distinct from maternal effect inheritance.

Analysis of the hereditary transmission of mutant alleles of chloroplast and mitochondrial DNA has been difficult. First, the function of these organelles is dependent on gene products of both nuclear and organelle DNA. Second, the number of organelles contributed to each progeny often exceeds one. If many chloroplasts and/or mitochondria are contributed and only one or a few of them contain a mutant gene, the corresponding mutant phenotype may not be revealed. Analysis is thus much more involved than for Mendelian characters.

In this section, we shall discuss examples of inheritance patterns related to these organelles. In a future chapter (see Chapter 11) we will provide a more detailed analysis of the molecular genetics of these organelles.

Chloroplasts: Variegation in Four O'Clock Plants

In 1908, Carl Correns (one of the rediscoverers of Mendel's work) provided the earliest example of inheritance linked to chloroplast transmission. Correns discovered a variety of the four o'clock plant, *Mirabilis jalapa*, which had branches with either white, green, or variegated leaves. As shown in Table 7.1, inheritance in all possible combinations of crosses is determined strictly by the phenotype of the ovule source. For example, if the seeds (representing the progeny) were derived from ovules on branches with green leaves, all progeny plants bore only green leaves, regardless of the phenotype of the source of pollen.

Correns concluded that inheritance was through the cytoplasm of the maternal parent because the pollen, which, in contrast to the ovule, contributes little or no cytoplasm to the zygote, had no influence on the progeny phenotypes.

Iojap in Maize

A phenotype similar to *Mirabilis* but with a different pattern of inheritance has been analyzed in maize by Marcus M. Rhoades. In this case the expression of green, colorless, or green-and-colorless striped leaves is not only controlled by the cytoplasm, but, in addition, is influenced by a nuclear gene. This locus is called **iojap (ij)**. The wild-type allele is designated *Ij*. Plants that are homozygous for the mutation *(ij/ij)* have green-and-white striped leaves. However, when reciprocal crosses are made between plants with striped leaves and plants with green leaves, the results are seen to vary, depending on which parent is mutant (Figure 7–3). If the female is striped *(ij/ij)* and the male is green *(Ij/Ij)*, plants with colorless leaves, striped leaves, and green leaves are observed as progeny. If the male parent is striped *(ij/ij)* and the female is green *(Ij/Ij)*, only green plants are produced! In both types of cross, all offspring have identical genotypes *(Ij/ij)*. We may conclude that although a nuclear gene seems to be involved, the inheritance pattern is strictly influenced maternally.

We can understand this pattern better by examining the offspring resulting from self-fertilization of these heterozygous plants (Figure 7–3). The striped plant gives rise to progeny with colorless, striped, and green leaves, regardless of the genotype of the progeny. Green plants give rise to both green and striped progeny in a 3:1 ratio. Results of these self-fertilizations substantiate that the mutant chloroplasts are transmitted solely through the female cytoplasm, regardless of the plant's nuclear genotype.

TABLE 7.1	Offspring from crosses between flowers from various branches of variegated four o'clock plants		
		Location of Ovule	
Source of Pollen	*White branch*	*Green branch*	*Variegated branch*
White branch	White	Green	White, green, or variegated
Green branch	White	Green	White, green, or variegated
Variegated branch	White	Green	White, green, or variegated

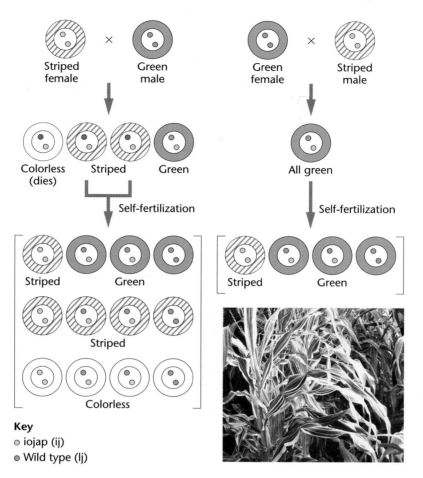

Figure 7–3 Maternal inheritance of striping in maize. Regardless of their genotype, offspring reflect the maternal phenotypes in their own appearance. Because color is due to chloroplasts in the leaves, inheritance is controlled by these organelles as they are passed through the maternal cytoplasm.

Key
- iojap (ij)
- Wild type (Ij)

Apparently, the nuclear genotype *ij/ij* somehow alters chloroplasts, which are then transmitted maternally. The colorless areas of the leaf are due to the lack of the green pigment chlorophyll in chloroplasts originally induced by the *ij* allele. Once acquired, chlorophyll-deficient (colorless) chloroplasts are transmitted through the maternal cytoplasm, establishing the phenotypes of leaves of progeny plants.

Chlamydomonas *Mutations*

The unicellular green alga *Chlamydomonas reinhardi* has provided an excellent system for the investigation of plastid and mitochondrial inheritance. The organism is eukaryotic and contains a single large chloroplast as well as numerous mitochondria. Matings are followed by meiosis, and the various stages of the life cycle are easily studied in culture in the laboratory. The first cytoplasmic mutant, *streptomycin resistance (sr)*, was reported in 1954 by Ruth Sager. Although *Chlamydomonas'* two mating types—*mt+* and *mt-*—appear to make equal cytoplasmic contributions to the zygote, Sager determined that the *sr* phenotype is transmitted only through the *mt+* parent.

Since this discovery, a number of other Chlamydomonas mutations (including an acetate requirement as well as resistance to or dependence on a variety of

bacterial antibiotics) have been discovered that show a similar uniparental inheritance pattern. These mutations have been linked to the transmission of the chloroplast, and their study has extended our knowledge of chloroplast inheritance.

Following fertilization, the single chloroplasts of the two mating types fuse. After the resulting zygote has undergone meiosis, it is apparent that the genetic information of the chloroplasts of progeny cells is derived only from the *mt+* parent. The genetic information present in the *mt-* chloroplast has degenerated!

Furthermore, studies suggested that these chloroplast mutations, representing numerous gene sites, form a single circular linkage group (a closed loop of DNA)—the first such extranuclear unit to be established in a eukaryotic organism. It is also apparent that the linkage groups from the two mating types are capable of undergoing recombination following zygote formation. All available evidence supports the hypothesis that this linkage group consists of DNA residing in the chloroplast.

Mitochondria: poky *in* Neurospora

Mitochondria, like chloroplasts, play a critical role in cellular bioenergetics, and they contain a distinctive genetic system. Mutants affecting mitochondrial function have been discovered and studied. As with chloroplast mutants, these are transmitted through the cytoplasm and result in non-Mendelian inheritance patterns. Mitochondrial genetic systems have now been extensively characterized.

In 1952, Mary B. and Hershel K. Mitchell discovered a slow-growing mutant strain of the mold *Neurospora crassa* and called it **poky**. (It is also designated *mi-1,* for *maternal inheritance.*) Studies have shown slow growth to be associated with impaired mitochondrial function related specifically to certain cytochromes that are essential to electron transport. The results of genetic crosses between wild-type and *poky* strains suggest that the trait is maternally inherited. If the female parent is *poky* and the male parent is wild type, all progeny colonies are *poky*. The reciprocal cross produces normal colonies.

Studies with *poky* mutants have taken advantage of a unique phase in the life cycle of fungi. Occasionally, hyphae from separate mycelia fuse with one another, giving rise to structures containing two or more nuclei in a common cytoplasm. If the hyphae contain nuclei of different genotypes, the structure is called a **heterokaryon**. The cytoplasm now contains mitochondria derived from both initial mycelia. A heterokaryon may give rise to haploid spores (or **conidia**) that produce new mycelia whose phenotypes may be determined.

Heterokaryons produced by the fusion of *poky* and wild-type hyphae initially show normal rates of growth and respiration. However, mycelia produced following conidia formation from these heterokaryons become progressively more abnormal until they show the *poky* phenotype. This occurs in spite of the presumed presence of both wild-type and *poky* mitochondria in the cytoplasm of the hyphae.

To explain the initial growth and respiration pattern, it is assumed that the wild-type mitochondria support the respiratory needs of the hyphae. The subsequent expression of the *poky* phenotype suggests that the presence of the *poky* mitochondria may somehow prevent or depress the function of these wild-type mitochondria. Perhaps the *poky* mitochondria replicate more rapidly and "wash out" or dilute wild-type mitochondria numerically. Another possibility is that *poky* mitochondria produce a substance that inactivates the wild-type organelle or interferes with the replication of its DNA (mtDNA). As a result of this type of interaction, *poky* is an example of what is referred to as a **suppressive mutation**. This general phenomenon is characteristic of many other suspected mitochondrial mutations of *Neurospora* and yeast.

Petite *in* Saccharomyces

Another extensive study of mitochondrial mutations has been performed with the yeast *Saccharomyces cerevisiae*. The first such mutation, **petite**, was described by Boris Ephrussi and his co-workers in 1956. The mutant is so named because of the small size of the yeast colonies.

Many independent **petite mutations** have since been discovered and studied. They all have a common characteristic: deficiency in cellular respiration involving abnormal electron transport. Fortunately, this organism is a facultative anaerobe and can grow by fermenting glucose through glycolysis. Thus, although colonies are small, the organism may survive the loss of mitochondrial function by generating energy anaerobically.

The complex genetics of *petite* mutations is diagrammed in Figure 7–4. A small proportion of these mutants exhibit Mendelian inheritance and are called **segregational *petites***, indicating that they are the result of nuclear mutations. The remainder demonstrate cytoplasmic transmission, producing one of two effects in matings. The **neutral *petites***, when crossed to wild type, produce meiotic products (ascospores) that give rise only to wild-type, normal-sized colonies. The same pattern continues if progeny of this cross are backcrossed to neutral *petites*. This is because the majority of neutrals lack mtDNA completely or have lost a substantial portion of it. Thus, the wild-type cell is the effective source of normal mitochondria capable of reproduction.

A third type, the **suppressive *petites***, behaves similarly to *poky* in *Neurospora*. Crosses between mutant and wild type give rise to mutant diploid zygotes, which, upon undergoing meiosis, yield all mutant cells. Under these conditions, the *petite* mutation behaves "dominantly" and seems to suppress the function of the wild-type mitochondria. Suppressive *petites* also have deletions of mtDNA, but they are not nearly as extensive as deletions in the neutral petites.

Suppressiveness remains unexplained. Two major hypotheses have been advanced. One explanation suggests that the mutant (or deleted) mtDNA replicates more rapidly, and thus mutant mitochondria "take over" or dominate the phenotype by numbers alone. The second explanation suggests that recombination occurs between the mutant and wild-type mtDNA, introducing errors into or disrupting the normal mtDNA. It is not yet clear which, if either, of these explanations is correct.

Mitochondrial DNA and Human Diseases

The DNA found in mitochondria (**mtDNA**) has been extensively characterized in a variety of organisms, including humans. Human mtDNA, which is circular, contains 16,569 base pairs and is strictly inherited maternally. The mitochondrial gene products include:

13 proteins, required for oxidative respiration

22 transfer RNAs (tRNAs), required for translation of proteins

2 ribosomal RNAs (rRNAs), required for translation of proteins

Because cellular respiration is such an intricate metabolic process, disruption of *any* mitochondrial gene by mutation is likely to have a severe impact on an organ-

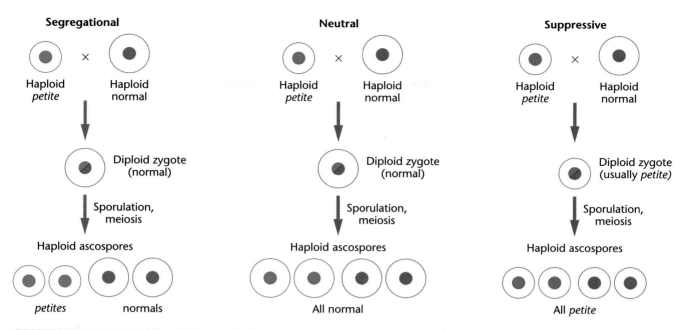

Segregational

Haploid *petite* × Haploid normal

Diploid zygote (normal)

Sporulation, meiosis

Haploid ascospores

petites   normals

Neutral

Haploid *petite* × Haploid normal

Diploid zygote (normal)

Sporulation, meiosis

Haploid ascospores

All normal

Suppressive

Haploid *petite* × Haploid normal

Diploid zygote (usually *petite*)

Sporulation, meiosis

Haploid ascospores

All *petite*

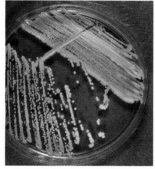

■ Figure 7–4 The outcome of crosses involving the three types of *petite* mutations affecting mitochondrial function in the yeast *Saccharomyces cerevisiae*. The photograph shows normal yeast colonies.

ism. We have seen this in our discussion of *petite*, which would be a lethal mutation were it not for yeast's ability to respire anaerobically. Thus, mtDNA would seem to be particularly vulnerable to mutations.

On the other hand, a zygote receives a large number of organelles through the egg, so if only one of them contains a mutation, its impact may be minimized because there will be many more nonmutant mitochondria that will function normally. During early development, cell division disperses the initial population of mitochondria present in the zygote to newly formed cells, in which these organelles reproduce autonomously. Therefore, adults will exhibit cells with a variable mixture of normal and abnormal organelles, should a deleterious mutation arise or already be present in the initial population of organelles. This is a condition called **heteroplasmy**.

In order for a human disorder to be attributable to genetically altered mitochondria, several criteria must be met.

1. Inheritance must exhibit a maternal rather than a Mendelian pattern.

2. The disorder must reflect a deficiency in the bioenergetic function of the organelle.

3. A specific genetic mutation in one of the mitochondrial genes must be documented.

Thus far, several cases are known that demonstrate these characteristics. **Myoclonic epilepsy and ragged red fiber disease (MERRF)** demonstrates a pedigree consistent with maternal inheritance. Only offspring of affected mothers inherit the disorder, while the offspring of affected fathers are all normal. Individuals with this rare disorder express deafness and dementia in addition to seizures. Both muscle fibers and mitochondria are abnormal in appearance. Such aberrant mitochondria appear to be devoid of internal cristae, as is evident in Figure 7–5. Upon analysis of mtDNA, a mutation has been found in one of the mitochondrial genes encoding a transfer RNA. This genetic alteration apparently interferes with the process of translation within the organelle. Presumably, the deficiency in efficient mitochondrial function is related to the manifestations of this disorder.

A second disorder, **Leber's hereditary optic neuropathy (LHON)**, also exhibits maternal inheritance as well as mtDNA lesions. The disorder is characterized by sudden bilateral blindness. The average age of vision loss is 27, but onset is quite variable. Four different mutations have been identified, all of which disrupt normal oxidative phosphorylation. Over 50 percent of cases are due to a mutation at a specific position in the mitochondrial gene encoding a subunit of NADH dehydrogenase. The amino acid arginine is converted to histidine. In large families,

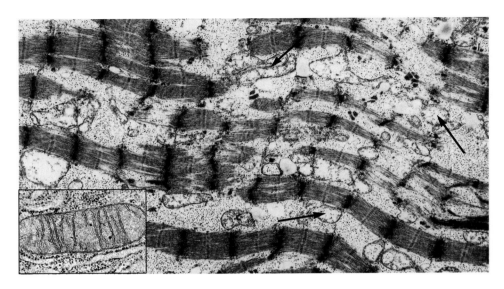

■ Figure 7–5   An electron micrograph of muscle tissue from an individual afflicted with MERRF, displaying abnormal mitochondria (see arrows) characterizing this human disorder. The photo insert at the bottom left shows a normal mitochondrion for comparison.

this mutation is transmitted to all maternal relatives. It is also of interest to note that in many instances of LHON, there is no family history. It appears that a significant number of cases may result from "new" mutations.

In a third disorder, **Kearns–Sayre syndrome (KSS)**, severely affected individuals lose their vision, undergo hearing loss, and display heart conditions. The genetic basis of KSS involves deletions at various positions within mtDNA. Many KSS patients are symptom-free as children but display progressive symptoms as adults. Analysis has revealed that the proportion of mtDNAs that reveal deletions increases as the severity of symptoms increases.

The study of hereditary, mitochondrial-based disorders provides insights into the importance and genetic basis of this organelle during normal development as well as the relationship between mitochondrial function and neuromuscular disorders. Furthermore, such study has suggested a hypothesis for aging based on the progressive accumulation of mtDNA mutations and the accompanying loss of mitochondrial function.

Infectious Heredity

There are numerous examples of cytoplasmically transmitted phenotypes in eukaryotes that are due to an invading microorganism or particle. The foreign invader coexists in a symbiotic relationship, is usually passed

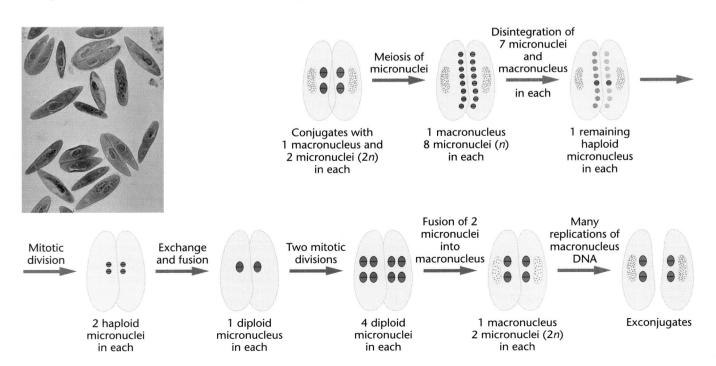

■ Figure 7–6   Genetic events occurring during conjugation in *Paramecium*. The photographic insert shows numerous pair of organisms undergoing conjugation.

through the maternal ooplasm to progeny cells or organisms, and confers a specific phenotype that may be studied. We shall consider several examples illustrating this phenomenon.

Kappa in Paramecium

First described by Tracy Sonneborn, certain strains of *Paramecium aurelia* are called **Killers** because they release a cytoplasmic substance called **paramecin** that is toxic and sometimes lethal to sensitive strains. This substance is produced by particles called **kappa** that replicate in the Killer cytoplasm. They contain DNA and protein, and depend for their maintenance on a dominant nuclear gene *K*. One cell may contain 100 to 200 such particles.

As illustrated in Figure 7–6 (on page 146), paramecia are diploid protozoans that can undergo sexual exchange of genetic information through the process of **conjugation**. In some instances, cytoplasmic exchange also occurs. There are a variety of ways in which the *K* gene and kappa can be transmitted.

Paramecium aurelia contain two diploid micronuclei. Early in conjugation, both micronuclei in each mating pair undergo meiosis, resulting in eight haploid nuclei. However, seven of these degenerate, and the remaining one undergoes a single mitotic division. Each cell then donates one of the two haploid nuclei to the other, re-creating the diploid condition in both cells. After the exchange, the two resulting cells, called **exconjugates**, are of identical genotype.

In a similar process involving only a single cell, **autogamy** occurs. Following meiosis of both micronuclei, eight micronuclei are formed. Seven of them degenerate and one survives. This nucleus divides, and the resulting nuclei fuse to re-create the diploid condition. If the original cell was heterozygous, autogamy results in homozygosity because the newly formed diploid nucleus was derived solely from a single haploid meiotic product. In a population of cells that were originally heterozygous, half of the new cells express one allele and half express the other allele.

Figure 7–7 illustrates the results of crosses between *KK* and *kk* cells, without and with cytoplasmic ex-

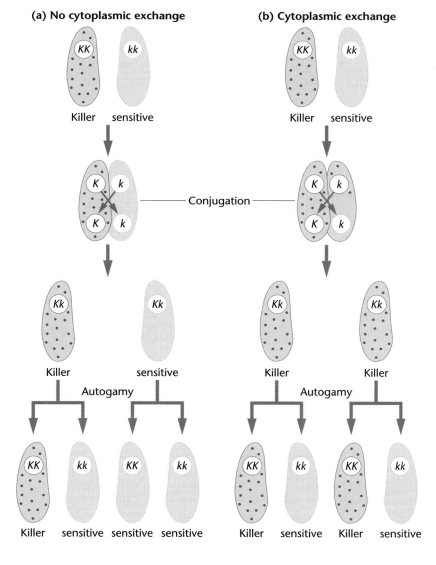

■ Figure 7–7 Results of crosses between *Killer* (*KK*) and *sensitive* (*kk*) strains of *Paramecium*, with and without cytoplasmic exchange during conjugation. The kappa particles (dots) are maintained only when a *K* allele is also present.

change. When no cytoplasmic exchange occurs, even though the resultant cells may be *Kk* (or *KK* following autogamy), they remain sensitive if no kappa particles are transmitted. When exchange occurs, the cells become Killers provided that the kappa particles are supported by at least one dominant *K* allele.

Kappa particles are bacterialike and may contain inactive viruses. One theory holds that these viruses of kappa may become activated. When they do, they produce the toxic products that are released and kill sensitive strains.

Infective Particles in Drosophila

Two examples of similar phenomena are known in *Drosophila*: **CO₂ sensitivity** and **sex-ratio**. In the former, flies that would normally recover from CO_2 anesthetization instead become permanently paralyzed and are killed by CO_2. Sensitive mothers pass this trait to all offspring. Furthermore, extracts of sensitive flies induce the trait when injected into resistant flies. Phillip L'Heritier has postulated that sensitivity is due to the presence of a virus, **sigma**. The particle has been visualized under the electron microscope; it is smaller than kappa. Attempts to transfer the virus to other insects have been unsuccessful, demonstrating that specific genes support the presence of sigma in *Drosophila*.

A second example of infective particles comes from the study of *Drosophila bifasciata*. A small number of these flies were found to produce predominantly female offspring if reared at 21°C or lower. This condition, designated sex-ratio, was shown to be transmitted to daughters but not to the low percentage of males produced. This phenomenon was subsequently investigated in *Drosophila willistoni*. In these flies, the injection of ooplasm from sex-ratio females into normal females induced the condition. This observation suggests that an extranuclear element is responsible for the sex-ratio phenotype. The agent has now been isolated and shown to be a protozoan. While the protozoan has been found in both males and females, it is lethal primarily to developing male larvae. There is now some evidence that a virus harbored by the protozoan may be responsible for producing a male-lethal toxin.

Genomic Imprinting

One of the major exceptions to the assumptions underlying the laws of Mendelian inheritance involves the case where genetic expression varies, depending on the parental origin of the chromosome carrying a particular gene. The phenomenon is called **genomic** (or **parental**) **imprinting**. It appears that, in some species, certain regions of chromosomes and the genes contained within them somehow retain a memory, or an "imprint," of their parental origin that influences their genetic expression. As a result, specific genes either are expressed or remain genetically silent—i.e., are not expressed, based on their imprint.

For example, in Huntington disease in humans, as pointed out previously, early onset of the disease most often occurs when the mutant gene is inherited from the father. In myotonic dystrophy, just the opposite is true. In this disorder, when early onset is observed, the affected offspring usually inherits the gene from the mother.

The imprinting step is thought to occur before or during gamete formation, leading to differentially marked genes (or chromosome regions) in sperm-forming versus egg-forming tissues. The process is clearly different from mutation because in the cases cited above, the imprint can be reversed in succeeding generations as these autosomal genes pass from mother to son to granddaughter, and so on.

Another example of imprinting involves the inactivation of one of the X chromosomes in mammalian females. As we will discuss in Chapter 8, a mechanism of **dosage compensation** exists whereby the random inactivation of either the paternal or maternal X chromosome occurs during embryonic development. In mice, however, prior to the development of the embryo proper, imprinting occurs in tissues such that the X chromosome of paternal origin is genetically inactivated in all cells while the genes on the maternal X chromosome remain genetically active. As embryonic development is subsequently initiated, the imprint is "released" and random inactivation of either the paternal or the maternal X chromosome can occur.

In 1991, more specific information became available when it was established that three specific mouse genes undergo imprinting. One of them is the gene encoding insulinlike growth factor II (*Igf2*). A mouse that carries two nonmutant alleles of this gene is normal in size, whereas a mouse that carries two mutant alleles lacks a growth factor and is a dwarf. The size of a heterozygous mouse (one allele normal and one mutant—Figure 7–8) depends on the parental origin of the normal allele. The mouse is normal in size if the normal allele came from the father, but dwarf if the normal allele came from the mother. From this it can be deduced that the normal *Igf2* gene is imprinted to function poorly during the course of egg production in females but functions normally when it has passed through sperm-producing tissue in males.

Imprinting continues to depend on whether the gene passes through sperm-producing or egg-forming tissue leading to the next generation. For example, a heterozygous normal-sized male (above) will donate a "normal-functioning" wild-type allele to half his offspring, which will counteract a mutant allele received from the mother.

In humans, two distinct genetic disorders are thought to be caused by differential imprinting of the same region of chromosome 15 (15q1). In both cases, the disorders appear to be due to an identical deletion of this region in one member of the chromosome 15 pair. The first disor-

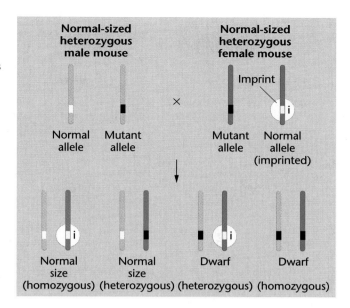

■ Figure 7–8 The effect of imprinting on the *Igf2* gene in the mouse, which produces dwarf mice in the homozygous condition. Heterozygous offspring that receive the normal allele from their father are normal in size. Heterozygotes that receive the normal allele from their mother, which has been imprinted (**i**), are dwarf.

GENETICS, TECHNOLOGY, AND SOCIETY

Mitochondrial DNA and the Mystery of the Romanovs

By most accounts, Nicholas II, the last Tsar of Russia, was a substandard monarch. He was accused of bungling during the Russo-Japanese War of 1904–1905, and his regime was plagued by corruption and incompetence. Even so, he probably didn't deserve the fate that befell him and his family one summer night in 1918. As we shall see, a full understanding of that event has relied on, of all things, mitochondrial DNA.

After being forced to abdicate in 1917, ending 300 years of Romanov rule, Tsar Nicholas and the imperial family were banished to Ekaterinburg in western Siberia. There, it was believed, they would be out of reach of the fiercely anti-imperialist Bolsheviks, who were then fighting to gain control of the country. But the Bolsheviks eventually caught up with the Romanovs. On a July night in 1918, Tsar Nicholas, Tsarina Alexandra, (grand-daughter of Queen Victoria), their five children (Olga, 22, Tatiana, 21, Marie, 19, Anastasia, 17, and Alexis, 13), their family doctor, and three of their servants were awakened and brought to a downstairs room of the house where they were being held prisoner. There they were made to form a double row against the wall, presumably so that a photograph could be taken. Instead, 11 men with

revolvers burst into the room and opened fire. After exhausting their ammunition, they proceeded to bayonet the bodies and smash their faces in with rifle butts. The corpses were then hauled away and flung down a mineshaft, only to be pulled out two days later and dumped into a shallow grave, doused with sulfuric acid, and covered over. Here they rested for more than 60 years.

An air of mystery soon developed around the demise of the Romanovs. Did all of the children of Nicholas and Alexandra die with their parents that bloody night, or did the youngest daughter, Anastasia, get away? Over the years, the possibility that Anastasia miraculously escaped execution has inspired countless books, a Hollywood movie, a ballet, a Broadway play, and, most recently, an animated movie. In all of these retellings, Anastasia re-emerges to claim her birthright as the only surviving member of the Romanovs.

Adding to the puzzle was one Anna Anderson. Two years after being dragged from a Berlin canal after a suicide attempt in 1920, she began claiming to be the Grand Duchess Anastasia. Despite a history of mental instability and a curious inability to speak Russian, she managed to convince many people.

The unraveling of the mystery began in 1979, when a Siberian geologist and a Moscow filmmaker discovered four skulls they believed to belong to the

Tsar's family. It wasn't until the summer of 1991, after the arrival of *glasnost,* that exhumation began. Altogether almost 1000 bone fragments were recovered, which were re-assembled into nine skeletons, five females and four males. Based on measurements of the bones and computer-assisted superimposition of the skulls onto photographs, the remains were tentatively identified as belonging to the murdered Romanovs. But there were still two missing bodies, one of the daughters (believed to be Anastasia) and the boy Alexis.

The next step in authenticating the remains involved studies of DNA that were conducted by Pavel Ivanov, the leading Russian forensic DNA analysis, in collaboration with Peter Gill of the British Forensic Science Service. Their goals were to establish family relationships among the remains, and then to determine, by comparisons with living relatives, whether the family group was in fact the Romanovs. They froze bone fragments from the nine skeletons in liquid nitrogen, ground them into a fine powder, and extracted small amounts of DNA, which comprised both nuclear DNA and mitochondrial DNA (mtDNA). Genomic DNA typing of each skeleton confirmed the familial relationships and showed that indeed one of the princesses and Alexis were missing. Final proof that the bones belonged to the Romanovs awaited analysis of mtDNA, however.

(continues)

(continued)

Mitochondrial DNA is ideal for forensic studies for several reasons. Since all the mitochondria in a human cell are descended from the mitochondria present in the egg, mtDNA is transmitted strictly from mother to offspring, never from father to offspring. Therefore, mtDNA sequences can be used to trace maternal lineages without the complicating effects of meiotic crossing over, which recombines maternal and paternal nuclear genes every generation. In addition, mtDNA is small (16,600 base pairs) and present in 500 to 1000 of copies per cell, so it is much easier to recover intact than nuclear DNA.

Ivanov's group amplified two highly variable regions from the mtDNA isolated from all nine bone samples and determined the nucleotide sequences of these regions. By comparing these sequences with living relatives of the Romanovs, they hoped to establish the identity of the Yekaterinburg remains once and for all. The sequences from Tsarina Alexandra were an exact match with those from Prince Philip of England, who is her grandnephew, verifying her identity. Authentication was more complicated for the Tsar, however.

The Tsar's sequences were compared with those from the only two living relatives who could be persuaded to participate in the study, Countess Xenia Cheremeteff-Sfiri (his great-grandniece) and James George Alexander Bannerman Carnegie, third Duke of Fife (a more distant relative, descended from a line of women stretching back to the Tsar's grandmother). These comparisons produced a surprise. At position 16169 of the mtDNA, the Tsar seemed to have either one or another base, a C or a T. The Countess and the Duke, in contrast, both had only T. The conclusion was that Tsar Nicholas had two different populations of mitochondria in his cells, each with a different base at position 16169 of its DNA. This condition, called *heteroplasmy,* is now believed to occur in 10 to 20 percent of people.

This ambiguity between the presumed Tsar and his two living maternal relatives cast doubt on the identification of the remains. Fortunately, the Russian government granted a request to analyze the remains of the Tsar's younger brother, Grand Duke Georgij Romanov, who died in 1899 of tuberculosis. The Grand Duke's mtDNA was found to have the same heteroplasmic variant at position 16169, either a C or a T. It was concluded that the Yekaterinburg bones were of the doomed imperial family. With years of controversy finally resolved, the now-authenticated remains of Tsar Nicholas II, Tsarina Alexandra, and three of their daughters are to be buried in the St. Peter and Paul Cathedral in St. Petersburg on July 17, 1998, 80 years to the day after they were murdered.

This DNA analysis did not solve the mystery of the fate of Anastasia, however. Did she die with her parents, sister, and brother in 1918? Or is it possible that Anna Anderson was telling the truth, that she was the escaped duchess? In a separate study, Anna Anderson's nuclear and mtDNA was recovered from intestinal tissue preserved from an operation five years before her death in 1984. Analysis of this DNA proved that she was not Anastasia, but rather a Polish peasant named Franziska Schanzkowska.

If Anastasia's remains were not among those of her parents and sisters and if Anna Anderson was an imposter, what *did* happen to Anastasia? Most evidence suggests that Anastasia and her brother Alexis were not found with the others in the mass grave because their bodies were burned over the grave site two days after the killings and the ashes scattered, never to be found again. Not exactly a Hollywood ending.

References

Gibbons, A. 1998. Calibrating the mitochondrial clock. *Science* 279: 28-29.

Ivanov, P., et al. 1996. Mitochondrial DNA sequence heteroplasmy in the Grand Duke of Russia Georgij Romanov establishes the authenticity of the remains of Tsar Nicholas II. *Nature Genet.* 12: 417-420.

Masse, R. (1996). *The Romanovs: The final chapter*. New York: Ballantine Books.

Stoneking, M., et. al. 1995. Establishing the identity of Anna Anderson Manahan. *Nature Genet.* 9: 9-10.

der, **Prader–Willi syndrome (PWS)**, results when only an undeleted maternal chromosome remains. If only an undeleted paternal chromosome remains, an entirely different disorder, **Angelman syndrome (AS)**, results.

Prader–Willi syndrome is clearly different from AS. In the former, mental retardation is noted, as well as a severe eating disorder marked by an uncontrollable appetite, obesity, and diabetes. Angelman syndrome, on the other hand, leads to distinct clinical manifestations involving behavior as well as mental retardation. One can conclude that region 15q1 is imprinted differently in male versus female gametes, and that both a maternal and a paternal region are required for normal development.

The area of genomic imprinting has received a great deal of recent research attention; many questions remain unanswered. It is not known how many genes are subject to imprinting, nor its developmental role. While it appears that regions of chromosomes rather than specific genes are imprinted, the molecular mechanism of imprinting is still a matter for conjecture. It has been hypothesized that **DNA methylation** may be involved. In vertebrates, methyl groups can be added to the carbon atom at position 5 in cytosine (see Chapter 9) as a result of the activity of the enzyme **DNA methyl transferase**. Methyl groups are added when the dinucleotide CpG or groups of CpG units (called **CpG islands**) are present along a DNA chain. DNA methylation is a reasonable mechanism for establishing a molecular imprint, since there is some evidence that a high level of methylation can inhibit gene activity and that active genes (or their regulatory sequences) are often undermethylated. Additionally, variation in methylation has been noted in the mouse genes that undergo imprinting. Whatever the cause of this phenomenon, it is a fascinating topic and one that will receive significant attention in future research studies.

Chapter Summary

1. Patterns of inheritance sometimes vary from that expected of biparental transmission of nuclear genes. In such instances, phenotypes most often appear to result from genetic information transmitted through the egg.

2. Maternal effect patterns result when nuclear gene products controlled by the maternal genotype of the egg influence early development. *Ephestia* pigmentation and coiling in snails are examples.

3. Organelle heredity is based on the genotypes of chloroplast and mitochondrial DNA as these organelles are transmitted through the egg to offspring.

4. Chloroplast mutations affect the photosynthetic capabilities of plants, while mitochondrial mutations may affect cells that are highly dependent on ATP gener-

ated through cellular respiration. The resulting mutants display phenotypes related to the loss of function of these organelles.

5. Another form of extranuclear inheritance is due to the transmission of infectious microorganisms, which establish symbiotic relationships with their host cells. Kappa particles and CO_2 sensitivity and sex-ratio determinants are examples.

6. Genomic imprinting is a phenomenon whereby transmission expression of certain genes varies, depending on which parent transmitted the gene.

INSIGHTS
and
SOLUTIONS

1. Analyze the theoretical pedigree below and determine the most consistent interpretation of how the trait is inherited. Are there any inconsistencies in the various individuals?

Solution:

The trait is passed from all male parents to all but one offspring and is never passed maternally. Individual IV-7 (a female) is the only exception.

2. Can the above explanation be attributed to a gene on the Y chromosome? Defend your answer.

Solution:

No, since male parents pass the trait to their daughters as well as to their sons.

3. Is the above case an example of a paternal effect or of paternal inheritance?

Solution:

It has all the earmarks of paternal inheritance, because males pass the trait to all of their offspring. To assess whether the trait is due to a paternal effect (resulting from a nuclear gene in the male gamete), analysis of further matings would be needed.

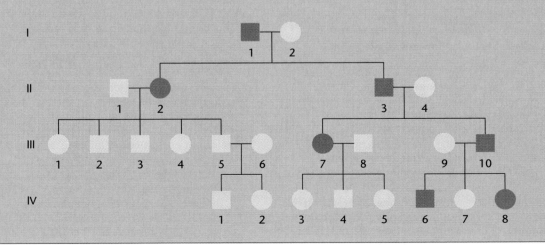

Key Terms

Angelman syndrome (AS), 150
autogamy, 147
CO_2 sensitivity, 148
conidia, 144
conjugation, 147
CpG islands, 150
DNA methylation, 150
DNA methyl transferase, 150
dosage compensation, 148
exconjugate, 147
genomic imprinting, 139
heterokaryon, 144

heteroplasmy, 145
iojap (ij), 142
kappa, 146
Kearns-Sayre syndrome (KSS), 146
Killer strain, 147
Leber's hereditary optic neuropathy (LHON), 145
maternal effect, 139
maternal inheritance, 141
mitochondrial DNA (mtDNA), 144
myoclonic epilepsy and ragged red fiber disease (MERRF), 145

neutral *petite*, 144
non-Mendelian inheritance, 139
paramecin, 147
petite mutation, 144
poky mutation, 144
Prader–Willi syndrome (PWS), 150
segregational *petite*, 144
sex-ratio, 148
sigma, 148
suppressive mutation, 144
suppressive *petite*, 144

Problems and Discussion Questions

1. What genetic criteria distinguish a case of extranuclear inheritance from a case of Mendelian autosomal inheritance? from a case of X-linked inheritance?

2. In *Limnaea*, what results would be expected in a cross between a *Dd* dextrally coiled and a *Dd* sinistrally coiled snail, assuming cross-fertilization occurs as shown in Figure 7–2? What results would occur if the *Dd* dextral snail produced only eggs and the *Dd* sinistral snail produced only sperm?

3. Streptomycin resistance in *Chlamydomonas* may result from a mutation in a chloroplast gene or in a nuclear gene. What phenotypic results would occur in a cross between a member of an *mt*+ strain resistant in both genes and a member of an *mt*− strain sensitive to the antibiotic? What results would occur in the reciprocal cross?

4. A plant may have green, white, or green-and-white (variegated) leaves on its branches owing to a mutation in the chloroplast (which produces the white leaves). Predict the results of the following crosses:

	Ovule Source		Pollen Source
(a)	Green branch	×	White branch
(b)	White branch	×	Green branch
(c)	Variegated branch	×	Green branch
(d)	Green branch	×	Variegated branch

5. In diploid yeast strains, sporulation and subsequent meiosis can produce haploid ascospores. These may fuse to reestablish diploid cells. When ascospores from a segregational *petite* strain fuse with those of a normal wild-type strain, the diploid zygotes are all normal. However, following meiosis, ascospores are 1/2 *petite* and 1/2 normal. Is the segregational *petite* phenotype inherited as a dominant or a recessive gene?

6. Predict the results of a cross between ascospores from a segregational *petite* strain and a neutral *petite* strain. Indicate the phenotype of the zygote and the ascospores it may subsequently produce.

7. Described below are the results of three crosses between strains of *Paramecium*. Determine the genotypes of the parental strains.

   (a) Killer × sensitive ⟶ 1/2 Killer: 1/2 sensitive
   (b) Killer × sensitive ⟶ all Killer
   (c) Killer × sensitive ⟶ 3/4 Killer: 1/4 sensitive

8. *Chlamydomonas*, a eukaryotic green alga, is sensitive to the antibiotic erythromycin that inhibits protein synthesis in prokaryotes.

   (a) Explain why.
   (b) There are two mating types in this alga, *mt*+ and *mt*−. If an *mt*+ cell sensitive to the antibiotic is crossed with an *mt*− cell that is resistant, all progeny cells are sensitive. The reciprocal cross (*mt*+ resistant and *mt*− sensitive) yields all resistant progeny cells. Assuming that the mutation for resistance is in chloroplast DNA, what can be concluded?

9. In *Limnaea*, a cross in which the snail contributing the eggs was dextral but of unknown genotype mated with another snail of unknown genotype and phenotype. All F_1 offspring exhibited dextral coiling. Ten of the F_1 snails were allowed to undergo self-fertilization. One-half produced only dextrally coiled offspring, while the other half produced only sinistrally coiled offspring. What were the genotypes of the original parents?

10. In *Drosophila subobscura*, the presence of a recessive gene called *grandchildless (gs)* causes the offspring of homozygous females, but not homozygous males, to be sterile. Can you offer an explanation as to why females but not males are affected by the mutant gene?

11. A male mouse from a true-breeding (hypothetical) strain of hyperactive animals is crossed to a female mouse from a true-breeding lethargic strain (also hypothetical). All of the progeny are lethargic. When the F1 males and females are mated, all of the progeny are lethargic. What is the best explanation for this data? Propose a cross to test your explanation.

12. The specification of the anterior–posterior axis in a *Drosophila* embryo is controlled by gene products that are synthesized and loaded into the maturing oocyte during oogenesis. Mutations in the genes that code for these proteins result in abnormalities of axis during embryogenesis. Describe how the resultant inheritance patterns involving mutations such as these vary from the results of crosses involving biparental (Mendelian) inheritance and from organelle heredity?

Selected Readings

Bogorad, L. 1981. Chloroplasts. *J. Cell Biol.* 91:256s–70s.

Cattanach, B. M., and Jones, J. 1994. Genetic imprinting in the mouse: implications for gene regulation. *J. Inherit. Metab. Dis.* 17:403–20.

Cohen, S. 1973. Mitochondria and chloroplasts revisited. *Am. Sci.* 61:437–45.

Ekstrom, T. 1994. Parental imprinting and the *Igf2* gene. *Horm. Res.* 42:176–81.

Freeman, G., and Lundelius, J. W. 1982. The developmental genetics of dextrality and sinistrality in the gastropod *Lymnaea peregra. Wilhelm Roux Arch.* 191:69–83.

Gillham, N. W. 1978. *Organelle hereditary.* New York: Raven Press.

Goodenough, U., and Levine, R. P. 1970. The genetic activity of mitochondria and chloroplasts. *Sci. Am.* (Nov.) 223:22–29.

Grivell, L. A. 1983. Mitochondrial DNA. *Sci. Am.* (March) 248:78–89.

Lander, E. S., et al. 1990. Mitochondrial diseases: Gene mapping and gene therapy. *Cell* 61:925–26.

Levine, R. P., and Goodenough, U. 1970. The genetics of photosynthesis and of the chloroplast in *Chlamydomonas reinhardi. Annu. Rev. Genet.* 4:397–408.

Margulis, L. 1970. *Origin of eukaryotic cells.* New Haven, CT: Yale University Press.

Mitchell, M. B., and Mitchell, H. K. 1952. A case of maternal inheritance in *Neurospora crassa. Proc. Natl. Acad. Sci. USA* 38:442–49.

Nusslein-Volhard, C. 1996. Gradients that organize embryo development. *Sci. Am.* (Aug.) 275:54–61.

Preer, J. R. 1971. Extrachromosomal inheritance: Hereditary symbionts, mitochondria, chloroplasts. *Annu. Rev. Genet.* 5:361–406.

Rosing, H. S., et al. 1985. Maternally inherited mitochondrial myopathy and myoclonic epilepsy. *Ann. Neurol.* 17:228–37.

Sager, R. 1965. Genes outside the chromosomes. *Sci. Am.* (Jan.) 212:70–79.

———. 1985. Chloroplast genetics. *BioEssays* 3:180–84.

Slonimski, P. 1982. *Mitochondrial genes.* Cold Spring Harbor, NY: Cold Spring Harbor Laboratory.

Sonneborn, T. M. 1959. Kappa and related particles in *Paramecium. Adv. Virus Res.* 6:229–356.

Strathern, J. N., et al., eds. 1982. The molecular biology of the yeast *Saccharomyces: Life cycle and inheritance.* Cold Spring Harbor, NY: Cold Spring Harbor Laboratory.

Sturtevant, A. H. 1923. Inheritance of the direction of coiling in *Limnaea. Science* 58:269–70.

Surani, M. A. 1994. Genomic imprinting: control of gene expression by epigenetic inheritance. *Curr. Opin. Cell Biol.* 6:390–95.

Tzagoloff, A. 1982. *Mitochondria.* New York: Plenum Press.

Wallace, D.C. 1988. Familial mitochondrial encephalomyopathy (MERRF): Genetic, pathophysiological and biochemical characterization of a mitochondrial DNA disease. *Cell* 55:601–10.

———. 1992. Mitochondrial genetics: A paradigm for aging and degenerative diseases. *Science* 256:628–32.

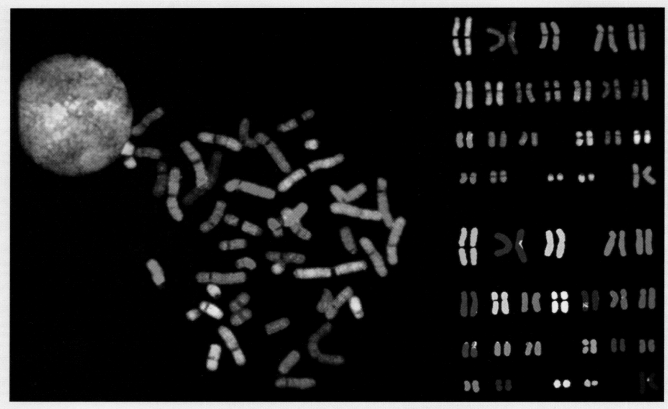

Spectral karyotyping of human chromosomes utilizing differentially labeled "painting" probes.

CHAPTER OUTLINE

CHAPTER
8
Chromosome Variation and Sex Determination

Chapter Concepts

Genetic information of a diploid organism is delicately balanced in both content and location within the genome. A change in chromosome number or in the arrangement of a chromosome region often results in phenotypic variation or disruption of development of an organism. Because the chromosome is the unit of transmission in meiosis, such variations are passed to offspring in a predictable manner, resulting in many interesting genetic situations. As a result of studying variation in the sex chromosomes, valuable insights have been gained into the mode of sex determination in many organisms.

Up to this point in the text, we have emphasized how mutations and the resulting alleles affect an organism's phenotype, and how traits are passed from parents to offspring according to Mendelian principles. In this chapter, we shall look at phenotypic variation occurring as a result of changes in the genetic material that are more substantial than alterations of individual genes. These involve modifications at the level of the chromosome.

Although members of diploid species normally contain precisely two haploid chromosome sets, many cases are known in which some variation from this pattern occurs. Modifications include variations in the number of individual chromosomes as well as rearrangements of the genetic material either within or among chromosomes. Taken together, such changes are called **chromosome mutations** or **chromosome aberrations**, to distinguish such genetic alterations from gene mutations. Because it is the chromosome, and not the gene, that is transmitted according to Mendelian laws, chromosome aberrations are transmitted to offspring in a predictable manner, resulting in many interesting examples of heritable phenotypic variation.

We shall begin this chapter with the analysis of variation in the number of sex chromosomes in both humans and *Drosophila*. As a result of these studies, valuable insights have been gained into how sex is determined. We shall also pursue several other topics related to the genetic function of sex chromosomes.

Variation in Chromosome Number: An Overview

Before embarking on a discussion of variations involving the number of chromosomes in organisms and sex determination, it is useful to establish the terminology that describes such changes. Variation in chromosome number ranges from the addition or loss of one or more chromosomes to the addition of one or more haploid sets of chromosomes. When an organism gains or loses one or more chromosomes, but not a complete set, the condition of **aneuploidy** is created. The loss of a single chromosome creates a condition called **monosomy**. The gain of one chromosome to an otherwise diploid genome results in **trisomy**. These are contrasted with the condition of **euploidy**, where complete haploid sets of chromosomes are found. If there are three or more sets, the more general term **polyploidy** is applicable. Those with three sets are called **triploid**; those with four sets are **tetraploid**, and so on. Table 8.1 provides a useful organizational framework for you to follow as we discuss each of these categories and the subsets within them.

Chromosome Composition and Sex Determination in Humans

In our discussion of X-linkage in Chapter 4, we pointed out that, as part of the diploid chromosome

TABLE 8.1	Terminology for variation in chromosome numbers

Term	Explanation
Aneuploidy	$2n$ plus or minus chromosomes
Monosomy	$2n - 1$
Trisomy	$2n + 1$
Tetrasomy, pentasomy, etc.	$2n + 2, 2n + 3$, etc.
Euploidy	Multiples of n
Diploidy	$2n$
Polyploidy	$3n, 4n, 5n, \ldots$
Triploidy	$3n$
Tetraploidy, pentaploidy, etc.	$4n, 5n$, etc.
Autopolyploidy	Multiples of the same genome
Allopolyploidy (Amphidiploidy)	Multiples of different genomes

composition of both humans and *Drosophila*, females possess two X chromosomes while males possess one X and one Y chromosome. This observation might lead us to conclude that the Y chromosome causes maleness in both species; however, this is not necessarily the case. Perhaps the lack of a second X chromosome somehow causes maleness, and the Y plays no role in sex determination. Perhaps the presence of two X chromosomes causes femaleness, and the Y plays no role. The evidence that clarified which explanation was correct awaited the study of variations in the sex chromosome composition of both humans and flies. As we shall see, the first explanation, where the Y determines maleness, is valid in humans but not in *Drosophila*.

Klinefelter and Turner Syndromes

Around 1940 it was observed that two human abnormalities, the **Klinefelter** and **Turner syndromes**,* are characterized by aberrant sexual development. Individuals with Klinefelter syndrome [Figure 8–1(a)] have genitalia and internal ducts that are usually male, but their testes are underdeveloped and fail to produce sperm. Although masculine development occurs, feminine sexual development is not entirely suppressed. Slight enlargement of the breasts is common, for example. Ambiguous sexual characteristics may lead to difficulties in social adjustment.

In Turner syndrome [Figure 8–1(b)], the affected individual has female external genitalia and internal ducts, but the ovaries are rudimentary. Other charac-

teristic abnormalities include short stature (usually under 5 feet), a webbed neck, and a broad, shieldlike chest.

In 1959, the karyotypes of individuals with these syndromes were determined to be abnormal with respect to the sex chromosomes. Individuals with Klinefelter syndrome most often are trisomic and have an XXY complement in addition to 44 autosomes. People with this karyotype are designated **47,XXY**. Individuals with Turner syndrome are monosomic, having only 45 chromosomes, including just a single X chromosome; they are designated **45,X**. Karyotypes of both conditions are shown in Figure 8–1. Note the convention used in designating chromosome compositions. The number indicates how many chromosomes are present, and the information after the comma designates the deviation from the normal diploid content. Both conditions result from nondisjunction of the sex chromosomes during meiosis.

These karyotypes and their corresponding sexual phenotypes allow us to conclude that the Y chromosome determines maleness in humans. In its absence, the sex of the individual is female, even if only a single X chromosome is present. The presence of the Y chromosome in the individual with Klinefelter syndrome is sufficient to determine maleness, even though its expression is not complete. Similarly, in the absence of a Y chromosome, as in the case of individuals with Turner syndrome, no masculinization occurs.

Klinefelter syndrome occurs in about 2 of every 1000 male births. The karyotypes **48,XXXY**, **48,XXYY**, **49,XXXXY**, and **49,XXYY** are similar phenotypically to 47,XXY, but manifestations are often more severe in individuals with a greater number of X chromosomes.

Karyotypes other than 45,X also lead to Turner syndrome. These include individuals with two apparent cell lines, each exhibiting different karyotypes. Such individuals are called **mosaics**, which result from mitotic errors during early development. The most common chromosome combinations are 45,X/46,XY and 45,X/46,XX. Turner syndrome is observed in only about 1 in 3000 female births, a frequency much lower than that for Klinefelter syndrome. One explanation for this difference is the observation that a substantial majority of 45,X fetuses die *in utero* and are aborted spontaneously.

47,XXX Syndrome

The presence of three X chromosomes along with a normal set of autosomes **(47,XXX)** results in female differentiation. This syndrome, which is estimated to occur in about 1 of 1200 female births, is highly variable in expression. Frequently, 47,XXX women are perfectly normal. In other cases, underdeveloped secondary sex character-

---

*Although the possessive form of the names of most syndromes (eponyms) is sometimes used (e.g., Klinefelter's), the current preference is to use the nonpossessive form for syndromes.

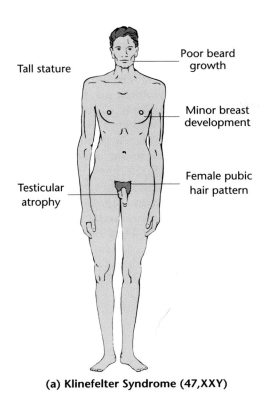

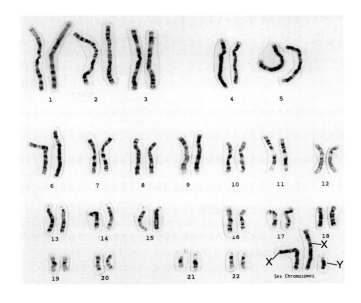

Tall stature

Poor beard growth

Minor breast development

Testicular atrophy

Female pubic hair pattern

(a) Klinefelter Syndrome (47,XXY)

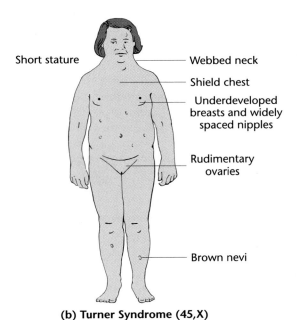

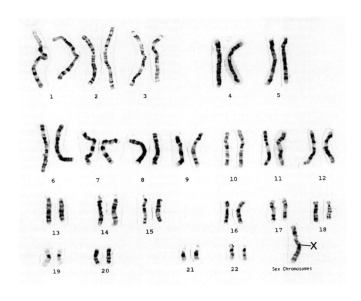

Short stature

Webbed neck

Shield chest

Underdeveloped breasts and widely spaced nipples

Rudimentary ovaries

Brown nevi

(b) Turner Syndrome (45,X)

■ Figure 8–1   The karyotypes and depictions of the characteristics of (a) Klinefelter syndrome (47,XXY) and (b) Turner syndrome (45,X). The sex chromosomes are labeled.

istics, sterility, and mental retardation may occur. In rare instances, **48,XXXX** and **49,XXXXX** karyotypes have been reported. The syndromes associated with these karyotypes are similar to but more pronounced than the 47,XXX. Thus, the presence of additional X chromosomes appears to disrupt the delicate balance of genetic information essential to normal female development.

47,XYY Condition

Another human trisomy, **47,XYY**, has been discovered and intensively investigated. Studies of this condition, in which the only deviation from diploidy is the presence of an additional Y chromosome to the normal male karyotype, have led to an interesting controversy.

In 1965, Patricia Jacobs discovered that 9 of 315 males in a Scottish maximum security prison had the 47,XYY karyotype. These males were significantly above average in height and had been involved in criminal acts with serious social consequence. Of the 9 males studied, 7 were of subnormal intelligence, and all suffered personality disorders. In several other studies, similar findings were obtained.

Because of these investigations, the phenotype and frequency of the 47,XYY condition in criminal and noncriminal populations have been examined more extensively. Above-average height and subnormal intelligence have been generally substantiated, and the frequency of males displaying this karyotype is indeed higher in penal and mental institutions compared with unincarcerated males. The possible correlation between this chromosome composition and antisocial and criminal behavior has been of considerable interest. A particularly relevant question involves the characteristics displayed by XYY males who are not incarcerated. The only nearly constant association is that such individuals are over 6 feet tall! Because it is now clear that many XYY males do not exhibit any form of antisocial behavior and lead normal lives, we must conclude that there is no absolute correlation between the extra Y chromosome and behavior.

The Human Y Chromosome and Male Development

Before turning to other types of chromosome variation and their effects in humans and other organisms, it is appropriate to review the information available concerning how the Y chromosome results in male development in mammals. Studies in humans have been instrumental in increasing our understanding of this extensive group of animals. We have previously alluded to the fact that this chromosome shares only limited homology with loci on the X chromosome, but it does carry genetic information that controls sexual development.

Therefore, some region of the Y chromosome contains genetic information that is responsible for the **testis-determining factor (TDF)**, a product that somehow triggers the undifferentiated gonadal tissue of the embryo to form testes. In its absence, female development occurs. Research has focused on just what constitutes the region and the product.

It is now clear that a small part of the human Y chromosome contains a gene called **SRY (sex-determining region Y)**. In rare individuals, sex chromosome compositions do not match their expected sexual phenotype. Evidence proving that *SRY* is indeed the gene responsible has relied on the molecular geneticist's ability to identify the presence or absence of DNA sequences in these particular individuals.

There are human males who demonstrate two X and no Y chromosomes. However, they have attached to one of their Xs the region of the Y containing *SRY*. There are also females who have only one X but also have one Y chromosome. Their Y is missing the *SRY* region. These observations argue strongly in favor of the role of *SRY* in male development.

The final proof that this gene is responsible for causing maleness involves an experiment using **transgenic mice**, in which a similar region, *Sry*, has been identified. Such animals arise from fertilized eggs that have foreign DNA injected into them and incorporated into the genetic composition of the developing embryo. When DNA containing only *Sry* is injected into normal XX eggs, most mice develop into males!

Thus, the relevant gene has been identified. It is present in all mammals thus far examined, having been conserved throughout evolution. How the product of this gene triggers the embryonic gonadal tissue to develop into testes rather than ovaries is now a reasonable question to ask, and one that is amenable to investigation.

The X Chromosome and Dosage Compensation

The presence of two X chromosomes in normal human females and only one X in normal human males is unique compared with the equal numbers of autosomes present in the cells of both sexes. On theoretical grounds alone, it is possible to speculate that this situation should create a genetic dosage problem between males and females. Since females have two copies of the X and males have only one, there is the potential for expressing twice as much gene product for all X-linked genes. In this section, we will describe research findings regarding X-linked gene expression that demonstrate the existence of a genetic **dosage compensation** mechanism.

Barr Bodies and Dosage Compensation

Murray L. Barr and Ewart G. Bertram's experiments with female cats, and Keith Moore and Barr's subsequent study with humans, demonstrate a genetic mechanism in mammals that compensates for X-chromosome dosage disparities. Barr and Bertram observed a darkly staining body in interphase nerve cells of female cats. They found that this structure was absent in similar cells of males. In human females, this body can be easily demonstrated in cells derived from the epithelial cells of the cheek or in fibroblasts, but not in similar male cells. This highly condensed structure, about 1 mm in diameter, lies against the nuclear envelope of interphase cells (Figure 8–2). It stains positively in the Feulgen reaction for DNA.

Current experimental evidence strongly suggests that this body, called a **sex chromatin body** or often a **Barr body**, is an inactivated X chromosome. Ohno was the first to suggest that the Barr body arises from one of the

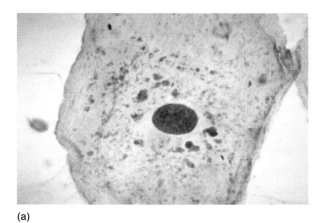

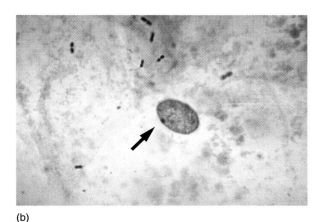

(a)

(b)

■ Figure 8–2 Photomicrographs comparing (a) a cheek epithelial cell nucleus that fails to reveal a Barr body with (b) one that demonstrates a Barr body (see arrow). This structure, also called a sex chromatin body, represents an inactivated X chromosome.

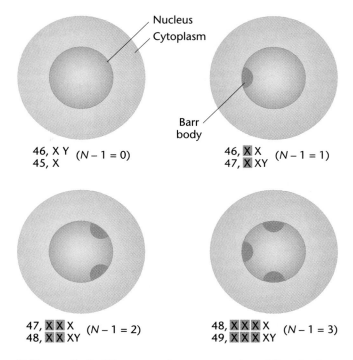

■ Figure 8–3 Diagrammatic representation of Barr body occurrence in various human karyotypes, where all X chromosomes except one ($N − 1$) are inactivated.

two X chromosomes. This hypothesis is attractive because it provides a mechanism for dosage compensation. If one of the two X chromosomes is inactive in the cells of females, the dosage of genetic information that may be expressed in males and females is equivalent. Convincing but indirect evidence for this hypothesis comes from the study of the sex chromosome syndromes described earlier in this chapter. Regardless of how many X chromosomes exist, all but one of them appear to be inactivated and can be seen as Barr bodies. For example, no Barr bodies are seen in Turner 45,X females; one is seen in Klinefelter 47,XXY males; two in 47,XXX females; three in 48,XXXX females; and so on (Figure 8–3). Therefore, the number of Barr bodies follows an $N − 1$ rule, where N is the total number of X chromosomes present.

The Lyon Hypothesis

The inactivation of one of the two X chromosomes raises numerous questions. In mammalian females, one X chromosome is of maternal origin while the other is of paternal origin. Which one is inactivated? Is the inactivation random? Is the same chromosome inactive in all somatic cells? In 1961, Mary Lyon and Liane Russell independently proposed a hypothesis that addresses these questions. They postulated that the inactivation of X chromosomes occurs randomly in somatic cells at a point early in embryonic development, and once inactivation has occurred, all progeny cells have the same X chromosome inactivated.

This explanation, which has come to be called the **Lyon hypothesis**, was initially based on observations of female mice that were heterozygous for sex-linked coat color genes. The pigmentation of these heterozygous females was mottled, with large patches of skin expressing the color allele on one X and other patches expressing the allele on the other X. Indeed, if one or the other of the two X chromosomes was inactive in adjacent patches of cells, such a phenotypic pattern would result. Similar mottling occurs in the black and orange patches of female tortoise-shell and calico cats. Such X-linked coat color patterns do not occur in male cats because all cells are hemizygous for only one X-linked coat color allele. A calico cat is shown in Figure 8–4.

The Lyon hypothesis is generally accepted as valid. One extension of the hypothesis is that *mammalian females are mosaics for all heterozygous X-linked alleles*. Some areas of the body express only the maternally derived alleles, and others express only the paternally derived alleles. Two especially interesting examples involve **red-green color blindness** and **anhidrotic ectodermal dysplasia**,

■ Figure 8–4 A calico cat, in which the random distribution of orange and black patches illustrates the Lyon hypothesis. The white patches are due to still another gene, distinguishing calicos from tortoiseshell cats.

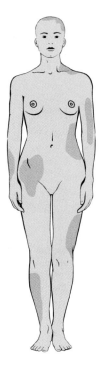

■ Figure 8–5 Depiction of the absence of sweat glands (shaded regions) in a female who is heterozygous for the X-linked condition anhidrotic ectodermal dysplasia.

both X-linked recessive disorders in humans. In the former case, hemizygous males are fully color-blind in all retinal cells. However, heterozygous females display mosaic retinas with patches of defective color perception and surrounding areas with normal color perception. Males who are hemizygous for anhidrotic ectodermal dysplasia show absence of teeth, sparse hair growth, and lack of sweat glands. The skin of heterozygous females reveals patterns of tissue with and without sweat glands (Figure 8–5). In both examples, random inactivation of one or the other X chromosome early in the development of heterozygous females has led to these occurrences.

The Mechanism of Inactivation

The least understood aspect of the Lyon hypothesis is the mechanism of chromosome inactivation in mammals. How are almost all genes of an entire chromosome inactivated? Recent investigations are beginning to clarify this issue. A single region of the human X chromosome, called the **X-inactivation center** (**XIC**), is the major control unit. Genetic expression of this region occurs only on the X chromosome that is inactivated. The constant association of expression and X chromosome inactivation support the conclusion that this region is an important genetic component in the process.

A gene, *XIST* (*X-inactive specific transcript*), is now believed to represent the critical locus within the XIC. A comparable region (Xic) and gene (*Xist*) exist in the mouse. Recently, in 1996, a research group led by Graeme Penny provided the most convincing evidence thus far that expression (transcription) of *Xist* is the critical event in chromosome inactivation. These researchers were able to inactivate the gene by introducing a targeted deletion into it that eliminated inactivation of the chromosome bearing this alteration. An interesting question remains unanswered. In cells with more than two chromosomes, what sort of "counting" mechanism determines that all but one X chromosome are inactivated? Another question is how inactivation is stably maintained in progeny cells. Whatever the answers to these questions, the discovery of these inactivation centers is exciting and pushes us closer to an understanding of how dosage compensation is accomplished in mammals.

Chromosome Composition and Sex Determination in *Drosophila*

Because males and females in *Drosophila* have the identical sex chromosome composition as humans, we might assume that the Y also causes maleness in these flies. However, the elegant work of Calvin Bridges in 1916 showed this not to be true. He studied flies with quite varied chromosome compositions, leading him to the conclusion that the Y chromosome is not involved in sex determination in this organism. Instead, Bridges proposed that the X chromosomes and autosomes together play a critical role in sex determination.

Bridges' work can be divided into two phases: (1) a study of offspring resulting from *nondisjunction* of the X chromosomes during meiosis in females, and (2) subsequent work with progeny of triploid (3*n*) females. **Nondisjunction** is the failure of paired chromosomes to segregate or separate during the anaphase stage of the first or second meiotic division. The result is the production of two abnormal gametes: One of these contains an extra chromosome (*n* + 1), whereas the other lacks a chromosome (*n* − 1). Fertilization of such gametes produces (2*n* + 1) or (2*n* − 1) aneuploid zygotes. As a result, in addition to the normal complement of six autosomes, the resulting flies had either an XXY or an X0 sex chromosome composition. (The zero signifies that the second chromosome is absent.) The XXY flies were normal females, and the X0 flies were sterile males. The presence of the Y chromosome in the XXY flies did not cause maleness, and its absence in the X0 flies did not produce femaleness. From these data, Bridges concluded that the Y chromosome in *Drosophila* lacks male-determining factors, but apparently contains genetic information essential to male fertility because the X0 males were sterile.

Bridges was able to clarify the mode of sex determination in *Drosophila* by studying the progeny of triploid females (3*n*), which have three copies each of the haploid complement of chromosomes. These females apparently originate from rare diploid eggs fertilized by normal haploid sperm. Triploid females have heavy-set bodies, coarse bristles, and coarse eyes and may be fertile. Because of the presence of an odd number of each chromosome (3), a wide range of chromosome compositions is distributed into gametes. These give rise to offspring with a variety of abnormal chromosome constitutions. A correlation between the sexual morphology, chromosome composition, and Bridges' interpretation is shown in Figure 8–6. *Drosophila* has a haploid number of 4, thereby displaying three pairs of autosomes in addition to its sex chromosomes.

Bridges realized that the critical factor in determining sex is *the ratio of X chromosomes to the number of haploid sets of autosomes present*. Normal (2X:2A) and triploid (3X:3A) females each have a ratio equal to 1.0, and both are fertile. As the ratio exceeds unity (3X:2A, or 1.5, for example), what was originally called a **superfemale** is produced. Because this female is rather weak and infertile and has lowered viability, this type is now more appropriately called a **metafemale**.

Normal (XY:2A) and sterile (X0:2A) males each have a ratio of 1:2, or 0.5. When the ratio decreases to 1:3, or 0.33, as in the case of an XY:3A male, infertile **metamales** result. Other flies recovered by Bridges in these studies contained an X:A ratio intermediate between 0.5 and 1.0. These flies were generally larger, and they exhibited a variety of morphological abnormalities and

rudimentary bisexual gonads and genitalia. They were invariably sterile and were designated as **intersexes**.

These results indicate that in *Drosophila*, male-determining factors are not localized on the sex chromosomes, but are instead found on the autosomes. Some female-determining factors, however, are localized on the X chromosomes. Thus, with respect to primary sex determination, male gametes containing one of each autosome plus a Y chromosome result in male offspring not because of the presence of the Y chromosome, but because of the lack of an X chromosome. This mode of sex determination is explained by the **genic balance theory**. Bridges proposed that a threshold for maleness is reached when the X:A ratio is 1:2 (X:2A), but that the presence of an additional X (XX:2A) alters this balance and results in female differentiation.

Several mutant genes have now been identified that are involved in sex determination in *Drosophila*. The recessive autosomal gene *transformer* (*tra*), discovered by Alfred H. Sturtevant, is especially interesting. Homozygous mutant females are transformed into sterile males, but males are unaffected when homozygous for *tra*.

Another gene, *Sex-lethal* (*Sxl*), has been shown to play a critical role in sex determination, serving as a "master switch" for the activation of a group of at least four other regulatory genes. Activation of the *Sxl* gene, located on the X chromosome, is essential to subsequent female development, and relies on a ratio of X chromosomes to sets of autosomes that equals 1.0. In the absence of activation, resulting from an X:A ratio of 0.5, male development occurs.

The *Sxl* locus is part of a hierarchy of gene expression and exerts control over still other genes, including the *tra* gene (discussed above) and the *dsx* (*doublesex*) gene. The wild-type allele of *tra* is activated by the product of *Sxl* only in females, which in turn influences the expression of *dsx*. Ultimately, different forms of the *dsx* gene product are formed that account for sexual dimorphism.

Aneuploidy

We turn now to a consideration of variations in the number of autosomes and the genetic consequence of such changes. The most common examples of aneuploidy, in which an organism has a chromosome number other than an exact multiple of the haploid set, are cases in which a single chromosome is either added to or lost from a normal diploid set. Such circumstances can arise as a result of primary or secondary nondisjunction (Figure 8–7). The loss of one chromosome produces a 2*n* − 1 complement and is called monosomy; the gain of one chromosome produces a 2*n* + 1 complement and is described as trisomy. The 2*n* + 2 and 2*n* + 3 conditions are called **tetrasomy** and **pentasomy**, indicating that an individual has four or five copies of a chromosome in an otherwise diploid genome.

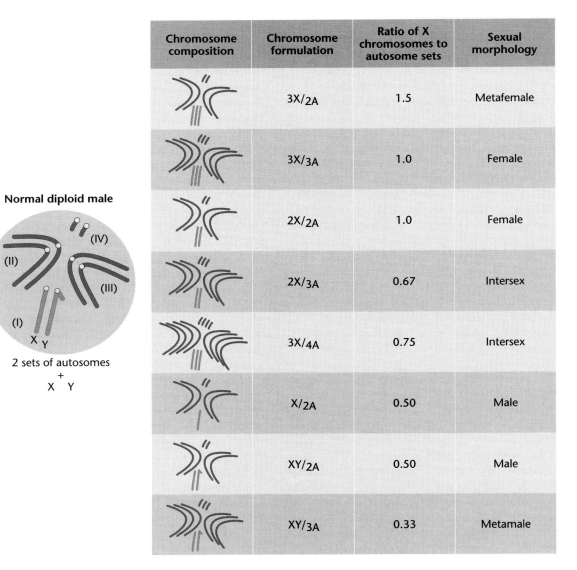

Chromosome composition	Chromosome formulation	Ratio of X chromosomes to autosome sets	Sexual morphology
	3X/2A	1.5	Metafemale
	3X/3A	1.0	Female
	2X/2A	1.0	Female
	2X/3A	0.67	Intersex
	3X/4A	0.75	Intersex
	X/2A	0.50	Male
	XY/2A	0.50	Male
	XY/3A	0.33	Metamale

Normal diploid male

2 sets of autosomes
+
X   Y

■ Figure 8–6   Chromosome compositions, the ratios of X chromosomes to sets of autosomes, and the resultant sexual morphology in *Drosophila melanogaster*. The normal diploid male chromosome composition is shown as a reference on the left. It contains two sets of the three autosomes plus an X and Y chromosome (XY/2A).

Monosomy

Although monosomy for one of the sex chromosomes is fairly common, monosomy for one of the autosomes is not usually tolerated in animals. In *Drosophila*, flies that are monosomic for the very small chromosome 4—a condition referred to as **Haplo-IV**—develop more slowly, exhibit reduced body size, and have impaired viability. Monosomy for the larger chromosomes 2 and 3 is apparently lethal, because such flies have never been recovered.

The failure of monosomic individuals to survive in many animal species is at first quite puzzling, since at least a single copy of every gene is present in the remaining homolog. However, if just one of those genes is represented by a lethal allele, the unpaired chromosome condition leads to the death of the organism. This occurs because monosomy unmasks recessive lethals

that are tolerated in heterozygotes carrying the corresponding wild-type alleles.

Aneuploidy is better tolerated in the plant kingdom. Monosomy for autosomal chromosomes has been observed in maize, tobacco, the evening primrose *Oenothera*, and the Jimson weed *Datura*, among other plants. Nevertheless, such monosomic plants are usually less viable than their diploid derivatives. Haploid pollen grains, which undergo extensive development before participating in fertilization, are particularly sensitive to the lack of one chromosome and are seldom viable.

Partial Monosomy: Cri-du-Chat Syndrome

In humans, autosomal monosomy has not been reported beyond birth. Individuals with such chromosome complements are undoubtedly conceived, but

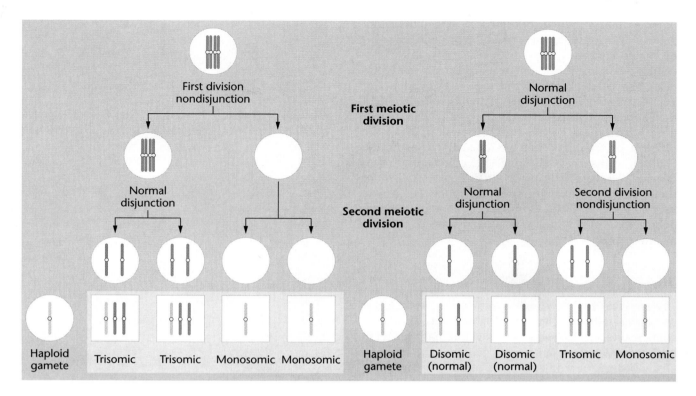

■ Figure 8–7   Diagram illustrating nondisjunction during the first and second meiotic divisions. In both cases, some gametes are formed that either contain two members of a specific chromosome or lack it altogether. Following fertilization by a normal haploid gamete, monosomic, disomic (normal), or trisomic zygotes result.

none apparently survive embryonic and fetal development. There are, however, examples of survivors with **partial monosomy**, in which only part of one chromosome is lost. These cases are also referred to as **segmental deletions**. One such case was first reported by Jérôme LeJeune in 1963 when he described the clinical symptoms of the **cri-du-chat (cry of the cat) syndrome**. This syndrome is associated with the loss of much of the short arm of chromosome 5 (Figure 8–8). Thus, the genetic constitution may be designated as **46,5p–**, meaning that such an individual has all 46 chromosomes but that some of the p arm (the short or petite arm) is missing.

Infants with this syndrome may exhibit anatomic malformations, including gastrointestinal and cardiac complications, and are often mentally retarded. Abnormal development of the glottis and larynx is characteristic

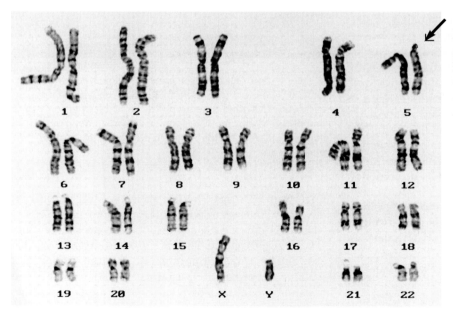

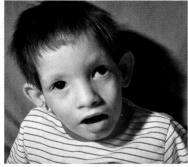

■ Figure 8–8   A representative karyotype and photograph of a child exhibiting cri-du-chat syndrome (46,5p–). In the karyotype, the arrow identifies the nearly complete absence of the short arm of one member of the chromosome 5 homologs.

of individuals with this syndrome. As a result, the infant has a cry similar to the meowing of a cat, giving the syndrome its name.

Since 1963, hundreds of cases of cri-du-chat syndrome have been reported worldwide. An incidence of 1 in 50,000 live births has been estimated. The length of the short arm that is deleted varies somewhat; longer deletions appear to have a greater impact on the physical, psychomotor, and mental skill levels of those children who survive. Although the effects of the syndrome are severe, many individuals achieve a level of social development in the trainable range. Those who receive home care and early special schooling are ambulatory, develop self-care skills, and learn to communicate verbally.

Trisomy

In general, the effects of trisomy parallel those of monosomy. However, the addition of an extra chromosome produces somewhat more viable individuals in both animal and plant species than does the loss of a chromosome. In animals, this is often true, provided that the chromosome involved is relatively small. However, the addition of a large autosome to the diploid complement in both *Drosophila* and humans has severe effects and is usually lethal during development.

In plants, trisomic individuals are viable, but their phenotype may be altered. A classical example involves the Jimson weed *Datura*, whose diploid number is 24. Twelve primary trisomic conditions are possible, and examples of each one have been recovered. Each trisomy alters the phenotype of the plant's capsule sufficiently (Figure 8–9) to produce a unique phenotype. These capsule phenotypes were first thought to be caused by mutations in one or more genes.

Down Syndrome

The only human autosomal trisomy in which a significant number of individuals survive longer than a year past birth was discovered in 1866 by Langdon Down. The condition is now known to result from trisomy of chromosome 21, one of the G group[*] (Figure 8–10), and is called **Down syndrome** or simply **trisomy 21** (and is designated **47,21+**). This trisomy is found in approximately 3 infants in every 2000 live births.

The overt phenotype of these individuals is so similar that they bear a striking resemblance to one another. They display a prominent epicanthic fold in the corner of the eye and are characteristically short. They may have small, round heads; protruding, furrowed tongues, which cause the mouth to remain partially open; and short, broad hands with fingers showing characteristic palm and

*On the basis of size and centromere placement, human chromosomes are divided into seven groups: A (1–3); B (4–5); C (6–12); D (13–15); E (16–18); F (19–20); and G (21–22).

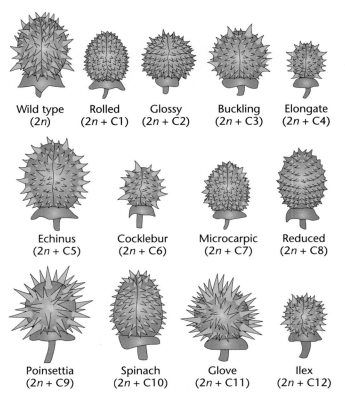

Wild type (2*n*) Rolled (2*n* + C1) Glossy (2*n* + C2) Buckling (2*n* + C3) Elongate (2*n* + C4)

Echinus (2*n* + C5) Cocklebur (2*n* + C6) Microcarpic (2*n* + C7) Reduced (2*n* + C8)

Poinsettia (2*n* + C9) Spinach (2*n* + C10) Glove (2*n* + C11) Ilex (2*n* + C12)

■ Figure 8–9 Drawings of capsule phenotypes of the fruits of *Datura stramonium*. In comparison with wild type, each of the 12 other phenotypes is the result of trisomy of one of the 12 chromosomes that are characteristic of the haploid genome. The photograph illustrates the normal plant.

fingerprint patterns. Physical, psychomotor, and mental development is retarded, and IQ is seldom above 70.

Down children have a shortened life expectancy, and few survive to age 50. They are prone to respiratory disease and heart malformations and show an incidence of leukemia approximately 15 times as high as that of the normal population. However, careful medical scrutiny

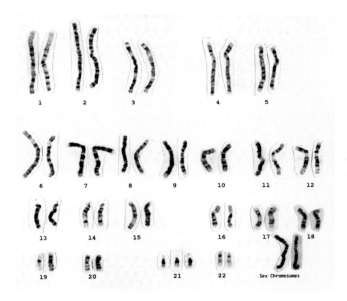

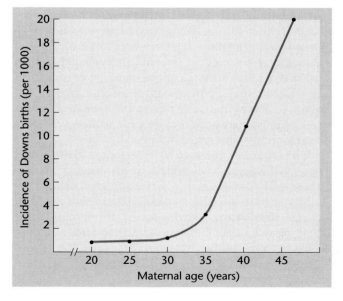

■ Figure 8–10 The karyotype and a photograph of a child with Down syndrome. In the karyotype, three members of the G-group chromosome 21 are present, creating the 47,21+ condition.

and treatment throughout their lives has extended their survival significantly. A striking observation is that death of older Down syndrome adults is frequently due to Alzheimer's disease.

One way in which this trisomic condition may originate is through nondisjunction of chromosome 21 during meiosis. Failure of paired homologs to disjoin during anaphase I or of chromatids to disjoin during anaphase II can result in male or female gametes with the $n + 1$ chromosome composition. Following fertilization with a normal gamete, the trisomic condition is created. Chromosome analysis has shown that while the additional chromosome may be derived from either the mother or father, the ovum is most often the source.

Before the development of techniques that distinguish paternal from maternal homologs, this conclusion was supported by other indirect evidence derived from studies of the age of mothers giving birth to Down infants. Figure 8–11 shows an analysis of the distribution of maternal age and the incidence of Down syndrome newborns. The frequency of Down births increases dramatically as the age of the mother increases. While the frequency is about 1 in 1000 at maternal age 30, a tenfold increase to a frequency of 1 in 100 is noted at age 40. The frequency increases still further to about 1 in 50 at age 45.

While the nondisjunctional event that produces Down syndrome seems more likely to occur during oogenesis in women between the ages of 35 and 45, we do not know with certainty why this is so. However, one observation may be relevant. In human females, all primary oocytes have been formed by birth. Therefore, once ovulation begins, each succeeding ovum has been arrested in meiosis for about a month longer than the one preceding it. Women 30 or 40 years old produce ova that are significantly older and arrested longer than those they ovulated 10 or 20 years previously. However, it is not yet known whether ovum age is the cause

of the increased incidence of nondisjunction leading to Down syndrome.

These statistics pose a serious problem for the woman who becomes pregnant late in her reproductive years. **Genetic counseling** early in such pregnancies serves two purposes. First, it informs the parents about the probability that their child will be affected and educates them about Down syndrome. Although some individuals with Down syndrome must be institutionalized, others benefit greatly from special education programs and may be cared for at home. Further, these children are noted for their affectionate, loving natures. Second, a genetic counselor may recommend a prenatal diagnostic technique such as **amniocentesis** or **chorionic villus sampling (CVS)**. These techniques require

■ Figure 8–11 Incidence of Down syndrome births contrasted with maternal age.

the removal and culture of fetal cells. The karyotype of the fetus may then be determined by cytogenetic analysis. If the fetus is diagnosed as having Down syndrome, a therapeutic abortion is one option the parents may consider.

Because Down syndrome appears to be caused by a random error—nondisjunction of chromosome 21 during maternal or paternal meiosis—the disorder is not expected to be inherited. Nevertheless, Down syndrome occasionally runs in families. This condition, **familial Down syndrome**, involves a **translocation** of chromosome 21, another type of chromosomal aberration, which we will discuss later in this chapter.

Viability in Human Aneuploid Conditions

The reduced viability of individuals with recognized monosomic and trisomic conditions is evident. Only two other trisomies in humans survive to term. Both **Patau** and **Edwards syndromes, 47,13+** and **47,18+**, respectively, result in severe malformations and early lethality. Figure 8–12 illustrates the abnormal karyotype and the many defects characterizing Edwards infants.

Such observations lead us to believe that many other aneuploid conditions arise, but that the affected fetuses do not survive to term. This observation has been confirmed by karyotypic analysis of spontaneously aborted fetuses. These studies have revealed some rather striking statistics. At least 15 to 20 percent of all conceptions terminate in spontaneous abortion (some estimates are considerably higher)! About 30 percent of all spontaneous abortuses demonstrate some form of chromosomal anomaly, and approximately 90 percent of all chromosomal anomalies are terminated prior to birth as a result of spontaneous abortion.

A large percentage of spontaneous abortuses demonstrating chromosomal abnormalities are aneuploids. The aneuploid with highest incidence among abortuses is the 45,X condition, which produces an infant with Turner syndrome if the fetus survives to term.

An extensive review of this subject by David H. Carr also reveals that a significant percentage of abortuses are trisomic for one of the chromosome groups. Trisomies for every human chromosome have been recovered. Monosomies were seldomly found, however, even though nondisjunction should produce $n - 1$ gametes with a frequency equal to $n + 1$ gametes. This finding suggests that gametes lacking a single chromosome are functionally impaired to a serious degree or that the embryo dies so early in its development that recovery occurs infrequently. Various forms of polyploidy and other miscellaneous chromosomal anomalies were also found in Carr's study.

These observations support the hypothesis that normal embryonic development requires a precise diploid complement of chromosomes to maintain a delicate equilibrium in the expression of genetic information.

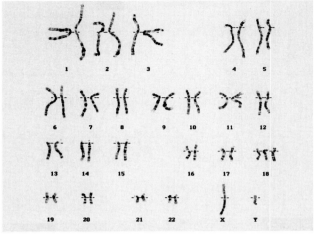

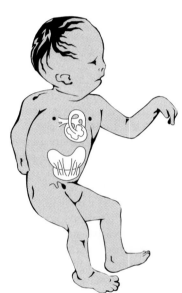

Growth failure
Mental retardation
Open skull sutures at birth
High arched eyebrows
Low set deformed ears
Short sternum
Ventricular septal defect
Flexion deformities of fingers
Abnormal kidneys
Persistent ductus arteriosus
Deformity of hips
Prominent external genitalia
Muscular hypertonus
Prominent heel
Dorsal flexion of big toes

■ Figure 8–12   The karyotype and phenotypic depiction of an infant with Edwards syndrome. Three members of the E-group chromosome 18 are present, creating the 47,18+ condition.

The prenatal mortality of most aneuploids provides a barrier against the introduction of a general form of genetic anomalies into the human population.

Polyploidy and Its Origins

The term polyploidy describes instances in which more than two multiples of the haploid chromosome set are found. The naming of polyploids is based on the number of sets of chromosomes found: A triploid has $3n$ chromosomes; a tetraploid has $4n$; a **pentaploid**, $5n$; and so forth. Several general statements may be made about polyploidy. This condition is relatively infrequent in most animal species, but is well known in lizards, amphibians, and fish. It is much more common in plant species. Odd numbers of chromosome sets are not usually maintained reliably from generation to generation, because a polyploid organism with an uneven number of homologs usually does not produce genetically balanced gametes. For this reason, triploids, pentaploids, and so on, are not usually found in plant species that depend solely upon sexual reproduction for propagation.

Polyploidy can originate in two ways: (1) The addition of one or more extra sets of chromosomes, identical to the normal haploid complement of the same species, results in **autopolyploidy**; and (2) the combination of chromosome sets from different species may occur as a consequence of hybridization and results in **allopolyploidy**. Thus, the distinction between auto- and allopolyploidy is based on the genetic origin of the extra chromosome sets.

In our discussion of polyploidy, we will use the following symbols to clarify the origin of additional chromosome sets. For example, if *A* represents the haploid set of chromosomes of any organism, then

$$A = a_1 + a_2 + a_3 + a_4 + \cdots + a_n$$

where a_1, a_2, and so on, represent individual chromosomes and where *n* is the haploid number. A normal diploid organism is represented simply as *AA*.

Autopolyploidy

In **autopolyploidy**, each additional set of chromosomes is identical to the parent species. Therefore, triploids are represented as *AAA*, tetraploids are *AAAA*, and so forth.

Autotriploids may arise in several ways. A failure of all chromosomes to segregate during meiotic divisions may produce a diploid gamete. If such a gamete is fertilized by a haploid gamete, a zygote with three sets of chromosomes is produced. Or, rarely, two sperm may fertilize an ovum, resulting in a triploid zygote. Triploids can also be produced under experimental conditions by crossing diploids with tetraploids. Diploid organisms produce gametes with *n* chromosomes, while tetraploids produce *2n* gametes. Upon fertilization, the desired triploid is produced.

Tetraploid cells may be produced experimentally from diploid cells by applying cold or heat shock during meiosis or by applying colchicine to somatic cells undergoing mitosis. **Colchicine**, an alkyloid derived from the autumn crocus, interferes with spindle formation; thus, replicated chromosomes cannot be separated at anaphase and migrate to opposite poles. When colchicine is removed, the cell may reenter interphase. When the paired sister chromatids separate and uncoil, the nucleus contains twice the diploid number of chromosomes and is effectively *4n*. This process is illustrated in Figure 8–13.

In general, autopolyploid plants are larger than their diploid relatives. Often, the flower and fruit are increased in size. This increase seems to be due to larger cell size rather than greater cell number. Although autopolyploids do not contain new or unique information compared with the diploid relative, such varieties may be of greater commercial value. Economically important triploid plants include several potato species of the genus *Solanum*, Winesap apples, commercial bananas, seedless watermelons, and the cultivated tiger lily *Lilium tigrinum*. These plants are propagated asexually. Diploid bananas contain hard seeds, but the commercial, triploid, "seedless" variety has edible seeds. Tetraploid alfalfa, coffee, peanuts, and McIntosh apples are also of economic value because they are larger or grow more vigorously than their diploid or triploid counterparts. The commercial strawberry is an octaploid. These observations attest to the importance of autopolyploidy in domesticated plants.

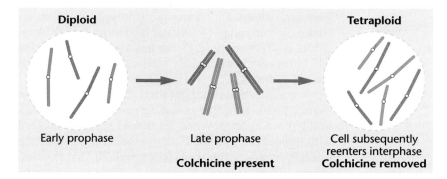

■ Figure 8–13 The potential involvement of colchicine in doubling the chromosome number, as occurs during the production of an autotetraploid. Two pairs of homologous chromosomes are followed. While each chromosome has replicated its DNA earlier during interphase, the chromosomes do not appear as double structures until late prophase. When anaphase fails to occur, the chromosome number doubles if the cell reenters interphase.

Allopolyploidy

Polyploidy may also result from hybridization of two closely related species. If a haploid ovum from a species with chromosome sets *AA* is fertilized by sperm from a species with sets *BB*, the resulting hybrid is *AB*, where $A = a_1, a_2, a_3, \ldots, a_n$ and $B = b_1, b_2, b_3, \ldots, b_n$. The hybrid may be sterile because of its inability to produce viable gametes. More often, this occurs because some or all of the *a* and *b* chromosomes cannot synapse in meiosis, since they lack sufficient homology to one another. As a result, unbalanced genetic conditions occur. If, however, the new *AB* genetic combination undergoes a natural or induced chromosomal doubling, a fertile *AABB* tetraploid is produced. These events are illustrated in Figure 8–14. Since this polyploid contains the equivalent of four haploid genomes and because the hybrid contains unique genetic information compared with either parent, such an organism is called an **allotetraploid** (from the Greek word *allo*, meaning other or different). An equivalent term, **amphidiploid**, is also used to describe this situation in which a hybrid organism contains two complete diploid genomes. This latter term is preferred when the original species are known.

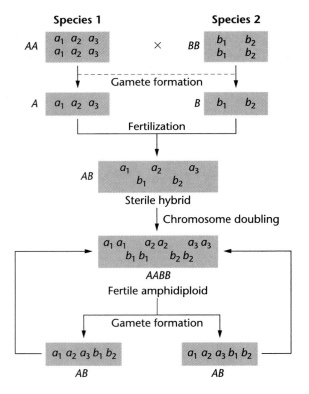

■ Figure 8–14   The origin and propagation of an allotetraploid. Species 1 contains genome *A,* consisting of three nonhomologous chromosomes, a_1, a_2, and a_3. Species 2 contains genome *B,* consisting of two nonhomologous chromosomes, b_1 and b_2. Following fertilization between members of the two species and chromosome doubling, a fertile allotetraploid containing four complete haploid genomes (*AABB*) is formed.

■ Figure 8–15   The amphidiploid form of *Gossypium*, the cultivated cotton plant.

A classical example of allotetraploidy in plants is the cultivated species of American cotton, *Gossypium* (Figure 8–15). This species has 26 pairs of chromosomes: 13 are large and 13 are much smaller. When it was discovered that Old World cotton had only 13 pairs of large chromosomes, allopolyploidy was suspected. After an examination of wild American cotton revealed 13 pairs of small chromosomes, this speculation was strengthened. J. O. Beasley was able to reconstruct the origin of cultivated cotton experimentally. He crossed the Old World strain with the wild American strain and then treated the hybrid with colchicine to double the chromosome number. The result of these treatments was a fertile amphidiploid variety of cotton. It contained 26 pairs of chromosomes and characteristics similar to the cultivated variety.

Amphidipoids often exhibit characteristics of both parental species. Successful commercial hybridization has been performed using the grasses wheat and rye. Wheat (genus *Triticum*) has a basic haploid genome of 7 chromosomes. Cultivated autopolyploids exist, including tetraploid ($4n = 28$) and hexaploid ($6n = 42$) varieties. Rye (genus *Secale*) also has a genome consisting of 7 chromosomes. The only cultivated species of rye is diploid ($2n = 14$).

Using a technique that parallels the steps outlined in Figure 8–14, geneticists have produced various allopolyploid hybrids. When tetraploid wheat is crossed with diploid rye and the F_1 is treated with colchicine, a hexaploid variety ($6n = 42$) is derived. The hybrid, designated *Triticale* (see Figure 1–9), represents a new genus. Fertile hybrid varieties derived from various wheat and rye species may be crossed together or backcrossed. These crosses have created many variations of the genus *Triticale*.

The hybrid plants demonstrate characteristics of both wheat and rye. For example, certain hybrids combine the high protein content of wheat with the high content of the amino acid lysine of rye. The lysine content is low in wheat and thus is a limiting nutritional factor. Wheat is considered a high-yielding grain, whereas rye is noted for its versatility of growth in unfavorable environments. *Triticale* species, combining both traits, have the potential for significantly increasing grain production. Programs designed to improve crops through hybridization have long been underway in several underdeveloped countries of the world, as discussed in Chapter 1.

Recall that in Chapter 6 we discussed the use of **somatic cell hybrids** to map human genes. This technique has also been applied to the production of allopolyploid plants (Figure 8–16). Cells from the developing leaves of plants can be isolated and treated to remove their cell wall, resulting in **protoplasts**. These altered cells can be maintained in culture and stimulated to fuse with protoplasts from other plants, producing somatic cell hybrids.

When cells from different plant species are fused in this way, hybrid cells can be produced with unique chromosome combinations. Since protoplasts can be in-

duced to divide and differentiate into stems that develop leaves, the potential for producing allopolyploids is available in the research laboratory. In some cases, entire plants may be derived from cultured protoplasts. If only stems and leaves are produced, these may be grafted onto the stem of another plant. If flowers are formed, fertilization may yield mature seeds, which, upon germination, yield an allopolyploid plant.

Variation in Chromosome Structure and Arrangement: An Overview

The second general class of chromosome aberrations includes structural changes that delete, add, or rearrange substantial portions of one or more chromosomes. Included in this category are **deletions** and **duplications** of genes or part of a chromosome and rearrangements of genetic material in which a chromosome segment is either **inverted**, exchanged with a segment of nonhomologous chromosome, or merely

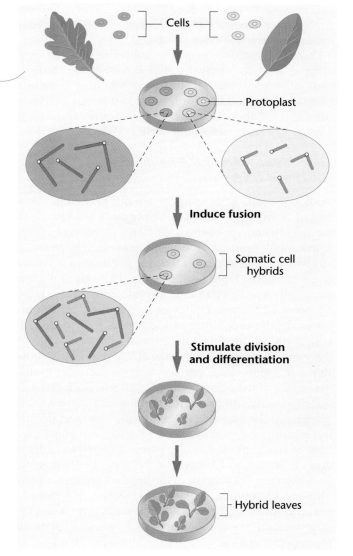

■ Figure 8–16 Application of the somatic cell hybridization technique in the production of an amphidiploid. Cells from the leaves of two species of plants are removed and cultured. The cell walls are digested away and the resultant protoplasts are induced to undergo cell fusion. The hybrid cell is selected and stimulated to divide and differentiate, as illustrated in the photograph. An amphidiploid has a complete set of chromosomes from each parental cell type and displays phenotypic characteristics of each. Two pairs of chromosomes from each species are depicted.

transferred to one, leading to a **translocation**. Before discussing these aberrations, we will present several general statements pertaining to them.

In most instances, these changes are due to one or more breaks along the axis of a chromosome, followed by either the loss or rearrangement of some genetic material. Chromosomes can break spontaneously, but the rate of breakage may increase in cells exposed to chemicals or radiation. Although the ends of chromosomes, the **telomeres**, do not readily fuse with ends of "broken" chromosomes or with other telomeres, the ends produced at points of breakage are "sticky" and can rejoin other broken ends. If a breakage and rejoining event does not merely reestablish the original relationship, and if the alteration occurs in germ plasm, the gametes will contain the structural rearrangement, which will be heritable.

If the aberration is found in one homolog but not the other, unusual but characteristic pairing configurations are formed during meiotic synapsis. These patterns are useful in identifying the type of change that has occurred. If no loss or gain of genetic material occurs, individuals bearing the aberration "heterozygously" are likely to be unaffected phenotypically. However, the unusual pairing arrangements often lead to gametes that are duplicated or deficient for some chromosomal regions. When this occurs, the offspring of "carriers" of certain aberrations often have an increased probability of demonstrating phenotypic manifestations.

Deletions

A deletion refers to the condition whereby a portion of a chromosome is lost. (It is also sometimes called a **deficiency**.) The deletion may be very small and involve just part of a gene, or it may be larger and encompass a few genes or more. The missing region may include either end of the chromosome, or it may be an internal portion of the chromosome. These situations are called **terminal** and **intercalary deletions**, respectively. Both result from one or more breaks in the chromosome. During synapsis between a chromosome with a large intercalary deletion and a normal complete homolog, the unpaired region of the normal homolog must "buckle out." Such a configuration is called a **deficiency loop** or **compensation loop**. Such a loop can be visualized in the large polytene chromosomes derived from insect salivary gland cells. Polytene chromosomes, discussed in detail in Chapter 2, display a banding pattern that is characteristic for each chromosome, making them useful in cytological studies. The formation of such a loop in *Drosophila* polytene chromosomes is diagrammed in Figure 8–17.

Duplications

When any part of the genetic material—a single locus or a large piece of a chromosome—is present more

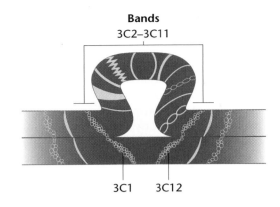

Bands
3C2–3C11

3C1   3C12

■ Figure 8–17   Deletion loop formed in salivary chromosomes of the X chromosome of *Drosophila melanogaster*. The deletion encompasses bands 3C2 through 3C11.

than once in the genome, it is called a duplication. As in deletions, pairing in heterozygotes may produce a compensation loop. Duplications may arise as the result of unequal crossing over between synapsed chromosomes during meiosis (Figure 8–18) or through a replication error prior to meiosis. In the former case, both a duplication and a deletion are produced.

Three interesting aspects of duplications will be considered. First, they may result in gene redundancy. Second, as with deletions, duplications may produce phenotypic variation. Third, according to one convincing theory, duplications have also been an important source of genetic variability during evolution.

Gene Redundancy and Amplification— Ribosomal RNA Genes

Although many gene products are not needed in every cell of an organism, other gene products are known to be essential components of all cells. For example, ribosomal RNA must be present in abundance in order to support protein synthesis. The more metabolically active a cell is, the higher is the demand for this molecule. We might hypothesize that a single copy of the gene encoding rRNA may be inadequate in many cells. Studies using the technique of molecular hybridization, which allows the determination of the percentage of the genome coding for specific RNA sequences, show that our hypothesis is correct! Indeed, there are multiple copies of the genes coding for rRNA. Such DNA is called **rDNA**, and the general phenomenon is called **gene redundancy**. For example, in the common intestinal bacterium *Escherichia coli (E. coli)*, about 0.4 percent of the haploid genome consists of rDNA. This is equivalent to 5 to 10 copies of the gene. In *Drosophila melanogaster*, 0.3 percent of the haploid genome, equivalent to 130 copies, consists of rDNA. Although the presence of multiple copies of the same gene is not restricted to those coding for rRNA, we will focus on them in this section.

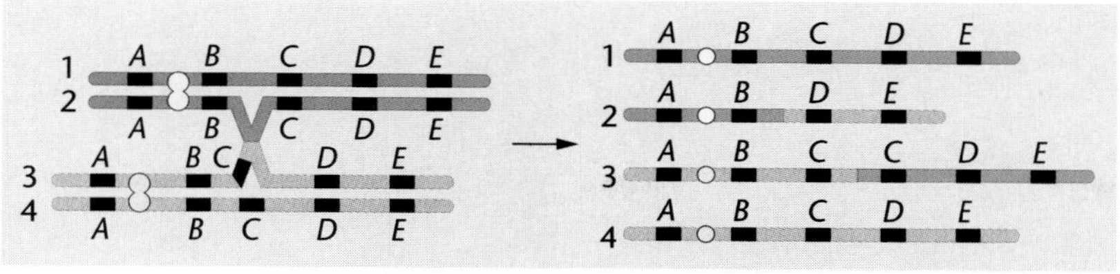

■ Figure 8–18 The origin of duplicated and deficient regions of chromosomes as a result of unequal crossing over. The tetrad at the left is mispaired during synapsis. A single crossover between chromatids 2 and 3 results in deficient and duplicated chromosomal regions (see chromosomes 2 and 3, respectively, on the right). The two chromosomes that are not involved in the crossover event remain normal in their gene sequence and content.

In some cells, particularly oocytes, even the normal redundancy of rDNA may be insufficient to provide adequate amounts of rRNA and ribosomes. Oocytes store abundant nutrients in the ooplasm for use by the embryo during early development. In addition, more ribosomes are included in the oocytes than in any other cell type. By considering how the amphibian *Xenopus laevis* acquires this abundance of ribosomes, we will see a second way in which the amount of rRNA is increased. This phenomenon is referred to as **gene amplification**.

The genes that code for rRNA are located in an area of the chromosome known as the **nucleolar organizer region (NOR)**. The NOR is intimately associated with the nucleolus, which is a processing center for ribosome production. Molecular hybridization analysis has shown that each NOR in the frog *Xenopus* contains the equivalent of 400 redundant gene copies coding for rRNA. Even this number of genes is apparently inadequate to synthesize the vast amount of ribosomes that must accumulate in the amphibian oocyte to support development following fertilization.

To further amplify the number of rRNA genes, the rDNA is selectively replicated, and each new set of genes is released from its template. Because each new copy is equivalent to an NOR, multiple small nucleoli are formed around each NOR in the oocyte. As many as 1500 of these "micronucleoli" have been observed in a single oocyte. If we multiply the number of micronucleoli (1500) by the number of gene copies in each NOR (400), we see that amplification in *Xenopus* oocytes can result in over half a million gene copies! If each copy is transcribed only 20 times during the maturation of the oocyte, in theory, sufficient copies of rRNA may be produced to result in well over 1 million ribosomes.

The **Bar** *Eye Mutation in* **Drosophila**

Duplications may cause phenotypic variation that at first might appear to be caused by a simple gene mutation. The *Bar* eye phenotype in *Drosophila* is a classical example. Instead of the normal oval eye shape, *Bar*-eyed flies have narrow, slitlike eyes. This phenotype is inherited in the same way as a dominant X-linked mutation.

In the early 1920s, Alfred H. Sturtevant and Thomas H. Morgan discovered and investigated this "mutation." As illustrated in Figure 8–19, normal wild-type females (B^+/B^+) have about 800 facets in each eye. Heterozygous females (B/B^+) have about 350 facets, while homozygous females (B/B) average only about 70 facets. Females are occasionally recovered with even fewer facets and are designated as *double Bar* (B^D/B^+).

About ten years later, Calvin Bridges and Herman J. Muller compared the polytene X chromosome banding pattern of the *Bar* fly with that of the wild-type fly. Such chromosomes contain specific banding patterns that have been well categorized into regions. Their studies reveal that one copy of region 16A of the X chromosome is present in wild-type flies and that this region is duplicated in *Bar* flies and triplicated in *double Bar* flies. These observations provide evidence that the Bar phenotype is not the result of a simple chemical change in the gene, but is instead a duplication.

The Role of Gene Duplication in Evolution

One of the most intriguing aspects of the study of evolution is the consideration of the mechanisms for genetic variation. The origin of unique gene products present in more recently evolved organisms but absent in ancestral forms is a topic of particular interest. In other words, how do "new" genes arise?

In 1970, Susumo Ohno published a provocative monograph, *Evolution by Gene Duplication,* in which he suggested that gene duplication is essential to the origin of new genes during evolution. Ohno's thesis is based on the supposition that the gene products of unique genes, present as only a single copy in the genome, are indispensable to the survival of members of any species during evolution. Therefore, unique genes are not free to accumulate mutations sufficient to alter their primary function and give rise to new genes.

However, if an essential gene were to become duplicated in a germ cell, the new copy would be inherited in all future generations. Because it is an extra copy, major mutational changes in it are tolerated because the original gene still provides the genetic information for its essential function. The duplicated copy is now free to

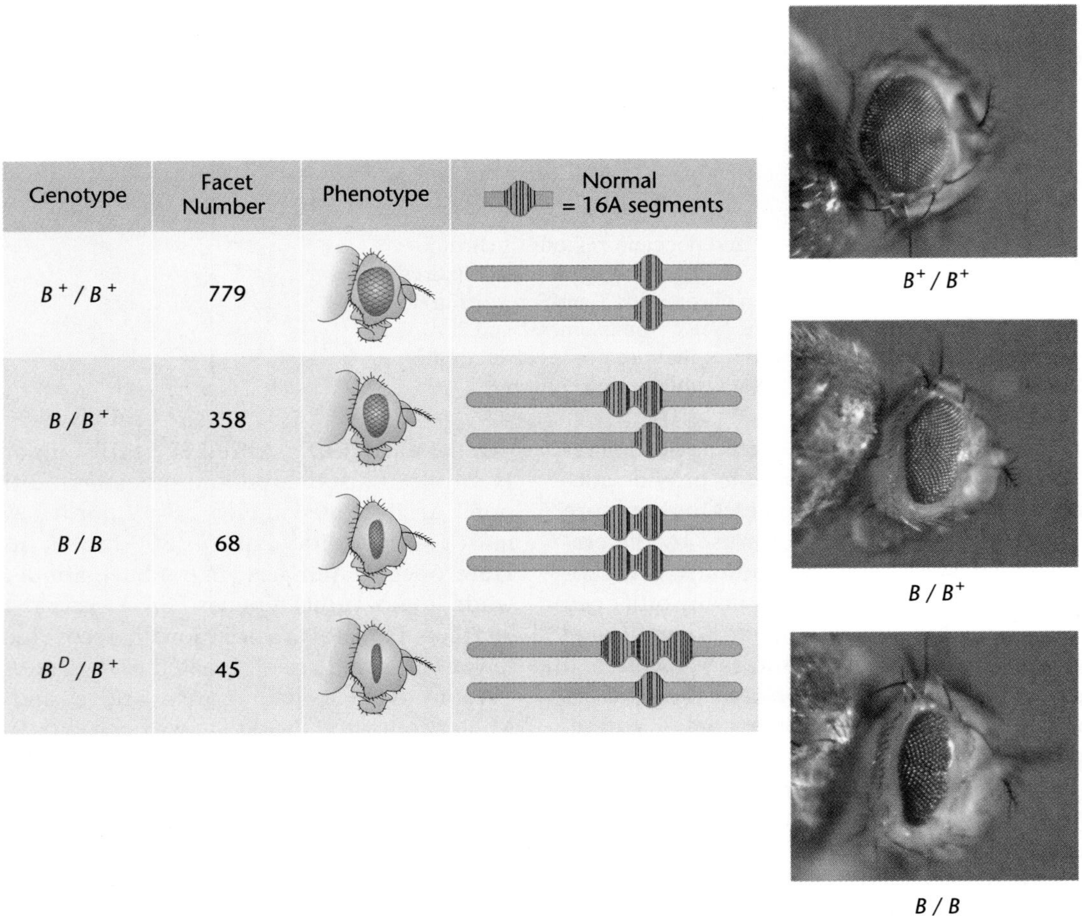

Genotype	Facet Number	Phenotype	Normal = 16A segments	
B^+ / B^+	779			
B / B^+	358			
B / B	68			
B^D / B^+	45			

■ Figure 8–19   Summary of the duplication genotypes and resultant *Bar* eye phenotypes. Photographs illustrate the actual *Bar* eye phenotypes compared to wild type (*B+/B+*).

undergo large numbers of mutational changes within the organisms of any species over long periods of time. Over short intervals, the new genetic information may be of no practical advantage. However, over long periods of evolutionary time, the gene may change sufficiently so that its product assumes a divergent role in the cell. The new function may impart an "adaptive" advantage to organisms carrying such unique genetic information, enhancing their fitness. Ohno outlined a mechanism through which sustained genetic variability may have originated.

Ohno's thesis is supported by the discovery of genes that have a substantial amount of their DNA sequence in common, but whose gene products are distinct. For example, trypsin and chymotrypsin fit this description, as do myoglobin and hemoglobin. The DNA sequence is so similar (homologous) in each case that one may conclude that members of each pair of genes arose from a common ancestral gene through duplication. During evolution, the related genes diverged sufficiently that their products became unique.

Other support includes the presence of **gene families**, groups of contiguous genes whose products perform the same function. Again, members of a family show DNA sequence homology sufficient to conclude that they share a common origin. The various types of human hemoglobin polypeptide chains are examples. Other examples include the immunologically important **T-cell receptors** and antigens encoded by the **major histocompatibility complex** (see Chapter 19). As more and more genes and protein molecules are sequenced, the list grows.

Inversions

The **inversion**, another class of structural variation, is a type of chromosomal aberration in which a segment of a chromosome is turned around 180° within a chromosome. An inversion does not involve a loss of genetic information, but simply rearranges the linear gene sequence. An inversion requires breaks at two points along the length of the chromosome and subsequent reinsertion of the inverted segment. Figure 8–20 illustrates how an inversion might arise. By forming a chromosomal loop prior to breakage, the newly created "sticky ends" are brought close together and rejoined.

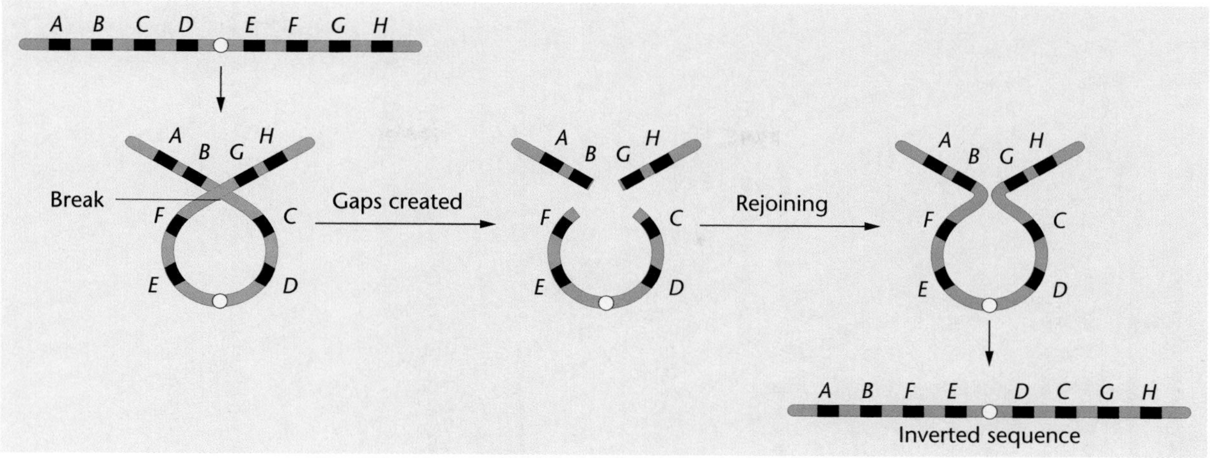

■ Figure 8–20   One possible origin of a pericentric inversion.

The inverted segment may be short or quite long and may or may not include the centromere. If the centromere is not part of the rearranged chromosome segment, the inversion is said to be **paracentric**. If the centromere is part of the inverted segment, the term **pericentric** describes the inversion, which is the type illustrated in Figure 8–20.

Although inversions may seem to have a minimal impact on the genetic material, their consequences are of interest to geneticists. Organisms that are heterozygous for inversions may produce aberrant gametes that have a major impact on their offspring.

Consequences of Inversions During Gamete Formation

If only one member of a homologous pair of chromosomes has an inverted segment, normal linear synapsis during meiosis is not possible. Organisms with one inverted chromosome and one noninverted homolog are called **inversion heterozygotes**. Pairing between two such chromosomes in meiosis may be accomplished only if they form an **inversion loop**.

If crossing over does not occur within the inverted segment of the inversion loop, the homologs will segregate and result in two normal and two inverted chromatids that are distributed into gametes. However, if crossing over occurs within the inversion loop, abnormal chromatids are produced. The effects of single exchange events within a paracentric inversion is diagrammed in Figure 8–21(a).

As in any meiotic tetrad, a single crossover between nonsister chromatids produces two parental chromatids and two recombinant chromatids. One recombinant chromatid is **dicentric** (two centromeres), and one recombinant chromatid is **acentric** (lacking a centromere). Both contain duplications and deletions of chromosome segments as well. During anaphase, an acentric chromatid moves randomly to one pole or the other or may be lost, while a dicentric chromatid is

pulled in two directions. This polarized movement produces **dicentric bridges** that are cytologically recognizable. A dicentric chromatid will usually break at some point so that part of the chromatid goes into one gamete and part into another gamete during the reduction divisions. Therefore, gametes containing either recombinant chromatid are deficient in genetic material. When such a gamete participates in fertilization, the zygote most often develops abnormally, if at all.

A similar chromosomal imbalance is produced as a result of a crossover event between a chromatid bearing a pericentric inversion and its noninverted homolog, as shown in Figure 8–21(b). The recombinant chromatids that are directly involved in the exchange have duplications and deletions. However, no acentric or dicentric chromatids are produced. Gametes receiving these chromatids also produce inviable embryos following their participation in fertilization.

Because fertilization events involving these aberrant chromosomes do not produce viable offspring, it *appears* as if the inversion suppresses crossing over because crossover gametes are not recovered in the offspring. *Actually*, in inversion heterozygotes, the inversion has the effect of *suppressing the recovery of crossover products when chromosome exchange occurs within the inverted region*. If crossing over always occurred within a paracentric or pericentric inversion, 50 percent of the gametes would be ineffective. The viability of the resulting zygotes is therefore greatly diminished. Furthermore, up to one-half of the viable gametes have the inverted chromosome, and the inversion will be perpetuated within the species. The cycle will be repeated continuously during meiosis in future generations.

Translocations

Translocation, as the name implies, involves the movement of a segment of a chromosome to a new place in the genome. Translocation may occur within a single

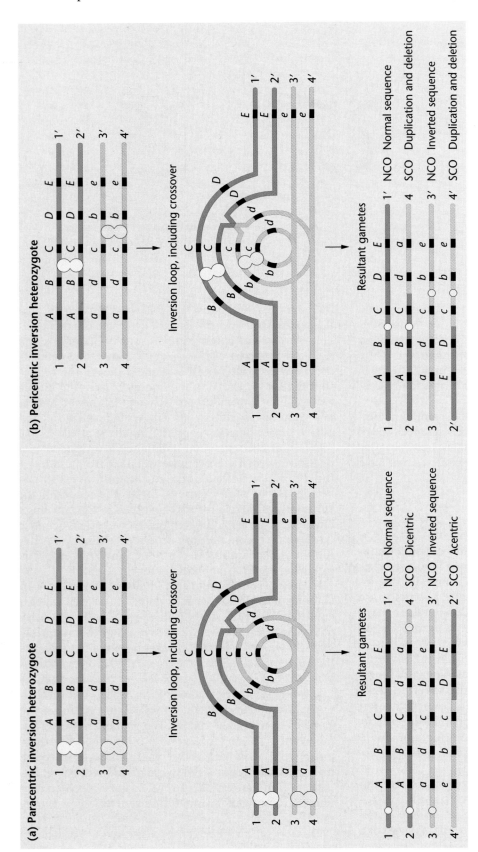

Figure 8–21 The effects of a single crossover with an inversion loop in cases involving (a) a paracentric inversion and (b) a pericentric inversion. In (a), two altered chromosomes are produced, one that is acentric and one that is dicentric. Both chromosomes also contain duplicated and deficient regions. In (b), two altered chromosomes are produced, both with duplicated and deficient regions.

chromosome or between nonhomologous chromosomes. The exchange of segments between two nonhomologous chromosomes is a type of structural variation called a **reciprocal translocation**. The possible origin of a relatively simple reciprocal exchange is illustrated in Figure 8–22(a). The least complex way for this event to occur is for two nonhomologous chromosome arms to come close to each other so that an exchange between them is facilitated. For this type of translocation, only two breaks are required, one on each chromosome. If the exchange includes internal chromosome segments, four breaks are required, two on each chromosome.

The genetic consequences of reciprocal translocations are, in several instances, similar to those of inversions.

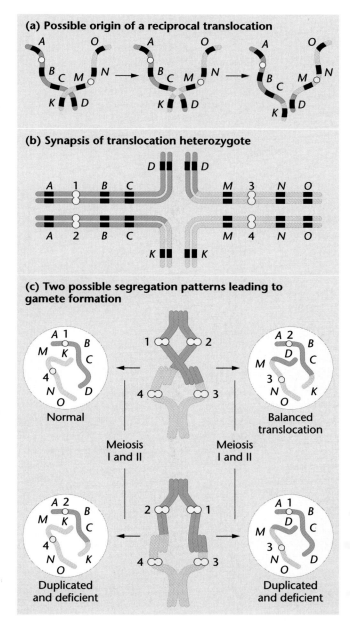

(a) Possible origin of a reciprocal translocation

(b) Synapsis of translocation heterozygote

(c) Two possible segregation patterns leading to gamete formation

Normal

Balanced translocation

Meiosis I and II

Meiosis I and II

Duplicated and deficient

Duplicated and deficient

■ Figure 8–22 Origin, synapsis, and gamete formation in a reciprocal translocation.

For example, genetic information is not lost or gained. Rather, there is only a rearrangement of genetic material. The presence of a translocation does not, therefore, directly alter viability of individuals bearing it.

Homologs that are heterozygous for a reciprocal translocation undergo unorthodox synapsis during meiosis. As shown in Figure 8–22(b), pairing results in a crosslike configuration. As with inversions, genetically unbalanced gametes are also produced as a result of this unusual alignment during meiosis. In the case of translocations, however, aberrant gametes are not necessarily the result of crossing over. To see how unbalanced gametes are produced, focus on the homologous centromeres in Figures 8–22(b) and 8–22(c). According to the principle of independent assortment, the chromosome containing centromere 1 will migrate randomly toward one pole of the spindle during the first meiotic anaphase; it will travel with *either* the chromosome having centromere 3 or the chromosome having centromere 4. The chromosome with centromere 2 will move to the other pole along with *either* the chromosome containing centromere 3 *or* centromere 4. This results in four potential meiotic products. The 1,4 combination contains chromosomes that are not involved in the translocation. The 2,3 combination, however, contains translocated chromosomes. These contain a complete complement of genetic information and are balanced. The other two potential products, the 1,3 and 2,4 combinations, contain chromosomes displaying duplicated and deleted segments.

When incorporated into gametes, the resultant meiotic products are genetically unbalanced. If they participate in fertilization, lethality often results. As few as 50 percent of the progeny of parents that are heterozygous for a reciprocal translocation may survive. This condition, called **semisterility**, has an impact on the reproductive fitness of organisms, thus playing a role in evolution. Furthermore, in humans, such an unbalanced condition results in partial monosomy or trisomy, leading to a variety of birth defects.

Translocations in Humans: Familial Down Syndrome

Research performed since 1959 has revealed numerous translocations in members of the human population. One common type of translocation involves breaks at the extreme ends of the short arms of two nonhomologous acrocentric chromosomes. These small segments are lost, and the larger segments fuse at their centromeric region. This type of translocation produces a new, large submetacentric or metacentric chromosome, often called a **Robertsonian translocation**.

Just such a translocation accounts for cases in which Down syndrome is inherited or familial. Earlier in this chapter we pointed out that most instances of Down syndrome are due to trisomy 21. This chromosome composition results from nondisjunction during meiosis of one parent. Trisomy accounts for over 95 percent of all cases of Down syndrome. In such instances, the chance of the same parents producing a second afflicted child is extremely low. However, in the remaining families with a Down child, the syndrome occurs in a much higher frequency over several generations.

Cytogenetic studies of the parents and their offspring from these unusual cases explain the cause of familial Down syndrome. Analysis reveals that one of the parents contains a **14/21 D/G translocation** (Figure 8–23). That is, one parent has the majority of the G-group chromosome 21 translocated to one end of the D-group chromosome 14. This individual is normal even though

he or she has only 45 chromosomes. Following meiosis, one-fourth of the individual's gametes will have two copies of chromosome 21: a normal chromosome and a second copy translocated to chromosome 14. When such a gamete is fertilized by a standard haploid gamete, the resulting zygote has 46 chromosomes but three copies of chromosome 21. These individuals exhibit Down syndrome. Other potential surviving offspring contain either the standard diploid genome (without a translocation) or the balanced translocation like the parent. Both cases result in normal individuals. Knowledge of translocations has allowed geneticists to resolve the seeming paradox of an inherited trisomic phenotype in an individual with an apparent diploid number of chromosomes.

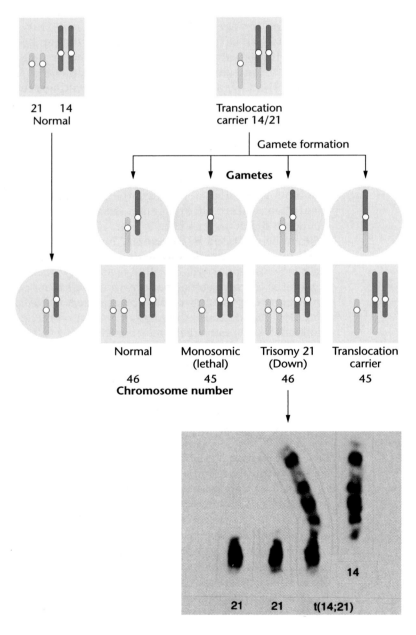

■ Figure 8–23 Illustration of chromosomal involvement in familial Down syndrome. The photograph illustrates the relevant chromosomes from a trisomy 21 offspring produced by a translocation carrier parent.

Fragile Sites in Humans

We conclude this chapter by briefly discussing the results of an interesting discovery made around 1970 during observations of metaphase chromosomes prepared following human cell culture. In cells derived from certain individuals, a specific area along one of the chromosomes failed to stain, giving the appearance of a gap. In other individuals whose chromosomes displayed or expressed such morphology, the gap appeared in different positions within the set of chromosomes. Such areas eventually became known as **fragile sites**, since they were considered susceptible to chromosome breakage when cultured in the absence of certain chemicals such as folic acid, which is normally present in the medium. Fragile sites were at first considered curiosities, until a strong association was subsequently shown to exist between one of the sites and a form of mental retardation.

The cause of the fragility at these sites is unknown. Because they appear to represent points along the chromosome that are susceptible to breakage, these sites may indicate regions where the chromatin is not tightly coiled. It should be noted that almost all studies of fragile sites have been carried out *in vitro*, and it is unknown how this phenotype is expressed *in vivo*.

Fragile X Syndrome (Martin–Bell Syndrome)

Most fragile sites do not appear to be associated with any clinical syndrome. However, individuals bearing a folate-sensitive site on the X chromosome (Figure 8–24) exhibit the **fragile X syndrome** (or **Martin-Bell syndrome**), the most common form of inherited mental retardation. This syndrome affects about 1 in 1250 males and 1 in 2500 females. Because it is a dominant trait, females carrying only one fragile X chromosome can be mentally retarded. Fortunately, the trait is not fully expressed, as only about 30 percent of fragile X females

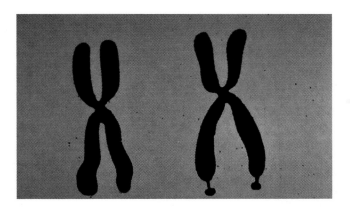

■ Figure 8–24 A normal human X chromosome (left) contrasted with a fragile X chromosome. The "gap" (at the bottom of the chromosome) is associated with the fragile X syndrome, including mental retardation.

are retarded, whereas about 80 percent of fragile X-bearing males are mentally retarded. In addition to mental retardation, affected males have characteristic long, narrow faces with protruding chins, enlarged ears, and increased testicular size.

A gene that spans the fragile site may be responsible for this syndrome. This gene, known as *FMR-1*, is one of a growing number of genes in which a sequence of three nucleotides is repeated many times, expanding the size of one end of the gene. This phenomenon, called **trinucleotide repeats**, has been recognized in other human disorders, including Huntington disease. The number of repeats varies immensely within the human population, and this number correlates directly with expression of fragile X syndrome. Normal individuals have up to 50 repeats, whereas those with 50 to 200 repeats are considered to be "carriers" of the disorder. Above 200 repeats leads to expression of the syndrome.

The fragile X breakpoint is contained in this variable-length region, and full expression of the phenotype depends on both an increase in length and chemical modifications of the DNA nucleotides in this gene segment. The *FMR-1* gene is known to be expressed in the brain, but the exact nature and function of the encoded protein awaits further research. The gene may encode a DNA-binding protein based on the analysis of its nucleotide sequence. Although other genes may turn out to be involved in fragile X syndrome, both the structure and the function of the FMR-1 gene make it the leading candidate to provide insights into the molecular basis of this common form of mental retardation.

Fragile X syndrome illustrates an important phenomenon known as **genetic anticipation**, whereby expression of a malady becomes more severe in successive generations. In the case of this disorder, the number of trinucleotide repeats increases as the affected X chromosome passes through the gametes to offspring. Although it is not yet clear how an expansion of these repeats affects the severity, and often the onset, of phenotypic expression, the correlation is extremely strong and is not limited to fragile X syndrome. Similar findings have been made in the genes responsible for **Huntington disease** (see Chapter 4) and **myotonic dystrophy**, the most common form of adult muscular dystrophy. In both human disorders, trinucleotide repeats are also involved. We will return to a discussion of these repeats in Chapter 15.

The FHIT Gene and Human Lung Cancer

A second significant association between a fragile site and a human condition was reported in 1996 by Carlo Croce, Kay Huebner, and their colleagues, who showed that a gene located on chromosome 3, *FHIT* (standing for *f*ragile *hi*stidine *t*riad), is altered in cells taken from tumors recovered from individuals with lung cancer. In

80 percent of the tumors and cell lines derived from them, an abnormal *FHIT* gene was exhibited. This gene is part of the fragile area of the autosome designated *FRA3B*, which has also been linked to cancer of the esophagus, colon, and stomach.

FHIT may be more fragile in some people than in others, leading to a susceptibility to lung cancer following environmental insults to lung tissue by agents such as cigarette smoke. The research team found a variety of mutant sequences of the gene in cells derived from the tumors, in which the gene had apparently been broken and then fused back together incorrectly. These sequences potentially lead to the synthesis of an altered gene product or to no product being synthesized.

These are very recent findings, and ones that will undoubtedly yield many more significant details in the future. The impact of the discovery is indeed great. An explanation at the molecular genetic level may be provided that explains why some smokers are at risk for lung cancers, other smokers are not, and why some people who have never smoked nevertheless develop lung cancer. Furthermore, the potential for **genetic screening** is clearly on the scientific horizon.

Chapter Summary

1. Investigations into the uniqueness of each organism's chromosomal constitution have further enhanced our understanding of genetic variation. Alterations of the precise diploid content of chromosomes are referred to as chromosomal aberrations or chromosomal mutations.

2. In humans, the study of individuals with altered sex chromosome compositions has established that the Y chromosome is responsible for male differentiation. The absence of the Y leads to female differentiation. Similar studies in *Drosophila* have excluded the Y in such a role, instead demonstrating that a balance between the number of X chromosomes and sets of autosomes is the critical factor.

3. Dosage compensation mechanisms limit the expression of X-linked genes in females who have two X chromosomes, as compared to males who have only one X. In mammals, compensation is achieved as a result of the inactivation of either the maternal or paternal X early in development. This process results in the formation of Barr bodies in female somatic cells.

4. The Lyon hypothesis states that, early in development, inactivation is random between the maternal and paternal X. All subsequent progeny cells inactivate the same X as their progenitor cell. Mammalian females thus develop as genetic mosaics with respect to their expression of heterozygous X-linked alleles.

5. Deviations from the diploid chromosomal number, or mutations in the structure of the chromosome, are inherited in predictable Mendelian fashion; they often result in inviable organisms or substantial changes in the phenotype.

6. Aneuploidy is the gain or loss of one or more chromosomes from the diploid content, resulting in conditions of monosomy, trisomy, tetrasomy, and so on. Studies of monosomy, as in Turner syndrome, and trisomy, as in Down syndrome, have increased our understanding of the delicate genetic balance that must exist in order for normal development to occur.

7. When complete sets of chromosomes are added to the diploid genome, polyploidy is created. These sets may have identical or diverse genetic origin, creating either autopolyploidy or allopolyploidy, respectively.

8. Large segments of the chromosome may be modified by deletions or duplications. Deletions may produce serious conditions such as the cri-du-chat syndrome in humans, while duplications may be particularly important as a source of redundant or unique genes.

9. Inversions and translocations, while altering the gene order along chromosomes, cause no loss of genetic information or deleterious effects initially. However, individuals who are heterozygous for inversions or translocations may produce genetically abnormal gametes following meiosis, often causing lethality.

10. Fragile sites in human mitotic chromosomes have sparked research interest because one such site on the X chromosome is associated with fragile X syndrome, the most common form of inherited mental retardation. Another fragile site, located on chromosome 3, has been linked to lung cancer.

Key Terms

acentric chromosome, 173
allopolyploidy, 167
allotetraploid, 168
amniocentesis, 165
amphidiploid, 168
aneuploidy, 155
anhidrotic ectodermal dysplasia, 159
autopolyploidy, 167
autotriploid, 167
Barr body, 158
chorionic villus sampling (CVS), 165
chromosome aberration, 155
chromosome mutation, 155
colchicine, 167
color blindness, red-green, 159
compensation loop, 170
cri-du-chat (cry of the cat) syn-
      drome, 163
deficiency, 170
deficiency loop, 170
deletion, 169
dicentric bridge, 173
dicentric chromosome, 173
dosage compensation, 158
Down syndrome, 164
duplication, 169
Edwards syndrome, 166
euploidy, 155
familial Down syndrome, 166
fragile site, 177
fragile X syndrome (Martin-Bell
      syndrome), 177
gene amplification, 171
gene family, 172
gene redundancy, 170

genetic anticipation, 177
genetic counseling, 165
genetic screening, 178
genic balance theory, 161
Haplo-IV, 162
Huntington disease, 177
intercalary deletion, 170
intersex, 161
inversion, 169
inversion heterozygote, 173
inversion loop, 173
Klinefelter syndrome, 156
Lyon hypothesis, 159
major histocompatibility
      complex, 172
metafemale, 161
metamale, 161
monosomy, 155
mosaic, 156
myotonic dystrophy, 177
nondisjunction, 161
nucleolar organizer region
      (NOR), 171
paracentric inversion, 173
partial monosomy, 163
Patau syndrome, 166
pentaploid, 166
pentasomy, 161
pericentric inversion, 173
polyploidy, 155
protoplast, 169
rDNA, 170
rearrangement, 170
reciprocal translocation, 175
Robertson translocation, 175

segmental deletion, 163
semisterility, 175
sex chromatin body, 158
sex-determining region Y
      (SRY), 158
somatic cell hybrid, 169
T-cell receptor, 172
telomere, 170
terminal deletion, 170
testis-determining factor (TDF), 158
tetraploid, 155
tetrasomy, 161
transgenic mice, 158
translocation, 170
trinucleotide repeats, 177
triploid, 155
trisomy, 155
trisomy 21, 164
Turner syndrome, 156
X-inactivation center (XIC), 160
XIST (*X-inactive specific
      transcript*), 160
14/21 D/G translocation, 176
45,X, 156
46,5p–, 163
47,XXX, 156
47,XXY, 156
47,XYY, 157
47,13+, 166
47,18+, 166
47,21+, 164
48,XXXY, 156
48,XXYY, 156
49,XXXXY, 156
49,XXXYY, 156

INSIGHTS and SOLUTIONS

1. In a cross using maize involving three genes, *a,* *b,* and *c,* a heterozygote (*abc/+++*) was test-crossed (to *abc/abc*). Even though the three genes were separated by at best five map units, only two phenotypes were recovered: *abc* and *+++*. Additionally, the cross produced significantly fewer viable plants than expected. Can you propose why no other phenotypes were recovered and why the viability was reduced?

Solution:

One of the two chromosomes may contain an inversion that overlaps all three genes, effectively precluding the recovery of any "crossover" offspring. If this is a paracentric inversion and the genes are clearly separated (assuring that a significant number of crossovers will occur between them), then numerous acentric and dicentric chromosomes will be formed, resulting in the observed reduction in viability.

2. A male *Drosophila* from a wild-type stock was discovered to have only 7 chromosomes, whereas the normal 2*n* number is 8. Close examination revealed that one member of chromosome IV (the smallest chromosome) was attached to (translocated to) the distal end of chromosome II and was missing its centromere, thus accounting for the reduction in chromosome number.

(a) Diagram all members of chromosomes II and IV during synapsis in meiosis I.

Solution:

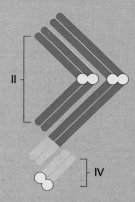

(b) If this male mates with a female with a normal chromosome composition who is homozygous

for the recessive chromosome IV mutation *eyeless (ey),* what chromosome compositions will occur in the offspring regarding chromosomes II and IV?

Solution:

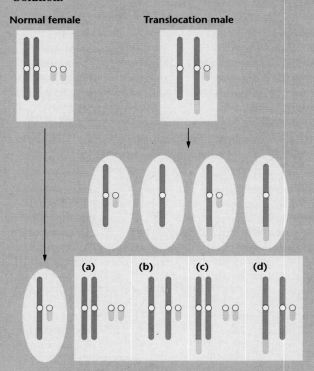

Normal female Translocation male

(c) What phenotypic ratio will result regarding the presence of eyes, assuming all abnormal chromosome compositions survive?

Solution:

(a)—normal (heterozygous)

(b)—eyeless (monosomic, contains chromosome IV from mother)

(c)—normal (heterozygous)

(d)—normal (heterozygous)

The final ratio is 3/4 normal:1/4 eyeless.

Problems and Discussion Questions

1. Contrast the mechanisms underlying sex determination in humans compared to *Drosophila*. Describe the evidence in support of your answer.

2. What is a Barr body? Indicate the expected number of Barr bodies in the following individuals: Klinefelter syndrome; Turner syndrome; 47,XYY; 47,XXX; 47,XXXX.

3. Define the Lyon hypothesis. Relate the effect of this hypothesis to the human retina of a female heterozygous for the X-linked *color-blindness* gene.

4. Considering the information presented in Chapter 4 about lethal genes, speculate as to why the vast majority of human 45,X conceptions fail to survive to birth.

5. It has been suggested that any male-determining genes contained on the Y chromosome in humans should not be located in the limited region that synapses with the X chromosome during meiosis. What might be the outcome if such genes were located in this region?

6. Define and distinguish between the following pairs of terms:

   aneuploidy/euploidy

   monosomy/trisomy

   autopolyploidy/allopolyploidy

   paracentric inversion/pericentric inversion

7. What evidence suggests that Down syndrome is more often the result of nondisjunction during oogenesis rather than during spermatogenesis?

8. What conclusions have been drawn about human aneuploidy as a result of karyotypic analyses of abortuses?

9. Discuss the role of polyploidy in evolution. How is its role different in animal and plant species?

10. What experimental techniques may be used to convert diploid plants to tetraploids?

11. Discuss the origin of cultivated American cotton.

12. For a species with a diploid number of 18, indicate how many chromosomes will be present in the somatic nuclei of individuals who are haploid, triploid, tetraploid, trisomic, and monosomic.

13. Discuss the possible mechanisms involved in the production of deletions, duplications, inversions, and translocations.

14. Contrast the synaptic configurations of homologous pairs of chromosomes in which one member is normal and the other member has sustained a deletion, a duplication, or an inversion.

15. In a cross in *Drosophila*, a female that was heterozygous for the autosomally linked genes *a, b, c, d,* and *e* (*abcde*/+++++) was test-crossed to a male that was homozygous for all recessive alleles. Even though the distance between each of the above loci was at least 3 map units, only four phenotypes were recovered:

Phenotype	No. of Flies
+ + + + +	440
a b c d e	460
+ + + + e	48
a b c d +	52
Total	1000

Why are many expected crossover phenotypes missing? Can any of these loci be mapped from the data given here? If so, determine map distances.

16. Discuss Ohno's hypothesis of the role of gene duplication in the process of evolution.

17. A human female with Turner syndrome also expresses the recessive sex-linked trait hemophilia, as her father did. Which parent underwent nondisjunction during meiosis, giving rise to the gamete responsible for the syndrome?

18. The primrose *Primula kewensis* has 36 chromosomes that are similar in appearance to the chromosomes in two related species, *Primula floribunda* ($2n = 18$) and *Primula verticillata* ($2n = 18$). How could *P. kewensis* arise from these species? How would you describe *P. kewensis* in genetic terms?

19. *Drosophila* may be trisomic for chromosome 4 and remain fertile. The recessive mutation *bent (b)* is located on chromosome 4. Predict the F_1 results of crossing:

   trisomic *bent (b/b/b)* × normal disomic *(B/B)*

20. Cat breeders are aware that kittens expressing the X-linked calico coat pattern are almost invariably females. Why?

21. A couple, looking ahead to a family, were aware that through the past three generations on the male's side, a substantial number of stillbirths had occurred and several malformed babies had been born who died early in childhood. The wife had studied genetics and urged her husband to visit a genetic counseling clinic, where a complete karyotype-banding analysis was performed. Although it was found that he had a normal complement of 46 chromosomes, analysis revealed that one member of the chromosome 1 pair contained an inversion covering 70 percent of its length. The homolog of chromosome 1 and all other chromosomes were normal.

   (a) How would you explain the high incidence of past stillbirths?

(b) What would you predict about the probability of malformations in this couple's children?

(c) What other advice would you give the couple, if they were determined to have children?

22. The *SRY* gene (described in this chapter) is located on the Y chromosome very close to the region that pairs with the X chromosome during meiosis in a male. Given this information, propose a model to explain the generation of unusual males who have two X chromosomes (with an *SRY*-containing piece of the Y attached to one X chromosome).

23. The Xg cell-surface antigen is coded for by a gene located on the X chromosome. There is no equivalent on the Y chromosome. Two codominant alleles of this gene have been identified: Xg^1 and Xg^2. A woman of genotype Xg^2Xg^2 marries a man of genotype Xg^1 and they produce a Klinefelter son of genotype Xg^1Xg^2Y. Using proper genetic terminology, explain briefly how this individual was generated. In which parent and in which meiotic division did the error occur?

24. The marine echiurid worm *Bonellia viridis* is an extreme example of the environment's influence on sex determination. Undifferentiated larvae either remain free-swimming and differentiate into females or they settle on the proboscis of an adult female and become males. If larvae that have been on a female proboscis for a short period are removed and placed in seawater, they develop as intersexes. If larvae are forced to develop in an aquarium where pieces of probosces have been placed, they develop into males. Contrast this mode of sexual differentiation with that of mammals. Suggest further experimentation to elucidate the mechanism of sex determination in *Bonellia*.

25. Shown below are four graphs that plot the percentage of males occurring in various reptile groups versus the temperature that fertilized eggs encounter during early development. Interpret these data as they relate to reptilian sex determination.

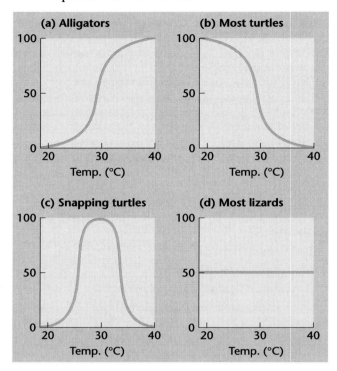

Selected Readings

Barr, M. L. 1966. The significance of sex chromatin. *Int. Rev. Cytol.* 19:35–39.

Beasley, J. O. 1942. Meiotic chromosome behavior in species, species hybrids, haploids, and induced polyploids of *Gossypium. Genetics* 27:25–54.

Blakeslee, A. F. 1934. New jimson weeds from old chromosomes. *J. Hered.* 25:80–108.

Borgaonker, D. S. 1989. *Chromosome variation in man: A catalogue of chromosomal variants and anomalies,* 5th ed. New York: Alan R. Liss.

Boue, A. 1985. Cytogenetics of pregnancy wastage. *Adv. Hum. Genet.* 14:1–58.

Burgio, G. R., et. al., eds. 1981. *Trisomy 21.* New York: Springer-Verlag

Carr, D. H. 1971. Genetic basis of abortion. *Annu. Rev. Genet.* 5:65–80.

Court-Brown, W. M. 1968. Males with an XYY sex chromosome complement. *J. Med. Genet.* 5:341–59.

Croce, C. M. 1996. The *FHIT* gene at 3p14.2 is abnormal in lung cancer. *Cell* 85:17–26.

Davidson, R., Nitowski, H., and Childs, B. 1963. Demonstration of two populations of cells in human females heterozygous for glucose-6-phosphate dehydrogenase variants. *Proc. Natl. Acad. Sci. USA* 50:481–91.

DeArce, M. A., and Kearns, A. 1984. The fragile X syndrome: The patients and their chromosomes. *J. Med. Genet.* 21:84–91.

Erickson, J. D. 1976. The secondary sex ratio of the United States, 1969–71: Association with race, parental ages, birth order, paternal education and legitimacy. *Ann. Hum. Genet. (London)* 40:205–12.

————. 1979. Paternal age and Down syndrome. *Am. J. Hum. Genet.* 31:489–97.

Gorman, M., Kuroda, M., and Baker, B. S. 1993. Regulation of sex-specific binding of maleness dosage compensation protein to the male X chromosome in *Drosophila. Cell* 72:39–49.

Hassold, T. J., et. al. 1980. Effect of maternal age on autosomal trisomies. *Ann. Hum. Genet. (London)* 44:29–36.

Hassold, T., and Jacobs, P. A. 1984. Trisomy in man. *Annu. Rev. Genet.* 18:69–98.

Head, G., May, R., and Pendleton, L. 1987. Environmental determination of sex in the reptiles. *Nature* 329:198–99.

Hecht, F. 1988. Enigmatic fragile sites on human chromosomes. *Trends Genet.* 4:121–22.

Hook, E. B. 1973. Behavioral implications of the human XYY genotype. *Science* 179:139-50.

Hulse, J. H., and Spurgeon, D. 1974. Triticale. *Sci. Am.* (Aug.) 231:72–81.

Jacobs, P. A., et al. 1974. A cytogenetic survey of 11,680 newborn infants. *Ann. Hum. Genet.* 37:359–76.

Kaiser, P. 1984. Pericentric inversions: Problems and significance for clinical genetics. *Hum. Genet.* 68:1–47.

Koopman, P., et al. 1991. Male development of chromosomally female mice transgenic for *Sry*. *Nature* 351:117–121.

Lewis, W. H., ed. 1980. *Polyploidy: Biological relevance.* New York: Plenum Press.

Lyon, M. F. 1988. X-chromosome inactivation and the location and expression of X-linked genes. *Am. J. Hum. Genet.* 42:8–16.

———. 1962. Sex chromatin and gene action in the mammalian X chromosome. *Am. J. Hum. Genet.* 14:135–48.

———. 1988. X-chromosome inactivation and the location and expression of X-linked genes. *Am. J. Hum. Genet.* 42:8–16.

———. 1993. Epigenetic inheritance in mammals. *Trends Genet.* 9:123–28.

Mantell, S. H., Mathews, J. A., and McKee, R. A. 1985. *Principles of plant biotechnology: An introduction to genetic engineering in plants.* Oxford: Blackwell.

McLaren, A. 1988. Sex determination in mammals. *Trends Genet.* 4:153–57.

McMillen, M. M. 1979. Differential mortality by sex in fetal and neonatal deaths. *Science.* 204:89–91.

Obe, G., and Basler, A. 1987. *Cytogenetics: Basic and applied aspects.* New York: Springer-Verlag.

Ohno, S. 1970. *Evolution by gene duplication.* New York: Springer-Verlag.

Oostra, B. A., and Verkerk, A. J. 1992. The fragile X syndrome: Isolation of the *FMR-1* gene and characterization of the fragile X mutation. *Chromosoma* 101:381–87.

Page, D. C., et al. 1987. The sex-determining region of the human Y chromosome encodes a finger protein. *Cell* 51:1091–1104.

Patterson, D. 1987. The causes of Down syndrome. *Sci. Am.* (Aug.) 257:52–61.

Penny, G. D., et al. 1996. Requirement for *Xist* in X chromosome inactivation. *Nature* 379:131–37.

Shepard, J., et al. 1983. Genetic transfer in plants through interspecific protoplast fusion. *Science* 21:683–88.

Simmonds, N. W. ed. 1976. *Evolution of crop plants.* London: Longman.

Sutherland, G. 1985. The enigma of the fragile X chromosome. *Trends Genet.* 1:108–11.

Strickberger, M. W. 1996. *Evolution,* 2nd ed. Boston: Jones and Bartlett.

Taylor, A. I. 1968. 1968. Autosomal trisomy syndromes: A detailed study of 27 cases of Edwards syndrome and 27 cases of Patau syndrome. *J. Med. Genet.* 5:227–52.

Tjio, J. H., and Levan, A. 1956. The chromosome number of man. *Hereditas* 42:1–6.

Wilkins, L. E., Brown, J. X., and Wolf, B. 1980. Psychomotor development in 65 home-reared children with cri-du-chat syndrome. *J. Pediatr.* 97:401–5.

Witkin, H. A., et. al. 1976. Criminality in XYY and XXY men. *Science* 193:547–55.

Yunis, J. J., ed. 1977. *New chromosomal syndromes.* Orlando, FL:Academic Press.

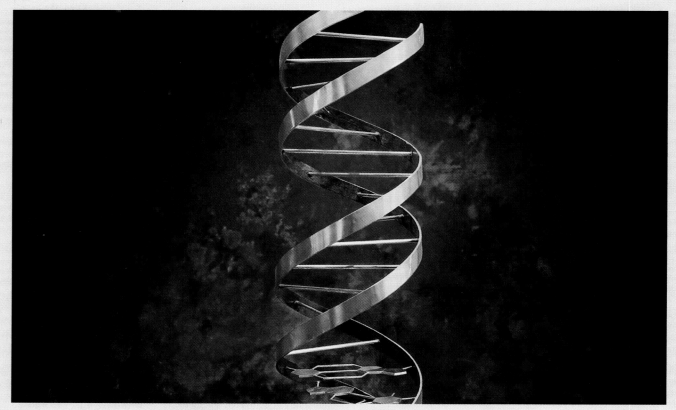

Bronze sculpture of double-helical DNA.

CHAPTER OUTLINE

CHAPTER
9

DNA—The Physical Basis of Life

Chapter Concepts

With few exceptions, the nucleic acid DNA serves as the genetic material in every living thing. The structure of DNA provides the chemical basis for storing and expressing genetic information within cells, as well as transmitting it to future generations. The molecule takes the form of a double-stranded helix united by hydrogen bonds formed between complementary nucleotides. In some viruses, RNA serves as the genetic material.

Earlier in the text, we discussed the presence of genes on chromosomes that control phenotypic traits and the way in which the chromosomes are transmitted through gametes to future offspring. Logically, there must be some form of information contained in genes, which, when passed to a new generation, influences the form and characteristics of the offspring; this is called the **genetic information**. We might also conclude that this same information in some way directs the many complex processes leading to the adult form.

Until 1944 it was not clear what chemical component of the chromosome makes up genes and constitutes the genetic material. Because chromosomes were known to have both a nucleic acid and a protein component, both were considered candidates. In 1944, however, there emerged direct experimental evidence that the nucleic acid DNA serves as the informational basis for the process of heredity.

Once the importance of DNA in genetic processes was realized, work was intensified with the hope of discerning not only the structural basis of this molecule but also the relationship of its structure to its function. Between 1944 and 1953, many scientists sought information that might answer the most significant and intriguing question in the history of biology: How does DNA serve as the genetic basis for the living process? The answer was believed to depend strongly on the chemical structure of the DNA molecule, given the complex but orderly functions ascribed to it.

These efforts were rewarded in 1953 when James Watson and Francis Crick set forth their hypothesis for the double-helical nature of DNA. The assumption that the molecule's functions would be clarified more easily once its general structure was determined proved to be correct. This chapter initially reviews the evidence that DNA is the genetic material and then discusses the elucidation of its structure.

Characteristics of the Genetic Material

The genetic material has several characteristics: **replication, storage of information, expression of that information**, and **variation by mutation**. "Replication" of the genetic material is one facet of cell division, a fundamental property of all living organisms. Once the genetic material of cells has been replicated, it must then be partitioned equally into daughter cells.

The characteristic of "storage" may be viewed as a repository of genetic information present in cells of an organism that may or may not be expressed. It is clear that while most cells contain a complete complement of DNA, at any point in time they express only a part of this genetic potential. For example, bacteria turn many genes on only in response to specific environmental conditions, and turn them off when such conditions change. In vertebrates, skin cells may display active melanin genes but never activate their hemoglobin

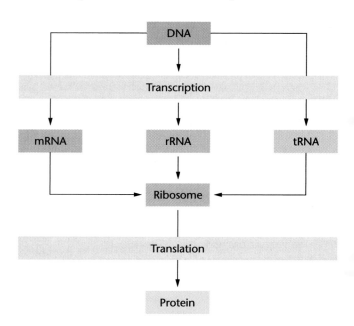

■ Figure 9–1   A simplified view of information flow involving DNA, RNA, and proteins within cells.

genes; digestive cells activate many genes specific to their function, but do not activate their melanin genes.

"Expression" of the stored genetic information is a complex process and is the basis for the concept of **information flow** within the cell. Figure 9–1 shows a simplified illustration of this concept. The initial event is the **transcription** of DNA, resulting in the synthesis of three types of RNA molecules: **messenger RNA (mRNA), transfer RNA (tRNA),** and **ribosomal RNA (rRNA).** Of these, mRNAs are translated into proteins. Each type of mRNA is the product of a specific gene and leads to the synthesis of a different protein. **Translation** occurs in conjunction with ribosomes and involves tRNA, which adapts the chemical information in mRNA to the amino acids that make up proteins. Collectively, these processes serve as the foundation for the **central dogma of molecular genetics**: "DNA makes RNA, which makes proteins."

The genetic material is also the source of "variability" among organisms through the process of mutation. If a mutation—a change in the chemical composition of DNA—occurs, the alteration will be reflected during transcription and translation, affecting the specified protein. If a mutation is present in gametes, it will be passed to future generations and, with time, may become distributed in the population. Genetic variation, which also includes rearrangements within and between chromosomes (see Chapter 8), provides the raw material for the process of evolution.

The Genetic Material: Early Studies

The idea that genetic material is physically transmitted from parent to offspring has been accepted for as long as the concept of inheritance has existed. Beginning in the late nineteenth century, research into the structure of biomolecules progressed considerably, setting the stage for the description of the genetic material in chemical terms. Although proteins and nucleic acid were both considered major candidates for the role of the genetic material, many geneticists, until the 1940s, favored proteins. This is not surprising, since a diversity of proteins was known to be abundant in cells, and much more was known about protein chemistry.

DNA was first studied in 1868 by a Swiss chemist, Friedrick Miescher. He was able to isolate cell nuclei and derive an acid substance containing DNA that he called **nuclein**. As investigations progressed, however, DNA, which was shown to be present in chromosomes, seemed to lack the chemical diversity necessary to store extensive genetic information. This conclusion was based largely on Phoebus A. Levene's observations in 1910 that DNA contained approximately equal amounts of four quite similar molecules called **nucleotides**. Levene postulated incorrectly that identical groups of these four components were repeated over and over, which was the basis of his **tetranucleotide hypothesis** for DNA structure. Attention was thus directed away from DNA, favoring proteins, much by default. However, in the 1940s, Erwin Chargaff showed that Levene's proposal was incorrect when he demonstrated that most organisms do not contain precisely equal proportions of the four nucleotides. We will see later that the structure of DNA accounts for Chargaff's observations.

Evidence Favoring DNA in Bacteria and Bacteriophages

The 1944 publication by Oswald Avery, Colin MacLeod, and Maclyn McCarty concerning the chemical identity of a "transforming principle" in bacteria marked the initial event leading to the acceptance of DNA as the genetic material. Along with the subsequent findings of other research teams, this work constituted direct experimental proof that in the organisms studied, DNA, and not protein, is the biomolecule responsible for heredity. This period marked the beginning of an era of discovery in biology that has revolutionized our understanding of life on earth.

The initial evidence implicating DNA as the genetic material was derived from studies of prokaryotic bacteria and the viruses that infect them, called **bacteriophages**. The reasons for their use will become apparent as the experiments are studied. Primarily, bacteria and their viruses are capable of rapid growth because they complete life cycles in hours. They also may be manipulated experimentally, and mutations may easily be induced and selected.

Transformation Studies

The research that provided the foundation for Avery, MacLeod, and McCarty's work was initiated in 1927 by Frederick Griffith, a medical officer in the British Ministry of Health. He performed experiments with several different strains of the bacterium *Diplococcus pneumoniae*.* Some were **virulent strains**, which cause pneumonia in certain vertebrates (notably humans and mice), while others were **avirulent strains**, which do not cause illness.

The difference in virulence is related to the polysaccharide capsule of the bacterium. Virulent strains have this capsule, whereas avirulent strains do not. The nonencapsulated bacteria are readily engulfed and destroyed by phagocytic cells in the animal's circulatory system. Virulent bacteria, which possess the polysaccharide coat, are not easily engulfed; they multiply and cause pneumonia.

The presence or absence of the capsule is the basis for a visible difference between colonies of virulent and avirulent strains. Encapsulated bacteria form a **smooth**, shiny-surfaced colony (**S**) when grown on an agar culture plate; nonencapsulated strains produce **rough** colonies (**R**) (Figure 9–2). Thus, virulent and avirulent strains may be distinguished easily by standard microbiological culture techniques.

Each strain of *Diplococcus* may be one of dozens of different types called **serotypes**. The specificity of the serotype is due to the detailed chemical structure of the polysaccharide constituent of the thick, slimy capsule. Serotypes are identified by immunological techniques and are usually designated by Roman numerals. Griffith used the avirulent type II*R* and the virulent type III*S* in his critical experiments. Table 9.1 summarizes the characteristics of these strains.

Griffith knew from the work of others that only living virulent cells would produce pneumonia in mice. If heat-killed virulent bacteria are injected into mice, no pneumonia results, just as living avirulent bacteria fail to produce the disease. Griffith's critical experiment (Figure 9–2) involved an injection into mice of living II*R* (avirulent) cells combined with heat-killed III*S* (virulent) cells. Since neither cell type caused death in mice when injected alone, Griffith expected that the double injection would not kill the mice. But, after five days, all the mice that received double injections were dead. Analysis of the blood of the dead mice revealed large numbers of living type III*S* bacteria!

As far as could be determined, these III*S* bacteria were identical to the III*S* strain from which the heat-killed cell preparation had been made. Control mice, injected only with living avirulent II*R* bacteria, did not develop pneumonia and remained healthy. This finding suggested strongly that the occurrence of the living III*S* bacteria in the dead mice was not caused by faulty technique or contamination, but that some interaction between the two types of injected bacteria had occurred.

Griffith concluded that the heat-killed III*S* bacteria were responsible for converting live avirulent II*R* cells into virulent III*S* cells. Calling the phenomenon **transformation**, he suggested that the **transforming principle** might be some part of the polysaccharide capsule or some compound required for capsule synthesis, although the capsule alone did not cause pneumonia. To use Griffith's term, the transforming principle from the dead III*S* cells served as a "pabulum" for the II*R* cells.

Griffith's work led other physicians and bacteriologists to explore the phenomenon of transformation. By 1931, Henry Dawson and his coworkers showed that transformation could occur *in vitro* (in a test tube containing only bacterial cells); that is, injection into mice was not necessary for transformation to occur. By 1933, Lionel J. Alloway had refined the *in vitro* experiments by using extracts from *S* cells added to living *R* cells. The soluble filtrate from the heat-killed *S* cells was as effective in inducing transformation as were the intact cells! Alloway and others did not view transformation as a genetic event, but rather as a physiological modification of some sort. Nevertheless, the experimental evidence that a chemical substance was responsible for transformation was quite convincing.

Then, in 1944, after ten years of work, Avery, MacLeod, and McCarty published their results in what is now regarded as a classic paper in the field of molecular genetics. They reported that they had obtained the transforming principle in a highly purified state, and that beyond reasonable doubt it was DNA. The details of their work are outlined in Figure 9–3.

These researchers began their isolation procedure with large quantities (50–75 liters) of liquid cultures of type III*S* virulent cells. The cells were centrifuged, collected, and heat-killed. Following various chemical treatments, a soluble filtrate was derived from these cells that retained the ability to induce transformation of type II*R* avirulent cells.

Further testing established beyond doubt that the transforming principle was DNA. The soluble filtrate was treated with a protein-digesting enzyme, called a **protease**, and an RNA-digesting enzyme, called **ribonuclease**. Such treatment destroyed the activity of any remaining protein and RNA. Nevertheless, transforming activity still remained. The conclusion was drawn that neither protein nor RNA was responsible

TABLE 9.1	Strains of *Diplococcus pneumoniae* used by Frederick Griffith in his original transformation experiments		
Serotype	**Colony Morphology**	**Capsule**	**Virulence**
II*R*	Rough	Absent	Avirulent
III*S*	Smooth	Present	Virulent

*Note that this organism is now designated *Streptococcus pneumonia*.

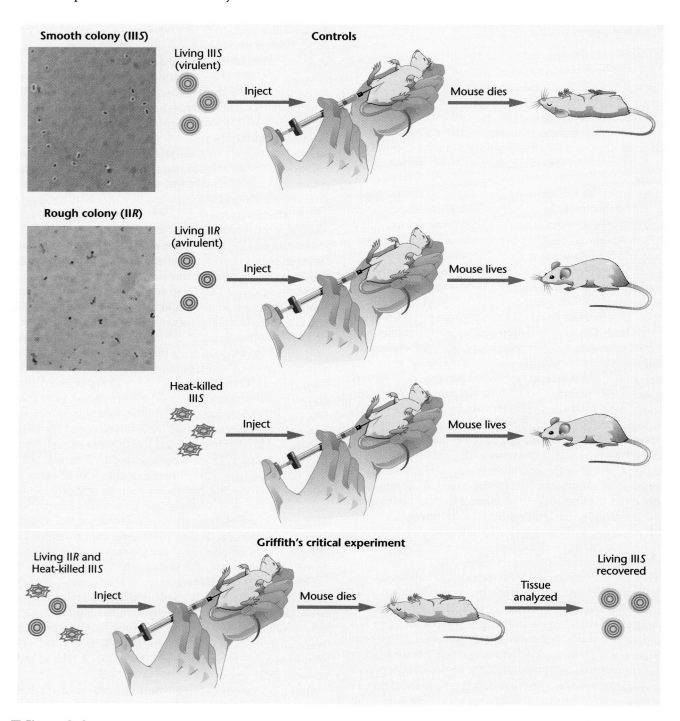

Smooth colony (III*S*)

Controls

Living III*S*
(virulent)

Inject

Mouse dies

Rough colony (II*R*)

Living II*R*
(avirulent)

Inject

Mouse lives

Heat-killed
III*S*

Inject

Mouse lives

Griffith's critical experiment

Living II*R* and
Heat-killed III*S*

Inject

Mouse dies

Tissue
analyzed

Living III*S*
recovered

■ Figure 9–2   Summary of Griffith's transformation experiment. The photographs show bacterial colonies containing cells with capsules (type III*S*) and without capsules (type II*R*).

for transformation. The final confirmation came with experiments using crude samples of the DNA-digesting enzyme **deoxyribonuclease**, which was isolated from dog and rabbit sera. Digestion with this enzyme was shown to destroy transforming activity present in the filtrate. There could be no doubt that the active transforming principle in these experiments was DNA!

The great amount of work, the confirmation and reconfirmation of the conclusions drawn, and the logic of the experimental design involved in the research of these three scientists are truly impressive. The conclusion of their 1944 publication, however, can be stated very simply: "The evidence presented supports the belief that a nucleic acid of the desoxyribose* type is the fundamental unit of the transforming principle of *Pneumococcus* Type III."

*Desoxyribose is now spelled deoxyribose.

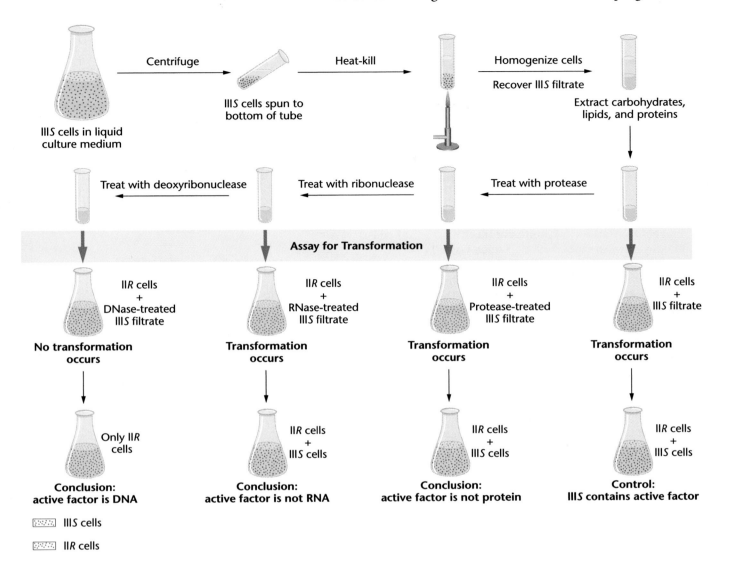

■ Figure 9–3   Summary of Avery, MacLeod, and McCarty's experiment demonstrating that DNA is the transforming principle.

Avery and his coworkers recognized the genetic and biochemical implications of their work. They suggested that the transforming principle interacts with the II*R* cell and gives rise to a coordinated series of enzymatic reactions that culminates in the synthesis of the type III*S* capsular polysaccharide. They emphasized that, once transformation occurs, the capsular polysaccharide is produced in successive generations. Transformation is therefore heritable, and the process affects the genetic material.

Transformation, discussed in more detail in Chapter 16, has been shown to occur in *Hemophilus influenzae*, *Bacillus subtilis*, *Shigella paradysenteriae*, and *Escherichia coli*, among many other microorganisms. Transformation of numerous genetic traits other than colony morphology has been demonstrated, including ones involving resistance to antibiotics and the ability to metabolize various nutrients. These observations further strengthened the belief that transformation by

DNA is primarily a genetic event, rather than simply a physiological change. This idea is pursued again in the "Insights and Solutions" section at the end of this chapter.

The Hershey–Chase Experiment

The second major piece of evidence supporting DNA as the genetic material was provided by the study of the bacterial virus T2. This virus, also called a bacteriophage or just a **phage**, has as its host the bacterium *Escherichia coli* and consists of a protein coat surrounding a core of DNA. The phage's external structure is composed of a hexagonal head plus a tail. The life cycle of bacteriophages such as T2 is shown in Figure 9–4.

In 1952, Alfred Hershey and Martha Chase published the results of experiments designed to clarify the events leading to phage reproduction. Several of the experiments clearly established the independent functions of phage protein and nucleic acid in the reproduction process associated with the bacterial cell. Hershey and Chase knew from existing data that:

1. T2 phages consist of approximately 50 percent protein and 50 percent DNA.

2. Infection is initiated by adsorption of the phage by its tail fibers to the bacterial cell.

3. The production of new viruses occurs within the bacterial cell.

It appeared that some molecular component of the phage, DNA and/or protein, enters the bacterial cell and directs viral reproduction. Which was it?

Hershey and Chase used radioisotopes to follow the molecular components of phages during infection. Both ^{32}P and ^{35}S, radioactive forms of phosphorus and sulfur, were used. Because DNA contains phosphorus but not sulfur, ^{32}P effectively labels DNA. Because proteins contain sulfur but not phosphorus, ^{35}S labels protein. *This is a key point in the experiment.* If *E. coli* cells are first grown in the presence of ^{32}P *or* ^{35}S and then infected with T2 viruses, the progeny phage will have *either* a labeled DNA core *or* a labeled protein coat, respectively. These radioactive phages may be isolated and used to infect unlabeled bacteria (Figure 9–5).

When labeled phage and unlabeled bacteria are mixed, an adsorption complex is formed as the phages attach their tail fibers to the bacterial wall. These complexes were isolated and subjected to a high shear force by placing them in a blender. This force strips off the attached phages, which may then be analyzed separately (Figure 9–5). By tracing the radioisotopes, Hershey and Chase were able to demonstrate that most of the ^{32}P-labeled DNA had been transferred into the bacterial cell following adsorption; on the other hand, most of the ^{35}S-labeled protein remained outside the bacterial cell and was recovered in the phage "ghosts" (empty phage coats) after the blender treatment. Following this separation, the bacterial cells, which now contained viral DNA, were eventually lysed as new phages were produced. These progeny contained ^{32}P, but not ^{35}S.

Hershey and Chase interpreted these results as indicating that the protein of the phage coat remains outside the host cell and is not involved in the production of new phages. On the other hand, and most important, phage DNA enters the host cell and directs phage multiplication. Hershey and Chase had demonstrated that in phage T2, DNA, not protein, is the genetic material.

These experiments, along with those of Avery and his colleagues, provided convincing evidence to most geneticists that DNA is the molecule responsible for heredity. This conclusion has since served as the cornerstone of the field of molecular genetics.

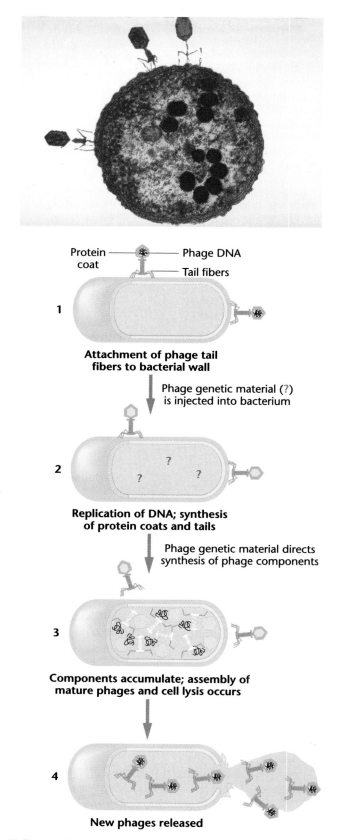

Protein coat — Phage DNA
Tail fibers

1 Attachment of phage tail fibers to bacterial wall

Phage genetic material (?) is injected into bacterium

2 Replication of DNA; synthesis of protein coats and tails

Phage genetic material directs synthesis of phage components

3 Components accumulate; assembly of mature phages and cell lysis occurs

4 New phages released

■ Figure 9–4   The life cycle of a T-even bacteriophage. The electron micrograph shows *E. coli* during infection by phage T2.

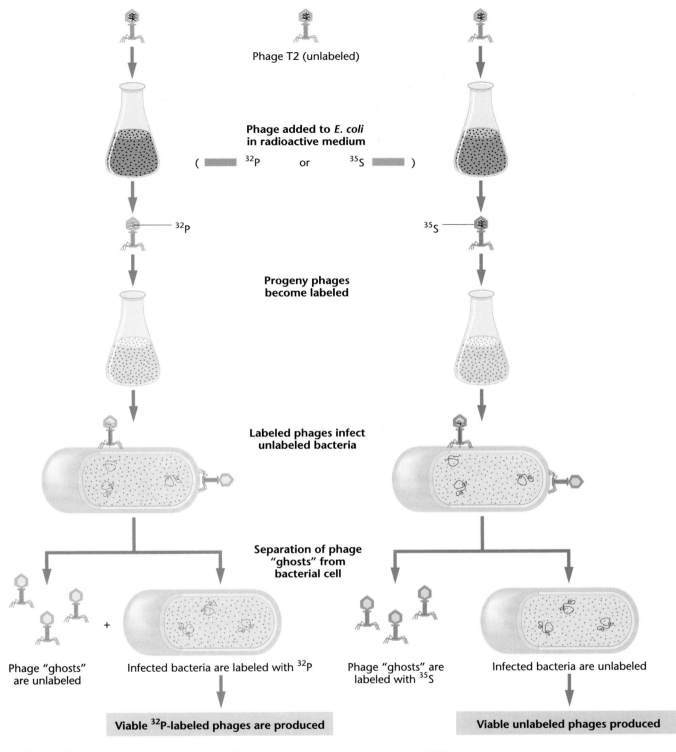

Phage T2 (unlabeled)

**Phage added to *E. coli*
in radioactive medium**

(▬▬▬ ^{32}P   or ^{35}S ▬▬▬)

^{32}P

^{35}S

**Progeny phages
become labeled**

**Labeled phages infect
unlabeled bacteria**

**Separation of phage
"ghosts" from
bacterial cell**

+

Phage "ghosts"
are unlabeled

Infected bacteria are labeled with ^{32}P

Phage "ghosts" are
labeled with ^{35}S

Infected bacteria are unlabeled

Viable ^{32}P-labeled phages are produced

Viable unlabeled phages produced

■ Figure 9–5   Summary of the Hershey–Chase experiment demonstrating that DNA, and not protein, is responsible for directing the reproduction of phage T2 during the infection of *E. coli*.

Transfection Experiments

During the eight years following the publication of the Hershey–Chase experiment, additional research with the same organisms provided even more solid proof that DNA is the genetic material. In 1957, several reports demonstrated that if *E. coli* were treated with the enzyme **lysozyme**, the outer wall of the cell could be removed without destroying the bacterium. Enzymatically treated cells are naked, so to speak, and contain only the cell membrane as the outer boundary of the cell. Such structures are called **protoplasts** (or **spheroplasts**). John Spizizen and Dean Fraser reported independently that by using protoplasts, they were able to initiate phage multiplication with disrupted T2 particles. That is, provided that protoplasts are used, it is not necessary for a virus to be intact in order for infection to occur.

Similar but refined experiments were reported in 1960 using only the DNA purified from bacteriophages. This process of infection by only the viral nucleic acid, called **transfection**, proves conclusively that phage DNA alone contains all the necessary information for production of mature viruses. Thus, the evidence that DNA serves as the genetic material in all organisms was further strengthened, even though all direct evidence had been obtained from bacterial and viral studies.

Observations Favoring DNA in Eukaryotes

Around 1950, most eukaryotic organisms were not amenable to the types of experiments that demonstrated DNA as the genetic material in bacteria and viruses. Nevertheless, it was generally assumed that the genetic material would be a universal substance and also serve this role in eukaryotes. Support for this assumption relied initially on many diverse circumstantial observations. Together, they provided support that DNA serves as the genetic material in eukaryotes. We shall look at several such observations.

Distribution of DNA

The genetic material should be found where it functions—in the nucleus as part of chromosomes. Both DNA and protein fit this criterion. However, protein is also abundant in the cytoplasm, while DNA is not. Both mitochondria and chloroplasts are known to perform genetic functions, and DNA is also present in these organelles. Thus, DNA is found only where primary genetic function is known to occur. Protein, however, is found everywhere in the cell. These observations are consistent with the interpretation favoring DNA over protein as the genetic material.

Because it had been established earlier that chromosomes within the nucleus contain the genetic material, a correlation was expected between the ploidy (n, $2n$, etc.) of cells and the quantity of the molecule that functions as the genetic material. Meaningful comparisons may be made between gametes (sperm and eggs) and somatic or body cells. The latter are recognized as being diploid ($2n$) and containing twice the number of chromosomes as gametes, which are haploid (n).

Table 9.2 compares the amount of DNA found in haploid sperm and the diploid nucleated precursors of red

TABLE 9.2	DNA content of haploid versus diploid cells of various species (in picograms)	
Organism	*n*	*2n*
Human	3.25	7.30
Chicken	1.26	2.49
Trout	2.67	5.79
Carp	1.65	3.49
Shad	0.91	1.97

Note: Sperm (n) and nucleated precursors to red blood cells ($2n$) were used to contrast ploidy levels.

blood cells from a variety of organisms. There is a close correlation between the amount of DNA and the number of sets of chromosomes. No consistent correlation can be observed between gametes and diploid cells for proteins, thus favoring DNA over proteins as the genetic material of eukaryotes.

Mutagenesis

Ultraviolet (UV) light is one of a number of agents capable of inducing mutations in the genetic material. Simple organisms such as yeast and other fungi can be irradiated with various wavelengths of UV light, and the effectiveness of each wavelength may be measured by the number of mutations it induces. When the data are plotted, an **action spectrum** of UV light as a mutagenic agent is obtained. This action spectrum may then be compared with the **absorption spectrum** of any molecule suspected to be the genetic material (Figure 9–6). The molecule serving as the genetic material is expected to absorb at the wavelengths found to be mutagenic.

UV light is most mutagenic at the wavelength (λ) of 260 nanometers (nm). Both DNA and RNA absorb UV light most strongly at 260 nm. On the other hand, protein absorbs most strongly at 280 nm, yet no significant mutagenic effects are observed at this wavelength. This indirect evidence supports the idea that a nucleic acid is the genetic material and tends to exclude protein.

Direct Evidence Favoring DNA in Eukaryotes

Although the circumstantial evidence described above does not constitute direct proof that DNA is the genetic material in eukaryotes, these observations spurred researchers to forge ahead, basing their work on this hypothesis. Today, there is no doubt of the validity of this conclusion. DNA *is* the genetic material in eukaryotes.

The strongest evidence that DNA is the genetic material is provided by molecular analysis utilizing **recombinant DNA technology**. In this procedure, segments of eukaryotic DNA corresponding to specific genes are

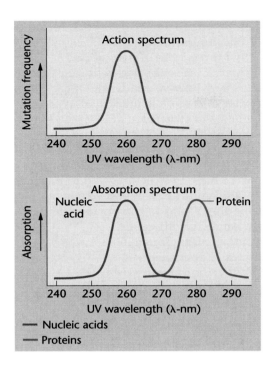

Nucleic acids
Proteins

■ Figure 9–6   The "action spectrum" determining the most effective mutagenic UV wavelength contrasted with the "absorption spectrum" of nucleic acids versus proteins in the UV range of the electromagnetic spectrum.

isolated and literally spliced into bacterial DNA. Such a complex can be inserted into a bacterial cell and its genetic expression monitored. If a eukaryotic gene is introduced, the presence of the corresponding eukaryotic protein product demonstrates directly that this DNA is present and functional in the bacterial cell. This has been shown to be the case in countless instances. For example, the products of the human genes that specify insulin and the interferon are produced by bacteria following insertion of the human genes that encode these proteins. As the bacterium divides, the eukaryotic DNA is replicated along with the host DNA and is distributed to the daughter cells, which also express the human genes and synthesize the corresponding proteins.

Work in the laboratory of Beatrice Mintz and others has further strengthened this evidence. This research has demonstrated that DNA encoding the human β-globin gene, when microinjected into a fertilized mouse egg, is later present and expressed in the adult mouse tissue and can be transmitted to that mouse's progeny! Such mice are examples of **transgenic organisms**. More recent work has introduced DNA representing the growth hormone gene from *rats* into fertilized *mouse* eggs. About one-third of the resultant animals grew to twice the size of normal mice, indicating that the foreign DNA was present *and* functional. Subsequent generations of mice received the gene and also grew to a large size.

RNA as the Genetic Material

Some viruses contain an RNA core rather than one composed of DNA. In these viruses, it appears that RNA must serve as the genetic material—an exception to the general rule that DNA performs this function. In 1956, it was demonstrated that when purified RNA from **tobacco mosaic virus (TMV)** is spread on tobacco leaves, the characteristic lesions caused by viral infection appear later on the leaves. It was concluded that RNA is the genetic material of this virus.

In 1965 and 1966, Norman Pace and Sol Spiegelman demonstrated further that RNA from the phage Qb could be isolated and replicated *in vitro*. Replication was dependent on an enzyme, **RNA replicase**, which was isolated from host *E. coli* cells following normal infection. When the RNA replicated *in vitro* was added to *E. coli* protoplasts, infection and viral multiplication (transfection) occurred. Thus, RNA synthesized in a test tube can serve as the genetic material in these phages by directing the production of all components necessary for viral replication.

Finally, one other group of RNA-containing viruses bears mentioning. These are the **retroviruses**, which replicate in an unusual way. Their RNA serves as a template for the synthesis of the complementary DNA molecule! The process, designated as **reverse transcription**, occurs under the direction of an RNA-dependent DNA polymerase enzyme called **reverse transcriptase**. Because the genetic material can be represented by this DNA intermediate, it can be incorporated into the genome of the host cell. Once it is present, transcription yields retroviral RNA chromosomes if the DNA is expressed. Retroviruses include the human immunodeficiency virus (HIV), which causes AIDS, and many RNA tumor viruses.

The Structure of DNA

Having established that DNA is the genetic material in all living organisms (except certain viruses), we turn now to a consideration of the structure of this nucleic acid. In 1953, James Watson and Francis Crick proposed that the structure of DNA is in the form of a double helix. Their proposal was published in a short paper in the journal *Nature*, reprinted in its entirety (see p. 201). In a sense, this publication constituted the finish of a highly competitive scientific race to obtain what some consider to be the most significant finding in the history of biology. This "race," as recounted in Watson's book *The Double Helix*, demonstrates the human interaction, genius, frailty, and intensity involved in the scientific effort that eventually led to the elucidation of DNA structure.

The data available to Watson and Crick, crucial to the development of their proposal, came primarily from two sources: (1) base composition analysis of hydrolyzed samples of DNA and (2) X-ray diffraction studies of DNA. The analytical success of Watson and Crick may be attributed to model building that conformed to the existing data. If the correct solution to the structure of DNA may be viewed as a puzzle, Watson and Crick, working in the Cavendish Laboratory in Cambridge, England, were the first to put together all of the pieces successfully.

Nucleic Acid Chemistry

Before turning to this work, a brief introduction to nucleic acid chemistry is in order. This chemical information was well known to Watson and Crick during their investigation and served as the basis of their model building.

DNA is a nucleic acid, and nucleotides are the building blocks of all nucleic acid molecules. Sometimes called mononucleotides, these structural units consist of three essential components: a **nitrogenous base**, a **pentose sugar** (a five-carbon sugar), and a **phosphate group**. There are two kinds of nitrogenous bases: the

nine-membered, double-ringed **purines** and the six-membered, single-ringed **pyrimidines**. Two types of purines and three types of pyrimidines are commonly found in nucleic acids. The two purines are **adenine** and **guanine**, abbreviated **A** and **G**. The three pyrimidines are **cytosine**, **thymine**, and **uracil**, abbreviated **C, T,** and **U**. The chemical structures of A, G, C, T, and U are shown in Figure 9–7(a). Both DNA and RNA contain A, C, and G; only DNA contains the base T, whereas only RNA contains the base U. Each nitrogen or carbon atom of the ring structures of purines and pyrimidines is designated by an unprimed number. Note that corresponding atoms in the purine and pyrimidine rings are numbered differently.

The pentose sugars found in nucleic acids give them their names. **Ribonucleic acids (RNA)** contain **ribose**, while **deoxyribonucleic acids (DNA)** contain **deoxyribose**. Figure 9–7(b) shows the structures for these two pentose sugars. Each carbon atom is distinguished by a number with a prime sign (e.g., C-1′, C-

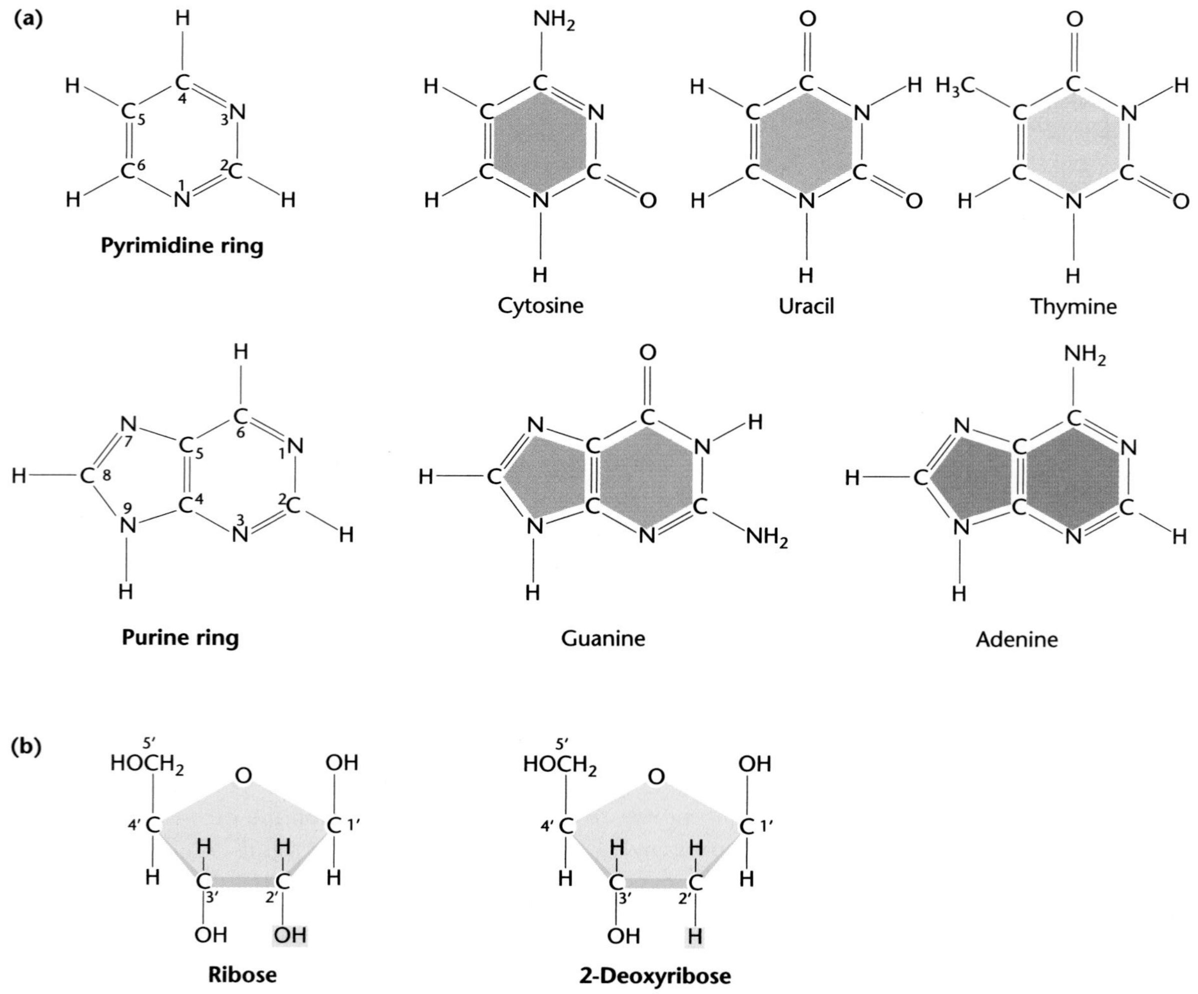

■ Figure 9–7 (a) Chemical structures of the pyrimidines and purines that serve as the nitrogenous bases in RNA and DNA. (b) Chemical ring structures of ribose and 2-deoxyribose, which serve as the pentose sugars in RNA and DNA, respectively.

2', etc.). As you can see, deoxyribose has a hydrogen atom rather than a hydroxyl group at the C-2' position compared with ribose. The presence of a hydroxyl group at the C-2' position thus distinguishes RNA from DNA.

If a molecule is composed of a purine or pyrimidine base and a ribose or deoxyribose sugar, the chemical unit is called a **nucleoside**. If a phosphate group is added to the nucleoside, the molecule is now called a **nucleotide**. Nucleosides and nucleotides are named according to the specific nitrogenous base (A, T, G, C, or U) that is part of the molecule. The structure of a nucleotide and the nomenclature used in naming DNA nucleotides and nucleosides are as shown in Figure 9–8.

The bonding between the components of a nucleotide is highly specific. The C-1' atom of the sugar is involved in the chemical linkage to the nitrogenous base. If the base is a purine, the N-9 atom is covalently bonded to the sugar. If the base is a pyrimidine, the bonding involves the N-1 atom. In a nucleotide, the phosphate group may be bonded to the C-2', C-3', or C-5' atom of the sugar. The C-5'-phosphate configuration is shown in Figure 9–8. It is by far the most prevalent one in biological systems and the one found in DNA and RNA.

Nucleotides are also described by the term **nucleoside monophosphate (NMP)**. The addition of one or two phosphate groups results in **nucleoside diphosphates (NDP)** and **triphosphates (NTP)**, as illustrated in Figure 9–9. The triphosphate form is significant because it serves as the precursor molecule during nucleic acid synthesis within the cell. Additionally, the triphosphates **adenosine triphosphate (ATP)** and **guanosine triphosphate (GTP)** are important in the bioenergetics of cells because of the large amount of energy involved in the addition or removal of the terminal phosphate group. The hydrolysis of ATP or GTP to ADP or GDP and inorganic phosphate (P_i) is accompanied by the release of a large amount of energy in the cell. When these chemical conversions are coupled to other reactions, the energy produced may be used to drive them. As a result, ATP and GTP are involved in many cellular activities.

The linkage between two mononucleotides consists of a phosphate group linked to two sugars. A **phosphodiester bond** is formed, because phosphoric acid has been joined to two alcohols (the hydroxyl groups on the two sugars) by an ester linkage on both sides. Figure 9–10 shows the resultant phosphodiester bond in DNA.

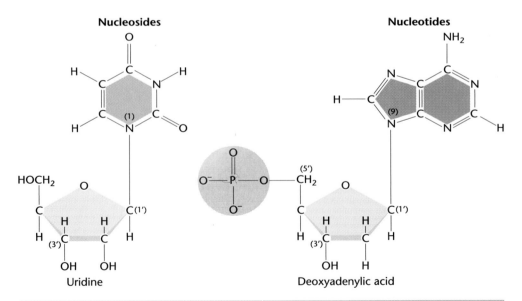

Ribonucleosides	Ribonucleotides
Adenosine	Adenylic acid
Cytidine	Cytidylic acid
Guanosine	Guanylic acid
Uridine	Uridylic acid
Deoxyribonucleosides	**Deoxyribonucleotides**
Deoxyadenosine	Deoxyadenylic acid
Deoxycytidine	Deoxycytidylic acid
Deoxyguanosine	Deoxyguanylic acid
Deoxythymidine	Deoxythymidylic acid

■ Figure 9–8 The structures and names of the nucleosides and nucleotides of DNA and RNA.

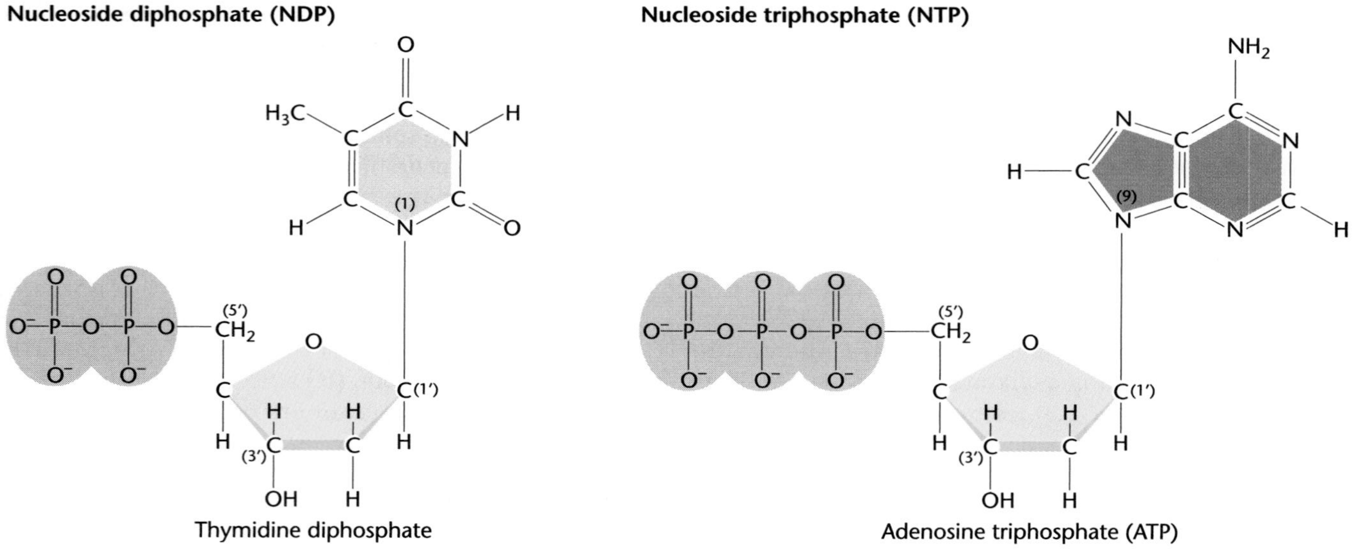

Nucleoside diphosphate (NDP)

Nucleoside triphosphate (NTP)

Thymidine diphosphate

Adenosine triphosphate (ATP)

■ Figure 9–9 The basic structures of nucleoside diphosphates and triphosphates, as illustrated by thymidine diphosphate and deoxyadenosine triphosphate.

Each bond has a C-3′ end and a C-5′ end. The joining of two nucleotides forms a dinucleotide; of three nucleotides, a trinucleotide; and so forth. Short chains consisting of fewer than 20 nucleotides are called **oligonucleotides**. Still longer chains are referred to as **polynucleotides**.

Long polynucleotide chains account for the large molecular weight and explain the most important property of DNA—storage of vast quantities of genetic information. If each nucleotide position in this long chain may be occupied by any one of four nucleotides, extraordinary variation is possible. For example, a polynu-

cleotide that is only 1000 nucleotides in length may be arranged 4^{1000} different ways, each one different from all other possible sequences. This potential variation in molecular structure is essential if DNA is to serve the function of storing the vast amounts of chemical information necessary to direct cellular activities.

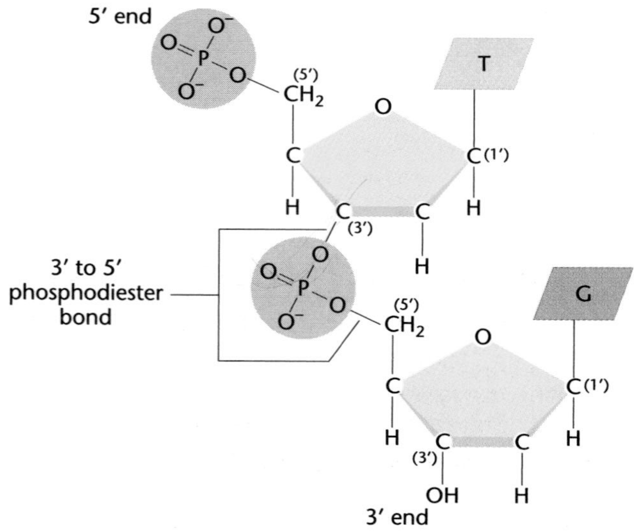

Base Composition Studies

Between 1949 and 1953, Erwin Chargaff and his colleagues used chromatographic methods to separate the four bases in DNA samples from various organisms. Quantitative methods were then used to determine the amounts of the four nitrogenous bases from each source. Table 9.3(a) provides some of Chargaff's original data. Parts (b) and (c) show more recently derived information that reinforces Chargaff's findings. As we shall see, interpretation of Chargaff's data was critical to the successful model of DNA put forward by Watson and Crick.

On the basis of these data, the following conclusions may be drawn.

1. The amount of adenine residues is proportional to the amount of thymine residues in DNA (columns 1, 2, and 5). Also, the amount of guanine residues is proportional to the amount of cytosine residues (columns 3, 4, and 6).

2. Based on the above proportionality, the sum of the purines (A + G) equals the sum of the pyrimidines (C + T), as shown in column 7.

3. The percentage of G + C does not necessarily equal the percentage of A + T. As we can see, this ratio

■ Figure 9–10 The linkage of two nucleotides by the formation of a C-3′ to C-5′ (3′-5′) phosphodiester bond, producing a dinucleotide.

TABLE 9.3	DNA base composition data

(a) Chargaff's data*

	Molar proportions[a]			
	1	*2*	*3*	*4*
Source	*A*	*T*	*G*	*C*
Ox thymus	26	25	21	16
Ox spleen	25	24	20	15
Yeast	24	25	14	13
Avian tubercle bacilli	12	11	28	26
Human sperm	29	31	18	18

(c) G + C content in several organisms

Organism	*% G + C*
Phage T2	36.0
Drosophila	45.0
Maize	49.1
Euglena	53.5
Neurospora	53.7

(b) Base compositions of DNAs from various sources

	Base composition				Base ratio			A + T/G + C ratio
	1	*2*	*3*	*4*	*5*	*6*	*7*	*8*
Source	*A*	*T*	*G*	*C*	*A/T*	*G/C*	*(A + G)/(C + T)*	*(A + T)/(C + G)*
Human	30.9	29.4	19.9	19.8	1.05	1.00	1.04	1.52
Sea urchin	32.8	32.1	17.7	17.3	1.02	1.02	1.02	1.58
E. coli	24.7	23.6	26.0	25.7	1.04	1.01	1.03	0.93
Sarcina lutea	13.4	12.4	37.1	37.1	1.08	1.00	1.04	0.35
T7 bacteriophage	26.0	26.0	24.0	24.0	1.00	1.00	1.00	1.08

Source: From Chargaff, 1950.

[a]Moles of nitrogenous constituent per mole of P (often, the recovery was less than 100 percent).

varies greatly in different organisms, as shown in column 8.

These conclusions indicate a definite pattern of base composition of DNA molecules. The data served as the initial clue to "the puzzle." Additionally, they directly refute Levene's tetranucleotide hypothesis, which stated that all four bases are present in equal amounts.

X-Ray Diffraction Analysis

When fibers of a DNA molecule are subjected to X-ray bombardment, these rays are scattered according to the molecule's atomic structure. The pattern of scatter may be captured as spots on photographic film and analyzed, particularly for the overall shape of and regularities within the molecule. This process, **X-ray diffraction analysis**, was applied successfully to the study of protein structure by Linus Pauling and other chemists. The technique had been attempted on DNA as early as 1938 by William Astbury. By 1947, he had detected a periodicity within the structure of the molecule of 3.4 angstroms (Å)*, which suggested to him that the bases were stacked like coins on top of one another.

Between 1950 and 1953, Rosalind Franklin, working in the laboratory of Maurice Wilkins, obtained im-

proved X-ray data from more purified samples of DNA (Figure 9–11). Her work confirmed the 3.4-Å periodicity seen by Astbury and suggested that the structure of DNA was some sort of helix. However, she did not propose a definitive model. Pauling had analyzed the work of Astbury and others and proposed incorrectly that DNA is a triple helix.

The Watson–Crick Model

Watson and Crick published their analysis of DNA structure in 1953.** By building models under the constraints of the information just discussed, they proposed the double-helical form of DNA as shown in Figure 9–12(a). This model has the following major features.

1. Two long polynucleotide chains are coiled around a central axis, forming a right-handed double helix.

2. The two chains are antiparallel; that is, their C-5'-to-C-3' orientations run in opposite directions.

3. The bases of both chains are flat structures, lying perpendicular to the axis; they are "stacked" on one another, 3.4 Å (0.34 nm) apart, and are located on the inside of the structure.

---

*Today, measurement in nanometers (nm) is favored (1 nm = 10 Å).
**Reproduced on page 201.

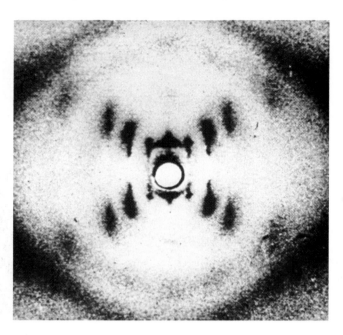

■ Figure 9–11 An X-ray diffraction photograph of the B form of crystallized DNA. The dark patterns at the top and bottom provide an estimate of the periodicity of nitrogenous bases, which are 3.4 Å apart. The central pattern is indicative of the molecule's helical structure.

4. The nitrogenous bases of opposite chains are paired to one another as the result of the formation of **hydrogen bonds** (described in the following discussion); in DNA, only A═T and G≡C pairs are allowed.

5. Each complete turn of the helix is 34 Å (3.4 nm) long; thus, ten bases exist in each chain per turn.

6. In any segment of the molecule, alternating larger **major grooves** and smaller **minor grooves** are apparent along the axis.

7. The double helix measures 20 Å (2.0 nm) in diameter.

The nature of base pairing (point 4 above) is the most genetically significant feature of the model. Before discussing it in detail, several other important features warrant emphasis. First, the antiparallel nature of the two chains is a key part of the double-helix model. While one chain runs in the 5′-to-3′ orientation (what seems right side up to us), the other chain is in the 3′-to-5′ orientation (and thus appears upside down). This is illustrated in Figure 9–12(b). Given the constraints of the bond angles of the various nucleotide components, the double helix could not be constructed easily if both chains ran parallel to one another.

The key to the model proposed by Watson and Crick is the specificity of base pairing. Chargaff's data had suggested that the amounts of A equaled T and that G equaled C. Watson and Crick realized that if A pairs with T and C pairs with G, thus accounting for these proportions, the members of each such base pair form

hydrogen bonds (Figure 9–13), providing the chemical stability necessary to hold the two chains together. Arranged in this way, both major and minor grooves become apparent along the axis. Further, a purine (A or G) opposite a pyrimidine (T or C) on each "rung of the spiral staircase" of the proposed helix accounts for the 20-Å (2-nm) diameter suggested by X-ray diffraction studies.

The specific A═T and G≡C base pairing is the basis for the concept of **complementarity**. This term describes the chemical affinity provided by the hydrogen bonds between the bases. As we will see, this concept is very important in the processes of DNA replication and gene expression.

It is appropriate to inquire into the nature of a hydrogen bond, and whether is it strong enough to stabilize the helix. A hydrogen bond is a very weak electrostatic attraction between a covalently bonded hydrogen atom and an atom with an unshared electron pair. The hydrogen atom assumes a partial positive charge, while the unshared electron pair—characteristic of covalently bonded oxygen and nitrogen atoms—assumes a partial negative charge. These opposite charges are responsible for the weak chemical attractions. As oriented in the double helix, adenine forms two hydrogen bonds with thymine, and guanine forms three hydrogen bonds with cytosine. Although two or three hydrogen bonds taken alone are very weak, two or three thousand bonds in tandem (which would be found in two long polynucleotide chains) are capable of providing great stability to the helix.

Still another stabilizing factor is the arrangement of sugars and bases along the axis. In the Watson–Crick model, the **hydrophobic** (or "water-fearing") nitrogenous bases are stacked almost horizontally on the interior of the axis, thus shielded from water. The **hydrophilic** (or "water-loving") sugar–phosphate backbone is on the outside of the axis, where both components may interact with water. These molecular arrangements provide significant chemical stabilization to the helix.

A more recent and accurate analysis of the form of DNA that served as the basis for the Watson–Crick model has revealed a minor structural difference. A precise measurement of the number of base pairs (bp) per turn has demonstrated a value of 10.4 rather than the 10.0 predicted by Watson and Crick. Where, in the classical model, each base pair is rotated 36° around the helical axis relative to the adjacent base pair, the new finding requires a rotation of 34.6°. Thus, there are slightly more than 10 base pairs per turn.

The Watson–Crick model had an immediate effect on the emerging discipline of molecular biology. Even in their initial 1953 article, the authors noted, "It has not escaped our notice that the specific pairing we have postulated immediately suggests a possible copying mechanism for the genetic material." Two months later,

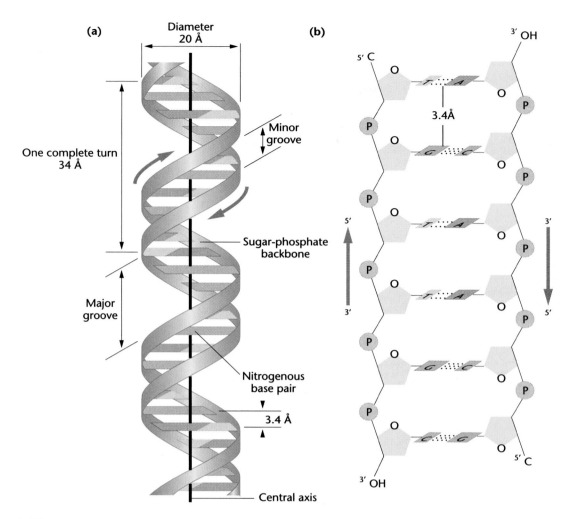

■ Figure 9–12   (a) A schematic representation of the DNA double helix as proposed by Watson and Crick. The ribbonlike strands constitute the sugar–phosphate backbones, and the horizontal rungs constitute the nitrogenous base pairs, of which there are 10 per complete turn. The major and minor grooves are apparent. A solid vertical bar representing the central axis has been placed through the center of the helix. (b) A representation of the antiparallel nature of the two strands of the helix.

in a second article in *Nature*, Watson and Crick pursued this idea, suggesting a specific mode of replication of DNA–the **semiconservative model.** The second article also alluded to two new concepts: (1) the storage of genetic information in the sequence of the bases and (2) the mutation or genetic change that would result from alteration of the bases. These ideas have received vast amounts of experimental support since 1953 and are now universally accepted.

The "synthesis" of ideas by Watson and Crick was highly significant with regard to subsequent studies of genetics and biology. The nature of the gene and its role in genetic mechanisms could now be viewed and studied in biochemical terms. Recognition of their work, along with that of Wilkins, led to their receipt of the Nobel Prize in Physiology and Medicine in 1962. This was one of many such awards bestowed for work in the field of genetics.

Other Forms of DNA

Under different conditions of isolation, several conformational forms of DNA have been recognized. At the time Watson and Crick performed their analysis, two forms—**A-DNA** and **B-DNA**—were known. Watson and Crick's analysis was based on X-ray studies of the B form by Franklin, which is present under aqueous, low-salt conditions and is believed to be the biologically significant conformation.

While DNA studies around 1950 relied on the use of X-ray diffraction, more recent investigations have been performed using **single-crystal X-ray analysis**. The earlier studies achieved limited resolution of about 5 Å, but single crystals diffract X-rays at about 1 Å, near atomic resolution. As a result, every atom is "visible," and much greater structural detail is available during analysis.

Adenine-thymine base pair

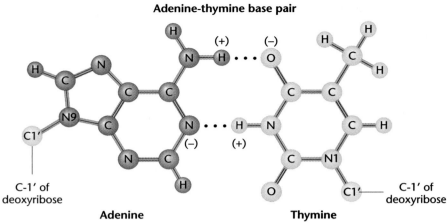

C-1' of deoxyribose

Adenine

C-1' of deoxyribose

Thymine

Guanine-cytosine base pair

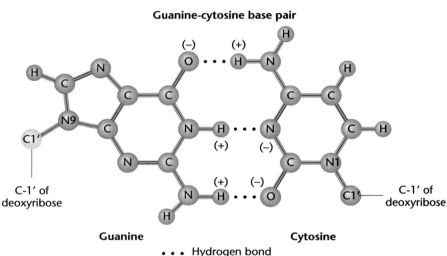

C-1' of deoxyribose

Guanine

C-1' of deoxyribose

Cytosine

• • • Hydrogen bond

■ Figure 9–13 Ball-and-stick models of A═T and G≡C base pairs. The rows of dots (· · ·) represent the hydrogen bonds that form.

Using these modern techniques, the A form of DNA has now been scrutinized. A-DNA is prevalent under high-salt or dehydration conditions. In comparison to B-DNA (Figure 9–14), A-DNA is slightly more compact, with 11 base pairs in each complete turn of the helix, which is 23 Å (2.3 mm) in diameter. While it is also a right-handed helix, the orientation of the bases is somewhat different. They are tilted and displaced laterally in relation to the axis of the helix. As a result of these differences, the appearance of the major and minor grooves is modified compared with those in B-DNA. It seems doubtful that A-DNA occurs under biological conditions.

Other forms of DNA (e.g., C-, D-, and E-DNA) are now known. Still another form, called **Z-DNA**, was discovered by Andrew Wang, Alexander Rich, and their colleagues in 1979, when they examined a small synthetic DNA fragment containing only C–G base pairs. Z-DNA takes on the rather remarkable configuration of a left-handed double helix (Figure 9–14). Like A- and B-DNA, Z-DNA consists of two antiparallel chains held together by Watson–Crick base pairs. Beyond these characteristics, Z-DNA is quite different. The left-handed helix is 18 Å (1.8 nm) in diameter, contains

12 base pairs per turn, and assumes a zigzag conformation (hence its name). The major groove present in B-DNA is nearly eliminated in Z-DNA.

Speculation has abounded over the possibility that regions of Z-DNA exist in the chromosomes of living organisms. The unique helical arrangement could provide an important recognition point for the interaction with other molecules. However, it is still not clear whether Z-DNA occurs *in vivo*.

The Structure of RNA

The second type of nucleic acid is the ribonucleic acid, or RNA. The structure of these molecules is similar to DNA, with several important exceptions. Although RNA also has as its building blocks nucleotides linked into polynucleotide chains, the sugar ribose replaces deoxyribose and the nitrogenous base uracil replaces thymine. Another important difference is that most RNA is single-stranded, although there are two important exceptions. First, RNA molecules sometimes fold back on themselves to form double-stranded regions of complementary base pairs. Second, some animal

Molecular Structure of Nucleic Acids:
A Structure for Deoxyribose Nucleic Acid

We wish to suggest a structure for the salt of deoxyribose nucleic acid (D. N. A.). This structure has novel features which are of considerable biological interest. A structure for nucleic acid has already been proposed by Pauling and Corey.[1] They kindly made their manuscript available to us in advance of publication. Their model consists of three intertwined chains, with the phosphates near the fibre axis, and the bases on the outside. In our opinion, this structure is unsatisfactory for two reasons: (1) We believe that the material which gives the X-ray diagrams is the salt, not the free acid. Without the acidic hydrogen atoms it is not clear what forces would hold the structure together, especially as the negatively charged phosphates near the axis will repel each other. (2) Some of the van der Waals distances appear to be too small.

Another three-chain structure has also been suggested by Fraser (in the press). In his model the phosphates are on the outside and the bases on the inside, linked together by hydrogen bonds. This structure as described is rather ill-defined, and for this reason we shall not comment on it.

We wish to put forward a radically different structure for the salt of deoxyribose nucleic acid. This structure has two helical chains each coiled round the same axis. We have made the usual chemical assumptions, namely, that each chain consists of phosphate diester groups joining β-D-deoxyribofuranose residues with $3',5'$ linkages. The two chains (but not their bases) are related by a dyad perpendicular to the fibre axis. Both chains follow right-handed helices, but owing to the dyad the sequences of the atoms in the two chains run in opposite directions. Each chain loosely resembles Furberg's[2] model No. 1; that is, the bases are on the inside of the helix and the phosphates on the outside. The configuration of the sugar and the atoms near it is close to Furberg's "standard configuration," the sugar being roughly perpendicular to the attached base. There is a residue on each chain every 3.4 Å in the z-direction. We have assumed an angle of 36° between adjacent residues in the same chain, so that the structure repeats after 10 residues on each chain, that is, after 34 Å. The distance

of a phosphates atom from the fibre axis is 10 Å. As the phosphates are on the outside, cations have easy access to them.

The structure is an open one, and its water content is rather high. At lower water content we would expect the bases to tilt so that the structure could become more compact.

The novel feature of the structure is the manner in which the two chains are held together by the purine and pyrimidine bases. The planes of the bases are perpendicular to the fibre axis. They are joined together in pairs, a single base from one chain being hydrogen-bonded to a single base from the other chain, so that the two lie side by side with identical z-co-ordinates. One of the pair must be a purine and the other a pyrimidine for bonding to occur. The hydrogen bonds are made as follows: purine position 1 to pyrimidine position 1; purine position 6 to pyrimidine position 6.

If it is assumed that the bases only occur in the structure in the most plausible tautomeric forms (that is, with the keto rather than the enol configuration) it is found that only specific pairs of bases can bond together. These pairs are: adenine (purine) with thymine (pyrimidine), and guanine (purine) with cytosine (pyrimidine).

In other words, if an adenine forms one member of a pair, on either chain, then on these assumptions the other member must be thymine; similarly for guanine and cytosine. The sequence of bases on a single chain does not appear to be restricted in any way. However, if only specific pairs of bases can be formed, it follows that if the sequence of bases on one chain is given, then the sequence on the other chain is automatically determined.

It has been found experimentally[3,4] that the ratio of the amounts of adenine to thymine, and the ratio of guanine to cytosine, are always very close to unity for deoxyribose nucleic acid.

It is probably impossible to build this structure with a ribose sugar in place of deoxyribose, as the extra oxygen atom would make too close a van der Waals contact.

The previously published X-ray data[5,6] on deoxyribose nucleic acid are insufficient for a rigorous test of our structure. So far as we can tell, it is

roughly compatible with the experimental data, but it must be regarded as unproved until it has been checked against more exact results. Some of these are given in the following communications. We were not aware of the details of the results presented there when we devised our structure, which rests mainly though not entirely on published experimental data and stereochemical arguments.

It has not escaped our notice that the specific pairing we have postulated immediately suggests a possible copying mechanism for the genetic material.

Full details of the structure, including the conditions assumed in building it, together with a set of co-ordinates for the atoms, will be published elsewhere.

We are much indebted to Dr. Jerry Donohue for constant advice and criticism, especially on interatomic distances. We have also been stimulated by a knowledge of the general nature of the unpublished experimental results and ideas of Dr. M. H. F. Wilkins, Dr. R. E. Franklin and their co-workers at King's College, London. One of us (J. D. W.) has been aided by a fellowship from the National Foundation for Infantile Paralysis.

J. D. Watson
F. H. C. Crick
Medical Research Council Unit for the Study of the Molecular Structure of Biological Systems, Cavendish Laboratory, Cambridge, England

[1]Pauling L., and Corey, R. B., *Nature,* 171, 346 (1953); *Proc. U.S. Nat. Acad. Sci.,* 39, 84 (1953).

[2]Furberg, S., *Acta Chem. Scand.,* 6, 634 (1952).

[3]Chargaff, E., for references see Zamenhof, S., Brawerman, G., and Chargaff, E., *Biochim. et Biophys. Acta,* 9, 402 (1952).

[4]Wyatt, G. R., *J. Gen. Physiol.,* 36, 201 (1952).

[5]Astbury, W. T., *Symp. Soc. Exp. Biol. 1, Nucleic Acid,* 66 (Camb. Univ. Press, 1947).

[6]Wilkins, M. H. F., and Randall, J. T., *Biochim. et Biophys. Acta,* 10, 192 (1953).

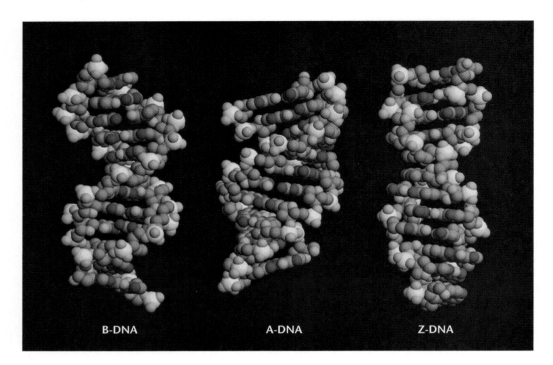

B-DNA A-DNA Z-DNA

■ Figure 9–14 Space-filling models of the B-, A-, and Z-forms of DNA.

viruses that have RNA as their genetic material contain double-stranded helices.

At least three classes of cellular RNA molecules function during the expression of genetic information: ribosomal RNA (rRNA), messenger RNA (mRNA), and transfer RNA (tRNA). These molecules all originate as complementary copies of one of the two strands of DNA segments during the process of transcription. That is, their nucleotide sequence is complementary to the deoxyribonucleotide sequence of DNA, which serves as the template for their synthesis. Uracil replaces thymine in RNA and is complementary to adenine during transcription.

Each class of RNA may be characterized by its size, sedimentation behavior in a centrifugal field, and genetic function. Sedimentation behavior depends on a molecule's density, mass, and shape, and its measure is called the **Svedberg coefficient (S)**. Table 9.4 relates the S values, molecular weights, and approximate number of nucleotides of the major forms of RNA. While higher S values almost always designate molecules of greater molecular weight, the correlation is not direct;

that is, a twofold increase in molecular weight does not lead to a twofold increase in S. This is because the size and the shape of the molecule impact on its rate of sedimentation (S). As you can see, wide variation exists in the size of the three classes of RNA.

Ribosomal RNA is generally the largest of these molecules (as is generally reflected in its S values) and usually constitutes about 80 percent of all RNA in the cell. The various forms of rRNA found in prokaryotes and eukaryotes differ distinctly in size. The values in Table 9.4 are based on eukaryotic molecules. Ribosomal RNAs are important structural components of **ribosomes**, which function as a nonspecific workbench during the synthesis of proteins during the process of translation.

Messenger RNA molecules carry genetic information from the DNA of the gene to the ribosome, where translation occurs. They vary considerably in length, which is partly a reflection of the variation in the size of the gene serving as the template for transcription of mRNA species. Note that the values shown in Table 9.4 are only estimates, particularly at their upper limits.

TABLE 9.4	Sedimentation coefficients, molecular weights, and number of nucleotides for various RNAs			
RNA Type	*Abbreviation*	*Svedberg Coefficient (S)*	*Molecular Weight*	*Number of Nucleotides*
Ribosomal RNA	rRNA	5S	35,000	120
		5.8S	47,000	160
		18S	700,000	1900
		28S	1,800,000	4800
Transfer RNA	tRNA	5S	23,000–30,000	75–90
Messenger RNA	mRNA	5S	25,000–1,000,000	100–10,000

Precursors of many mRNAs, called **primary transcripts**, may demonstrate values considerably higher.

Transfer RNA, the smallest class of RNA molecules, carries amino acids to the ribosome during translation. Because more than one tRNA molecule interacts simultaneously with the ribosome, the molecule's smaller size facilitates these interactions.

We will discuss the functions of the three classes of RNA in much greater detail in Chapter 12. In addition, as we proceed through the text, we will encounter other unique RNAs that perform various roles. For example, **small nuclear RNA (snRNA)** participates in processing mRNAs (Chapter 12). **Telomerase RNA** is involved in DNA replication at the ends of chromosomes (Chapter 10). And, **antisense RNA** is involved in gene regulation (Chapter 13). Our purpose in this section has been to contrast the structure of DNA, which stores genetic information, with that of RNA, which most often functions in the expression of that information.

Hydrogen Bonds and the Analysis of Nucleic Acids

The unique nature of the hydrogen bond imparts an interesting and important set of qualities to the chemical behavior of nucleic acids under both laboratory and physiological conditions. For example, if DNA is isolated and subjected to slow heating, the double helix is denatured and unwinds. If a mixture of single strands that are complementary to each other are slowly cooled, they will reassociate, reforming the helix. In the laboratory, these transformations may be "tracked" by monitoring the absorption of UV light (or optical density) at 260 nm (OD_{260}), using a spectrophotometer.

During unwinding, the viscosity of DNA decreases and UV absorption increases. A melting profile, in which OD_{260} is plotted against temperature, is shown for two DNA molecules in Figure 9–15. The midpoint of each curve is called the **melting temperature (T_m)**, where 50 percent of the strands are unwound. The molecule with a higher T_m has a higher percentage of G≡C base pairs than A=T base pairs compared to the molecule with the lower T_m, since G≡C pairs share three hydrogen bonds compared to two present between A=T pairs.

Molecular Hybridization Techniques

The property of denaturation/renaturation of nucleic acids is the basis for one of the most powerful and useful techniques in molecular genetics—**molecular hybridization**. Provided that a reasonable degree of base complementarity exists and under the proper temperature conditions, two nucleic acid strands from different sources will rejoin. As a result, molecular hybridization is possible between DNA strands from different species and between DNA and RNA strands. For example, an RNA molecule will hybridize with the seg-

ment of DNA from which it was transcribed or with a DNA molecule from a different species, provided that its nucleotide sequence is nearly the same.

The technique can even be performed using the DNA present in cytological preparations as the "target" for hybrid formation. This process is called **in situ molecular hybridization**. Mitotic or interphase cells are first fixed to slides and then subjected to hybridization conditions. Single-stranded DNA or RNA is added, and hybridization is monitored. The nucleic acid that is added may either be radioactive or contain a fluorescent label to allow its detection. In the former case, the technique of autoradiography may be used.

In Figure 9–16, the use of a fluorescent label is illustrated. A "probe," consisting of a short fragment of DNA that is complementary to DNA present in the centromere regions of chromosomes has been hybridized. Fluorescence occurs only in the centromere regions, thus identifying each one along its chromosome. Because fluorescence is used, the technique is known by the acronym **FISH (fluorescent in situ hybridization)**. The use of this technique in identification of chromosomal locations housing other specific genetic information has been a valuable addition to the repertoire of experimental techniques available to geneticists.

Reassociation Kinetics and Repetitive DNA

One extension of molecular hybridization procedures is the analysis of the *rate of reassociation* of complementary single strands of DNA. This technique, called **reassociation kinetics,** was first refined and studied by Roy Britten and David Kohne.

The DNA used in such studies is first fragmented into small pieces as a result of shearing forces introduced dur-

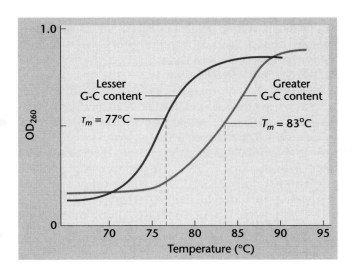

■ Figure 9–15   Comparison of the increase in UV absorbance with an increase in temperature for two DNA molecules with different G≡C contents. The molecule with a melting point (T_m) of 83°C has a greater G≡C content than the molecule with a T_m of 77°C.

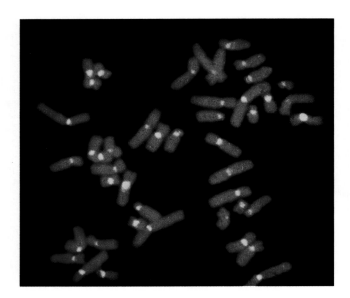

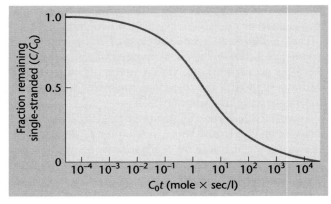

Figure 9–17 The ideal time course for reassociation of DNA (C/C_0) when, at time zero, all DNA consists of unique fragments of single-stranded complements. Note that the abscissa (C_0t) is plotted logarithmically.

Figure 9–16 *In situ* hybridization of human metaphase chromosomes using a fluorescent technique (FISH). The probe, specific to centromeric DNA, produces a yellow fluorescence signal indicating hybridization. The red fluorescence is produced by propidium iodide counterstaining of chromosomal DNA.

ing its isolation. The resultant DNA fragments have an average size of several hundred base pairs. These fragments are then dissociated into single strands by heating. The temperature is then lowered, and reassociation is monitored. During reassociation, pieces of single-stranded DNA collide randomly. If they are complementary, a stable double strand is formed; if not, they separate and are free to encounter other DNA fragments. The process continues until all possible matches are made.

The results of such an experiment are presented in Figure 9–17. The percentage of reassociation of DNA fragments is plotted against a logarithmic scale of normalized time, a function referred to as C_0t. In this term C_0 is equal to the initial concentration of DNA single strands and t is equal to time.

A great deal of information can be obtained from studies comparing the reassociation of DNA of different organisms. For example, we may compare the point in the reaction when one-half of the DNA is present as double-stranded fragments. This point is called the **half reaction time**. Provided that all DNA fragments contain unique nucleotide sequences and all are about the same size, $C_0t_{1/2}$ varies directly with the total length of the DNA.

In Figure 9–18, DNAs from three phage or bacterial sources are compared, each with a different genome size. As genome size increases, the curves obtained have a similar shape but are shifted farther and farther to the right, indicative of an extended reassociation time. Reassociation occurs more slowly in larger genomes because it takes longer for initial matches to be made if there are greater numbers of unique DNA fragments. This is so because collisions are random;

more sequences present will result in greater numbers of mismatches before each correct match is made.

When reassociation kinetics of DNA from eukaryotic organisms with much larger genome sizes were first studied, a surprising observation was made. Rather than exhibiting a reduced rate of reassociation, the data revealed that *some* of the DNA segments reassociate even more rapidly than those derived from *E. coli*! The remainder of the DNA, as expected because of its greater size and complexity, takes longer to reassociate.

For example, Britten and Kohne examined DNA derived from calf thymus tissue. Based on their observations (Figure 9–19), they hypothesized correctly that the rapidly reassociating fraction might represent repetitive sequences present many times in the calf genome. This interpretation would explain why these DNA segments reassociate so rapidly. Multiple copies of the same sequence are much more likely to make matches, thus reassociating more quickly than single copies. On the other hand, the remaining DNA segments consist of unique nucleotide sequences present only once in the genome. Because

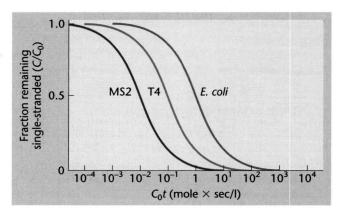

Figure 9–18 Comparison of the reassociation rate (C/C_0) of DNA derived from phage MS-2, phage T4, and *E. coli*. The genome of T4 is larger than MS-2, and that of *E. coli* is larger than T4.

there are many more of these unique sequences in calf thymus than in *E. coli*, their reassociation takes longer.

The copies present many times in the genome are referred to collectively as **repetitive DNA**. Repetitive DNA is prevalent in eukaryotic genomes and is important to our understanding of how genetic information is organized in chromosomes. Careful study has shown that there are various levels of repetition. Cases are known where short DNA sequences are repeated over a million times, where longer sequences are repeated only a few times, and where intermediate levels of sequence redundancy are present. The discovery of repetitive DNA was one of the first clues that much of the DNA in eukaryotes is not contained in genes that encode proteins! We will return to the topic of repetitive DNA in Chapter 11.

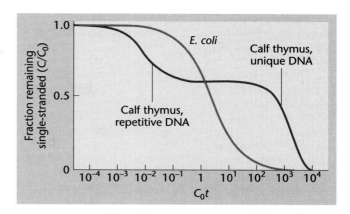

■ Figure 9–19 C_0t curve of calf thymus DNA compared with *E. coli*. The repetitive fraction of calf DNA reassociates more quickly than that of *E. coli*, while the more complex unique calf DNA takes longer to reassociate than that of *E. coli*.

Chapter Summary

1. The existence of a genetic material capable of replication, storage, expression, and mutation is deducible from the observed patterns of inheritance in organisms. Both proteins and nucleic acids were initially considered as possible candidates for the genetic material.

2. Transformation studies, as well as experiments using bacteria infected with bacteriophages, strongly suggested that DNA is the genetic material for bacteria and most viruses.

3. Initially, only circumstantial observations supported the concept of DNA controlling inheritance in eukaryotes. These included the distribution of DNA in the cell, quantitative analysis of DNA, and UV-induced mutagenesis. More recent recombinant DNA techniques, as well as experiments with transgenic mice, have provided direct experimental evidence that the eukaryote genetic material is DNA.

4. In some viruses, RNA serves as the genetic material. These include bacteriophages as well as some plant and animal viruses.

5. By the 1950s, many scientists sought to determine the structure of DNA. These efforts culminated in 1953 with Watson and Crick's proposal. Based on base-pairing information and X-ray diffraction data, they constructed a model, the key features of which include two antiparallel polynucleotide chains held together in a right-handed double helix by the hydrogen bonds formed between complementary bases. To date, the basic tenets of this double helix have held true.

6. RNA varies from DNA by virtue of most often being single-stranded, containing uracil rather than thymine, and having ribose rather than deoxyribose as its constituent sugar.

7. The unique nature of the hydrogen bond allows double-stranded nucleic acids to be dissociated and reassociated experimentally, leading to the analytical techniques of molecular hybridization and the study of reassociation kinetics.

8. The study of reassociation kinetics has led to the discovery of repetitive DNA sequences characteristic of eukaryotes in contrast to unique DNA sequences.

Key Terms

INSIGHTS and SOLUTIONS

In contrast to the preceding chapters, this chapter does not emphasize genetic problem solving. Instead, it recounts some of the initial experimental analysis that served as the cornerstone of modern genetics. Quite fittingly, then, our Insights and Solutions section shifts its emphasis to experimental rationale and analytical thinking, an approach that will continue through the remainder of the text whenever appropriate.

1. Based strictly on the transformation analysis of Avery, MacLeod, and McCarty, what objection might be made to the conclusion that DNA is the genetic material? What other conclusion might be considered?

Solution: Based solely on their results, it may be concluded that DNA is essential for transformation. However, DNA might have been a substance that caused capsular formation by converting nonencapsulated cells *directly* to ones with a capsule. That is, DNA may simply have played a catalytic role in capsular synthesis, leading to cells displaying smooth type III colonies.

2. What observations argue against this objection?

Solution: First, transformed cells pass the trait on to their progeny cells, thus supporting the conclusion that DNA is responsible for heredity, not for the direct production of polysaccharide coats.

Second, subsequent transformation studies over the next five years showed that other traits, such as antibiotic resistance, could be transformed. Therefore, the transforming factor has a broad general effect, not one specific to polysaccharide synthesis.

3. If RNA were the universal genetic material, how would this have affected the Avery experiment and the Hershey–Chase experiment?

Solution: In the Avery experiment, RNase rather than DNase would have eliminated transformation. Had this occurred, Avery and his colleagues would have concluded that RNA was the transforming factor. Hershey and Chase would have received identical results, since ^{32}P would also label RNA, but not protein.

4. Sea urchin DNA, which is double-stranded, was shown to contain 17.5 percent of its bases in the form of cytosine (C). What percentages of the other three bases are expected to be present in this DNA?

Solution: The amount of C equals G, so guanine is also present as 17.5 percent. The remaining bases, A and T, are present in equal amounts and together they represent the rest of the bases (100 − 35). Therefore, A = T = 65/2 = 32.5 percent.

Problems and Discussion Questions

1. The functions ascribed to the genetic material are replication, expression, storage, and mutation. What does each of these terms mean?

2. Discuss the reasons why proteins were generally favored over DNA as the genetic material before 1940. What was the role of the tetranucleotide hypothesis in this controversy?

3. Contrast the various contributions made to an understanding of transformation by Griffith with those of Avery and his coworkers.

4. Why were ^{32}P and ^{35}S chosen for use in the Hershey–Chase experiment? Discuss the rationale and conclusions of this experiment.

5. What observations are consistent with DNA serving as the genetic material in eukaryotes? List and discuss. What direct evidence exists?

6. What are the exceptions to the general rule that DNA is the genetic material in all organisms? What evidence supports these exceptions?

7. Draw the chemical structure of the three components of a nucleotide and then link the three together. What atoms are removed from the structures when the linkages are formed?

8. Adenine may also be named 6-amino purine. How would you name the other four nitrogenous bases using this alternate system? (—O is oxy, and —CH_3 is methyl.)

9. Describe the various characteristics of the Watson–Crick double-helix model for DNA.

10. What evidence did Watson and Crick have at their disposal in 1953? What was their approach in arriving at the structure of DNA?

11. Had Chargaff's data from a single source indicated the following, what might Watson and Crick have concluded?

	A	T	C	G
%	29	19	21	31

Why would this conclusion be contradictory to Wilkins' and Franklin's data?

12. List three main differences between DNA and RNA.

13. What is the chemical basis of molecular hybridization?

14. When the entire genome is utilized, why does unique sequence DNA take longer to reassociate than repetitive DNA sequences?

15. A genetics student was asked to draw the chemical structure of an adenine- and thymine-containing dinucleotide derived from DNA. The student made more than six major errors. His answer is shown below. One of them is circled, numbered 1A, and explained. Find five others. Circle them, number them 2A–6A, and briefly explain each, following the example.

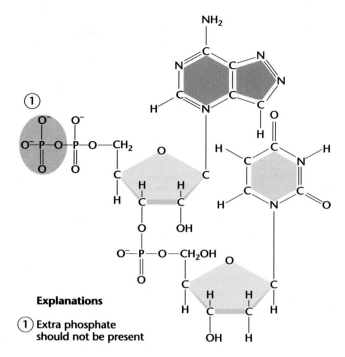

Explanations

① Extra phosphate should not be present

16. Newsdate: March 1, 2005. A unique creature has been discovered during exploration of outer space. Recently, its genetic material has been isolated and analyzed. This material is similar in some ways to DNA in its chemical makeup. It contains in abundance the four-carbon sugar erythrose and a molar equivalent of phosphate groups. Additionally, it contains six nitrogenous bases: adenine (A), guanine (G), thymine (T), cytosine (C), hypoxanthine (H), and xanthine (X). These bases exist in the following relative proportions:

$$A = T = H \text{ and } C = G = X$$

X-ray diffraction studies have established a regularity to the molecule and a constant diameter of about 30 Å. Together, these data have suggested a model for the structure of this molecule. (a) Propose a general model of this molecule. Describe it briefly. (b) What base-pairing properties must exist for H and for X in the model? (c) Given the constant diameter of 30 Å, do you think that H and X are either (1) both purines or both pyrimidines or (2) one is a purine and one is a pyrimidine?

17. A primitive eukaryote was discovered that displayed a unique nucleic acid as its genetic material. Analysis revealed the following:

(i) X-ray diffraction studies display a general pattern similar to DNA, but with somewhat different dimensions and more irregularity.

(ii) A major hyperchromic shift is evident upon heating and monitoring UV absorption at 260 nm.

(iii) Base composition analysis reveals four bases in the following proportions:

Adenine—8% Guanine—37%

Xanthine—37% Hypoxanthine—18%

(iv) About 75% of the sugars are deoxyribose while 25% are ribose.

Attempt to solve the structure of this molecule by postulating a model that is consistent with the above observations.

18. One of the most common spontaneous lesions that occurs in DNA under physiological conditions is the hydrolysis of the amino group of cytosine, converting it to uracil. What would be the effect of a uracil replacing cytosine in the DNA?

19. In some organisms, cytosine is methylated at carbon 5 of the pyrimidine ring after it is incorporated into DNA. If a 5-methyl cytosine is hydrolyzed as described in the previous problem, what base will be generated?

Selected Readings

Avery, O. T., MacLeod, C. M., and McCarty, M. 1944. Studies on the chemical nature of the substance inducing transformation of pneumococcal types. Induction of transformation by a desoxyribonucleic acid fraction isolated from pneumococcus type III. *J. Exp. Med.* 79:137–58. (Reprinted in Taylor, J. H. 1965. *Selected papers in molecular genetics.* Orlando, FL: Academic Press.)

Britten, R. J., and Kohne, D. E. 1970. Repeated segments of DNA. *Sci. Am.* (Apr.) 222:24–31.

Chargaff, E. 1950. Chemical specificity of nucleic acids and mechanism for their enzymatic degradation. *Experientia* 6:201–9.

Dawson, M. H. 1930. The transformation of pneumococcal types: I. The interconvertibility of type-specific *S. pneumococci. J. Exp. Med.* 51:123–47

DeRobertis, E. M., and Gurdon, J. B. 1979. Gene transplantation and the analysis of development. *Sci. Am.* (Dec.) 241:74–82.

Dickerson, R. E. 1983. The DNA helix and how it is read. *Sci. Am.* (June) 249:94–111.

Dickerson, R. E., et al. 1982. The anatomy of A-, B-, and Z-DNA. *Science* 216:475–85.

Dubos, R. J. 1976. *The professor, the institute and DNA: Oswald T. Avery, his life and scientific achievements.* New York: Rockefeller University Press.

Felsenfeld, G. 1985. DNA. *Sci. Am.* (Oct.) 253:58–78.

Fraenkel-Conrat, H., and Singer, B. 1957. Virus reconstruction: II. Combination of protein and nucleic acid from different strains. *Biochem. Biophys. Acta* 24:530–48 (Reprinted in Taylor, J.H. 1965. *Selected papers in molecular genetics*, Orlando, FL: Academic Press.)

Franklin, R. E., and Gosling, R. G. 1953. Molecular configuration in sodium thymonucleate. *Nature* 171:740–41.

Griffith, F. 1928. The significance of pneumococcal types. *J. Hyg.* 27:113–59.

Guthrie, G. D., and Sinsheimer, R. L. 1960. Infection of protoplasts of *Escherichia coli* by subviral particles. *J. Mol. Biol.* 2:297–305.

Hershey, A. D., and Chase, M. 1952. Independent functions of viral protein and nucleic acid and in growth of bacteriophage. *J. Gen. Physiol.* 36:39–56. (Reprinted in Taylor, J. H. 1965. *Selected papers in molecular genetics.* Orlando, FL: Academic Press.)

Judson, H. 1979. *The eighth day of creation: Makers of the revolution in biology.* New York: Simon & Schuster.

Levene, P. A., and Simms, H. S. 1926. Nucleic acid structure as determined by electrometric titration data. *J. Biol. Chem.* 70:327–41.

McCarty, M. 1980. Reminiscences of the early days of transformation. *Annu. Rev. Genet.* 14:1–16.

————. 1985. *The transforming principle: Discovering that genes are made of DNA.* New York: W. W. Norton.

Olby, R. 1974. *The path to the double helix.* Seattle: University of Washington Press.

Palmiter, R. D., and Brinster, R. L. 1985. Transgenic mice. *Cell* 41:343–45.

Pauling, L., and Corey, R. B. 1953. A proposed structure for the nucleic acids. *Proc. Natl. Acad. Sci. USA* 39:84–97.

Rich, A., Nordheim, A. and Wang, A. H.-J. 1984. The chemistry and biology of left-handed Z-DNA. *Annu Rev. Biochem.* 53:791–846.

Spizizen, J. 1957. Infection of protoplasts by disrupted T2 viruses. *Proc. Natl. Acad. Sci. USA* 43:694–701.

Stent, G. S., ed. 1981. *The double helix: Text, commentary review, and original papers.* New York: W. W. Norton.

Stewart, T. A., Wagner, E. F., and Mintz, B. 1982. Human β-globin gene sequences injected into mouse eggs, retained in adults, and transmitted to progeny. *Science* 217:1046–48.

Varmus, H. 1988. Retroviruses. *Science* 240:1427–35.

Watson, J. D. 1968. *The double helix.* New York: Atheneum.

Watson, J. D., and Crick, F. C. 1953a. Molecular structure of nucleic acids. A structure for deoxyribose nucleic acids. *Nature* 171:737–38.

Watson, J. D., and Crick, F. C. 1953b. Genetic implications of the structure of deoxyribose nucleic acid. *Nature* 171:964.

Weinberg, R. A. 1985. The molecules of life. *Sci. Am.* (Oct.) 253:48–57.

Wilkins, M. H. F., Stokes, A. R., and Wilson., H. R. 1953. Molecular structure of desoxypentose nucleic acids. *Nature* 171:738–40.

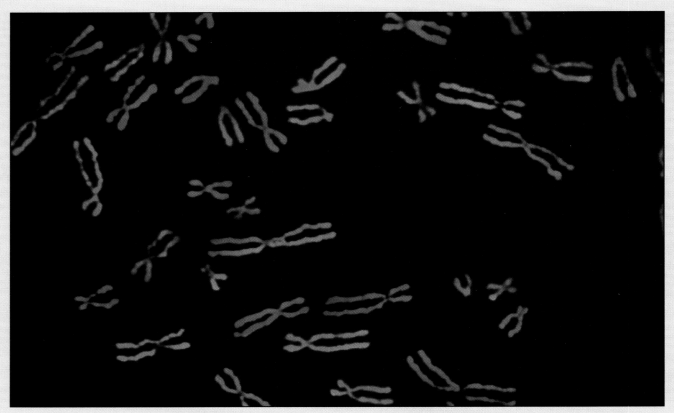

Human metaphase chromosomes, each composed of two sister chromatids joined at a common centromere.

CHAPTER OUTLINE

DNA—Replication and Synthesis

Chapter Concepts

Genetic continuity between parental and progeny cells is made possible by semiconservative replication of DNA, as predicted by the Watson–Crick model. Each strand of the parent helix serves as a template for the production of its complement. Synthesis of DNA is a complex but orderly process orchestrated by a myriad of enzymes and other molecules. Together, they function with great fidelity to polymerize nucleotides into polynucleotide chains. Other enzymes interact with DNA, leading to genetic recombination.

Following Watson and Crick's proposal for the structure of DNA, scientists focused their attention on how this molecule is replicated. This process is an essential function of the genetic material and must be executed precisely if genetic continuity between cells is to be maintained following cell division. This is an enormous, complex task. Consider for a second that in the human genome, some three billion (10^9) base pairs exist within the 23 chromosomes. To duplicate a molecule of this size faithfully requires a mechanism of extreme precision. Even an error rate of only 10^{-6} (one in a million) will still create errors (3000), obviously an excessive number during each replication cycle. While it is not error-free, an extremely accurate system of DNA replication has evolved in all organisms.

As Watson and Crick wrote in their 1953 paper, the model of the double helix provided them the initial insight into how replication could occur. This mode, called **semiconservative replication**, has since received strong experimental support from studies of viruses, prokaryotes, and eukaryotes.

Once the general mode of replication was made clear, research was intensified to determine the precise details of DNA synthesis. What has since been discovered is that numerous enzymes and other proteins are needed to copy a DNA helix. Because of the complexity of the chemical events during synthesis, this subject remains an extremely active area of research.

In this chapter, we will discuss the general mode of replication as well as the specific details of the synthesis of DNA. The research leading to this knowledge is still another link in our understanding of life processes at the molecular level.

The Mode of DNA Replication

It was apparent to Watson and Crick that because of the arrangement and nature of the nitrogenous bases, each strand of a DNA double helix could serve as a template for the synthesis of its complement (Figure 10–1). They proposed that if the helix were unwound, each nucleotide along the two parent strands would have an affinity for its complementary nucleotide. As we learned in Chapter 9, the complementarity is due to the potential hydrogen bonds that can be formed. If thymidylic acid (T) were present, it would "attract" adenylic acid (A); if guanidylic acid (G) were present, it would "attract" cytidylic acid (C); likewise, A would attract T, and C would attract G. If these nucleotides were then linked covalently into polynucleotide chains along both templates, the result would be the production of two new but identical double strands of DNA. This concept is illustrated in Figure 10–1. Each replicated DNA molecule would consist of one "old" and one "new" strand; hence the reason for the name semiconservative replication.

Two other possible modes of replication also rely on the parental strands as a template. In **conservative**

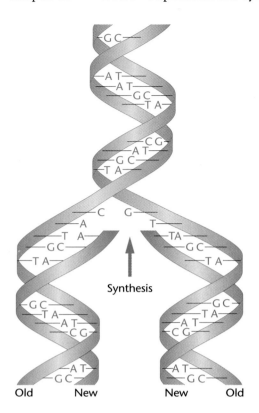

Synthesis

Old   New   New   Old

■ Figure 10–1 General model of semiconservative replication of DNA.

replication, synthesis of complementary polynucleotide chains occurs as described above. Following synthesis, however, the two newly created strands are brought together, and the parental strands reassociate. The original helix is thus "conserved."

In the second alternative mode, called **dispersive replication**, the parental strands are seen to be dispersed into two new double helices following replication. Each strand would then consist of both old and new DNA. This mode would involve cleavage of the parental strands during replication. It is the most complex of the three possibilities and is therefore least likely.

The Meselson–Stahl Experiment

In 1958, Matthew Meselson and Franklin Stahl published the results of an experiment providing strong evidence that semiconservative replication is used by cells to produce new DNA molecules. *Escherichia coli* cells were grown for many generations in a medium where $^{15}NH_4Cl$ (ammonium chloride) was the only nitrogen source. A "heavy" isotope of nitrogen, ^{15}N, contains one more neutron than the naturally occurring ^{14}N isotope. Unlike "radioactive" isotopes, ^{15}N is stable and thus does not decay (i.e., it is not radioactive). After many generations, all nitrogen-containing molecules, including the nitrogenous bases of DNA, contained the heavier isotope in the *E. coli* cells. DNA containing ^{15}N may be distinguished from ^{14}N-contain-

ing DNA by the use of **sedimentation equilibrium centrifugation**, in which samples are "forced" by centrifugation through a density gradient of a heavy metal salt such as cesium chloride. The more dense ^{15}N-DNA will reach equilibrium in the gradient at a point closer to the bottom (where the density is greater) than will ^{14}N-DNA.

In this experiment, uniformly labeled ^{15}N cells were transferred to a medium containing only $^{14}NH_4Cl$. Thus, all new synthesis of DNA during replication contained only the "lighter" isotope of nitrogen. The time of transfer to the new medium was taken as time zero ($t = 0$). The *E. coli* cells were allowed to replicate during several generations, with cell samples removed after each replication cycle. From each sample, DNA was isolated and subjected to sedimentation equilibrium centrifugation. The results are depicted in Figure 10–2.

After one generation, the isolated DNA was all present in a single band of intermediate density—the expected result for semiconservative replication. Each replicated molecule would be composed of one new ^{14}N-strand and one old ^{15}N-strand, as seen in Figure 10–3. This result effectively ruled out the conservative replication mode, in which two distinct bands are predicted to occur.

After two cell divisions, DNA samples showed two density bands: One was intermediate and the other was lighter, corresponding to the ^{14}N position in the gradient. Similar results occurred after a third generation, except that the proportion of the ^{14}N-band increased. If replication were dispersive, all subsequent generations after $t = 0$ would demonstrate DNA of an intermediate density. In each subsequent generation the ratio $^{14}N/^{15}N$ would increase, and the hybrid band would become lighter and lighter, eventually approaching the ^{14}N-band. Since this result was not observed, the dispersive mode was ruled out. Thus, the results of the Meselson–Stahl experiment provided strong support for the semiconservative mode of DNA replication, as postulated by Watson and Crick.

Semiconservative Replication in Eukaryotes

In 1957, the year before the work of Meselson and his colleagues was published, evidence was presented by J. Herbert Taylor, Philip Woods, and Walter Hughes that semiconservative replication also occurs in a eukaryotic organism. They experimented with root tips of the broad bean *Vicia faba*, which are an excellent source of dividing cells. These researchers examined the chromosomes of these cells following replication of DNA. They were able to monitor the process of replication by labeling DNA with ^3H-thymidine, a radioactive precursor of DNA, and then performing autoradiography.

The technique of **autoradiography** is a cytological procedure that allows the location of an isotope to be identified within the cell. In this procedure, a photographic emulsion is placed over a section of cellular

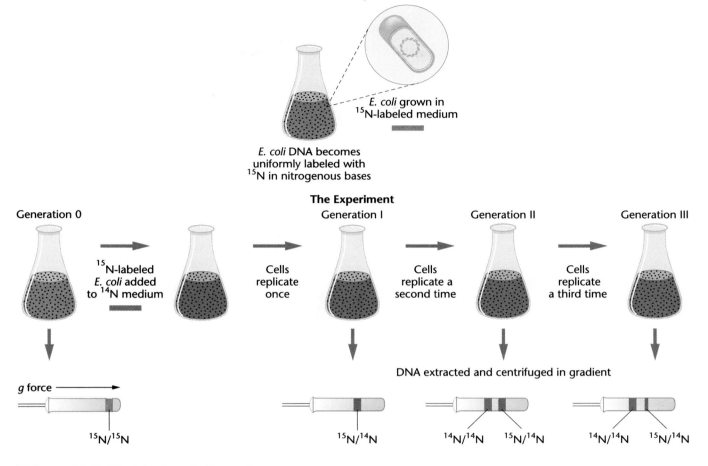

Figure 10–2 The Meselson–Stahl experiment.

material (root tips in this experiment), and the preparation is stored in the dark. The slide is then developed, much as photographic film is processed. Because the radioisotope emits energy, the emulsion turns black at the approximate point of emission following development. The end result is the presence of dark spots or "grains" on the surface of the section, identifying within the cell the location of newly synthesized DNA.

Root tips were grown for approximately one generation in the presence of the radioisotope and then placed in unlabeled medium, where cell division continued. At the conclusion of each generation, cultures were arrested at metaphase by the addition of colchicine (a chemical derived from the crocus plant, which poisons the mitotic spindle fibers), and chromosomes were examined by autoradiography. Figure 10–4 illustrates

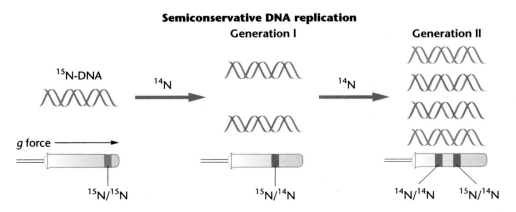

Figure 10–3 The expected results of two generations of semiconservative replication in the Meselson–Stahl experiment.

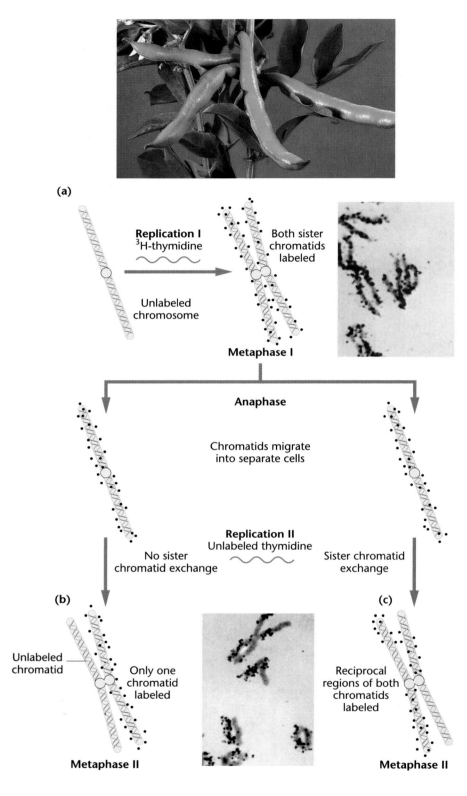

(a)

Replication I
³H-thymidine

Unlabeled
chromosome

Both sister
chromatids
labeled

Metaphase I

Anaphase

Chromatids migrate
into separate cells

Replication II
Unlabeled thymidine

No sister
chromatid exchange

Sister chromatid
exchange

(b)

Unlabeled
chromatid

Only one
chromatid
labeled

Metaphase II

(c)

Reciprocal
regions of both
chromatids
labeled

Metaphase II

■ Figure 10–4 Depiction of the experiment by Taylor, Woods, and Hughes demonstrating the semiconservative mode of replication of DNA in root tips of *Vicia faba*. The plant is shown in the top photograph. (a) An unlabeled chromatid proceeds through the cell cycle in the presence of ³H-thymidine. As it enters mitosis, both sister chromatids of the chromosome are labeled, as shown by autoradiography (black grains). (b) After a second generation of replication, this time in the absence of ³H-thymidine, only one chromatid of each chromosome is expected to be surrounded by grains. Except where a reciprocal exchange has occurred between sister chromatids (c), the expectation was upheld. The two photographs are of the actual autoradiograms generated in the expeiment.

replication of a single chromosome over two division cycles as well as the distribution of grains. In this experiment, labeled thymidine is found only in association with chromatids that contain newly synthesized DNA.

The results are compatible with the semiconservative mode of replication. After the first replication cycle, radioactivity is detected over both sister chromatids. This finding is expected because each chromatid will contain one "new" radioactive DNA strand and one "old" unlabeled strand. After the second replication cycle, *which also takes place in unlabeled medium*, only one of the two new sister chromatids should be radioactive because half of the parent strands are unlabeled. With only minor exceptions of **sister chromatid exchange** (see Chapter 6), this result was observed.

Together, the Meselson–Stahl experiment and the experiment by Taylor, Woods, and Hughes soon led to the general acceptance of the semiconservative mode of replication. The same conclusion has been reached in studies with other organisms. These experiments also strongly supported Watson and Crick's proposal for the double-helix model of DNA.

Replication Origins and Forks

The semiconservative mode of replication represents the general pattern by which DNA is duplicated. Before turning to the details of how DNA is actually synthesized, we will mention several other topics that are relevant to the complete description of semiconservative replication. The first concerns the **origin of replication** along any chromosome. Where along the chromosome is DNA replicated initially? Is each point of origin random or located in a specific region along the chromosome? And, is there only a single origin, or more than one point where DNA synthesis begins? As we address these questions, we shall define the length of DNA that is replicated following the initiation of synthesis at a single origin as a unit called the **replicon**.

A second issue involves the direction of replication. Once it begins, does it move in a single direction or in both directions away from the origin? This consideration distinguishes between **unidirectional** and **bidirectional replication**, respectively. At each point of replication, the strands of the helix must unwind, creating what is called a **replication fork**. Bidirectional replication creates two such forks, which move apart in opposite directions away from the origin.

The evidence is reasonably clear regarding these topics. In bacteria and most bacterial viruses, which have only a single circular chromosome, there is one specific region where replication is initiated. In *E. coli* this region, called *oriC*, has been located. It consists of 245 base pairs, but only a small number are actually essential to the initiation of DNA synthesis. Since there is only a single point of origin of DNA synthesis, in bacteriophages and bacteria the entire chromosome constitutes one replicon.

In *E. coli*, replication is bidirectional from *ori C*, proceeding in both directions until the entire chromosome is replicated, as illustrated in Figure 10–5. Two replication forks are established that move away from the origin in opposite directions and eventually merge as semiconservative replication of the entire chromosome is completed. The termination region of the chromosome is called *ter*.

■ Figure 10–5
Bidirectional replication of the *E. coli* chromosome. The arrows identify the advancing replication forks. The micrograph is of a bacterial chromosome in the process of replication.

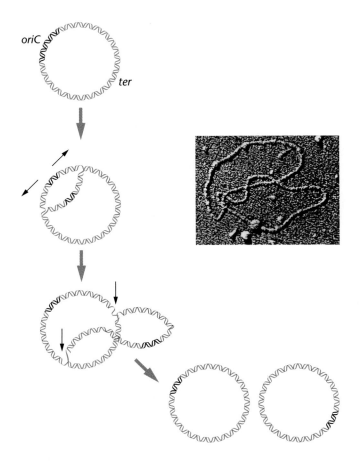

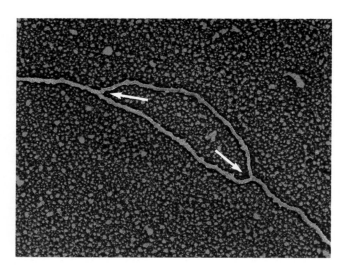

■ **Figure 10–6** Transmission electron micrograph of human DNA from a HeLa cell, illustrating dual replication forks (see arrows) and the replication bubble that characterizes DNA replication within a single replicon.

Compared to bacteria, as described above, a major difference exists in eukaryotes. While replication is bidirectional, creating two replication forks at each point of initiation of replication (Figure 10–6), there are multiple origins along each chromosome. As a result, during the S phase of interphase, numerous replicating events occur along each chromosome. Eventually, the numerous replication forks merge, completing replication of the entire chromosome. The presence of multiple replicons is undoubtedly related to the much greater length and complexity of a single eukaryotic chromosome compared to one from a bacterium. With many replicons active simultaneously, replication can be completed much more quickly.

Synthesis of DNA in Microorganisms

The determination that replication is semiconservative and bidirectional indicates only the *pattern* of DNA duplication and the association of finished strands with one another once synthesis is completed. A much more

complex issue is how the *actual synthesis* of long complementary polynucleotide chains occurs on a DNA template. As in most studies of molecular biology, this question was first approached by using microorganisms. Research began about the same time as the Meselson–Stahl work, and even today this topic is an active area of investigation. What is most apparent in this research is the tremendous chemical complexity of the biological synthesis of DNA.

DNA Polymerase I

Studies of the enzymology of DNA replication were first reported by Arthur Kornberg and colleagues in 1957. They isolated an enzyme from *E. coli* that was able to direct DNA synthesis in a cell-free (*in vitro*) system. The enzyme is now called **DNA polymerase I**, since it was the first of several to be isolated. Kornberg determined the following major requirements for *in vitro* DNA synthesis under the direction of the enzyme.

1. All four deoxyribonucleoside triphosphates (dATP, dCTP, dGTP, dTTP = dNTP)*
2. Template DNA

If any one of the four deoxyribonucleoside triphosphates was omitted from the reaction, no synthesis occurred. If derivatives of these precursor molecules other than the nucleoside triphosphate were used (nucleotides or nucleoside diphosphates), synthesis did not occur. If no template DNA was added, synthesis of DNA occurred but was reduced greatly. Template-dependent synthesis directed by Kornberg's enzyme appeared to be exactly the type required for semiconservative replication. The reaction is summarized in Figure 10–7. The enzyme has since been shown to consist of a single polypeptide containing 928 amino acids.

Fidelity of Synthesis

Having shown how DNA was synthesized, Kornberg sought to demonstrate the accuracy, or fidelity, with which the enzyme had replicated the DNA template.

*dNTP designates the deoxyribose forms of the four nucleoside triphosphates; in a similar way, dNMP refers to the monophosphate forms.

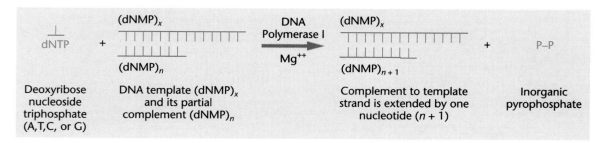

■ **Figure 10–7** The chemical reaction catalyzed by DNA polymerase I. During each step, a single nucleotide is added to the growing complement of the DNA template, using a nucleoside triphosphate as the substrate. The release of inorganic pyrophosphate drives the reaction energetically.

Deoxyribose nucleoside triphosphate (A,T,C, or G) + DNA template (dNMP)$_x$ and its partial complement (dNMP)$_n$ → Complement to template strand is extended by one nucleotide ($n+1$) + Inorganic pyrophosphate

Because the nucleotide sequences of the template and the product could not be determined in 1957, he had to rely initially on several indirect methods.

One of Kornberg's approaches was to compare the nitrogenous base compositions of the DNA template with those of the recovered DNA product. Table 10.1 shows Kornberg's base composition analysis of three DNA templates. These may be compared with the DNA product synthesized in each case. Within experimental error, the base composition of each product agreed with the template DNAs used. These data, along with other types of comparisons of template and product, suggested that the templates were replicated faithfully.

Synthesis of Biologically Active DNA

Despite Kornberg's extensive work, not all researchers were convinced that DNA polymerase I was the enzyme that replicates DNA within cells (*in vivo*). The primary reservations involved observations that the *in vitro* rate of synthesis was much slower than the *in vivo* rate, that the enzyme was much more effective replicating single-stranded DNA than double-stranded DNA, and that the enzyme appeared to be able to *degrade* DNA as well as to *synthesize* it.

Faced with the uncertainty of the true cellular function of DNA polymerase I, Kornberg pursued another approach. He reasoned that if the enzyme could be used to synthesize **biologically active DNA** *in vitro*, then DNA polymerase I must be the major catalyzing force for DNA synthesis within the cell. The term *biological activity* means that the DNA synthesized is capable of supporting metabolic activities and directing reproduction of the organism from which it was originally duplicated.

In 1967, Mehran Goulian, Kornberg, and Robert Sinsheimer showed that the DNA of the small bacteriophage φX174 could be completely copied by DNA polymerase I *in vitro*, and that the new product could be isolated and used to infect *E. coli*. This resulted in the production of mature phages under the direction of the synthetic DNA, thus demonstrating biological activity!

This demonstration of biological activity was viewed as a precise assessment of faithful copying. If even a single error had occurred to alter the base sequence of any of the 5386 nucleotides constituting the φX174 chromosome, the change might easily have caused a mutation that would prohibit the production of viable phages.

DNA Polymerases II and III

Although DNA synthesized under the direction of polymerase I demonstrated biological activity, a more serious reservation about the enzyme's true biological role was raised in 1969. Peter DeLucia and John Cairns reported the discovery of a mutant strain of *E. coli* that was deficient in polymerase I activity. The mutation was designated *polA1*. In the absence of the functional enzyme, this mutant strain of *E. coli* still duplicated its DNA and reproduced successfully! Other properties of the mutation led DeLucia and Cairns to conclude that in the absence of polymerase I, these cells are highly deficient in their ability to "repair" DNA. For example, the mutant strain is highly sensitive to ultraviolet light and radiation, both of which damage DNA and are therefore mutagenic. Nonmutant bacteria are able to repair a great deal of UV-induced damage.

These observations led to two conclusions:

1. There must be at least one other enzyme present in *E. coli* cells that is responsible for replicating DNA *in vivo*.

2. DNA polymerase I may serve only a secondary function *in vivo*. This function is now believed by Kornberg and others to be critical to the *fidelity* of DNA synthesis, but this enzyme is not one that actually synthesizes the complementary strand.

To date, two unique DNA polymerases have been isolated from cells lacking polymerase I activity. These two enzymes have also been isolated from normal cells that also contain polymerase I.

The characteristics of these two enzymes, called **DNA polymerases II and III**, are contrasted with DNA polymerase I in Table 10.2. As is evident from that information, all three share several characteristics. While none can initiate DNA synthesis on a template, all can elongate an existing DNA strand, called a **primer**. As we shall see, RNA is also an adequate primer and is, in fact, what is utilized initially!

Another important characteristic is the direction of synthesis. Elongation occurs by polymerizing nucleotides in the 5′-to-3′ direction. That is, each new nucleotide is added at its 5′-phosphate end to the 3′-OH

TABLE 10.1	Base composition of the DNA template and the product of replication in Kornberg's early work

Organism	Template or Product	%A	%T	%G	%C
T2	Template	32.7	33.0	16.8	17.5
	Product	33.2	32.1	17.2	17.5
E. coli	Template	25.0	24.3	24.5	26.2
	Product	26.1	25.1	24.3	24.5
Calf	Template	28.9	26.7	22.8	21.6
	Product	28.7	27.7	21.8	21.8

Source: Kornberg, 1960.

TABLE 10.2	Comparative properties of the three bacterial DNA polymerases

Properties	I	II	III
Initiation of chain synthesis	–	–	–
5′-3′ polymerization	+	+	+
3′-5′ exonuclease activity	+	+	+
5′-3′ exonuclease activity	+	–	–
Molecules of polymerase/cell	400	?	15

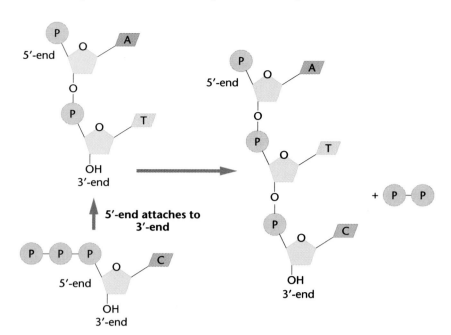

■ Figure 10–8
Demonstration of 5'-to-3' synthesis of DNA.

of the growing polynucleotide. Each addition creates a new exposed 3'-OH group on the sugar, which can then participate in the next reaction. Synthesis occurring in this way is illustrated in Figure 10–8.

The DNA polymerase enzymes are all large, complex proteins that exhibit molecular weights in excess of 100,000 daltons. All three possess 3'-to-5' **exonuclease activity**. This means that they can polymerize in one direction and then reverse directions and excise nucleotides just added. As we will see, this activity provides a capacity to proofread and remove an incorrect nucleotide.

What, then, are the roles of the three polymerases *in vivo*? Polymerase III is considered to be the enzyme responsible for the polymerization essential to replication. Its 3'-to-5' exonuclease activity also allows it to proofread, excise, and then correct base pairs created in error during polymerization. As we will soon see, gaps are a natural occurrence on one of the two strands during replication as RNA primers are removed. It is believed that polymerase I, originally studied by Kornberg, is responsible for removing the primer as well as for the synthesis that fills these gaps. Its exonuclease activity also allows proofreading to occur during this process. Alternatively, another enzyme, RNase, has been discovered that may actually remove the RNA primer prior to the activation of DNA polymerase I activity.

Polymerase II appears to be involved in repair synthesis of DNA that has been damaged by external forces, such as ultraviolet light. It is encoded by a gene that may be activated by disruption of DNA synthesis at the replication fork.

We end this section by emphasizing the complexity of the DNA polymerase III molecule. Its active form, called a **holoenzyme**, consists of two sets (a dimer) of ten separate polypeptide chains (see Table 10.3) and exhibits a molecular weight in excess of 600,000 dal-

tons. The largest subunit, α, has a molecular weight of 140,000 daltons and, along with two other subunits (ε and θ), constitutes the "core" enzyme responsible for the polymerization activity of the holoenzyme. The α subunit is responsible for nucleotide polymerization on the template strands, whereas the ε subunit of the core enzyme possesses the 3'–5' exonuclease activity.

A second group of five subunits (γ, δ, δ', χ, and ψ) forms what is called the γ complex, which is involved in "loading" the enzyme onto the template at the replication fork. This enzymatic function is energy-requiring and dependent on the hydrolysis of ATP. As we will discuss in more detail below, dimers of the β subunit serve as donut-shaped clamps that prevent the core enzyme from falling off the template during polymerization. Finally, the τ subunit functions to hold together the two core polymerases

TABLE 10.3	Subunits constituting DNA polymerase III holoenzyme	
Subunit	*Function*	*Groupings*
α	5'–3' polymerization	"Core" enzyme: elongates poly-nucleotide chain and proofreads
ε	3'–5' exonuclease	
θ	??	
γ	Loads enzyme on template (serves as clamp loader)	γ complex
δ		
δ'		
χ		
ψ		
β	Sliding clamp structure (processivity factor)	
τ	Dimerizes core complex	

at the replication fork. Together with several other proteins at the replication fork, a complex nearly as large as a ribosome is present. This molecular entity is referred to as a **replisome**. Further details will be forthcoming later concerning the function of DNA polymerase III.

DNA Synthesis: A Model

We have thus far established that replication is semiconservative and bidirectional along a single replicon in bacteria and many viruses. Also, we know that synthesis is in the 5'-to-3' mode under the direction of DNA polymerase III, creating two replication forks. These move in opposite directions away from the origin of synthesis. We are now ready to pursue several other topics that describe how DNA synthesis is accomplished at a replication fork. In combination with the information provided above, we shall establish a coherent model of DNA replication that takes into account the following additional points.

1. A mechanism must exist by which the helix is initially unwound (or denatured) and stabilized in this "open" configuration so that synthesis may proceed along both strands.

2. As unwinding and subsequent DNA synthesis proceeds, increased coiling creates tension farther down the helix, which must be reduced.

3. A primer of some sort must be synthesized so that polymerization can commence under the direction of DNA polymerase III.

4. Once the RNA primers have been synthesized, DNA polymerase III commences synthesis of the complement of both strands of the parent molecule. Because the strands are antiparallel, continuous synthesis in the direction in which the replication fork moves is possible along only one of the two strands. On the other strand, synthesis is discontinuous in the opposite direction.

5. The RNA primers must be removed prior to completion of replication. The gaps that are temporarily created must be filled with DNA that is complementary to the template at each location.

6. The newly synthesized DNA strand that fills each temporary gap must be ligated to the adjacent strand of DNA.

As we consider the above points, Figures 10–9, 10–10, and 10–11 will be used to illustrate how each issue is resolved. Figure 10–12 will summarize the model of DNA synthesis.

Unwinding the DNA Helix

As discussed earlier, there is only a single origin along the circular chromosome of most bacteria and viruses at which DNA synthesis is initiated. This region of the *E.*

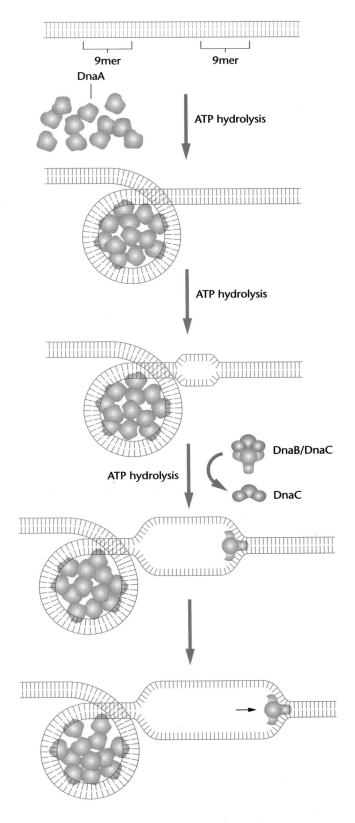

■ Figure 10–9 Helical unwinding of DNA during replication as accomplished by DnaA, DnaB, and DnaC proteins. Initial binding of many monomers of dnaA occurs at DNA sites containing repeating sequences of 9 nucleotides, called 9mers. Not illustrated are 13mers that are also involved.

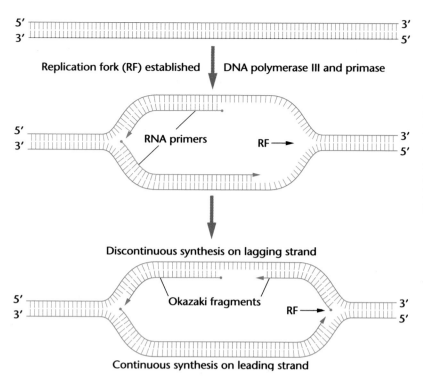

Replication fork (RF) established | DNA polymerase III and primase

RNA primers

RF →

Discontinuous synthesis on lagging strand

Okazaki fragments

RF →

Continuous synthesis on leading strand

■ Figure 10–10 **Illustration** of the opposite polarity of DNA synthesis (shown in blue) along the two strands necessitated by the requirement of 5′-to-3′ synthesis of DNA polymerase III. On the lagging strand, synthesis must be discontinuous, resulting in the production of Okazaki fragments. On the leading strand, synthesis is continuous. RNA primers are utilized and shown in green.

coli chromosome, *oriC*, has been particularly well studied. Called oriC, the origin of replication consists of 245 base pairs characterized by the presence of repeating sequences of 9 and 13 bases (called **9mers** and **13mers**). One particular protein, called **DnaA** (because it is encoded by the gene called dnaA), is responsible for the initial step in unwinding the helix. A number of subunits of the DnaA protein bind to each of several 9mers. This step is essential in facilitating the subsequent binding of **DnaB** and **DnaC** proteins that further open and destabilize the helix (Figure 10–9). Proteins such as these that require the energy normally supplied by the hydrolysis of ATP in order to break hydrogen bonds and denature the double helix are called **helicases**.

As unwinding proceeds, other proteins, called **single-stranded binding proteins (SSBPs)**, stabilize this conformation and a coiling tension is created ahead of the replication fork. Oftentimes, various forms of **supercoiling** occur. These may take the form of added twists and turns of the DNA in circular molecules, much like the coiling that would be created in a rubber band by holding one end and twisting the other. Such supercoiling can be relaxed by the action of an enzyme, **DNA gyrase**, a member of a larger group referred to as **DNA topoisomerases**. Depending on the form of the enzyme involved, either single- or double-stranded "cuts" are made by DNA gyrase. The enzyme also catalyzes localized manipulations of DNA strands that have the effect of "undoing" the twists and knots created during supercoiling. The strands are then resealed. To drive these various reactions, the energy released during ATP hydrolysis is required.

In combination with the polymerase complex, these proteins comprise an array of molecules that partici-

pate in DNA synthesis and are part of what we have previously called the replisome.

Initiation of Synthesis

Once a small portion of the helix is unwound, initiation of synthesis may occur. As mentioned previously, DNA polymerase III requires a free 3′ end as part of a primer in order to elongate a polynucleotide chain. This prompted researchers to investigate how the first nucleotide could be added, as no free 3′-hydroxyl group is initially present. There is now evidence that RNA is involved as the primer in initiating DNA synthesis.

It is thought that a short segment of RNA, complementary to DNA, is first synthesized on the DNA template. The RNA, about 5 to 15 nucleotides long, is made under the direction of a form of RNA polymerase called **primase**. The RNA polymerase does not require a free 3′ end to initiate synthesis. It is to this short segment of RNA that DNA polymerase III begins to add 5′-deoxyribonucleotides (Figure 10–10). After DNA synthesis occurs at an area adjacent to the RNA primer, the RNA segment is clipped off and replaced with DNA. Both steps are thought to be performed by DNA polymerase I. RNA priming has been recognized in viruses, bacteria, and several eukaryotic organisms and is thought to be a universal phenomenon.

Continuous and Discontinuous DNA Synthesis

We must now reconsider the fact that the two strands of a double helix are antiparallel to each other. One runs in the 5′ to 3′ direction, while the other has the opposite

3' to 5' polarity. Because DNA polymerase III synthesizes DNA in only the 5'–3' direction, simultaneous synthesis of antiparallel strands along an advancing replication fork occurs in one direction along one strand and in the opposite direction on the other.

As the strands unwind and the replication fork progresses down the helix, only one of the two DNA strands can serve as a template for **continuous DNA synthesis**. This is called the **leading strand**. As the fork progresses, many points of initiation are necessary on the opposite, or **lagging strand**, resulting in **discontinuous DNA synthesis** (Figure 10–10).

Evidence in support of discontinuous DNA synthesis was first provided by Reiji and Tuneko Okazaki and their colleagues. They discovered that when bacteriophage DNA is replicated in *E. coli,* some of the newly formed DNA that is hydrogen-bonded to the template strand is present as small fragments containing 1000 to 2000 nucleotides. RNA primers are part of each such fragment. These pieces, called **Okazaki fragments**, must then be joined enzymatically. As synthesis proceeds, the Okazaki fragments of low molecular weight are indeed converted into longer and longer DNA strands of higher molecular weight.

Discontinuous synthesis of DNA requires the removal of the RNA primer as well as an enzyme that can unite the smaller products into the longer continuous molecules that represent the lagging strand. While DNA polymerase I removes the primer and replaces the missing nucleotides, **DNA ligase** has been shown to be capable of catalyzing the formation of the phosphodiester bond, the last step in sealing the gap existing between discontinuously synthesized strands. The evidence that DNA ligase does perform this function during DNA synthesis is strengthened by the observation of a ligase-deficient mutant strain (*lig*) of *E. coli*. In this strain, Okazaki fragments accumulate in particularly large amounts. Apparently, they are not joined adequately.

Discontinuous synthesis on the lagging strand, as described above, is characteristic of both bacteria and eukaryotic cells.

Concurrent Synthesis on the Leading and Lagging Strands

Given the above model, the question has arisen as to how the holoenzyme of DNA polymerase III accomplishes synthesis on both the leading and lagging strands. Are the events totally separate, involving two copies of the enzyme? The available evidence suggests that this is not the case. Instead, it is believed that a mechanism exists to allow simultaneous synthesis at the replication fork on both strands, with nucleotide polymerization occurring on each template strand under the direction of one of the two monomers making up the dimeric holoenzyme As Figure 10–11 illustrates, if the lagging strand forms a loop, it is possible for simultaneous synthesis to occur. In keeping with our knowledge of Okazaki fragments, after each 100 to 200 base pairs of synthesis, the lagging-strand duplex is released by the enzyme. A new loop is then formed with the lagging template strand, and the process is repeated. Looping inverts the orientation of the template, but not the direction of actual synthesis on the lagging strand, which is always in the 5' to 3' direction.

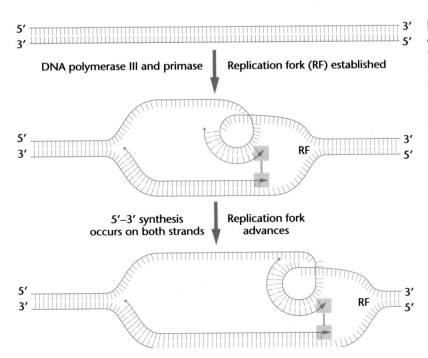

■ Figure 10–11 Illustration of how concurrent DNA synthesis may be achieved on both the leading and lagging strands at a single replication fork. The lagging template strand is "looped" in order to invert the physical direction of synthesis, but not the biochemical direction.

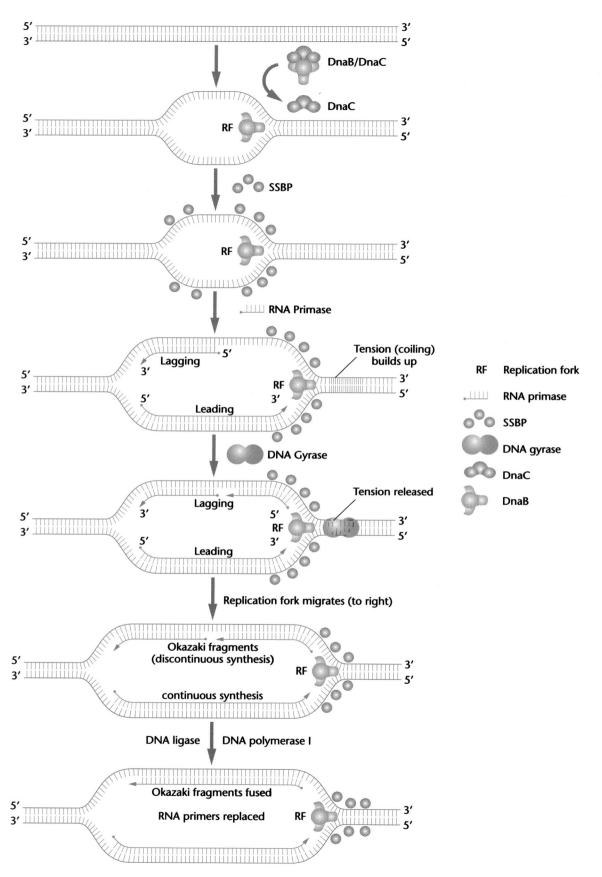

■ Figure 10–12 **A model of DNA synthesis involving the many enzymes and proteins essential to the process, as described in the text and indicated in the key.**

Proofreading

The underpinning of semiconservative replication is the synthesis of a new strand of DNA that is precisely complementary to the template strand at each nucleotide position. Although the action of DNA polymerases is very accurate, synthesis is not perfect. Occasionally, a noncomplementary nucleotide is inserted erroneously. To compensate for such inaccuracies, both polymerases I and III possess **3′ to 5′ exonuclease activity**. They are capable of detecting, pausing, and excising a mismatched nucleotide (in the 3′ to 5′ direction). Once the mismatched nucleotide is removed, 5′ to 3′ synthesis can again proceed.

This process is called **exonuclease proofreading** and serves to increase the fidelity of synthesis. In the case of the holoenzyme form of DNA polymerase III, the ε subunit is responsible for enhancing the proofreading step. Strains of *E. coli* have been isolated in which a mutation has occurred that renders the ε subunit nonfunctional. The error rate (the mutation rate) during DNA synthesis is increased in these strains as a result.

Summary of DNA Synthesis

To summarize the process of DNA synthesis at the level of the replication fork in bacteria and bacteriophages, the following steps provide the biochemical basis for semiconservative replication. These steps are illustrated in Figure 10–12.

1. Unwinding proteins (called helicases) denature the helix at the origin of synthesis, creating the replication fork. Other proteins (SSBPs) stabilize the newly created single strands of denatured DNA.

2. As the replication fork moves away from the origin, increased coiling of the helix occurs ahead of the fork. The tension thus created is diminished by the enzymatic action of the topoisomerase enzyme DNA gyrase. This topoisomerase cuts DNA strands, allowing uncoiling to occur, and then reseals the DNA strands.

3. Initiation of DNA synthesis occurs concurrently on both template strands and involves an RNA primer synthesized under the direction of a unique RNA polymerase called primase. The resultant RNAs are complementary to their DNA templates.

4. DNA polymerase III polymerizes complementary DNA strands by simultaneously elongating the existing primer in the 5′ to 3′ direction.

5. As the replication fork moves away from the origin, synthesis is continuous on the leading strand, but is discontinuous on the lagging strand, producing short polynucleotides called Okazaki fragments.

6. The RNA primers are subsequently removed and the resulting gaps are filled with DNA under the direction of DNA polymerase I. The final gaps between the existing DNA and the replacement DNA are closed by DNA ligase.

7. Along the lagging strand, the Okazaki fragments are joined by DNA ligase.

8. As synthesis proceeds along both the leading and lagging strands, proofreading by DNA polymerase III occurs, and semiconservative replication is achieved.

Because the investigation of DNA synthesis is still an extremely active area of research, this model will no doubt be extended in the future. In the meantime, it provides a summary of DNA synthesis against which genetic phenomena can be interpreted.

Genetic Control of Replication

Much of what we know and have outlined in the previous section concerning the details of DNA replication in viruses and bacteria has been based on the genetic analysis of the process. For example, we have already discussed the *polA1* mutation, the study of which revealed that DNA polymerase I is not the major enzyme responsible for replication. Many other mutations have been isolated that interrupt or seriously impair some aspect of replication, such as the ligase-deficient and the proofreading-deficient mutations mentioned previously. These can be studied most easily if they are **temperature-sensitive mutations** (one type of **conditional mutation**). Such mutations are expressed under one condition (in this case, at a restrictive temperature), but are not expressed under a separate condition (in this case, the permissive temperature). As a result, a mutation that would otherwise be lethal may be maintained at the permissive temperature, while the genetic effects may be investigated by shifting mutant cells to the restrictive temperature. Investigation of such temperature-sensitive mutants has the potential to provide insights into the product and the associated function of the normal, nonmutant gene.

As shown in Table 10.4, the enzyme product or its general role in replication has been ascertained for a variety of genes in *E. coli*. For example, numerous mutations in genes specifying the subunits of polymerases I, II, and III have been isolated. Genes have also been identified that encode products involved in specification of the origin of synthesis, helix unwinding and stabilization, initiation and priming, relaxation of supercoiling, repair, and ligation. The discovery of such a large group of genes attests to the complexity of the process of replication, even in the relatively simple prokaryote. This complexity is not unexpected, given the enormous quantity of DNA that must be unerringly replicated in a very brief time. As we will see, the process is even more involved and therefore more difficult to investigate in eukaryotes.

TABLE 10.4	A list of various *E. coli* mutant genes and their products or role in replication

Mutant Gene	Enzyme or Role
polA	DNA polymerase I
polB	DNA polymerase II
dnaE,N,Q,X,Z	DNA polymerase III subunits
dnaG	Primase
dnaA,I,P	Initiation
dnaB,C	Helicase at *oriC*
oriC	Origin of replication
gyrA,B	Gyrase subunits
lig	Ligase
rep	Helicase
ssb	Single-stranded binding proteins
rpoB	RNA polymerase subunit

Eukaryotic DNA Synthesis

Most features of DNA synthesis found in microorganisms are now thought to apply also to eukaryotic systems. However, because eukaryotic cells contain about 50 times as much DNA per cell as prokaryotes and because the DNA of eukaryotes is complexed with a variety of proteins, eukaryotes face many problems during synthesis of DNA that are not encountered by prokaryotes. Also, these factors plus the general complexity of the eukaryotic cell compared with the prokaryotic cell have made analysis much more difficult.

Nevertheless, much has been learned. As we will see in detail in the next chapter, the major proteins that complex with eukaryotic DNA are positively charged **histones** that bond to the negatively charged phosphates, forming a repeating structure called the **nucleosome**. Prior to initiation of DNA synthesis at the replication fork, these histones must be dissociated from the DNA. Following replication, both daughter strands reassociate with histones to form the nucleosome structures characteristic of chromatin.

Eukaryotic cells have been found to contain five kinds of DNA polymerases, called α, β, γ, δ, and ε. It appears that the α and δ forms are the major enzymes involved in the replication of nuclear DNA. The β form may be involved in DNA repair, while the γ form is the only DNA polymerase found in mitochondria. Presumably, it is unique in its function within that organelle. Undoubtedly, there is a comparable but unique form of this enzyme associated with DNA replication in chloroplasts of plant cells. The δ form appears to function on the lagging strand during nuclear DNA replication. DNA polymerases of eukaryotes have the same fundamental requirements for DNA synthesis as bacterial and viral systems: four deoxyribonucleoside triphosphates, a template, and a primer.

Data and observations derived from autoradiographic and electron microscopic studies have pro-

vided many other insights. As mentioned earlier, eukaryotic DNA synthesis is bidirectional, creating two replication forks from each point of origin. As would be expected because of the larger amount of DNA in eukaryotes compared to prokaryotes, many more points of origin are present. In mammals, there are about 25,000 replicons present in the genome, each consisting of an average of 100,000 to 200,000 base pairs (100–200 kb). In *Drosophila*, there are about 3500 replicons per genome, with an average size of 40 kb.

To accommodate the increased number of replicons, many more DNA polymerase molecules are present in eukaryotic cells than are found in bacteria. While *E. coli* has about 15 copies of DNA polymerase III per cell, there may be up to 50,000 copies of the α form of DNA polymerase in eukaryotic cells. As a result, most replicons may be replicated at approximately the same time during the S phase of interphase.

Synthesis of DNA is also slower in eukaryotes. In *E. coli*, 100 kb are added to a growing chain per minute, while eukaryotic synthesis ranges from only 0.5 to 5 kb per minute. Nevertheless, *E. coli* requires 20 to 40 minutes to replicate its chromosome, while many cells of *Drosophila*, each with 40 times more DNA, accomplish the same task in only 3 minutes! The presence of smaller replicons in eukaryotes compensates for the slower rate of DNA synthesis in eukaryotes compared to that in prokaryotes.

Even though DNA synthesis within each replicon is slower in eukaryotes, most general aspects of chain elongation are thought to be similar. The actions of a variety of proteins—DNA helicase and SSBPs—are believed to modulate strand separation that precedes RNA priming by a primase enzyme. On the lagging strand, synthesis is discontinuous, resulting in Okazaki fragments that are subsequently linked together by DNA ligase. These fragments are about 10 times as small (100–150 nucleotides) in eukaryotes as in prokaryotes. While synthesis may occur continuously on the leading strand, it is not clear whether or not it can proceed uninterrupted for the full length of a replicon. This possibility has led to a description of synthesis on the leading strand as being **semidiscontinuous.**

DNA Synthesis at the Ends of Linear Chromosomes

One final aspect of eukaryotic versus prokaryotic DNA synthesis involves the fact that eukaryotic chromosomes are linear compared to the circular forms displayed by bacteria and most bacteriophages. A special problem is encountered during DNA replication at the "ends" of linear molecules. Such ends are part of each telomeric region of each chromosome.

While synthesis can proceed normally to the end of the leading DNA strand, a problem is encountered on the lagging DNA strand, as illustrated in Figure 10–13. The difficulty arises as the RNA primer is removed from the

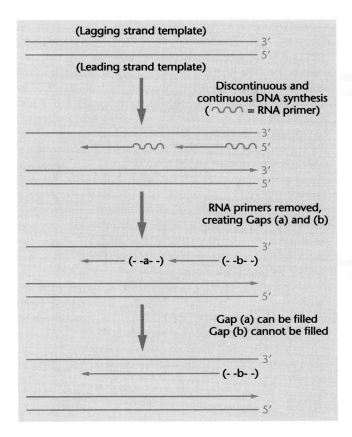

■ Figure 10–13   The difficulty encountered during the replication of the ends of linear chromosomes. A gap is left following synthesis on the lagging strand.

Telomerase adds several copies of the 6-nucleotide repeat to the 3′-end of the lagging strand (using 5′-to-3′ synthesis). These repeats appear to be capable of forming a "hairpin loop," which is stabilized by unorthodox hydrogen bonding between guanine residues (G–G). This creates a free 3′-OH end that, following removal of the RNA primer, can serve as a substrate for DNA polymerase I to fill in the gap. If the hairpin is then cleaved off, the potential loss of DNA is averted.

Further investigation of the *Tetrahymena* telomerase enzyme, isolated and studied extensively by Elizabeth Blackburn and Carol Greider, has yielded an extraordinary finding. This enzyme adds the same TTGGGG sequence to other DNA termini. Blackburn and Greider have now established how the enzyme works. They have discovered that the enzyme contains a short piece of RNA that is essential to its catalytic activity! The functional enzyme is thus a **ribonucleoprotein**. The RNA component encodes the sequences added by the enzyme, serving as its own template. The RNA contains

lagging strand. The newly created gap is normally filled by adding a nucleotide to the existing 3′-OH group provided during discontinuous synthesis (this would normally be present to the right of the gap in Figure 10–13). However, there is no strand present to provide the 3′-OH group because this is the end of the chromosome! As a result, each successive round of synthesis will theoretically shorten the chromosome by the length of the RNA primer. Because this is such a significant problem, we can predict that a molecular solution would have been forthcoming early in evolution, and be shared by all eukaryotes. Indeed, this appears to be the case.

In bacteria and viruses that contain circular DNA molecules, this is not a problem, since no free ends are encountered during replication. In eukaryotes, the discovery of a unique enzyme, **telomerase**, has helped us understand how some organisms solve this problem. In the ciliated protozoan *Tetrahymena*, all telomeres terminate in the sequence 5′-TTGGGG-3′. The enzyme is capable of adding repeats of TTGGGG to the ends of molecules that already contain this sequence. This process, now known to occur in other organisms under the direction of a similar enzyme, solves the problem of completing replication without leaving a gap and shortening the chromosome following each replication (Figure 10–14).

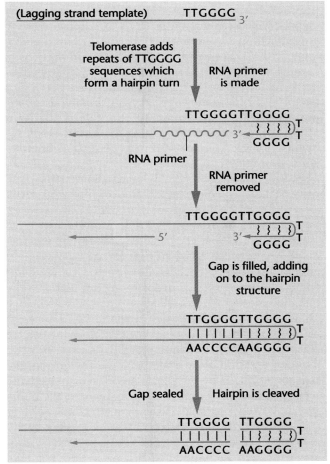

■ Figure 10–14   Diagram of the predicted solution to the theoretical problem posed in Figure 10–13. The enzyme telomerase directs synthesis of the TTGGGG repeated sequence, resulting in the formation of a hairpin structure. As described in the text, this process averts the creation of a gap during replication of the ends of linear chromosomes.

159 bases, including the sequence 5'-CCCCAA-3', which is complementary to the sequence whose synthesis it directs. Analogous enzyme functions have now been found in still other single-celled organisms. The RNA-containing telomerase enzyme behaves like the retroviral enzyme, reverse transcriptase, in that it synthesizes the DNA complement of an RNA template. In this case, the template is much shorter, and the enzyme supplies its own!

The analysis of telomeric DNA sequences, discussed in more detail in Chapter 11, has shown them to be highly conserved throughout evolution. Such conservation reflects not only the critical function of telomeres, but also the unusual nature of their DNA replication.

DNA Recombination

We conclude this chapter by returning to a topic discussed in Chapter 6—**genetic recombination**. There, it was pointed out that the process of crossing over depends on breakage and rejoining of the DNA strands between homologs. Now that we have discussed the chemistry and replication of DNA, it is appropriate to consider how recombination occurs at the molecular level. In general, the following information pertains to genetic exchange between any two homologous double-stranded DNA molecules, whether they be viral or bacterial chromosomes or eukaryotic homologs during meiosis. Genetic exchange at equivalent positions along two chromosomes with substantial DNA sequence homology is referred to as **general** or **homologous recombination**.

Several models are available to explain crossing over, but they all share certain common features. First, all are based on the initial proposals put forth independently by Robin Holliday and Harold L. K. Whitehouse in 1964. They also depend on the complementarity between DNA strands for their precision of exchange. Finally, each model relies on a series of enzymatic processes in order to accomplish genetic recombination.

One such model is illustrated in Figure 10–15. It begins with two paired DNA duplexes or homologs (a), each of which has a single-stranded nick introduced (b) at an identical position by an endonuclease. The ends of the strands produced by these cuts are then displaced and subsequently pair with their complements on the opposite duplex (c). A ligase then seals the loose ends (d), creating hybrid duplexes called **heteroduplex DNA molecules**. The exchange creates a cross-bridged or **Holliday structure**. The position of this cross-bridge can then move down the chromosomes as a result of a process called **branch migration** (e). This occurs as a result of a zipperlike action as hydrogen bonds are broken and then re-formed between complementary bases of the displaced strands of each duplex. This migration yields an increased length of heteroduplex DNA on both homologs.

If the duplexes separate (f) and the bottom portions rotate 180° (g), an intermediate planar structure called a **chi form** is created. If the two strands on opposite homologs previously uninvolved in the exchange are now nicked by an endonuclease (h) and ligation occurs (i), recombinant duplexes are created. Note that the arrangement of alleles is altered as a result of recombination.

Evidence supporting the above model includes the electron microscopic visualization of chi-form planar molecules from bacteria where four duplex arms are joined at a single point of exchange (Figure 10–15). Additionally, the discovery in E. coli of the **RecA protein** provides important evidence. This molecule promotes the exchange of reciprocal single-stranded DNA molecules as must occur in step (c) of the model. Further, the RecA protein enhances the hydrogen bond formation during strand displacement, thus initiating heteroduplex formation. Finally, many other enzymes that are essential to the nicking and ligation process have been discovered and investigated. Mutations that prevent genetic recombination have been found in a number of genes in viruses and bacteria. These are thought to represent genes, the products of which play an essential role in this process.

Gene Conversion

A modification of the above model has helped us to understand better a unique genetic phenomenon known as **gene conversion**. Initially found in yeast by Carl Lindegren and in *Neurospora* by Mary Mitchell, gene conversion is characterized by a genetic exchange ratio involving two closely linked genes that is *nonreciprocal*. If one were to cross two *Neurospora* strains each bearing a separate mutation (a+ X + b), a *reciprocal* recombination event between the genes would yield spore pairs of the ++ and *ab* genotypes. However, a nonreciprocal exchange yields one pair without the other. Working with pyridoxine mutants, Mitchell observed several asci containing spore patterns displaying the ++ genotype, but not the reciprocal product (*ab*). Because the frequency of these events was higher than the predicted mutation rate and thus could not be accounted for by that phenomenon, they were called **gene conversions**. They were so named because it appeared that one allele had somehow been "converted" to another during an event in which genetic exchange also occurred. Similar findings are apparent in the study of other fungi as well.

This phenomenon is now considered to be a consequence of the process of DNA recombination, as discussed in the previous section. During this process, heteroduplex formation is most often accompanied by mismatched bases. These appear as "errors" to the enzyme system that is capable of correcting these mismatches. It is during this repair process that the conversion step occurs. We will return to the topic of DNA repair (such as this) in Chapter 12.

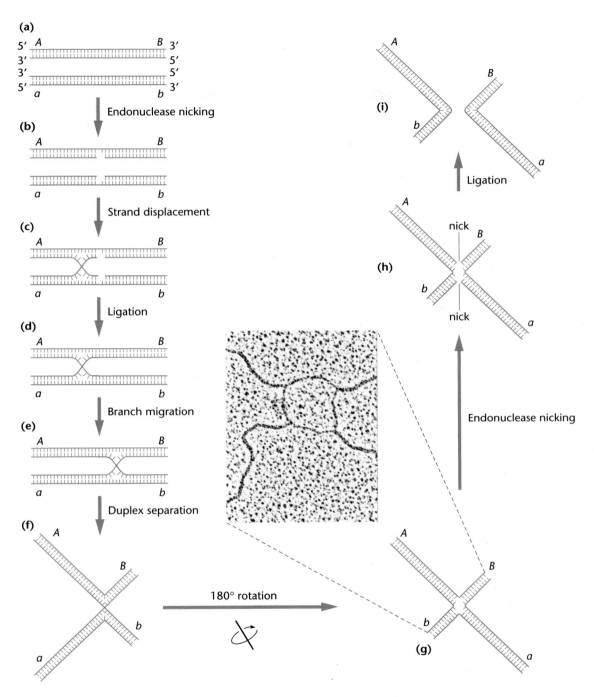

■ Figure 10–15   Depiction of how genetic recombination occurs as a result of the break-age and rejoining of heterologous DNA strands. Each stage is described in the text. The electron micrograph shows a Holliday Junction similar to the diagram in (g) that has been partially denatured to clearly show the path of the four DNA strands.

Gene conversion events have helped to explain other puzzling genetic phenomena in fungi. For example, when mutant and wild-type alleles of a single gene are studied in a cross, asci should yield equal numbers of mutant and wild-type spores. However, exceptional asci with 3:1 or 1:3 ratios are sometimes observed. These ratios are more easily understood when interpreted in terms of gene conversion. The phenomenon has also been detected during mitotic events in fungi, as well as during the study of unique compound chromosomes in *Drosophila*.

Telomerase: The Key to Immortality?

Humans, like all multicellular organisms, are programmed to grow old and to die. As we age, our immune systems become less efficient, wound healing is impaired, tissues and organs lose resilience. It has always been a mystery why we go through these age-related declines, and why each species has a characteristic, finite lifespan. Why do we grow old? Can we reverse this march to mortality? Some recent discoveries suggest that the answers to these questions may lie at the ends of our chromosomes.

The study of human aging begins with a study of human cells, growing in culture dishes. Like the organisms from which the cells are taken, cells in culture have a finite lifespan. This "replicative senescence" was noted over 30 years ago by Hayflick. He reported that normal human fibroblasts lose their ability to grow and divide after about 50 cell divisions. These senescent cells remain metabolically active, but can no longer proliferate. Eventually, they die. Although it not known whether cellular senescence causes organismal aging directly, the evidence is suggestive. For example, cells from young people go through more divisions in culture than cells from older people: Human fetal cells divide 60 to 80 times before undergoing senescence, whereas cells from older adults divide only 10 to 20 times. In addition, cells from short-lifespan species stop growing after fewer divisions than cells from longer-lifespan species: Mouse cells divide 10 to 15 times in culture, but tortoise cells will undergo over 100 divisions. Moreover, cells from patients with genetic premature aging syndromes (such as Werner syndrome) undergo fewer divisions in culture than cells from normal patients.

Another characteristic of aging cells, both in culture and in whole organisms, is that their telomeres become shorter. *Telomeres* are the tips of linear chromosomes and consist of several thousand repeats of a short DNA sequence (TTGGG in humans). Telomeres help preserve the structural integrity of chromosomes by protecting their ends from degrading or from fusing to other chromosomes. Telomeres are created and maintained by *telomerase*—a remarkable RNA-containing enzyme that adds telomeric

DNA sequences onto the ends of linear chromosomes. Telomerase also solves the "end-replication" problem—which asserts that linear DNA molecules will become shorter at each replication because DNA polymerase cannot synthesize new DNA at the 3′ ends of each parent strand. By adding numerous telomeric repeat sequences onto the 3′ ends of chromosomes, telomerase prevents chromosomes from shrinking into oblivion. Unfortunately, normal aging cells contain little if any telomerase. As a result, telomere length decreases by about 100 base pairs every time a normal cell divides in culture. It is thought that telomere shortening may act as a clock that counts cell divisions and instructs the cell to stop dividing.

Could telomere shortening decrease the lifespan of cells and ultimately of organisms? If so, could an increase in telomere length grant us perpetual youth and vitality? A recent study suggests that it may be possible to reverse senescence by artificially increasing the amount of telomerase in our cells. When the investigators introduced cloned telomerase genes into normal human cells in culture, telomeres lengthened by thousands of base pairs and the cells continued to grow long past their senescence point. These observations confirm that telomere length can act as a cellular clock. In addition, they suggest that some of the atrophy of tissues that accompanies old age may someday be reversed by introducing telomerase genes. However, before we rush out to buy telomerase pills, we need to consider a possible consequence of cellular immortality—cancer.

Although normal cells undergo senescence after a specific number of cell divisions, cancer cells do not. It is thought that cancers arise after several genetic mutations accumulate in a cell. These mutations disrupt the normal checks and balances that control cell growth and division. It seems logical that successful cancer cells also need to stop the normal aging clock. If their telomeres become shorter after each cell division, tumor cells will eventually succumb to aging and cease growth. However, if they synthesize telomerase, they will arrest the ticking of the senescence clock and become immortal. In keeping with this idea, it has been found that 80 to 90 percent of human tumor cells contain telomerase

activity, and have stable telomeres. Although there is currently some debate about whether the presence of telomerase is a prerequisite for—or simply a consequence of—cell transformation, it is possible that the acquisition of telomerase activity may be an important step in the development of a cancer cell. Therefore, any attempt to increase telomerase activity in normal cells carries the risk of enhancing the development of tumors.

An attractive possibility is that telomerase may be an ideal target for anti-cancer drugs. Treatments that inhibit telomerase might destroy cancer cells by allowing their telomeres to shorten, thereby forcing the cells into senescence. Because most normal human cells do not express telomerase, anti-telomerase drugs might be specific for tumor cells and hence less toxic than most current anti-cancer drugs. Although we do not know yet whether this approach will work in animals, it appears to work in cultured tumor cells. Tumor cells that are treated with an anti-telomerase agent lose telomeric sequences and arrest growth after about 25 cell divisions.

Before anti-telomerase drugs can be developed and used on humans, several questions must be answered. Is telomerase synthesized and required by some normal human cells (such as lymphocytes and macrophages)? If so, anti-telomerase drugs may be unacceptably toxic. Could some cancer cells compensate for the loss of telomerase by using other telomere-lengthening mechanisms (such as recombination)? If so, anti-telomerase drugs may be doomed to fail. Even if we inhibit telomerase activity in tumor cells, could the cells undergo multiple divisions before reaching senescence, and still damage the host?

Will telomerase allow us to arrest cancers and to reverse the descent into old age? Time will tell.

References

Bodnar, A.G., et al. 1998. Extension of life-span by introduction of telomerase into normal human cells. *Science* 279:349–352.

de Lange, T. 1998. Telomeres and senescence: ending the debate. *Science* 279:334–335.

Greider, C. W. 1998. Telomerase activity, cell proliferation, and cancer. *Proc. Natl. Acad. Sci. USA.* 95:90–92.

Chapter Summary

1. In theory, three modes of DNA replication are possible: semiconservative, conservative, and dispersive. Though all three rely on base complementarity, semiconservative replication is the most straightforward and was predicted.

2. In 1958, Meselson and Stahl resolved this problem in favor of semiconservative replication in *E. coli*, showing that newly synthesized DNA consists of one old strand and one new strand. Taylor, Woods, and Hughes used root tips of the broad bean to demonstrate semiconservative replication in eukaryotes.

3. During the same period, Kornberg isolated DNA polymerase I from *E. coli* and demonstrated it to be an enzyme capable of *in vitro* DNA synthesis, provided that a template and precursor nucleoside triphosphates were supplied.

4. The subsequent discovery of the *polA1* mutant strain of *E. coli*, capable of DNA replication in spite of its lack of polymerase I activity, cast doubt on this enzyme's *in vivo* replicative function. DNA polymerases II and III were then isolated. Polymerase III has been identified as the enzyme responsible for DNA replication *in vivo*.

5. During the process of DNA synthesis, the double helix unwinds, forming a replication fork where synthesis begins. Proteins stabilize the unwound helix and assist in relaxing the coiling tension created ahead of the replication activity.

6. Synthesis is initiated at specific sites along each template strand by RNA primase, which results in a short segment of RNA that provides a suitable 3′ end, upon which DNA polymerase III can begin polymerization.

7. Because of the antiparallel nature of the double helix, polymerase III synthesizes DNA continuously on the leading strand in a 5′-to-3′ direction. On the opposite strand, called the lagging strand, synthesis results in short Okazaki fragments that are later joined by DNA ligase.

8. DNA polymerase I removes and replaces the RNA primer with DNA, which is joined to the adjacent polynucleotide by DNA ligase.

9. The isolation of numerous phage and bacterial mutant genes affecting many of the molecules involved in the replication of DNA has helped to define the complex genetic control of the entire process.

10. DNA replication in eukaryotes is similar to, but more complex than, replication in prokaryotes. For example, replication at the ends (telomeres) of linear molecules poses a special problem, which can be solved by a unique RNA-containing enzyme called telomerase.

11. Homologous recombination between genetic molecules relies on a series of enzymes that can cut, realign, and reseal DNA strands. The phenomenon of gene conversion, in which one allele appears to be converted to another, may best be explained in terms of mismatch repair synthesis during these exchanges.

Key Terms

autoradiography, 212
bidirectional replication, 215
biologically active DNA, 217
branch migration, 226
chi form, 226
conditional mutation, 223
conservative replication, 211
continuous DNA synthesis, 221
discontinuous DNA synthesis, 221
dispersive replication, 212
DnaA protein, 220
DnaB protein, 220
DnaC protein, 220
DNA gyrase, 220
DNA helicase, 220
DNA ligase, 221
DNA polymerase, 216
DNA topoisomerase, 220

exonuclease activity, 218
exonuclease proofreading, 223
gene conversion, 226
genetic recombination, 226
heteroduplex DNA molecule, 226
histone, 224
Holliday structure, 226
holoenzyme, 218
homologous recombination, 226
lagging strand, 221
leading strand, 221
nucleosome, 224
Okazaki fragment, 221
oriC, 215
origin of replication, 215
primase, 220
primer, 217
RecA protein, 226

replication fork, 215
replicon, 215
replisome, 219
sedimentation equilibrium centrifugation, 212
semiconservative replication, 211
semidiscontinuous, 224
single-stranded binding protein (SSBP), 220
sister chromatid exchange, 215
supercoiling, 220
telomerase, 225
temperature-sensitive mutation, 223
ter, 215
unidirectional replication, 215
9mer, 220
13mer, 220

INSIGHTS
and
SOLUTIONS

1. Predict the theoretical results of conservative and dispersive models of DNA synthesis using the conditions of the Meselson–Stahl experiment. Follow the results through two generations of replication after cells have been shifted to ^{14}N-containing medium, using the following sedimentation pattern:

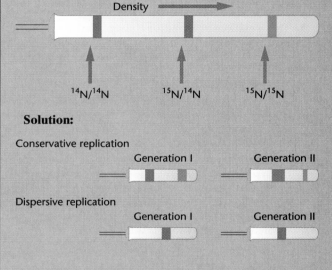

Density

^{14}N/^{14}N ^{15}N/^{14}N ^{15}N/^{15}N

Solution:

Conservative replication

Generation I   Generation II

Dispersive replication

Generation I   Generation II

2. Mutations in the *dnaA* gene of *E. coli* are lethal and can only be studied following the isolation of condi-

tional, temperature-sensitive mutations. Such mutant strains grow nicely and replicate their DNA at the permissive temperature of 18°C, but they do not grow or replicate their DNA at the restrictive temperature of 37°C. Two observations were useful in determining the function of the *dnaA* gene product. First, *in vitro* studies using DNA templates that have been nicked (opened) do not require the DnaA protein. Second, if intact cells are grown at 18°C and then shifted to 37°C, DNA synthesis continues at this temperature until one round of replication is completed, and DNA synthesis stops. What do these observations suggest about the role of the dnaA gene product?

Solution: These observations suggest that *in vivo* the DnaA protein is essential to the initiation of DNA synthesis. At 18°C (the permissive temperature) the mutation is not expressed and DNA synthesis begins. Following the shift to the restrictive temperature, DNA synthesis already initiated continues, but no new synthesis can begin. Because the dnaA protein is not required for synthesis of "nicked" DNA, this observation suggests that the protein functions during initiation by interacting with the intact helix and somehow facilitating the localized denaturing necessary for synthesis to proceed. In fact, both conclusions are valid.

Problems and Discussion Questions

1. Compare conservative, semiconservative, and dispersive modes of DNA replication.

2. In the Meselson–Stahl experiment, which of the three modes of replication could be ruled out after one round of replication? After two rounds?

3. Predict the results of the experiment by Taylor, Woods, and Hughes if replication were (a) conservative and (b) dispersive.

4. What are the requirements for the *in vitro* synthesis of DNA under the direction of DNA polymerase I?

5. What is meant by "biologically active" DNA?

6. Why was the phage φX174 chosen for the experiment demonstrating biological activity?

7. What was the significance of the *polA1* mutation during the study of DNA synthesis?

8. Summarize the properties and functions of polymerase I, II, and III.

9. List the proteins that unwind and stabilize DNA during *in vivo* DNA synthesis. How do they function?

10. Define and indicate the significance of (a) Okazaki fragments, (b) DNA ligase, and (c) primer RNA during DNA synthesis.

11. Outline the current model for DNA synthesis.

12. Why should DNA synthesis be more complex in eukaryotes than in bacteria? How is DNA synthesis similar in the two types of organisms?

13. Describe why DNA synthesis is "problematic" at the telomeres. What is unique about the enzyme that makes synthesis possible?

14. Define gene conversion and describe how this phenomenon is related to genetic recombination.

15. (a) Many of the gene products that are involved in DNA synthesis were initially defined by studying mutant *E. coli* strains that could not synthesize DNA. The α subunit of DNA pol III is responsible for the 5′ to 3′ polymerase (chain elongation) function and is coded for by the *dnaE* gene. What would be the resulting phenotype of a mutation in this gene that inactivates the α subunit? How is it possible to maintain such a strain?

(b) The ε subunit of DNA pol III contains the 3′ to 5′ exonuclease activity and is coded for by the gene *dnaQ*. If this gene is mutated, what would be the resulting phenotype?

16. Suppose that *E. coli* synthesizes DNA at a rate of 100,000 nucleotides per minute and takes 40 minutes to replicate its chromosome.

(a) How many base pairs are present in the entire *E. coli* chromosome?

(b) What is the physical length of the chromosome in its helical configuration; that is, what is the circumference of the circular chromosome?

Selected Readings

Blackburn, E. H. 1991. Structure and function of telomeres. *Nature* 350:569–572.

DeLucia, P., and Cairns, J. 1969. Isolation of an *E. coli* strain with a mutation affecting DNA polymerase. *Nature* 224:1164–66.

Dressler, D., and Potter, H. 1982. Molecular mechanisms in genetic recombination. *Annu. Rev. Biochem.* 51:727–61.

Greider, C. W., and Blackburn, E. H. 1989. A telomeric sequence in the RNA of *Tetrahymena* telomerase required for telomere repeat synthesis. *Nature* 337:331–36.

———. 1996. Telomeres, telomerase, and cancer. *Sci. Am.* (Feb.) 274:92–97.

Herendeen, D. R., and Kelly, T. J. 1996. DNA polymerase III: Running rings around the fork. *Cell* 84:5–8.

Holliday, R. 1964. A mechanism for gene conversion in fungi. *Genet. Res.* 5:282–304.

Huberman, J. C. 1987. Eukaryotic DNA replication: A complex picture partially clarified. *Cell* 48:7–8.

Kornberg, A. 1960. Biological synthesis of DNA. *Science* 131:1503–8.

———. 1974. *DNA synthesis.* New York: W. H. Freeman.

———. 1979. Aspects of DNA replication. *Cold Spring Harbor Symp. Quant. Biol.* 43:1–10.

———, and Baker, T. A. 1992. *DNA replication,* 2nd ed. New York: W. H. Freeman.

Lindegren, C. C. 1953. Gene conversion in *Saccharomyces. J. Genet.* 51:625–37.

Lodish, H., et al. 1995. *Molecular cell biology,* 3rd ed. New York: Scientific American Books.

Meselson, M., and Stahl, F. W. 1958. The replication of DNA in *Escherichia coli. Proc. Natl. Acad. Sci. USA* 44:671–82.

Mitchell, M. B. 1955. Aberrant recombination of pyridoxine mutants of *Neurospora. Proc. Natl. Acad. Sci. USA* 41:215–20.

Radding, C. M. 1978. Genetic recombination: Strand transfer and mismatch repair. *Annu. Rev. Biochem.* 47:847–80.

Radman, M., and Wagner, R. 1988. The high fidelity of DNA duplication. *Sci. Am.* (Aug.) 259:40–46.

Stahl, F. W. 1979. *Genetic recombination: Thinking about it in phage and fungi.* New York: W. H. Freeman.

———. 1987. Genetic recombination. *Sci. Am.* (Feb.) 256:90–101.

Taylor, J. H., Woods, P. S., and Hughes, W. C. 1957. The organization and duplication of chromosomes revealed by autoradiographic studies using tritium-labeled thymidine. *Proc. Natl. Acad. Sci. USA* 48:122–28.

Wang, J. C. 1982. DNA topoisomerases. *Sci. Am.* (July) 247:94–108.

Watson, J. D., et al. 1987. *Molecular biology of the gene, Vol. 1. General principles,* 4th ed. Menlo Park, CA: Benjamin/Cummings.

Whitehouse, H. L. K. 1982. *Genetic recombination: Understanding the mechanisms.* New York: Wiley.

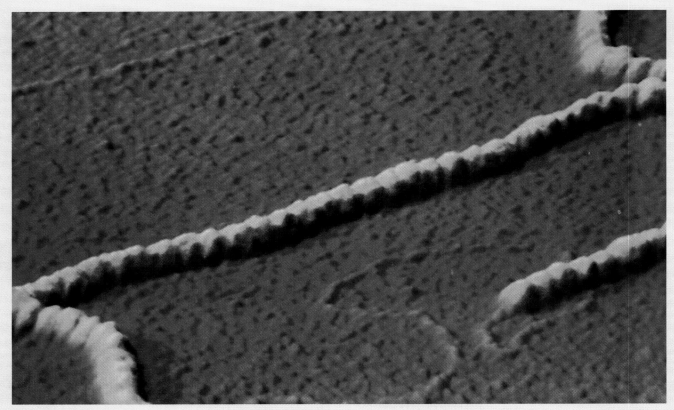

Chromatin fibers viewed using a scanning transmission electron microscope (STEM).

CHAPTER OUTLINE

CHAPTER
11

Organization of DNA in Chromosomes and Genes

Chapter Concepts

DNA is organized in a manner that is consistent with the complexity of the host structure with which it is associated. Viruses, bacteria, mitochondria, and chloroplasts contain a shorter, often circular, DNA molecule that is relatively free of proteins. Eukaryotic cells contain greater amounts of DNA organized into nucleosomes and present as chromatin fibers. This increase in complexity is related to the larger amount of genetic information present as well as the greater complexity associated with their genetic function. In eukaryotes, genes are organized in various ways, ranging from single copies to families of related genes repeated in tandem arrays. The eukaryotic genome contains large amounts of noncoding DNA, some of which interrupts the coding portions of genes.

Once it was understood that DNA is the genetic material, it became very important to determine how DNA is organized in genes and chromosomes. There has been much interest in these topics because knowledge of the organization of the determination of the genetic material and associated molecules will undoubtedly provide valuable insights into other aspects of genetics. For example, how the genetic information is stored, expressed, and regulated must be related to the organization of the genetic molecule, DNA. In eukaryotes, how the chromatin fibers characteristic of interphase are condensed into chromosome structures visible during mitosis and meiosis is also of great interest.

In this chapter, we will first provide a survey of the various ways in which DNA is organized into chromosomes, including examples from viruses, bacteria, and eukaryotes. The genetic material has been studied using numerous approaches, including molecular analysis and direct visualization by light and electron microscopy.

We will then turn to consideration of how genes are organized within the genome of the organism. Such organization is least complex in bacteriophages and bacteria, in which the chromosome consists largely of an array of contiguous genes. Each gene consists of an uninterrupted linear array of nucleotides, most of which either encode the amino acid sequences of proteins or are involved in gene regulation.

In comparison, a large portion of many eukaryotic genes consists of (1) noncoding DNA sequences called **introns** and (2) flanking regions of the gene, whose coded information does not become part of the final RNA transcript. There are also extensive regions of the genome called intergenic noncoding areas, some of which consists of repetitive DNA.

Beyond this information, we are learning much more about the organization of genes within the genome. We will review some of these topics and expand upon them with particular emphasis on the organization of multigene families. These topics represent some of the most exciting and unexpected discoveries made in the field of genetics.

Viral and Bacterial Chromosomes

In comparison with eukaryotes, the chromosomes of viruses and bacteria are much less complicated. They usually consist of a single nucleic acid molecule, largely devoid of associated proteins. Contained within the single chromosome of viruses and bacteria is much less genetic information than in the multiple chromosomes comprising the genome of eukaryotes. These characteristics have greatly simplified analysis, providing a fairly comprehensive view of viral and bacterial chromosomes.

The chromosomes of viruses consist of a nucleic acid molecule—either DNA or RNA—which can be either single- or double-stranded. They may exist as circular structures (closed loops), or they may take the form of linear molecules. The single-stranded DNA of the **φX174 bacteriophage** and the double-stranded DNA of the **polyoma virus** are ring-shaped molecules within the protein coat of the mature virus and within the host cell. The **bacteriophage lambda (λ),** on the other hand, possesses a linear double-stranded DNA molecule prior to infection, which forms a closed loop upon infection of the host cell. Still other viruses, such as the **T-even series of bacteriophages,** have linear, double-stranded chromosomes of DNA that do not form circles inside the bacterial host. Thus, circularity is not an absolute requirement for replication in some viruses.

Viral nucleic acid molecules have been visualized with the electron microscope. Figure 11–1 shows a mature bacteriophage lambda with its double-stranded DNA molecule in the circular configuration. One feature shared by viruses, bacteria, and eukaryotic cells is the ability to package an exceedingly long DNA molecule into a relatively small volume. In phage λ, the DNA is 17 μm long and must fit into the phage head, which is less than 0.1 μm on any side. As you examine Figure 11–1, note that the phage is magnified six times as much as the DNA molecule.

Table 11.1 compares the length of the chromosomes of several viruses to the size of their head. In each case, a similar packaging feat must be accomplished. The dimensions given for phage T2 may be compared with the micrograph of both the DNA and viral particle shown in Figure 11–2. Seldom does the space available in the head of a virus exceed the chromosome volume by more than a factor of 2. In many cases, almost all space is filled, indicating nearly perfect packing. Once it is packed within the head, the genetic material is functionally inert until it is released into a host cell.

Bacterial chromosomes are also relatively simple in form compared with those of eukaryotic cells. They always consist of a double-stranded DNA molecule, compacted into a structure sometimes referred to as the **nucleoid.** *Escherichia coli,* the most extensively studied bacterium, has a large circular chromosome, measuring approximately 1200 μm (1.2 mm) in length. When the cell is gently lysed and the chromosome re-

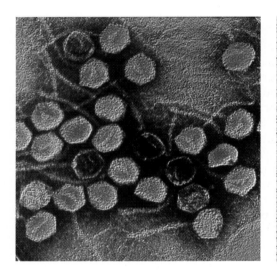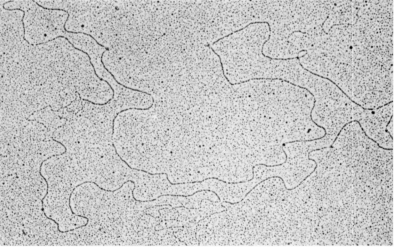

■ Figure 11–1   Electron micrographs of (a) phage λ and (b) the DNA isolated from it. The chromosome is 17 μm long.

| TABLE 11.1 | The genetic material of representative viruses and bacteria |

			Nucleic Acid		Overall Size of Viral Head or Bacteria (μm)
	Organism	Type	SS or DS[a]	Length (μm)	
Viruses	φX174	DNA	SS	2.0	0.025 × 0.025
	Tobacco mosaic virus	RNA	SS	3.3	0.30 × 0.02
	Lambda phage	DNA	DS	17.0	0.07 × 0.07
	T2 phage	DNA	DS	52.0	0.07 × 0.10
Bacteria	*Hemophilus influenzae*	DNA	DS	832.0	1.00 × 0.30
	Escherichia coli	DNA	DS	1200.0	2.00 × 0.50

[a]SS = single-stranded; DS = double-stranded.

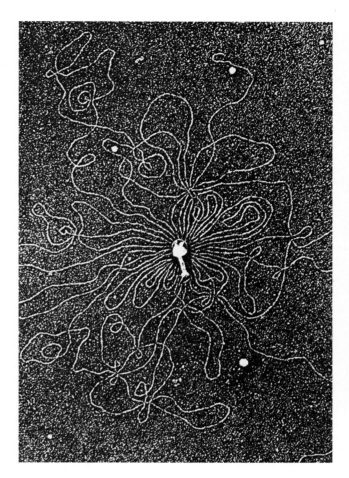

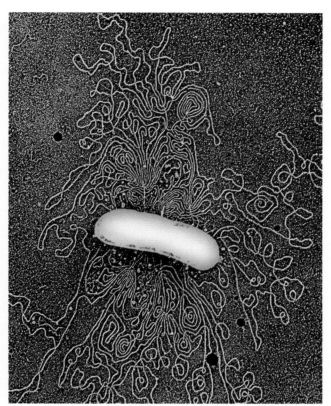

■ Figure 11–3   Electron micrograph of the bacterium *Escherichia coli*, which has had its DNA released by osmotic shock. The chromosome is 1200 μm long.

■ Figure 11–2   Electron micrograph of bacteriophage T2, which has had its DNA released by osmotic shock. The chromosome is 52 μm long.

■ Figure 11–3   Electron micrograph of the bacterium *Escherichia coli*, which has had its DNA released by osmotic shock. The chromosome is 1200 μm long.

leased, it can be visualized under the electron microscope (Figure 11–3).

This DNA is found to be associated with several types of **DNA-binding proteins,** including those called **HU** and **H.** These proteins are small but abundant in the cell and contain a high percentage of positively charged amino acids that can bond ionically to the negative charges of the phosphate groups in DNA. As we will soon see, these proteins resemble structurally similar molecules called **histones** that are found associated with eukaryotic DNA. Unlike the tightly packed chromosome of a virus, the bacterial chromosome is not functionally inert. In spite of the compacted condition of the bacterial chromosome, replication and transcription occur readily.

Mitochondrial and Chloroplast DNA

Numerous observations have demonstrated that both **mitochondria** and **chloroplasts** contain their own genetic information. This was suggested by the discovery of mutations in yeast, other fungi, and plants that alter the function of these organelles (see Chapter 7). Transmission of these mutations did not always demonstrate biparental inheritance patterns characteristic of nuclear genes. Instead, a **uniparental mode of inheritance** was often observed. Because the origin of both mitochondria and chloroplasts is also often uniparental, these observations suggested that these organelles might house their own DNA that influences their function.

Thus, geneticists set out to look for more direct evidence of DNA in these organelles. Electron microscopists not only documented the presence of DNA in both organelles, they also saw DNA in a form quite unlike that seen in the chromosomes of eukaryotic cells. This DNA looked remarkably similar to that seen in viruses and bacteria! As we shall see, this similarity, among other observations, led to the idea that both mitochondria and chloroplasts are derived from primitive organisms that were free-living and much like bacteria.

Molecular Organization and Function of Mitochondrial DNA

Extensive information is now available about the molecular aspects of **mitochondrial DNA (mtDNA)** and

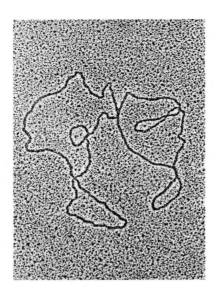

■ Figure 11–4   Electron micrograph of mitochondrial DNA (mtDNA) derived from *Xenopus laevis*.

related gene function. In most eukaryotes, mtDNA (Figure 11–4) is a circular duplex that replicates semiconservatively and is free of the chromosomal proteins characteristic of eukaryotic DNA. In size, mtDNA differs among organisms. This can be seen by examining the information presented in Table 11.2. In a variety of animals, mtDNA consists of about 16,000 to 18,000 (16–18 kb). In vertebrates, there are 5 to 10 molecules per organelle. A considerably greater amount of DNA is present in plant mitochondria, where 100 kb is not unusual.

Several general statements can now be made concerning mtDNA. There appear to be few or no gene repetitions, and replication is dependent upon enzymes encoded by nuclear DNA. Genes have been identified that encode products essential to transcription and translation, including ribosomal RNAs, transfer RNAs, and numerous products essential to the cellular respiratory functions of the organelles. However, genetic function within mitochondria is also dependent on nuclear-encoded gene products.

TABLE 11.2	The size of mtDNA in different organisms
Organism	*Size in Kilobases*
Human	16.6
Mouse	16.2
Xenopus (frog)	18.4
Drosophila (fruit fly)	18.4
Saccharomyces (yeast)	84.0
Pisum sativum (pea)	110.0

Molecular Organization and Function of Chloroplast DNA

There is now available a substantial amount of molecular information about **chloroplast DNA (cpDNA)** and about the genetic function of this organelle. Chloroplasts, like mitochondria, contain their own genetic system distinct from that found in the nucleus and cytoplasm. As in mitochondria, the molecular components needed for transcription and translation are derived from both nuclear and the organellar genetic information. While chloroplast DNA is much larger than mitochondrial DNA, it is nevertheless very similar to that found in prokaryotic cells.

DNA isolated from chloroplasts (Figure 11–5), is found to be circular, double-stranded, replicated semiconservatively, and free of the associated proteins characteristic of eukaryotic DNA. Compared with nuclear DNA of the same organism, it invariably shows a different buoyant density and base composition.

In *Chlamydomonas*, there are about 75 copies of the chloroplast DNA molecule per organelle. Each copy consists of a length of DNA that contains 195 kb, almost twice the size of a T-even bacteriophage genome. In higher plants such as the sweet pea, multiple copies of the DNA molecule are present in each organelle, but the molecule is considerably smaller than that in *Chlamydomonas*, consisting of 134 kb.

Based on the observation that mitochondrial and chloroplast DNA and their genetic apparatus are similar to their counterparts in bacteria, a theory of the origin of these organelles has been proposed. Championed by Lynn Margulis and others, this hypothesis, called the **endosymbiont theory,** states that mitochondria and chloroplasts may have originated as distinct bacterialike particles that became incorporated into primitive eukaryotic cells. In the coevolution of

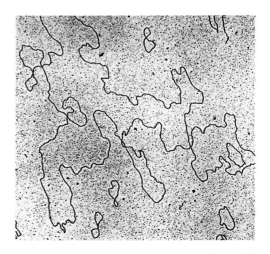

■ Figure 11–5   Electron micrograph of chloroplast DNA derived from corn.

this symbiotic relationship, the particles lost their ability to function independently, and the eukaryotic host cell became dependent on them. While there are many unanswered questions, the basic tenets of this theory are difficult to dispute.

Organization of DNA in Chromatin

The structure and organization of the genetic material in eukaryotic cells is much more intricate than in viruses or bacteria. This complexity is due to the greater amount of DNA per chromosome and the presence of large numbers of proteins associated with DNA in eukaryotes. For example, while DNA in the *E. coli* chromosome is 1200 μm long, the DNA in human chromosomes ranges from 14,000 to 73,000 μm in length. In a single human nucleus, all 46 chromosomes contain sufficient DNA to extend almost 2 m! This genetic material, along with its associated proteins, is contained within a nucleus that usually measures about 5 μm in diameter.

For many reasons the intricacy of the organization of eukaryotic DNA is to be expected. It parallels the structural and biochemical diversity of the many types of eukaryotic cells present in a multicellular organism. Different cells assume specific functions based upon highly specific biochemical activity. While all cells carry a full genetic complement, different cells activate different sets of genes. A highly ordered regulatory system governing the readout of this information must exist if dissimilar cells performing different functions are present. Such a system must in some way be imposed on or related to the molecular structure of the genetic material.

Although bacteria can reproduce themselves, they never exhibit a complex process similar to mitosis. As described in Chapter 2, eukaryotic cells exhibit a highly organized cell cycle. During interphase, the genetic material and associated proteins are uncoiled and dispersed throughout the nucleus in a condition referred to as **chromatin.** When mitosis begins, the chromatin condenses greatly, and during prophase it is compressed into recognizable chromosomes. This condensation represents a contraction in length of some 10,000 times for each chromatin fiber and is the basis of the **folded-fiber model** of chromosome structure (see Figure 2–14). This highly regular condensation-uncoiling cycle poses special organizational problems in eukaryotic genetic material.

Early studies of the structure of eukaryotic genetic material concentrated on intact chromosomes, preferably large ones, because of the limitations of light microscopy. Discussion of two such structures—polytene and lampbrush chromosomes—was presented in Chapter 2. Subsequently, new techniques for biochemical

analysis, as well as the examination of relatively intact eukaryotic chromatin and mitotic figures under the electron microscope, greatly enhanced our understanding of chromosome structure.

As established earlier, the genetic material of viruses and bacteria consists of essentially naked strands of DNA or RNA. In eukaryotic chromatin, a substantial amount of protein is associated with the chromosomal DNA in all phases of the eukaryotic cell cycle. The associated proteins are divided into basic, positively charged **histones** and less positively charged **nonhistones**.

Nucleosome Structure

The general model for chromatin structure is based on the assumption that chromatin fibers, composed of DNA and protein, undergo extensive coiling and folding as they are condensed within the cell nucleus. Of the proteins associated with DNA, the histones clearly play the most essential structural role. Histones contain large amounts of the positively charged amino acids lysine and arginine, making it possible for them to bond electrostatically to the negatively charged phosphate groups of nucleotides. Recall that a similar interaction has been proposed for several bacterial proteins. There are five main types of histones (Table 11.3).

X-ray diffraction studies confirm that histones play an important role in chromatin structure. Chromatin produces regularly spaced diffraction rings, suggesting that repeating structural units occur along the chromatin axis. If the histone molecules are chemically removed from chromatin, the regularity of this diffraction pattern is disrupted.

The basic model for chromatin structure was worked out in the mid-1970s. Several observations are relevant to the development of this model:

1. Digestion of chromatin by certain endonucleases, such as micrococcal nuclease, yields DNA fragments that are approximately 200 base pairs in length, or multiples thereof. This demonstrates that enzymatic digestion is not random; if it were we would expect a wide range of fragment sizes. Thus, chromatin consists of some type of repeating unit, each of which is protected from enzymatic cleavage, except where

TABLE 11.3	Categories and properties of histone proteins	
Histone Type	*Lysine-Arginine Content*	*Molecular Weight (daltons)*
H1	Lysine-rich	23,000
H2A	Slightly lysine-rich	14,000
H2B	Slightly lysine-rich	13,800
H3	Arginine-rich	15,300
H4	Arginine-rich	11,300

any two units are joined. It is the area between adjacent units that is attacked and cleaved by the nuclease.

2. Electron microscopic observations of chromatin have revealed that chromatin fibers are composed of linear arrays of spherical particles (Figure 11–6). Discovered by Ada and Donald Olins, the particles occur regularly along the axis of a chromatin strand and resemble beads on a string. These particles are now referred to as **v-bodies** or **nucleosomes** (v is the Greek letter nu). These findings conform nicely to the earlier proposal, which suggests the existence of repeating units.

3. Results of the study of precise interactions of histone molecules and DNA in the nucleosomes constituting chromatin show that histones H2A, H2B, H3, and H4 occur as two types of tetramers: $(H2A)_2$-$(H2B)_2$ and $(H3)_2$-$(H4)_2$. This suggests that each repeating nucleosome unit consists of one of each tetramer. This octamer interacts with about 200 base pairs of DNA. Such a structure is consistent with previous observations and provides the basis for a model that explains the interaction of histone and DNA in chromatin.

4. When nuclease digestion time is extended, DNA is removed from both the entering and exiting strands, creating a **nucleosome core particle** consisting of 146 base pairs. This number is fairly constant in other organisms studied. The DNA lost in this prolonged digestion is responsible for linking nucleosomes together. This **linker DNA** is associated with histone H1.

5. On the basis of the above information as well as X-ray and neutron-scattering analysis of crystallized core particles by John T. Finch, Aaron Klug, and others, a detailed model of the nucleosome was put forward (Figure 11–7). In this model, the 146-base-pair DNA core exists as a secondary helix surrounding the octamer of histones. The coiled DNA does not quite complete two full turns. The entire nucleosome is ellipsoidal in shape, measuring about 110 Å (11 nm) in its greatest dimension.

The extensive investigation of nucleosomes now provides the basis for predicting how the packaging of chromatin within the nucleus occurs. In its most extended state under the electron microscope, the chromatin fiber is about 110 Å (11 nm) in diameter, a size consistent with the longer dimension of the ellipsoid nucleosome. It is believed that chromatin fibers consist of long strings of repeating nucleosomes [(Figure 11.7(b)].

The 110-Å nucleosome string is further folded into a thicker fiber, sometimes called a **solenoid**, which is 300 Å (30 nm) in diameter. This fiber appears to consist of numerous nucleosomes coiled closely together.

In the overall transition from a fully extended chromatin fiber to the extremely condensed status of the mitotic chromosome, a packing ratio (the ratio of DNA length to the length of the structure containing it) of about 500 must be achieved. Our model accounts for a ratio of only about 50. Obviously, the larger fiber can be further bent, coiled, and packed as even greater condensation occurs during the formation of a mitotic chromosome.

Heterochromatin

All recent evidence supports the concept that the DNA of each eukaryotic chromosome consists of one continuous double-helical fiber along its entire length. A continuous fiber is the basis of the **unineme model of DNA** within a chromosome. This concept, along with our knowledge of nucleosomes, might lead you to suspect that the chromosome would demonstrate structural uniformity along its length. However, in the early part of this century, it was observed that some parts of the chromosome remain condensed and stain deeply during interphase, but most do not. The terms **heterochromatin** and **euchromatin** were coined in 1928 to describe the parts of chromosomes that remain tightly coiled and condensed and those that are uncoiled, respectively.

Subsequent investigation has revealed a number of characteristics of heterochromatin that distinguish it from euchromatin. The condensed heterochromatic areas are genetically inactive because either they lack genes or they contain genes that are repressed. Also, heterochromatin replicates later during the S phase of

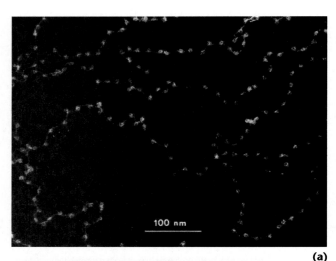

(a)

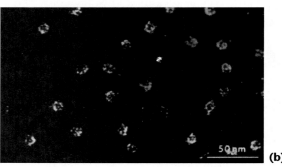

100 nm

50 nm

(b)

■ Figure 11–6 (a) Dark-field electron micrograph of nucleosomes present in chromatin derived from a chicken erythrocyte nucleus. (b) Dark-field electron micrograph of nucleosomes produced by micrococcal nuclease digestion.

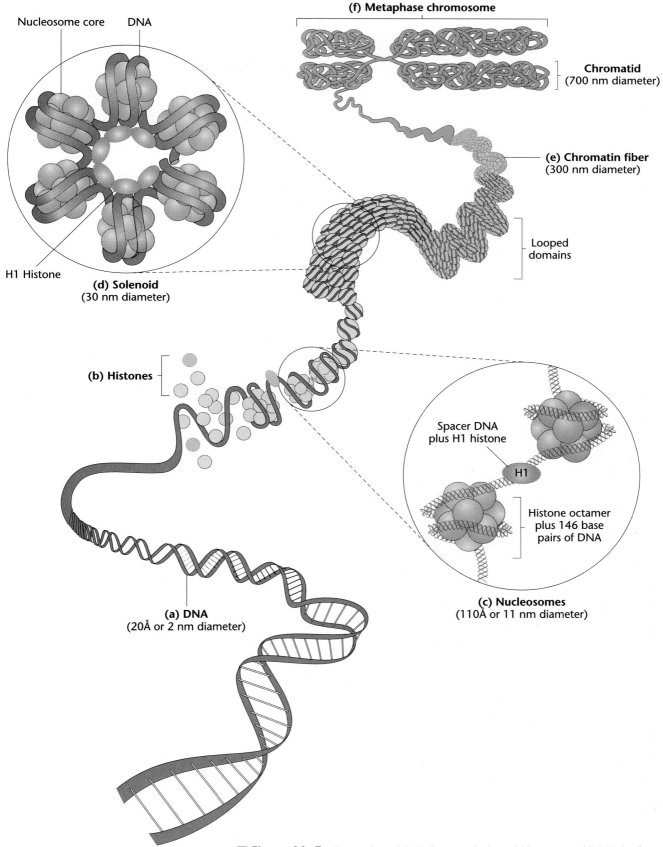

Nucleosome core   DNA

H1 Histone

(d) Solenoid
(30 nm diameter)

(b) Histones

(a) DNA
(20Å or 2 nm diameter)

(f) Metaphase chromosome

Chromatid
(700 nm diameter)

(e) Chromatin fiber
(300 nm diameter)

Looped
domains

Spacer DNA
plus H1 histone

H1

Histone octamer
plus 146 base
pairs of DNA

(c) Nucleosomes
(110Å or 11 nm diameter)

■ Figure 11–7 General model of the association of histones and DNA in the nucleosome, illustrating the way in which the chromatin fiber may be coiled into a more condensed structure, ultimately producing a mitotic chromosome.

the cell cycle than does euchromatin. The discovery of heterochromatin provided the first clues that parts of eukaryotic chromosomes do not always encode proteins. Instead, some chromosome regions are thought to be involved in the maintenance of the chromosome's structural integrity and in other functions such as chromosome movement during cell division.

Heterochromatin is characteristic of the genetic material of eukaryotes. Early cytological studies showed that areas of the **centromeres** are composed of heterochromatin. The ends of chromosomes, called **telomeres,** are also heterochromatic. In some cases, whole chromosomes are heterochromatic. Such is the case with the mammalian Y chromosome, which for the most part is genetically inert. And, as we discussed in Chapter 8, the inactivated X chromosome in mammalian females is condensed into an inert Barr body. In some species, such as mealy bugs, all chromosomes of one entire haploid set are heterochromatic.

When certain heterochromatic areas from one chromosome are translocated to a new site on the same or another nonhomologous chromosome, genetically active areas sometimes become genetically inert if they now lie adjacent to the translocated heterochromatin. This influence on existing euchromatin is one example of what is more generally referred to as a **position effect.** That is, the position of a gene or groups of genes relative to other genetic material may affect their expression. We first introduced this term in Chapter 8.

Chromosome Banding

Until about 1970, mitotic chromosomes viewed under the light microscope could be distinguished only by their relative sizes and the positions of their centromeres. Even in organisms with a low haploid number, two or more chromosomes are often indistinguishable from one another. However, around 1970, differential staining along the longitudinal axis of mitotic chromosomes was made possible by new cytological procedures. These methods are now called **chromosome banding techniques** because the staining patterns resemble the bands of polytene chromosomes.

One of the first chromosome banding techniques was devised by Mary Lou Pardue and Joe Gall. They found that if chromosome preparations of mice were heat-denatured and then treated with Giemsa stain, a unique staining pattern emerged. The centromeric regions of mitotic chromosomes preferentially took up the stain! This cytological technique stains a specific area of the chromosome composed of heterochromatin. A diagram of the mouse karyotype treated in this way is shown in Figure 11–8. The staining pattern is referred to as **C-banding.** Note that all of the mouse chromosomes are telocentric, where staining is apparent.

Still other chromosome banding techniques were developed about the same time. A group of Swedish researchers, led by Tobjorn Caspersson, used a technique

■ Figure 11–8   Karyotype of a male mouse (top) and a female mouse (bottom), where chromosome preparations were processed to demonstrate C-bands. The centromeres stain intensely.

that provided even greater staining differentiation of metaphase chromosomes. They used fluorescent dyes that bind to nucleoprotein complexes and produce unique banding patterns. When the chromosomes are treated with the fluorochrome quinacrine mustard and viewed under a fluorescent microscope, precise patterns of differential brightness are seen. Each of the 23 human chromosome pairs can be distinguished by this technique. The bands produced by this method are called **Q-bands.**

Another banding technique produces a staining pattern nearly identical with the Q-bands. This method, which produces **G-bands,** involves the digestion of the mitotic chromosomes in cytological preparations with the proteolytic enzyme trypsin, followed by Giemsa staining. Figure 11–9 illustrates G-banding in the human karyotype. Another technique results in the reverse G-band staining pattern called an **R-band** pattern. In 1971, a meeting was held in Paris to establish the nomenclature for these various patterns in humans. Still other banding techniques are now available.

Research is underway to elucidate the molecular mechanisms involved in producing these banding patterns. The variety of staining reactions under different conditions reflects the heterogeneity and complexity of chromosome composition.

Organization of the Eukaryotic Genome

Having established the general view of the chromatin fiber, we now turn to a consideration of the organization of the **genome** of eukaryotic organisms. The genome is

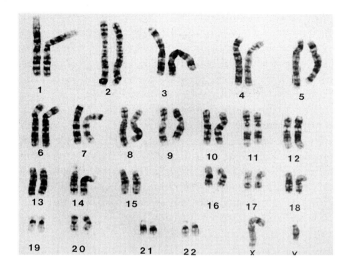

■ Figure 11–9   G-banded karyotype of a normal human male. Chromosomes were derived from cells in metaphase.

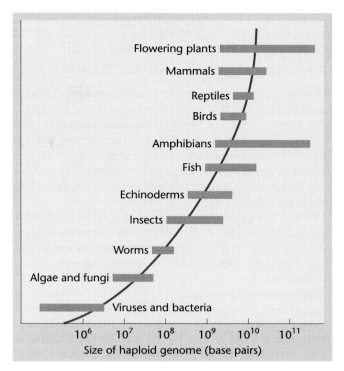

■ Figure 11–10   The range of DNA content of the haploid genome of many representative groups of organisms.

defined as the complete haploid set of chromosomes constituting the genetic material of an organism.

Prior to 1970, little was known about eukaryotic gene structure and organization at the molecular level. Nevertheless, many geneticists believed that when eukaryotic genes could be examined directly, they might well be different from those of viruses and bacteria. Multicellular eukaryotes, it was reasoned, have several unique requirements that distinguish them from phages and bacteria. Primarily, cell differentiation, tissue organization, and coordinated development and function depend on the regulation of genetic expression. Do these requirements also depend on different modes of gene structure and organization?

When the technologies of DNA analysis were developed (Chapter 17), the answer to this question began to emerge. The eukaryotic gene is far more complex than we could ever have imagined. In this section we review many findings related to gene structure and the organization of the eukaryotic genome and the structure of genes.

Eukaryotic Genomes and the C-Value Paradox

The *amount* of DNA contained in the haploid genome of a species is called the **C value.** When such values were determined for a large variety of eukaryotic organisms (Figure 11–10), several trends were apparent. The most notable trends are that:

1. Eukaryotes contain substantially more DNA in their genomes than viruses or prokaryotes and exhibit a wide variation among different groups.
2. Evolutionary progression has been accompanied by increased amounts of DNA. In many major phylogenetic groups, more evolutionarily advanced forms generally contain more DNA than do less advanced forms.

It might be argued that increasing *C* values are simply the result of a greater need for increased amounts and varieties of gene products in more complex organisms. However, several observations make this explanation unacceptable. First, the increases are very dramatic. Whereas viruses and bacteria contain 10^4 to 10^6 nucleotide pairs in their single chromosomes, eukaryotes contain 10^7 to 10^{11} nucleotide pairs in each set of chromosomes. Does a eukaryote require 10^4 (10,000) times as many genes as the most complex bacterium, as these figures might suggest? Second, closely related organisms with the same degree of complexity in body form and tissue and organ types often vary tenfold or more in DNA content. Third, the DNA contents of amphibians and flowering plants, which varies as much as 100-fold within their taxonomic classes, are often substantially greater than that of other, more recently evolved eukaryotes. Despite this observation, there is no correlation between increased genome size and morphological complexity in these groups.

Therefore, it is doubtful that the development of greater complexity during evolution can account for the amount of DNA found in eukaryotic genomes. This conclusion is the basis of what has been called the **C-value paradox:** *Excess DNA is present that does not seem to be essential to the development or evolutionary progression of eukaryotes.* Is this excess DNA vital to the organisms carrying it in their genomes, or is it simply excess DNA that has somehow accumulated during evolution and has no function?

Repetitive DNA: Centromeres and Telomeres

Careful analysis of the eukaryotic genome is beginning to unravel the mystery surrounding the *C*-value paradox. For example, in Chapter 9 we established that noncoding sequences are prevalent in the form of various types of **repetitive DNA.** As we will see below, an understanding of the molecular organization of repetitive DNA provides important insights into the organization of the genome.

We shall begin our discussion with a consideration of the centromere, the region of the chromosome that attaches to one or more spindle fibers and moves to one of the poles during the anaphase stages of mitosis and meiosis (see Chapter 2).

Separation of chromatids is essential to the fidelity of chromosome distribution during mitosis and meiosis. Most estimates of infidelity during mitosis are exceedingly low: 1×10^{-5} to 1×10^{-6}, or 1 error per 100,000 to 1,000,000 cell divisions. As a result, it has been generally assumed that analysis of the DNA sequence of centromeric regions will provide insights into the rather remarkable features of this chromosomal region. This DNA region is designated the **CEN.**

Analysis of the CEN regions of the yeast *Saccharomyces cerevisiae* chromosomes provided the basis for a model system first described by John Carbon and Louis Clarke. Because each centromere serves an identical function, it is not surprising that all CENs were found to be remarkably similar in their organization. The CEN region of yeast chromosomes consists of about 225 base pairs, which can be divided into three regions (Figure 11–11). The first and third regions (I and III) are relatively short and highly conserved, consisting of only 8 and 26 base pairs, respectively. Region II, which is larger (80–85 base

pairs) and extremely A=T rich (up to 94 percent), varies in sequence among different chromosomes.

Mutational analysis suggests that regions I and II are less critical to centromere function than region III. Mutations in the former regions are often tolerated, but mutations in region III usually disrupt centromere function. The DNA sequence of this region may be essential to binding to spindle fibers.

The amount of DNA associated with centromeres of multicellular eukaryotes is much more extensive than in yeast. For example, highly repetitive **"satellite"** DNA sequences are localized in the centromere regions of mice. Such sequences, absent from yeast but characteristic of most multicellular organisms, vary considerably in size. For example, the 10-base-pair sequence AATAACATAG is tandemly repeated many times in the centromeres of all four chromosomes of *Drosophila.* In humans, one of the most recognized satellite DNA sequences is the **alphoid family.** Found mainly in the centromere regions, alphoid sequences, each about 170 base pairs in length, are present in tandem arrays of up to 1 million base pairs.

The role of this highly repetitive DNA in centromere function remains unclear; it is known that the sequences are not transcribed. There is some sequence variation among members of the alphoid family. The number of repeats is specific to each human chromosome. Perhaps there is a shorter sequence common to the alphoid family that, like yeast CEN DNA, is absolutely critical to centromere function.

The second important structure of chromosomes is the **telomere,** found at the ends of linear chromosomes. The function of telomeres is to provide stability to the chromosome by rendering chromosome ends generally inert in interactions with other chromosome ends. In contrast to broken chromosomes, whose ends may rejoin other such ends, telomere regions do not fuse with one another or with broken ends. It is thought that some aspect of the molecular structure of telomeres

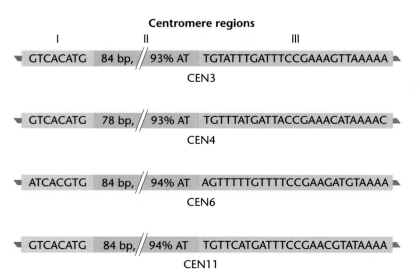

Centromere regions

I   II   III

GTCACATG   84 bp, // 93% AT   TGTATTTGATTTCCGAAAGTTAAAAA
CEN3

GTCACATG   78 bp, // 93% AT   TGTTTATGATTACCGAAACATAAAAC
CEN4

ATCACGTG   84 bp, // 94% AT   AGTTTTTGTTTTCCGAAGATGTAAAA
CEN6

GTCACATG   84 bp, // 94% AT   TGTTCATGATTTCCGAACGTATAAAA
CEN11

■ Figure 11–11 Nucleotide sequence information derived from DNA of the three major centromere regions of chromosomes 3, 4, 6, and 11 of yeast.

must be unique compared with most other chromosome regions.

As with centromeres, the problem was first approached by investigating the smaller chromosomes of simple eukaryotes, such as protozoans and yeast. The idea that all telomeres in a given species might share a common nucleotide sequence has now been borne out.

Two types of telomere sequences have been discovered. The first type, called simply **telomeric DNA sequences,** consists of short tandem repeats. It is this group that contributes to the stability and integrity of the chromosome. In the ciliate *Tetrahymena,* over 50 tandem repeats of the hexanucleotide sequence GGGGTT occur. In another ciliate, *Oxytricha,* GGGGTTTT is repeated many times. In humans, the sequence GGGATT is repeated numerous times. The analysis of telomeric DNA sequences has shown them to be highly conserved over long periods of evolutionary time, reflecting the critical role they play in maintaining the integrity of chromosomes.

The second type, **telomere-associated sequences,** are also repetitive and are found both adjacent to and within the telomere. These sequences vary from one another in different organisms, and their significance remains unknown.

As discussed in Chapter 10, replication of the telomere requires a unique RNA-containing enzyme **telomerase.** In its absence, the DNA at the ends of chromosomes becomes shorter during each replication. Because single-celled eukaryotes are immortalized cells, telomerase is critical for the survival of such species. In multicellular organisms, such as humans, telomerase is essential in germline cells, but is inactive in most all types of somatic cells. Chromosome shortening is considered part of the natural process of cell aging, serving as an internal clock. In human cancer cells, which have become immortalized, the transition to malignancy appears to require the activation of telomerase in order to overcome the normal senescence associated with chromosome shortening.

Repetitive DNA: SINEs, LINEs, and VNTRs

A brief review of other prominent categories of repetitive DNA sheds more light on our understanding of the eukaryotic genome. In this discussion, we will also account for more of the genomic DNA that fails to encode proteins.

In addition to highly repetitive DNA, which constitutes about 5 percent of the human genome (and 10 percent of the mouse genome), a second category, **moderately** (or **middle**) **repetitive DNA,** is fairly well characterized. Because a great deal is being learned about the human genome, we will use our own species to illustrate the part played by this category of DNA in genome organization.

Moderately repetitive DNA consists of either interspersed or tandemly repeated sequences. That which is interspersed is either short or long. **Short interspersed elements,** called **SINEs,** are less than 500 base pairs long and are present as many as a million times.

The best-characterized human SINE is a set of closely related sequences called the ***Alu* family** (the name is based on the presence of DNA sequences recognized by the restriction endonuclease Alu). Members of this DNA family, also found in other mammals, are 200 to 300 base pairs long and are present 700,000 to 900,000 times throughout the genome, both between and within genes. In humans, this family encompasses almost 10 percent of the entire genome.

Alu sequences are particularly interesting, although their function, if any, is yet undefined. Members of the *Alu* family are sometimes transcribed. The role of this RNA is unclear, but in some cases it may be related to the observation that *Alu* elements are potentially transposable within the genome. *Alu* sequences are thought to have arisen from an RNA element whose DNA complement was dispersed throughout the genome as a result of the activity of reverse transcriptase (an enzyme that synthesizes DNA on an RNA template).

Long interspersed elements (LINEs) represent another category of moderately repetitive DNA. In humans, the most prominent example is a family of LINEs designated **LI.** Members of this sequence family are around 6400 base pairs long and are present between 3000 and 40,000 times, according to different estimates. Their 5′ end is highly variable. Their role within the genome has yet to be defined.

Moderately repetitive DNA also may be clustered as tandem repeats rather than interspersed throughout the genome. In some cases, this category includes functional genes present in multiple copies. For example, in humans, many of the copies of genes encoding ribosomal RNA (rRNA) are clustered on the p arm of the acrocentric chromosomes 13, 14, 15, 21, and 22. We will return momentarily to discuss this arrangement of tandemly repeated genes within the genome.

A second type of tandem repeat includes those called **variable-number tandem repeats (VNTRs).** The repeating DNA sequence or VNTRs may be 15 to 100 base pairs long. The number of tandem copies of any specific sequence varies in individuals, creating localized regions of 1000 to 5000 base pairs in length. As we will see in Chapter 17, such regions in humans are the basis for the forensic technique referred to as **DNA fingerprinting.** VNTRs may be found within and between genes.

Together, the various forms of moderately repetitive DNA comprise up to 30 percent of the human genome. However, over 60 percent of the genome is still unaccounted for. Such observations are not uncommon. While the proportion of the genome consisting of repetitive DNA varies among organisms, one feature seems to be shared: *Only a small part of the genome codes for proteins.* For example, the 20,000 to 30,000 genes encoding proteins in sea urchins occupy less than 10 percent of the genome. In *Drosophila,* only 5 to 10 percent of the genome is occupied by genes coding for proteins. In humans, it appears that only 1 to 2 percent of the genome encodes proteins. We must conclude that although there is sufficient DNA to code for over 1 million genes in many eukaryotic organisms, *the vast majority of DNA does not represent genes that encode proteins.*

Eukaryotic Gene Structure

We turn now to a brief review of the structure of the eukaryotic gene. As this discussion unfolds, you should refer to Figure 11–12, which presents a model of the eukaryotic gene.

An important insight has been derived from a comparison of mRNA molecules and the DNA of the genes from which they are derived. It has been found that many internal base sequences of genes, referred to as **intervening sequences,** or simply **introns** (for short), are not represented in "mature" mRNAs that are translated into proteins. The remaining areas of each gene that are ultimately translated into the amino acid sequence of the encoded protein are called **exons.** While the entire gene is transcribed into a large precursor mRNA, which is part of a pool of other such molecules called **heterogeneous nuclear RNA (hnRNA),** the intron regions of the initial transcript are excised and the exon regions of the transcript are spliced back together before translation. We will pursue the topic of introns and exons in greater detail in Chapter 12.

Given this information about internal gene sequences, we can ask whether the nucleotide sequences found in introns solve the C-value paradox. That is, do introns account for the remainder of the noncoding sequences beyond those of repetitive DNA? The answer is emphatically no! As shown in Table 11.4, the presence of introns does substantially increase the size of the average eukaryotic gene. However, introns account for no more than 10 percent of the noncoding DNA in the genome. We have yet to account for a great deal of noncoding DNA.

Aside from introns, we shall review briefly the presence and nature of the noncoding flanking regions to complete our discussion of the eukaryotic gene. It is now clear that the 5′ region upstream from the coding

| | | Minimum | |
Gene	mRNA Size	Transcript Length	Ratio
Rabbit β-globin	589	1295	2.2
Mouse β^{maj}-globin	620	1382	2.2
Mouse β^{min}-globin	575	1275	2.2
Mouse α-globin	585	850	1.5
Rat insulin I	443	562	1.3
Rat insulin II	443	1061	2.4
Chick ovalbumin	1859	7500	4.0
Chick ovomucoid	883	5600	6.3
Chick lysozyme	620	3700	6.0

TABLE 11.4 A comparison of mRNAs and the initial transcripts from which they are derived

sequence is critical to efficient transcription. Closest to the coding sequence is the **promoter** with its **TATA box,** where RNA polymerase binds before initiation of transcription. The TATA box is so named because a base sequence of that nature has been conserved in most eukaryotic genes. Farther upstream is the **CCAAT box,** which is involved somehow in the regulation of transcription. In numerous cases, an **enhancer** region, which modulates transcription, also is present. Enhancers also may be found within the coding sequence or as part of the 3′ downstream region, beyond the coding sequence. Note also that in Figure 11–12 there are upstream (5′) and downstream (3′) noncoding sequences that are part of the transcriptional unit. These are part of the primary transcript of a gene, but they are trimmed off as the mRNA matures.

Since flanking regions are critical to specific gene function, their discovery has extended our knowledge of gene structure considerably.

Multigene Families: The α- and β-Globin Genes

Another aspect of the organization of the eukaryotic genome includes the distribution of related genes. Called **multigene families,** members share DNA sequence homology, and their gene products are functionally related. Often, but not always, they are found together in a single location along a chromosome. The

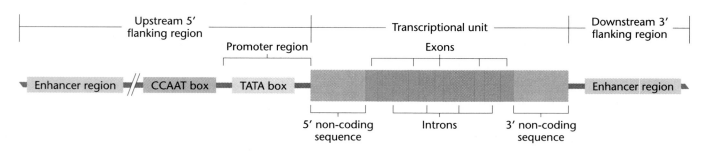

■ Figure 11–12 Modern concept of the eukaryotic gene.

examination of multigene families further provides valuable insights into the *C*-value paradox.

We will first examine cases where groups of genes encode very similar, but not identical, polypeptide chains that become part of proteins that are very closely related in function. The globin gene families, responsible for encoding the various polypeptides that are part of hemoglobin molecules, are examples that provide many insights into the organization of the genome. Then, we will consider the case of identical genes that are tandemly repeated, one after the other, as represented by ribosomal RNA genes.

The human **alpha-** and **beta-globin gene families** are two of the most extensive and best characterized groups. The alpha family resides on the short arm of chromosome 16 and contains five genes, whereas the beta family resides on the short arm of chromosome 11 and contains six genes. Members of both families show homology to one another, but not nearly to the extent that members within the same family exhibit homology to one another. Members of both families encode globin polypeptides that combine into a single tetrameric molecule, which interacts with a heme group that can reversibly bind to oxygen. And, within each group of genes, members are coordinately turned on and off during the embryonic, fetal, and adult stages of develop-ment in the precise order in which they occur along the chromosome within each family.

The alpha family (Figure 11–13) spans more than 30,000 nucleotide pairs (30 kb) and contains five genes: the zeta (ζ) gene, expressed only in the embryonic stage, two nonfunctional **pseudogenes,** and two copies of the alpha (α) gene. The α2 gene is expressed during the fetal stage, whereas the α1 gene is expressed during the adult stage. Pseudogenes, found in both families, are similar in sequence to the other genes in their family, but contain significant nucleotide substitutions, deletions, and duplications that prevent their transcription. The first pseudogene in the alpha family shows greater similarity to the zeta gene, while the second shows greater similarity to the alpha gene. Pseudogenes are designated by the prefix (Ψ) (the Greek letter psi), followed by the symbol of the gene they most resemble. Thus, the designation Ψα1 indicates a pseudogene of the first alpha gene.

Examination of the organization of the members of the alpha family as well as their intragenic distribution of introns and exons [Figure 11–13(a)] reveals several interesting features. First, only a small portion of the chromosomal region housing the entire alpha family is occupied by the three functional genes. The region consists almost entirely of intergenic regions. Second, the

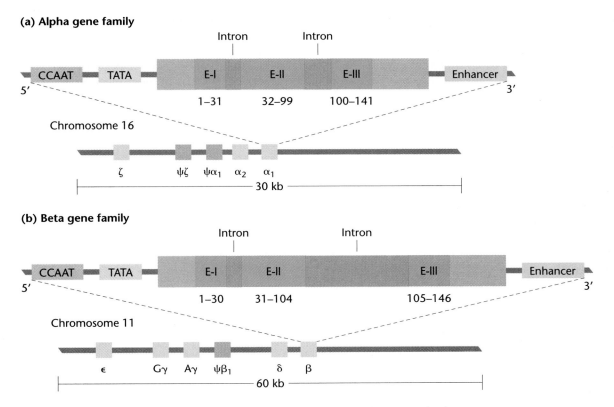

(a) Alpha gene family

(b) Beta gene family

■ Figure 11–13 The comparative arrangement of the human alpha and beta gene families, depicted in parts (a) and (b), respectively. The internal organization of the functional genes from each family are also shown. All genes contain three exons (E-I, E-II, and E-III) and two introns, as well as 5′ and 3′ noncoding regions. The arabic numbers designate the amino acid positions encoded by each exon.

functional genes in this family contain two introns at precisely the same positions within the genes. The nucleotide sequences within corresponding exons are nearly identical when the zeta and alpha genes are compared. Both encode polypeptide chains that are 141 amino acids long. However, the intron sequences are extremely divergent, even though they are about the same size. Significantly, a high percentage of the nucleotide sequence within each gene is contained in these noncoding introns. The above information bears on both the structural organization of a gene family and our earlier discussion of the *C*-value paradox.

The beta-globin family in humans is even more extensive, containing six genes (including one pseudogene) and occupying a 60-kb sequence of DNA [Figure 11–13(b)]. As with the alpha family, the sequence of genes along the chromosome parallels the order of expression during development.

The five functional genes encode products that are 146 amino acids long, and each contains two similarly sized introns at exactly the same positions. The second intron is significantly larger than its counterpart in the functional alpha genes. The introns and their locations within the beta gene are shown in Figure 11–13(b).

The flanking regions of each gene have also been studied. Both a TATA box (29 or 30 base pairs upstream from the start codon) and a CCAAT region (70 to 78 base pairs upstream) have been observed. Downstream, globin enhancers have been identified.

Several observations of the β-globin genes pertain to the *C*-value paradox. Only about 5 percent of the 60-kb region consists of coding sequences. The remaining 95 percent includes the introns, flanking regions, and spacer DNA found between genes. Of this percentage, only about 11 percent consists of introns. The remainder of the DNA serves no known function and consists of sequences that are not found elsewhere in the genome.

Genome Analysis

We conclude this chapter with a brief overview of what is unquestionably one of the most exciting areas of current investigation in the field of genetics: *characterization of the DNA sequences constituting the comparative genomes of a variety of model species, including humans.* The organisms that are being studied have been selected based on the rich genetic background available for each of them. As a result, knowledge of DNA sequences from these genomes will provide the basis for confirming, fine-tuning, and integrating previous genetic findings as well as opening new lines of investigation based on previously undiscovered information. Additionally, the organisms represent many stages of evolution, thus promising to provide valuable insights into this important biological process.

The Genome Project

When the technology to rapidly clone and sequence DNA was developed, as will be described in Chapter 17, it was only logical to consider sequencing the entire genome of organisms, including humans. With regard to our own species, discussion about the feasibility began in the mid-1980s. By 1990, the goals were established, an international effort was agreed upon, and the project was launched. As Table 11.5 reveals, six previously well-studied organisms, in addition to humans, have been included in the formal study. In the United States, approximately $200 million in government funding was provided during each of the first five years, primarily by the National Institutes of Health and the Department of Energy in support of the project, which is scheduled for completion about the year 2003.

In addition to the seven organisms studied under the auspices of the Genome Project, Table 11.5 includes information for two additional organisms whose genomes have also been completely sequenced as a result of sep-

TABLE 11.5	A comparison of the genome size and estimated number of genes in organisms whose genome has been or is being sequenced		

Organism/Type	Genome Size Base Pairs	Estimated Number of Genes
Mycoplasma genitalium (bacterium)[a]	0.58×10^6	470
Hemophilus influenzae (bacterium)[a]	1.83×10^6	1,743
Escherichia coli (bacterium)	4.67×10^6	4,288
Saccharomyces cerevisiae (yeast)	1.20×10^7	5,885
Arabidopsis thaliana (plant)	1.0×10^8	25,000
Caenorhabditis elegans (roundworm)	1.0×10^8	13,000
Drosophila melanogaster (fruit fly)	1.2×10^8	10,000
Mus musculus (mouse)	3.0×10^9	60–100,000
Homo sapiens (human)	3.2×10^9	60–100,000

[a]Organisms not part of the formal Genome Project.

arate research initiatives—*Mycoplasma genitalium* and *Hemophilus influenzae*. Examination of the information involving the nine organisms reveals an interesting comparison of the genome sizes and the estimated number of genes in each organism. The number of nucleotide pairs ranges from about one-half million (0.5 Mb) in *Mycoplasma* to 3.2 billion (3200 Mb) in humans. The number of genes ranges from a low of 482 in the above bacterium to 80,000 in humans. It is fascinating to dissect this information. The complexity of the three different bacterial forms is clearly related to an increase in genome size and gene number. Yeast, as the prototypic single-celled eukaryote, has further expanded its genome size and gene number. The plant and the two invertebrates have almost ten times more DNA than yeast, but only two to four times as many genes! The two mammals (the mouse and humans) reveal about a 30-fold jump in DNA content compared to the invertebrates, but less that a tenfold increase in gene number.

When these figures for all nine organisms are compared, the most apparent finding involves the density of coding regions in each genome. For the three bacteria (*Mycoplasma, Hemophilus,* and *E. coli*) there are about 1000 genes per Mb unit of DNA. In the unicellular yeast, the density decreases to less than one-half this value. There is a further progressive decline in gene density as one compares the model plant species (*Arabidopsis*), to the roundworm (*Caenorhabditis*), to the fruit fly (*Drosophila*), and then to mice and humans, which are roughly equivalent, averaging only about 25 genes per Mb of DNA. These findings represent a definite evolutionary trend, where more advanced genomes have acquired extensive regions of DNA that fail to encode gene products. In *Mycoplasma* and *Hemophilus,* 85 to 90 percent of all DNA constitutes functional gene areas, whereas in humans it is doubtful if as much as 10 percent will be found to represent functional coding regions.

Chapter Summary

1. Knowledge of the organization of the molecular components forming chromosomes is essential to the understanding of the function of the genetic material. Largely devoid of associated proteins, bacteriophage and bacterial chromosomes are naked DNA molecules present in a form equivalent to the Watson–Crick model.

2. Mitochondria and chloroplasts contain DNA that encodes products essential to their biological function. This DNA is remarkably similar in form and appearance to bacterial and phage DNA, lending support to the endosymbiont theory, which suggests that these organelles were once free-living organisms.

3. The eukaryotic chromatin fiber is a nucleoprotein organized into repeating units called nucleosomes. Composed of about 200 base pairs of DNA and an octamer of four types of histones, the nucleosome is important in facilitating the conversion of the extensive chromatin fiber characteristic of interphase into the highly condensed chromosome seen in mitosis.

4. The structural heterogeneity of the chromosome axis has been established as a result of both biochemical and cytological investigation. Heterochromatin, prematurely condensed in interphase, is genetically inert. The centromeric and telomeric regions, the Y chromosome, and the Barr body are examples.

5. DNA analysis has revealed highly conserved, unique nucleotide sequences in both the heterochromatic centromere and telomere regions. These sequences distinguish these structures from other parts of the chromosome and provide the basis for the critical features imparted to the chromosome by these regions.

6. The *C*-value paradox, made apparent during the study of the organization of eukaryotic DNA, suggests that eukaryotic organisms contain much more DNA than is necessary to encode the gene products essential to the development and normal functions of an organism.

7. Detailed analysis of eukaryotic gene structure has revealed numerous categories of noncoding DNA sequences. Introns, internal noncoding sequences, are represented in the initial mRNA molecules or heterogeneous RNA, but never appear in the mature mRNAs from which proteins are synthesized.

8. The flanking regions of genes, particularly those giving rise to the 5′-initiation point of mRNA, have also been analyzed. Three upstream regions that appear to be essential to efficient transcription are the TATA box, the CCAAT box, and the enhancer element.

9. Multigene families are composed of structurally related genes, usually clustered together, encoding functionally similar products. The human alpha- and beta-globin families are two examples.

10. A massive research initiative is now under way to provide comparative genome sequencing information for a variety of organisms, from bacteria to humans. When complete, the data promise to provide comprehensive insights into the organization of DNA as it serves as the genetic material.

Key Terms

INSIGHTS
and
SOLUTIONS

A previously undiscovered single-celled organism was found living at a great depth on the ocean floor. Its nucleus contained only a single linear chromosome containing 7×10^6 nucleotide pairs of DNA coalesced with three types of histonelike proteins. Consider the following experimental observations.

1. A short micrococcal nuclease digestion yielded DNA fractions consisting of 700, 1400, and 2100 base pairs. Predict what these fractions represent. What conclusions can be drawn?

Solution: The chromatin fiber may consist of a variation of nucleosomes containing 700 base pairs of DNA. The 1400 and 2100 base-pair fractions represent two and three nucleosomes, respectively, linked together. Enzymatic digestion has been incomplete in the latter two fractions.

2. Analysis of individual nucleosomes revealed that each unit contained one copy of each protein, and that the short linker DNA contained no protein bound to it. If the entire chromosome consists of nucleosomes, how many are there, and how many total proteins are needed to form them?

Solution: Because the chromosome contains 7×10^6 base pairs of DNA, the number of nucleosomes, each containing 7×10^2 base pairs, is equal to

$$\frac{7 \times 10^6}{7 \times 10^2} = 10^4 \text{ nucleosomes}$$

The chromosome thus contains 10^4 copies of each of the three proteins, for a total of 3×10^4 molecules.

3. Analysis then revealed the above organism's DNA to be a double helix similar to the Watson–Crick model, but containing 20 base pairs per complete turn of the right-handed helix. The physical size of the nucleosome was exactly double the volume occupied by that found in all other known eukaryotes, by virtue of increasing the distance along the fiber axis by a factor of 2. Compare the degree of compaction of this organism's nucleosome to that found in other eukaryotes.

Solution: The unique organism compacts a length of DNA consisting of 35 complete turns of the helix (700 base pairs per nucleosome/20 base pairs per turn) into each nucleosome. The normal eukaryote compacts a length of DNA consisting of 20 complete turns of the helix (200 base pairs per nucleosome/10 base pairs per turn) into a nucleosome one-half the volume of that in the unique organism. The degree of compaction is therefore less in the unique organism.

Problems and Discussion Questions

1. Compare and contrast the chemical nature, size, and form assumed by the genetic material of viruses and bacteria.

2. Contrast the DNA associated with mitochondria and chloroplasts.

3. Describe the sequence of research findings leading to the model of chromatin structure. What is the molecular composition and arrangement of the nucleosome?

4. Provide a comprehensive definition of heterochromatin and list as many examples as you can.

5. Mammals contain a diploid genome consisting of at least 10^9 base pairs. If this amount of DNA is present as chromatin fibers in which each group of 200 base pairs of DNA is combined with 9 histones into a nucleosome, and each group of 5 nucleosomes is further condensed into a structure called a **solenoid,** achieving a final packing ratio of 50, determine:

   (a) The total number of nucleosomes in all fibers

   (b) The total number of solenoids in all fibers

   (c) The total number of histone molecules combined with DNA in the diploid genome

   (d) The combined length of all fibers

6. Assume that a viral DNA molecule is in the form of a 50-mm-long, circular rod of a uniform 20-Å diameter. If this molecule is contained within a viral head that is a sphere with a diameter of 0.08 mm, will the DNA molecule fit into the viral head, assuming complete flexibility of the molecule? Justify your answer mathematically.

7. Describe what is meant by exon shuffling.

8. The β-globin gene family consists of 60 kb of DNA, yet only 5 percent of the DNA encodes β-globin gene products. Account for as much of the remaining 95 percent of the DNA as you can.

9. Describe the rRNA gene family in eukaryotes. What accounts for the major difference in the size of tandem repeats in different organisms?

10. "In the 1990s, the *C*-value paradox is not so paradoxical." Agree or disagree with this statement and support your opinion.

Selected Readings

Blackburn, E. H. 1990. Telomeres: Structure and synthesis. *J. Biol. Chem.* 265:5919–21.

———. 1991. Structure and function of telomeres. *Nature* 350:569–73.

Bloom, K., Hill, A., and Yeh, E. 1986. Structural analysis of a yeast centromere. *BioEssay* 4:100–5.

Carbon, J. 1984. Yeast centromeres: Structure and function. *Cell* 37:352–53.

Chen, T. R., and Ruddle, F. H. 1971. Karyotype analysis utilizing differential stained constitutive heterochromatin of human and murine chromosomes. *Chromosoma* 34:51–72.

Collins, F. S. 1995. Ahead of schedule and under budget: The Genome Project passes its fifth birthday. *Proc. Natl. Acad. Sci. USA* 92:10821–23.

Fritsche, E. F., Lawn, R. M., and Maniatis, T. 1980. Molecular cloning and characterization of the human β-like globin gene cluster. *Cell* 19:959–72.

Gall, J. G. 1981. Chromosome structure and the *C*-value paradox. *J. Cell Biol.* 91:3s–14s.

Green, B. R., and Burton, H. 1970. *Acetabularia* chloroplast DNA: Electron microscopic visualization. *Science* 168:981–82.

Hewish, D. R., and Burgoyne, L. 1973. Chromatin sub-structure. The digestion of chromatin DNA at regularly spaced sites by a nuclear deoxyribonuclease. *Biochem. Biophys. Res. Commun.* 52:504–10.

Hsu, T. C. 1973. Longitudinal differentiation of chromosomes. *Annu. Rev. Genet.* 7:153–77.

Korenberg, J. R., and Rykowski, M. C. 1988. Human genome organization: *Alu*, LINEs and the molecular structure of metaphase chromosome bands. *Cell* 53:391–400.

Kornberg, R. D. 1975. Chromatin structure: A repeating unit of histones and DNA. *Science* 184:868–71.

Kornberg, R. D., and Klug, A. 1981. The nucleosome. *Sci. Am.* (Feb.) 244:52–64.

Maniatis, T., et al. 1980. The molecular genetics of human hemoglobins. *Annu. Rev. Genet.* 14:145–78.

Moyzis, R. K. 1991. The human telomere. *Sci. Am.* (Aug.) 265:48–55.

Olins, A. L., and Olins, D. E. 1974. Spheroid chromatin units (ν bodies). *Science* 183:330–32.

———. 1978. Nucleosomes: The structural quantum in chromosomes. *Am. Sci.* 66:704–11.

Schmid, C. W., and Jelinek, W. R. 1982. The *Alu* family of dispersed repetitive sequences. *Science* 216:1065–70.

Singer, M. F. 1982. SINEs and LINEs: Highly repeated short and long interspersed sequences in mammalian genomes. *Cell* 28:433–34.

van Holde, K. E. 1989. *Chromatin.* New York: Springer-Verlag.

Verma, R. S., ed. 1988. *Heterochromatin: Molecular and structural aspects.* Cambridge, England: Cambridge University Press.

Willard, H. F. 1990. Centromeres of mammalian chromosomes. *Trends Genet.* 6:410–16.

Yunis, J. J. 1981. Chromosomes and cancer: New nomenclature and future directions. *Hum. Pathol.* 12:494–503.

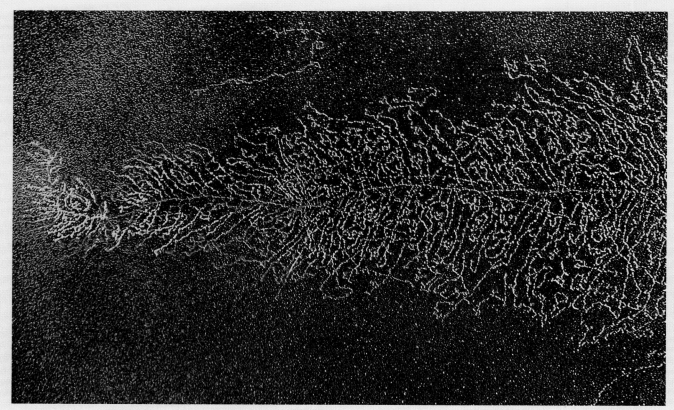

Electron micrograph visualizing the process of transcription.

CHAPTER OUTLINE

CHAPTER
12

Storage and Expression of Genetic Information

Chapter Concepts

Genetic information, stored in DNA and transferred to RNA during the process of transcription, is present as three-letter code words. Using four different letters, corresponding to the four ribonucleotides in RNA, the 64 possible triplet codons specify the 20 different amino acids found in proteins, as well as provide signals that initiate and terminate protein synthesis. This unique language is the basis of life as we know it.

In previous chapters, we have established that DNA is the genetic material and explored the structure and organization of DNA as it serves as the basis for heredity in all living things. This molecule provides the chemical basis for storage of genetic information. Although we have yet to discuss the details of the processes, we know that the DNA of an active gene is first transcribed into an RNA complement that is subsequently translated into the end product of the gene— a protein. Thus, the sequence of deoxyribonucleotides in DNA is first transferred to the complementary sequence of ribonucleotides in RNA, which then specifies the insertion of amino acids during protein synthesis. The end products of various genes are responsible for the normal and mutant phenotypes of organisms.

A fundamental question, then, is how an RNA molecule consisting of only four types of nucleotides (A, U, C, and G) can specify 20 amino acids. This question poses an intriguing theoretical problem. Although the earliest proposals were imaginative, it was not until ingenious analytical research was applied that the code was deciphered. It was established that the code is triplet in nature. Code words, or **codons**, consisting of three ribonucleotides specify the insertion of amino acids into a polypeptide chain during its synthesis. The work leading to these discoveries occurred most intensively in the late 1950s and early 1960s, one of the most exciting periods in the study of molecular genetics. This research revealed the intricacies of the specific chemical language that serves as the basis of all life on earth.

The Genetic Code: An Overview

Before considering the analytical approaches used in arriving at our current understanding of the genetic code, we shall provide a summary of the general features that characterize it.

1. The genetic code is written in linear form using the ribonucleotide bases that compose the letters in mRNA molecules. The ribonucleotide sequence is derived from its complement in DNA.

2. Each code word within the mRNA contains three letters. The code is referred to as a **triplet codon**, where each group of *three* ribonucleotides specifies *one* amino acid.

3. The code is **unambiguous**, meaning that each triplet specifies only one amino acid.

4. The code is **degenerate**, meaning that more than one triplet may specify a given amino acid. This is the case for 18 of the 20 amino acids.

5. The code is **ordered**. Degenerate codons for a given amino acid are grouped together, most often varying by only the third base.

6. The code contains "start" and "stop" punctuation signals. Certain triplets are necessary to **initiate** and to **terminate** translation.

7. No commas (or internal punctuation) are used in the code. Thus, the code is said to be **commaless**. Once translation of mRNA begins, each three ri-

bonucleotides are read in turn, one after the other.

8. The code is **nonoverlapping**. Once translation commences, any single ribonucleotide at a specific location within the mRNA is part of only one triplet.

9. The code is almost **universal**. With only minor exceptions, a single coding dictionary is used by almost all viruses, prokaryotes, and eukaryotes.

Early Thinking About the Code

Before it became clear that mRNA serves as an intermediate in transferring genetic information from DNA to proteins, it was thought that DNA itself might encode the synthesis of proteins directly. The central question, whether DNA or RNA houses the code, was how only four letters—the four nucleotides—could specify 20 words—the amino acids. Once mRNA was discovered, it was clear that even though genetic information is stored in DNA, the code that is translated into proteins resides in RNA.

As to the size of the code, in the early 1960s, Sidney Brenner argued that it must be a triplet because three-letter words represent the minimal use of four letters to specify 20 amino acids. For example, four nucleotides, taken two at a time, provide only 16 unique code words (4^2). Although a triplet code provides 64 words (4^3)—clearly more than the 20 needed—a three-letter code is much simpler than a four-letter code, by which 256 words (4^4) would be specified. Brenner also argued that the code was nonoverlapping.

By 1960, many significant questions had been posed, and the scene was clearly set for the design of experimentation to decipher the code. In 1961, several noteworthy events occurred. First, François Jacob and Jacques Monod formally postulated the existence of messenger RNA (mRNA) as an intermediate between DNA and protein. This suggested how investigators might design experiments aimed at deciphering the code.

Second, research by Francis Crick and his colleagues provided some of the earliest experimental evidence concerning the nature of the code. Studying phage T4 and the unique type of mutation called a **frameshift**, they provided support for the triplet nature of the code.

Frameshift mutations occur as a result of the addition or deletion of one or more nucleotides in the gene and subsequently the mRNA transcribed by it. The gain or the loss of one or more letters shifts the frame of reading during translation. Crick and his colleagues found that the gain or loss of one or two nucleotides caused such a mutation, but when three nucleotides were involved, the frame of reading was reestablished (Figure 12–1). This would not occur if the code consisted of anything other than a triplet. This work also suggested that most triplet codes were not blank, but rather encode amino acids, supporting the concept of a degenerate code.

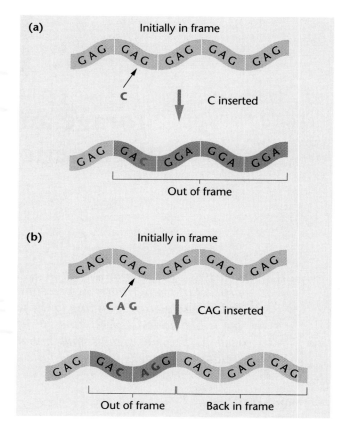

■ Figure 12–1 Schematic diagram of the effect of frameshift mutations on a DNA molecule repeating the triplet sequence GAG. In part (a) the insertion of a single nucleotide has shifted all subsequent reading frames. In part (b) the insertion of three nucleotides changes only two reading frames, but all remaining frames are retained in their original sequence.

Deciphering the Code: Initial Studies

In 1961, Marshall Nirenberg and J. Heinrich Matthaei published results that characterized the first specific coding sequences. These results served as a cornerstone for the complete analysis of the code. Nirenberg and Matthaei's success depended on the use of two experimental tools: a **cell-free protein-synthesizing system** and an enzyme, **polynucleotide phosphorylase**, which allowed the production of synthetic mRNAs. These mRNAs served as templates for polypeptide synthesis in the cell-free system.

In the cell-free (*in vitro*) system, amino acids can be incorporated into polypeptide chains. This *in vitro* mixture, as might be expected, must contain the essential factors for protein synthesis in the cell: ribosomes, tRNAs, amino acids, and other molecules essential to translation. In order to trace protein synthesis, one or more of the amino acids must be radioactive. Finally, an mRNA must be added, which serves as the template to be translated.

In 1961 mRNA had yet to be isolated. However, the use of the enzyme polynucleotide phosphorylase allowed artificial synthesis of RNA templates, which could be added to the cell-free system. This enzyme, isolated from bacteria, catalyzes the reaction shown in Figure 12–2. Discovered in 1955 by Marianne Grunberg-Manago and Severo Ochoa, the enzyme functions metabolically in bacterial cells to degrade RNA. However, *in vitro*, with high concentrations of ribonucleoside diphosphates, the reaction can be "forced" in the opposite direction to synthesize RNA, as illustrated.

In contrast to RNA polymerase, polynucleotide phosphorylase requires no DNA template. As a result, ribonucleotides are assembled at random, according to the relative concentration of the four ribonucleoside diphosphates added to the reaction mixtures. *This point is absolutely critical to understanding the work of Nirenberg and others in the ensuing discussion.*

Taken together, the cell-free system for protein synthesis and the availability of synthetic mRNAs provides a means of deciphering the ribonucleotide composition of various triplets encoding specific amino acids.

Nirenberg and Matthaei's Homopolymers

In their initial experiments, Nirenberg and Matthaei synthesized **RNA homopolymers**, each consisting of only one ribonucleotide. Therefore, the mRNA added to the *in vitro* system was either UUUUUU. . . , AAAAAA. . . , CCCCCC. . . , or GGGGGG. . . . In testing each mRNA, they were able to determine which, if any, amino acids were incorporated into newly synthesized proteins. They determined this by labeling one of the 20 amino acids added to the *in vitro* system and conducting a series of experiments, each with a different amino acid made radioactive.

For example, consider one of the initial experiments, in which ^{14}C-phenylalanine was used (Table 12.1). From these and related experiments, Nirenberg and Matthaei concluded that the message poly U (polyuridylic acid) directs only the incorporation of phenylalanine into the

TABLE 12.1 Incorporation of ^{14}C-phenylalanine into protein

Artificial mRNA	Radioactivity (counts/min)
None	44
Poly U	39,800
Poly A	50
Poly C	38

Source: After Nirenberg and Matthaei, 1961.

homopolymer polyphenylalanine. Assuming a triplet code, they had determined the first specific codon assignment: UUU codes for phenylalanine.

In the same way, they quickly found that AAA codes for lysine and CCC codes for proline. Poly G did not serve as an adequate template, probably because the molecule folds back on itself. Thus, the assignment for GGG had to await other approaches. Note that the specific triplet codon assignments were possible only because of the use of homopolymers. This method yields only the composition of triplets, not their sequence. However, three U's, C's, or A's can have only one possible sequence (i.e., UUU, CCC, and AAA).

The Use of Mixed Copolymers

With these techniques in hand, Nirenberg and Matthaei, as well as Ochoa and coworkers, turned to the use of **RNA heteropolymers**. In this technique, two or more different ribonucleoside diphosphates are added in combination to form the message. These researchers reasoned that if the relative proportion of each type of ribonucleoside diphosphate is known, the frequency of any particular triplet codon occurring in the synthetic mRNA can be predicted. If the mRNA is then added to the cell-free system and the percentage of any particular amino acid present in the new protein is ascertained, correlations may be made and composition assignments predicted.

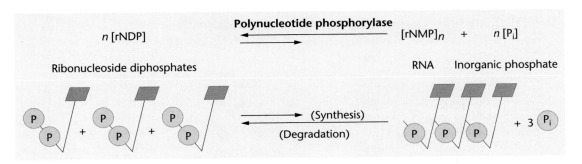

■ Figure 12–2 The reaction catalyzed by the enzyme polynucleotide phosphorylase. Note that the equilibrium of the reaction favors the degradation of RNA but can be "forced" in the direction favoring synthesis.

This concept is illustrated in Figure 12–3. Suppose that A and C are added in a ratio of 1A:5C. Now, the insertion of a ribonucleotide at any position along the RNA molecule during its synthesis is determined by the ratio of A:C. Therefore, there is a 1/6 possibility for an A and a 5/6 chance for a C to occupy each position. On this basis, we can calculate the frequency of any given triplet appearing in the message.

For AAA, the frequency is $(1/6)^3$, or about 0.4 percent. For AAC, ACA, and CAA, the frequencies are identical—that is, $(1/6)^2(5/6)$, or about 2.3 percent for each of the three different sequences. Together, all three 2A:1C triplets account for 6.9 percent of the total three-letter sequences. In the same way, each of three 1A:2C triplets accounts for $(1/6)(5/6)^2$, or 11.6 percent each (for a total of 34.8 percent). CCC is represented by $(5/6)^3$, or 57.9 percent of the triplets.

By examining the percentages of any given amino acid incorporated into the protein synthesized under the direction of this message, it is possible to propose probable base composition assignments (Figure 12–3). Because proline appears 69 percent of the time and because 69 percent is close to 57.9 percent + 11.6 percent, we can deduce that proline is coded by CCC and by one triplet of the 2C:1A variety. Histidine, at 14 percent, is probably coded by one 2C:1A (about 11 percent) and one 1C:2A (about 2 percent). Threonine, at 12 percent, is likely coded by only one 2C:1A. Asparagine and glutamine each appear to be coded by one of the 1C:2A triplets, and lysine appears to be coded by AAA.

Using as many as all four ribonucleotides to construct the mRNA, many similar experiments were conducted. Although determination of the *composition* of triplet code words corresponding to all 20 amino acids represented a very significant breakthrough, *specific sequences* of triplets were still unknown. Their determination awaited still other approaches.

The Triplet Binding Technique

It was not long before more advanced techniques were developed. In 1964 Nirenberg and Philip Leder developed the **triplet binding assay**, which led to specific assignments of triplets. The technique took advantage of the observation that ribosomes, when presented with an RNA sequence as short as three ribonucleotides, will bind to it and attract the correct charged tRNA (the tRNA–amino acid complex) corresponding to the triplet code. For example, if ribosomes are presented with an RNA triplet UUU and tRNA^{phe}, a complex will

Possible compositions	Probability of occurrence of any triplet	Possible triplets	Final %
3A	$(1/6)^3 = 1/216 = 0.4\%$	AAA	0.4
1C:2A	$(1/6)^2(5/6) = 5/216 = 2.3\%$	AAC ACA CAA	3 x 2.3 = 6.9
2C:1A	$(1/6)(5/6)^2 = 25/216 = 11.6\%$	ACC CAC CCA	3 x 11.6 = 34.8
3C	$(5/6)^3 = 125/216 = 57.9\%$	CCC	57.9
			100.0%

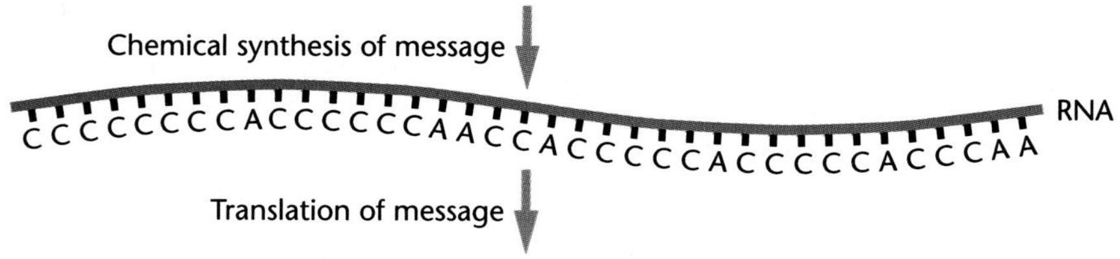

Chemical synthesis of message

CCCCCCCACCCCCCAACCACCCCCACCCCCACCCAA   RNA

Translation of message

Percentage of amino acids in protein		Probable base composition assignments
Lysine	1%	AAA
Glutamine	2%	1C:2A
Asparagine	2%	1C:2A
Threonine	12%	2C:1A
Histidine	14%	2C:1A, 1C:2A
Proline	69%	CCC, 2C:1A

■ Figure 12–3   Results and interpretation of a mixed copolymer experiment in which a ratio of 1A:5C is used (1/6A:5/6C).

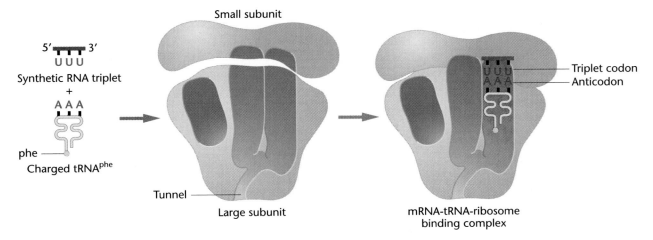

■ Figure 12–4   Molecular components used in the triplet binding assay.

form that is similar to what actually occurs *in vivo*. The triplet acts like a codon in mRNA and attracts the complementary sequence called the **anticodon** found within tRNA (Figure 12–4). Anticodons are those triplet sequences in tRNAs that are complementary to the codons of mRNA. Although it was not yet feasible to synthesize long stretches of RNA chemically, triplets of known sequence could be constructed in the laboratory to serve as templates.

All that was needed now was a method to determine which tRNA–amino acid complex is bound to the RNA–ribosome complex. The test system that was devised is quite simple. The amino acid to be tested is made radioactive, and a charged tRNA is produced. Because code compositions were known, it was possible to narrow the decision as to which amino acids should be tested for each specific triplet.

The radioactive charged tRNA, the RNA triplet, and ribosomes are incubated together on a nitrocellulose filter, which retains the larger ribosomes but not the other individual components, such as RNA triplets or charged tRNA. These components are much smaller and pass through the pores of the filter. If no radioactivity is retained on the filter, an incorrect amino acid has been tested. If radioactivity remains on the filter, it is retained because the charged tRNA has bound to the triplet associated with the ribosome. In such a case, a specific codon assignment may be made.

Work proceeded in several laboratories, and in many cases clear-cut, unambiguous results were obtained. Table 12.2, for example, shows 26 unique triplets assigned to 9 different amino acids. However, in some cases the degree of binding was insufficient, and unambiguous assignments were not possible. Eventually, about 50 of the 64 triplets were assigned. The binding technique was a major innovation in deciphering the code. Based on these specific assignments of triplets to amino acids, two major conclusions were drawn. The genetic code is **degenerate;** that is, one amino acid may be specified by more than one triplet. The code is also **unambiguous;**

TABLE 12.2	Amino acid assignments to specific trinucleotides derived from the triplet binding assay
Trinucleotide	**Amino Acid**
UGU UGC	Cysteine
GAA GAG	Glutamic acid
AUU AUG AUA	Isoleucine
UUA UUG CUU	Leucine
CUC CUA CUG	Leucine
AAA AAG	Lysine
AUG	Methionine
UUU UUC	Phenylalanine
CCU CCC	Proline
CCG CCA	Proline
UCU UCC	Serine
UCA UCG	Serine

that is, a single triplet specifies only one amino acid. As we shall see later in this chapter, these conclusions have been upheld with only minor exceptions.

The Use of Repeating Copolymers

Still another innovative technique used to decipher the genetic code was developed by Gobind Khorana, who was able to chemically synthesize long repeating RNA sequences that could be used in the cell-free protein synthesizing system. First, he created shorter sequences (e.g., di-, tri-, and tetranucleotides), which were then replicated many times and finally joined enzymatically to form the long polynucleotides.

As illustrated in Figure 12–5, a dinucleotide made in this way is converted to a message with two repeating triplets. A trinucleotide is converted to one with three potential triplets, depending on the point at which initiation occurs. Similarly, a tetranucleotide creates four repeating triplets.

When these synthetic mRNAs are added to a cell-free system, the predicted number of different amino acids incorporated is upheld. Several examples are shown in

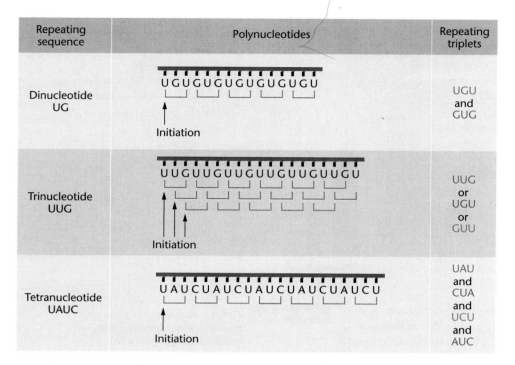

Repeating sequence	Polynucleotides	Repeating triplets
Dinucleotide UG	U G U G U G U G U G U G U G U Initiation	UGU and GUG
Trinucleotide UUG	U U G U U G U U G U U G U U G U U G U Initiation	UUG or UGU or GUU
Tetranucleotide UAUC	U A U C U A U C U A U C U A U C U A U C U Initiation	UAU and CUA and UCU and AUC

■ Figure 12–5 The conversion of di-, tri-, and tetranucleotides into repeating copolymers, as devised by Khorana. The triplet codons produced in each case are shown.

Table 12.3. When such data are combined with conclusions drawn from other approaches (composition assignment, triplet bindings), specific assignments become possible.

TABLE 12.3	Amino acids incorporated using repeated synthetic copolymers of RNA	
Repeating Copolymer	*Codons Produced*	*Amino Acids in Polypeptides*
UG	UGU GUG	Cysteine Valine
AC	ACA CAC	Threonine Histidine
UUC	UUC UCU CUU	Phenylalanine Serine Leucine
AUC	AUC UCA CAU	Isoleucine Serine Histidine
UAUC	UAU CUA UCU AUC	Tyrosine Leucine Serine Isoleucine
GAUA	GAU AGA UAG AUA	None

From such interpretations, Khorana reaffirmed triplets already deciphered and filled in gaps left by other approaches. For example, the use of two tetranucleotide sequences, GAUA and GUAA, suggested that at least two triplets are termination signals. This conclusion was drawn because neither of these sequences directed the incorporation of any amino acids into a polypeptide (Table 12.3). Because there are no triplets common to both messages (repeating GAUA and GUAA sequences), it was predicted that each repeating sequence contains at least one triplet that terminates protein synthesis. As we shall see, there are three such triplets, one of which is included in poly-(GAUA), accounting for lack of polypeptides.

The Coding Dictionary

The various techniques applied to decipher the genetic code have yielded a dictionary of 61 triplet codon–amino acid assignments. The remaining three triplets are termination signals, and do not specify any amino acid. Figure 12–6 designates the assignments in a particularly illustrative form first suggested by Francis Crick.

Degeneracy, Wobble, and Order in the Code

A general pattern of coding assignments becomes apparent when this presentation of the genetic code is inspected. Most evident is that the code is degenerate. That is, almost all amino acids are specified by two, three, or four triplets. Three amino acids (serine, arginine, and leucine) are each encoded by six different

Second position

First position (5′-end)	U	C	A	G	Third position (3′-end)
U	UUU *phe*	UCU *ser*	UAU *tyr*	UGU *cys*	U
	UUC	UCC	UAC	UGC	C
	UUA *leu*	UCA	UAA *Stop*	UGA *Stop*	A
	UUG	UCG	UAG *Stop*	UGG *trp*	G
C	CUU *leu*	CCU *pro*	CAU *his*	CGU *arg*	U
	CUC	CCC	CAC	CGC	C
	CUA	CCA	CAA *gln*	CGA	A
	CUG	CCG	CAG	CGG	G
A	AUU *ile*	ACU *thr*	AAU *asn*	AGU *ser*	U
	AUC	ACC	AAC	AGC	C
	AUA	ACA	AAA *lys*	AGA *arg*	A
	AUG *met*	ACG	AAG	AGG	G
G	GUU *val*	GCU *ala*	GAU *asp*	GGU *gly*	U
	GUC	GCC	GAC	GGC	C
	GUA	GCA	GAA *glu*	GGA	A
	GUG	GCG	GAG	GGG	G

Initiation Termination

■ Figure 12–6 The coding dictionary. AUG encodes methionine, which initiates most polypeptide chains. All other amino acids except tryptophan, which is encoded by UGG, are represented by two to six triplets. The triplets UAA, UAG, and UGA are termination signals and do not encode any amino acids.

codons. Only tryptophan and methionine are encoded by single triplets.

What is also evident is the pattern of degeneracy. Most often in a set of codons specifying the same amino acid, the first two letters are the same, with only the third differing. This interesting pattern prompted Crick in 1966 to postulate the **wobble hypothesis**.

Crick proposed that the first two ribonucleotides of triplet codes are more critical than the third member in attracting the correct tRNA. Based on this proposal, Crick suggested that hydrogen bonding at the third position of the codon–anticodon interactions need not adhere as specifically as the first two members according to the base-pairing rules. After examining the coding dictionary carefully, he proposed a new set of base-pairing rules at the third position of the codon.

This relaxed base-pairing requirement, or "wobble," allows the anticodon of a single tRNA species to pair with more than one triplet in mRNA. The code's degeneracy often allows substitution at the position of the third base without changing the amino acid. Consistent with the wobble hypothesis and coding assignments, it appears that U at the third position of the anticodon of tRNA may pair with A or G at the third position of the triplet in mRNA, and that G may likewise pair with U or C. Inosine, one of the modified bases found in tRNA, may pair with C, U, or A. Applying these wobble rules, a minimum of 30 different tRNA species is necessary to accommodate the 61 triplets specifying an amino acid. If nothing more, wobble can be considered an economy measure, provided that fidelity of translation is not compromised.

More recently, another observation has become apparent in the pattern of triplet sequences and their corresponding amino acids, leading to the description referred to as an **ordered code**. Chemically similar amino acids often share one or two "middle" bases in the different triplets that encode them. For example, U or C are often present in the second position of triplets that specify hydrophobic amino acids, including valine and alanine, among others. Charged amino acids, such as lysine and aspartic acid, or hydrophilic amino acids, such as glycine and serine, most often are specified by triplet codons with A or G in the second position. The chemical properties of amino acids will be discussed in more detail in Chapter 14. The end result of an "ordered" code is that it buffers the potential effect of mutation on protein function. While many mutations of the second base of triplet codons result in a change of one amino acid to another, the change is often to an amino acid with similar chemical properties. In such cases, protein function may not be noticeably altered.

Initiation and Termination

Initiation of protein synthesis is a highly specific process. In bacteria, the initial amino acid inserted into all polypeptide chains is a modified form of methionine—***N*-formylmethionine (fmet)**. Only one codon, AUG, codes for methionine. It is called the **initiator codon**, and it provides the "start signal" when it appears as the initial triplet in mRNA. However, when AUG appears internally in mRNA, unformylated methionine is inserted into the polypeptide chain. Rarely, still another triplet, GUG, specifies methionine during initiation. It is not clear why this occurs, since GUG normally encodes valine.

In bacteria, either the formyl group is removed from the initial methionine upon completion of synthesis of a protein, or the entire formylmethionine residue is removed. In eukaryotes, methionine is also the initial amino acid during polypeptide synthesis; however, it is not formylated.

As mentioned in the preceding section, three other triplets (UAA, UAG, and UGA) serve as punctuation signals and do not code for any amino acid. They are not recognized by a tRNA molecule, and termination of translation occurs when they are encountered. Mutations that produce any of the three triplets internally within a gene also result in termination. As a result, only a partial polypeptide is synthesized, since it is released from the ribosome prematurely. When such a change occurs, it is called a **nonsense mutation**.

Confirmation of Code Studies: Phage MS2

The aspects of the genetic code discussed so far yield a fairly complete picture. The code is triplet in nature, degenerate, unambiguous, and commaless, but it contains punctuation with respect to start and stop signals. These individual principles have been confirmed by detailed analysis of the RNA-containing bacteriophage MS2 by Walter Fiers and coworkers.

MS2 is a bacteriophage that infects *E. coli*. Its nucleic acid (RNA) contains only about 3500 ribonucleotides, making up only three genes. These genes specify a coat protein, an RNA-directed replicase, and a maturation protein (the A protein). This simple system of a small genome and few gene products allowed Fiers and his colleagues to sequence the genes and their products. The amino acid sequence of the coat protein was completed in 1970, and the nucleotide sequence of the gene and a number of nucleotides on each end of it were reported in 1972.

The coat protein contains 129 amino acids, and the gene contains 387 nucleotides, as expected for a triplet code (129×3). Each amino acid and triplet corresponds in linear sequence to the correct codon in the RNA code word dictionary, providing direct proof of the **colinear relationship** between nucleotide sequence and amino acid sequence. The codon for the first amino acid is preceded by AUG, the common initiator codon; and the codon for the last amino acid is succeeded by two consecutive termination codons, UAA and UAG.

By 1976, the other two genes and their protein products had been sequenced, providing similar confirmation. The analysis showed clearly that the genetic code as established in bacterial systems is identical in this virus. We shall now consider briefly some other evidence suggesting that the code is also identical in eukaryotes.

Universality of the Code?

Between 1960 and 1978, it was generally assumed that the genetic code would be found to be universal, applying similarly to viruses, bacteria, and eukaryotes. Certainly, the nature of mRNA and the translation machinery seemed to be very similar in these organisms. For example, cell-free systems derived from bacteria could translate eukaryotic mRNAs. Poly U was shown to stimulate translation of polyphenylalanine in cell-free systems when the components were derived from eukaryotes. Many recent studies involving recombinant DNA technology (see Chapter 17) have revealed that eukaryotic genes can be inserted into bacterial cells and transcribed and translated. Within eukaryotes, mRNAs from mice and rabbits have been injected into amphibian eggs and translated efficiently. For the many eukaryotic genes that have been sequenced, notably those for hemoglobin molecules, the amino acid sequence of the encoded proteins adheres to the coding dictionary established from bacterial studies.

However, several 1979 reports on the coding properties of DNA derived from yeast and human mitochondria (mtDNA) altered the principle of universality of the genetic language. Since then, mtDNA has been examined in many other organisms.

Mitochondria contain unique DNA, and transcription and translation occur within these organelles. Cloned mtDNA fragments were sequenced and compared with the amino acid sequences of various mitochondrial proteins, revealing several exceptions to the coding dictionary (Table 12.4). Most surprising was that the codon UGA, normally causing termination in mRNA, specifies the insertion of tryptophan during translation of mRNA originating in yeast and human mitochondria. In human mitochondria, AUA, which normally specifies isoleucine, directs the internal insertion of methionine.

TABLE 12.4	Exceptions to the universal code		
Triplet	*Normal Code Word*	*Altered Code Word*	*Source*
UGA	termination	trp	Human and yeast mitochondria Mycoplasma
CUA	leu	thr	Yeast mitochondria
AUA	ile	met	Human mitochondria
AGA AGG	arg	termination	Human mitochondria
UAA	termination	gln	Paramecium Tetrahymena Stylonychia
UAG	termination	gln	Paramecium

In yeast mitochondria, threonine is inserted instead of leucine when CUA is encountered in mRNA.

More recently, in 1985, several other exceptions to the standard coding dictionary were discovered. These and other altered codes are also summarized in Table 12.4. Such alterations have been observed in the bacterium *Mycoplasma capricolum*, as well as in the protozoan ciliates *Paramecium*, *Tetrahymena,* and *Stylonychia*. As shown, each change converts one of the termination codons to glutamine or tryptophan. These changes are significant because both a prokaryote and several eukaryotes are involved, representing distinct species that have evolved over a long period of time.

Note the apparent pattern in several of the altered codon assignments. The change in coding capacity involves only a shift in recognition of the third, or wobble, position. For example, AUA specifies isoleucine during translation in the cytoplasm and methionine in the mitochondrion. In cytoplasmic translation, methionine is specified by AUG. In a similar way, UGA calls for termination in the cytoplasmic system but tryptophan in the mitochondrion. In the cytoplasm, tryptophan is specified by UGG. Although it has been suggested that such changes in codon recognition may represent an evolutionary trend toward reducing the number of tRNAs needed in mitochondria, the significance of these findings is not yet clear. It is known that only 22 tRNA species are encoded in human mitochondria. However, until still other examples are revealed, the differences must be considered as exceptions to the previously established general coding rules.

Expression of Genetic Information: An Overview

Even while the genetic code was being studied, it was quite clear that proteins were the end products of most genes. Thus, while some geneticists were attempting to elucidate the code, other research efforts were directed toward the nature of **genetic expression**. The central question was how DNA, a nucleic acid, is able to specify a protein composed of amino acids. Put still another way, how is information transferred from DNA to protein? We shall return in the next chapter to examine the evidence supporting the conclusion that proteins are specified by genes, as well as to discuss protein structure and function. Here we shall emphasize the concept of **information flow** as it occurs from DNA to protein. This topic was extremely interesting and exciting in the early 1960s, and it remains no less so as we move toward the end of this century.

Genetic information, stored in DNA, was shown to be transferred to RNA during the initial stage of gene expression. The process by which RNA molecules are synthesized on a DNA template is called **transcription**. The ribonucleotide sequence of RNA, written in a genetic code, is then capable of directing the process of **translation**. During translation, polypeptide chains—the precursors of proteins—are synthesized. Protein synthesis is dependent on a series of **transfer RNA (tRNA)** molecules, which serve as adaptors between the codons of mRNA and the amino acids specified by them. In addition, the process occurs only in conjunction with an intricate cellular organelle, the **ribosome**.

The processes of transcription and translation are complex molecular events. Like the replication of DNA, both rely heavily on base-pairing affinities between complementary nucleotides. The initial transfer from DNA to mRNA produces a molecule that is complementary to the gene sequence of one of the two strands of the double helix. Each triplet codon of the mRNA is complementary to the anticodon region of its appropriate tRNA, and the corresponding amino acid, attached to the charged tRNA, is inserted correctly into the polypeptide chain during translation. In the following sections, we will describe in detail how these processes were discovered and how they are executed.

Transcription: RNA Synthesis

The idea that RNA is involved as an intermediate molecule in the process of information flow between DNA and protein is suggested by the following observations.

1. DNA is for the most part associated with chromosomes in the nucleus of the eukaryotic cell. However, protein synthesis occurs in association with ribosomes located outside the nucleus in the cytoplasm. Therefore, DNA does not participate directly in protein synthesis.

2. RNA is synthesized in the nucleus of eukaryotic cells, where DNA is found, and is chemically similar to DNA.

3. Following its synthesis, most RNA migrates to the cytoplasm, where protein synthesis occurs.

4. The amount of RNA is generally proportional to the amount of protein in a cell.

Collectively, these observations suggest that genetic information, stored in DNA, is transferred to an RNA intermediate, which directs the synthesis of proteins. As with most new ideas in molecular genetics, the initial supporting experimental evidence was based on studies of bacteria and their phages. It was clearly established that: (1) during initial infection RNA synthesis preceded phage protein synthesis; and (2) the RNA is complementary to phage DNA.

The results of these experiments agree with the concept of a **messenger RNA (mRNA)** being made on a

DNA template and then directing the synthesis of specific proteins in association with ribosomes. This concept was formally proposed by François Jacob and Jacques Monod in 1961 as part of a model for gene regulation in bacteria. Since then, mRNA has been isolated and studied thoroughly. There is no longer any question about its role in genetic processes.

RNA Polymerase

In order to prove that RNA may be synthesized on a DNA template, it was necessary to demonstrate that there is an enzyme capable of directing this synthesis. By 1959, several investigators, including Samuel Weiss, had independently discovered such a molecule from rat liver. Called **RNA polymerase**, the enzyme has the same general requirements as DNA polymerase, the major exception being that the ribose rather than the deoxyribose form of the sugar is present in each nucleotide that is inserted into the new RNA molecule. The overall reaction that summarizes the synthesis of RNA on a DNA template may be expressed as

$$n(\text{NTP}) \xrightarrow[\text{enzyme}]{\text{DNA}} (\text{NMP})_n + n(\text{PP}_i)$$

As the equation reveals, nucleoside triphosphates (NTPs) serve as substrates for the enzyme. It catalyzes the polymerization of nucleoside monophosphates (NMPs), or nucleotides, into a polynucleotide chain $(\text{NMP})_n$. Nucleotides are linked during synthesis by 5'-3' phosphodiester bonds. The energy created by cleaving the triphosphate precursor into the monophosphate form drives the reaction and inorganic phosphates (PP_i) are generated.

A second equation summarizes the sequential addition of each ribonucleotide as the process of transcription is in progress:

$$(\text{NMP})_n + \text{NTP} \xrightarrow[\text{enzyme}]{\text{DNA}} (\text{NMP})_{n+1} + \text{PP}_i$$

As this equation shows, each step of transcription involves the addition of one ribonucleotide (NMP) to the growing polyribonucleotide chain $(\text{NMP})_{n+1}$, using a nucleoside triphosphate (NTP) as the precursor.

RNA polymerase from *E. coli* has been extensively characterized and shown to consist of subunits designated α_2, β, β', and σ. The active form of the enzyme $\alpha_2\beta\beta'\sigma$ has a molecular weight of almost 500,000 daltons. Of these subunits, it is the β and the β' **polypeptides** that provide the catalytic basis and active site for transcription. As we will see, the σ **(sigma) subunit** plays a regulatory function involving the initiation of RNA transcription.

While there is only a single form of the enzyme in *E. coli*, three different forms of RNA polymerase are involved in the transcription of the various types of RNA synthesized in eukaryotes (Table 12.5). The

TABLE 12.5   RNA polymerases in eukaryotes

Form	Product	Location
I	rRNA	Nucleolus
II	mRNA	Nucleoplasm
III	5S rRNA	Nucleoplasm
	tRNA	Nucleoplasm

three eukaryotic polymerases all consist of a greater number of polypeptide subunits than the bacterial form of the enzyme.

The Sigma Subunit, Promoters, and Template Binding

Transcription results in the synthesis of a single-stranded RNA molecule that is complementary to a region along one of the two strands of the DNA double helix. For the purpose of future discussion, the strand that is transcribed is called the **template strand** and its complement is referred to as the **partner strand**. The initial step of transcription is referred to as **template binding**, where RNA polymerase interacts physically with DNA (Figure 12–7). The site of initial binding in bacteria is achieved as a result of the recognition by the sigma subunit (σ) of the holoenzyme of specific DNA sequences called **promoters**.

These regions are located in the 5'- upstream region (to the left in the illustration) from the point of initial transcription of a gene. It is believed that the enzyme "explores" a length of DNA until the promoter region is recognized, and a bound complex results. Once this occurs, the helix is denatured or unwound locally, making the DNA template accessible to the action of the enzyme. The polymerase binds to about 60 nucleotides, 40 of which are upstream from the point of initiation.

The importance of promoter sequences cannot be overemphasized. They govern the efficiency of initiation of transcription by directing RNA polymerase to the proper starting point. Both strong promoters and weak promoters are recognized, leading to a variation of initiation from once every 1 to 2 seconds to only once every 10 to 20 minutes. Mutations in promoter sequences are known to reduce the initiation of gene expression severely. And, as we shall see later in this chapter, a consensus sequence rich in adenine and thymine residues (called the TATA box) is also present, but farther upstream (at the 5' end of the gene), in almost all eukaryotic genes.

A **consensus sequence** is one that is present in similar positions in the genes of evolutionarily diverse organisms. The high degree of conservation through evolution attests to the critical nature of consensus sequences. In the case of promoters, the important role of such sequences in transcription is apparent.

(a) Transcription components

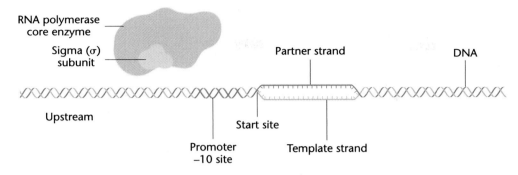

(b) Template binding and initiation of transcription

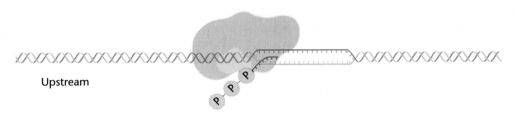

(c) Chain elongation

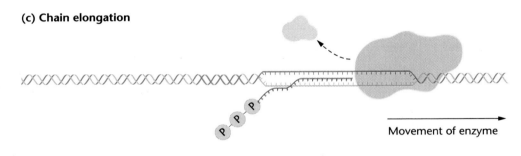

Movement of enzyme

■ Figure 12–7 Schematic representation of the early stages of transcription in prokary-otes, including (a) the transcription components; (b) template binding and initiation involving the sigma subunit of RNA polymerase; and (c) chain elongation, after sigma has dissociated from the transcription complex.

The Synthesis of RNA

Once the promoter has been recognized and bound by the enzyme complex, RNA polymerase catalyzes the insertion of the first 5′-ribonucleotide, which is complementary to the first nucleotide at the start site of the DNA template strand. Unlike DNA synthesis, no primer is required. Subsequent ribonucleotide complements are inserted and linked together by phosphodiester bonds as RNA polymerization proceeds. This process, called **chain elongation** [Figure 12–7(c)], continues in the 5′-to-3′ direction, creating a temporary DNA/RNA duplex whose chains run antiparallel to one another.

After a few ribonucleotides have been added to the growing RNA chain, the σ subunit is dissociated from the holoenzyme, and elongation proceeds under the direction of the core enzyme. In *E. coli* this process proceeds at the rate of about 50 nucleotides/second at 37°C.

Eventually, the enzyme traverses the entire gene and encounters a specific DNA sequence that, when transcribed as part of the RNA, serves as a termination signal. The RNA folds back on itself, forming a stem and loop (sometimes called a hairpin) that interacts with RNA polymerase. The enzyme pauses, and the RNA is disssociated from the DNA template. In many cases, termination is facilitated by the **termination factor, rho (ρ)**. How rho works is not yet clear, although there are DNA sequences at the ends of genes that appear to be related to its function.

Once the transcribed RNA molecule is released from the DNA template, the core enzyme dissociates. Under the direction of RNA polymerase, an RNA molecule is synthesized that is precisely complementary to a DNA sequence representing the template strand of a gene. Wherever an A, T, C, or G residue existed, a corresponding U, A, G, or C residue has been incorporated into the RNA molecule. The significance of this synthesis is enormous, because it is the initial step in the process of information flow within the cell.

Transcription in Eukaryotes

Much of our knowledge of transcription has been derived from studies of prokaryotes. Many general aspects of the mechanics of these processes are similar in eukaryotes. There are, however, numerous notable differences, several of which will be discussed below. We will first summarize some of the major differences and then expand on several of these in greater detail.

1. Transcription in eukaryotes occurs within the nucleus under the direction of three separate forms of RNA polymerase (Table 12.5). Unlike prokaryotes, the RNA transcript is not free to associate with ribosomes prior to the completion of transcription. For the mRNA to be translated, it must move out into the cytoplasm.

2. Initiation and regulation of transcription involve a more extensive interaction between upstream DNA sequences and protein factors involved in stimulating and initiating transcription. There are, in addition to promoters, other control units called enhancers that may be located at positions other than in the 5′-regulatory region upstream from the point of initiation of transcription.

3. Maturation of eukaryotic mRNA from the primary transcript involves many complex stages referred to generally as "processing." Two initial steps involve the addition of a 5′ cap and a 3′ tail to most transcripts destined to become mRNAs.

4. Most notably, extensive modifications occur to the internal sequence of nucleotides of eukaryotic RNA transcripts that serve eventually as mRNAs. The initial (or primary) transcripts are much larger than those that are eventually translated. Sometimes called **pre-mRNAs**, they are part of a group of molecules found only in the nucleus—a group referred to collectively as **heterogeneous nuclear RNA (hnRNA)**. Such RNA molecules are of variable size and complexed with proteins, forming **heterogeneous ribonucleoprotein particles (hnRNPs)**. Only about 25 percent of hnRNA molecules are converted to mRNA. Those that are converted have substantial amounts of their ribonucleotide sequence excised, while the remaining segments are spliced back together prior to translation. This phenomenon has given rise to the concept of **split genes** and **splicing** in eukaryotes.

Eukaryotic Promoters, Enhancers, and Transcription Factors

Recognition of certain highly specific DNA regions by RNA polymerase is the basis of orderly genetic function in all cells. Both the polymerases and promoters leading to template binding have been found to be more complex in eukaryotes. RNA polymerase exists in three forms (Table 12.5), and each is larger and more intricate than the prokaryotic counterpart of the enzyme. Each eukaryotic polymerase consists of two large subunits and 10 to 15 smaller subunits. In regard to the initial template binding step and promoter regions, most is known about polymerase II, which transcribes all mRNAs in eukaryotes.

There are at least three *cis*-acting elements of a eukaryotic gene that function in the efficient initiation of transcription by polymerase II. The use of the term *cis* is drawn from organic chemistry nomenclature, meaning "next to, or on the same side as," in contrast to being "across from, or *trans* to" other functional groups. In molecular genetics, then, *cis* elements are part of the same DNA molecule.

The first such element is called the Goldberg–Hogness or TATA box, which is found about 25 nucleotide pairs upstream (–25) from the start point of transcription. The consensus sequence is a heptanucleotide consisting solely of A and T residues. The sequence and function are analogous to that found in the –10 promoter region of prokaryotic genes. Because such a region is common to most eukaryotic genes, the TATA box is thought to be nonspecific and simply have the responsibility for fixing the site of initiation of transcription by facilitating the denaturation of the helix. Such a conclusion is supported by the fact that A$=$T base pairs are less stable than G$\equiv$C pairs.

The second *cis*-acting element of interest is found farther upstream from the TATA box. In different genes studied, positions anywhere from 5 to 500 nucleotides upstream appear to modulate transcription. These regions of DNA have been located on the basis of the effects of their deletion on transcription. Their loss appears to reduce *in vivo* transcription drastically. Some regions, such as those associated with globin and the viral SV40 genes, are about 50 to 100 nucleotides from the TATA sequence. Others, such as those associated with the sea urchin H2A gene and the *Drosophila* glue protein gene, are 200 to 500 nucleotides upstream. Because the sequence CCAAT is frequently part of these regions, they are sometimes called CCAAT boxes.

The third element is represented by DNA regions called **enhancers**. While their location may vary, enhancers often may be found even farther upstream than the regions discussed above, or even downstream or within the gene. They have the effect of modulating transcription from a distance. Although they may not participate directly in template binding, they are essential to highly efficient initiation of transcription. We will return to a discussion of these elements in Chapter 13.

Complementing the more extensive *cis*-acting regulatory sequences associated with eukaryotic genes are various **trans-acting factors** that serve in the role of facilitating template binding, and, therefore, the initiation of transcription. These are proteins referred to as **transcription factors**. They are essential because RNA polymerase II cannot bind directly to eukaryotic promoter sites and initiate transcription without their presence. The transcription factors involved with human RNA polymerase II-binding are well characterized and designated **TFIIA**, **TFIIB**, etc. Each may consist of protein subunits. Those that bind directly to the TATA box sequence are sometimes called **TATA-binding proteins (TBPs)**. For example, one such TBP is part of **TFIID**, which is responsible for initial binding to the TATA box sequence. The TFIID consists of about ten subunits. Once initial binding to DNA occurs, at least seven other transcription factors bind sequentially to TFIID, forming an extensive pre-initiation complex, which is then bound by RNA polymerase II.

Transcription factors with similar activity have been discovered in a variety of eukaryotes, including *Drosophila* and yeast. The nucleotide sequences in all organisms studied demonstrate a high degree of conservation. These factors appear to supplant the role of the sigma factor in the prokaryotic enzyme. We will return in Chapter 13 to a consideration of the role of other, more specific transcription factors in eukaryotic gene regulation, as well as a discussion of the various DNA-binding domains that characterize some of these polypeptides.

Heterogeneous Nuclear RNA and Its Processing: Caps and Tails

Insights into other regions of DNA that do not encode proteins directly have come from the study of RNA. This research has provided detailed knowledge of eukaryotic gene structure. The genetic code is written in the ribonucleotide sequence of mRNA. This information originated, of course, in the template strand of DNA, where complementary sequences of deoxyribonucleotides exist. In bacteria, the relationship between DNA and RNA appears to be quite direct. The DNA base sequence is transcribed into an mRNA sequence, which is then translated into an amino acid sequence according to the genetic code.

In eukaryotes, however, the situation is much more complex than in bacteria. It has been found that many internal base sequences of a gene may never appear in the mature mRNA that is translated. Other modifications occur at the beginning and the end of the mRNA

prior to translation. These findings have made it clear that in eukaryotes, complex processing of mRNA occurs before it is transported to the cytoplasm to participate in translation.

By 1970, accumulating evidence showed that eukaryotic mRNA is transcribed initially as a much larger precursor molecule than that which is translated. This notion was based on the observation by James Darnell and coworkers of **heterogeneous nuclear RNA (hnRNA)** in mammalian cells. Heterogeneous RNA, complexed with an abundant variety of proteins (creating **hnRNPs—heterogeneous nuclear ribonucleoproteins**), is large but of variable size (up to 107 daltons). It is found only in the nucleus. Importantly, hnRNA was found to contain nucleotide sequences common to the smaller mRNA molecules present in the cytoplasm. Because of this observation, it was proposed that the initial transcript of a gene results in a large RNA molecule that must first be processed in the nucleus before it appears in the cytoplasm as a mature mRNA molecule. The various processing steps, to be discussed below, are summarized in Figure 12–8.

The initial **posttranscriptional modification** of eukaryotic RNA transcripts destined to become mRNAs involves the 5′-end of these molecules, where a **7-methylguanosine (7mG) cap** is added. The cap appears to be important to the subsequent processing within the nucleus, perhaps by protecting the 5′-end of the molecule from nuclease attack. Subsequently, this cap may be involved in transport of mature mRNAs across the nuclear membrane into the cytoplasm. The cap is fairly complex and distinguished by a unique 5′-5′ bonding between the cap and the initial ribonucleotide of the RNA. Some eukaryotes also contain a methyl group ($—CH3$) on the 2′-carbon of the ribose sugars of the first two ribonucleotides of the RNA.

A subsequent discovery provided further insights into the processing of RNA transcripts during the maturation of mRNA. Both hnRNAs and mRNAs have been found to contain at their 3′-end a stretch of as many as 250 adenylic acid residues. Such **poly-A sequences** are added after the 5′-7mG cap has been added. The 3′-end of the initial transcript is first cleaved enzymatically at a point some 10 to 35 ribonucleotides from a highly conserved AAUAAA sequence. Then polyadenylation occurs by virtue of sequential addition of adenylic acid residues. Poly A has now been found at the 3′-end of almost all mRNAs studied in a variety of eukaryotic organisms. The exceptions seem to be the products of histone genes.

The importance of the AAUAAA sequence and the 3′-tail became apparent when mutations in the AAUAAA sequence were investigated. Cells bearing such mutations cannot add the poly A sequence and, in the absence of this tail, the RNA transcripts are rapidly degraded. Therefore, the poly A tail is critical if an RNA transcript is to be further processed and transported to the cytoplasm.

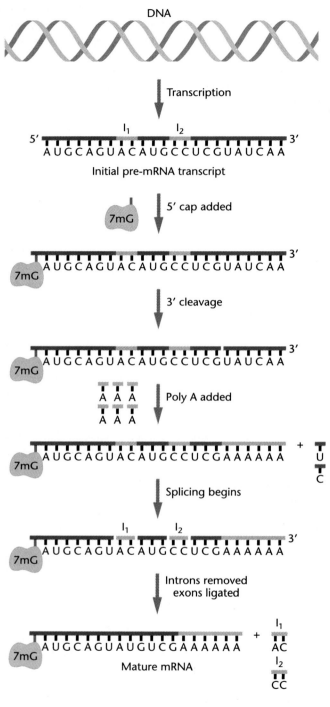

■ Figure 12–8 The conversion in eukaryotes of heterogeneous nuclear RNA (hnRNA) to messenger (mRNA), which contains a 5' cap and a 3' poly A tail.

Intervening Sequences and Split Genes

One of the most exciting discoveries in the history of molecular genetics occurred in 1977. At this time, direct evidence was provided by Susan Berget, Philip Sharp, and others that the genes of certain animal viruses contain internal nucleotide sequences that are not expressed in the amino acid sequence of the proteins they

encode. That is, certain internal sequences in DNA do not always appear in the mature mRNA that is translated into a protein.

Such nucleotide segments have been called **intervening sequences**, contained within split genes. Those DNA sequences whose complements are not present in the final mRNA product are also called **introns** ("int" for intervening), and those retained and expressed are called **exons** ("ex" for expressed). Removal of introns occurs as a result of an excision and rejoining process referred to as **splicing**.

Similar discoveries were soon to be made in eukaryotes. Two approaches have been most fruitful. The first involves molecular hybridization of purified, functionally mature mRNAs along with DNA containing the gene specifying that message. When hybridization occurs between nucleic acids that are not perfectly complementary, **heteroduplexes** are formed that may be visualized with the electron microscope. As illustrated in Figure 12–9, introns present in DNA but absent in mRNA must loop out and remain unpaired. This figure shows an electron micrograph and an interpretive drawing derived from hybridization between the template strand of the gene encoding chicken ovalbumin and the mature RNA prior to its translation into that

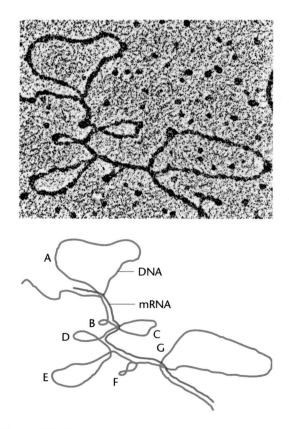

■ Figure 12–9 An electron micrograph and an interpretive drawing of the hybrid molecule formed between the template DNA strand of the ovalbumin gene and the mature ovalbumin mRNA. Seven DNA introns, A–G, produce unpaired loops.

protein. There are seven introns (A–G) whose sequences are present in DNA but not in the final mRNA.

The second approach provides more specific information. It involves a comparison of nucleotide sequences of DNA with those of mRNA and amino acid sequences. Such an approach allows the precise identification of all intervening sequences.

Thus far, a large number of genes from diverse eukaryotes have been shown to contain introns. One of the first so identified was the **beta-globin gene** in mice and rabbits, as studied independently by Philip Leder and Richard Flavell. The mouse gene contains an intron 550 nucleotides long, beginning immediately after the codon specifying the 104th amino acid. In the rabbit (Figure 12–10), there is an intron of 580 base pairs near the codon for the 110th amino acid. Additionally, a second intron of about 120 nucleotides exists earlier in both genes. Similar introns have been found in the beta-globin gene in all mammals examined.

Several genes, notably those coding for histones and interferon, appear to contain no introns. However, intervening sequences have been identified in most other eukaryotic genes that have been examined.

As pointed out above, a more extensive set of introns has been located in the **ovalbumin gene** of chickens. The gene has been extensively characterized by Bert O'Malley in the United States and by Pierre Chambon in France. As shown in Figure 12–10, the gene contains seven introns. Notice that the majority of the gene's DNA sequence is "silent," being composed of introns. The initial RNA transcript is four times the length of the mature mRNA. You should compare the information on the ovalbumin gene presented in Figures 12–9 and 12–10. Can you match the unpaired loops in Figure 12–9 with the sequence of introns specified in Figure 12–10?

The list of genes containing intervening sequences is growing rapidly. In fact, few eukaryote genes seem to be without introns. An extreme example of the number of introns in a single gene is that found in one of the chicken genes, *pro-α-2(1) collagen*. One of several genes coding for a subunit of this connective tissue protein, pro-α-2(1) collagen contains about 50 introns. The precision with which cutting and splicing occur must be extraordinary if errors are not to be introduced into the mature mRNA.

Splicing Mechanisms: Autocatalytic RNAs

The discovery of split genes represents one of the most exciting genetic findings in molecular genetics. As a result, intensive investigation has been directed toward elucidation of the mechanism by which introns of RNA are excised and exons are spliced back together. A great deal of progress has already been made. Interestingly, it appears that somewhat different mechanisms exist for different types of RNA as well as for RNAs produced in mitochondria and chloroplasts.

Based on splicing mechanisms, there are several groups of introns. One group (type I) is illustrated by the primary transcript of rRNAs derived from the ciliate protozoan *Tetrahymena*. This transcript contains an intron that must be removed in order for the mature rRNA to be formed. Surprisingly, the enzymatic activity necessary for removal of the intron is inherent in the intron itself. This self-excision process is illustrated in Figure 12–11. This amazing discovery, made in 1982 by Thomas Cech and his colleagues, revealed that RNA can demonstrate autonomous catalytic properties. As a result, the RNAs that are capable of splicing themselves are sometimes called **ribozymes**.

Chemically, two transesterification reactions occur, the first involving an interaction between guanosine, which acts as a cofactor in the reaction, and the primary transcript [Figure 12–11(a)]. The 3′-OH group of guanosine is transferred to the nucleotide adjacent to the 5′-end of the intron [Figure 12–11(b)]. The second reaction involves the interaction between the 3′-OH group of the left-hand exon and the phosphate on the 3′-end of the right-hand exon [Figure 12–11(c)]. The intron is spliced out and the two exon regions are ligated, leading to the mature rRNA [Figure 12–11(d)].

Self-exision of group I introns, as described above, is now known to apply to pre-rRNAs from other protozoans. Self-excision also seems to govern the removal of introns present in the primary mRNA and tRNA transcripts produced in the mitochondria and chloroplasts. These are referred to as group II introns. Like group I molecules, splicing involves two autocatalytic reactions leading to the excision of introns. However, guanosine is not involved as a cofactor.

Splicing Mechanisms: The Spliceosome

Another major group of introns is found in nuclear-derived transcripts representing mRNAs. Introns in nuclear-derived mRNA, in comparison to other RNAs

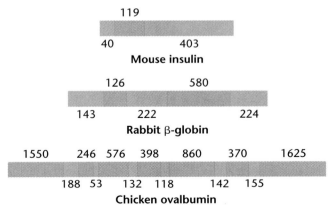

■ Figure 12–10 Intervening sequences in three eukaryotic genes. The numbers indicate nucleotides present in various intron and exon regions.

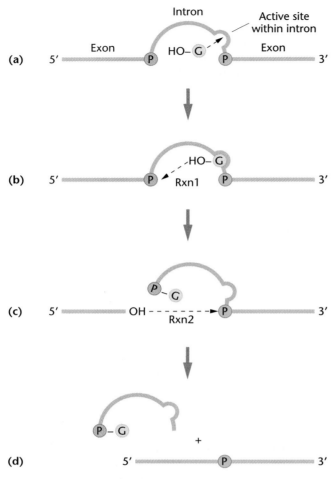

■ Figure 12–11 Splicing mechanism involved with the excisions of group I introns removed from the primary transcript of pre-rRNA. The process is one of self-excision, involving guanosine as a cofactor.

The U1 snRNA bears a nucleotide sequence that is homologous to that of the 5′-end of the intron. Base pairing resulting from this homology promotes binding that represents the initial step in the formation of the spliceosome. Following the addition of the other snurps (U2, U4, U5, U6), splicing commences.

As with group I intron splicing, two reactions, described as transesterifications, are involved. The first involves a 2′-OH group from an adenine (A) residue present within the intron and the phosphate group attached to the 3′-end of exon 1 [Figure 12–12(a)]. The position of the A residue represents what is called the **branch point** of the spliceosome structure. An intermediate set of structures is formed and the second reaction occurs. This involves the —OH group, which has been added to the 3′-end of the exon 1 [depicted on the left in Figure 12–12(b)]. Following excision of the intron and ligation of exon 1 and 2, a characteristic lariat-like structure remains [Figure 12–12(c)]. The intron has been removed and the exons spliced together.

The finding of "genes in pieces," as split genes have been described, raises many interesting questions and has provided great insights into the organization of eukaryotic genes. In addition, the processing involved in splicing represents a potential regulatory step during gene expression. For example, several cases are known

thus far discussed, can be much larger—up to 20,000 nucleotides—and they are more plentiful. Their removal appears to require a much more complex mechanism, which has been more difficult to define.

Many clues are now emerging. First, the nucleotide sequence around different introns is often similar. Most begin at the 5′-end with a GU dinucleotide sequence and terminate at the 3′-end with an AG dinucleotide sequence. These, as well as other consensus sequences shared by introns, attract specific molecules that form a complex essential to splicing. Such a complex, called a **spliceosome**, has been identified in extracts of yeast as well as mammalian cells. It is very large, being 40S in yeast and 60S in mammals. One group of components of these complexes consists of a unique set of **small nuclear RNAs (snRNAs)**. These RNAs are usually 100 to 200 nucleotides long. When complexed with proteins in the spliceosome they are called **small nuclear ribonucleoproteins (snRNPs, or snurps)**. Because they are rich in uridine residues, the snRNAs have been arbitrarily designated U1, U2, . . . , U6, etc.

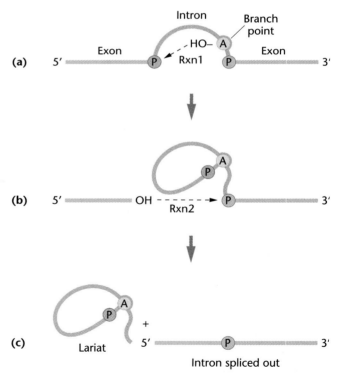

■ Figure 12–12 Splicing mechanism involving introns removed from pre-mRNAs. Excision is dependent on snRNAs and associated proteins (forming snurps), which are part of the structure referred to as the spliceosome. The lariat structure that is formed is characteristic of this mechanism.

in which introns present in pre-mRNAs derived from the same gene are spliced in more than one way, thereby yielding different collections of exons in the mature mRNA. This process, referred to as **alternative splicing**, yields a group of mRNAs that, upon translation, result in a series of related proteins called **isoforms**. A growing number of examples are now found in organisms ranging from viruses to *Drosophila* to humans. Alternative splicing of pre-mRNAs provides the basis for producing related proteins from a single gene.

RNA Editing

In the late 1980s, still another quite unexpected form of posttranscriptional RNA processing was discovered. In this case, the nucleotide sequence of a pre-mRNA is actually changed prior to translation. This form of processing is referred to as **RNA editing**. As a result of the process, the ribonucleotide sequence of the mature RNA differs from the sequence encoded in the exons of the DNA from which the RNA was transcribed.

There are two main types of RNA editing: **substitution editing**, in which the identities of individual nucleotide bases are altered; and **insertion/deletion editing**, in which nucleotides are added to or subtracted from the total number of bases. Substitution editing is used in some nuclear-derived eukaryotic RNAs and is highly prevalent in mitochondrial and chloroplast RNAs transcribed in plants. *Trypanosoma*, a parasite which causes African sleeping sickness, and its relatives use extensive insertion/deletion editing in mitochondrial RNAs. The number of uridines added to an individual transcript can make up more than 60 percent of the coding sequence, usually forming the initiation codon and placing the rest of the sequence into the proper reading frame. *Physarum polycephalum*, a slime mold, uses both substitution and insertion/deletion editing for its mitochondrial mRNAs.

Insertion/deletion editing in trypanosomes is directed by **gRNA (guide RNA)** templates which are also transcribed from the mitochondrial genome. These small RNAs are complementary to the edited region of the final, edited mRNAs. They base-pair with the pre-edited mRNAs to direct the editing machinery to make the correct changes.

Substitutional editing changes in RNAs are difficult to envision because they occur without any form of template information, characteristic of all other forms of information flow in the cell. The mechanism by which such specific nucleotides are targeted and very precise modifications are achieved remains a mystery. Further, it is not yet clear just how extensive the phenomenon will prove to be. The finding that a single C-to-U RNA editing step is an important part of the developmental regulation of the apolipoprotein of mammals has undoubtedly spurred interest. A second system, the synthesis of subunits constituting the glutamate receptor channels (GluR) in mammalian brain

tissue, is also affected by RNA editing. In this case, adenosine (A)-to-inosine (I) editing occurs in pre-mRNAs prior to their translation, where I is read as guanosine (G) during translation. A specific enzyme, double-stranded **RNA adenosine deaminase**, has been implicated in the editing. These changes alter the physiological parameters of the receptor containing "edited" subunits.

Findings such as these in mammals have established that RNA editing provides still another important mechanism of posttranscriptional modification, and that this process is not restricted to evolutionarily less advanced genetic systems such as mitochondria. It thus seems likely that RNA editing will be found to be more widespread as studies progress. Further, the process may have important implications in regulation of genetic expression.

Translation: Components Necessary for Protein Synthesis

Translation of mRNA is the biological polymerization of amino acids into polypeptide chains. The process, alluded to in our earlier discussion of the genetic code, occurs only in association with ribosomes. The central question in translation is how triplet ribonucleotides of mRNA direct specific amino acids into their correct positions in the polypeptide. The question was answered once transfer RNA (tRNA) was discovered. This class of molecules serves as adaptors between specific triplet codons in mRNA and their corresponding amino acids. Such an adaptor role of tRNA was postulated by Crick in 1957.

In association with a ribosome, mRNA presents a triplet codon that calls for a specific amino acid. Because a specific tRNA molecule contains within its composition three consecutive ribonucleotides complementary to the codon, they are called the **anticodon** and can base-pair with the codon. Another region of this tRNA is covalently bonded to the amino acid called for. Inside the ribosome, hydrogen bonding of tRNAs to mRNA holds the amino acid in proximity so that a peptide bond can be formed. This process occurs over and over as mRNA runs through the ribosome and amino acids are polymerized into a polypeptide.

In our discussion of translation, we will first consider the structure of the ribosome and transfer RNA, two of the major components essential for protein synthesis.

Ribosomal Structure

Because of its essential role in the expression of genetic information, the ribosome has been analyzed extensively. One bacterial cell contains about 10,000 of these structures, while a eukaryotic cell contains many times more. Electron microscopy has revealed that the bacterial ribosome is about 250 Å in its largest diameter and consists of a larger and a smaller subunit. Both subunits consist of one or more molecules of **rRNA** and an array of **ribosomal proteins**.

The specific differences between prokaryotic and eukaryotic ribosomes are summarized in Figure 12–13. The subunit and rRNA components are most easily isolated and characterized on the basis of their sedimentation behavior (their rate of migration) in sucrose gradients (see Chapter 9). The two subunits associated with each other constitute a **monosome**. In prokaryotes the monosome is a 70S particle, and in eukaryotes it is approximately 80S. Sedimentation coefficients, which reflect the variable rate of migration of different-sized particles and molecules when centrifuged, are not additive. For example, the 70S monosome consists of a 50S and a 30S subunit, and the 80S monosome consists of a 60S and a 40S subunit.

The larger subunit in prokaryotes consists of a 23S RNA molecule, a 5S rRNA molecule, and 32 ribosomal proteins. In the eukaryotic equivalent, a 28S rRNA molecule is accompanied by a 5.8S and 5S rRNA molecule and about 50 proteins. In the smaller prokaryotic subunits, a 16S rRNA component and 21 proteins are found. In the eukaryotic equivalent, an 18S rRNA component and about 33 proteins are found. The approximate molecular weights and number of nucleotides of these components are shown in Figure 12–13.

Molecular hybridization studies have established the degree of redundancy of the genes coding for the rRNA components. The *E. coli* genome contains seven copies of a single sequence that codes for all three components—23S, 16S, and 5S. The initial transcript of these genes produces a 30S RNA molecule that is enzymatically cleaved into these smaller components.

In eukaryotes, many more copies of a sequence encoding the 28S and 18S components are present. In *Drosophila*, approximately 120 copies per haploid genome are each transcribed into a molecule of about 34S. This is processed to the 28S, 18S, and 5.8S rRNA species. In *X. laevis*, over 500 copies per haploid genome are present. In mammalian cells, the initial

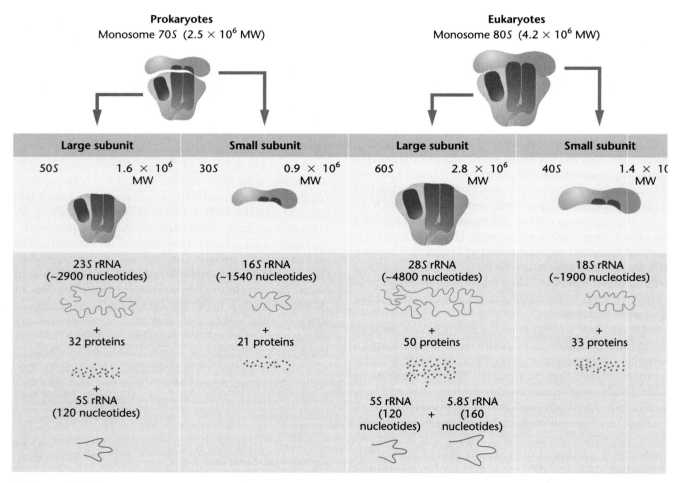

■ Figure 12–13   A comparison of the components in the prokaryotic and eukaryotic ribosome.

transcript is 45S. The rRNA genes are part of the moderately repetitive DNA fraction and are present in clusters at various chromosomal sites. Each cluster consists of **tandem repeats**, with each unit separated by a noncoding **spacer DNA** sequence. In humans, these gene clusters have been localized on the ends of the short arms of chromosomes 13, 14, 15, 21, and 22.

The unique 5S rRNA component of eukaryotes is not part of the larger transcript, as it is in *E. coli.* Instead, genes coding for this ribosomal component are distinct and located separately. In humans, a gene cluster encoding them has been located on chromosome 1.

Despite the detailed knowledge available on the structure and genetic origin of the ribosomal components, a complete understanding of the function of these components has eluded geneticists. This is not surprising, because the ribosome is perhaps the most intricate of all cellular organelles. In bacteria, the monosome has a combined molecular weight of 2.5 million daltons!

tRNA Structure

Because of its small size and stability in the cell, tRNA is the best characterized RNA molecule. It is composed of only 75 to 90 nucleotides, displaying a nearly identical structure in bacteria and eukaryotes. In both types of organisms, tRNAs are transcribed as larger precursors, which are cleaved into mature 4S tRNA molecules. In *E. coli*, for example, tRNA[tyr] (the superscript identifies the specific amino acid that binds a particular tRNA) is composed of 77 nucleotides, yet its precursor contains 126 nucleotides.

In 1965, Robert Holley and his coworkers reported the complete sequence of tRNA[ala] isolated from yeast. Of great interest was the finding that there are a number of nucleotides unique to tRNA. Each is a modification of one of the four nitrogenous bases expected in RNA (G, C, A, and U). These include **inosinic acid**, which contains the purine **hypoxanthine, ribothymidylic acid,** and **pseudouridine**, among others. These modified structures, sometimes referred to as *unusual, rare*, or *odd bases*, are created posttranscriptionally. That is, the unmodified base (A, U, C, or G) is produced during transcription, and then enzymatic reactions catalyze the base modifications.

Holley's sequence analysis led him to propose the two-dimensional **cloverleaf model** of transfer RNA. It had been known that tRNA demonstrates secondary structure due to base pairing. Holley discovered that he could arrange the linear model in such a way that several stretches of base pairing would result. This arrangement created a series of paired stems and unpaired loops resembling the shape of a cloverleaf. Loops consistently contained modified bases, which do not generally form base pairs. Holley's model is shown in Figure 12–14.

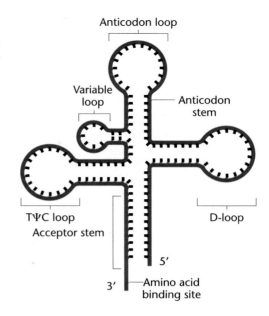

■ Figure 12–14 The two-dimensional cloverleaf model of transfer RNA.

Because the triplets GCU, GCC, and GCA specify alanine, Holley looked for the theoretical anticodon of his tRNA[ala] molecule. He found it in the form of CGI, at the top loop of the cloverleaf. Recall from Crick's wobble hypothesis that I (inosinic acid) is predicted to pair with U, C, or A. Thus, the **anticodon loop** was established.

As other tRNA species were examined, numerous constant features were observed. First, at the 3′ end, all tRNAs contain the sequence . . . **pCpCpA-5′**. It is to the terminal adenosine residue that the amino acid is covalently bonded during charging. At the 5′ terminus, all tRNAs contain . . . **pG**. Additionally, the lengths of various stems and loops are very similar. All tRNAs examined also contain an anticodon complementary to the known amino acid code for which it is specific, which are present in the same position of the cloverleaf as well.

Because the cloverleaf model was predicted strictly on the basis of nucleotide sequence, there was great interest in X-ray crystallographic examination of tRNA, which reveals three-dimensional structure. By 1974, Alexander Rich and his coworkers in the United States, as well as J. Roberts, B. Clark, Aaron Klug, and their colleagues in England, had been successful in crystallizing tRNA and performing X-ray crystallography at a resolution of 3 Å. At such resolution, the pattern formed by individual nucleotides is discernible.

As a result of these studies, a complete three-dimensional model is now available (Figure 12–15). The model reveals tRNA to exhibit a reversed L-shape. At one end of the L is the anticodon loop and stem, and at the other end is the 3′-acceptor region where the amino acid is bound. It has been speculated that the shapes of

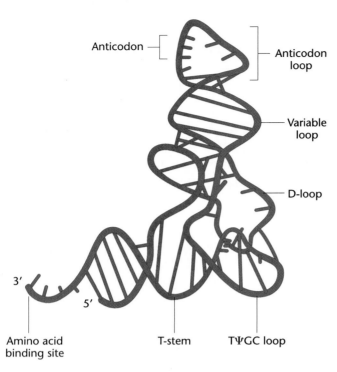

■ Figure 12–15 A three-dimensional model of transfer RNA.

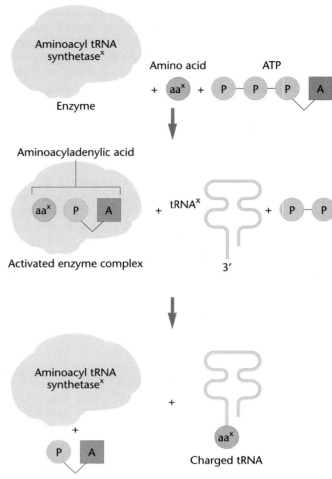

■ Figure 12–16 Steps involved in charging tRNA. The "X" denotes that for each amino acid (aa), only the corresponding tRNA and specific aminoacyl synthetase enzyme are involved in the charging process.

the intervening loops may be recognized by specific enzymes responsible for adding the amino acid to tRNA, the subject to which we now turn our attention.

Charging tRNA

Before translation can proceed, the tRNA molecules must be linked chemically to their respective amino acids. This activation process, called **charging**, occurs under the direction of enzymes called **aminoacyl tRNA synthetases**. Because there are 20 amino acids, one can reason that there must be a minimum of 20 different tRNA molecules, one for each amino acid. Because each charging event must be highly specific, there must also be a unique synthetase enzyme for each reaction. As we know from our earlier discussion, there are 61 triplet codes. Thus, in theory, there could be 61 specific tRNAs and enzymes. However, based on the ability of the third member of a triplet code to "wobble," fewer are actually needed. It is now thought that there are at least 32 different tRNAs; it is also believed that there are only 20 synthetases, one for each amino acid, regardless of the greater number of corresponding tRNAs.

The charging process is outlined in Figure 12–16. In the initial catalysis step, the amino acid is converted to an activated form, reacting with ATP to form an **aminoacyladenylic acid**. A covalent linkage is formed between the 5′-phosphate group and the carboxyl end

of the amino acid. This molecule remains associated with the enzyme, forming an activated complex, which then reacts with a specific tRNA molecule. In this second step, the amino acid is transferred to this tRNA and bonded covalently to the adenine residue at the 3′ end. The charged tRNA may participate directly in protein synthesis. Aminoacyl tRNA synthetases are highly specific enzymes because they recognize only one amino acid and only a subset of corresponding tRNAs called **isoaccepting tRNAs**. This is a crucial point if fidelity of translation is to be maintained. The basis for this binding has sometimes been referred to as the **second genetic code**, although this is not a particularly apt description.

Translation: The Process

In a way similar to transcription, the process of translation can be best described by breaking it into discrete steps. Be aware, however, that translation is a dynamic, ongoing process. Correlate the following discussion with the step-by-step characterization of the process in Figure 12–17. Many of the protein factors and their roles in translation are summarized in Table 12.6.

Initiation (Steps 1–3 of Figure 12–17)

Recall that ribosomes serve as a nonspecific workbench for the translation process. Most ribosomes, when they are not involved in translation, are dissociated into their large and small subunits. Initiation of translation in *E. coli* involves the small subunit, an mRNA molecule, a specific initiator tRNA, GTP, Mg^{2+}, and at least three proteinaceous **initiation factors (IFs)**. The initiator molecules are not part of the ribosome, but are required to enhance the binding affinity of the translational components. In prokaryotes, the initiation code of mRNA, AUG, calls for the modified amino acid **formylmethionine**.

The small ribosomal subunit binds to several initiation proteins, and this complex in turn binds to mRNA. In bacteria, this binding involves a sequence of up to six ribonucleotides (AGGAGG), which precedes the initial AUG codon. This sequence (containing only purines and called the **Shine–Dalgarno sequence**) base-pairs with a region of the 16S rRNA of the small ribosomal subunit.

Another initiation protein then facilitates the binding of charged formylmethionyl-tRNA to the small subunit in response to the AUG triplet. This step "sets" the reading frame so that all subsequent groups of three ribonucleotides are translated accurately. This aggregate represents the **initiation complex**, which then combines with the large ribosomal subunit. In this process, a molecule of GTP is hydrolyzed, providing the required energy, and the initiation factors are released.

Elongation (Steps 4–9 of Figure 12–17)

Once both subunits of the ribosome are assembled with the mRNA, binding sites for two charged tRNA molecules are formed. These are designated the **P**, or **peptidyl**, and the **A**, or **aminoacyl, sites**. The initiator tRNA binds to the P site, provided that the AUG triplet is in the corresponding position of the small subunit. The sequence of the second triplet in mRNA dictates which charged tRNA molecule will become positioned at the A site. Once it is present, **peptidyl transferase** catalyzes the formation of the peptide bond, which links the two amino acids together. This enzyme is part of the large subunit of the ribosome. At the same time, the covalent bond between the amino acid and the tRNA occupying the P site is hydrolyzed (broken). The product of this reaction is a dipeptide, which is attached to the tRNA at the A site. The step in which the growing polypeptide chain increases in length by one amino acid at a time is called **elongation**.

Before elongation can be repeated, the tRNA attached to the P site, which is now uncharged, must be released from the large subunit. The uncharged tRNA is thought to move transiently through a third site on the ribosome called the **E site** (E stands for "exit"—Because the E site is occupied only briefly, we have not included it in every frame of Figure 12–17). The entire **mRNA-tRNA-aa$_2$-aa$_1$** complex now shifts in the direction of the P site by a distance of three nucleotides. This event requires several protein **elongation factors (EFs)** as well as the energy derived from hydrolysis of GTP (Table 12.6). The result is that the third triplet of mRNA is now in a position to direct another specific charged tRNA into the A site. One simple way to distinguish the two sites in your mind is to remember that, *following the shift*, the P site contains a tRNA attached to a peptide chain (P for peptide), while the A site contains a tRNA with an amino acid attached (A for amino acid).

The sequence of elongation is repeated over and over. An additional amino acid is added to the growing polypeptide chain each time the mRNA advances by one triplet through the ribosome. Once a reasonably sized polypeptide chain is assembled (30 amino acids, by one estimate), it begins to emerge from the base of the large subunit of the ribosome, as illustrated in step 9 of Figure 12–17. A **tunnel** exists within the subunit, through which the elongating polypeptide works its way out of the ribosome.

The efficiency of the process is remarkably high; the observed error rate is only 10^{-4}. An incorrect amino acid will thus occur once in every 20 polypeptides of an average length of 500 amino acids! In *E. coli*, elongation occurs at a rate of about 15 amino acids per second at 37°C. The process can be likened to a tape moving through a tape recorder. As the tape moves, sequential sound is emitted from the recorder. Likewise, as mRNA moves, a growing polypeptide is produced by the ribosome.

Termination (Steps 10–11 of Figure 12–17)

The termination of protein synthesis is signaled by one or more of three triplet codes: UAG, UAA, or UGA. These codons do not specify an amino acid, nor do they direct tRNA into the A site. The finished polypeptide is therefore still attached to the terminal tRNA at the P site. The termination codon signals the action of GTP-dependent **release factors** (Table 12.6), which cleave the polypeptide chain from the terminal tRNA. Once this cleavage occurs, the tRNA is released from the ribosome, which then dissociates into its subunits. If a termination codon should appear in the middle of an mRNA molecule as a result of mutation, the same process occurs, and the polypeptide chain is terminated prematurely.

Polyribosomes

As elongation proceeds and the initial portion of mRNA has passed through the ribosome, the message is free to associate with another small subunit to form another initiation complex. This process can be repeated several times with a single mRNA and results in what are called **polyribosomes** or just **polysomes**.

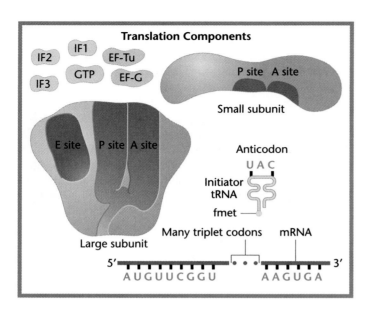

Translation Components

IF2 IF1 EF-Tu
IF3 GTP EF-G

P site A site

Small subunit

E site P site A site

Large subunit

Anticodon

U A C

Initiator tRNA

fmet

Many triplet codons   mRNA

5′ —————————————— 3′
A U G U U C G G U   A A G U G A

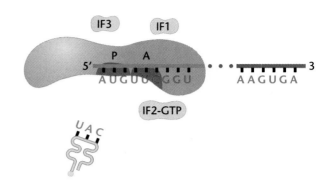

IF3   IF1

P A

5′ ————— • • • ———— 3′
A U G U U C G G U   A A G U G A

IF2-GTP

U A C

1. mRNA binds to small subunit along with initiation factors (IF1, 2, 3)

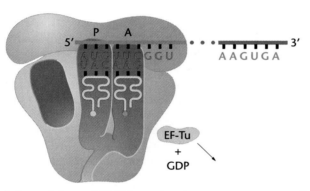

P   A

5′ ——————— • • • —————— 3′
A U G U U C G G U   A A G U G A
U A C A A G

EF-Tu
+
GDP

4. Second charged tRNA has entered A site, facilitated by EF-Tu

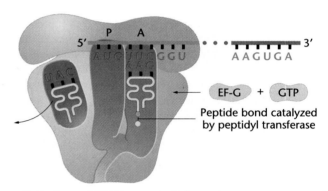

P   A

5′ ——————— • • • —————— 3′
A U G U U C G G U   A A G U G A
A A C

EF-G + GTP

Peptide bond catalyzed by peptidyl transferase

5. Uncharged tRNA moves to E-site and then out of ribosome; dipeptide bond forms

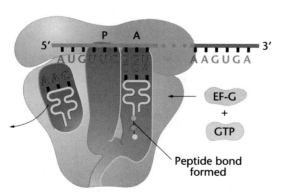

P   A

5′ ——————— • • • —————— 3′
A U G U U C G G U   A A G U G A
A A C   C X

EF-G
+
GTP

Peptide bond formed

8. Tripeptide formed; second elongation step completed; uncharged tRNA moves to E site

Many elongation steps

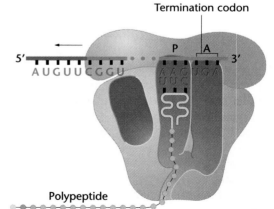

Termination codon

P   A

5′ ——————— • • • —————— 3′
A U G U U C G G U   A A G U G A
U U C

Polypeptide

9. Polypeptide chain synthesized and exiting ribosome

■ Figure 12–17 Schematic representation of the process of translation, depicting the steps involved in protein synthesis.

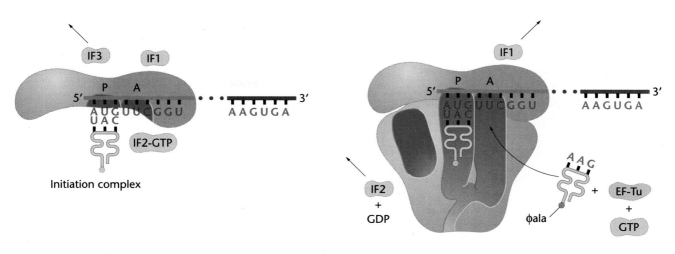

2. Initiator tRNA^fmet binds to mRNA codon in P site; IF3 released

3. Large subunit binds to complex; IF1 and IF2 released; EF-Tu binds to tRNA, facilitating entry into A site

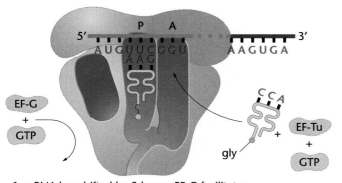

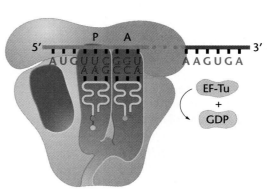

6. mRNA has shifted by 3 bases; EF-G facilitates translocation step; first elongation step completed

7. Third charged tRNA has entered A site, facilitated by EF-Tu

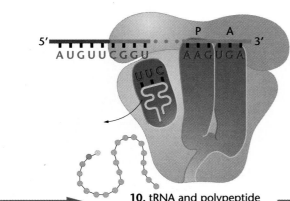

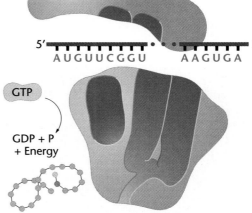

10. tRNA and polypeptide chain released

11. GTP-dependent termination factors activated; components separate; polypeptide folds into protein

TABLE 12.6	Various protein factors involved during translation in *E. coli*	
Process	**Factor**	**Role**
Initiation of translation	IF1	Stabilizes 30S subunit
	IF2	Binds fmet-tRNA to 30S-mRNA complex
	IF3	Binds 30S subunit to mRNA; dissociates monosomes into subunits following termination
Elongation of polypeptide	EF-Tu	Binds GTP; brings aminoacyl-tRNA to the A site of ribosome
	EF-Ts	Generates active EF-Tu
	EF-G	Stimulates translocation; GTP-dependent
Termination of translation and release of polypeptide	RF1	Catalyzes release of the polypeptide chain from tRNA and dissociation of the translocation complex; specific for UAA and UAG termination codons.
	RF2	Behaves like RF1; specific for UGA and UAA codons
	RF3	Stimulates RF1 and RF2

Polyribosomes can be isolated and analyzed following a gentle lysis of cells. Figure 12–18 illustrates these complexes as seen under the electron microscope; in part (a) of this figure, note the presence of mRNA between the individual ribosomes. The micrograph in part (b) is even more remarkable because it shows the polypeptide chains emerging from the ribosomes during translation. The formation of polysome complexes represents an efficient use of the components available for protein synthesis during a unit time.

To complete the analogy with tapes (mRNA) and tape recorders (ribosomes), in polysome complexes one tape would be played simultaneously, but at any given moment the transcripts (polypeptides) would all be at different stages of completion.

Translation in Eukaryotes

The general features of the model just presented were initially derived from investigations of the translation process in bacteria. During that discussion, we pointed out one of the main differences between translation in prokaryotes and eukaryotes. In the latter group, translation occurs on ribosomes that are larger and whose rRNA and protein components are more complex than prokaryotes (see Figure 12–13).

In addition, several other notable differences need to be mentioned. Eukaryotic mRNAs are much longer lived than their prokaryotic counterparts. Most exist for hours rather than minutes prior to their degradation by nucleases in the cell. Thus eukaryotic mRNAs are available much longer to orchestrate protein synthesis.

Two aspects that involve the initiation of translation are different in eukaryotes. First, as we discussed during our consideration of mRNA maturation, the 5′ end is "capped" with a 7-methylguanosine residue. The presence of this cap, absent in prokaryotes, is essential to efficient translation. RNAs lacking the cap are translated poorly. Additionally, most eukaryotic mRNAs contain a short recognition sequence that surrounds the initiating AUG codon—**5′-ACCAUGG**. . . . Named after Marilyn Kozak who discovered it, the **Kozak sequence**

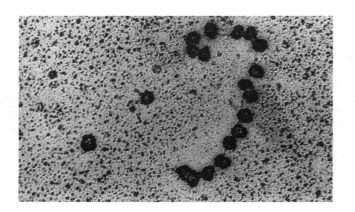
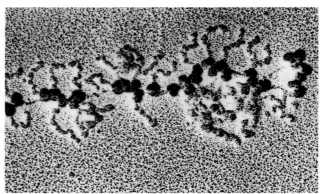

■ Figure 12–18 Polyribosomes visualized under the electron microscope. Those in part (a) were derived from rabbit reticulocytes engaged in the translation of hemoglobin mRNA and in part (b) from giant salivary gland cells of the midgefly, *Chironomus thummi*. In part (b), the nascent polypeptide chain is apparent as it emerges from each ribosome. Its length increases as translation proceeds from left (5′) to right (3′) along the mRNA.

GENETICS, TECHNOLOGY, AND SOCIETY

Antisense Oligonucleotides: Attacking the Messenger

Standard chemotherapies for diseases such as cancer and HIV are often accompanied by toxic side effects. Conventional therapeutic drugs target both normal and diseased cells, with diseased or infected cells being only slightly more susceptible than the patient's normal cells. Scientists have long wished for a magic bullet that could seek out and destroy the virus or the cancer cell, leaving normal cells alive and healthy. Over the last decade, one particularly promising candidate has emerged—the *antisense oligonucleotide.*

Antisense therapies have arisen from an understanding of the molecular biology of gene expression. Gene expression is a two-step process. First, a single-stranded messenger RNA (mRNA) is copied from one strand of the duplex DNA molecule. Second, the mRNA is transported to the cytoplasm, complexed with ribosomes, and its genetic information is translated into the amino acid sequence of a polypeptide.

Normally, a gene is transcribed into RNA from only one strand of the DNA duplex, referred to as the template. The resulting RNA is known as sense RNA. However, it is sometimes possible for the other DNA strand to be copied into RNA. The RNA produced by transcription of the "wrong" strand of DNA is called antisense RNA. As with complementary strands of DNA, complementary strands of RNA can form double-stranded molecules. Naturally occurring antisense RNAs have been detected in both bacteria and eukaryotes, where they may play regulatory roles in gene expression.

The formation of duplex structures between a sense and an antisense RNA may affect the sense RNA in one of several different ways. If the antisense RNA hybridizes to the 5′ end of the sense RNA, it may physically block ribosome binding and hence inhibit translation. If other portions of the sense RNA are hybridized, translation will also be inhibited, even if ribosome binding is achieved. Further, the presence of antisense RNA may trigger the degradation of the sense RNA, because double-stranded RNA molecules are attacked by intracellular ribonucleases. In each case, gene expression is blocked.

What makes the antisense approach so exciting is its potential specificity. Scientists can design antisense RNA (or DNA) molecules of known nucleotide sequence and then synthesize large amounts of these nucleic acids *in vitro.* Usually, oligonucleotides (up to 20 nucleotides) are utilized. It is then theoretically possible to treat cells with synthetic antisense oligonucleotides, to have the oligonucleotides enter the cell and bind precise target mRNAs, and to turn off the synthesis of one specific protein. If that protein is necessary for viral reproduction or cancer cell growth (but is not necessary in normal cells), the antisense oligonucleotide should have only therapeutic effects.

In the last few years, laboratory tests of antisense drugs have been so promising that several clinical trials are now in progress. For example, the ability of antisense oligonucleotides to inhibit replication of human cytomegalovirus (CMV) is being tested. CMV is a common virus that is found in most people. Although it causes few problems in people with normal immune systems, it can cause serious symptoms in people with conditions (such as AIDS) that impair the immune system. Up to 40 percent of AIDS patients develop retinitis or blindness as a result of CMV infections of the eye. In 1998, Isis Pharmaceuticals and CIBA Vision reported that introduction of antisense CMV oligonucleotides into the eyes of AIDS patients with CMV-induced retinitis significantly delayed disease progression compared to an untreated group. Although it is not known how the oligonucleotides are operating in these clinical trials, laboratory studies suggest that these antisense oligonucleotides trigger the destruction of CMV mRNA and interfere with the absorption of virus to the surface of host cells.

Antisense oligonucleotides may also act as anti-inflammatory drugs. Clinical trials are in progress to test antisense compounds in the treatment of Crohn disease, rheumatoid arthritis, psoriasis, ulcerative colitis, and transplant rejection. One interesting antisense oligonucleotide—now in clinical trials conducted by Isis and Boehringer Ingelheim—binds to the mRNA that encodes ICAM-1. ICAM-1 is a cell surface glycoprotein that helps activate immune system and inflammatory cells. It is often overexpressed in tissues that suffer extreme inflammatory responses. It is hoped that antisense oligonucleotides might temper inflammatory responses by reducing expression of ICAM-1. Results of a recent clinical trial suggest that antisense ICAM-1 may be an effective treatment for Crohn disease. Crohn disease is a form of inflammatory bowel disease that affects about 200,000 people in the United States. It is a debilitating, chronic disease, and treatments such as steroids and immunosuppressive drugs are often ineffective and have toxic side effects. In one clinical trial, almost 50 percent of Crohn patients went into remission after treatment with antisense ICAM-1, whereas none of the placebo group did so. In addition, one-third of the treated patients were well enough to stop steroid treatments by the end of the trial period. In this trial, the antisense ICAM-1 drugs appeared to be well tolerated and safe.

Some interesting clinical trials are in progress to test antisense oligonucleotides as treatments for some types of cancer. These oligonucleotides are designed to reduce the synthesis of proteins that are either overexpressed in cancer cells or are present as mutant forms. These antisense drugs will be evaluated as treatments for tumors of the ovary, prostate, breast, brain, colon and lung. Other antisense drugs in the pipeline are designed to attack the hepatitis B, hepatitis C, human papilloma, and AIDS viruses.

If antisense therapeutics pass the scrutiny of these scientific and clinical trials, we may indeed have acquired a magic molecular bullet to use in the battle against a variety of diseases.

References

Guru, T. 1995. Antisense has growing pains. *Science* 270: 575–77.

Roush, W. 1997. Antisense aims for a renaissance. *Science* 276: 1192–93.

Sharma, H. W. and Narayanann, R. 1995. The therapeutic potential of antisense oligonucleotides. *BioEssays* 17:1055–63.

appears to function during initation in the same way that the Shine–Dalgarno sequence functions in prokaryotic mRNA. Both greatly facilitate the initial binding of mRNA to the small subunit of the ribosome.

The second factor related to initiation of translation involves the insertion of the first amino acid. Initiation of eukaryotic translation does not require the amino acid formylmethionine. However, as in prokaryotes, the AUG triplet is essential to the formation of the transla-tional complex, and a unique transfer RNA ($tRNA_i{}^{met}$) is used during initiation.

Finally, protein factors similar to those in prokaryotes guide initiation, elongation, and termination of transla-tion in eukaryotes. Many of these eukaryotic factors are clearly homologous to their counterparts in prokary-otes. However, there may be a greater number of fac-tors required during each of these steps, and they may be somewhat more complex in prokaryotes.

Chapter Summary

1. The genetic code, stored in DNA, is transferred to RNA, where it is used to direct the synthesis of polypeptide chains. It is degenerate, unambiguous, nonoverlapping, and commaless.

2. The complete coding dictionary, determined using various experimental approaches, reveals that of the 64 possible codons, 61 encode the 20 amino acids found in proteins, while three triplets terminate trans-lation. One of these 61 is the initiation codon and specifies methionine.

3. The observed pattern of degeneracy often involves only the third letter of a triplet series, leading Crick to propose the wobble hypothesis.

4. Confirmation for the coding dictionary, including codons for initiation and termination, was obtained by comparing the complete nucleotide sequence of phage MS2 with the amino acid sequence of a corresponding protein. Other findings support the belief that, with only minor exceptions, the code is universal for all organisms.

5. Transcription and translation—RNA and protein biosyn-thesis, respectively—are the fundamental processes es-sential to the expression of genetic information.

6. The processes of transcription and translation, like DNA replication, can be subdivided into the stages of initiation, elongation, and termination. Both processes rely on base-pairing affinities between complemen-tary nucleotides.

7. Transcription describes the synthesis, under the direc-tion of RNA polymerase, of a strand of RNA that is complementary to a DNA template.

8. Translation is the complex energy-requiring process in-volving charged tRNA molecules, numerous proteins, ribosomes, and mRNA. Transfer RNA (tRNA) serves as the adaptor molecule between an mRNA triplet and the appropriate amino acid. The ribosome serves as the workbench for translation.

9. The processes of transcription and translation are more complex in eukaryotes than in prokaryotes. The primary transcript in eukaryotes must be modified in various ways, including (1) the addition of a cap and poly A tail and (2) the removal, through splicing, of intervening sequences, or introns. RNA editing of pre-mRNA prior to its translation also occurs in some organisms.

Key Terms

alternative splicing, 267
amber (UAG), 256
aminoacyl site (A site), 271
aminoacyl tRNA synthetase, 270
aminoacyladenylic acid, 270
anticodon, 255
antisense oligonucleotides, 275
β and β′ polypeptides, 260
beta-globin gene, 265
CCAAT box, 262
cell-free protein-synthisizing system, 252
chain elongation, 261
charging (tRNA), 270
cloverleaf model, 269

codon, 251
colinear relationship, 258
commaless code, 251
consensus sequence, 260
degenerate code, 251
double-stranded RNA adenosine deaminase, 267
elongation, 271
elongation factor, 271
enhancer, 262
exit site (E site), 271
exon, 264
formylmethionine, 271
frameshift mutation, 252
genetic expression, 259

guide RNA (gRNA), 267
Goldberg–Hogness (TATA) box, 262
heteroduplex, 264
heterogeneous nuclear RNA (hnRNA), 262
heterogeneous ribonucleoprotein particle (hnRNP), 262
hypoxanthine, 269
information flow, 259
initiation complex, 271
initiation factor, 271
initiation of translation, 251
initiator codon, 257
inosinic acid, 269

INSIGHTS *and* SOLUTIONS

1. Had evolution seized on 6 bases (three complementary base pairs) rather than 4 bases within the structure of DNA, calculate how many triplet codons would be possible. Would 6 bases accommodate a two-letter code, assuming 20 amino acids and start and stop codons?

Solution: Six things taken three at a time will produce 6^3 or 216 triplet codes. If the code was a doublet, there would be 6^2 or 36 two-letter codes, more than enough to accommodate 20 amino acids and start–stop punctuation.

2. In a heteropolymer experiment using 1/2 C:1/4 A:1/4 G, how many different triplets will occur in the synthetic RNA molecule? How frequently will the most frequent triplet occur?

Solution: There will be 3^3 or 27 triplets produced. The most frequent will be CCC, present $(1/2)^3$ or 1/8 of the time.

3. In a regular copolymer experiment, where UUAC is repeated over and over, how many different triplets will occur in the synthetic RNA, and how many amino acids will occur in the polypeptide when this RNA is translated? Be sure to consult Figure 12–6.

Solution: The synthetic RNA will repeat four triplets—UUA, UAC, ACU, and CUU—over and over.

Because both UUA and CUU encode leucine, while ACU and UAC encode threonine and tyrosine, respectively, polypeptides synthesized under the directions of such an RNA contain three amino acids in the repeating sequence leu-leu-thr-tyr.

4. Actinomycin D inhibits DNA-dependent RNA synthesis. This antibiotic is added to a bacterial culture in which a specific protein is being monitored. Compared to a control culture, translation of the protein declines over a period of 20 minutes, until no further protein is made. Explain these results.

Solution: The mRNA, which is the basis for the translation of the protein, has a lifetime of about 20 minutes. When actinomycin D is added, transcription is inhibited and no new mRNAs are made. Those already present support the translation of the protein for up to 20 minutes.

Problems and Discussion Questions

1. Crick, Barnett, Brenner, and Watts-Tobin, in their studies of frameshift mutations, found that either 3 (+)'s or 3 (−)'s restored the correct reading frame. If the code were a sextuplet (consisting of six nucleotides), would the reading frame be restored by either of the above combinations?

2. In a mixed copolymer experiment using polynucleotide phosphorylase, 3/4 G:1/4 C was added to form the synthetic message. The amino acid composition of the ensuing protein was determined:

Glycine	36/64	(56%)
Alanine	12/64	(19%)
Arginine	12/64	(19%)
Proline	4/64	(6%)

From this information:

(a) Indicate the percentage (or fraction) of the time each possible triplet will occur in the message.

(b) Determine a complete set of base composition assignments for all amino acids present.

(c) Considering the wobble hypothesis, predict as many specific triplet assignments as possible.

3. In a mixed copolymer experiment, messengers were created with either 4/5 C:1/5 A or 4/5 A:1/5 C. These messages yielded proteins with the following amino acid compositions. Using these data, predict the most specific coding composition for each amino acid.

_4/5C:1/5A_		_4/5A:1/5C_	
Proline	63.0%	Proline	3.5%
Histidine	13.0%	Histidine	3.0%
Threonine	16.0%	Threonine	16.6%
Glutamine	3.0%	Glutamine	13.0%
Asparagine	3.0%	Asparagine	13.0%
Lysine	0.5%	Lysine	50.0%
	98.5%		98.5%

4. In a coding experiment using repeating copolymers (as shown in Table 12.3), the following data were obtained:

Copolymer	Condons Produced	Amino Acids in Polypeptide
AG	AGA, GAG	Arg, Glu
AAG	AGA, AAG, GAA	Lys, Arg, Glu

AGG is known to code for arginine. Taking into account the wobble hypothesis, assign each of the four remaining different triplet codes to its correct amino acid.

5. In the triplet binding technique, radioactivity remains on the filter when the amino acid corresponding to the triplet is labeled. Explain the basis of this technique.

6. In studies of the amino acid sequence of wild-type and mutant forms of tryptophan synthetase in _E. coli_, the following changes have been observed:

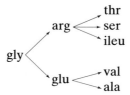

Determine a set of triplet codes in which a single nucleotide change produces each amino acid change.

7. Define and differentiate between transcription and translation. Where do these processes fit into the central dogma of molecular genetics?

8. List all of the molecular constituents present in a functional polyribosome.

9. Contrast the roles of tRNA and mRNA during translation and list all enzymes that participate in the transcription and translation process.

10. Francis Crick proposed the "adaptor hypothesis" for the function of tRNA. Why did he choose that description?

11. What molecule bears the codon? the anticodon?

12. The α chain of eukaryotic hemoglobin is composed of 141 amino acids. What is the minimum number of nucleotides in an mRNA coding for this protein chain? Assuming that each nucleotide is 0.34 nm long in the mRNA, how many triplet codes can at one time occupy space in a ribosome that is 20 nm in diameter?

13. Summarize the steps involved in charging tRNAs with their appropriate amino acids.

14. In 1962, F. Chapeville and others reported an experiment in which they isolated radioactive ^{14}C-cysteinyl-tRNA^{cys} (charged tRNA^{cys} + cysteine). They then removed the sulfur group from the cysteine, creating alanyl-tRNA^{cys} (charged tRNA^{cys} + alanine). When alanyl-tRNAcys was added to a synthetic mRNA calling for cysteine but not alanine, a polypeptide chain was synthesized containing alanine. What can you conclude from this experiment?

15. A short RNA molecule was isolated that demonstrated a hyperchrome shift indicating secondary structure. Its sequence was determined to be AGGCGCCGACUCUACU.

(a) Predict a two-dimensional model for this molecule.

(b) What DNA sequence would give rise to this RNA molecule through transcription?

(c) If the molecule were a tRNA fragment containing a CGA anticodon, what would the corresponding codon be?

(d) If the molecule were an internal part of a message, what amino acid sequence would result from it following translation? (Refer to the code chart in Figure 12–6.)

16. Given below is the sequence of a short wild-type protein as well as various mutant forms of the protein.

(a) Describe the general type of mutation in each case.

(b) Indicate the specific base-pair change that occurred for each mutation. Assume these mutations resulted from single base-pair changes. (There may be more than one correct answer.)

(c) A fourth mutation is identified. The strain carrying this mutation produces abnormally low amounts of wild-type protein. In what part of the gene is this mutation likely to be located?

Wild-type:	met	trp	tyr	arg	gly	ser	pro	thr
Mutant #1:	met	trp						
Mutant #2:	met	cys	ile	val	val	leu	gln	
Mutant #3:	met	trp	his	arg	gly	ser	pro	thr

Selected Readings

Alberts, B., et al. 1994. *Molecular biology of the cell*, 3rd ed. New York: Garland.

Barrell, B. G., Banker, A. T., and Drouin, J. 1979. A different genetic code in human mitochondria. *Nature* 282:189–94.

Brenner, S. 1989. *Molecular biology: A selection of papers.* Orlando, FL: Academic Press.

———, Jacob, F., and Meselson, M. 1961. An unstable intermediate carrying information from genes to ribosomes for protein synthesis. *Nature* 190:575–80.

Cech, T. R. 1986. RNA as an enzyme. *Sci. Am.* (Nov.) 255(5):64–75.

———. 1987. The chemistry of self-splicing RNA and RNA enzymes. *Science* 236:1532–39.

Chambon, P. 1981. Split genes. *Sci. Am.* (May) 244:60–71.

Cold Spring Harbor Laboratory. 1966. The genetic code. *Cold Spring Harb. Symp.*, vol. 31.

Crick, F. H. C. 1962. The genetic code. *Sci. Am.* (Oct.) 207:66–77.

———. 1966. The genetic code: III. *Sci. Am.* (Oct.) 215:55–63.

———. 1979. Split genes and RNA splicing. *Science* 204:264–71.

Darnell, J. E. 1983. The processing of RNA. *Sci. Am.* (Oct.) 249:90–100.

———. 1985. RNA. *Sci. Am.* (Oct.) 253:68–87.

Dickerson, R. E. 1983. The DNA helix and how it is read. *Sci. Am.* (Dec.) 249:94–111.

Fiers, W., et al. 1976. Complete nucleotide sequence of bacteriophage MS2 RNA: Primary and secondary structure of the replicase gene. *Nature* 260:500–7.

Hamkalo, B. 1985. Visualizing transcription in chromosomes. *Trends Genet.* 1:255–60.

Holley, R. W., et al. 1965. Structure of a ribonucleic acid. *Science* 147:1462–65.

Judson, H. E. 1979. *The eighth day of creation.* New York: Simon & Schuster.

Kable, M.L., et. al. 1996. RNA editing: A mechanism for gRNA-specifed uridylate insertion into precursor mRNA. *Science* 273:1189–95.

Khorana, H.G. 1967. Polynucleotide synthesis and the genetic code. *Harvey Lect.* 62:79–105.

Lake, J. A. 1981. The ribosome. *Sci. Am.* (Aug.) 245:84–97.

Lodish, H., et al. 1995. Molecular cell biology, 2nd ed. New York: Scientific American Books.

Maniatis, T., and Reed, R. 1987. The role of small nuclear ribonucleoprotein particles in pre-mRNA splicing. *Nature* 325:673–78.

Miller, O. L., Hamkalo, B., and Thomas, C. 1970. Visualization of bacterial genes in action. *Science* 169:392–95.

Nirenberg, M. W. 1963. The genetic code: II. *Sci. Am.* (Mar.) 190:80–94.

———, and Matthaei, H. 1961. The dependence of cell-free protein synthesis in *E. coli* upon naturally occurring or synthetic polyribosomes. *Proc. Natl. Acad. Sci. USA* 47:1588–602.

Nomura, M. 1984. The control of ribosome synthesis. *Sci. Am.* (Jan.) 250:102–14.

O'Malley, B., et al. 1979. A comparison of the sequence organization of the chicken ovalbumin and ovomucoid genes. In *Eucaryotic gene regulation*, ed. R. Axel et al., pp. 281–99. Orlando, FL: Academic Press.

Padgett, R. A., et al. 1986. Splicing of messenger RNA precursors. *Annu. Rev. Biochem.* 55:1119–50.

Reed, R., and Maniatis, T. 1985. Intron sequences involved in lariat formation during pre-mRNA splicing. *Cell* 41:95–105.

Rich, A., and Houkim, S. 1978. The three-dimensional structure of transfer RNA. *Sci. Am.* (Jan.) 238:52–62.

Rich, A., Warner, J. R., and Goodman, H. M. 1963. The structure and function of polyribosomes. *Cold Spring Harbor Symp. Quant. Biol.* 28:269–85.

Sharp, P. A. 1987. Splicing of messenger RNA precursors. *Science* 235:766–71.

———. 1994. Nobel Lecture: Split genes and RNA splicing. *Cell* 77:805–15.

Steitz, J. A. 1988. Snurps. *Sci. Am.* (June) 258(6):56–63.

Stryer, L. 1995. *Biochemistry*, 4th ed. New York: W. H. Freeman.

Volkin, E., Astrachan, L., and Countryman, J. L. 1958. Metabolism of RNA phosphorus in *E. coli* infected with bacteriophage T7. *Virology* 6:545–55.

Watson, J. D. 1963. Involvement of RNA in the synthesis of proteins. *Science* 140:17–26.

———, et al. 1987. *Molecular biology of the gene*, 4th ed. Menlo Park, CA: Benjamin-Cummings.

Zubay, G. L., and Marmur, J. 1973. *Papers in biochemical genetics*, 2nd ed. New York: Holt, Rinehart and Winston.

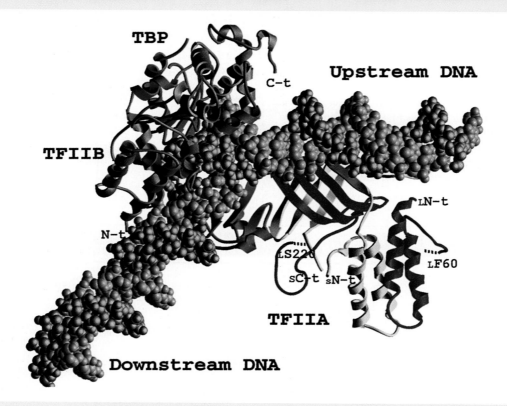

Molecular complex formed between TATA box-binding protein (TBP), transcription factors TFIIA and TFIIB, and TATA box DNA during the initiation of transcription.

CHAPTER OUTLINE

CHAPTER
13

Regulation of Gene Expression

Chapter Concepts

Expression of genetic information is dependent on regulatory mechanisms that either activate or repress the transcription of genes. Transcription is modulated by the interaction of various regulatory molecules with DNA sequences, most often located upstream from affected genes. Genetic regulation in eukaryotes also occurs during posttranscriptional events.

In previous chapters, we established how DNA is organized into genes, how genes store genetic information, and how this information is expressed. We now consider one of the most fundamental issues in molecular genetics: *How is genetic expression regulated?* A functional bacterial cell contains thousands of proteins present at widely different concentrations, yet each is encoded by a single gene. Synthesis of bacterial gene products changes dramatically in response to environmental conditions. Bacterial cells normally synthesize the enzymes to metabolize lactose only when it is present in the environment and other, more readily metabolized carbon sources are not. In the absence of lactose, the enzymes needed to metabolize this sugar are also absent. These observations lend support to the idea that gene action is regulated.

In eukaryotic organisms, cells of the pancreas do not make retinal pigment, and retinal cells do not make insulin. In multicellular eukaryotes, genetic regulation is at the heart of cellular differentiation (Chapter 20). Differential gene expression serves as the basis for phenotypic specialization at the cellular and tissue levels in both plants and animals.

In this chapter, we will discuss examples of gene regulation in bacteria, bacteriophages, and eukaryotes. Pivotal to this discussion is the idea that all cells in an organism contain a complete set of genetic information characteristic of that species. Regulation is not accomplished by eliminating unused genetic information; instead, mechanisms have evolved to control the expression of genes. Some of these mechanisms, particularly in bacterial systems and their phages, have been extensively characterized.

Genetic Regulation in Prokaryotes: An Overview

Regulation of gene expression has been studied extensively in prokaryotes, particularly in *Escherichia coli*. Highly efficient mechanisms have evolved that turn genes on and off, depending on the cell's metabolic needs in particular environments. Detailed analysis of proteins in *E. coli* has shown that for the more than 4000 polypeptide chains encoded by the genome, there is a vast range of concentration of gene products. Some proteins may be present in as few as 5 to 10 molecules per cell, whereas others, such as ribosomal proteins and the many proteins involved in the glycolytic pathway, are present in as many as 100,000 copies per cell.

While a basal level of most gene products exists, this level can be altered in response to chemical signals from the environment. Enzymes produced in response to such signals are referred to as **inducible**. In contrast, other proteins are produced continuously, regardless of the chemical makeup of the environment, and are called **constitutive**.

Studies have also revealed cases where the presence of a specific molecule causes inhibition of genetic expression. This is often the case for molecules that are the end products of biosynthetic pathways. Amino acids can be synthesized by bacterial cells, or, if available, they can be taken up from the environment. If a specific amino acid is present in the environment, it is inefficient for the cell to synthesize the enzymes necessary for the production of that amino acid. In this case, the presence of the amino acid represses transcription of mRNA for the appropriate biosynthetic enzymes, and it is an example of a **repressible** system of gene regulation.

Gene regulation, whether inducible or repressible, may be under **negative** or **positive control**. Under negative control, gene expression occurs unless it is shut off by a **repressor** protein. In positive-control systems, gene expression occurs only in the presence of an **active regulator** protein. Some control systems contain elements of both positive and negative control. Examples discussed in the following sections will help distinguish among these mechanisms.

Lactose Metabolism in *E. coli*: An Inducible System

In prokaryotes, structural genes tend to be organized in clusters controlled from a single regulatory site. This site is linked to the gene cluster it controls and is known as a *cis*-acting element. The *cis*-acting sites are usually located upstream from the gene cluster they regulate. Usually, the genes in the cluster have related functions, and they encode the enzymes in a metabolic pathway. Interactions at the regulatory site involve the binding of molecules that control transcription of the gene cluster. Such molecules are called *trans*-acting elements. Actions at the regulatory site determine whether the genes are expressed or not, and thus whether the corresponding enzymes or other protein products are present or not. Binding of a *trans*-acting element at a *cis*-acting site can regulate the gene cluster positively (by turning genes in the cluster on) or negatively (by turning the genes off). In this section, we will discuss how such bacterial gene clusters are coordinately regulated.

The gene cluster involved in the metabolism of lactose in *E. coli* has been studied extensively. Beginning in 1946 with the studies of Jacques Monod, and with significant contributions by Joshua Lederberg, François Jacob, and André Lwoff, genetic and biochemical evidence provided clear insights into the way in which the genes responsible for lactose metabolism are regulated. In the presence of lactose, the concentration of the enzymes responsible for its metabolism increases rapidly from 5 to 10 molecules to thousands per cell. The enzymes are **inducible**, and lactose serves as the **inducer**.

The lactose gene cluster (Figure 13–1) contains a regulatory gene, a control site, and three adjacent structural genes which encode the enzymes involved in lactose metabolism. The three structural genes are transcribed into a single **polycistronic mRNA**. Clustered structural genes transcribed as a single mRNA and their adjacent control regions are called **operons**. Together, the genes in the *lac* operon function in a coordinated fashion and provide a rapid response to the presence or absence of lactose in the environment.

Structural Genes of the lac Operon

Genes that code for the primary structure of enzymes are called **structural genes**. The *lacZ* gene encodes **β-galactosidase**, an enzyme that converts the disaccharide lactose to the monosaccharides glucose and galactose (Figure 13–2). This conversion is essential if lactose is to serve as the primary energy source in glycolysis.

The *lacY* gene codes for **β-galactoside permease**, a membrane-bound protein which assists the transport of lactose into the bacterial cell. The third gene in the cluster, *lacA*, codes for **transacetylase**, an enzyme whose role is still not completely clear.

Studies on these genes utilized *lac⁻* mutants that eliminated the function of one or the other enzyme. Mutants of *lacZ⁻* or *lacY⁻* are unable to utilize lactose as an energy source. Mapping studies by Lederberg established that all three genes are closely linked to one another in the order *Z–Y–A* (Figure 13–1).

Regulatory Mutations in the lac Operon

How does lactose activate structural genes and induce the synthesis of the enzymes for lactose metabolism? The answer to this question required the study of a second class of mutations, called **constitutive mutants**. In these mutants the enzymes are produced whether or not lactose is present. These mutations served as the basis for studies that defined the regulatory scheme for lactose metabolism.

The constitutive mutant *lacI⁻* maps at a site adjacent to, but independent from, the locus of the structural genes, and has its own promoter and termination region. A second class of constitutive mutants mapped to a locus immediately adjacent to the structural genes. This class is designated *lacOᶜ* and represents the **operator region** of the *lac* operon.

The Operon Model: Negative Control

In 1961, Jacob and Monod proposed a scheme of negative control of regulation called the **operon model**, whereby a group of genes is regulated and expressed together as a unit. In their specific model [Figure 13–3(a)], the operon consists of the *Z*, *Y*, and *A* structural genes as well as the adjacent sequences of DNA referred to as the operator region. They argued that the *lacI* gene regulates the transcription of the structural genes by producing a **repressor molecule**. The repressor

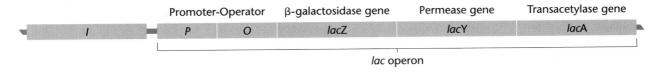

| Promoter-Operator | β-galactosidase gene | Permease gene | Transacetylase gene |

| *I* | *P* | *O* | *lacZ* | *lacY* | *lacA* |

lac operon

Figure 13–1 Organization of the gene cluster and regulatory units involved in the control of lactose metabolism.

was hypothesized to be **allosteric**, meaning that the molecule interacts reversibly with another molecule, causing both a conformational change in three-dimensional shape and a change in chemical activity.

Jacob and Monod suggested that the repressor normally interacts with the DNA sequence of the operator region. When it does so, it inhibits the action of RNA polymerase, effectively repressing the transcription of the structural genes [Figure 13–3(b)]. However, when lactose is present, this disaccharide binds to the repressor, causing the allosteric conformational change. This change alters the binding site of the repressor, rendering it incapable of interacting with operator DNA [Figure 13–3(c)]. In the absence of the repressor–operator interaction, RNA polymerase transcribes the structural genes, and the enzymes necessary for lactose metabolism are produced. Since transcription occurs only when the repressor fails to bind to the operator region, **negative control** is exerted.

The operon model uses these potential molecular interactions to explain the efficient regulation of the structural genes for lactose metabolism. In the absence

of lactose, the enzymes encoded by the genes are not needed and so are repressed. When lactose is present, it indirectly induces the activation of the genes by binding with the repressor. If all lactose is metabolized, none is available to bind to the repressor, which is again free to bind to operator DNA and repress transcription.

Both *lacI⁻* and *lacOᶜ* constitutive mutations interfere with these molecular interactions, allowing continuous transcription of the structural genes. In the case of *lacI⁻* mutants, the repressor product is altered and cannot bind to the operator region, so the structural genes are always transcribed. In the case of *lacOᶜ* mutants, the nucleotide sequence of the operator DNA is altered and will not bind with a normal repressor molecule. The result is the same: Structural genes are always transcribed. Both types of constitutive mutations are illustrated diagrammatically in Figures 13–3(d) and 13–3(e).

Genetic Proof of the Operon Model

The operon model makes predictions that can be tested using genetic methods. The major assumptions to be tested are as follows: (1) The *lacI* gene produces a diffusible cellular product; (2) the *lacO* region does not encode a gene product; and (3) the *lacO* region must be adjacent to the *lac* structural genes in order to regulate transcription.

Bacteria are naturally haploid organisms. However, partially diploid strains can be developed that allow the assessment of these assumptions. In such cases, the entire host chromosome is present along with an extra copy of one or a few genes of choice. The extra genes are inserted as part of a plasmid called the **F factor**, designated F′ in Table 13.1. It is possible to construct heterozygous genotypes where a *lacI⁺* gene has been introduced into a *lacI⁻* host, or where a *lacO⁺* region is added to a *lacOᶜ* host. The Jacob–Monod operon model predicts that adding a *lacI⁺* gene to a *lacI⁻* cell should restore inducibility, because a normal repressor would again be produced. Adding a *lacO⁺* region to a *lacOᶜ* cell should have no effect on constitutive enzyme production, since regulation depends on a *lacO⁺* region in the host DNA, immediately adjacent to the structural genes.

Results of these experiments are shown in Table 13.1, where *Z* represents all three structural genes. In the cases just described, the Jacob–Monod model is upheld (part B of Table 13.1). Part C shows the reverse experiments, where either a *lacI⁻* gene or a *lacOᶜ* region is

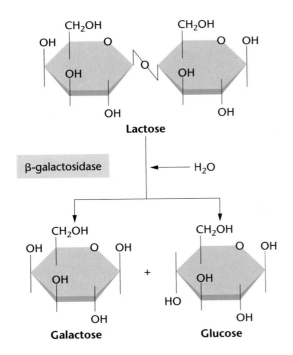

Lactose

β-galactosidase ◄—— H_2O

Galactose   Glucose

Figure 13–2 The catabolic conversion of the disaccharide lactose into its monosaccharide units, galactose and glucose.

(a) Components

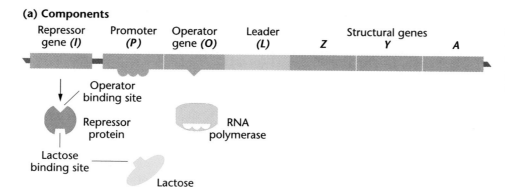

(b) $I^+ O^+ Z^+ Y^+ A^+$ (wild type) – No lactose present – Repressed

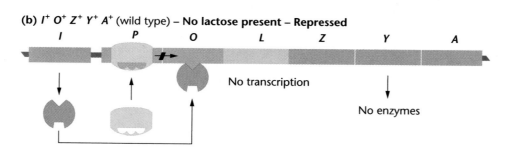

(c) $I^+ O^+ Z^+ Y^+ A^+$ (wild type) – Lactose present – Induced

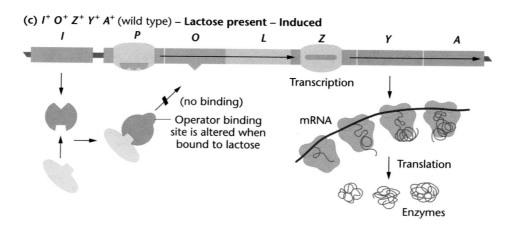

■ Figure 13–3 The components involved in the regulation of the *lac* operon and their interaction under various genotypic conditions, as described in the text.

added to cells of normal inducible genotypes. The model predicts that inducibility will be maintained in these partial diploids, and it is.

Another prediction of the operon model is that certain mutations in the *lacI* gene should have the opposite effect of *lacI⁻* mutants. That is, instead of being constitutive by failing to interact with the operator, mutant repressor molecules should be produced that cannot interact with the inducer, lactose. As a result, the repressor would remain bound to the operator sequence, and the structural genes would be repressed permanently. In this case, the presence of an additional *lacI⁺* gene would have little or no effect on repression.

As shown in part D of Table 13.1, such a mutation, *lacIˢ*, was discovered in which the operon is "superrepressed." Adding a *lacI⁺* gene to these cells does not re-

lieve repression of gene activity. These observations again provide support for the operon model.

Isolation and Crystallographic Analysis of the lac Repressor

The most direct proof for the role of repressors in operon regulation was the isolation and characterization of the repressor molecule predicted by the model. This was a difficult task because each *E. coli* cell contains only about 10 repressor molecules. In spite of the enormity of the task, Walter Gilbert and Benno Müller-Hill reported its isolation in 1966. The functional repressor is a tetramer of four polypeptides (MW 38,000). It has two binding sites, one for lactose and one for the DNA sequences of the operator region. Confir-

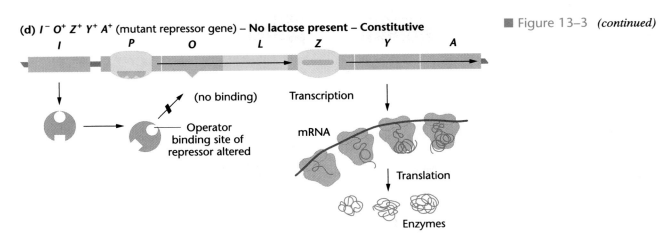

(d) $I^-\ O^+\ Z^+\ Y^+\ A^+$ (mutant repressor gene) – **No lactose present – Constitutive**

■ Figure 13–3 *(continued)*

(e) $I^+\ O^c\ Z^+\ Y^+\ A^+$ (mutant operator gene) – **No lactose present – Constitutive**

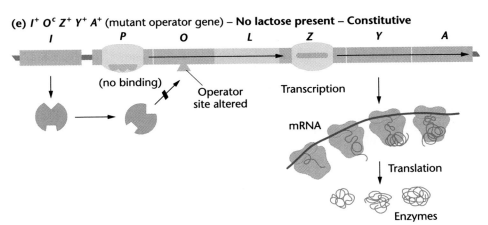

TABLE 13.1	A comparison of gene activity (+ or -) in the presence or absence of lactose for various *E. coli* genotypes

	Genotype	Presence of β-Galactosidase Activity	
		Lactose Present	Lactose Absent
A.	$I^+O^+Z^+$	+	−
	$I^+O^+Z^-$	−	+
	$I^-O^+Z^+$	+	+
	$I^+O^cZ^+$	+	+
B.	$I^-O^+Z^+/F'I^+$	+	−
	$I^+O^cZ^+/F'O^+$	+	+
C.	$I^+O^+Z^+/F'I^-$	+	−
	$I^+O^+Z^+/F'O^c$	+	−
D.	$I^sO^+Z^+$	−	−
	$I^sO^+Z^+/F'I^+$	−	−

Note: In parts B to D, most genotypes are partially diploid, containing an F factor plus attached genes (F′).

mation of its role was provided by showing that the repressor binds to DNA containing a *lacO⁺* gene but not to DNA with a *lacOᶜ* gene. Furthermore, repressor binding activity could not be demonstrated among the proteins isolated from *lacI⁻* cells, as predicted.

Recently, in 1996, Mitchell Lewis, Ponzy Lu, and their colleagues succeeded in determining the crystal structure of the lac repressor as well as the structure of the repressor bound to IPTG and to operator DNA. The four monomers present in the repressor tetramer are identical, consisting of 360 amino acids. In the crystallographic studies, each tetramer binds to two symmetrical operator DNA helices. Binding by the repressor distorts the conformation of DNA, causing it to bend away from the repressor. A generalized image of this binding is illustrated in Figure 13–4. In this depiction, an auxiliary operator region located 82 base pairs upstream from the *lacZ* gene is also bound, creating a "repression loop" of DNA.

These studies have also defined the three-dimensional conformational changes that accompany the allosteric transitions occurring during the interactions with the inducer molecules. Many of the related findings demonstrate that the molecular mechanisms giving rise

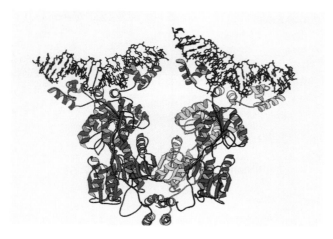

■ Figure 13–4 Model of the *lac* repressor bound to DNA constituting the operator region of the *lac* operon, based on crystallographic analysis.

to the transitions are prototypes for similar transitions known to occur in eukaryotic systems.

The crystallographic studies bring us to a new level of understanding in the regulatory process occurring within the *lac* operon. Remarkably, this current information has confirmed at the level of molecular visualization so many of the things that Jacob and Monod were able to predict based strictly on genetic grounds during their 1965 Nobel Prize-winning research.

The CAP Protein: Positive Control of the lac Operon

When the *lac* repressor is bound to the inducer, the *lac* operon is activated and RNA polymerase transcribes the structural genes. This process is initiated as a result of the binding that occurs between the polymerase and the nucleotide sequence of the **promoter region**, found upstream (the 5′ end) from the initial coding sequences. Within the *lac* operon, the promoter is found between the *I* gene and the operator region (O^c)— (see Figure 13–1). Careful examination has revealed that, in this case, polymerase binding is not very efficient unless another protein is present to facilitate the process. This protein is called the **catabolite-activating protein (CAP)**.

The discovery of CAP and the region within the promoter where it binds (the **CAP-binding site**) occurred as a result of investigation of a most interesting observation. Even when cellular conditions are inducible (in the presence of lactose), *transcription of the operon is inhibited if glucose is present*. This inhibition, called **catabolite repression**, is a reflection of the greater simplicity with which glucose may be metabolized in comparison to lactose. The cells "prefer" glucose, and if present, they do not activate the *lac* operon, even when lactose is present.

Regulation of the *lac* operon by CAP and the role of glucose in catabolite repression is summarized in Figure 13–5. In the absence of glucose, and under inducible conditions, CAP exerts positive control by binding to the CAP site, facilitating RNA polymerase binding at the promoter, and thus transcription. Therefore, for maximal transcription, repressor must be bound by lactose, and CAP must be bound to the CAP-binding site.

Regulation of the *lac* operon by catabolite repression results in efficient energy utilization, because the presence of glucose will override the need for the metabolism of lactose, should it also be available to the cell. Catabolite repression involving CAP has also been observed for other inducible operons, including those that control the metabolism of galactose and arabinose.

Tryptophan Metabolism in *E. coli*: A Repressible Gene System

Although induction had been known for some time, it was not until 1953 that Monod and his coworkers discovered the phenomenon of **enzyme repression**. Wild-type *E. coli* are capable of producing the enzymes required for the biosynthesis of amino acids and other essential biomolecules. Monod focused his studies on the amino acid tryptophan and the enzyme **tryptophan synthetase**. He discovered that if tryptophan is present in sufficient quantities in the growth medium, the enzymes necessary for its synthesis are **repressed**. Energetically, such enzyme repression is highly economical to the cell, because synthesis is unnecessary in the presence of adequate amounts of tryptophan.

Further investigation showed that enzymes encoded by five contiguous genes on the *E. coli* chromosome are involved in tryptophan biosynthesis. These genes are part of an operon (the *trp* operon). In the presence of tryptophan, all genes are coordinately repressed, and none of the enzymes is synthesized. Because of the similarity between repression by tryptophan and the induction of enzymes in lactose metabolism, Jacob and Monod proposed a model of gene regulation analogous to the *lac* system.

To account for repression, they suggested that *an inactive repressor is normally made* that alone cannot bind to the operator region within the operon. However, the repressor can interact with tryptophan when it is present. Binding with tryptophan activates the repressor, allowing it to bind to the operator, repressing transcription. In this case, when the end product of this biosynthetic pathway is present, further transcription of tryptophan enzymes is shut off. Since the regulatory complex inhibits transcription of the operon, this repressible system, like the *lac* system, is under negative control. Because tryptophan participates in repression, it is referred to as a **corepressor** in this regulatory scheme.

Genetics of the trp Operon

The model of a repressible operon is supported by genetic evidence. Two distinct categories of constitutive mutations were isolated. The first class, *trpR*⁻, maps at a

(a) In the absence of glucose, cAMP levels increase and no repression occurs

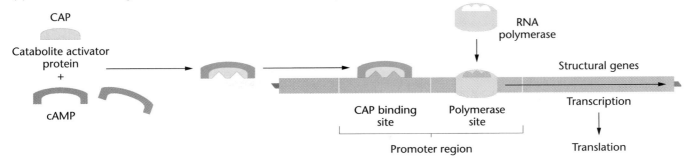

(b) In the presence of glucose, cAMP levels decrease and catabolite repression occurs

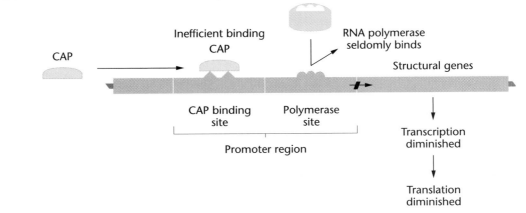

■ Figure 13–5 Catabolite repression. In the absence of glucose, cAMP levels increase, resulting in cAMP-CAP binding and the ensuing stimulation of transcription.

distance from the structural genes. This locus encodes the repressor protein. The *trpR⁻* mutants either inhibit the interaction of a mutant repressor with tryptophan or inhibit repressor formation entirely. In the absence of repressor, the operon is transcriptionally active. The presence of a *trpR⁺* gene restores repression, and the operon is transcriptionally inactive.

A second constitutive mutant behaves like the *lacOᶜ* mutation in the *lac* operon. It is found immediately adjacent to the structural genes in the *trp* operon. Addition of a wild-type gene in mutant cells does not restore enzyme repression. This is predictable if the mutation represents an operator region that can no longer interact with the repressor–tryptophan complex.

The entire *trp* operon has now been well defined, as shown in Figure 13–6. Five contiguous structural genes (*trpE, trpD, trpC, trpB,* and *trpA*) are transcribed as a polycistronic message that directs translation of the enzymes for the biosynthesis of tryptophan. As in the *lac* operon, a promoter region (*trpP*) is the binding site for RNA polymerase, and an operator region (*trpO*) binds the repressor. In the absence of binding, transcription is initiated within the overlapping *trpP/trpO* region and proceeds along a leader sequence. Transcription is initiated 162 nucleotides prior to the first structural gene (*trpE*). The leader sequence contains a regulatory se-

quence, called an **attenuator**. This regulatory unit is an integral part of the control mechanism of this operon.

Attenuation

Even in the presence of tryptophan, when repression should be complete, transcription is often initiated. As a result, repression in this system is said to be "weak" in comparison to that which occurs in the *lac* operon. While strong versus weak repression may seem difficult to understand at first, such phenomena are not unexpected. Molecular binding between the protein repressor and operator DNA is a dynamic process subject to chemical equilibria. If there is a high degree of affinity between the two, repression at equilibrium will be quite effective. If there is less affinity between the operator and the repressor, repression will be weak.

The process of **attenuation**, as studied extensively by Charles Yanofsky and his colleagues, represents a second level of regulation of the *trp* operon. In this context, *attenuate* means to reduce in amount. In the presence of tryptophan, mRNA synthesis may be initiated, but is frequently terminated at a point about 140 nucleotides along the transcript. This point is identified in Figure 13–6(a) as the **attenuator**. In the absence of tryptophan, attenuation is overcome and transcription

(a) The tryptophan operon

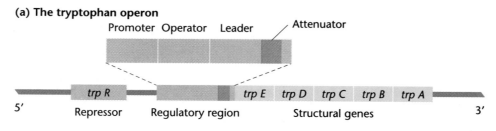

(b) Activation of structural genes in absence of tryptophan

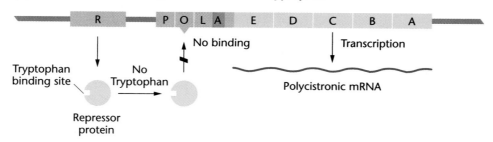

(c) Repression of structural genes in presence of tryptophan

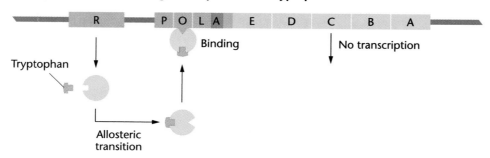

■ **Figure 13–6** (a) The components involved in the regulation of the tryptophan operon. (b) Regulatory conditions involving either activation or (c) repression of the structural genes. In the absence of tryptophan, an inactive repressor is made that cannot bind to the operator (O), thus allowing transcription to proceed. In the presence of tryptophan, it binds to the repressor, causing an allosteric transition to occur. This complex binds to the operator region, resulting in repression of the operon.

proceeds, leading to the production of enzymes necessary for tryptophan biosynthesis.

Attenuation involves folding of the RNA transcribed from the leader sequence. During attenuation (when tryptophan is abundant), the structure of the RNA transcribed *from the leader sequence* mimics that found at the end of mRNA molecules, forming a "hairpin loop." This configuration leads to the premature termination of transcription, reducing the amount of mRNA produced. A model for the mechanism of attenuation and how it is overcome in the absence of tryptophan has been proposed by Yanofsky. As in the studies of Jacob and Monod, the prediction of molecular events was initially based on genetic evidence, including the existence of mutations within the leader sequence that abolish attenuation.

Genetic Regulation in Phage Lambda: Lysogeny or Lysis?

Our understanding of genetic regulation at the transcriptional level has benefitted from studies of bacteriophage lambda as well as from studies of operons in

bacteria. Lambda DNA contains about 45,000 base pairs, enough to encode 35 to 40 genes. Following infection of *E. coli*, phage lambda can follow either a **lysogenic** or a **lytic** pathway. In the lysogenic pathway, phage DNA is integrated into the bacterial genome and is almost totally repressed; in the lytic pathway, the phage DNA is transcribed, and viral reproduction ensues.

In the lysogenic pathway, the genes responsible for phage reproduction and lysis are turned off by the **λ repressor protein**. Since it is produced by one of the virus's own genes, *c*I, we will refer subsequently to this protein as the ***c*I repressor**. If the lytic pathway is followed, the expression of the *c*I gene is repressed by a second protein, **Cro**, which is produced by the *cro* gene. Both the *c*I repressor and the Cro protein have been isolated and characterized. Their interaction with λ DNA has also been determined. The operator regions that bind the repressor are shown in Figure 13–7.

The *c*I repressor was isolated and characterized by Mark Ptashne in 1967. It is a protein consisting of 236 amino acids, and it controls the rate at which classes of λ mRNA are made. The Cro protein consists of 66 amino acids. When repressor is produced, it recognizes two different operator regions in λ DNA, O_L and O_R,

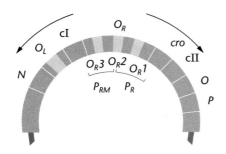

■ Figure 13–7 The regulatory sites controlling lysogeny and lysis in phage λ.

present on either side (left or right) of the cI gene. When repressor is bound to these regions, two sets of early genes are repressed, causing the remainder of the λ genes to be turned off. In this case, the repressor behaves as a negative control element. During binding, the cI repressor stimulates transcription of the cI gene and acts as a positive control element, enhancing cI transcription by a factor of ten.

When no repressor is bound to O_L and O_R, transcription initiation at promoters (P_{RM} and P_R) proceeds, resulting in the production of two proteins, N and Cro. The N protein functions as an antiterminator, allowing complete transcription of genes essential to reproduction and lysis. The Cro protein acts as a repressor of cI gene transcription. Mutational analysis supports this model. Mutations in either O_L or O_R prevent repressor binding and abolish the potential for lysogeny. cro^- mutations abolish the potential for lysis. These proteins served as models for the binding of proteins to DNA. Both Cro and cI repressors form dimers that bind in the major groove of the DNA helix. DNA-binding proteins play important roles in controlling gene expression in both prokaryotes and eukaryotes.

Genetic Regulation in Eukaryotes: An Overview

Following the discovery and characterization of bacterial operons, many studies were initiated to locate comparable regulatory systems in eukaryotes. To date, a similar system has been found only in yeast and the nematode *C. elegans*. Almost 40 years of research has established that operons are not common in multicellular eukaryotes. As we shall see, however, this is not surprising given the complexity of gene expression in eukaryotes.

Each eukaryotic cell contains a much greater amount of genetic information, and this DNA is complexed with histones and other proteins to form chromatin. Genetic information in eukaryotes is carried on many chromosomes, rather than one, and these chromosomes are enclosed within a nuclear membrane. In eukaryotes, the process of transcription is spatially and temporally separated from that of translation, and involves

three classes of RNA polymerases. Transcripts of eukaryotic genes are processed, cleaved, and realigned before transport to the cytoplasm, and many transcribed sequences never leave the nucleus. In addition, during the development of multicellular organisms, differentiation and changes in gene expression are often influenced by cellular interactions and external signals such as hormones.

Regulation of eukaryotic gene expression can potentially occur at many levels (Figure 13–8). These include (1) transcriptional control, (2) processing of the pre-mRNA, (3) transport to the cytoplasm, (4) stability of the mRNA, (5) selecting which mRNAs are translated, and (6) posttranslational modification of the protein product. Most eukaryotic genes are regulated, in part, at the transcriptional level. In the following sections, emphasis is placed on transcriptional control, although other levels of control will also be discussed. There are two main components of transcriptional control of gene expression: short DNA sequences that serve as recognition sites, and regulatory proteins that bind to these sites.

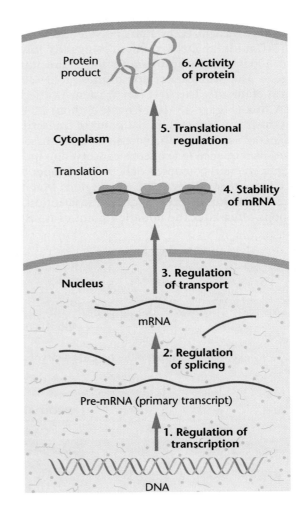

■ Figure 13–8 Various levels of regulation that are possible during genetic expression.

Regulatory Elements and Eukaryotic Genes

The internal structure of eukaryotic genes (see Chapter 12) includes regulatory sequences adjacent to genes that control transcription. These sequences are of two types: **promoters** and **enhancers** (Figure 13–9). These regulatory elements can be on either side of a gene, or at some distance from the gene. When they are adjacent to the structural genes, they are called *cis* regulators, as opposed to *trans* regulators that are not adjacent to the genes that they regulate.

Promoters

Promoters, which have counterparts in bacteria, consist of nucleotide sequences that serve as the recognition point for RNA polymerase binding. They represent the region necessary to *initiate* transcription and are located immediately adjacent to the genes they regulate. Promoter regions are usually several hundred nucleotides in length.

Eukaryotic promoters require the binding of a number of protein factors to initiate transcription. Promoters that are recognized by RNA polymerase II consist of short modular DNA sequences usually located within 100 base pairs upstream (in the 5′ direction) of the gene. The promoter region of most genes shares several elements. The first is a sequence called the **TATA box** (Figure 13–10). Located about 25 to 30 bases upstream from the initial point of transcription (designated as –25 to –30), it consists of an 8-base-pair consensus sequence (a sequence conserved in most or all genes studied) composed only of T=A pairs, often flanked on either side by G≡C-rich regions. Mutations in the TATA box severely reduce transcription, and deletions often alter the initiation point of transcription.

Many promoters contain other components; one of these is called the **CAAT box**. Its consensus sequence is CAAT or CCAAT, and it frequently appears in the region –70 to –80 base pairs from the start site. Mutational analysis suggests that the CAAT box is critical to the promoter's ability to facilitate transcription. Mutations on either side of this element have little or no effect on transcription, whereas mutations within the CAAT sequence lower the rate of transcription dramatically. A third modular element of some promoters is called the GC box, which has the consensus sequence GGGCGG and is often found at about position –110.

Enhancers

In addition to the promoter region, transcription of most, if not all, eukaryotic genes is regulated by additional DNA sequences called **enhancers**. These regions interact with regulatory proteins and can increase the efficiency of transcription initiation or activate the promoter. Thus, there is some analogy between enhancers and operator regions in prokaryotes. However, enhancers appear to be much more complex in both structure and function. Enhancers can be distinguished from promoters by several characteristics.

1. The position of the enhancer need not be fixed; it can be placed upstream, downstream, or within the gene it regulates.

2. Its orientation can be inverted without significant effect on its action.

3. If an enhancer is moved to another location in the genome, or if an unrelated gene is placed near an enhancer, transcription of the adjacent gene is enhanced.

Most eukaryotic genes are under the control of enhancers. In the immunoglobulin heavy-chain genes, an enhancer is located in an intron between two coding regions. In this case, the enhancer is located

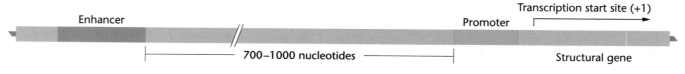

■ Figure 13–9 Organization of the typical eukaryotic gene and transcriptional control regions.

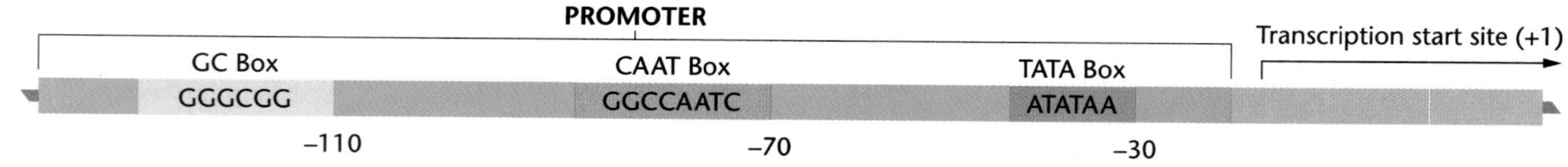

■ Figure 13–10 The modular regulatory elements and their location within the promoter region in eukaryotic genes, including the TATA box, the CAAT box, and the GC box.

within the gene it regulates. Downstream enhancers are found in the human β-globin gene, and in chickens, an enhancer is located between the β-globin and the ε-globin genes. In yeast, regulatory sequences similar to enhancers, called **upstream activator sequences (UAS)**, can function upstream at variable distances (see below). They differ from enhancers in that they cannot function downstream of the transcription start point.

The most intriguing question about enhancers is how they are able to exert control over transcription at a great distance from either promoters or the transcriptional start site. As we will discuss in a later section, transcription factors bind to enhancers and alter the configuration of chromatin by bending or looping the DNA to bring distant enhancers and promoters into direct contact in order to form complexes with transcription factors and polymerases. In the new configuration, transcription is stimulated to a higher level, increasing the overall rate of RNA synthesis.

Transcription Factors and Gene Regulation

As in prokaryotes, transcriptional control in eukaryotes involves interaction between DNA sequences adjacent to genes and DNA-binding proteins. Regulatory sequences adjacent to a wide range of genes have been identified, mapped, and sequenced. Proteins have been identified that are not part of the RNA polymerase molecule itself, but are needed for the initiation of transcription. These are called **transcription factors**. These proteins control where, when, and how genes are expressed.

Transcription factors are modular structures with at least two functional domains: one that binds to DNA sequences present in promoters and enhancers (**DNA-binding domain**), and another that activates transcription via protein–protein interaction (**trans-activating domain**). Each domain consists of a specific sequence of amino acids within the protein that is responsible for the binding function.

Genetic Analysis of Transcription Factors: GAL4

One of the first model systems used to study eukaryotic genetic regulation involves the set of genes in yeast that encode enzymes essential for galactose metabolism. Expression of the genes encoding these enzymes is regulated by the presence or absence of galactose. In the absence of galactose, these genes are not transcribed. If galactose is added to the growth medium, immediate transcription of the genes commences, and the mRNA concentration of these transcripts increases by a thousandfold. A mutation in one of the genes (*GAL4*) prevents the activation of these genes in the presence of galactose, indicating that transcription is genetically regulated.

Transcription of the structural genes for galactose metabolism is controlled by DNA sequences called **UAS$_G$s** (upstream activating sequence of galactose genes). The GAL4 protein binds to the UAS$_G$ regions and activates transcription of the genes for galactose metabolism, establishing that the *GAL4* gene encodes a transcription factor.

The *GAL4* gene product is a protein of 881 amino acids that has been the subject of extensive analysis and experimentation. This protein includes: (1) a DNA-binding domain that recognizes and binds to sequences in the UAS$_G$s; and (2) a *trans*-activating domain, which is also essential for the activation of transcription. Results of several experiments suggest strongly that, in addition to UAS$_G$ binding, gene activation requires direct interaction between the activating domain of the transcription factor and other proteins. A general model involving these components and an adjacent promoter and its accompanying gene is depicted in Figure 13–11.

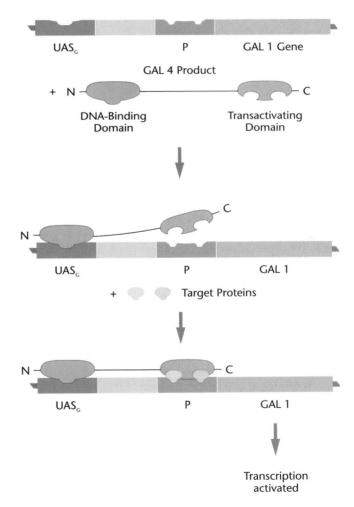

■ Figure 13–11   Illustration of the regulation of a gene involved in galactose metabolism (*GAL1*) by the *GAL4* gene product. This regulatory protein contains a DNA-binding domain and a *trans*-activating domain, as described in the text.

How this interaction activates transcription is an important question. There are several possibilities, including stabilization of the binding between the promoter DNA and RNA polymerase, increasing the rate at which the double-stranded DNA within the transcribed region is unwound, or attracting and stabilizing other factors which bind to the promoter or to the RNA polymerase. An attractive candidate for an activator target is **TFIID**, a complex of several proteins that binds to the TATA sequence of promoters and is required for transcription. TFIID consists of the **TATA-binding protein (TBP)**, and about ten other proteins called **TBP-associated factors**, or TAFs (some of which are described in later sections).

Structural Motifs of Transcription Factors

The domains of eukaryotic transcription factors take on several forms. The **DNA-binding domains** have distinctive three-dimensional structural patterns or **motifs**. There are three major types of these structural motifs: **helix–turn–helix (HTH)**, **zinc finger**, and **leucine zippers**. This classification is not exhaustive, and other new groups will undoubtedly be established as new factors are characterized.

The first DNA-binding domain to be discovered was the helix–turn–helix (HTH) motif. In prokaryotes, HTH motifs have been identified in the *cro* repressor (in the DNA-binding domain), the *lac* repressor, the *trp* repressor, and other proteins (Figure 13–12). Studies indicate that the HTH motif is present in many prokaryotic DNA-binding proteins. This motif is characterized by its geometric conformation rather than a distinctive amino acid sequence. Two adjacent a helices separated by a "turn" of several amino acids enables the protein to bind to DNA (and after which the motif is named). Unlike several of the other DNA-binding motifs, the HTH pattern cannot fold or function alone, but is always part of a larger DNA-binding domain.

The potential for forming helix–turn–helix geometry has been recognized in distinct regions of a large number of eukaryotic genes known to regulate developmental processes. Present almost universally in eukaryotic organisms and called the **homeobox**, a stretch of 180 base pairs specifies a 60-amino-acid **homeodomain** sequence that can form a helix–turn–

helix structure. Of the 60 amino acids, many are basic (arginine and lysine), and a conserved sequence is found among these many genes. We shall discuss homeobox-containing genes in Chapter 20 because of their significance to developmental processes.

Zinc fingers are one of the major structural families of eukaryotic transcription factors, and they are involved in many aspects of gene regulation. Zinc fingers were originally discovered in the *Xenopus* transcription factor TFIIIA. This structural motif has now been identified in proto-oncogenes (Chapter 19), genes that regulate development in *Drosophila* (the Krüppel gene, see Chapter 20), in proteins whose synthesis is induced by growth factors and differentiation signals, and in transcription factors. There are several types of zinc finger proteins, each with a distinctive structural pattern.

One such zinc finger protein contains clusters of two cysteine and two histidine residues at repeating intervals (Figure 13–13). The interspersed cysteine and histidine residues covalently bind zinc atoms, folding the amino acids into loops (these are the zinc "fingers"). Each finger consists of approximately 23 amino acids, with a loop of 12 to 14 amino acids between the Cys and His residues, and a linker between loops consisting of 7 or 8 amino acids. The amino acids in the loop interact with and bind to specific DNA sequences. Studies have shown that zinc fingers bind in the major groove of the DNA helix and wrap at least part way around the DNA. Within the major groove, the zinc finger makes contact with a set of DNA bases and may form hydrogen bonds with the bases, especially in G-rich strands. The number of fingers in a zinc finger transcription factor varies from 2 to 13, as does the length of the DNA-binding sequence.

The third type of domain is represented by the **leucine zipper**. First seen as a stretch of 35 amino acids in a nuclear protein in rat liver, four leucine residues are spaced 7 amino acids apart, flanked by basic amino acids. The leucine-rich regions form a helix with leucine residues protruding at every other turn. When two such molecules dimerize (Figure 13–14), the leucine residues "zip" together. The dimer contains two alpha-helical regions adjacent to the zipper, which bind to phosphate residues and specific bases in DNA, making the dimer look like a pair of scissors.

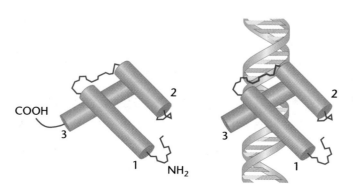

■ Figure 13–12 A helix–turn–helix or homeodomain, where three different planes of the a helix are established and bind in the grooves of DNA double helix.

(a) (b)

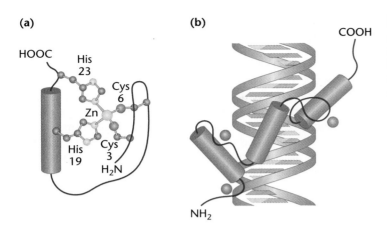

■ **Figure 13–13** A zinc finger, where cysteine and histidine residues bind to a Zn²⁺ ion, looping the amino acid chain out into a fingerlike configuration.

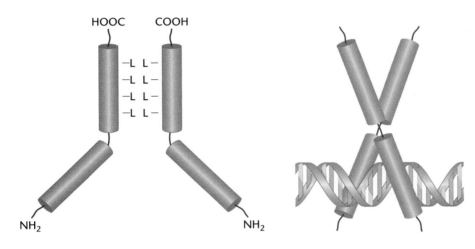

■ **Figure 13–14** A leucine zipper, where dimers result from leucine residues at every other turn of the α-helix in facing stretches of one or two polypeptide chains. When the α-helical regions form a leucine zipper, the regions beyond the zipper form a Y-shaped region that grips the DNA in a scissors-like configuration.

In addition to domains that bind DNA, recall that transcription factors contain domains that activate transcription. These regions can occupy from 30 to 100 amino acids and are distinct from the DNA-binding domains. These stretches of amino acids interact with other transcription factors (such as those that bind to the TATA sequence) or directly with the RNA polymerase.

Overall, the picture of transcriptional regulation in eukaryotes is somewhat complex, but a number of generalizations can be drawn. The structural organization of chromatin and alterations in chromatin structure to allow binding of transcription factors are considered to be the primary levels of regulation. The regulation of transcription by protein factors is largely positive, although transcription repressors are now being recognized as important components of gene regulation. The binding of one or more factors at promoter regions is a prerequisite to transcriptional activation of a locus. Promoter and enhancer sequences are recognized and bound by transcription factors.

Assembling the Transcription Complex

Some insights into the transcriptional apparatus of type II genes (those transcribed by RNA polymerase II) are now available which suggest that a series of transcriptional factors (e.g., TFIID) are assembled at the promoter in a specific order. To initiate formation of the apparatus, the TFIID complex binds to the TATA promoter via its TATA-binding protein (TBP) (Figure 13–15). About 20 base pairs of DNA are involved in binding the TBP, and the other subunits then bind to the growing complex. TFIID responds to contact with

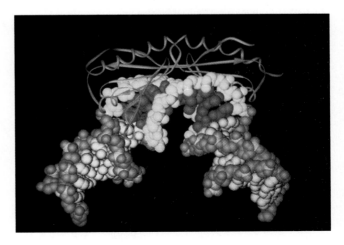

■ **Figure 13–15** Molecular complex formed between the TATA box-binding protein (green) and the 8-bp TATA box (red), itself part of DNA (yellow and blue).

activator proteins by conformational changes that expedite the binding of other transcription factors (e.g., TFIIA, TFIIB) and RNA polymerase.

Unlike the situation in prokaryotes, where polymerase binds directly to the DNA promoter region, eukaryotic polymerases bind to transcription factor proteins that are in turn bound to the DNA promoter region (Figure 13–16). At this point, transcription of the DNA downstream may ensue at a minimal *basal* level. The final stage involves the achievement of the *induced* state, where transcription is stimulated above the basal level. This state is not yet well defined, but involves other areas of the promoter region, enhancers, and numerous transcription factors, which control the assembly of the transcription complex and the rate at which RNA polymerase initiates transcription. Factors bound to enhancers at a distance from the site are thought to interact with the transcription complex, looping out the DNA that separates the enhancer from the complex.

Genomic Alterations and Gene Expression: DNA Methylation

Alteration of chromatin conformation is one of several ways in which gene expression can be regulated. One type of change in chromatin that plays a role in gene regulation involves adding or removing methyl groups to the bases in DNA. The DNA of most eukaryotic organisms is modified after replication by the enzyme-mediated addition of methyl groups to bases and sugars. **DNA methylation** most often involves cytosine. Approximately 5 percent of the cytosine residues are methylated in the genome of any given eukaryotic species.

The ability of base methylation to alter gene expression is known from studies on the *lac* operon in *E. coli*. Methylation of DNA in the operator region, even at a single cytosine residue, can cause a marked change in the affinity of the repressor for the operator. Methylation of cytosine occurs at the 5′ position (Figure 13–17), causing the methyl group to protrude into the major

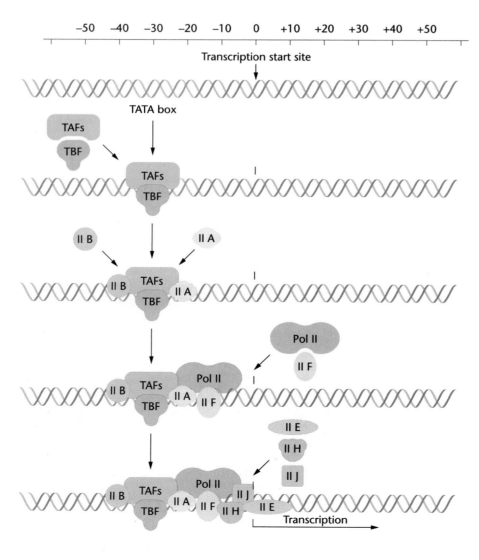

■ Figure 13–16 The assembly of transcription factors required for the initiation of transcription by RNA polymerase II, as described in the text. RNA polymerase II (pol II) binds to transcription factors through protein–protein interactions.

Comparison of cytosine that is methylated at the number 5 carbon with 5-azacytosine, where a nitrogen atom is substituted for the number 5 carbon atom. As a result, azacytosine cannot be methylated.

groove of the DNA helix, where it can alter the binding of proteins to the DNA.

Methylation occurs most often in the cytosine of CG doublets in DNA, usually in both strands:

$$5'\text{-}{}^{m}CpG\text{---}3'$$

$$3'\text{---}GpC^{m}\text{-}5'$$

Analysis of the methylation of a given gene in different tissues shows that, in general, if a gene is expressed, it is not methylated, or has a low level of methylation.

Evidence for the role of methylation as a factor in the regulation of eukaryotic gene expression is somewhat indirect and is based on a number of observations. As indicated above, an inverse relationship exists between the degree of methylation and the degree of expression. That is, low amounts of methylation are associated with high levels of gene expression, and high levels of methylation are associated with low levels of gene expression. In mammalian females the inactivated X chromosome, which is almost totally inactive in gene expression, has a higher level of methylation than does the active X chromosome. Within the inactive X, those regions that escape inactivation have much lower levels of methylation than those seen in adjacent, inactive regions.

Second, methylation patterns are tissue-specific and, once established, are heritable for all cells of that tissue. Perhaps the strongest evidence for the role of methylation in gene expression comes from studies using base analogs. The nucleotide 5'-azacytidine is incorporated into DNA in place of cytidine and cannot be methylated (Figure 13–17), causing undermethylation of sites where it is incorporated. Incorporation of 5'-azacytidine causes changes in the pattern of gene expression and can stimulate expression of alleles on inactivated X chromosomes.

As stated above, the available evidence indicates that the absence of methyl groups in DNA is related to increases in gene expression. Methylation cannot, however, be regarded as a universal mechanism for gene regulation, because methylation is not a general phenomenon in eukaryotes. In *Drosophila*, for example, there is no methylation of DNA. Thus, methylation may represent only one of a number of ways in which gene expression can be regulated by genomic changes.

Gene Regulation by Steroid Hormones

We end our consideration of eukaryotic gene regulation at the level of transcription with a short discussion of how **steroid hormones** affect their target cells. Our knowledge of this system provides a nice recap of the various aspects of eukaryotic regulation discussed thus far. Steroid hormones are used to regulate growth and development and to maintain homeostasis. The major sex hormones are all steroids, as is vitamin D, and the homeostatic adrenal hormones that regulate glucose metabolism and mineral utilization.

The general scheme of hormone action is illustrated in Figure 13–18. Hormones enter the cell by passing through the plasma membrane and binding to a specific **hormone receptor protein** in the cytoplasm. The receptor–hormone complex is translocated to the nucleus and activates transcription of one or more specific genes.

Hormone receptors and the DNA sequences to which they bind have been studied intensively over the last decade. All receptors analyzed to date have three functional domains: a variable N-terminus domain, unique to each receptor; a short, highly conserved central domain that binds to DNA; and a C-terminus domain that binds to the hormone. The DNA-binding central domain contains two zinc fingers. The zinc fingers in hormone receptors bind to specific DNA sequences known as **hormone-responsive elements (HREs)**.

In general, HREs share some characteristics with promoters and enhancers. They are composed of short consensus sequences that are related but not always identical. HREs are often located several hundred bases upstream from the transcription start site and may be present in multiple copies. Often, HREs are present within promoter or enhancer sequences.

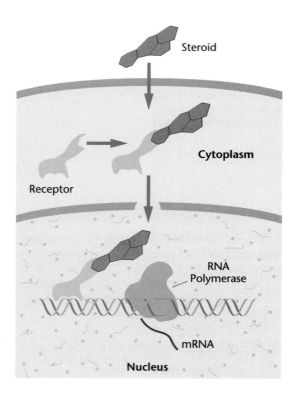

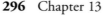 Figure 13–18 Stages of steroid hormone effects on gene expression. Steroids in the circulating system pass through the plasma membranes of target cells and bind to cytoplasmic receptor proteins. The steroid–receptor complex moves to the nucleus and binds to DNA at hormone receptor elements, stimulating transcription of steroid-induced genes.

While binding of the receptor to the HRE is necessary for activation of a specific gene, it may not be sufficient on its own to cause activation. Binding of the receptor to the HRE may serve simply to facilitate the interaction of other transcription factors by altering the chromatin structure in the HRE and adjacent regions. Such an alteration may make other binding sites, including the promoter, available to bind transcription factors and RNA polymerase II, resulting in the initiation of transcription.

In some ways, then, steroid hormone regulation of gene transcription in eukaryotes is similar to the positive-control systems present found in some of the operons in prokaryotes. An external effector binds to a cytoplasmic receptor, changes the configuration of the receptor, and moves the receptor to the DNA, where it acts as a transcription factor. In other ways, however, the regulation of gene expression is quite different in that the DNA-binding sequence is often at a great distance from the regulated gene, and that alterations in chromatin structure mediate the action of the receptor.

Postranscriptional Regulation of Gene Expression

We end our discussion of eukaryotic gene regulation by considering the possibilities for regulation *after* transcription has occurred. These are referred to as **posttranscriptional modes of regulation**. Many potential levels exist. For example, eukaryotic nuclear RNA transcripts are chemically modified prior to translation: mRNA is altered by the addition of a 5′-cap and a poly-A tail at the 3′-end, noncoding introns are removed, and the remaining exons are spliced together. Each of these processing steps offers several possibilities for regulation. Additionally, the general stability of a given mRNA may vary, thus determining the length of time it is present to direct translation.

Alternative Processing Pathways for mRNA

One of the best documented posttranscriptional mechanisms involves alternative splicing of mRNAs. Alternative splicing can generate different forms of a protein, giving rise to a situation where expression of one gene can give rise to a family of related proteins. Figure 13–19 illustrates an example where the polypeptide products derived from a single type of pre-mRNA are distinct from one another. The initial bovine pre-mRNA transcript is processed into one or two **preprotachykinin mRNAs (PPT mRNAs)**. The precursor mRNA molecule potentially includes the genetic information specifying two neuropeptides called P and K. These two peptides are members of the family of sensory neurotransmitters referred to as **tachykinins**, and they are believed to play different physiological roles. While the P neuropeptide is restricted largely to tissues of the nervous system, the K neuropeptide is found more predominantly in the intestine and thyroid.

The RNA sequences for both neuropeptides are derived from the same gene. However, processing of the initial RNA transcript can occur in two different ways. In one case, exclusion of the K-exon during processing results in the α-PPT mRNA, which upon translation yields neuropeptide P but not K. Conversely, processing that includes both the P and K exons yields β-PPT mRNA, which upon translation results in the synthesis of both the P and K neuropeptides. Analysis of the relative levels of the two types of RNA has demonstrated striking differences between tissues. In nervous system tissues, α-PPT mRNA predominates by as much as a threefold factor, while β-PPT mRNA is the predominant type in the thyroid and intestine.

Given the existence of alternative splicing, how many different polypeptides can be derived from the same pre-mRNA? Work on alpha-tropomyosin has provided a partial answer to this question. In rats, the alpha-

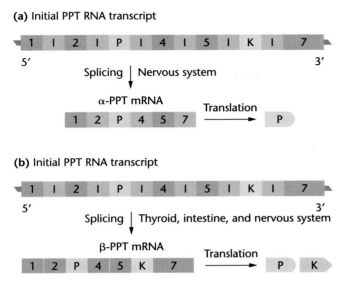

(a) Initial PPT RNA transcript

Splicing | Nervous system

α-PPT mRNA

Translation → P

(b) Initial PPT RNA transcript

Splicing | Thyroid, intestine, and nervous system

β-PPT mRNA

Translation → P K

■ Figure 13–19
Diagrammatic representation of the alternative splicing of the initial RNA transcript of the preprotachykinin gene (PPT). Introns are unshaded and are labeled I. Exons are either numbered or designated by the letter P or K. (a) When the K exon is excluded, α-PPT mRNA is produced, and only the P neuropeptide is synthesized. (b) Inclusion of P and K exons leads to β-PPT mRNA, which upon translation yields both the P and K tachykinin neuropeptides.

tropomyosin gene contains a total of 14 exons, six of which make up three pairs that are spliced alternatively. Only one member of each pair ends up in the finished mRNA, never both. Alternative splicing of this pre-mRNA results in ten different forms of alpha-tropomyosin, many of which are expressed in a tissue-specific manner. Another gene, for the muscle form of troponin T, a protein that regulates calcium needed for contraction, produces 64 known forms of the protein from a single pre-mRNA by alternative splicing.

Chapter Summary

1. A system of genetic regulation must exist if the complete genome is not to be continuously active in transcription throughout the life of every cell of all species. Highly refined mechanisms have evolved that regulate transcription, maintaining genetic efficiency.

2. Both inducible and repressible operons, illustrated by the *lac* and *trp* gene complexes, respectively, have been documented and studied in bacteria. They involve genes of a regulatory nature in addition to structural genes that code for the enzymes of the system. Both operons are under the negative control of a repressor molecule.

3. The catabolite-activating protein (CAP) facilitates the binding of RNA polymerase to the promoter in the *lac* and other operons. Catabolite repression, a mechanism that represses the operon when glucose is present, has evolved presumably because glucose can be utilized more efficiently than lactose.

4. An additional regulatory step, referred to as attenuation, has been studied in the *trp* operon, whereby transcription begins, but is interrupted under appropriate cellular conditions.

5. Repression of transcription in phage lambda has been shown to be due to the product of one of its own genes, *cI*. Another gene product, Cro, acts to inhibit the transcription of *cI* and promotes the lytic pathway when the activity of cII protein is reduced during the presence of glucose.

6. Transcription in eukaryotes is controlled by regulatory DNA sequences known as promoters and enhancers. Regulatory sequences, including the TATA box and the CCAAT box, are elements of promoters found near the transcriptional starting point. Enhancer elements, which appear to control the degree of transcription, can be located before, after, or within the gene expressed.

7. Transcription factors are proteins that bind to DNA recognition sequences within the promoters and enhancers and activate transcription through protein–protein interaction. Such factors display various types of motifs that are related to their potential binding to DNA.

8. Alteration of chromatin conformation is one way in which gene expression can be regulated, as illustrated by the degree of methylation of cytosine residues in DNA.

9. Gene expression in eukaryotes can be regulated by steroid hormones originating outside the cell. Such signals are transduced by proteins that shuttle from the cytoplasm to the nucleus and regulate patterns of transcription.

10. Several types of posttranscriptional control of gene expression are possible in eukaryotes. One such mechanism is alternate processing of a single class of pre-mRNAs to generate different mRNA species.

Key Terms

INSIGHTS and SOLUTIONS

1. A theoretical operon (*theo*) in *E. coli* contains several structural genes encoding enzymes that are involved sequentially in the biosynthesis of an amino acid. Unlike the *lac* operon, where the repressor gene is separate from the operon, the gene encoding the regulator molecule is contained within the *theo* operon. When the end product (the amino acid) is present, it combines with the regulator molecule, and this complex binds to the operator, repressing the operon. In the absence of the amino acid, the regulatory molecule fails to bind to the operator, and transcription proceeds.

Characterize this operon; then consider the following mutations as well as the situation in which the wild-type gene is present along with the mutant gene in partially diploid cells (F′). In each case, will the operon be active or inactive in transcription, assuming that the mutation affects the regulation of the *theo* operon? Compare each response to the equivalent situation in the *lac* operon.

  (a) Mutation in the operator gene

  (b) Mutation in the promoter region

  (c) Mutation in the regulator gene

Solution: The operon is under negative control and is repressible. The regulatory molecule, when bound to the amino acid, binds to the operator region and inhibits gene expression.

(a) As in the *lac* operon, a mutation in the *theo* operator gene inhibits binding with the repressor complex, and transcription occurs constitutively. The presence of an F′ plasmid bearing the wild-type allele would have no effect.

(b) A mutation in the *theo* promoter region would no doubt inhibit binding to RNA polymerase and therefore inhibit transcription. This would also happen in the lac operon. A wild-type allele present in an F′ plasmid would have no effect.

(c) A mutation in the *theo* regulator gene, as in the *lac* system, may inhibit either its binding to the repressor or its binding to the operator gene. In both cases, transcription will be constitutive because the *theo* system is repressible. Both cases result in the failure of the regulator to bind to the operator, allowing transcription to proceed. In the *lac* system, failure to bind the corepressor lactose would permanently repress the system. The addition of a wild-type allele would restore repressibility, provided that this gene was transcribed constitutively.

Problems and Discussion Questions

1. Contrast the need for the enzymes involved in the metabolism of lactose and tryptophan in bacteria in the presence and absence of lactose and tryptophan, respectively.
2. Contrast positive and negative control systems.
3. Contrast the role of the repressor in an inducible system and in a repressible system.
4. For the following *lac* genotypes, predict whether the structural genes *ZYA* are constitutive, permanently repressed, or inducible in the presence of lactose.

Genotype	Constitutive	Repressed	Inducible
$I^+O^+Z^+$			X
$I^-O^+Z^+$			
$I^+O^cZ^+$			
$I^-O^+Z^+/F'I^+$			
$I^+O^cZ^+/F'O^+$			
$I^sO^+Z^+$			
$I^sO^+Z^+/F'I^+$			

5. For the genotypes and condition (lactose present or absent) in the table below, predict whether functional enzymes are made, nonfunctional enzymes are made, or no enzymes are made.
6. In a theoretical operon, genes *a, b, c,* and *d* represent the repressor gene, the promoter sequence, the operator gene, and the structural gene, but not necessarily in that order. This operon is concerned with the metabolism of a theoretical molecule (tm). From the data given immediately to the right, first decide if the operon is inducible or repressible. Then assign *a, b, c,* and *d* to the four parts of the operon (AE = active enzyme, IE = inactive enzyme, NE = no enzyme).

Genotype	tm Present	tm Absent
$A^+B^+C^+D^+$	AE	NE
$A^-B^+C^+D^+$	AE	AE
$A^+B^-C^+D^+$	NE	NE
$A^+B^+C^-D^+$	IE	NE
$A^+B^+C^+D^-$	AE	AE
$A^-B^+C^+D^+/F'A^+B^+C^+D^+$	AE	AE
$A^+B^-C^+D^+/F'A^+B^+C^+D^+$	AE	NE
$A^+B^+C^-D^+/F'A^+B^+C^+D^+$	AE + IE	NE
$A^+B^+C^+D^-/F'A^+B^+C^+D^+$	AE	NE

7. If the *cI* gene of phage λ contained a mutation with effects similar to the *lacI⁻* mutation in *E. coli*, what would be the result?
8. Why is gene regulation assumed to be more complex in a multicellular eukaryote than in a prokaryote? Why is the study of this phenomenon in eukaryotes more difficult?
9. List and define the levels of gene regulation discussed in this chapter.
10. Distinguish between the regulatory elements referred to as promoters and enhancers.
11. Describe the role of transcription factors in the regulation of gene expression.
12. How are genes regulated by steroid hormones?
13. Contrast the modification of chromatin structure and posttranscriptional modes of gene regulation in eukaryotes.
14. Contrast and compare gene regulation in prokaryotes and eukaryotes.

Genotype	Condition	Functional Enzyme Made	Nonfunctional Enzyme Made	No Enzyme Made
$I^+O^+Z^+$	No lactose	—	—	X
$I^+O^cZ^+$	Lactose			
$I^-O^+Z^-$	No lactose			
$I^-O^+Z^-$	Lactose			
$I^-O^+Z^+/F'I^+$	No lactose			
$I^+O^cZ^+/F'O^+$	Lactose			
$I^+O^+Z^-/F'I^+O^+Z^+$	Lactose			
$I^-O^+Z^-/F'I^+O^+Z^+$	No lactose			
$I^sO^+Z^+/F'O^+$	No lactose			
$I^+O^cZ^+/F'O^+Z^+$	Lactose			

Selected Readings

Beato, M. 1989. Gene regulation by steroid hormones. *Cell* 56:335–44.

Berget, S. M. 1995. Exon recognition in vertebrate splicing. *J. Biol. Chem.* 270:2411–14.

Bernstein, B. E., Hoffman, R. C., and Klevit, R. F. 1994. Sequence-specific DNA recognition by Cys2, His2 zinc fingers. *Ann. N.Y. Acad. Sci.* 726:92–102.

Blumenthal, T. 1995. Trans-splicing and polycistronic transcription in *Caenorhabditis elegans. Trends Genet.* 11:132–36.

Busch, S. J., and Sassone-Corsi, P. 1990. Dimers, leucine zippers and DNA-binding domains. *Trends Genet.* 6:36–40.

Dynan, W. S. 1988. Modularity in promoters and enhancers. *Cell* 58:1–4.

Gilbert, W., and Müller-Hill, B. 1966. Isolation of the *lac* repressor. *Proc. Natl. Acad. Sci. USA* 56:1891–98.

———. 1967. The lac operator in DNA. *Proc. Natl. Acad. Sci. USA* 58:2415–21.

Gilbert, W., and Ptashne, M. 1970. Genetic repressors, *Sci. Am.* (June) 222:36–44.

Hayes, J. J., and Wolffe, A. P. 1992. The interaction of transcription factors with nucleosomal DNA. *BioEssays* 14:597–603.

Hershey, A. D., ed. 1971. *The bacteriophage lambda.* Cold Spring Harbor, NY: Cold Spring Harbor Laboratory.

Jacob, F., and Monod, J. 1961. Genetic regulatory mechanisms in the synthesis of proteins. *J. Mol. Biol.* 3:318–56.

Jacobson, R. H., and Tjian, R. 1996. Transcription factor IIA: A structure with multiple functions. *Science* 272:830–36.

Johnson, A. D., et al. 1981. λ repressor and *cro*-components of an efficient molecular switch. *Nature* 294:217–23.

Kakidani, H., and Ptashne, M. 1988. *GAL4* activates gene expression in mammalian cells. *Cell* 52:161–67.

Lewis, M., et al. 1996. Crystal structure of the lactose operon repressor and its complexes with DNA and inducer. *Science* 271:1247–54.

Littlewood, T. D., and Evans, G. I. 1994. Transcription factors 2: Helix-loop-helix. *Protein Profile* 1:639–709.

Lopez, H. J. 1995. Developmental role of transcription factor isoforms generated by alternate splicing. *Dev. Bio.* 172:396–411.

Maniatis, T., Goodbourn, S., and Fischer, J. A. 1987. Regulation of inducible and tissue-specific expression. *Science* 236:1237–45.

———, and Ptashne, M. 1976. A DNA operator-repressor system. *Sci. Am.* (Jan.) 234:64–76.

McCarthy, J. E., and Brimacombe, R. 1994. Prokaryotic translation: The interactive pathway leading to initiation. *Trends Genet.* 10:402–7.

McKnight, S. L. 1995. Transcription revisited: A commentary on the 1995 Cold Spring Harbor Meeting, Mechanisms of Eukaryotic Transcription. *Genes Dev.* 10:367–381.

Meehan, R., et. al. 1992. Transcriptional repression by methylation of CpG. *J. Cell Sci.* Suppl. 16:9–14.

Miller, J. H., and Reznikoff, W. S. 1978. *The operon.* Cold Spring Harbor, NY: Cold Spring Harbor Laboratory.

Mitchell, P. J., and Tjian, R. 1989. Transcriptional regulation in mammalian cells by sequence-specific DNA binding proteins. *Science* 245:371–78.

Müller-Hill, B. 1996. *The lac operon: A short history of a genetic paradigm.* Hawthorne, NY: Walter de Gruyter.

Pieler, T., and Bellefroid, E. 1994. Perspectives on zinc finger protein function and evolution—an update. *Mol. Biol. Rep.* 20:1–8.

Ptashne, M. 1986. *A genetic switch.* Palo Alto, CA: Blackwell Scientific.

———, and Gann, A. A. F. 1990. Activators and targets. *Nature* 346:329–31.

Ptashne, M., Johnson, A. D., and Pabo, C. O. 1982. A genetic switch in a bacterial virus. *Sci. Am.* (Nov.) 247:128–40.

Stringer, K. F., Ingles, C. J., and Greenblatt, J. 1990. Direct and selective binding of an acidic transcriptional activation domain to the TATA-box factor TFIID. *Nature* 345:783–86.

Struhl, K. 1993. Yeast transcription factors. *Curr. Opin. Cell Biol.* 5:513–20.

Tate, P., and Bird, A. 1993. Effects of DNA methylation on DNA-binding proteins and gene expression. *Curr. Opin. Genet. Dev.* 3:226–31.

Yankofsky, C. 1981. Attenuation in the control of expression of bacterial operons. *Nature* 289:751–58.

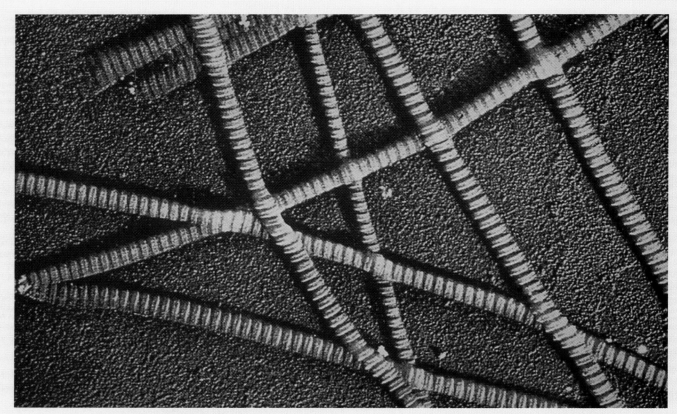

An electron micrograph of collagen fibers.

CHAPTER OUTLINE

CHAPTER
14

Proteins:
The End Product of Genes

Chapter Concepts

The end products of most gene expression are polypeptide chains. They achieve a three-dimensional conformation based on their primary amino acid sequences, often interacting with other such chains to create functional protein molecules. The function of any protein is closely tied to its structure, which can be disrupted by mutation, leading to a distinctive phenotypic effect.

In Chapter 12 we established that there is a genetic code that stores information in the form of triplet nucleotides in DNA, and that this information can be expressed through the orderly processes of transcription and translation. The final product of gene expression, in almost all instances, is a polypeptide chain consisting of a linear series of amino acids, whose sequence has been prescribed by the genetic code. In this chapter, we will review the evidence that confirmed that proteins are the end products of genes. Then we will discuss briefly the various levels of protein structure, diversity, and function. This information provides an important foundation in our understanding of how mutations, which arise in DNA, can result in the variety of phenotypic effects observed in organisms.

Garrod and Bateson:
Inborn Errors of Metabolism

The first insight into the role of proteins in genetic processes was provided by observations made by Sir Archibald Garrod and William Bateson early in this century. Garrod was born into an English family of medical scientists. His father was a physician with a strong interest in the chemical basis of rheumatoid arthritis, and his eldest brother was a leading zoologist in London. Thus, it is not surprising that, as a practicing physician, Garrod became interested in several human disorders that seemed to be inherited. Although he studied albinism

and cystinuria, we will describe in some detail his most significant investigation, that of the disorder **alkaptonuria**. Individuals afflicted with this disorder cannot metabolize the alkapton 2,5-dihydroxyphenylacetic acid, also known as **homogentisic acid**. As a result, an important metabolic pathway (Figure 14–1) is blocked. Homogentisic acid accumulates in cells and tissues and is excreted in the urine. The molecule's oxidation products are black and thus are easily detectable in the diapers of newborns and in urine samples. The products tend to accumulate in cartilaginous areas, causing a darkening of the ears and nose. In joints, this deposition leads to a benign arthritic condition. This rare disease is not serious, but it persists throughout an individual's life.

Garrod studied alkaptonuria by increasing dietary protein or adding to the diet the amino acids phenylalanine or tyrosine, which are chemically related to homogentisic acid. Under such conditions, homogentisic acid levels increase in the urine of alkaptonurics but not in unaffected individuals. Garrod concluded that normal individuals can break down, or catabolize, this alkapton but that afflicted individuals cannot. By studying the patterns of inheritance of the disorder, Garrod further concluded that alkaptonuria is inherited as a simple recessive trait.

On the basis of these conclusions, Garrod hypothesized that hereditary information controls chemical reactions in the body and that the inherited disorders he studied are the result of alternative modes of metabolism. While *genes* and *enzymes* were not familiar terms in Garrod's time, he used the corresponding concepts

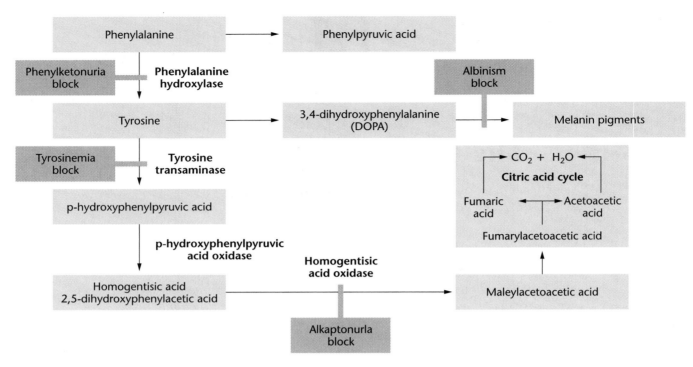

Figure 14–1 Metabolic pathway involving phenylalanine and tyrosine. Metabolic blocks resulting from mutations lead to the disorders phenylketonuria, alkaptonuria, albinism, and tyrosinemia.

of *unit factors* and *ferments*. Garrod published his initial observations in 1902.

Only a few geneticists, including Bateson, were familiar with or referred to Garrod's work. His ideas fit nicely with Bateson's belief that inherited conditions are caused by the lack of some critical substance. In 1909, Bateson published *Mendel's Principles of Heredity*, in which he linked ferments with heredity. However, for almost 30 years, most geneticists failed to see the relationship between genes and enzymes. Garrod and Bateson, like Mendel, were ahead of their time.

Phenylketonuria

Described first in 1934, **phenylketonuria (PKU)** may result in mental retardation and is inherited as an autosomal recessive disease. Afflicted individuals have still another step blocked in the metabolic pathway just discussed. They are unable to convert the amino acid phenylalanine to the amino acid tyrosine (see Figure 14–1). Phenylalanine and tyrosine differ by only a single hydroxl group (—OH), present in tyrosine, but absent in phenylalanine. The reaction is catalyzed by the enzyme **phenylalanine hydroxylase**, which is inactive in affected individuals and active at about a 30 percent level in heterozygotes. The enzyme functions in the liver. While the normal blood level of phenylalanine is about 1 mg/100 ml, phenylketonurics show a level as high as 50 mg/100 ml.

As phenylalanine accumulates, it may be converted to phenylpyruvic acid and, subsequently, other derivatives.

These are less efficiently resorbed by the kidney and tend to spill into the urine more quickly than phenylalanine. Both phenylalanine and derivatives enter the cerebrospinal fluid, resulting in elevated levels in the brain. The presence of these substances during early development is thought to cause mental retardation.

Retardation can be prevented by PKU screening of newborns. When the condition is detected in the analysis of an infant's blood, a strict dietary regimen is instituted. A low-phenylalanine diet can reduce such by-products as phenylpyruvic acid, and abnormalities characterizing the disease can be diminished. Screening of newborns is done routinely in almost every state in this country. Phenylketonuria occurs in approximately 1 in 11,000 births.

Knowledge of inherited metabolic disorders such as alkaptonuria and phenylketonuria has caused a revolution in medical thinking and practice. No longer is human disease attributed solely to the action of invading microorganisms, viruses, or parasites. We know now that literally thousands of abnormal physiological conditions are caused by errors in metabolism that are the result of mutant genes. These human biochemical disorders are far-ranging and include all classes of organic biomolecules.

The One-Gene:One-Enzyme Hypothesis

In two separate investigations beginning in 1933, George Beadle provided the first convincing experimental evidence that genes are directly responsible for the synthesis of enzymes. The first investigation, conducted in collaboration with Boris Ephrussi, involved *Drosophila* eye pigments. Encouraged by these findings, Beadle then joined with Edward Tatum to investigate nutritional mutations in the pink bread mold *Neurospora crassa*. The latter investigation led to the **one-gene:one-enzyme hypothesis**.

Beadle and Tatum: Neurospora Mutants

In the early 1940s, Beadle and Tatum chose to work with the organism *Neurospora crassa* because much was known about its biochemistry, and mutations could be induced and isolated with relative ease. By inducing mutations, they produced strains that had genetic blocks of reactions essential to the growth of the organism.

Beadle and Tatum knew that *Neurospora* can manufacture nearly everything necessary for normal development. For example, using rudimentary carbon and nitrogen sources, this organism can synthesize 9 water-soluble vitamins, 20 amino acids, numerous carotenoid pigments, and purines and pyrimidines. Beadle and Tatum irradiated asexual conidia (spores) with X-rays to increase the frequency of mutations and allowed them to progress through a sexual cycle, forming haploid ascospores. These were grown on "complete" medium containing all necessary growth factors (vitamins, amino acids, etc.). Under such growth conditions, a mutant strain would be able to grow by virtue of supplements present in the enriched complete medium. All cultures were then transferred to minimal medium. If growth occurred on minimal medium, the organisms were able to synthesize all necessary growth factors themselves, and it was concluded that the culture did not contain a mutation. If no growth occurred, then the strain being tested contained a nutritional mutation, and the only task remaining was to determine its type. Both cases are illustrated in Figure 14–2(a).

Many thousands of individual spores derived by this procedure were isolated and grown on complete medium. In subsequent tests on minimal medium, many cultures failed to grow, indicating that a nutritional mutation had been induced. To identify the mutant type, the mutant strains were tested on a series of different minimal media, each containing a group of vitamins, amino acids, purines, or pyrimidines until a supplement that permitted growth was found [Figure 14–2(b)]. Once the general category of supplements was determined, individual components were tested [Figure 14–2(c)]. Beadle and Tatum reasoned that the specific supplement that restores growth is the molecule that the mutant strain could *not* synthesize. In the case of our illustration in Figure 14–2, the amino acid tyrosine restored growth and represents the molecule whose synthesis was interrupted.

The first mutant strain isolated by Beadle and Tatum required vitamin B-6 (pyridoxin) in the medium, and the second one required vitamin B-1 (thiamine). Using the same procedure, they eventually isolated and studied hundreds of mutants.

The findings derived from testing over 80,000 spores convinced Beadle and Tatum that genetics and biochemical events have much in common. It seemed likely that each nutritional mutation caused the loss of the enzymatic activity that facilitates an essential reaction in wild-type organisms. It also appeared that a mutation could be found for nearly any enzymatically controlled reaction. Beadle and Tatum had thus provided sound experimental evidence that **one gene specifies one enzyme**, an idea alluded to over 30 years earlier by Garrod and Bateson. With modifications, this concept was to become a major principle of genetics.

Genes and Enzymes: Analysis of Biochemical Pathways

The one-gene:one-enzyme concept and its attendant methods have been used over the years to work out many details of metabolism in *Neurospora, Escherichia coli*, and a number of other microorganisms. One of the first metabolic pathways to be investigated in detail was that leading to the synthesis of the amino acid arginine in *Neurospora*. By studying seven mutant strains, each requiring arginine for growth (*arg⁻*), Adrian Srb and Norman Horowitz were able to ascertain a partial biochemical pathway leading to the synthesis of this molecule. The rationale followed in their work illustrates how genetic analysis can be used to establish biochemical information.

Srb and Horowitz tested each mutant strain's ability to reestablish growth if either **citrulline** or **ornithine**, two compounds with close chemical similarity to arginine, was used as a supplement to minimal medium. If either was able to substitute for arginine, Srb and Horowitz reasoned that it must be involved in the biosynthetic pathway of arginine. In fact, they found that both molecules could be substituted in one or more strains.

Of the seven mutant strains, four of them (*arg 1–4*) grew if supplied with either citrulline, ornithine, or arginine. Two of them (*arg 5–6*) grew if supplied with citrulline or arginine. One strain (*arg 7*) would grow only if arginine was supplied. Neither citrulline nor ornithine could substitute for it. From these experimental observations, the following pathway was deduced:

$$\text{Precursor} \xrightarrow[\text{Enzyme A}]{arg\ 1\text{--}4} \text{Ornithine} \xrightarrow[\text{Enzyme B}]{arg\ 5\text{--}6} \text{Citrulline} \xrightarrow[\text{Enzyme C}]{arg\ 7} \text{Arginine}$$

The reasoning supporting these conclusions is based on the following logic. If mutants *arg 4–7* can grow regardless of which of the three molecules is supplied as a supplement to minimal medium, the mutations preventing growth must cause a metabolic block that occurs *prior* to the involvement of ornithine, citrulline, or arginine in the pathway. When any one of these three molecules is

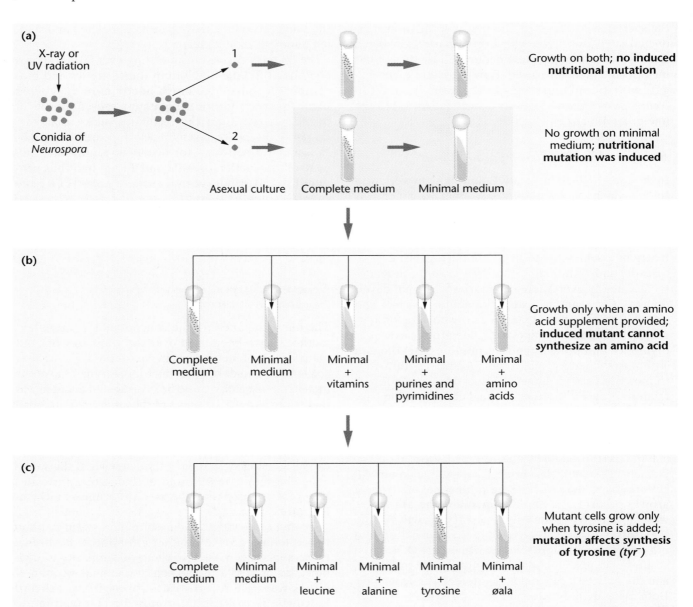

■ Figure 14–2 Induction, isolation, and characterization of a nutritional auxotrophic mutation in *Neurospora*. In (a), conidium 1 is not affected, but conidium 2 contains a mutation. In (b) and (c), the precise nature of the mutation is determined to involve the biosynthesis of tyrosine.

added, its presence bypasses the block. As a result, it can be concluded that both citrulline and ornithine are involved in the biosynthesis of arginine. However, the sequence of their participation in the pathway cannot be determined on the basis of these data.

On the other hand, the *arg 2* and *3* mutations grow if supplied citrulline but not if they are supplied only with ornithine. Therefore, ornithine must occur in the pathway *prior to* the block. Its presence will not overcome the block. Citrulline, however, does overcome the block, so it must be involved beyond the point of block-

age. Therefore, the conversion of ornithine to citrulline represents the correct sequence in the pathway.

Finally, it can be concluded that *arg 1* represents a mutation that prevents the conversion of citrulline to arginine. Neither ornithine nor citrulline can overcome the metabolic block because both participate earlier in the pathway.

Taken together, these reasons support the sequence of biosynthesis outlined here. Ever since Srb and Horowitz's work in 1944, the detailed pathway has been worked out and the enzymes controlling each step

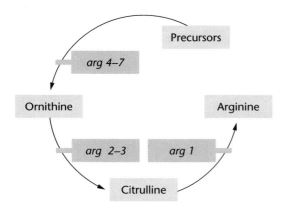

■ Figure 14–3 Abbreviated pathway resulting in the biosynthesis of arginine in *Neurospora*.

characterized. The metabolic pathway is shown in Figure 14–3.

The concept of one-gene:one-enzyme developed in the early 1940s was not accepted immediately by all geneticists. This is not surprising, because it was not yet clear how mutant enzymes could cause variation in many phenotypic traits. For example, Drosophila mutants demonstrated altered eye size, wing shape, wing vein pattern, and so on. Plants exhibited mutant varieties of seed texture, height, and fruit size. How an inactive, mutant enzyme could result in such phenotypes was puzzling to many geneticists. Another reason for their reluctance to accept this concept was the paucity of information then available in molecular genetics. It was not until 1944 that Avery, MacLeod, and McCarty showed DNA to be the transforming factor, and not until the early 1950s that most geneticists believed that DNA serves as the genetic material (see Chapter 9). However, by that time, the evidence was overwhelming in support of the concept of enzymes as gene products, which was accepted as valid. Nevertheless, the question of how DNA specifies the structure of enzymes remained unanswered at that time.

One-Gene:One-Protein/ One-Gene:One-Polypeptide

Two factors soon modified the one-gene:one-enzyme hypothesis. First, while nearly all enzymes are proteins, not all proteins are enzymes. As the study of biochemical genetics proceeded, it became clear that all proteins are specified by the information stored in genes, leading to a revision of the one-gene:one-enzyme idea to the more inclusive description, **one-gene:one-protein**. Second, proteins were often shown to have a subunit structure consisting of two or more **polypeptide chains**. This is the basis of the quaternary structure of proteins, which we will discuss later in this chapter. Because each distinct polypeptide chain is encoded by a separate gene, a more modern statement of Beadle and Tatum's basic principle is **one-gene:one-polypeptide chain**. The need to modify the original hypothesis became apparent during the analysis of hemoglobin structure in individuals afflicted with sickle-cell anemia.

Sickle-Cell Anemia

The first direct evidence that genes specify proteins other than enzymes came from work on mutant hemoglobin molecules derived from humans afflicted with the disorder **sickle-cell anemia**. Affected individuals contain erythrocytes which, under low oxygen tension, become elongated and curved because of the polymerization of hemoglobin. The "sickle" shape of these erythrocytes is in contrast to the normal biconcave disk shape of erythrocytes (Figure 14–4). Those afflicted with the disease suffer attacks when red blood cells aggregate on the venous side of capillary systems, where oxygen tension is very low. As a result, a variety of tissues may be deprived of oxygen and suffer severe damage. When this occurs, an individual is said to experience a sickle-cell crisis. If untreated, a crisis may be fatal. The kidneys, muscles, joints, brain, gastrointestinal tract, and lungs may be affected.

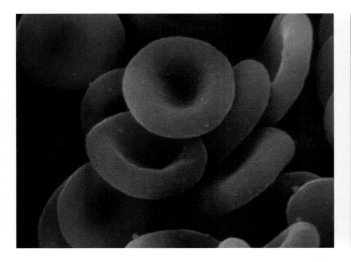

■ Figure 14–4   A comparison of erythrocytes derived from HbA (left) and HbS (right).

In addition to suffering crises, afflicted individuals are also anemic, because their erythrocytes are destroyed more rapidly than normal red blood cells. Compensatory physiological mechanisms include increased red cell production by bone marrow and accentuated heart action. These mechanisms lead to abnormal bone size and shape as well as dilation of the heart.

In 1949, James Neel and E. A. Beet demonstrated that the disease is inherited as a Mendelian trait. Pedigree analysis revealed three genotypes and phenotypes controlled by a single pair of alleles, Hb^A and Hb^S. Normal and affected individuals result from the homozygous genotypes Hb^A/Hb^S and Hb^S/Hb^S, respectively. The red blood cells of the heterozygote Hb^A/Hb^S, which exhibits the **sickle-cell trait** but not the disease, undergo much less sickling because over half of their hemoglobin is normal. Although they are less affected than homozygotes, such persons are carriers of the disorder.

In the same year, Linus Pauling and his coworkers provided the first insight into the molecular basis of the disease. They showed that hemoglobins isolated from diseased and normal individuals differ in their rates of electrophoretic migration. In this technique, charged molecules migrate in an electric field. If the net charge of two molecules is different, the rate of migration will vary. On this basis, Pauling and his colleagues concluded that a chemical difference exists between the two types of hemoglobin. The two molecules are now designated **HbA** and **HbS**.

Figure 14–5(a) illustrates the migration pattern of hemoglobin derived from individuals with all three possible genotypes when subjected to **starch gel electrophoresis**. The gel provides the supporting medium for the molecules during migration. In this experiment, samples are placed at a point of origin between the cathode (−) and the anode (+), and an electric field is applied. The migration pattern reveals that all molecules move toward the anode, indicating a net negative charge. However, HbA migrates farther than HbS, suggesting that its net negative charge is greater. The electrophoretic pattern of hemoglobin derived from carriers reveals the presence of both HbA and HbS, confirming their heterozygous genotype.

Pauling's findings suggested two possibilities. It was known that **hemoglobin** consists of four nonproteina-

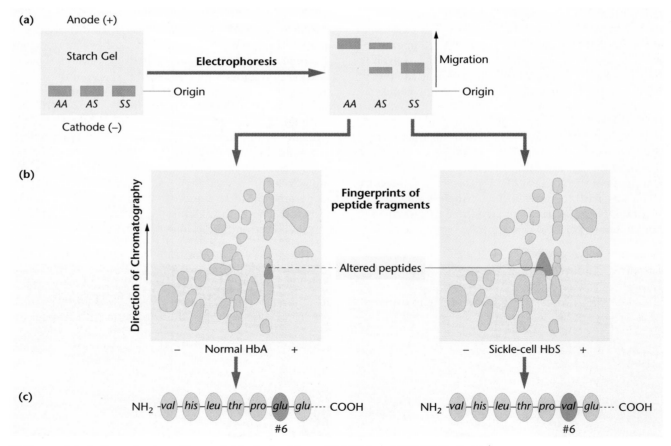

■ Figure 14–5 Investigation of hemoglobin derived from Hb^A/Hb^A and Hb^S/Hb^S individuals using electrophoresis, fingerprinting, and amino acid analysis. Hemoglobin from individuals with sickle-cell anemia: (a) migrates differently in an electrophoretic field; (b) shows an altered peptide in fingerprint analysis; and (c) shows an altered amino acid, valine, at the sixth position in the β chain. During electrophoresis, heterozygotes reveal both forms of hemoglobin.

ceous, iron-containing **heme groups** and a **globin portion** that contains four polypeptide chains. For example, human adult hemoglobin contains two identical alpha (α) chains of 141 amino acids and two identical beta (β) chains of 146 amino acids in its quaternary structure. The alteration in net charge in HbS could be due, theoretically, to a chemical change in either the heme or the globin portion of the molecule.

Work carried out between 1954 and 1957 by Vernon Ingram resolved this question. He demonstrated that the chemical change occurs in the primary structure of the globin portion of the hemoglobin molecule. Using the **fingerprinting technique**, Ingram showed that HbS differs in amino acid composition compared to HbA.

The fingerprinting technique involves enzymatic digestion of the protein into peptide fragments. The mixture is then placed on absorbent paper and exposed to an electric field, where migration occurs according to net charge. The paper is then turned at a right angle and placed in a solvent, where chromatographic action causes migration of the peptides in the second direction. The end result is a two-dimensional separation of the peptide fragments into a distinctive pattern of spots or "fingerprints." Ingram's work revealed that HbS and HbA differed by only a single peptide fragment [Figure 14–5(b)]. Further analysis then revealed a single amino acid change: Valine was substituted for glutamic acid at the sixth position of the β chain, accounting for the peptide difference [Figure 14–5(c)].

The significance of this discovery has been multifaceted. It clearly established that a single gene provides the genetic information for a single polypeptide chain. Studies of HbS also demonstrate that a mutation can affect the phenotype by directing a single amino acid substitution. Also, by providing the explanation for sickle-cell anemia, the concept of **molecular disease** was firmly established. Finally, this work led to a thorough study of human hemoglobins, which has provided valuable genetic insights.

In the United States, sickle-cell anemia is found almost exclusively in the black population. Extensive research is now underway to determine modes of treatment for those with the disease. It affects about one in every 625 black infants born in this country. Currently, about 50,000 to 75,000 individuals are afflicted. In about one of every 145 black married couples, both partners are heterozygous carriers, so each of their children has a 25 percent chance of having the disease.

Human Hemoglobins

Molecular analysis reveals that a variety of hemoglobin molecules are produced in humans. All are tetramers consisting of different combinations of seven distinct polypeptide chains, each encoded by a separate gene. In our discussion of sickle-cell anemia, we learned that HbA contains two **alpha (α)** and two **beta (β) chains.**

HbA represents about 98 percent of all hemoglobin found in an individual's erythrocytes after the age of 6 months. The remaining 2 percent consists of **HbA$_2$**, a minor adult component. This molecule contains two alpha and two **delta (δ) chains**. The delta chain is very similar to the beta chain, consisting of 146 amino acids, but varies somewhat in its amino acid sequence.

During embryonic and fetal development, quite a different set of hemoglobins is found. The earliest set to develop is called **Gower 1** and contains two **zeta (ζ) chains**, which are alphalike, and two **epsilon (ε) chains**, which are betalike. By 8 weeks of gestation, this embryonic form is gradually replaced by still another hemoglobin molecule with different chains. This molecule is called **HbF**, or **fetal hemoglobin**, and consists of two alpha chains and two **gamma (γ) chains**. These gamma chains are of nearly identical types and are designated $^G\gamma$ and $^A\gamma$. Both are betalike and differ from each other by only a single amino acid.

The nomenclature and sequence of appearance of the five tetramers described so far are summarized in Table 14.1. The genes coding for each of the seven chains have been mapped. Those coding for α and ζ are located on chromosome 16, while those coding for ε, $^G\gamma$, $^A\gamma$, δ, and β are located on chromosome 11. In each case, they are clustered together, constituting gene families.

Colinearity

Once it was established that genes specify the synthesis of polypeptide chains, the next logical question was how genetic information contained in the nucleotide sequence of a gene can be transferred to the amino acid sequence of a polypeptide chain. It seemed most likely that a **colinear relationship** would exist between the two molecules. That is, the order of nucleotides in the DNA of a gene would correlate directly with the order of amino acids in the corresponding polypeptide.

The experimental evidence in support of the concept of colinearity involves studies of the A subunit of the enzyme **tryptophan synthetase** in *E. coli*. Charles Yanofsky isolated many independent mutants that had lost activity of the enzyme. He was able to map these mutations and establish their location with respect to one another *within the gene*. Then he determined the location of the amino acid substitution within each mu-

TABLE 14.1   Chain compositions of human hemoglobins from conception to adulthood

Hemoglobin Type	Chain Composition
Embryonic-Gower 1	$\zeta_2\varepsilon_2$
Fetal-HbF	$\alpha_2{}^G\gamma_2$
	$\alpha_2{}^A\gamma_2$
Adult-HbA	$\alpha_2\beta_2$
Minor adult-HbA$_2$	$\alpha_2\delta_2$

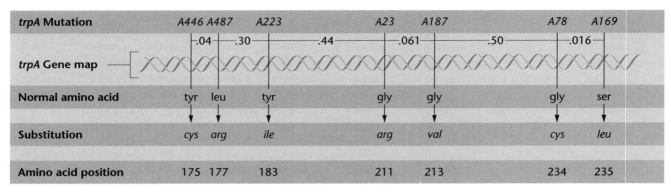

trpA Mutation	A446 A487	A223		A23	A187		A78	A169

■ Figure 14–6 Demonstration of colinearity between the genetic map of various *trp A* mutations in *E. coli* and the affected amino acids in the protein product. The numbers (.04, .30, etc.) shown between mutations represent linkage distances.

tant protein. When the two sets of data were compared, the colinear relationship was apparent. The location of each mutation in the *trp A* gene correlates with the position of the altered amino acid in the A polypeptide of tryptophan synthetase. This comparison is illustrated in Figure 14–6. Recall that we have already discussed the details of how information is transferred from DNA to protein (transcription and translation) in Chapter 12.

Protein Structure and Function

We turn now to a discussion of protein structure and function. How is it that these molecules play such a critical role in determining cellular activities? As we will see, the structure of proteins is intimately related to their functional diversity.

Protein Structure

First, we should differentiate between the terms **polypeptide** and **protein**. Both describe a molecule composed of amino acids. The molecules differ, however, in their state of assembly and functional capacity.

Polypeptides are the precursors of proteins. As assembled on the ribosome during translation, the molecule is called a *polypeptide*. When it is released from the ribosome following translation, the polypeptide folds up and assumes a higher order of structure. When this occurs, a three-dimensional conformation in space emerges. In many cases, several polypeptides interact to produce this conformation. Whether or not several polypeptides interact, the three-dimensional conformation is essential to the function of the molecule. When the final conformation is achieved, the function molecule is now appropriately called a *protein*.

The polypeptide chains of proteins, like nucleic acids, are linear nonbranched polymers. There are 20 amino acids that serve as the building blocks (subunits) of proteins. Each amino acid has a **carboxyl group**, an **amino group**, an **R (radical) group** (or side chain), and a **hydrogen atom** bonded covalently to a **central carbon**

atom. This central atom is often referred to as the α carbon atom. It is the R group that provides each amino acid with its chemical identity. Figure 14–7 illustrates the 20 different R groups, which show a variety of configurations and may be divided into four main classes: (1) **nonpolar (hydrophobic)**, (2) **polar (hydrophilic)**, (3) **negatively charged (acidic)**, and (4) **positively charged (basic)**. Because polypeptides are often long polymers, and because each position may be occupied by any one of 20 amino acids, each with unique chemical properties, an enormous variety of polypeptides is possible. For example, if an average polypeptide is composed of 200 amino acids (molecular weight about 20,000 daltons), 20^{200} different molecules, each with a unique sequence, can be created using the 20 building blocks!

Around 1900, the German chemist Emil Fischer determined the manner in which the amino acids are bonded together. He showed that the amino group of one amino acid can react with the carboxyl group of another amino acid during a dehydration reaction, releasing a molecule of H_2O. The resulting covalent bond is known as a **peptide bond** (Figure 14–8). Two amino acids linked together constitute a dipeptide, three a tripeptide, and so on. When more than ten amino acids are linked by peptide bonds, the chain is referred to as a polypeptide. Generally, no matter how long a polypeptide is, it will contain a free amino group at one end (the **N-terminus**) and a free carboxyl group at the other end (the **C-terminus**).

Four levels of **protein structure** are recognized: **primary (I°)**, **secondary (II°)**, **tertiary (III°)**, and **quaternary (IV°)**. The sequence of amino acids in the linear backbone of the polypeptide constitutes its primary structure. As we have seen in Chapter 12, this sequence

■ Figure 14–7 Chemical structures of the 20 amino acids found in living organisms, divided into four major categories. Biochemical abbreviations and designations (e.g., arg-A, trp-W, etc.) are shown in parentheses for each amino acid. The inset at the bottom right shows the common structural feature of all amino acids.

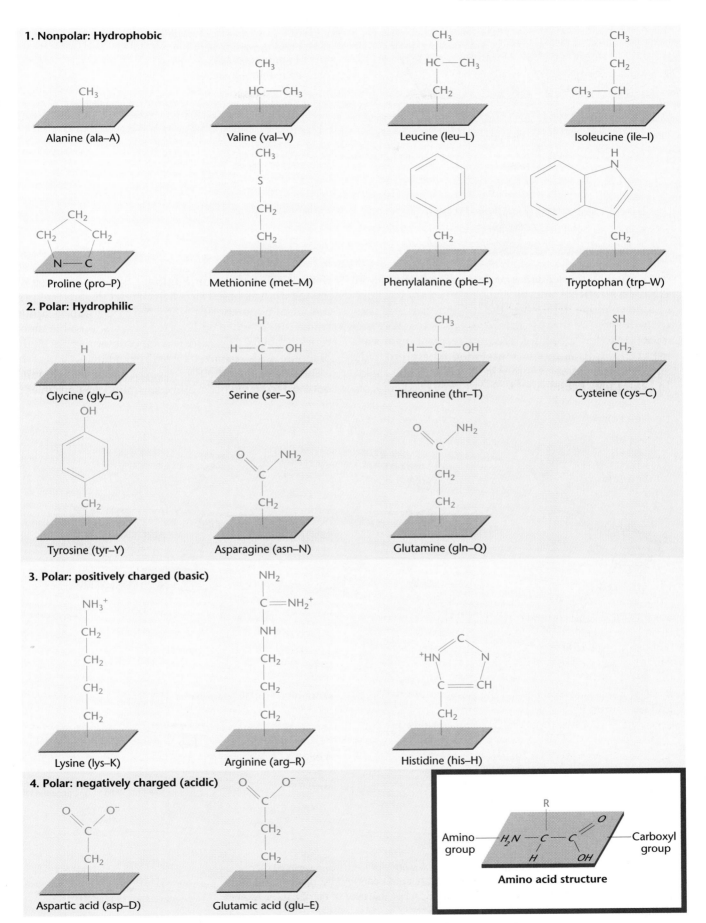

1. Nonpolar: Hydrophobic

Alanine (ala–A)

Valine (val–V)

Leucine (leu–L)

Isoleucine (ile–I)

Proline (pro–P)

Methionine (met–M)

Phenylalanine (phe–F)

Tryptophan (trp–W)

2. Polar: Hydrophilic

Glycine (gly–G)

Serine (ser–S)

Threonine (thr–T)

Cysteine (cys–C)

Tyrosine (tyr–Y)

Asparagine (asn–N)

Glutamine (gln–Q)

3. Polar: positively charged (basic)

Lysine (lys–K)

Arginine (arg–R)

Histidine (his–H)

4. Polar: negatively charged (acidic)

Aspartic acid (asp–D)

Glutamic acid (glu–E)

Amino group

Carboxyl group

Amino acid structure

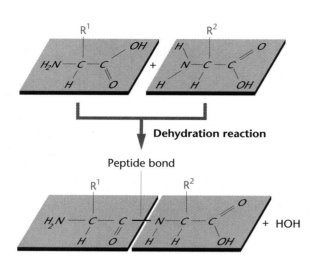

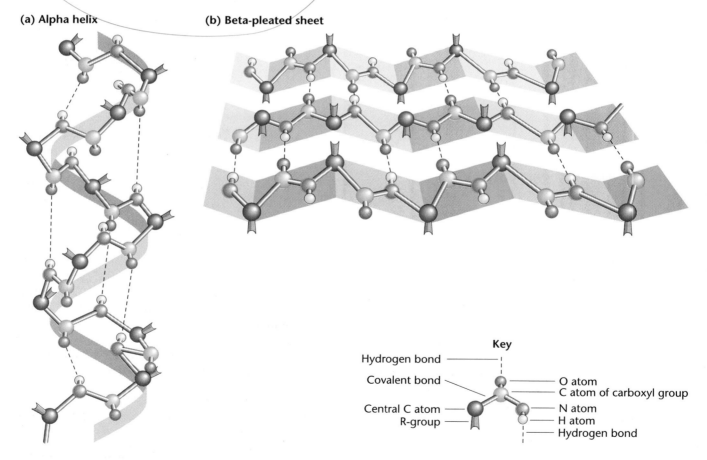

■ Figure 14–8 Peptide bond formation between two amino acids resulting from a dehydration reaction.

is specified by the sequence of deoxyribonucleotides in DNA via an mRNA intermediate. The primary structure of a polypeptide determines the specific characteristics of the higher orders of structure as a protein is formed.

The secondary structure refers to a regular or repeating configuration in space assumed by amino acids aligned closely to one another in the polypeptide chain. In 1951, Linus Pauling and Robert Corey predicted, on theoretical grounds, an **α (alpha) helix** as one type of secondary structure. The α-helix model [Figure 14–9(a)] has since been confirmed by X-ray crystallographic studies. It is rodlike and has the greatest possible theoretical stability. The helix is composed of a spiral chain of amino acids stabilized by hydrogen bonds.

The side chains (the R groups) of amino acids extend outward from the helix, and each amino acid occupies a

(a) Alpha helix

(b) Beta-pleated sheet

Key

Hydrogen bond ————
Covalent bond
Central C atom ————
R-group ————

O atom
C atom of carboxyl group
N atom
H atom
Hydrogen bond

■ Figure 14–9 (a) The right-handed α helix, which represents one form of secondary structure of a polypeptide chain. (b) the β-pleated-sheet configuration, an alternative form of secondary structure of polypeptide chains. For the sake of clarity, some atoms are not shown.

vertical distance of 1.5 Å in the helix. There are 3.6 amino acids per turn. While left-handed helices are theoretically possible, all natural proteins that demonstrate a helix are right-handed.

In 1951, Pauling and Corey proposed another secondary structure, the **β-pleated-sheet configuration**. In this model, a single polypeptide chain folds back on itself, or several chains run in either parallel or antiparallel fashion next to one another. Each such structure is stabilized by hydrogen bonds formed between atoms present on adjacent chains [Figure 14–9(b)]. A single zigzagging plane is formed in space with adjacent amino acids 3.5 Å apart.

As a general rule, most proteins demonstrate a mixture of α-helical and β-pleated-sheet structure. For example, globular proteins, most of which are round in shape and water-soluble, usually contain a core of β-pleated-sheet structure as well as many areas demonstrating α-helical structure. The more rigid structural proteins, many of which are water-insoluble, rely on more extensive β-pleated-sheet regions for their rigidity. For example, **fibroin**, a protein made by the silk moth, depends extensively on this form of secondary structure.

While the secondary structure describes the arrangement of amino acids within certain areas of a polypeptide chain, the tertiary structure of proteins defines the three-dimensional conformation of the entire chain in space. The molecule twists and turns and loops around itself in a very precise fashion, characteristic of the specific protein. Three aspects of this III° structure are most important in determining this conformation and in stabilizing the molecule:

1. Covalent disulfide bonds form between closely aligned cysteine residues to form the unique amino acid cystine.

2. Nearly all of the polar, hydrophilic R groups become located on the surface, where they interact with water.

3. The nonpolar, hydrophobic R groups are usually located on the inside of the molecule, where they interact with one another, avoiding interaction with water.

It is important to emphasize that the tertiary (three-dimensional) conformation achieved by any protein is the direct result of the primary (I°) structure of the polypeptide. The genetic code need only specify the sequence of amino acids in order to encode information leading to the functional tertiary structure of proteins. The three stabilizing factors listed above depend on the location of each amino acid relative to all others in the chain. As folding occurs, the most thermodynamically stable conformation possible results.

A model of the three-dimensional tertiary structure of the respiratory pigment **myoglobin** is shown in Figure 14–10. This III° level of organization is extremely important because the specific function of any protein is the direct result of its three-dimensional conformation.

The **quaternary level of organization** is characteristic of proteins composed of more than one polypeptide

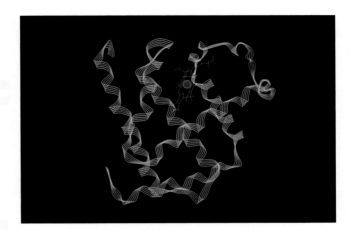

■ Figure 14–10 A ribbon drawing illustrating the tertiary (III°) level of protein structure in myoglobin, a respiratory pigment. The bound oxygen atom is shown in red.

chain. The IV° structure describes the conformation of the associated polypeptide chains in relation to one another. This type of protein is called **oligomeric**, and each chain is called a **protomer** or, less formally, a **subunit**. The individual protomers have native conformations that fit together in a specific complementary fashion. Hemoglobin, an oligomeric protein of four polypeptide chains, has been studied in great detail. Its IV° structure is shown in Figure 14–11. Many enzymes, including DNA and RNA polymerase, demonstrate IV° structure.

Chaperones and Protein Folding

Because of the relationship of the three-dimensional structure to protein function, it has been of great interest to confirm how polypeptide chains fold into their final conformation. For many years, it was thought that protein folding was a spontaneous process, whereby the molecule achieved maximum thermodynamic stability, based largely on the properties inherent in the amino acid sequence of the polypeptide chain(s) composing the protein as well as the interactions of the polypeptide subunits. However, numerous studies have shown that, for many proteins, folding is dependent upon members of a family of still other ubiquitous proteins called **chaperones**.

Chaperone proteins (sometimes called *molecular chaperones* or *chaperonins*) function to facilitate the folding of other proteins. Although the mechanism whereby they accomplish this role is still obscure, the interaction between chaperones and their target proteins is known not to be covalent. Furthermore, like enzymes, chaperones do not become part of the final product.

The initial discovery of chaperones was made in the 1970s, when they were first called **heat-shock proteins** because they were found to be induced in *Drosophila* following exposure of flies to elevated temperatures. Since then, this type of chaperone has been discovered

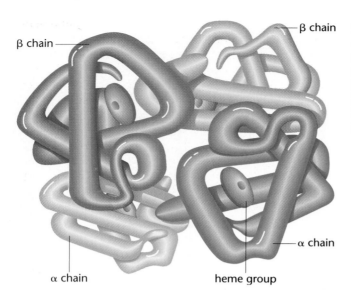

β chain

β chain

α chain

α chain

heme group

■ Figure 14–11 Schematic drawing illustrating the quaternary level (IV°) of protein structure as seen in hemoglobin. Four chains (2 alpha and 2 beta) interact with four heme groups to form the functional molecule.

in a variety of organisms, including bacteria, animals, and plants, where they are overproduced in response to heat shock, presumably to minimize the denaturation process that proteins undergo when heated.

In cells that are not heat-shocked, other types of chaperones appear to be required for the normal assembly of many proteins. To facilitate assembly, they may bind to and stabilize polypeptide chains in a conformation that is essential for subsequent folding. This binding is sometimes envisioned as a process that prevents incorrect folding by restricting certain potential conformations. In other cases, chaperones are believed to bind to proteins that must partially unwind in order to be transported across intracellular membranes. The binding is thought to prevent potential aggregation that could occur during this unwinding. Still another type of chaperone is known to function to maintain histone proteins in a conformation accessible to interaction with DNA during the assembly of nucleosomes (see Chapter 11).

Even though the molecular mechanism of chaperone action remains obscure and a topic of intense research effort, it is clear that these molecules are of great importance to protein structure and function. Their discovery has expanded our view of protein folding.

Posttranslational Modification and Protein Targeting

Before turning to a discussion of protein function, it is important to point out that polypeptide chains, like RNA, are often modified once they have been synthesized. This concept is broadly described as **posttranslational modification**. While many of these alterations are detailed biochemical transformations and beyond the scope of this discussion, you should be aware that they occur and that they are critical to the functional capability of the final protein product. Furthermore, many

such modifications are involved in the regulation of protein activity.

Several examples are presented below, the last of which will be discussed in slightly more detail.

1. **The N-terminus and C-terminus amino acids are usually removed or modified**. For example, the initial N-terminal formylmethionine residue in bacterial polypeptides is usually removed enzymatically. Often, the amino group of the initial methionine residue is removed, and the amino group of the N-terminal residue is acetylated in eukaryotic polypeptide chains.

2. **Individual amino acid residues are sometimes modified**. For example, phosphates may be added to the hydroxyl groups of certain amino acids such as tyrosine. Modifications such as this create negatively charged residues that may bond ionically with other molecules. The process of phosphorylation is extremely important in regulating many cellular activities as a result of the action of enzymes called **kinases**. In other proteins, methyl groups may be added enzymatically.

3. **Carbohydrate side chains are sometimes attached**. These are added covalently, producing **glycoproteins**, an important category of molecules that includes many antigenic determinants.

4. **Polypeptide chains may be trimmed**. For example, insulin is first translated into a longer molecule that is enzymatically trimmed to its final 51-amino-acid form.

5. **Signal sequences are removed**. At the N-terminal end of some proteins is found a sequence of up to 30 amino acids that plays an important role in directing the protein to the location in the cell where it functions. Therefore, this is called a **signal sequence**, and it determines the final destination of a protein in the cell. This process is called **protein targeting**.

For example, proteins whose fate involves secretion or which are to become part of the plasma membrane, are dependent on specific sequences for their initial transport into the lumen of the endoplasmic reticulum. While the signal sequence of various proteins with a common destination might differ in their primary amino acid sequence, they share many chemical properties. For example, those destined for secretion all contain a string of up to 15 hydrophobic amino acids preceded by a positively charged amino acid at the N-terminus of the signal sequence. Once transported, but prior to achieving their functional status as proteins, the signal sequence is enzymatically removed from these polypeptides.

6. **Polypeptide chains are often complexed with metals**. The tertiary and quaternary levels of protein structure often include and are dependent on metal atoms. The function of the protein is thus dependent on a molecular complex that includes both polypeptide chains and metal atoms. Hemoglobin, containing four iron atoms along with four polypeptide chains, is a good example.

These various types of posttranslational modifications are no doubt important in the conversion of newly translated polypeptide chains into their final three-dimensional conformation in space. Therefore, such modifications are critical in achieving the functional status specific to proteins.

Protein Function

Proteins are the most abundant macromolecules found in cells. As the end products of genes, they play diverse roles. For example, the respiratory pigments hemoglobin and **myoglobin** transport oxygen, which is essential for cellular metabolism. **Collagen** and **keratin** are examples of structural proteins associated with the skin, connective tissue, and hair of vertebrates. **Actin** and **myosin** are contractile proteins, found in abundance in

muscle tissue. Still other examples are the **immunoglobins**, which function in the immune systems of vertebrates; **transport proteins**, involved in movement of molecules across membranes; **hormones**, which regulate various types of chemical activity; and **histones**, which bind to DNA in eukaryotic organisms.

The largest group of proteins with a related function are the **enzymes**. These molecules specialize in catalyzing biological reactions. Enzymes increase the rate at which a chemical reaction reaches equilibrium, but they do not alter the end point of the chemical equilibrium. Their remarkable, highly specific catalytic properties largely determine the biomolecular nature of any cell type. The specific functions of many enzymes involved in the genetic and cellular processes of cells are described throughout the text.

Biological catalysis is a process whereby the **energy of activation** for a given reaction is lowered (Figure 14–12). The energy of activation is the increased kinetic energy state that molecules must usually reach before they can react with one another. This state can be attained as a result of elevated temperatures, but enzymes allow biological reactions to occur at lower physiological temperatures. In this way, enzymes make possible life as we know it.

The catalytic properties and specificity of an enzyme are determined by the chemical configuration of the molecule's **active site**. This site is associated with a crevice, a cleft, or a pit on the surface of the enzyme, which binds the reactants, or substrates, facilitating their interaction. Enzymatically catalyzed reactions control metabolic activities in the cell. Each reaction is either **catabolic** or **anabolic**. Catabolism is the degradation of large molecules into smaller, simpler ones with the release of chemical energy. Anabolism is the synthetic phase of metabolism, yielding nucleic acids, proteins, lipids, and carbohydrates. Metabolic pathways that serve the dual function of anabolism and catabolism are **amphibolic**.

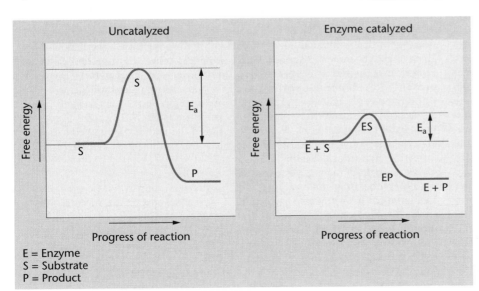

E = Enzyme
S = Substrate
P = Product

■ Figure 14–12 Energy requirements of an uncatalyzed versus an enzymatically catalyzed chemical reaction. The energy of activation (E_a) necessary to initiate the reaction is substantially lower as a result of the interaction of the enzyme and substrates.

GENETICS, TECHNOLOGY, AND SOCIETY

Mad Cows and Heresies: The Prion Story

Before "mad cow disease" hit the headlines in 1996, few people had heard of bovine spongiform encephalopathy (BSE) or its human cousin, Creutzfeldt–Jakob disease (CJD). Although a BSE epidemic had been raging in British cattle herds since the mid-1980s, no humans had shown signs of the disease and it was assumed that measures to control the epidemic would be adequate. That complacency changed on March 20, 1996, when the British government announced the appearance of 10 cases of a new variant of CJD (vCJD) and admitted that the victims may have contracted the disease by eating BSE-contaminated beef. The ensuing panic triggered political turmoil in Europe, a worldwide ban on British beef and beef products, and the near collapse of the $8.9 billion British beef industry. The European Union demanded the slaughter and incineration of 4.7 million British cattle, a campaign that would cost the government over $12 billion to compensate farmers, import milk, and buy cattle for new herds. Some people felt that the anxiety over BSE was greatly exaggerated, as it was unlikely that humans could catch BSE from cows. However, recent studies confirm that BSE and vCJD are so similar at the molecular and pathological levels that it is very likely they are the same disease. At least 20 people have now died of vCJD, and it is still not known how many other people who may have eaten the contaminated beef are also infected.

BSE and CJD are members of a group of neurological diseases known as spongiform encephalopathies. These diseases display long incubation times of up to 20 or 30 years. Infected brain tissue comes to resemble a sponge (hence "spongiform") and becomes riddled with proteinaceous deposits. Victims of the disease lose motor function, become demented, and eventually die. CJD cases arise spontaneously and randomly, at a rate of 1 per million per year worldwide. A less common form of CJD is inherited as an autosomal dominant condition. But the transmitted forms are the strangest. CJD can be transmitted through corneal or nervous tissue grafts, or by injection of growth hormone purified from human pituitary glands. Kuru, a CJD-like disease of the Fore people of New Guinea, was transmitted from person to person via ritualistic cannibalism. Transmissible spongiform encephalopathies in animals include scrapie (sheep and goats) and mad cow disease (BSE). Like Kuru, BSE and scrapie are passed from animal to animal by ingestion of diseased animal remains, particularly neural tissue. The epidemic of BSE in Britain occurred because diseased entrails of cows and sheep were processed and fed to cattle as a protein supplement. In 1988, the British government banned the use of cows and sheep as feed for other cows and sheep, and the epidemic of BSE is subsiding.

For many years, the spongiform encephalopathies defied analysis. The diseases are difficult to study, as they require injection of infected brain material into the brains of experimental animals and the diseases take months or years to develop. In addition, the infectious agent is apparently not a virion or bacterium, and infected animals do not develop antibodies to these mysterious agents. Scrapie infectious material is unaffected by radiation or nucleases that damage nucleic acids. However, it is destroyed by reagents that hydrolyze or modify proteins. In the early 1980s, American scientist Stanley Prusiner purified the infectious agent and concluded that it consisted only of protein. He proposed that scrapie was spread by an infectious protein particle that he called a "prion." His hypothesis was immediately dismissed by most scientists, as the idea of an infectious agent that contained no DNA or RNA as genetic material was heretical. However, Prusiner and others have presented evidence supporting the prion hypothesis, and the notion that disease can be transmitted by an infectious particle that contains no genetic material has gained acceptance. In fact the prion hypothesis has become so mainstream that Prusiner was awarded the 1997 Nobel Prize in Physiology and Medicine for his pioneering work on spongiform encephalopathies. Although some scientists disagree with the protein-only hypothesis, no one has yet shown that the infectious agent contains genetic material.

If prions are comprised only of protein, how do they cause disease? The answer may be as strange as the prion itself. The protein that makes up a prion (PrP) is also a normal protein which is synthesized in neurons and found in the brains of all adult animals. The difference between PrP in a normal brain and PrP in an infected brain lies in their protein secondary structures. Normal, noninfectious PrP folds into α-helices, whereas infectious prion PrP folds into β-pleated sheets. When a normal PrP molecule makes contact with an infectious PrP molecule, the normal protein is somehow unfolded and refolded into the abnormal PrP conformation. Once the normal PrP molecule has been transformed into an abnormal PrP molecule, it spreads its lethal conformation to neighboring PrP molecules, and the process takes off in a chain reaction. Normal PrP is a soluble protein that is easily destroyed by heat or enzymes that digest proteins. However, abnormal infectious PrP is insoluble in detergents and resists both heat and protease digestion. It forms deposits in the brain, causes death of neurons, and is nearly indestructible. Hence, spongiform encephalopathies can be considered diseases of protein secondary structure.

A great number of pressing questions still need to be addressed. Is BSE now out of the food chain? How many humans might be infected with BSE, and what is the incubation time for vCJD? How much infected material does a person need to ingest in order to develop disease, and can prions accumulate in the body over time? Do some people have a genetic susceptibility to spongiform encephalopathies, whereas others have resistance? Can prions exist in other parts of the body besides the brain and spinal cord? Can we develop diagnostic tests and therapies for BSE and vCJD? Have we heard the end of the mad cow disease story, or is it just the beginning?

References

Aguzzi, A., and Weissmann, C. 1997. Prion research: The next frontiers. *Nature* 389:795–98.

Almong, J., and Pattison, J. 1997. Human BSE. *Nature* 389:437-38.

Prusiner, S. B. The prion diseases. *Sci. Am.* (Jan.) 272:48-57.

Chapter Summary

1. Basic to Garrod's studies in the early 1900s of inborn human metabolic disorders such as cystinuria, albinism, and alkaptonuria was the concept that genes control the synthesis of specific metabolic products.

2. The investigation of nutritional requirements in *Neurospora* by Beadle and Tatum made it clear that mutations cause the loss of enzyme activity. Their work led to the concept of one gene:one enzyme.

3. The concept of the one-gene:one-enzyme hypothesis was later revised. Pauling and Ingram's investigations of hemoglobins from patients with sickle-cell anemia led to the discovery that one gene directs the synthesis of only one polypeptide chain.

4. Thorough investigations have revealed the existence of several major types of human hemoglobin molecules found in the embryo, fetus, and adult. Specific genes control each polypeptide chain constituting these various hemoglobin molecules.

5. The proposal suggesting that a gene's nucleotide sequence specifies in a colinear way the sequence of amino acids in a polypeptide chain was confirmed by Yanofsky's experiments involving mutations in the tryptophan synthetase gene in *E. coli.*

6. Proteins, the end products of genes, demonstrate four levels of structural organization that together provide the chemical basis for their three-dimensional conformation.

7. Polypeptide chains are often modified chemically once they have been synthesized, a process referred to as posttranslational modification. A signal sequence is often present at the N-terminal end of proteins that plays a role in directing the protein to its final destination in the cell. This process is called protein targeting.

8. Of the myriad functions performed by proteins, the most influential role is assumed by enzymes. These highly specific, cellular catalysts play a central role in the production of all classes of molecules in living systems.

Key Terms

actin, 315
active site, 315
alkaptonuria, 303
alpha (α) chain, 309
alpha (α) helix, 312
amino group, 310
amphibolic, 315
anabolic, 315
beta (β) chain, 309
β-pleated sheet, 313
carboxyl group, 310
catabolic, 315
central carbon atom, 310
chaperones, 313
citrulline, 305
colinear relationship, 309
C-terminus, 310
delta (δ) chain, 309
energy of activation, 315
enzyme, 315
epsilon (ε) chain, 309
fetal hemoglobin, 309
fibroin, 313
fingerprinting technique, 309
gamma (γ) chain, 309
glycoprotein, 314
Gower 1 hemoglobin, 309

HbA, 308
HbA₂, 309
HbF, 309
HbS, 308
heme group, 309
hemoglobin, 308
histone, 315
homogentisic acid, 303
hydrogen atom, 310
immunoglobin, 315
keratin, 315
kinase, 314
molecular disease, 309
myoglobin, 313
moysin, 315
negatively charged (acidic) R group, 310
nonpolar (hydrophobic) R group, 310
N-terminus, 310
oligomeric protein, 313
one-gene:one-enzyme hypothesis, 305
one-gene:one-polypeptide chain hypothesis, 307
ornithine, 305
peptide bond, 310

phenylalanine hydroxylase, 304
phenylketonuria (PKU), 304
polar (hydrophilic) R group, 310
polypeptide, 310
positively charged (basic) R group, 310
posttranslational modification, 314
primary (I°) protein structure, 310
prion, 316
protein, 310
protein targeting, 314
protomer, 313
quaternary (IV°) protein structure, 310
R (radical) group, 310
secondary (II°) protein structure, 310
sickle-cell anemia, 307
signal sequence, 314
starch gel electrophoresis, 308
subunit chain, 313
tertiary (III°) protein structure, 310
transport protein, 315
tryptophan synthetase, 309
zeta chain, 309

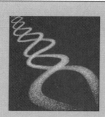

INSIGHTS
and
SOLUTIONS

1. The following growth responses were obtained using four mutant strains of *Neurospora* and the related compounds A, B, C, and D. None of the mutations grows on minimal medium. Draw all possible conclusions.

		Growth Product		
		C	D	B	A
Mutation	1	−	−	−	−
	2	−	+	+	+
	3	−	−	+	+
	4	−	−	+	−

Solution: First, nothing can be concluded about mutation 1, except that it is lacking some essential growth factor, perhaps even unrelated to the biochemical pathway represented by mutations 2–4; nor can anything be concluded about compound C. If it is involved in the pathway, it is a product synthesized prior to the synthesis of A, B, and D. We must now analyze these three compounds and the control of their synthesis by the enzymes encoded by genes 2, 3, and 4. Because product B allows growth in all three cases, it may be considered the "end product." It bypasses the block in all three instances. Using similar reasoning, product A precedes B in the pathway because it allows a bypass in two of the three steps. Product D precedes B, yielding the more complete solution:

$$C \ (?) \longrightarrow D \longrightarrow A \longrightarrow B$$

Now determine which mutations control which steps. Since mutation 2 can be alleviated by products D, B, and A, it must control a step prior to all three products, perhaps the direct conversion to D, although we can't be certain. Mutation 3 is alleviated by B and A, so its effect must precede them in the pathway. Thus, we will assign it as controlling the conversion of D to A. Likewise, we can assign mutation 4 to the conversion of A to B, leading to the more complete solution:

$$C \ (?) \xrightarrow{2 \ (?)} D \xrightarrow{3} A \xrightarrow{4} B$$

Problems and Discussion Questions

1. Discuss the potential difficulties involved in designing a diet to alleviate the symptoms of phenylketonuria.

2. The synthesis of flower pigments is known to be dependent upon enzymatically controlled biosynthetic pathways. In the crosses shown below, postulate the role of mutant genes and their products in producing the observed phenotypes.

   (a) P_1: white strain A × white strain B
      F_1: all purple
      F_2: 9/16 purple: 7/16 white

   (b) P_1: white × pink
      F_1: all purple
      F_2: 9/16 purple: 3/16 pink: 4/16 white

3. A series of mutations in the bacterium *Salmonella typhimurium* result in the requirement of either tryptophan or some related molecule in order for growth to occur. From the data shown below, suggest a biosynthetic pathway for tryptophan.

	Growth Supplement				
Mutation	Minimal Medium	Anthranilic Acid	Indole Glycerol Phosphate	Indole	Tryptophan
trp-8	−	+	+	+	+
trp-2	−	−	+	+	+
trp-3	−	−	−	+	+
trp-1	−	−	−	−	+

4. The study of biochemical mutants in organisms such as *Neurospora* has demonstrated that some pathways are branched. The data shown below illustrate the branched nature of the pathway resulting in the synthesis of thiamine. Why don't the data support a linear pathway? Can you postulate a branched pathway for the synthesis of thiamine in *Neurospora*?

Mutation	Growth Supplement			
	Minimal Medium	Pyrimidine	Thiazole	Thiamine
thi-1	−	−	+	+
thi-2	−	+	−	+
thi-3	−	−	−	+

P_1	F_1	F_2
(1) orange × red	all orange	3/4 orange 1/4 red
(2) orange × yellow	all orange	3/4 orange 1/4 yellow
(3) red × yellow	all orange	9/16 orange 4/16 yellow 3/16 red

5. Explain why the one-gene:one-enzyme concept is not considered completely accurate.

6. Using sickle-cell anemia as a basis, describe what is meant by a molecular or genetic disease. What are the similarities and dissimilarities between this type of a disorder and a disease caused by an invading microorganism?

7. Describe colinearity.

8. Define and compare the four levels of protein organization.

9. List as many different protein functions as you can, with an example of each.

10. How does an enzyme function? Why are enzymes essential for living organisms on earth?

11. Naturally occurring wood lily plants have orange flowers and are true-breeding. A genetics student discovered both red and yellow true-breeding variants and proceeded to hybridize these strains:

Being knowledgeable about transmission genetics, the student surmised that two gene pairs were involved; on this basis she proposed how these genes account for the various colors.

(a) What outcome(s) of the crosses suggested that two gene pairs were at work?

(b) Assuming she was correct and using the mutant gene symbols *y* for yellow and *r* for red, propose which genotypes give rise to which phenotypes.

(c) The student then devised several possible biochemical pathways for production of these pigments as well as which steps are enzymatically controlled by which genes. These are listed below. Which of these (one or more) is/are consistent with your proposal in part (b) and the original data? Assume that a mixture of both red and yellow pigment results in orange flowers. Defend your answer.

Selected Readings

Bartholome, K. 1979. Genetics and biochemistry of phenylketonuria-Present state. *Hum. Genet.* 51:241–45.

Bateson, W. 1909. *Mendel's principles of heredity.* Cambridge, England: Cambridge University Press.

Beadle, G. W. 1946. Genes and the chemistry of the organism. *Am. Sci.* 34:31–53.

Beadle, G. W., and Tatum, E. L. 1941. Genetic control of biochemical reactions in *Neurospora. Proc. Natl. Acad. Sci. USA* 27:499–506.

Beet, E. A. 1949. The genetics of the sickle-cell trait in a Bantu tribe. *Ann. Eugenics* 14:279–84.

Brenner, S. 1955. Tryptophan biosynthesis in *Salmonella typhimurium. Proc. Natl. Acad. Sci. USA* 41:862–63.

Dickerson, R. E., and Geis, I. 1983. *Hemoglobin: Structure, function, evolution, and pathology.* Menlo Park, CA: Benjamin/Cummings.

Doolittle, R. F. 1985. Proteins. *Sci. Am.* (Oct.) 253:88–99.

Ezzell, C. 1994. Evolutions: Molecular chaperones and protein folding. *J. NIH Res.* 6:103.

Garrod, A. E. 1902. The incidence of alkaptonuria: A study in chemical individuality. *Lancet* 2:1616–20.

————. 1909. *Inborn errors of metabolism.* London: Oxford University Press. (Reprinted 1963, Oxford University Press, London.)

Ingram, V. M. 1957. Gene mutations in human hemoglobin: The chemical difference between normal and sickle cell hemoglobin. *Nature* 180:326–28.

Koshland, D. E. 1973. Protein shape and control. *Sci. Am.* (Oct.) 229:52–64.

Neel, J. V. 1949. The inheritance of sickle-cell anemia. *Science* 110:64–66.

Pauling, L., et al. 1949. Sickle cell anemia, a molecular disease. *Science* 110:543–48.

Yanofsky, C., et al. 1967. The complete amino acid sequence of the tryptophan synthetase A protein and its colinear relationship with the genetic map of the A gene. *Proc. Natl. Acad. Sci. USA* 57:296–98.

Mutant erythrocytes derived from an individual with sickle-cell anemia.

CHAPTER OUTLINE

CHAPTER
15

DNA—Mutation, Repair, and Transposable Elements

Chapter Concepts

Aside from chromosomal mutations, the major basis of diversity among organisms, even those that are closely related, is genetic variation at the level of the gene. The origins of such variation are gene mutations, whereby coding sequences are altered as a result of substitution, addition, or deletion of one or more bases within those sequences. Such mutations are subject to elaborate repair mechanisms, make the study of genetics possible, and are the raw material on which evolution relies.

In Chapter 9, we defined the four characteristics or functions ascribed to the genetic information: replication, storage, expression, and variation by mutation. In a sense, mutation is a failure to store the genetic information faithfully. If a change occurs in the stored information, it may be reflected in the expression of that information and will be propagated following replication. Historically, the term *mutation* includes both chromosomal changes and changes within single genes. We have discussed the former alterations in Chapter 8, referring to them collectively as **chromosomal aberrations**. In this chapter, we will be concerned with **gene mutations**. A change may be a simple substitution of one nucleotide or may involve the insertion or deletion of one or more nucleotides within the normal sequence of DNA.

Mutations provide the basis for genetic studies. The resulting phenotypic variability provides the basis for the geneticist to identify and study the genes that control the traits that have been modified. Without the phenotypic variability that mutations provide, genetic analysis would be impossible. For example, if all pea plants displayed a uniform phenotype, Mendel would have had no basis for his experimentation. Because of the importance of mutations, great attention has been given to their origin, induction, and classification.

Certain organisms lend themselves to induction of mutations that can be detected easily and studied throughout reasonably short life cycles. Viruses, bacteria, fungi, fruit flies, other invertebrates, certain plants,

and mice fit these criteria. Thus, these organisms have been used in studying mutation and mutagenesis, and through other studies they have also contributed to more general aspects of genetic knowledge.

Classification of Mutations

There are various schemes by which mutations are classified. They are not mutually exclusive, but instead depend simply on which aspects of mutation are being investigated or discussed. In this section, we describe several distinctions that are used to classify mutations.

Spontaneous versus Induced Mutations

All mutations are described as either spontaneous or induced. While these two categories overlap to some degree, **spontaneous mutations** are those that arise in nature. No specific agents—other than natural forces—are associated with their occurrence, and they are generally assumed to be changes in the nucleotide sequences of genes that occur at random sites.

What causes spontaneous mutations? Although their origin is not fully understood, many such mutations can be linked to normal chemical processes or phenomena that result in rare errors. Such errors may cause alterations in the chemical structure of nitrogenous bases that are part of existing genes. They occur most often during the enzymatic process of DNA replication. It is generally agreed that any natural phenomenon that heightens chemical reactivity in cells will lead to more errors. For example, background radiation from cosmic and mineral sources and ultraviolet light from the sun are energy sources to which most organisms are exposed. As such, these energy sources may be factors leading to spontaneous mutations. Once an error is present in the genetic code, it may be reflected in the amino acid composition of the specified protein. If the changed amino acid is present in a part of the molecule critical to its structure or biochemical activity, a functional alteration *may* result.

In contrast to spontaneous mutations, those that arise as a result of the influence of any artificial factor are considered to be **induced mutations**. The earliest demonstration of the induction of mutation occurred in 1927, when Herman J. Muller reported that X-rays could cause mutations in *Drosophila*. In 1928, Lewis J. Stadler reported the same finding in barley. In addition to various forms of radiation, a wide spectrum of chemical agents is also known to be mutagenic, the aspects of which we will examine later in this chapter.

Gametic versus Somatic Mutations

Mutations may occur in either somatic or gametic cells of eukaryotes. Those that create recessive autosomal alleles in somatic cells are rarely of any consequence. Most such mutations are masked by their functional dominant counterparts. Somatic mutations will have a greater impact if they are dominant or if they are X-linked recessives, since such mutations are more likely to be immediately expressed. Similarly, the impact will be more noticeable if such somatic mutations occur early in development, where undifferentiated cells will give rise to several differentiated tissues or organs. Mutations occurring in adult tissues are most often masked by the thousands upon thousands of nonmutant cells performing the normal function.

There is a very important exception to the above description. If a mutation occurs that disrupts the control of cell proliferation, such a mutation may cause cells to divide uncontrollably, causing a cancerous condition. This topic will be pursued extensively in Chapter 19.

Mutations in gametes or gamete-forming tissues are part of the germ line and are of greater concern because they are transmitted to offspring. **Dominant autosomal mutations** will be expressed phenotypically in the first generation. **X-linked recessive mutations** arising in the gametes of a heterogametic female may be expressed in hemizygous male offspring. This will occur provided that the male offspring receives the affected X chromosome. Because of heterozygosity, the occurrence of an **autosomal recessive mutation** in the gametes of either males or females (even one resulting in a lethal allele) may go unnoticed for many generations,

until the resultant allele has become widespread in the population. The new allele will become evident only when a chance mating brings two copies of it together in the homozygous condition.

Other Categories of Mutation

Various types of mutations are classified on the basis of their effect on the organism. We will briefly review several of them again. Note that a single mutation may well fall into more than one category.

The most easily observed mutations are those affecting a **morphological trait**. For example, all of Mendel's pea characters and many genetic variations encountered in the study of *Drosophila* fit this designation. They cause obvious changes in morphology.

A second broad category of mutations includes those that exhibit **nutritional** or **biochemical variations** in phenotype. In bacteria and fungi, the inability to synthesize a particular amino acid or vitamin is an example of a typical nutritional mutation. In humans, sickle-cell anemia and hemophilia are examples of biochemical mutations. While such mutations in these organisms are not visible and do not always affect specific morphological characters, they can have a more general effect on the well-being and survival of the affected individual.

A third category consists of mutations that affect behavior patterns of an organism. For example, mating behavior or circadian rhythms of animals may be altered. The primary effect of **behavior mutations** is often difficult to discern. For example, the mating behavior of a fruit fly may be impaired if it cannot beat its wings. However, the defect may be in (1) the flight muscles, (2) the nerves leading to them, or (3) the brain, where the nerve impulses that initiate wing movements originate. The study of behavior and the genetic factors influencing it has benefited immensely from investigations of behavior mutations.

Another type of mutation may affect the regulation of genes. A regulatory gene may produce a product that controls the transcription of another gene. In other instances, a region of DNA either close to or far away from a gene may modulate its activity! In either case, **regulatory mutations** may disrupt normal regulatory processes and permanently activate or inactivate a gene. Our knowledge of genetic regulation depends on the study of mutations that disrupt this process.

Still another group consists of **lethal mutations**. Nutritional and biochemical mutations may also fall into this category. A mutant bacterium that cannot synthesize a specific amino acid it needs will be unable to grow and divide if plated on a medium lacking that amino acid. Various human biochemical disorders, such as Tay-Sachs disease and Huntington disease, are lethal at different points in the life cycle of humans.

Finally, any of the above categories can exist as **conditional mutations**. Even though a mutation is present in the genome of an organism, it may not be evident under certain conditions. Among the best examples are **temperature-sensitive mutations**, found in a variety of organisms. At certain "permissive" temperatures, a mutant gene product functions normally, only to lose its functional capability at a different, "restrictive" temperature. When shifted to a restrictive temperature, the impact of the mutation becomes apparent, even lethal, and is amenable to investigation. The study of conditional mutations has been extremely important in experimental genetics, particularly in understanding the function of genes essential to the viability of organisms.

Detection of Mutation

Before geneticists can study the mutational process directly or obtain mutant organisms for genetic investigations, they must be able to detect mutations. The ease and efficiency of detecting mutations in a particular organism has generally determined the organism's usefulness in genetic studies. In this section, we will use several examples to illustrate how mutations are detected.

Detection in Bacteria and Fungi

Detection of mutations is most efficient in haploid microorganisms such as bacteria and fungi. Detection depends on a selection system by which mutant cells can be isolated easily from nonmutant cells. The general principles are similar in bacteria and fungi. To illustrate, we will describe how nutritional mutations in the fungus *Neurospora crassa* are detected.

Neurospora is a pink mold that normally grows on bread. It may also be cultured in the laboratory. This eukaryotic mold is haploid in the vegetative phase of its life cycle. Thus, mutations may be detected without the complications generated by heterozygosity in diploid organisms. Wild-type *Neurospora* grows on a **minimal culture medium** of glucose, a few inorganic acids and salts, a nitrogen source such as ammonium nitrate, and the vitamin biotin. Induced nutritional mutants will not grow on minimal medium, but will grow on a supplemented or **complete medium** that also contains numerous amino acids, vitamins, nucleic acid derivatives, and so forth. Microorganisms that are nutritional wild types (requiring only minimal medium) are called **prototrophs**, while those mutants that require a specific supplement to the minimal medium are called **auxotrophs**.

Nutritional mutants may be detected and isolated by their failure to grow on minimal medium and their ability to grow on complete medium. The mutant cells cannot synthesize some essential compound that is absent in minimal medium but present in complete medium. Once a nutritional mutant is detected and isolated, the missing compound is determined by attempts to grow the mutant strain in a series of tubes, each containing minimal medium supplemented with a single compound. The auxotrophic mutation can be defined in this way.

In Chapter 14, this detection technique was used by Beadle and Tatum during their "one-gene:one-enzyme" work, where we discussed it in the context of isolation of mutant *Neurospora* strains (see Figure 14–2). Similar techniques have been useful in the analysis of microorganisms (Chapter 16). These have been particularly critical during the study of molecular genetics.

Detection in Drosophila

Muller, in his studies demonstrating that X-rays are mutagenic, developed a number of detection systems in *Drosophila melanogaster*. These systems can be used to estimate both spontaneous and induced rates of X-linked and autosomal recessive lethal mutations. We will consider the **attached-X procedure**, which assesses X-linked mutations.

Attached-X females have two X chromosomes attached to a single centromere and one Y chromosome, in addition to the normal diploid complement of autosomes. When attached-X females are mated to males with normal sex chromosomes (XY), four types of progeny result: triplo-X females that die, viable attached-X females, YY males that also die, and viable XY males. Figure 15–1 illustrates how P_1 males that have been treated with a mutagenic agent produce F_1 male offspring that express any induced X-linked recessive mutation.

The same approach works equally well whether or not a mutagenic agent has been used. When not, spontaneous mutations are expressed in the first generation. In detection techniques devised for recessive autosomal lethals in *Drosophila,* dominant marker mutations are followed through a series of three generations. Although these techniques are more cumbersome to perform, they are also fairly efficient.

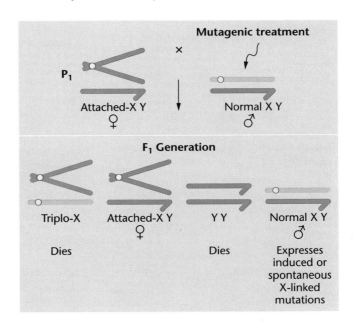

■ Figure 15–1 The attached-X method for detection of induced morphological mutations in *Drosophila.*

Detection in Plants

Genetic variation in plants is extensive. Mendel's peas, for example, were the basis for the fundamental postulates of transmission genetics. Studies of plants have also enhanced our understanding of gene interaction, polygenic inheritance, linkage, sex determination, chromosome rearrangements, and polyploidy. Many variations can be detected simply by observation. However, there are also techniques for the detection of biochemical mutations in plants. The first is the analysis of biochemical composition of plants. For example, the isolation of proteins from maize endosperm, hydrolysis of these proteins, and determination of their amino acid composition have revealed that the **opaque-2 mutant strain** contains significantly more lysine than other, nonmutant lines. Since lysine content is usually low in maize protein, this mutation significantly improves the nutritional value of maize. As a result of this discovery, plant geneticists and other specialists have analyzed the amino acid compositions of various strains of other grain crops, including rice, wheat, barley, and millet. The resulting information is useful in combating malnutrition diseases resulting from inadequate protein or the lack of essential amino acids in the diet.

Another detection technique involves tissue culture of plant cell lines in defined medium. The plant cells are handled as microorganisms, and resistance to herbicides or disease toxins may be determined by adding these compounds to the culture medium. Techniques associated with conditional lethal mutants are useful with plant cells in tissue culture. For example, temperature-sensitive mutations in plants are being explored, particularly in tobacco. These studies may add significantly to our understanding of plant growth, metabolism, and genetics.

Detection in Humans

Because humans are obviously not suitable experimental organisms, the techniques that have been developed for the detection of mutations in organisms such as *Drosophila* are not available to human geneticists. To determine the genetic basis for any human characteristic or disorder, geneticists first analyze a pedigree that traces the family history as far back as possible. If a trait is shown to be inherited, it is possible to predict whether the mutant allele is behaving as a dominant or a recessive and whether it is X-linked or autosomal. Dominant mutations are the simplest to detect. If they are present on the X chromosome, affected fathers pass the phenotypic trait to all their daughters. If dominant mutations are autosomal, approximately 50 percent of the offspring of an affected heterozygous individual are expected to show the trait. Figure 15–2 shows a pedigree illustrating the initial occurrence of an autosomal dominant allele for cataracts of the eye. The parents in generation I were unaffected, but one of three offspring (generation II) developed cataracts. This female, the proband, produced two children, of which the male child was affected. Of his six offspring, four of six were affected (generation IV). These obser-

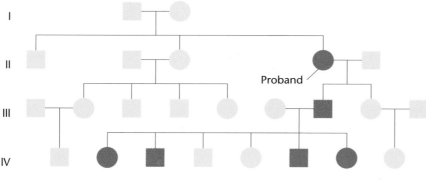

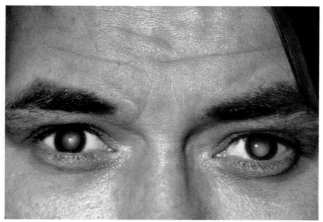

■ Figure 15–2 A hypothetical pedigree of inherited cataract of the eye, consistent with a dominant mode of inheritance. The photograph shows a male with bilateral cataracts.

vations are consistent with, but do not prove, an autosomal dominant mode of inheritance. However, the high percentage of affected offspring in generation IV favors this conclusion. Also, the unaffected daughter in this generation argues against X-linkage, because she received her X chromosome from her affected father. This conclusion is sound, provided that the mutant allele is completely penetrant.

X-linked recessive mutations may also be detected by pedigree analysis, as discussed in Chapter 4. The most famous case of an X-linked mutation in humans is that of **hemophilia**, which was found in the descendants of Queen Victoria. The recessive mutation for hemophilia has occurred many times in human populations, but the political consequences of the mutation that occurred in the royal family were sweeping. Inspection of the pedigree in Figure 15–3 leaves little doubt that Victoria was heterozygous (*Hh*) for the trait. Her father was not affected, and there is no reason to believe that her mother was a carrier, as Victoria was. Robert Massie's

Nicholas and Alexandra and Robert and Suzanne Massie's *Journey* (see this chapter's selected readings) provide fascinating reading on the topic of hemophilia.

It is also possible to detect autosomal recessive alleles. Because this type of mutation is "hidden" when it is heterozygous, it is not unusual for the trait to appear only intermittently through a number of generations. A mating between an affected individual and a homozygous normal individual will produce unaffected heterozygous carrier children. Matings between two carriers will produce, on the average, one-fourth affected offspring.

In addition to pedigree analysis, human cells may now be routinely cultured *in vitro*. This procedure has allowed the detection of many more mutations than any other form of analysis. Analysis of enzyme activity, protein migration in electrophoretic fields, and direct sequencing of DNA and proteins are among the techniques that have demonstrated wide genetic variation between individuals in human populations.

(a)

(b)

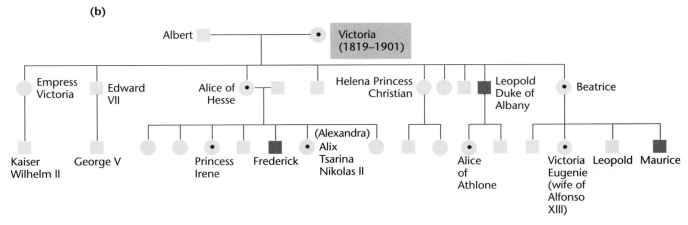

■ Figure 15–3 An abbreviated pedigree of hemophilia in the royal family descended from Queen Victoria. The pedigree is typical of the transmission of X-linked recessive traits. Circles with a dot in them indicate presumed female heterozygous carriers. The photograph shows Queen Victoria (seated front center) and some of her immediate family.

Spontaneous Mutation Rate

The types of detection systems just described allow geneticists to estimate mutation rates. It is of considerable interest to determine the rate of spontaneous mutation. Such information provides insights into evolution and provides the baseline for measuring the relative rate of experimentally induced mutation. Induction of mutation can only be ascertained when the induced rate clearly exceeds the spontaneous rate for the organism under study.

Examination of the spontaneous rate in a variety of organisms reveals many interesting points. First, the rate is exceedingly low for all organisms studied. Second, the rate is seen to vary considerably in different organisms. Third, even within the same species, the spontaneous mutation rate varies from gene to gene.

Viral and bacterial genes undergo spontaneous mutation on an average of about 1 in 100 million (10^{-8}) cell divisions. While *Neurospora* exhibits a similar rate, maize, *Drosophila*, and humans demonstrate a rate several orders of magnitude higher. The genes studied in these groups average between 1/1,000,000 and 1/100,000 (10^{-6} and 10^{-5}) mutations per gamete formed. Mouse genes are still another order of magnitude higher in their spontaneous mutation rate, 1/100,000 to 1/10,000 (10^{-5} to 10^{-4}). It is not clear why such a large variation occurs in mutation rate. The variation might reflect the relative efficiency of enzyme systems whose function is to repair errors created during replication. Repair systems will be discussed later in this chapter.

The Molecular Basis of Mutation

We will use a *simplified* definition of a gene in the description of the molecular basis of mutation. In this context, it is easiest to consider a gene as a linear sequence of nucleotide pairs representing stored chemical information. Based on the triplet nature of the genetic code, each sequence of three nucleotides specifies a single amino acid in the corresponding polypeptide. Any change that disrupts the coded information provides

sufficient basis for a mutation. The least complex change is the substitution of a single nucleotide. In Figure 15–4, such a change is compared with our own written language, using three-letter words to be consistent with the genetic code. A change of one letter can alter the meaning of the sentence, from "THE CAT SAW THE DOG" to "THE CAT SAW THE HOG" or "THE BAT SAW THE DOG," creating what is called *missense*. These are analogies to what are most appropriately referred to as **base substitutions** or **point mutations**. The mutation has turned accurate information that makes sense (at least to the cat and the dog) into various forms of inaccurate information referred to as *missense*.

Two other more formal terms are used to describe nucleotide substitutions. If a purine replaces a purine or a pyrimidine replaces a pyrimidine, a **transition** has occurred. If a purine and a pyrimidine are interchanged, a **transversion** has occurred.

A second type of change in the nucleotide sequence that may occur is the insertion or deletion of a single nucleotide at any point along the gene. As illustrated in Figure 15–4, the remainder of the three-letter (code) words become garbled, creating much more extensive missense. These examples are called **frameshift mutations** because the frame of reading has become altered.

The analogy in Figure 15–4 demonstrates that insertions and deletions have the potential to change all subsequent triplets in a gene. It is probable that, of the 64 possible triplets, one of many altered triplets will be either UAA, UAG, or UGA. These are termination codons (see Chapter 12). When one is encountered during translation, polypeptide synthesis is terminated. Obviously, the results of frameshift mutations can be very severe!

Tautomeric Shifts

In 1953, immediately after they had proposed a molecular structure of DNA, Watson and Crick published a paper in which they discussed the genetic implications of this structure. They recognized that the purines and

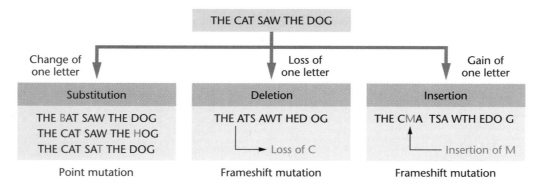

■ Figure 15–4 The impact of the substitution, deletion, or insertion of one letter in a sentence composed of three-letter words as analogies to point and frameshift mutations.

pyrimidines found in DNA could exist in **tautomeric forms**; that is, each can exist in alternative chemical forms, differing by only a single proton shift in the molecule. Watson and Crick suggested that **tautomeric shifts** could result in base-pair changes or mutations.

The most stable tautomers of the nitrogenous bases result in the standard hydrogen bonds that serve as the basis of the double helical model of DNA. The less frequently occurring tautomers are capable of hydrogen bonding with noncomplementary bases. However, the pairing is always between a pyrimidine and a purine. Figure 15–5 compares the normal base-pairing relationships with the rare unorthodox pairings. The biologically important unstable tautomers involve keto–enol pairs for thymine and guanine, and amino–imino pairs for cytosine and adenine.

The effect leading to mutation occurs during DNA replication when a rare tautomer in the template strand matches with a noncomplementary base. In the next round of replication, the "mismatched" members of the base pair are separated, and each specifies its normal complementary base. The end result is a transition mutation (see Figure 15–6).

Base Analogs

Base analogs, which are mutagenic chemicals, are molecules that may substitute for purines or pyrimidines during nucleic acid biosynthesis. The halogenated derivative of uracil in the number-5 position of the pyrimidine ring **5-bromouracil (5-BU)*** is a good example. Figure 15–7 compares the structure of this thymine analog with the structure of thymine. The presence of the bromine atom in place of the methyl group increases the probability that a tautomeric shift will occur. If 5-BU is incorporated into DNA in place of thymine and a tautomeric shift to the enol form occurs, 5-BU base-pairs with guanine. After one round of replication, an A=T to G≡C transition results. There are other base analogs that are mutagenic. One, **2-amino purine (2-AP)**, can serve successfully as an analog of adenine. In addition to its base-pairing affinity with thymine, 2-AP can also base-pair with cytosine. As such, transitions from A=T to G≡C may result following replication.

Because of the specificity by which base analogs such as 2-AP induce transition mutations, base analogs may also be used to induce reversion to the wild-type nucleotide sequence. This alteration is called **reverse mutation**. The process can occur spontaneously, but at a much lower rate.

Alkylating Agents

The sulfur-containing mustard gases were one of the first groups of chemical mutagens discovered. This discovery was made in studies involving chemical warfare

---

*If 5-BU is chemically linked to d-ribose, the nucleoside analog bromodeoxyuridine (BUdR) is formed.

(a) Standard base-pairing arrangements

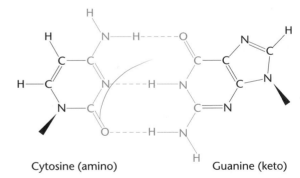

Thymine (keto)   Adenine (amino)   Cytosine (amino)   Guanine (keto)

(b) Anomalous base-pairing arrangements

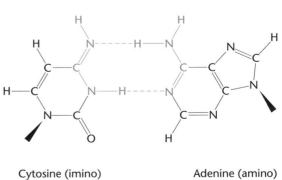

Thymine (enol)   Guanine (keto)   Cytosine (imino)   Adenine (amino)

■ Figure 15–5   The standard base-pairing relationships compared with anomalous arrangements occurring as a result of tautomeric shifts. The dense arrow head indicates the point of bonding to the pentose sugar.

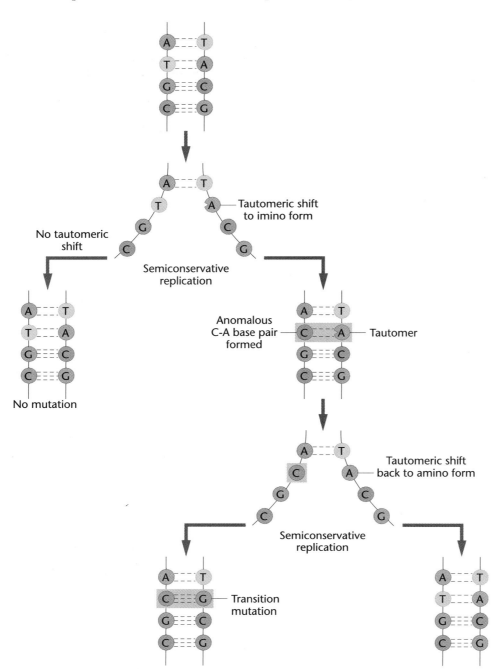

during World War II. Mustard gases are **alkylating agents**; that is, they donate an alkyl group such as CH₃— or CH₃—CH₂— to amino or keto groups in nucleotides. **Ethylmethane sulfonate (EMS)**, for example, alkylates the keto group in the number 6 position of guanine and in the number 4 position of thymine (Figure 15–8). As with base analogs, base-pairing affinities are altered and transition mutations result. In the case of **6-ethylguanine**, this molecule acts like a base analog of adenine, causing it to pair with thymine. Table 15.1 lists the chemical names and structures of several frequently used alkylating agents that are known to be mutagenic.

Acridine Dyes and Frameshift Mutations

Other chemical mutagens cause frameshift mutations, as originally depicted in Figure 15–4. These result from the addition or removal of one or more base pairs in the polynucleotide sequence of the gene. Inductions of frameshift mutations have been studied in detail with a group of aromatic molecules known as **acridine dyes**. The chemical structures of **proflavin**, the most widely studied acridine mutagen, and **acridine orange** are shown in Figure 15–9. Acridine dyes are of about the same dimension as a nitrogenous base pair and are

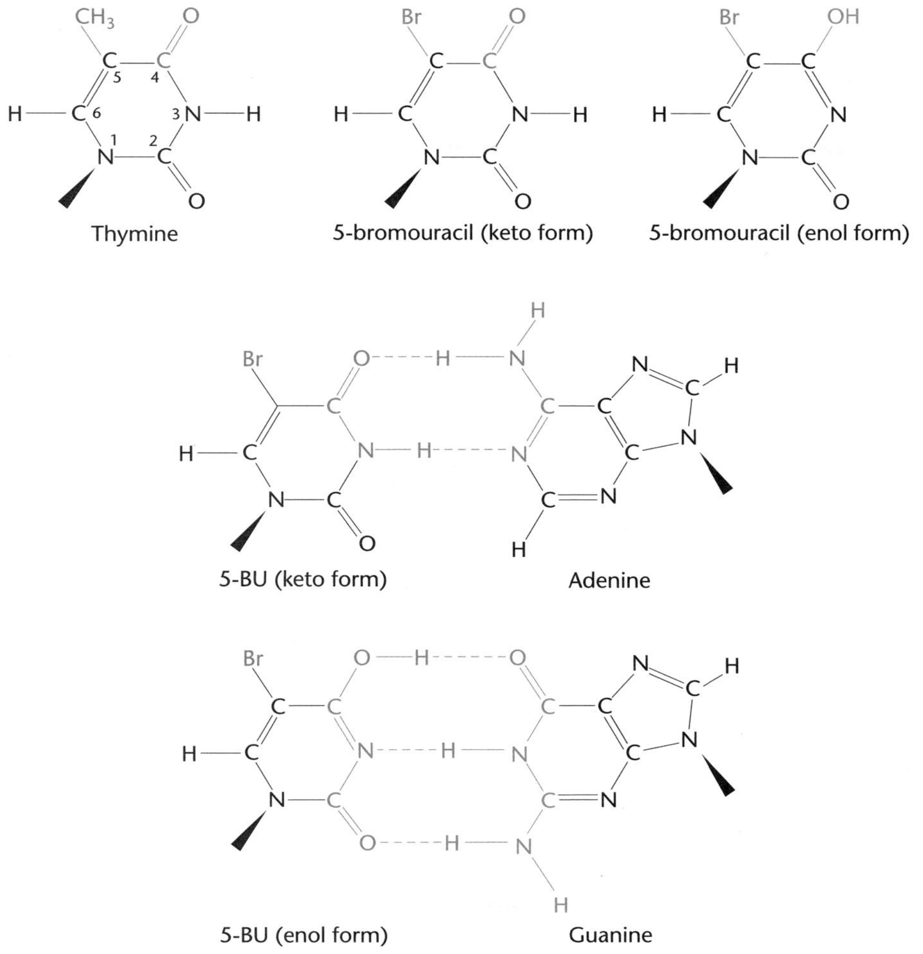

■ Figure 15–7 Similarity of 5-bromouracil (5-BU) structure to thymine structure. In the common keto form, 5-BU pairs normally with adenine, behaving as an analog. In the rare enol form, it pairs anomalously with guanine.

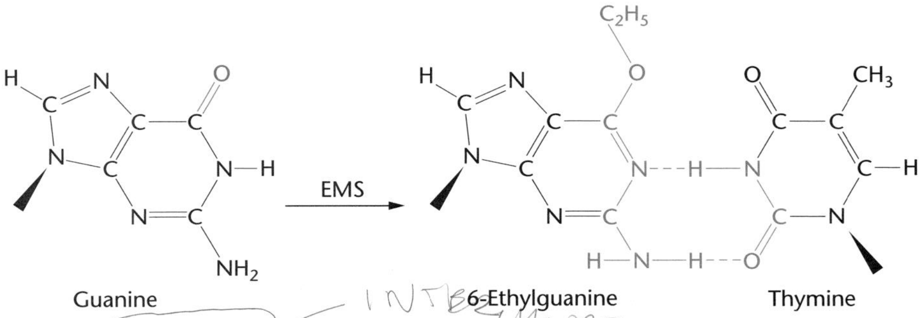

Thymine

5-bromouracil (keto form)

5-bromouracil (enol form)

5-BU (keto form)   Adenine

5-BU (enol form)   Guanine

■ Figure 15–8 Conversion of guanine to 6-ethylguanine by the alkylating agent ethylmethane sulfonate (EMS). 6-ethylguanine base-pairs with thymine.

Guanine   6-Ethylguanine   Thymine

known to intercalate or wedge between purines and pyrimidines of intact DNA. Intercalation of acridine dyes induces contortions in the DNA helix, causing deletions and insertions.

One model suggests that the resultant frameshift mutations are generated at gaps produced in DNA during replication, repair, or recombination. During these events, there is the possibility of slippage and improper base pairing of one strand with the other. The model suggests that intercalation of the acridine into an improperly base-paired region can extend the existence of these slippage structures. If so, the probability in-

creases that the mispaired configuration will exist when synthesis and rejoining occurs, thereby resulting in an addition or deletion of one or more bases from one of the strands.

Apurinic Sites and Other Lesions

Still another type of mutation involves the spontaneous loss of one of the nitrogenous bases in an intact double-helical DNA molecule. Most frequently, such an event involves either guanine or adenine. These sites, created by the "breaking" of the glycosidic bond linking the 1′-

TABLE 15.1	Alkylating agents	
Common Name or Symbol	**Chemical Name**	**Chemical Structure**
Mustard gas (sulfur)	Di-(2-chloroethyl) sulfide	$Cl-CH_2-CH_2-S-CH_2-CH_2-Cl$
EMS	Ethylmethane sulfonate	$CH_3-CH_2-O-\overset{\overset{\displaystyle O}{\|\|}}{\underset{\underset{\displaystyle O}{\|}}{S}}-CH_3$
EES	Ethylethane sulfonate	$CH_3-CH_2-O-\overset{\overset{\displaystyle O}{\|\|}}{\underset{\underset{\displaystyle O}{\|}}{S}}-CH_2-CH_3$

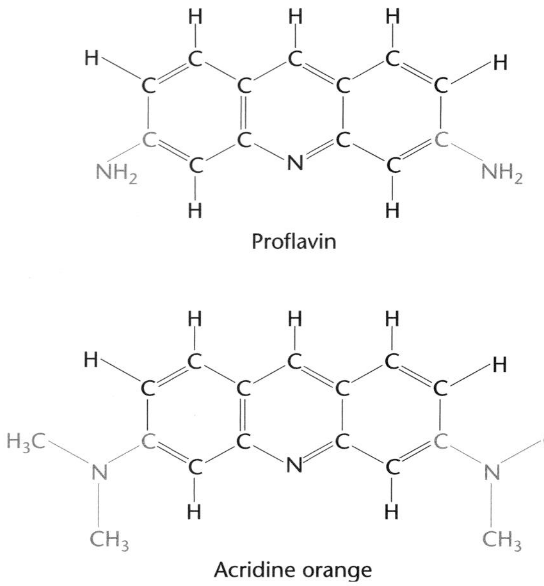

Proflavin

Acridine orange

■ Figure 15–9 Chemical structures of proflavin and acridine orange, which intercalate between bases within the DNA helix, causing frameshift mutations.

C of d-ribose and the 9-N of the purine ring, are called **apurinic sites (AP sites)**. It has been estimated that thousands of such spontaneous lesions are formed daily in the DNA of mammalian cells in culture.

The absence of a nitrogenous base at an AP site will alter the genetic code if the strand involved is transcribed and translated. If replication occurs, the AP site is an inadequate template and may cause replication to stall. If a nucleotide is inserted, it is frequently incorrect, causing still another mutation! Fortunately, as we will soon see, cells contain repair systems that often counteract and correct this type of lesion.

Several other types of lesions are known to be the source of some mutations. In the process of **deamination**, an amino group is converted to a keto group in cytosine and adenine. In these two cases, cytosine is converted to uracil and adenine is changed to hypoxanthine.

The major effect of these changes is to alter the base-pairing specificities of these two molecules during DNA replication. For example, cytosine normally pairs with guanine. Following its conversion to uracil, which pairs with adenine, the original G≡C pair is converted to an A=U pair and, following an additional replication, to an A=T pair. When adenine is deaminated, an original A=T pair is converted to a G≡C pair because hypoxanthine pairs naturally with cytosine. Nitrous acid is a known mutagen that is capable of inducing deamination of bases in DNA.

One other type of mutational lesion is the group caused by oxidation reactions. Active forms of oxygen radicals such as hydrogen peroxide (H_2O_2) and superoxides (O_2^-) can cause damage to DNA bases and result in mispairing during replication.

Ultraviolet Radiation and Thymine Dimers

In Chapter 9 we emphasized the fact that purines and pyrimidines absorb ultraviolet (UV) radiation most intensely at a wavelength of about 260 nm. This property has been used extensively in the detection and analysis of nucleic acids. In 1934, as a result of studies involving *Drosophila* eggs, it was discovered that UV radiation is mutagenic. By 1960, several studies concerning the *in vitro* effect on the components of nucleic acids had been completed, with the following conclusions. The major effect of UV radiation is on pyrimidines, where dimers are formed, particularly between two thymine residues (Figure 15–10). While cytosine-cytosine and thymine-cytosine dimers may also be formed, they are less prevalent. It is believed that the dimers distort the DNA conformation and inhibit normal replication, which seems to be responsible, at least in part, for the killing effects of UV radiation on microorganisms.

For this UV-induced lesion to be mutagenic and not lethal, cells must somehow overcome the inhibition of replication, even if it means inserting an incorrect nucleotide during synthesis. A system in bacteria has been discovered that, when activated, appears to allow the

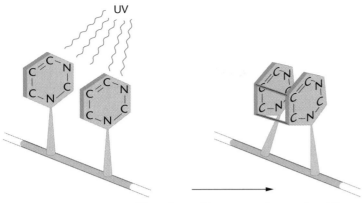

UV

Dimer formed between adjacent thymidine
residues along a DNA strand

■ Figure 15–10 Formation
of a thymine dimer induced
by UV radiation. Note the distortion of the thymine
residues that result.

"block" to be bypassed. The products of several genes
in *Escherichia coli*, including *lexA* and *recA*, somehow
allow the strict adherence of the insertion of complementary bases to be relaxed. While this decreases the fidelity during replication, the process allows the survival
of otherwise lethal effects of UV radiation.

This survival response is also likely to be activated by
other types of lesions that block replication. For example, the AP sites discussed above undoubtedly block
replication in a manner similar to pyrimidine dimers,
and they also induce the SOS response, allowing error-
prone replication to occur. We will return to the subject
of DNA repair, including the specific repair of UV-induced pyrimidine dimers, later in this chapter, where
we will treat the topic more thoroughly.

Mutations in Humans: Case Studies

In this section, we examine the results of two studies
that have investigated the actual gene sequence of various mutations affecting humans. The first provides an
interesting insight into the molecular basis of the **ABO
antigens**, originally presented in Chapter 4 as an example of multiple alleles. The second case involves the nature of the mutations that have led to the devastating
X-linked disorder **muscular dystrophy**.

The ABO system is based on a series of antigenic determinants found on erythrocytes and other cells, particularly epithelial types. As discussed in Chapter 4,
three alleles of a single gene exist, the product of which
is designed to modify the H substance. The modification
involves glycosyltransferase activity, converting the H
substance to either the A or B antigen, as a result of the
product of the I^A or I^B allele, respectively, or failing to
modify the H substance, as a result of the I^O allele.

Using recombinant DNA technology, the responsible
gene has been examined in 14 cases of varying ABO
status. Four consistent nucleotide substitutions were
found when the DNAs of the I^A and I^B alleles were
compared. It is assumed that the changes in the amino
acid sequence of the glycosyltransferase gene product

resulting from these substitutions leads to the different
modifications of the H substance.

The situation with the I^O allele is unique and interesting. Individuals who are homozygous for this allele
are type O, lack glycosyltransferase activity, and fail
to modify the H substance. Analysis of the DNA of
this allele shows one consistent change compared to
that of the other alleles: the deletion of a single nucleotide early in the coding sequence, causing a
frameshift mutation. A complete messenger RNA is
transcribed, but upon its translation, the frame of
reading shifts at the point of deletion and continues
out of frame for about 100 nucleotides before a
"stop" codon is encountered. At this point, premature
termination of the resulting polypeptide chain occurs,
producing a nonfunctional product.

These results provide a direct molecular explanation of
the ABO allele system and the basis for the biosynthesis
of the corresponding antigens. The molecular basis for
the antigenic phenotypes is the result of structural alterations, or mutations, of the nucleotide sequence of the
gene encoding the glycosyltransferase enzyme.

The second case of mutational analysis involves the
severe disorder muscular dystrophy. It is characterized
by progressive muscle degeneration, or myopathy, resulting in the death of affected individuals by early
adulthood. Because the condition is recessive and X-
linked, and because affected males do not reproduce,
females are rarely affected by the disorder. The incidence of 1/3500 live male births makes muscular dystrophy one of the most common life-shortening
hereditary disorders known. Two forms exist:
Duchenne muscular dystrophy (DMD) is more common and more severe than the allelic form, called
Becker muscular dystrophy (BMD).

The region containing the gene has been analyzed extensively and consists of over 2 million base pairs. In
unaffected individuals, transcription results in a messenger RNA containing about 14,000 bases (14 kb) that
is translated into the protein **dystrophin**, consisting of
3685 amino acids. This protein can be detected in most
cases of the less severe BMD, but is rarely found in

DMD. This has led to the hypothesis that most mutations that cause BMD do not alter the reading frame, but that most DMD mutations change the reading frame early in the gene, resulting in premature termination of dystrophin translation. This hypothesis is consistent with the observed differences in severity of these two forms.

In an extensive analysis of the DNA of 194 patients (160 DMD and 34 BMD), J. T. Den Dunnen and associates found that 128 of these mutations (66 percent) consisted of substantial deletions or duplications. Of 115 deletions, 17 occurred in BMD, and of 13 duplications, 1 was in BMD, with the remainder being found in DMD cases. In most cases, the above results were consistent with the "reading frame" hypothesis. With few exceptions, DMD mutations changed the frame of reading of exon areas. BMD mutations usually did not alter the reading frame. Perhaps the most noteworthy finding is the high percentage of deleterious mutations studied that represent the deletion or duplication of nucleotides within the gene. This observation reflects the fact that a mutation caused by a random single nucleotide substitution within a gene is more likely to be tolerated without the devastating effect of muscular dystrophy than the addition or loss of numerous nucleotides that may alter the frame of reading. There are three reasons for this.

1. A nucleotide substitution may not change the encoded amino acid because the code is degenerate.

2. If an amino acid substitution does result, the change may not be present at a location within the protein that is critical to its function.

3. Even if the altered amino acid is present at a critical region, it may still have little or no effect on the function of the protein. For example, an amino acid might be changed to another with nearly identical chemical properties or to one with very similar recognition properties, such as shape.

As a result, single-base substitutions may have little or no effect on protein function, or may simply reduce the efficiency but not eliminate the functional capacity of the gene product. It is clear that we cannot look at mutations with the oversimplified expectation that most of them are single-base substitutions. As more of them are analyzed directly, our picture of mutation will become increasingly clear.

Trinucleotide Repeats in Fragile-X Syndrome, Myotonic Dystrophy, and Huntington Disease

Between 1991 and 1993, molecular analysis of the DNA representing the genes responsible for various human disorders provided a remarkable set of observations. In the cases of three disparate genes, responsible for the X-linked **fragile-X syndrome** and the autosomal disorders **myotonic dystrophy** and **Huntington disease**, an intriguing similarity was discovered: Each gene was found to contain a unique trinucleotide DNA sequence repeated many times. While each repeated sequence is also present in the nonmutant (normal) allele of each gene, what characterizes each mutation is a significant, variable increase in the number of times the trinucleotide is repeated.

In several cases, a correlation has been found between the number of repeats and the onset of mutant gene expression. The greater the number, the earlier disease onset occurs. Of great interest and significance is the fact that the number of repeats may increase in each subsequent generation. This general phenomenon, which is called **genetic anticipation,** represents a unique form of mutation related to an instability of specific regions of the three genes. While other normal genes in humans, as well as many other species, are known to contain trinucleotide repeats, these repeats usually do not balloon in size, and the genes maintain normal function.

The gene responsible for fragile-X syndrome, *FMR-1*, may have several hundred to several thousand copies of the trinucleotide sequence CGG. Individuals with up to 50 copies are normal and do not display mental retardation associated with the syndrome. Individuals with 50 to 200 copies are considered "carriers." While they are normal, their offspring may contain even more copies and express the syndrome.

Myotonic dystrophy, the most common form of adult muscular dystrophy, contains multiple copies of the sequence CTG. Fewer than 35 copies results in normal gene expressions. Above this number, symptoms range from mild myopathy and cataracts to severe dystrophy, retardation, and even death in early childhood. Both severity and onset are correlated directly with the size of the repeated sequence.

Finally, the gene responsible for Huntington disease demonstrates a similar pattern. The trinucleotide CAG, present 10 to 35 times in normal individuals, increases significantly in diseased individuals. Recall from our discussion in Chapter 4 that the age of onset of this disease varies tremendously, most often being expressed in the mid- to late thirties. Much earlier onset occurs when very large numbers of copies are present. Interestingly, in still another disorder, **spinobulbar muscular atrophy (Kennedy disease)**, the gene contains repeated copies of the same triplet. However, diseased individuals usually contain only 30 to 60 copies of the CAG sequence.

The role of such repeated sequences in normal and mutant genes remains a mystery. While Huntington and Kennedy diseases contain the repeat within the coding portion of the gene, this is not the case in the other two disorders. In the gene responsible for fragile-X syndrome, the repeat is upstream (5′) in an area that is most often involved in regulating gene expression. In the case of myotonic dystrophy, the repeat is downstream (the 3′ end).

In addition to the role of the repeat in normal and mutant genes, the mechanism by which the sequence expands from generation to generation is of great interest. How such an instability during DNA replication affects only specific areas of certain genes is currently an important research topic. This instability seems to be more prevalent in humans than in many other organisms.

Detection of Mutagenicity: The Ames Test

There is particular concern about the possible mutagenic properties of any chemical that enters the human body, whether through the skin, digestive tract, or respiratory tract. For example, great attention has been given to residual materials of air and water pollution, second-hand smoke, food preservatives and additives, artificial sweeteners, herbicides, pesticides, and pharmaceutical products. While mutagenicity may be tested in various organisms, including *Drosophila*, mice, and cultured mammalian cells, the most common test involves bacteria and was devised by Bruce Ames.

The **Ames test** utilizes four tester strains of the bacterium *Salmonella typhimurium* that were selected for sensitivity and specificity for mutagenesis. One strain is used to detect specific base-pair substitutions, and the other three detect frameshift mutations. Each mutant strain requires histidine for growth (*his⁻*). The assay measures the frequency of reverse mutation, which yields wild-type (*his⁺*) bacteria. Greater sensitivity to mutagens occurs because these strains bear other mutations that eliminate the DNA excision repair system (discussed later in this chapter) and the lipopolysaccharide barrier that coats and protects the surface of the bacteria.

It is very interesting to note that many substances that enter the human body are relatively innocuous until activated metabolically to a more chemically reactive product. This usually occurs in the liver. Thus, the Ames test, which is performed *in vitro*, includes a step in which the test compound is incubated in the presence of a mammalian liver extract. Or, test compounds are actually injected into a mouse, which is later sacrificed and the liver removed. Extracts are then tested.

In the initial use of Ames testing in the 1970s, a large number of known carcinogens were examined. Over 80 percent of these were shown to be strong mutagens! This is not surprising, since transformation of cells to the malignant state undoubtedly occurs as a result of some alteration of DNA. Although a positive response as a mutagen does not prove the carcinogenic nature of a test compound, the Ames test is useful as a preliminary, screening device. It is used extensively in conjunction with the industrial and pharmaceutical development of chemical compounds.

Repair of DNA

Earlier in this chapter we established that DNA is vulnerable to various forms of errors and lesions that lead to gene mutations following replication. Living systems have evolved a variety of elaborate repair systems that are able to counteract many of the forms of DNA damage that lead to mutation. As we will see, such repair systems are essential to the survival of organisms on earth. For example, in humans the loss of just one type of repair system leads to a devastating, potentially life-shortening genetic disorder, xeroderma pigmentosum.

UV Radiation, Thymine Dimers, and Photoreactivation Repair

As established in Figure 15–10, UV light is mutagenic as a result of the creation of pryimidine dimers. The study of mutagenicity of UV radiation paved the way for the discovery of many naturally occurring forms of repair of DNA damage.

The first relevant discovery concerning UV repair in bacteria was made in 1949, when Albert Kelner observed the phenomenon of **photoreactivation repair.** He showed that the UV-induced damage to *E. coli* DNA could be partially reversed if, following irradiation, the cells were exposed briefly to light in the blue range of the visible spectrum. The photoreactivation repair process has subsequently been shown to be temperature dependent, thereby suggesting that the light-induced mechanism involves an enzymatically controlled chemical reaction. Visible light appears to induce the repair process of the DNA damaged by UV radiation.

Further studies of photoreactivation revealed that the process is due to a protein called the **photoreactivation enzyme (PRE).** This molecule may be isolated from extracts of *E. coli* cells. The enzyme's mode of action is to cleave the bonds between thymine dimers, thus reversing the effect of UV light on DNA [Figure 15–11(a)]. While the enzyme will associate with a dimer in the dark, it must absorb a photon of light to cleave the dimer.

The gene(s) encoding PRE have been preserved throughout evolution. Activity of this repair system has been detected in human cells in culture, as well as in other eukaryotes. The conservation of these genetic components over millions of years suggests that this repair system is extremely important to all organisms.

Excision Repair

Investigations in the early 1960s suggested that a repair system or systems that do not require light also exist in *E. coli*. Paul Howard-Flanders and co-workers isolated several independent mutants demonstrating increased sensitivity to UV radiation. One group was designated *uvr* (UV repair) and included the *uvrA, uvrB,* and *uvrC*

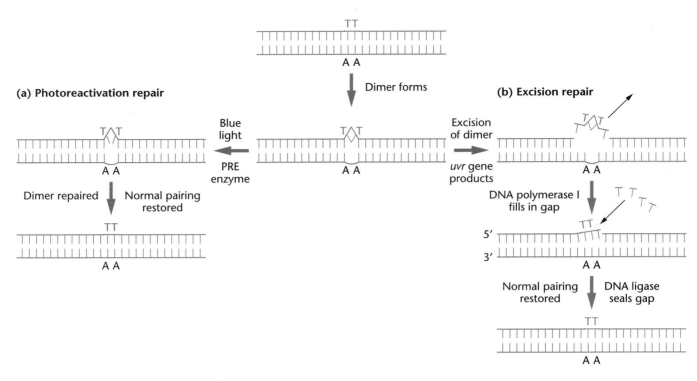

■ Figure 15–11 Contrasting diagrams of (a) photoreactivation repair and (b) excision repair of UV-induced thymine dimers. Actually, 12 bases are excised in prokaryotes and 28 are excised in eukaryotes during excision repair.

mutations. These genes and their protein products were subsequently shown to be involved in a process called **excision repair**. During this process, three steps have been shown to occur, as illustrated in Figure 15–11(b).

1. The distortion of the strand caused by the UV-induced dimer is recognized and enzymatically clipped out by a nuclease that cleaves the phosphodiester bonds. This "excision," which may include several nucleotides adjacent to the dimer as well, leaves a gap in the helix. The *uvr* gene products operate at this step.

2. **DNA polymerase I** fills this gap by inserting d-ribonucleotides complementary to those on the intact strand. The enzyme adds these bases to the 3′-OH end of the clipped DNA.

3. The joining enzyme **DNA ligase** seals the final "nick" that remains at the 3′-OH end of the last base inserted, closing the gap.

DNA polymerase I, the enzyme discovered by Kornberg, was once assumed to be the universal DNA-replication enzyme (see Chapter 10). The discovery of the *polA1* mutation demonstrated that this is not the case. *E. coli* cells carrying the *polA1* mutation lack functional polymerase I. However, replication of DNA takes place normally. Cells with this mutation are unusually sensitive to UV light. Apparently, such cells are unable to fill the gap created by the excision of the thymine dimers. This finding demonstrates the importance of excision repair mechanisms in counteracting the effects of UV light.

It has been shown subsequently that the process of excision repair can be activated in response to any damage to DNA that distorts the helix. For example, as we discussed earlier, the loss of a purine from d-ribose of one strand creates what is called an **apurinic site (AP site)**. The complementary pyrimidine on the opposite strand has nothing with which to form hydrogen bonds. Such a sugar with a missing base is recognized by an enzyme called **AP endonuclease**. The endonuclease makes a cut in the polynucleotide chain at the AP site. This creates the distortion that is recognized by the excision-repair system, which is then activated, leading ultimately to the correction of the error (the AP site).

Other enzymes recognize incorrect bases and stimulate repair. For example, one specific member of a group of enzymes called **DNA glycosylases** recognizes the presence of uracil when it is part of DNA. It cuts the glycosidic bond between the base and sugar, creating an AP site, which is then repaired as discussed above. Glycosylases are important repair components because, when created by deamination of cytosine, uracil will lead to a C≡G to T=A transition mutation after replication if it is not repaired.

In theory, excision repair may serve as the final step in a variety of repair processes, provided that the lesion or distortion in DNA may be recognized.

Proofreading and Mismatch Repair

As we pointed out in our discussion of DNA synthesis in Chapter 10, DNA polymerase III possesses a **proofreading** function. During polymerization, when an incorrect nucleotide is inserted, the enzyme complex has the potential to recognize the error and "reverse" it by cutting out the incorrect nucleotide and replacing it. In bacterial systems, proofreading is thought to increase fidelity during synthesis by two orders of magnitude. If initial mismatches occur in $1/10^5$ nucleotide pairs (a rate of 10^{-5}), proofreading decreases final mismatches to $1/10^7$ (a rate of 10^{-7}).

To cope with those errors that remain after proofreading, still another mechanism, called **mismatch repair,** may be activated. Proposed over 20 years ago by Robin Holliday, the molecular basis of this process is now well established. Like other DNA lesions, (1) the alteration or mismatch must be detected, (2) the incorrect nucleotide must be removed, and (3) replacement with the correct base must occur. But a special problem exists with correction of a mismatch. How does the repair system recognize which strand is correct (the template strand) and which contains the mismatched base (the newly synthesized strand)? How the repair system discriminates and recognizes the "new" nucleotide puzzled geneticists for decades. If the mismatch is recognized but no discrimination occurred and excision was random, half the time the strand bearing the correct base would be clipped out. The concept of strand discrimination by a repair enzyme is thus a critical step.

At least in some bacteria, including *E. coli*, this process has been elucidated and is based on the process of **DNA methylation**. These bacteria contain an enzyme, **adenine methylase**, that recognizes the DNA sequence

$$5' \ldots \text{G A T C} \ldots 3'$$
$$3' \ldots \text{C T A G} \ldots 5'$$

as a substrate. Upon recognition, a methyl group is added to each of the adenine residues. This modification is stable throughout the cell cycle.

Following a further round of replication, the newly synthesized strands remain temporarily unmethylated. It is at this point that the repair enzyme recognizes the mismatch and preferentially binds to the unmethylated strand. It is excised and the correct complement is inserted. Interestingly, the GATC sequence need only be within several thousand base pairs of the mismatch. A series of *E. coli* gene products, MutH, L, S, and U are involved in the discrimination step. Mutations in each result in strains that are deficient in mismatch repair.

Although it is agreed that mismatch repair undoubtedly also occurs in the DNA of higher organisms, the question of strand discrimination remains speculative in the absence of GATC methylation.

The SOS Response: Recombinational Repair

The final mode of repair to be discussed was discovered in an excision-defective strain of *E. coli* and first proposed by Miroslav Radman. Called **recombinational repair,** this system is thought to respond when damaged DNA has escaped repair and the damage disrupts the process of replication. Because this system "responds" to a signal of distress, so to speak (DNA damage), Radman initially referred to it as an **SOS response.** The cells that show this phenomenon are dependent on the product of a gene, *recA*, which is involved in several types of recombinational phenomena in *E. coli*.

When DNA bearing a lesion of some sort is being replicated, DNA polymerase at first stalls at and then skips over the distortion, creating a gap on one of the newly synthesized strands. To counteract this, the RecA protein directs a recombinational exchange process whereby this gap is filled as a result of the insertion of a segment that was initially present on the intact homologous strand. This creates a gap on the "donor" strand, which is filled by repair synthesis as replication proceeds.

Phil Hanawalt and Paul Howard-Flanders, among others, have established that as many as 20 different gene products are involved in this mode of repair. Of particular interest is the LexA protein product, which, when produced, serves to partially repress the transcription of the *recA* and *uvr* genes. However, when a RecA protein binds to single-stranded DNA in the area of a gap, this binding somehow activates a second function of the RecA protein—the ability to cleave the LexA repressor molecule, disrupting its regulatory capacity. The absence of a functional repressor molecule allows the activation of the *recA* and *uvr* genes, among others, leading to an increased production of the proteins for which they code. These products complete the repair process.

UV Radiation and Human Skin Cancer: Xeroderma Pigmentosum

The essential nature of any biological system can be assessed by examining the effects of the failure of the system. Regrettably, we can determine just how essential DNA repair is in humans by examining individuals who exhibit an inherited loss of function of one of the major repair systems. **Xeroderma pigmentosum (XP)**, a rare autosomal recessive disorder in humans, predisposes individuals to epidermal pigment abnormalities. Exposure to sunlight results in malignant skin tumors.

Figure 15–12 contrasts two XP individuals, one of whom was detected early and protected from sunlight. The condition is very severe and may be lethal. Because sunlight contains UV radiation, a causal relationship has been predicted between thymine dimer production and XP. It has also been of great interest to determine which of the three forms of repair processes that counteract the effects of UV-induced damage to DNA (if any) are operating in humans. It was suspected that XP individuals might lack one or more repair systems, which may cause them to be susceptible to UV-induced skin damage.

The modes of repair of UV-induced lesions have been investigated in human fibroblast cultures derived from XP and normal individuals. Fibroblasts are undifferentiated connective tissue cells. The results are varied and suggest that the XP phenotype may be caused by more than one mutant gene.

In 1968, James Cleaver showed that cells from XP patients were deficient in the **unscheduled DNA synthesis** elicited in normal cells by UV light. This assay is thought to represent activity of the excision repair system, suggesting that XP cells are deficient in this form of repair. In 1974, the presence of a **photoreactivation enzyme (PRE)** was established in normal human cells. Betsy Sutherland identified the enzyme first in leukocytes and subsequently in fibroblast cells. Sutherland demonstrated further that some XP cultures contain a lower PRE activity than control cultures. The activity in various XP cell strains ranges from 0 to 50 percent in cultures established from different patients.

The link between xeroderma pigmentosum and inadequate excision repair has been strengthened by the use of **somatic cell hybridization**, a technique we first introduced in Chapter 6. Cultured fibroblast cells from any two unrelated XP patients may be induced to fuse together, forming a heterokaryon in which the two nuclei share a common cytoplasm. Once fusion is achieved, excision repair, as assayed by unscheduled DNA synthesis, is assessed. Sometimes repair is reestablished in the heterokaryon. When this occurs, the two variants are said to demonstrate **complementation**. Alone, neither cell type demonstrates excision repair, but together in a heterokaryon the process occurs. In genetic terms, this is strong evidence that the two patients from whom the cells were derived have different genes affected that led to the disease. Complementation occurs because the heterokaryon has at least one normal copy of each gene.

Based on many studies, most patients have been divided into seven complementation groups, which suggests that at least seven different genes may be involved in excision repair. These genes or their protein products have now been identified. They are found on disparate regions of the genome, and a homologous gene for each has been identified in yeast. Approximately 20 percent of XP patients do not fall into any of the seven groups. They manifest similar symptoms, but their fibroblasts do not demonstrate defective excision repair. There is some evidence that they are less efficient at normal DNA replication.

The study of xeroderma pigmentosum established that normal individuals are susceptible to UV-induced damage of DNA by exposure to sunlight. However, this damage activates the repair systems that counteract it. We can expect that future work will clarify the precise role and mechanism of repair of UV-induced damage in humans.

High-Energy Radiation

Within the **electromagnetic spectrum**, energy varies inversely with wavelength (Figure 15–13). **X-rays, gamma rays,** and **cosmic rays** have even shorter wavelengths than UV light and are therefore more energetic. As a result, they are strong enough to penetrate deeply into tissues, causing ionization of the molecules encountered along the way. These sources of **ionizing radiation** could be predicted to be mutagenic, as established by Herman Muller and Lewis Stadler in the 1920s. Since

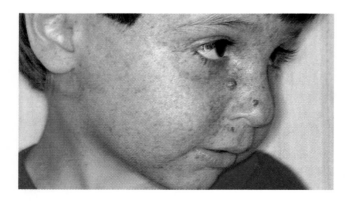
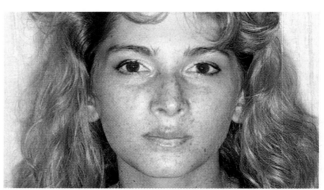

■ Figure 15–12   Xeroderma pigmentosum. The 4-year-old boy shows marked skin lesions induced by sunlight. Mottled redness (erythema) and irregular pigment changes that are a response to cellular injury are apparent. Two nodular cancers are present on his nose. The 18-year-old girl has been protected from sunlight since the diagnosis of xeroderma pigmentosum made in infancy. Several cancers have been removed.

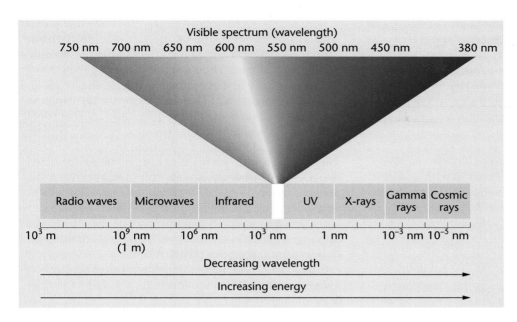

■ Figure 15–13 The components of the electromagnetic spectrum and their associated wavelengths.

that time, the effects of ionizing radiation, particularly X-rays, have been studied intensely.

As X-rays penetrate cells, electrons are ejected from the atoms of molecules encountered by the radiation. Stable molecules and atoms can be transformed into free radicals and reactive ions. Along the path of a **high-energy ray**, a trail of ions is left that can initiate a variety of chemical reactions. These reactions can affect the genetic material either directly or indirectly, altering the purines and pyrimidines in DNA and resulting in point mutations. Such ionizing radiation is also capable of breaking phosphodiester bonds, thus disrupting the physical integrity of chromosomes. This results in a variety of aberrations.

Figure 15–14 shows a plot of induced X-linked recessive lethal mutations versus the dose of X-rays administered. The graph shows a straight line that, if extrapolated, intersects near the zero axis. A linear relationship is evident between X-ray doses and the induction of mutation. For each doubling of dose, twice as many mutations are induced. Because the line intersects near the zero axis, this graph suggests that even very small doses of irradiation are mutagenic.

These observations may be interpreted in the form of the **target theory**, first proposed in 1924 by J. A. Crowther and F. Dessauer. The theory proposes that there are one or more sites, or targets, within cells and that a single event of irradiation at one site will bring about a damaging effect, or mutation. In a simple form, the target theory says that one "hit" of irradiation will cause one "event" or mutation, suggesting that the X-rays interact directly with the genetic material.

Two other observations concerning irradiation effects are of particular interest. First, in some organisms studied, the intensity of the dose (dose rate) administered seems to make little difference in mutagenic effect.

That is, a 100-roentgen exposure (a **roentgen** is a measure of energy dose), whether occurring in a single acute dose or cumulatively in many smaller chronic doses, seems to produce the same mutagenic effect. *Drosophila*, for example, shows this response. In mammals such as mice and humans, however, this is not the case. Repair of the damage seems to occur during the intervals between irradiation. As a result, several smaller doses are not as potent as a single large dose.

The second observation is that certain portions of the cell cycle are more susceptible to irradiation effects. As mentioned previously, in addition to a mutagenic effect, X-rays can also break chromosomes, resulting in terminal or intercalary deletions, translocations, inversions,

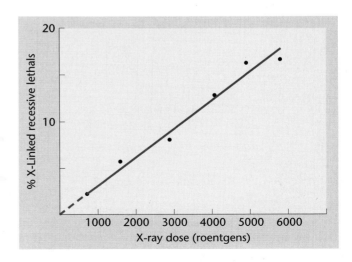

■ Figure 15–14 Plot of X-linked recessive mutations induced by increasing doses of X-rays. When extrapolated, the graph intersects the zero axis.

and general chromosome fragmentation. Damage occurs most readily when chromosomes are greatly condensed in mitosis. This property constitutes one of the reasons why radiation is used to treat human malignancy. Because tumorous cells undergo division more often than their nonmalignant counterparts, they are more susceptible to the immobilizing effect of radiation.

Site-Directed Mutagenesis

This section introduces a useful experimental technique, **site-directed mutagenesis**, that allows researchers to introduce a designed mutation at a prescribed site within a gene of interest. The technique relies on the availability of a cloned gene and utilizes a number of manipulations involving recombinant DNA technology, which will be introduced in Chapter 17. However, the underlying principles are based on information previously presented.

The goal of the technique is to alter one or more specific nucleotides within a gene in order to change a specific triplet codon. Upon transcription and translation, this change will cause the insertion of a "mutant" amino acid into the protein encoded by the original gene. Such designed mutations are particularly useful in studying the effects of genetic change on protein function.

The first step is to determine the nucleotide sequence of the gene being studied. This can be accomplished by DNA sequencing techniques, or it can be predicted if the amino acid sequence of the protein is known, by utilizing our knowledge of the genetic code.

The next step is to isolate the DNA of known sequence and obtain from it one of the two complementary strands. A decision is then made as to which nucleotides are to be changed. Then a small piece of DNA that is complementary to that region is synthesized chemically. It is complementary at all points, except in the triplet sequence (or sequences) that is to be altered. This triplet encodes the amino acid that will change in the protein.

As illustrated in Figure 15–15, this short piece of DNA is hybridized with the original parent strand, forming a partial duplex because of its complementarity along most of its length. If DNA polymerase and DNA ligase are then added to this hybrid complex, the short sequence is extended so that a duplex of the entire gene is formed. The two strands are perfectly complementary except at the point of alteration.

If this DNA undergoes semiconservative replication, two types of duplexes are formed: One is like the original, unaltered gene and the other contains the newly designed sequence. Using recombinant DNA technology, it is possible not only to complete the above manipulations, but also to allow the altered gene to be expressed so that large amounts of the desired protein are available for study. Or, the altered gene may be introduced into the genome of an organism and studied.

Knockout Genes and Transgenes

While a gene bearing a site-directed mutation may be investigated *in vitro,* the insertion of that gene into a living organism allows for the assessment of its *in vivo*

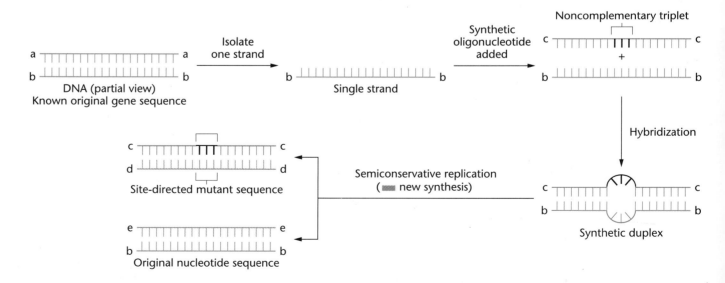

■ Figure 15–15 Site-directed mutagenesis, achieved by first obtaining a single strand of DNA from a gene of interest. This is hybridized with the complementary synthetic oligonucleotide containing one of the triplets altered so as to encode an amino acid of choice. Following semiconservative replication of the synthetic duplex, a different complementary triplet is present in one of the new duplexes. Upon transcription and translation, a mutant protein, "designed" in the laboratory, will be produced.

function. If the insertion process involves the *replacement* of the comparable gene of the organism from which it originated, the technique is referred to as **gene knockout**. The genetically altered organism is called a **knockout organism**, e.g., a "knockout mouse."

While specific gene replacement is difficult to achieve, it has been accomplished and is now fairly routine in research involving yeast and mice. In mice, the application of gene knockout techniques has been particularly fruitful in studies involving the genetic control of early development and behavior. Most often, a "loss-of-function" mutation replaces the normal gene, and the effects are investigated.

Furthermore, knockout mice now serve as models for studying human genetic disorders. In such cases, the comparable mouse gene that causes the human disorder is isolated, subjected to directed mutagenesis, and used to replace the nonmutant mouse gene. Cystic fibrosis and Duchenne muscular dystrophy have been investigated in this way. Applications of this general technology are discussed in more detail in Chapter 18.

If a gene is inserted into an organism *in addition* to its normal copies, it is called a **transgene**, and the organism is called a **transgenic organism** (e.g., a "transgenic plant"). In such cases, the gene may have undergone directed mutagenesis, or it may be a foreign gene isolated from another organism. This technology has been used more extensively than gene knockout, since it is easier to accomplish. This is so because specific replacement is not required.

The study of transgenic organisms is performed routinely in plants, *Drosophila*, mice, and a variety of other organisms. Of particular note is the potential provided in agricultural studies, as well as gene therapy techniques in our own species.

Transposable Genetic Elements

We conclude this chapter by introducing the phenomenon of **transposable genetic elements**. Sometimes referred to as **transposons**, these elements encompass a group of genetic units that are mobile. They can move or be "transposed" within the genome. Discussion of them in this chapter on mutation is appropriate because the movement of genetic units from one place in the genome to another often disrupts genetic function and results in phenotypic variation. As such, the impact of transposition of genetic units often fits into a broad definition of mutation.

As we shall see, transposable elements were first discovered almost 50 years ago in maize. However, even in the 1950s and 1960s, the idea that genetic information was not fixed within the genome of an organism was slow to find acceptance. Such a notion was quite alien to the classical interpretation of genes on chromosomes. It was not until other transposable elements

were discovered, and their molecular basis revealed, that the phenomenon was considered to be a bona-fide genetic process.

Insertion Sequences

Even though the presence of transposable elements in maize had been predicted earlier by Barbara McClintock, the first observation at the molecular level involved **insertion sequences (ISs)** in *E. coli*. Discovered in the early 1970s by a number of independent researchers, including Peter Starlinger and James Shapiro, insertion sequences were first visualized as a unique class of mutations affecting different genes in various bacterial strains. For example, the expression of a cluster of related genes involving galactose metabolism was repressed as a result of one such mutation.

This phenotypic effect was heritable, but was found not to be caused by a base-pair change characteristic of conventional gene mutations. Instead, it was shown that a short, specific DNA segment had been inserted into the bacterial chromosome at the beginning of the galactose gene cluster. When this segment was spontaneously excised from the bacterial chromosome, wild-type function was restored.

It was subsequently revealed that several other distinct DNA segments could behave in a similar fashion, inserting into the chromosome and affecting gene function. These DNA segments are relatively short, not exceeding 2000 base pairs [2 kilobases (kb)]. For example, the first insertion sequence to be characterized in *E. coli* was IS1. It is about 800 base pairs long, whereas IS2, 3, 4, and 5 are about 1250 to 1400 base pairs in length.

Analysis of the DNA sequences of most IS units reveals a feature that is important to their mobility. At each end, the nucleotide sequences consist of **inverted terminal repeats (ITRs)** of one another. Though Figure 15–16 shows this terminal repeating unit to consist of only a few nucleotides, many more are actually involved. For example, in *E. coli*, the IS1 termini contain about 20 nucleotide pairs, IS2 and IS3 about 40 pairs, and IS4 about 18 pairs. It seems likely that these terminal sequences are an integral part of the mechanism of insertion of IS units into DNA. That insertion of IS units

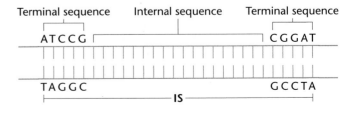

■ Figure 15–16 Diagrammatic representation of an insertion sequence (IS). The terminal sequences are perfect inverted repeats of one another.

is more likely to occur at certain DNA regions than others suggests that IS termini can recognize certain target sequences in the DNA during the process of insertion.

Careful investigation has revealed that IS units are present in the wild-type *E. coli* chromosome as well as in other autonomous segments of bacterial DNA called **plasmids**. Thus, their presence does not always result in mutation. In the *E. coli* chromosome, five or more copies of IS1, IS2, and IS3 are known to be present. The exact number of each varies, depending on the strain examined.

Bacterial Transposons

In addition to their potential mutational effects, IS units play an even more significant role in the formation and movement of the larger **transposon (Tn) elements**. Transposons in bacteria consist of IS units that contain within their internal DNA sequence genes whose functions are unrelated to the insertion process. Like IS units, Tn elements are mobile in both bacterial and viral chromosomes and in plasmids. The Tn elements provide a mechanism for movement of genetic information from place to place both within and between organisms. Transposons were first discovered to move between DNA molecules as a result of observations of antibiotic-resistant bacteria. In the mid-1960s, Susumu Mitsuhashi first suggested that genes responsible for resistance to several antibiotics were mobile and could move between bacterial plasmids and chromosomes.

Transposons have become the focus of increased interest, particularly because they have been found in organisms other than bacteria. Bacteriophages that demonstrate the ability to insert their genetic material into the host chromosome behave in a similar fashion. The **bacteriophage mu**, consisting of over 35,000 nucleotides, can insert its DNA at various places in the *E. coli* chromosome. Like IS units, if insertion occurs within a gene, mutant behavior at that locus results. Transposons have also been discovered in higher organisms, including yeast, corn, *Drosophila*, and humans.

The Ac-Ds *System in Maize*

About 20 years before the discovery of transposons in prokaryotic organisms, Barbara McClintock analyzed the genetic behavior of two mutations, ***Dissociation (Ds)*** and ***Activator (Ac)***, in corn plants (maize). Analysis involved an examination of the phenotypes of the kernels of maize. These observations were correlated with cytological examination of the maize chromosomes. Initially, McClintock determined that *Ds* is located on chromosome 9. Provided that *Ac* is also present in the genome, *Ds* has the effect of inducing breakage at a point on the chromosome adjacent to its location. If breakage occurs in somatic cells during their development, progeny cells often lose part of chromosome 9, causing a variety of phenotypic effects.

Subsequent analysis suggested to McClintock that both the *Ds* and *Ac* genes are sometimes moved, or *transposed*, to new locations in the genome. While *Ds* moves only if *Ac* is present, *Ac* is capable of autonomous movement. Where *Ds* comes to reside determines its genetic effect. Although it might cause chromosome breakage, it might instead inhibit gene expression. In cells where expression is inhibited, *Ds* might move again, releasing this inhibition. In these cases, the *Ds* element is believed to insert into a gene and subsequently to depart from it, causing changes in gene expression. McClintock concluded that the *Ds* and *Ac* genes are **transposable controlling elements**.

It was not until many years later that anything comparable to the controlling elements in maize was recognized in other organisms. When bacterial insertion sequences and transposons were discovered, many parallels were evident. Transposons and insertion sequences were seen to move into and out of chromosomes, to insert at different positions, and to affect gene expression at the point of insertion.

Several *Ac* and *Ds* elements have now been isolated and carefully analyzed. As a result of this information, the relationship between the two elements has been clarified. Several examples of these interactive influences are depicted in Figure 15–17. There is some evidence that the *Ac* element is a gene that encodes a **transposase enzyme**, essential to transposition of both *Ac* and *Ds* elements.

While the validity of McClintock's proposed mobile elements was questioned following her initial observations, molecular analysis has since verified her conclusions. For her work, Barbara McClintock was awarded the Nobel Prize in 1983.

Copia *and P Elements in* Drosophila

Transposable elements have been discovered in still other eukaryotic organisms, notably in yeast, *Drosophila*, and primates, including humans. In 1975, David Hogness and his colleagues, David Finnigan, Gerald Rubin, and Michael Young, identified a class of genes in *Drosophila melanogaster* which they designated as *copia*. These genes transcribe "copious" amounts of RNA (thus, their name). Present up to 30 times in the genome of cells, *copia* genes are nearly identical in nucleotide sequence. Mapping studies show that they are transposable to different chromosomal locations and are dispersed throughout the genome.

Copia **elements** appear to be only one of approximately 30 families of transposable elements in

(a)

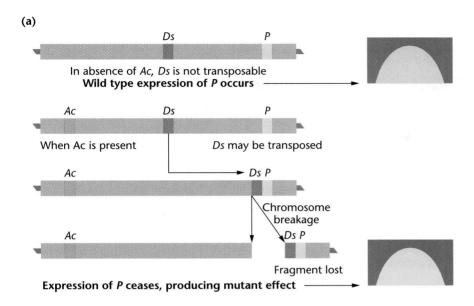

In absence of *Ac*, *Ds* is not transposable
Wild type expression of *P* occurs ⟶

When Ac is present   *Ds* may be transposed

⟶ Ds P

Chromosome breakage

Ds P

Fragment lost

Expression of *P* ceases, producing mutant effect ⟶

(b)

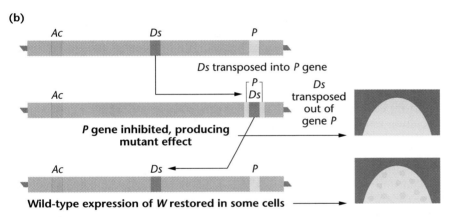

Ds transposed into *P* gene

Ds transposed out of gene *P*

P gene inhibited, producing mutant effect

Wild-type expression of *W* restored in some cells ⟶

■ Figure 15–17   Two consequences of the influence of the *Activator* (*Ac*) element on the *Dissociation* (*Ds*) element. In (a), *Ds* is transposed to a region adjacent to a theoretical gene *P*, which is responsible for pigment formation. Subsequent chromosomal breakage is induced, the *P*-bearing segment is lost, and mutant gene expression occurs, eliminating orangish pigment from the kernels. In (b), *Ds* is transposed to a region within the *P* gene, also causing mutant expression. *Ds* may "jump" out of the *P* gene in some cells, with the accompanying restoration of *P* gene activity, leading to spots of pigmentation on the kernels.

Drosophila, each of which is present 20 to 50 times in the genome. Together, these families constitute about 5 percent of the *Drosophila* genome and over half of the middle repetitive DNA of this organism. One estimate projects that 50 percent of all visible mutations in *Drosophila* are the result of the insertion of transposons into otherwise wild-type genes!

Despite the variability in DNA sequence between the members of different families, they share a common structural organization thought to be related to the insertion and excision processes of transposition. Each *copia* gene consists of approximately 5000 base pairs of DNA, including a long family-specific **direct terminal repeat (DTR)** sequence of 276 base pairs at each end. Within each repeat is a short **inverted terminal repeat (ITR)** of 17 base pairs. These features are illustrated in Figure 15–18. The DTR sequences are found in other transposons in other organisms but are not universal. However, the shorter ITR sequences are considered universal.

Insertion of *copia*, as with other transposable elements, appears to be dependent on ITR sequences and appears to occur at specific target sites in the genome. In general, eukaryotic transposons are strikingly similar to one another and share many features with those in bacteria.

Still another interesting category of transposable elements in *Drosophila* is the family called **P elements**. These were discovered while studying the phenomenon of **hybrid dysgenesis**, a condition that causes sterility, elevated mutation rate, and chromosome rearrangement in the offspring of crosses between certain strains of fruit flies.

Hybrid dysgenesis is caused by high rates of P-element transposition in the germ line, where these mobile DNA elements insert themselves into or near genes, thereby causing mutations. P elements are 2.9 kb long, with 31-bp terminal inverted repeats. The elements encode at least two proteins, one of which is the transposase enzyme that is required for transposition. The transposase is expressed only in the germ line, accounting for the tissue specificity of P-element transposition.

Mutations may arise from several kinds of insertional events. If a P element inserts into the coding region of a gene, it can destroy the normal gene product. If it inserts into the promoter region of a gene, it can affect the level of expression of the gene. Insertions into introns can affect splicing or cause premature termination of transcription.

Transposable Elements in Humans

Another class of mobile eukaryotic genetic units is the **Alu family**. Found in mammals, including humans, the *Alu* family consists of large numbers of short repetitive sequences, originally detected using reassociation kinetics. In humans, there are about 300,000 copies of this 200-to-300-base-pair sequence found interspersed throughout the genome. Because of their small size, they are members of a group of similar elements called **short interspersed elements (SINEs)**. These sequences represent much of the repetitive DNA in humans and constitute 3 percent of the total genome. The family has received its name because a high percentage of these sequences are cleaved by the restriction endonuclease *Alu I*.

The significance of the *Alu* family in genetic processes is based on the observations of their presence in the genomes of all primates as well as the fact that *Alu* sequences are represented in nuclear transcripts. These sequences are considered transposable based on several lines of evidence. Most important is the fact that they contain a 300-bp sequence flanked on either side by direct repeat sequences consisting of 7 to 20 bp. These flanking sequences are similar to those of bacterial insertion sequences that are related to the insertion process during transposition.

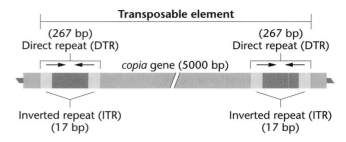

■ Figure 15–18 Structural organization of a *copia* transposable element in *Drosophila melanogaster*, illustrating the inverted terminal repeats.

rial insertion sequences that are related to the insertion process during transposition.

The potential mobility and mutagenic effect of these and other elements have far-reaching effects. A recent example involves a situation where a transposon has been "caught in the act." The case involves **hemophilia**, as investigated by Haig Kazazian and his colleagues. One cause of hemophilia is a defect in blood-clotting factor VIII, the product of an X-linked gene. Kazazian found two cases of male children where, sitting within this gene, there was a transposable element, much longer than *Alu* sequences, called a **LINE** (a long, interspersed nucleotide sequence). It is estimated that there are 100,000 LINEs of various sequences scattered throughout the human genome.

There has been great interest in determining if one of the mother's X chromosomes also contains this specific LINE. If so, the unaffected mother would be heterozygous and pass the LINE-containing chromosome to her son. The startling finding is that the LINE sequence is *not* present on either of her X chromosomes, but *was* detected on chromosome 22 of both parents. This suggests that this mobile element may have "moved" from one chromosome to another in the gamete-forming cells of the mother, prior to being transmitted to the son.

Many questions remain concerning this and other transposable elements. What is their origin? Were they once some sort of retrovirus? Exactly how do they move, and what has been their role during evolution? These and other questions will intrigue researchers for many years to come.

GENETICS, TECHNOLOGY, AND SOCIETY

Chernobyl's Legacy

On April 26, 1986, reactor #4 of the Chernobyl Nuclear Power Station overheated, exploded, and ejected massive amounts of radioactive material into the surrounding countryside and ultimately throughout the Northern Hemisphere. The accident killed 31 emergency workers and caused acute radiation sickness in over 200 others. In the nine days following the initial explosion, as the reactor's temperature approached meltdown, radioactive fission products continued to be discharged into the environment. High levels of radioactive iodine, xenon, strontium, and cesium were released into the atmosphere, along with over 5000 tons of vaporized boron carbide, dolomite, clay, sand, and lead. The plume of fallout traveled in a northwesterly direction through central Europe, reaching Finland and Sweden three days after the initial explosion. By May 5, the radioactive cloud reached the United Kingdom, and remnants of the cloud arrived in North America one week later. Millions of people in the Soviet Union and Europe were exposed to measurable amounts of radioactivity. People living within 30 km of Chernobyl were exposed to high levels of radioactivity prior to their evacuation 36 hours following the accident. Equally large were the exposures of some of the 600,000 military and civilian workers who were sent to Chernobyl to decontaminate the area and to encase the shattered reactor in a sarcophagus.

Chernobyl was the world's largest accidental release of radioactive material. In addition, it was a disaster that inflicted enormous economic and personal burdens on the Soviet people. The question that remains, over a decade after the accident, is whether Chernobyl's radioactive pollution directly threatens the long-term health of millions of people.

More is known about the effects of ionizing radiation on human health than any other agent (except perhaps cigarette smoking). Radiation refers to the energy that is emitted from a source, and includes heat, light, radio waves, microwaves, X-rays, and gamma rays from radioactive elements. Ionizing radiation (such as X-rays and gamma rays) is a subset of radiation that possesses sufficient energy to eject electrons from atoms. Ionizing radiation can damage any cellular component, and can alter nucleotides and induce strand breaks in DNA. These DNA lesions, if unrepaired by the cell, can lead to mutations or chromosomal translocations. It is this DNA-damaging capacity of ionizing radiation that makes it a mutagen.

We know from epidemiological studies that high doses of ionizing radiation increase the risk of developing certain cancers. Survivors of the atomic bomb blasts of Hiroshima and Nagasaki showed an increased incidence of leukemia within two years of the bombing. Leukemias peaked within 10 years and then declined. Breast cancers began to increase 10 years after exposure, as did cancers of the lung, thyroid, colon, ovary, stomach, and nervous system. However, the incidence of other cancers, such as those of the gall bladder, pancreas, uterus, and bone, did not increase due to radiation exposure. As ionizing radiation is known to induce DNA damage, offspring of A-bomb survivors were expected to show increases in birth defects. However, increases beyond rates expected in the unexposed population were not detectable.

The problem in extrapolating from Hiroshima to Chernobyl is the difference in dose. In A-bomb survivors, the rate of incidence of cancer increased among those exposed to at least 200 mSv (mSv=millisievert, a unit dose of absorbed radiation). It is estimated that people living in the most contaminated areas close to Chernobyl may have received about 50 mSv, and cleanup workers were exposed to approximately 250 mSv. Outside the Chernobyl area, radiation doses are estimated at 0.4–0.9 mSv in Germany and Finland, 0.01 mSv in the U.K., and 0.0006 mSv in the United States.

In order to put these numbers into context, the average dose for medical diagnostic procedures (such as chest and dental X-rays) is 0.39 mSv per year. A person's radiation exposure from natural background sources (cosmic rays, rocks, and radon gas) is about 2–3 mSv per year. Smokers expose themselves to an average of about 2.8 mSv extra per year from intake of naturally occurring radioactive materials in tobacco smoke. A whole-body X-ray dose of 3000 mSv or more can be fatal.

Of the 115,000 people evacuated from Chernobyl, it was estimated that 26 leukemias might appear above the 25 to 30 that would occur spontaneously. It has also been estimated that up to 17,000 additional cancers might occur in Europe, above the 123 million that would occur normally. At present, however, there have been no detectable increases in leukemia in contaminated areas of the Soviet Union, Finland, or Sweden, or in the 600,000 Chernobyl cleanup workers. It is still too early to detect increases in solid tumor incidence, which should only begin to appear 10 years after the accident.

Despite the lack of detectable increases in leukemias and solid tumors, one type of cancer does appear to be increasing in the Chernobyl area. The rate of childhood thyroid cancer has reached over 100 cases per million children per year, whereas normal rates would be expected to be between 0.5 and 3 cases per million children per year. The regions with the greatest contamination appear to have the highest rate of thyroid cancer. Although epidemiologists debate whether these increases are due to radiation or to increased reporting of cases, the scale of the increase, and the fact that radioactive isotopes of iodine comprised a significant portion of the Chernobyl fallout, make the link between thyroid cancer and Chernobyl fallout a plausible one.

The most profound effects of Chernobyl have been psychological. A recent study of Chernobyl cleanup workers found no increases in cancers, including leukemias or thyroid cancers. However, there was a 50 percent increase in suicides and a detectable increase in smoking and alcohol-related disease. This mirrors other studies showing that 45 percent of people living within 300 km of Chernobyl believe that they have a radiation-induced illness. These people may be correct, as health effects such as depression, sleep disturbance, hypertension, and altered perception have been documented. People suffer from the stress of relocation, loss of personal

(continues)

(continued)

and real property, and they lack a feeling of control over their radiation exposure or their future. Posttraumatic stress may be a greater threat to health than the actual radiation exposure from the accident. Psychological counseling for cleanup workers and for victims living in contaminated areas is nonexistent. Without solid information about their real risks, people feel that they live in constant danger and are simply awaiting the results of a cancer "lottery." Even if cancer rates and genetic defects do not increase dramatically, the indirect health effects from the Chernobyl Nuclear Power Plant explosion are, and continue to be, immense.

References

Anspaugh, L. R., Catlin, R. J., and Goldman, M. 1988. The global impact of the Chernobyl reactor accident. *Science* 242:1513–19.

Ginzberg, H. M. 1993. The psychological consequences of the Chernobyl accident—findings from the International Atomic Energy Agency study. *Public Health Rep.* 108:184–92.

Rahu, M., et al. 1997. The Estonian study of Chernobyl cleanup workers: II. Incidence of cancer and mortality. *Radiation Res.* 147:653–57.

Williams, D. 1994. Chernobyl, eight years on. *Nature* 371:556.

Chapter Summary

1. The phenomenon of mutation not only provides the basis for most of the inherent variation present in living organisms, but also serves as the working tool of the geneticist in studying and understanding the nature of genetic processes.

2. Mutations are distinguished by the tissues affected. Somatic mutations are those that may affect the individual but are not heritable. Mutations arising in gametes may produce new alleles that can be passed on to offspring and can enter the gene pool.

3. Another classification of mutations relies on their effect. Morphological mutations, for example, may be detected visibly. Other types include biochemical, lethal, conditional, and regulatory mutations, groups that are not mutually exclusive.

4. Spontaneous mutations may arise naturally as a result of rare chemical rearrangements of atoms, or tautomeric shifts, and as the result of errors occurring during DNA replication. Although spontaneous mutations are very rare, their rate of occurrence may be increased experimentally by a variety of mutagenic agents.

5. Mutagenic agents such as base analogs as well as alkylating and deaminating agents cause chemical changes in nucleotides that alter their base-pairing affinities. As a result, base substitution mutations arise following DNA replication. Mutagenicity of chemicals can be assessed using the Ames test.

6. Frameshift mutations, induced specifically by acridine dyes, arise when the addition or deletion of one or more nucleotides (but not multiples of three) occurs.

7. Direct analysis of DNA from individuals with specific ABO blood types and muscular dystrophy has been informative. Complete loss of function, as is the case in blood type O and the Duchenne form of muscular dystrophy, has occurred when deletions or duplications of nucleotides have shifted the reading frame.

8. Another form of mutation, discovered in several human disorders, includes unstable trinucleotide units that balloon in size during DNA replication and that are inherited in this form through successive generations. Increasing numbers of repeats often correlate with early onset and severity of disease.

9. Ultraviolet light and high-energy radiation from gamma, cosmic, or X-ray sources are also potent mutagenic agents. UV light induces the formation of pyrimidine dimers in DNA, while high-energy radiation causes the ionization of molecules in its path and is more penetrating than UV.

10. Various systems of genetically controlled DNA repair have been discovered, including photoreactivation and excision, mismatch, and recombinational repair. Loss of repair function by mutation in humans results in the severe disorder xeroderma pigmentosum.

11. Site-directed mutagenesis is a technique that allows researchers to create specific alterations in the nucleotide sequence of the DNA of genes.

12. Insertion sequences in bacteria and other transposable elements in eukaryotes have a profound effect on genetic expression, thus serving as a distinct category of mutagenic agents. A recent example involves a mutation that causes hemophilia in humans.

Key Terms

ABO antigen, 331
acridine dye, 328
acridine orange, 328
Activator (Ac) element, 340
adenine methylase, 335
alkylating agent, 328
Alu family, 342

Ames test, 332
AP endonuclease, 334
apurinic site, 330
attached-X procedure, 323
autosomal recessive mutation, 322
auxotroph, 323
base analog, 327

base substitution, 326
Becker muscular dystrophy (BMD), 331
behavior mutation, 322
biochemical variation, 322
chromosomal aberration, 321
complementation, 336

INSIGHTS *and* SOLUTIONS

1. How could you isolate a mutant strain of bacterial cells that is resistant to penicillin, an antibiotic that inhibits cell wall synthesis?

Solution:

Grow a culture of bacterial cells in liquid medium and plate the cells on agar medium to which penicillin has been added. Only penicillin-resistant cells will reproduce and form colonies. Each colony will, in all likelihood, represent a cloned group of cells with the identical mutation. Isolate members of each colony. To enhance the chance of such a mutation arising, you might want to add a mutagen to the liquid culture.

2. The base analog 2-amino purine (2-AP) substitutes for adenine during DNA replication, but it may base-pair with cytosine. The base analog 5-bromouracil (5-BU) substitutes for thymine, but it may base-pair with guanine. Follow the double-stranded trinucleotide sequence shown below through three rounds of replication, assuming that in the first round, both analogs are present and become incorporated wherever possible. In the second and third round of replication, they are removed. What final sequences occur?

Solution:

The solution to this problem is illustrated to the right.

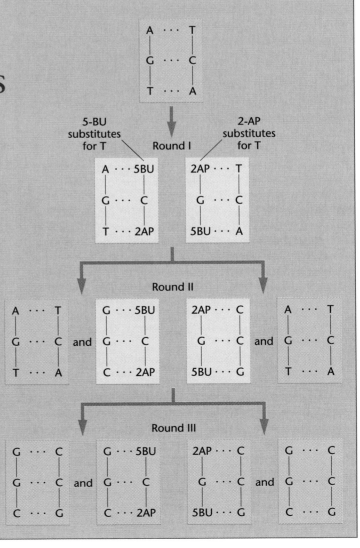

Problems and Discussion Questions

1. What is the difference between a chromosomal mutation and a gene mutation? Between a somatic and a gametic mutation?

2. Why do you suppose that a random mutation is more likely to be deleterious than beneficial?

3. Most mutations in a diploid organism are recessive. Why?

4. In *Drosophila*, induced mutations on chromosome 2 that are recessive lethals may be detected using a second chromosomal stock, *Curly, Lobe/Plum (Cy L/Pm)*. These alleles are all dominant and lethal in the homozygous condition. Detection is performed by crossing *Cy L/Pm* females to wild-type males that have been subjected to a mutagen. Three generations are required. In the F_1, *Cy L* males are selected and individually backcrossed to *Cy L/Pm* females. In the F_2 of each cross, flies expressing *Cy* and *L* are mated to produce a series of F_3 generations. Diagram these crosses and predict how an F_3 culture will vary if a recessive lethal was induced in the original test male compared to the case where no lethal mutation resulted. In which F_3 flies would a recessive morphological mutation be expressed?

5. Describe tautomerism and the way in which this chemical event may lead to mutation.

6. Acridine dyes induce frameshift mutations. Is such a mutation likely to be more detrimental than a point mutation, where a single pyrimidine or purine has been substituted? If so, why?

7. Contrast the various types of DNA repair mechanisms known to counteract the effects of UV light and other DNA damage. What is the role of visible light in photoreactivation?

8. Mammography is an accurate screening technique for the early detection of breast cancer in humans. Because this technique uses X-rays diagnostically, it has been highly controversial. Can you explain why? What reasons justify the use of X-rays for this type of medical diagnosis?

9. Presented below are theoretical findings from studies of heterokaryons formed from human xeroderma pigmentosum cell strains.

	XP1	XP2	XP3	XP4	XP5	XP6	XP7
XP1	0						
XP2	0	0					
XP3	0	0	0				
XP4	+	+	+	0			
XP5	+	+	+	+	0		
XP6	+	+	+	+	0	0	
XP7	+	+	+	+	0	0	0

Note: + = complementing; 0 = noncomplementing.

These data represent the occurrence of unscheduled DNA synthesis in the fused heterokaryon when neither of the strains alone showed synthesis. What does unscheduled DNA synthesis represent? Which strains fall into the same complementation groups? How many groups are revealed based on these limited data? How do we interpret the presence of these complementation groups?

10. If the human genome contains 100,000 genes, and the mutation rate at each of these loci is 5×10^{-5} per gamete formed, what is the average number of new mutations that exist in each individual? If the current population is 4.3 billion people, how many newly arisen mutations exist in the current populace?

11. Speculate as to how insertion sequences and other transposable elements disrupt genetic expression when inserted into wild-type genes.

12. Demonstrate your insights into both chromosomal and gene mutation by projecting yourself as one of the team of geneticists who immediately launched the study of the genetic effects of high-energy radiation on the surviving Japanese population following the atomic bomb attacks at Hiroshima and Nagasaki in 1945. Outline a comprehensive short-term and long-term study that would address this topic. Be sure to include stategies for considering the effects on both somatic and germ line tissues.

Selected Readings

Ames, B. N., McCann, J., and Yamasaki, E. 1975. Method for detecting carcinogens and mutagens with the *Salmonella*/mammalian microsome mutagenicity test. *Mut. Res.* 31:347–64.

Auerbach, C. 1978. Forty years of mutation research: A pilgrim's progress. *Heredity* 40:177–87.

Bates, G., and Lehrach, H. 1994. Trinucleotide repeat expansions and human genetic disease. *BioEssays* 16:277–83.

Beadle, G. W., and Tatum, E. L. 1945. *Neurospora* II. Methods of producing and detecting mutations concerned with nutritional requirements. *Am. J. Bot.* 32:678–86.

Berg, D., and Howe, M., eds. 1989. *Mobile* DNA. Washington, DC: American Society of Microbiology.

Carter, P. 1986. Site-directed mutagenesis. *Biochem. J.* 237:1–7.

Cleaver, J. E. 1968. Defective repair replication of DNA in xeroderma pigmentosum. *Nature* 218:652–56.

———, and Karentz, D. 1986. DNA repair in man: Regulation by a multiple gene family and its association with human disease. *Bioessays* 6:122–.

Cohen, S. N., and Shapiro, J. A. 1980. Transposable genetic elements. *Sci. Am.* (Feb.) 242:40–49.

Deering, R. A. 1962. Ultraviolet radiation and nucleic acids. *Sci. Am.* (Dec.) 207:135–44.

Den Dunnen, J. T., et al. 1989. Topography of the Duchenne muscular dystrophy (DMD) gene. *Am. J. Hum. Genet.* 45:835–47.

Devoret, R. 1979. Bacterial tests for potential carcinogens. *Sci. Am.* (Aug.) 241:40–49.

Federoff, N. V. 1984. Transposable genetic elements in maize. *Sci. Am.* (June) 250:85–98.

Friedberg, E.C., Walker, G.C., and Siede, W. 1995. *DNA repair and mutagenesis*. Washington, DC: ASM Press.

Hanawalt, P. C., and Haynes, R. H. 1967. The repair of DNA. *Sci. Am.* (Feb.) 216:36–43.

Haseltine, W. A. 1983. Ultraviolet light repair and mutagenesis revisited. *Cell* 33:13–17.

Howard-Flanders, P. 1981. Inducible repair of DNA. *Sci. Am.* (Nov.) 245:72–80.

Kelner, A. 1951. Revival by light. *Sci. Am.* (May) 184:22–25.

Knudson, A. G. 1979. Our load of mutations and its burden of disease. *Am. J. Hum. Genet.* 31:401–13.

Kraemer, F. H., et al. 1975. Genetic heterogeneity in xeroderma pigmentosum: Complementation groups and their relationship to DNA repair rates. *Proc. Natl. Acad. Sci. USA* 72:59–63.

Little, J. W., and Mount, D. W. 1982. The SOS regulatory system of *E. coli. Cell* 29:11–22.

Massie, R. 1967. *Nicholas and Alexandra*. New York: Atheneum.

———, and Massie, S. 1975. *Journey*. New York: Knopf.

McCann, J., Choi, E., Yamasaki, E., and Ames, B. 1975. Detection of carcinogens as mutagens in the *Salmonella*/microsome test: Assay of 300 chemicals. *Proc. Natl. Acad. Sci. USA* 72:5135–39.

Macdonald, M. E., et al. 1993. A novel gene containing a trinucleotide repeat that is expanded and unstable in Huntington's disease chromosome. *Cell* 72:971–80.

McKusick, V. A. 1965. The royal hemophilia. *Sci. Am.* (Aug.) 213:88–95.

Muller, H. J. 1927. Artificial transmutation of the gene. *Science* 66:84–87.

———. 1955. Radiation and human mutation. *Sci. Am.* (Nov.) 193:58–68.

O'Hare, K. 1985. The mechanism and control of P element transposition in *Drosophila. Trends Genet.* 1:250–54.

Osanna, N., Peterson, K. R., and Mount, D. W. 1986. Genetics of DNA repair in bacteria. *Trends Genet.* 2:55–58.

Radman, M., and Wagner, R. 1988. The high fidelity of DNA duplication. *Sci. Am.* (Aug.) 259(2):40–46.

Sherratt, D. J. (ed.)1995. *Mobile genetic elements*. New York: Oxford University Press.

Shortle, D., DiMario, D., and Nathans, D. 1981. Directed mutagenesis. *Annu. Rev. Genet.* 15:265–94.

Sigurbjornsson, B. 1971. Induced mutations in plants. *Sci. Am.* (Jan.) 224:86–95.

Stadler, L. J. 1928. Mutations in barley induced by X-rays and radium. *Science* 66:84–87.

Sutherland, B. M. 1981. Photoreactivation. *Bioscience* 31:439–44.

Tomlin, N. V., and Aprelikova, O. N. 1989. Uracil DNA glycosylases and DNA uracil repair. *Int. Rev. Cytol.* 114:81–124.

Vogel, F. 1992. Risk calculations for hereditary effects of ionizing radiation in humans. *Hum. Genet.* 89:127–46.

Wells, R. D. 1994. Molecular basis of genetic instability of triplet repeats. *J. Biol. Chem.* 271: 2875–78.

Wills, C. 1970. Genetic load. *Sci. Am.* (Mar.) 222:98–107.

Yamamoto, F., et al. 1990. Molecular genetic basis of the histo-blood group ABO system. *Nature* 345:229–33.

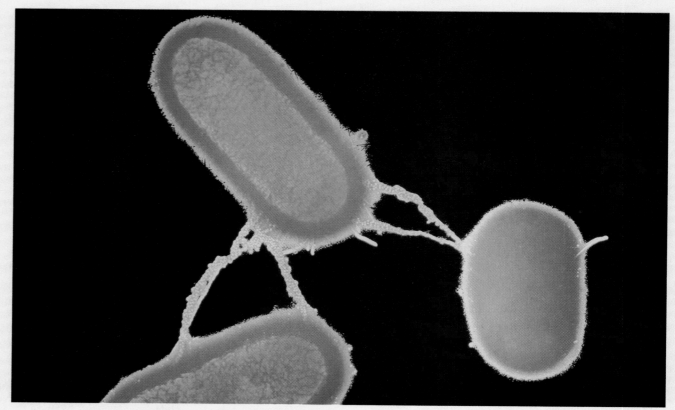

Transmission electron micrograph of conjugating E. coli.

CHAPTER OUTLINE

16

Genetics of Bacteria and Bacteriophages

Chapter Concepts

Bacteria and bacteriophages (bacterial viruses) have been the subject of extensive genetic analysis. They demonstrate mechanisms by which genetic recombination occurs, processes that may serve as the basis for genetic mapping. Bacteria contain extrachromosomal DNA in the form of plasmids. Both plasmids and bacteriophage DNA can be integrated into the bacterial chromosome.

The use of bacteria and bacteriophages has been essential to the accumulation of knowledge in many areas of genetic study. For example, much of what is known about molecular genetics, recobinational phenomena, and gene structure was initially derived from experimental work with these organisms. Their successful use in genetic studies is due to numerous factors. Both bacteria and their viruses have extremely short reproductive cycles. Hundreds of generations, giving rise to billions of genetically identical organisms, can be produced in short periods of time. It is also important that, like true-breeding strains in higher organisms, pure cultures of mutant strains of bacteria or the bacteriophages that infect them are available for genetic studies.

In this chapter, we will discuss a number of the historical developments that led to the extensive use of bacteria and bacteriophages in genetics and will examine in depth the processes by which these organisms undergo genetic recombination. These studies led to the production of extensive genetic maps of many bacteria and bacteriophages. We will focus on how mapping analysis is performed in microorganisms and viruses.

Bacterial Mutation and Growth

It has long been known that pure cultures of bacteria give rise to cells that exhibit heritable variation, particularly with respect to growth under unique environmen-

tal conditions. Prior to 1943, the source of this variation was hotly debated. The majority of bacteriologists believed that environmental factors induced changes in certain bacteria which led to their adaptation to the new conditions. For example, strains of *Escherichia. coli* are known to be sensitive to infection by the bacteriophage T1. Infection by the bacteriophage leads to reproduction of the virus at the expense of the bacterial cell, which is lysed as the new phages are released from the host cell. If a plate of *E. coli* is homogeneously sprayed with T1, almost all cells are lysed. Rare *E. coli* cells, however, survive infection and are not lysed. If these cells are isolated and established in pure culture, all descendants are *resistant* to T1 infection. The **adaptation hypothesis**, put forth to explain this type of observation, implied that the interaction of the phage and bacterium is essential to the acquisition of immunity. In other words, the phage "induced" resistance in the bacteria.

The occurrence of **spontaneous mutations** provided an alternative model to explain the origin of T1 resistance in *E. coli*. In 1943, Salvador Luria and Max Delbruck presented the first convincing evidence that bacteria, like eukaryotic organisms, are capable of spontaneous mutation. This experiment, referred to as the **fluctuation test**, marked the initiation of modern bacterial genetic study. Spontaneous mutation is now considered to be the primary source of genetic variation in bacteria.

Mutant cells that arise spontaneously in an otherwise pure culture can be isolated and established independently from the parent strain by using established selection techniques. As a result, mutations for almost any desired characteristic can now be induced and isolated. Because bacteria and viruses usually contain only a single chromosome and are therefore haploid, all mutations are expressed directly in the descendants of mutant cells, adding to the ease with which these microorganisms can be studied.

Bacteria are grown in a liquid culture medium or in a Petri dish on a semisolid agar surface. If the nutrient components of the growth medium are very simple and consist only of an organic carbon source (such as a glucose or lactose) and a variety of ions, including Na^+, K^+, Mg^{2+}, Ca^{2+}, and NH_4^+ present as inorganic salts, it is called **minimal medium**. In order to grow on such a medium, a bacterium must be able to synthesize all essential organic compounds (e.g., amino acids, purines, pyrimidines, sugars, vitamins, and fatty acids). A bacterium that can accomplish this remarkable biosynthetic feat—one that we ourselves cannot duplicate—is termed a **prototroph**. It is said to be wild type for all growth requirements. On the other hand, if a bacterium loses, through mutation, the ability to synthesize one or more organic components, it is said to be an **auxotroph**. For example, if it loses the ability to make histidine, then this amino acid must be added as a supplement to the minimal medium in order for growth to occur. The resulting bacterium is designated as a *his⁻* auxotroph, as opposed to its prototrophic *his⁺* counterpart.

In order to study mutant bacteria quantitatively, an inoculum of bacteria is placed in liquid culture medium. A characteristic growth pattern is exhibited, as illustrated in Figure 16–1. Initially, during the **lag phase**, growth is slow. Then, a period of rapid growth, called the **logarithmic (log) phase**, ensues. During this phase, cells divide many times with a fixed time interval between cell divisions, resulting in exponential growth. When a cell density of about 10^9 cells/ml is reached, nutrients and oxygen become limiting and cells cease dividing; at this point the cells enter the **stationary phase**. Because the doubling time during the log phase may be as short as 20 minutes, an initial inoculum of a few thousand cells can easily achieve maximum cell density during an overnight incubation.

Cells grown in liquid medium may be quantified by transferring them to semisolid medium in a Petri dish. Following incubation and many divisions, each cell gives rise to a visible colony on the surface of the medium. If the number of colonies is too great to count, then a series of successive dilutions (a technique called **serial dilution**) of the original liquid culture can be made and plated, until the colony number is reduced to the point where it can be counted (Figure 16–2). This technique allows one to calculate the number of bacteria present in the original culture.

For example, assume that the three dishes in Figure 16–2 represent serial dilutions of 10^{-3}, 10^{-4}, and 10^{-5} (left to right). We need only select the dish in which the number of colonies can be counted accurately. Because each colony arose from a single bacterium, the number of colonies multiplied by the dilution factor represents the number of bacteria in each milliliter of the initial inoculum used to start the serial dilutions. In our case, the dish farthest to the right has 15 colonies. The **dilution factor** for a 10^{-5} dilution is 10^5. Therefore, the initial number of bacteria is calculated to be 15×10^5 per milliliter.

Genetic Recombination in Bacteria: Conjugation

The development of techniques that allowed the identification and study of bacterial mutations led to detailed investigations of the arrangement of genes on bacterial chromosomes. In 1946, Joshua Lederberg and Edward Tatum initiated these studies. They showed that bacteria undergo **conjugation**, a process in which the genetic information from one bacterium is transferred to and recombined with that of another bacterium. Like meiotic crossing over in eukaryotes, genetic recombination in bacteria led to methodology for chromosome mapping.

Lederberg and Tatum's initial experiments were performed with two multiple-auxotroph strains of *E. coli* K12. Strain A required methionine and biotin in order to grow, while strain B required threonine, leucine, and thiamine (Figure 16–3). Neither strain would grow on minimal medium. The two strains were first grown separately in supplemented media, and then cells from both were mixed and grown together for several more generations. They were then plated on minimal medium. Any bacterial cells that grew on minimal medium were prototrophs. It was highly improbable that any of the cells that contained two or three mutant genes would undergo spontaneous mutation simultaneously at two or three locations, leading to wild-type

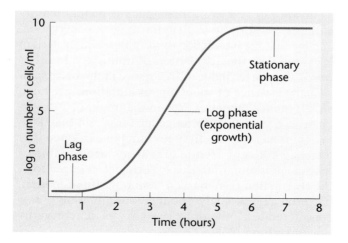

■ Figure 16–1 A typical bacterial population growth curve.

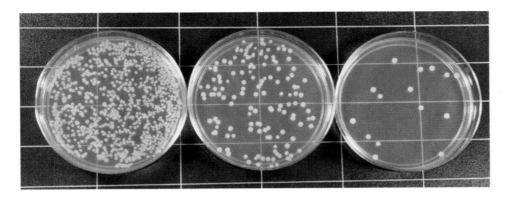

■ Figure 16–2 Results of the serial dilution technique and subsequent culture of bacteria. Each dilution varies by a factor of 10. Each colony, visible as a light circle, was derived from a single bacterial cell.

cells. Therefore, any prototrophs recovered were assumed to have arisen as a result of some form of genetic exchange and recombination between the two mutant strains.

In this experiment, prototrophs were recovered at a rate of $1/10^7$ (10^{-7}) cells plated. The controls for this experiment involved separate plating of cells from strains A and B on minimal medium. No prototrophs were recovered. Based on these observations, Lederberg and Tatum proposed that genetic exchange had occurred.

F^+ and F^- Bacteria

Many experiments designed to elucidate the genetic basis of conjugation soon followed. It became evident quickly that different strains of bacteria were involved in a unidirectional transfer of genetic material. That is, cells of one strain can serve as donors of parts of their chromosomes. Those that do are designated as **F^+ cells** (F stands for "fertility"). Recipient bacteria, which undergo genetic recombination by receiving the donor chromosome and exchanging part of it with part of their own, are designated as **F^- cells**. In our previous discussion (see Figure 16–3), one of the two strains (A or B) served as the donor (F^+), while the other served as the recipient (F^-).

It was established subsequently that cell contact is essential for chromosome transfer to occur. Support for this concept was provided by Bernard Davis, who designed a U-tube in which to grow F^+ and F^- cells (Figure 16–4). At the base of the tube was a sintered glass filter with a pore size that allowed passage of the liquid medium, but that was too small to allow the passage of bacteria. F^+ cells were placed on one side of the filter, and F^- cells were placed on the other side. The medium was moved back and forth across the filter so that the cells essentially shared a common medium during bacterial incubation. Samples from both sides of the tube were then plated on minimal medium, but no prototrophs were found. Davis concluded that *physical*

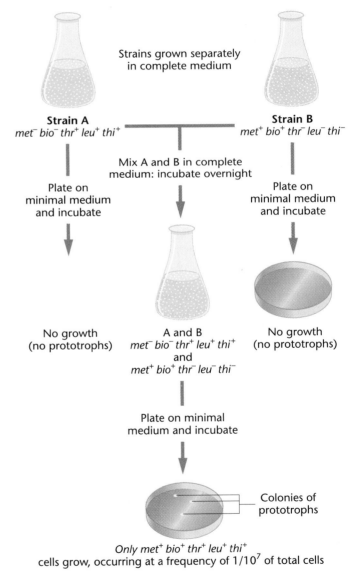

Strains grown separately in complete medium

Strain A
$met^- bio^- thr^+ leu^+ thi^+$

Strain B
$met^+ bio^+ thr^- leu^- thi^-$

Mix A and B in complete medium: incubate overnight

Plate on minimal medium and incubate

Plate on minimal medium and incubate

No growth (no prototrophs)

A and B
$met^- bio^- thr^+ leu^+ thi^+$
and
$met^+ bio^+ thr^- leu^- thi^-$

No growth (no prototrophs)

Plate on minimal medium and incubate

Colonies of prototrophs

Only $met^+ bio^+ thr^+ leu^+ thi^+$ cells grow, occurring at a frequency of $1/10^7$ of total cells

■ Figure 16–3 Genetic recombination involving two auxotrophic strains, producing prototrophs. Neither auxotroph will grow on minimal medium, but prototrophs will, allowing their recovery.

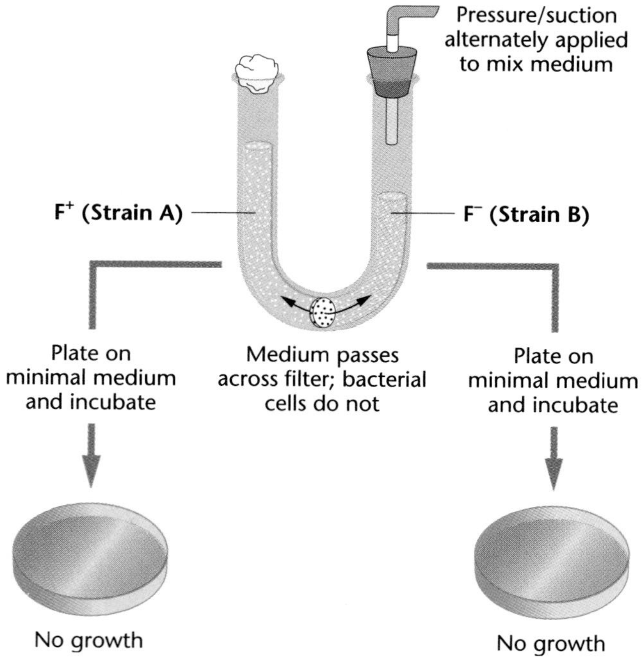

Pressure/suction alternately applied to mix medium

F⁺ (Strain A) — F⁻ (Strain B)

Plate on minimal medium and incubate

Medium passes across filter; bacterial cells do not

Plate on minimal medium and incubate

No growth

No growth

■ Figure 16–4 When strain A and strain B are grown in a common medium but are separated by a filter, no genetic recombination occurs, and no prototrophs are produced. This apparatus is called a Davis U-tube.

contact between cells of the two strains is essential to genetic recombination. This physical interaction is the initial step in the process of conjugation and is established by a structure called the **F sex pilus**. Bacteria often have many pili, which are tubular extensions of the cell. After contact has been initiated between mating pairs (Figure 16–5), transfer of the chromosome begins.

Evidence was provided subsequently that F⁺ cells contain an independent **fertility factor**, conferring their ability to donate part of their chromosome by passing one end of it through the pilus during conjugation. Experiments by the Lederbergs and by William Hayes and Luca Cavalli-Sforza established that, under certain conditions, donor ability can be lost by fertile cells. However, if these cells are then grown with fertile cells that retain donor ability, fertility is reestablished. These observations led to the hypothesis that an **F factor** responsible for donor ability exists, and that cells can lose and regain it.

This conclusion was further supported by the observation that, following conjugation and genetic recombination, recipient cells, originally F⁻, almost always become F⁺. Thus, in addition to the rare occurrence of gene transfer during conjugation, the F factor is passed to all recipient cells. On this basis, the initial crosses of the Lederberg and Tatum (Figure 16–3) may be designated:

STRAIN A		STRAIN B
F⁺	×	F⁻
DONOR		RECIPIENT

Confirmation of these proposals was forthcoming once the F factor was isolated. It has been shown to consist of a circular, double-stranded DNA molecule that is distinct from DNA constituting the bacterial chromosome and that contains about 100,000 nucleotide base pairs (100 kb). This amount of DNA is equivalent to about 2 percent of that making up the bacterial chromosome. Contained in the F factor DNA are 19 genes, the products of which are involved in the transfer of genetic information (*tra* genes), including those essential to the formation of the sex pilus.

As we soon shall see, the F factor is in reality an autonomous genetic unit referred to as a **plasmid**. However, in our historical coverage of its discovery, we will continue to refer to it as a "factor."

It is believed that the transfer of the F factor during conjugation involves separation of the two strands of the double helix making up the F factor and the movement of one of the two strands into the recipient. Both strands, one moving across the conjugation tube and one remaining in the donor cell, are replicated. The result is that both the donor *and* the recipient cells are F⁺. This process is diagrammed in Figure 16–6.

To summarize, an *E. coli* cell may or may not contain the F factor. When this factor is present, the cell is able to form a sex pilus and potentially serve as a donor of genetic information. During conjugation, a copy of the F factor is almost always transferred from the F⁺ cell to the F⁻ recipient, converting it to the F⁺ state. The question remains as to exactly why such a low proportion of

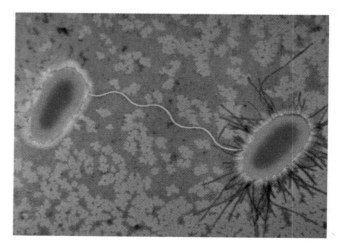

■ Figure 16–5 An electron micrograph revealing the sex pilus between two conjugating *E. coli* cells.

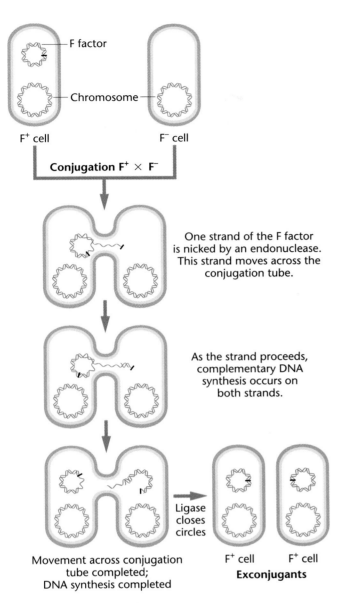

■ Figure 16–6   An F$^+$ × F$^-$ mating demonstrating how the recipient F$^-$ cell is converted to F$^+$. During conjugation, the DNA of the F factor is replicated with one new copy entering the recipient cell, converting it to F$^+$. The black bar has been added to the F factors to follow their rotation during replication.

(Figure labels:)
F factor
Chromosome
F$^+$ cell   F$^-$ cell
Conjugation F$^+$ × F$^-$
One strand of the F factor is nicked by an endonuclease. This strand moves across the conjugation tube.
As the strand proceeds, complementary DNA synthesis occurs on both strands.
Ligase closes circles
Movement across conjugation tube completed; DNA synthesis completed
F$^+$ cell   F$^+$ cell
Exconjugants

cells involved in these matings (10^{-7}) results in genetic recombination. As we will discuss, the answer awaited further experimentation.

Hfr Bacteria and Chromosome Mapping

Subsequent discoveries not only clarified how genetic recombination occurs, but also defined a mechanism by which the *E. coli* chromosome could be mapped. We shall first address chromosome mapping.

In 1950, Cavalli-Sforza treated an F$^+$ strain of *E. coli* K12 with nitrogen mustard, a chemical that is known to induce mutations. From these treated cells he recovered a genetically altered strain of donor bacteria that underwent recombination at a rate of $1/10^4$ (10^{-4}), 1000 times as frequently as the original F$^+$ strains. In 1953, Hayes isolated another strain that demonstrated a similar elevated frequency. Both strains were designated **Hfr**, meaning **high-frequency recombination**. Because Hfr cells behave as donors, they are a special class of F$^+$ cells.

Another important difference was noted between Hfr strains and the original F$^+$ strains. If the donor is an Hfr strain, recipient cells, while sometimes displaying genetic recombination, never become Hfr; that is, they remain F$^-$. In comparison, then,

$$F^+ \times F^- \rightarrow F^+ \text{ (low rate of recombination)}$$

$$Hfr \times F^- \rightarrow F^- \text{ (higher rate of recombination)}$$

Perhaps the most significant characteristic of Hfr strains is the *nature of recombination.* In any given strain, certain genes are more frequently recombined than others, and some not at all. This *nonrandom* pattern was shown to vary from Hfr strain to Hfr strain. While these results were puzzling, Hayes interpreted them to mean that some physiological alteration of the F factor had occurred, resulting in the production of Hfr strains of *E. coli.*

In the mid-1950s, experimentation by Ellie Wollman and François Jacob explained the difference between Hfr and F$^+$ and showed how Hfr strains allow genetic mapping of the *E. coli* chromosome. In their experiments, Hfr and F$^-$ strains with suitable marker genes were mixed and recombination of specific genes assayed at different times. To accomplish this, a culture containing a mixture of an Hfr and an F$^-$ strain was first incubated and samples removed at various intervals and placed in a blender. The shear forces created in the blender separated conjugating bacteria so that the transfer of the chromosome was terminated effectively. The cells were then assayed for genetic recombination.

This process, called the **interrupted mating technique**, demonstrated that specific genes of a given Hfr strain were transferred and recombined sooner than others. Figure 16–7 illustrates this point. During the first 8 minutes after the two strains were mixed, no genetic recombination could be detected. At about 10 minutes, recombination of the *azi* gene could be detected, but no transfer of the *ton^S*, *lac$^+$*, or *gal$^+$* genes was noted. By 15 minutes, 70 percent of the recombinants were *azi$^+$*; 30 percent were now also *ton^S*; but none was *lac$^+$* or *gal$^+$*. Within 20 minutes, the *lac$^+$* was found among the recombinants; and within 25 minutes, *gal$^+$* was also being transferred. Therefore, Wollman and Jacob had demonstrated *an ordered transfer of genes* that was correlated with the length of time conjugation was allowed to proceed.

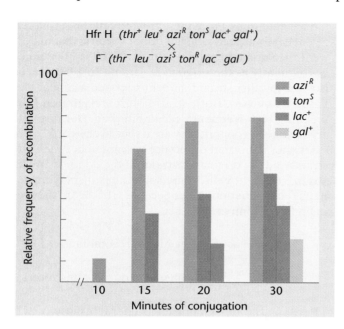

Hfr H *(thr⁺ leu⁺ aziᴿ tonˢ lac⁺ gal⁺)*
×
F⁻ *(thr⁻ leu⁻ aziˢ tonᴿ lac⁻ gal⁻)*

■ Figure 16–7 The progressive transfer during conjugation of various genes from a specific Hfr strain of *E. coli* to an F⁻ strain. Certain genes (*azi* and *ton*) are transferred sooner than others (*lac* and *gal*) and recombine more frequently. Others (*thr* and *leu*) are not transferred at all.

It appeared that the chromosome of the Hfr bacterium was transferred linearly and that the sequence and distance between the genes, as measured in minutes, could be determined from such experiments. This information served as the basis for the first genetic map of the *E. coli* chromosome (Figure 16–8). Apparently, conjugation does not usually last long enough to allow the entire chromosome to pass across to the recipient cell.

Wollman and Jacob then repeated the same type of experimentation with other Hfr strains, obtaining similar results with one important difference. While genes were always transferred linearly with time, as in their original experiment, which genes entered first and which followed later seemed to vary from Hfr strain to Hfr strain [Figure 16–9(a)]. When they reexamined the

■ Figure 16–8 A time map of the genes studied in the experiment depicted in Figure 16–7.

rate of entry of genes, and thus the different genetic maps for each strain, a definite pattern emerged. The major difference between each strain was simply the point of the origin (*O*) and the direction in which entry proceeded from that point [Figure 16–9(b)].

In order to explain these results, Wollman and Jacob postulated that the *E. coli* chromosome is circular (a closed circle, with no free ends). If the point of origin (*O*) varied from strain to strain, a different sequence of genes would be transferred in each case. But what determines *O*? They proposed that in various Hfr strains, the F factor integrates into the chromosome at different points. Its position determines the site of *O*. One such case is shown in Figure 16–10. During conjugation between an Hfr and an F⁻ cell, the position of the F factor determines the initial point of transfer. Those genes adjacent to *O* are transferred first. *The F factor becomes the last part that can be transferred.* However, conjugation rarely, if ever, lasts long enough to allow the entire chromosome to pass across the conjugation tube. This proposal explains why recipient cells, when mated with Hfr cells, remain F⁻.

The use of the interrupted mating technique with different Hfr strains has provided the basis for mapping the entire *E. coli* chromosome. Mapped in time units, strain K12 (or *E. coli* K12) is 100 minutes long. Over 900 genes have now been placed on the map.

Recombination in F⁺ × F⁻ Matings: A Reexamination

The above model of Hfr matings has helped geneticists to understand better how genetic recombination occurs during the F⁺ × F⁻ matings. Recall that recombination occurs much less frequently in them than in Hfr × F⁻ matings, and that random gene transfer is involved. The current belief is that when F⁺ and F⁻ cells are mixed, conjugation occurs readily and each F⁻ cell involved in conjugation receives a copy of the F factor, but no genetic recombination occurs. However, at a very low frequency in a population of F⁺ cells, the F factor integrates spontaneously at a random point into the bacterial chromosome. The integration converts each such F⁺ cell to the Hfr state. Therefore, in F⁺ × F⁻ crosses, the very low frequency of genetic recombination (10^{-7}) is attributed to the rare, newly formed Hfr cells, which then undergo conjugation with F⁻ cells. Because the point of integration is random, a nonspecific gene transfer ensues, leading to the low-frequency, random genetic recombination observed in the F⁺ × F⁻ matings.

The F′ State and Merozygotes

In 1959, during experiments with Hfr strains of *E. coli*, Edward Adelberg discovered that the F factor could lose its integrated status, causing reversion to the **F⁺ state** (Figure 16–11). When this occurs, the F factor frequently carries several adjacent bacterial genes along

(a)

Hfr strain	Order of transfer (Earliest)							(Latest)
H	thr – leu – azi – ton – pro – lac – gal – thi
1	leu – thr – thi – gal – lac – pro – ton – azi
2	pro – ton – azi – leu – thr – thi – gal – lac
7	ton – azi – leu – thr – thi – gal – lac – pro

(b)

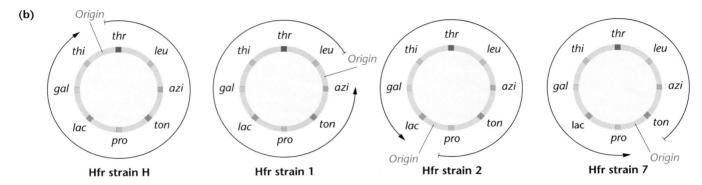

Hfr strain H   Hfr strain 1   Hfr strain 2   Hfr strain 7

■ Figure 16–9   (a) The order of gene transfer in four Hfr strains, suggesting that the *E. coli* chromosome is circular. (b) The point of the origin (*Origin*) of transfer is identified in each strain. Note that transfer can proceed in either direction, depending on the strain. The origin is determined by the point of integration into the chromosome of the F factor, and the direction of transfer is determined by the orientation of the F factor as it integrates.

with it. Adelberg labeled this condition F′ to distinguish it from F⁺ and Hfr. F′ is thus a special case of F⁺.

The presence of bacterial genes within a cytoplasmic F factor creates an interesting situation. An F′ bacterium behaves like an F⁺ cell, initiating conjugation with F⁻ cells. When this occurs, the F factor, along with the chromosomal genes, is transferred to the F⁻ cell. As a result, donor cell genes that are part of the F factor become duplicated in the recipient cell (because they are also present in the recipient cell's chromosome). This creates a partially diploid cell called a **merozygote**. Pure cultures of F′ merozygotes may be established. They have been extremely useful in the study of bacterial genetics, particularly in genetic regulation.

The Rec Proteins and Bacterial Recombination

Having established that a unidirectional transfer of chromosomal DNA occurs between bacteria, determining how the actual recombination event occurs in the recipient cell has been of great interest. Just how does the donor DNA replace the recipient DNA? As with many biological systems, the mechanism by which a biochemical event occurs is often deciphered as a result of isolating and studying mutations that interrupt the normal process. Major insights have been gained as a result of the discovery of mutations in a group of

genes named *rec*. The first mutant gene, *recA*, was found to diminish genetic recombination in *E. coli* 1000-fold, nearly eliminating it altogether. Other mutations, *recB, recC,* and *recD*, reduce recombination by over 100 times. Clearly, the normal, wild-type products of these genes play some role in the process.

By looking for a functional gene product that is present in normal cells but missing in mutant cells, several *rec* gene products have been isolated and found to play a role in the process by which DNA of the donor and recipient is recombined. The first is called the **RecA protein**.* The second is more complex and is called the **RecBCD protein**, consisting of polypeptides encoded by three genes. The roles of these proteins have now been elucidated *in vitro*, extending our knowledge of the process of recombination considerably.

The general scheme of **RecA-mediated recombination** is illustrated in Figure 16–12. The roles of these proteins have now been elucidated *in vitro*. As a result of this genetic research, our knowledge of the process of recombination has been extended considerably.

The RecA protein has a strong affinity *in vitro* for binding to one of the two strands that make up the double helix of DNA. As we saw in Figure 16–6, only one of the two stands that make up the donor DNA initially

*Note that the names of bacterial genes begin with lowercase letters and are italicized, while the corresponding gene products begin with capital letters and are not italicized, as illustrated by the *recA* gene and RecA protein.

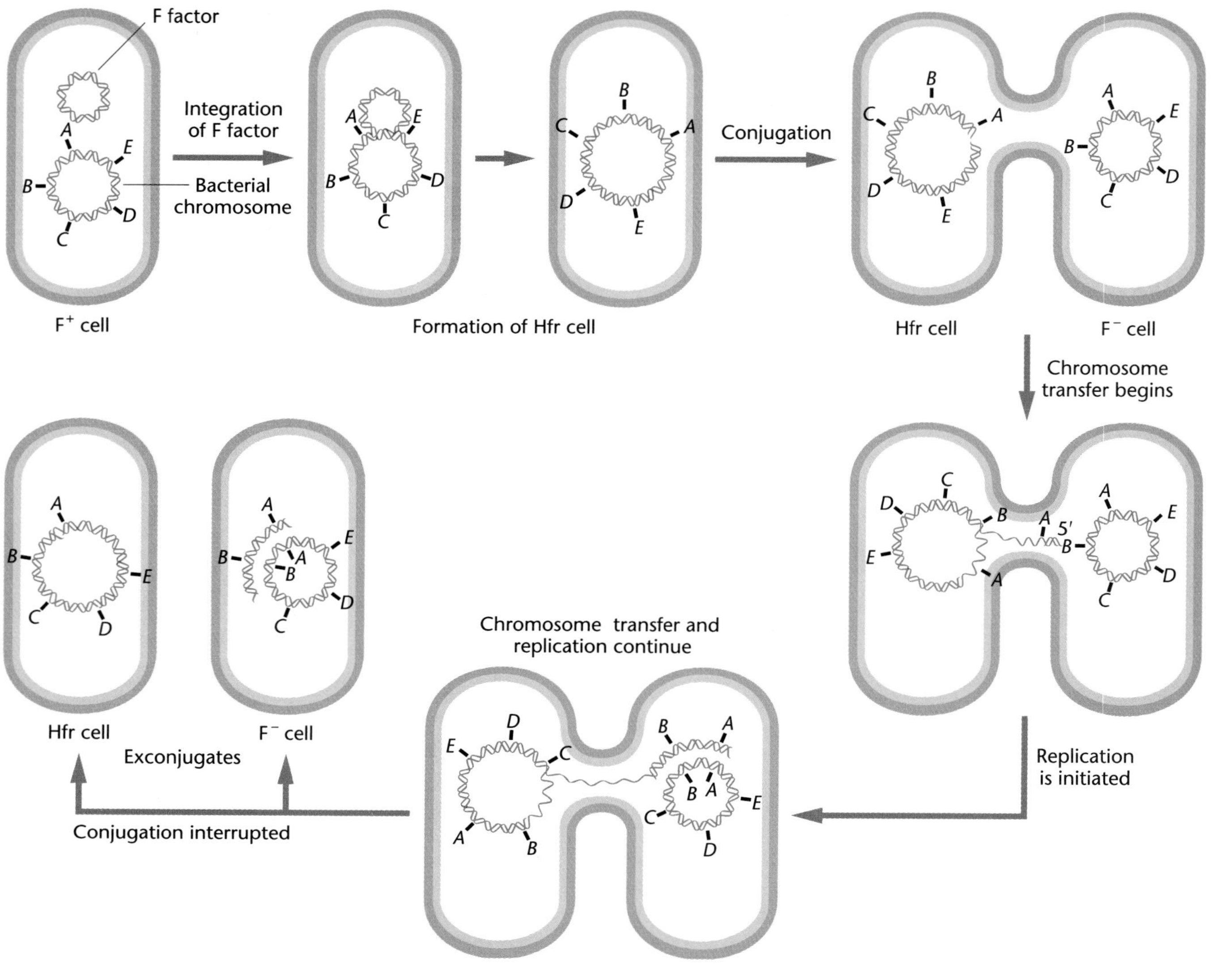

■ Figure 16–10 Conversion of F⁺ to an Hfr state occurs by the integration of the F factor into the bacterial chromosome. The point of integration determines the origin of transfer. During conjugation, the F factor is nicked by an enzyme. The factor is now on the end of the chromosome adjacent to the origin and is the last part to be transferred. Conjugation is usually interrupted prior to complete transfer. Above, only the *A* and *B* genes are transferred.

enters the recipient cell. The binding of RecA with this strand creates a DNA–protein complex that then invades and probes DNA of the recipient chromosome. The complex migrates along the chromosome until the region that is complementary to the ssDNA is located. When this homologous region, representing the same gene(s) as the invading DNA, is encountered, the invading DNA pairs with and eventually displaces its counterpart in the chromosome. The pairing and displacement, leading to genetic recombination, is facilitated by the RecA protein. The RecBCD enzyme is equally important to this process. It is thought to be essential to recombination in the role of "unwinding" and "cutting" DNA, which also facilitates the displacement of the recipient DNA with the invading molecule. This scheme will undoubtedly be clarified as further research

is conducted. A recent finding, for example, has shown that the pairing and displacement step, performed *in vivo* by RecA, can be accomplished *in vitro* by a 20-amino-acid peptide fragment derived from the parent molecule. This experimental observation pinpoints the region of the protein that is critical to its function.

Plasmids

When the F factor exists autonomously in the bacterial cytoplasm, this unit takes the form of a double-stranded closed circle (or loop) of DNA. These characteristics place the F factor in the more general category of the genetic structures called **plasmids**. These structures contain one or more genes and often contain

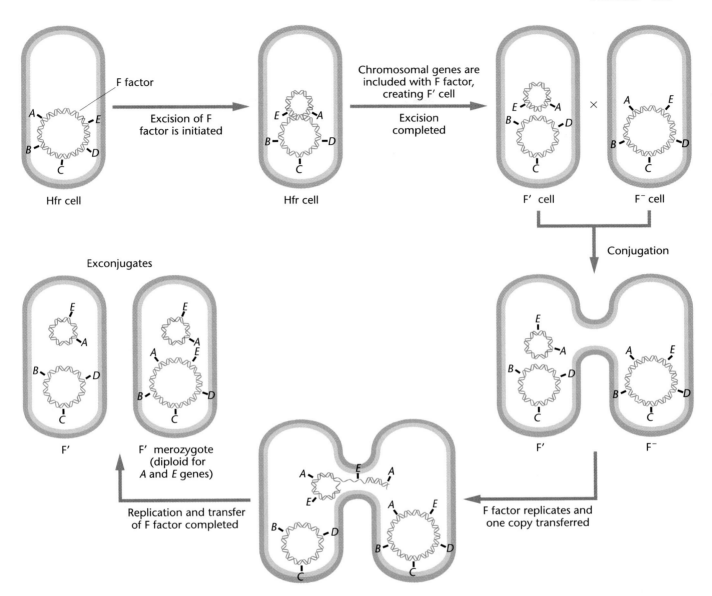

■ Figure 16–11 Conversion of an Hfr bacterium to F′ and its subsequent mating with an F⁻ cell. The conversion occurs when the F factor loses its integrated status. During excision from the chromosome, it carries with it one or more chromosomal genes (A and E). Following conjugation with such an F′ cell, an F⁻ recipient becomes partially diploid and is called a merozygote. It also becomes F′.

quite a few. Their replication depends on the same enzymes that replicate the chromosome of the cell, and they are distributed to daughter cells along with the host chromosome during cell division.

Plasmids are generally classified according to the genetic information specified by their DNA. The F factor confers fertility and contains genes that are essential for sex pilus formation. Other examples include the **R** and the **Col plasmids**. These plasmids confer multiple resistance to antibiotics and the ability to release toxic substances called **colicins**, respectively. An R plasmid is illustrated in Figure 16–13. It contains both r-determinants (for resistance) and RTFs, which facilitate plasmid transfer.

The R plasmid, which may be present in one to three copies per cell, is significant because it provides the means for rapid conferral of multiple drug resistance to cells that were previously sensitive to antibiotics. This phenomenon came to the attention of researchers in the late 1950s in Japan, when multiply resistant strains of the bacterium *Shigella* (which causes dysentery) were found in the guts of humans. *Shigella* had acquired this resistance following conjugation with *E. coli*, where the R factor originated. In such cases, antibiotic treatment kills off the sensitive cells, decreasing the competition that resistant cells encounter. The resistant cells then become the dominant "flora" of the gut.

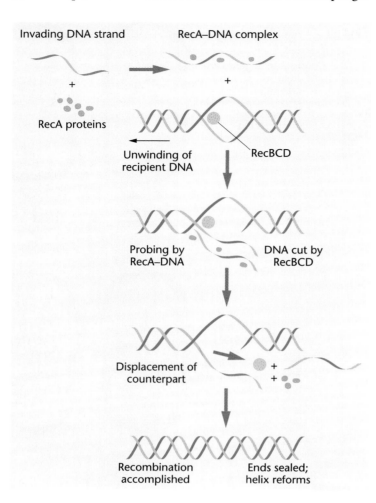

Invading DNA strand

RecA–DNA complex

+

RecA proteins

+

Unwinding of recipient DNA

RecBCD

Probing by RecA–DNA

DNA cut by RecBCD

Displacement of counterpart

+
+

Recombination accomplished

Ends sealed; helix reforms

■ Figure 16–12 A theoretical model illustrating the possible role of the RecA and RecBCD proteins in the process of genetic recombination between bacterial DNA molecules. The RecA protein has an affinity to bind to single-stranded regions of DNA and to facilitate the probing and displacement of the host DNA. The RecBCD enzyme unwinds and cuts the host DNA. One further round of DNA replication is needed to fully replace the host DNA.

Interest in plasmids has increased dramatically because of their role in the genetic technology referred to as **recombinant DNA research**. Specific genes from any source (e.g., a human cell) may be inserted into a plasmid, which may then be inserted into a bacterial cell. As the cell replicates its DNA and undergoes division, the foreign gene is treated like one of the cell's own. This is the major topic of Chapter 17.

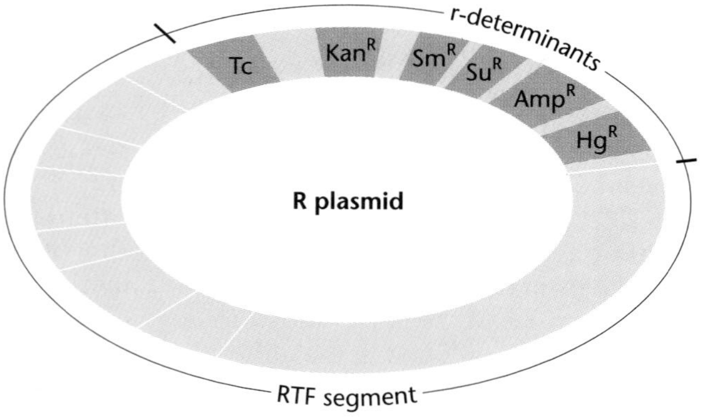

Bacterial Transformation

In Chapter 9 we described how the study of transformation in *Diplococcus pneumoniae* provided direct evidence that DNA is the genetic material. **Transformation**, like conjugation, provides a mechanism for the recombination of genetic information in certain bacteria. Transformation can also be used to map bacterial genes, although in a more limited way.

This process (Figure 16–14) consists of numerous steps that can be divided into two main categories: (1) entry of DNA into a recipient cell, and (2) recombination of the donor DNA with its homologous region in the recipient chromosome. In a population of bacterial cells, only those in a particular physiological state, referred to as **competence**, take up DNA. Entry apparently occurs at a limited number of receptor sites on the surface of the bacterial cell. The most effective length of transforming DNA is about 10,000 to 20,000 base pairs, an amount equal to about 1/200 of the *E. coli* chromosome. Passage across the cell wall and membrane is an active process that requires energy and spe-

r-determinants

Tc   Kan^R Sm^R Su^R

Amp^R

Hg^R

R plasmid

RTF segment

■ Figure 16–13 R plasmid containing resistance transfer factors (RTFs) and multiple r-determinants (Tc, tetracycline; Kan, kanamycin; Sm, streptomycin; Su, sulfonamide; Amp, ampicillin; Hg, mercury).

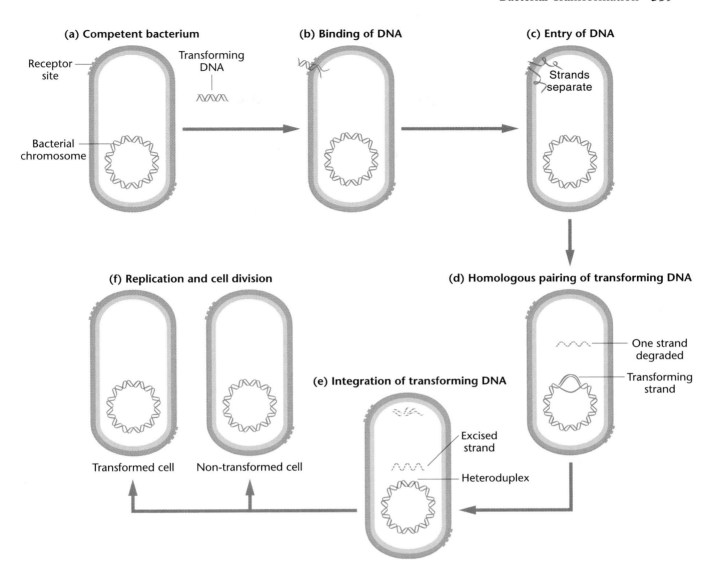

■ Figure 16–14 Proposed steps leading to transformation of a bacterial cell by exogenous DNA. Only one of the two strands of the entering DNA is involved in the transformation event, which is completed following cell division.

cific transport molecules. This model is supported by the fact that substances that inhibit energy production or protein synthesis in the recipient cell also inhibit the transformation process.

During the process of entry, one of the two strands of the double helix is digested by nucleases. The intact single strand of DNA then aligns with its complementary region of the bacterial chromosome. In a process involving several enzymes, the segment replaces its counterpart in the chromosome, which is excised and degraded (Figure 16–14).

For recombination to be detected, the transforming DNA must be derived from a different strain of bacteria that bears some genetic variation, such as a mutation. Once it is integrated into the chromosome, the recombinant region contains one host strand (present originally) and one mutant strand. Because these strands are from different sources, this helical region is

referred to as a **heteroduplex**. Following one round of semiconservative replication, one chromosome is restored to its identical configuration, and the other contains the mutant gene. Following cell division, one nonmutant (untransformed) cell and one mutant (transformed) cell are produced.

Transformation and Linked Genes

Because the effective size of transforming DNA is 10,000 to 20,000 base pairs, it contains a sufficient number of nucleotides to encode several genes. Genes that are adjacent or very close to one another on the bacterial chromosome may be carried on a single piece of DNA of this size. Because of this fact, a single event may result in the **cotransformation** of several genes simultaneously. Genes that are close enough to each other to be contransformed are said to be *linked*. In contrast to the use of the term *linkage* in eukaryotes to indicate all genes on a single chromosome, note that here linkage refers to the close proximity of genes.

If two genes are not linked, simultaneous transformation can occur only as a result of two independent events involving two distinct segments of DNA. As in double crossing over in eukaryotes, the probability of two independent events occurring simultaneously is equal to the product of the individual probabilities. Thus, the frequency of two unlinked genes being transformed simultaneously is much lower than if they are linked.

Subsequent studies have shown that, in addition to *Diplococcus pneumoniae*, a variety of bacteria readily undergo transformation (e.g., *Hemophilus influenzae*, *Bacillus subtilis*, *Shigella paradysenteriae*, and *E. coli*). Under certain conditions, studies have shown that relative distances between linked genes can be determined from transformation data. While analysis is more complex, such data are interpreted in a manner analogous to chromosome mapping in eukaryotes.

The Genetic Study of Bacteriophages

There is still a third mode involving genetic transfer between bacteria, called **transduction**. Since transduction is a process that is mediated by **bacteriophages**, we will delay its discussion temporarily until we have considered the genetics of bacteriophages and other viruses. As we will see, following infection, bacterial viruses are capable of either lysing the bacterial host or inserting their DNA into the host chromosome, where it may remain genetically quiescent. In some cases, it is when this integrated DNA is extricated from the host chromosome that transduction becomes possible.

Recombination has also been observed in bacterial viruses, but only when genetically distinguishable strains have simultaneously infected the same host cell. In this section, we will first review briefly the life cycle of a typical bacteriophage, one from the so-called **T-even phage series**. Then, we will discuss how these phages are studied during their infection of bacteria. Finally, we will contrast the two possible modes of behavior once the phage DNA is injected into the bacterial host. This information will serve as background for our discussion of transduction.

The T4 Phage Life Cycle

In Chapter 9 we discussed the Hershey–Chase experiment, which utilized bacteriophage T2 in providing support for the concept that DNA is the genetic material. Phage T4 is very similar and is typical of the family of T-even phages. It contains double-stranded DNA in a quantity sufficient to encode more than 150 average-sized genes. The genetic material is enclosed by an icosahedral protein coat (a polyhedron with 20 faces) making

up the head (or capsid) of the virus. This is connected to a tail that contains a collar and a contractile sheath that surrounds a central core. Six tail fibers protrude from the tail; their tips contain binding sites that specifically recognize unique areas of the external surface of the *E. coli* cell wall. The structure and assembly of the components of phage T4 are illustrated in Figure 6–15.

Binding of tail fibers to these specific areas of the cell wall is the first step of infection. Then, an ATP-driven contraction of the tail sheath causes the central core to penetrate the cell wall. The DNA in the head is extruded and then moves across the cell membrane into the bacterial cytoplasm. Within minutes, all synthesis of bacterial DNA, RNA, and protein is inhibited, and the production of viral macromolecules begins. This initiates the **latent period**, in which the production of viral components occurs prior to the assembly of virus particles.

RNA synthesis is initiated by the transcription of viral genes by the host cell's RNA polymerase. A plethora of early and late gene products are subsequently synthesized using the host cell's ribosomes. For example, about 20 different proteins are part of the protein capsid. Many others are structural components of the tail and its fibers. Additionally, numerous enzymes participate in assembly of mature bacteriophages, but are not themselves structural components. One late gene product, **lysozyme**, digests the bacterial cell wall, leading to the release of mature phages once they are assembled.

Prior to the synthesis of most viral gene products, semiconservative DNA replication begins, leading to a pool of viral DNA molecules that are available to be packaged into phage heads. It is at this time that recombination may occur between DNA molecules.

Assembly of mature viruses is a complex process that has been well studied by William Wood, Robert Edgar, and others. Three major independent pathways occur, leading to (1) DNA packaged into heads, (2) assembly of tails, and (3) synthesis and assembly of tail fibers. A mutation that interrupts any one of the three pathways does not inhibit the other two. As shown in Figure 16–15, the head is assembled and DNA is packaged into it. This complex then combines with the tail, and only then are the tail fibers added. Total construction is a combination of self-assembly and enzyme-directed processes. When approximately 200 viruses are assembled, the bacterial cell is ruptured by the action of lysozyme and the mature phages are released. If this occurs on a lawn of bacteria, the 200 phages will infect other bacterial cells and the process will be repeated over and over again. On the lawn, multiple infection cycles lead to a clear area where many bacteria have been lysed. Each such area is called a **plaque** and contains millions of genetic clones of the initial infecting T4 bacteriophage.

The Plaque Assay

The study of bacteriophages and other viruses has played a critical role in our understanding of molecular

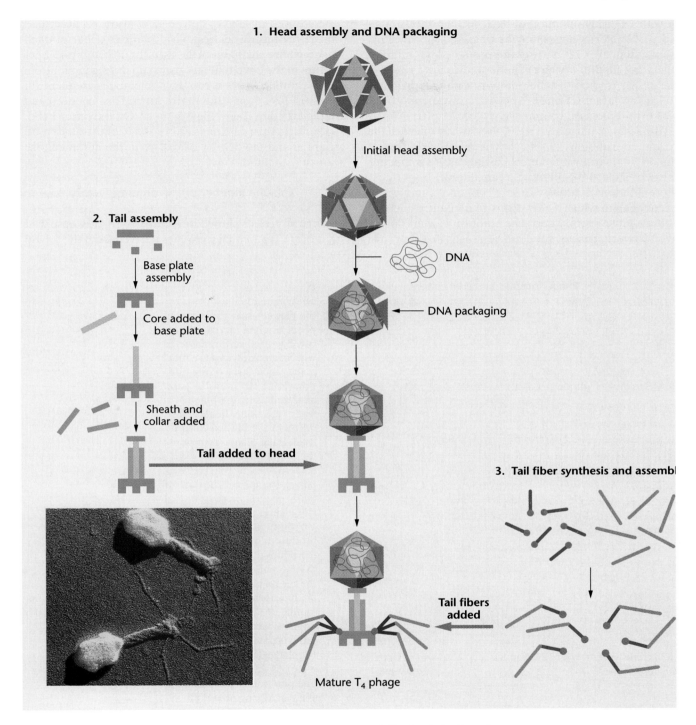

1. Head assembly and DNA packaging

Initial head assembly

DNA

DNA packaging

2. Tail assembly

Base plate
assembly

Core added to
base plate

Sheath and
collar added

Tail added to head

3. Tail fiber synthesis and assembly

**Tail fibers
added**

Mature T₄ phage

■ Figure 16–15   The structure and assembly of bacteriophage T4 following infection of
E. coli. Three separate pathways lead to the formation of an icosahedral head filled with
DNA, a tail containing a central core, and tail fibers. The tail is added to the head and then
tail fibers are added. Each step of each pathway is influenced by one or more phage genes.
The electron micrograph is of T4 phages.

genetics. During infection of bacteria, enormous quantities of bacteriophages may be obtained for investigation. Often, over 10^{10} viruses are produced per milliliter of culture medium. Many genetic studies have relied on the ability to quantify the number of phages produced following infection under specific culture conditions. The technique used routinely is called the **plaque assay**.

This assay is illustrated in Figure 16–16, where actual plaque morphology is also shown. A serial dilution of the original virally infected bacterial culture is first performed. Then, a 0.1-ml sample (an aliquot) from one or more dilutions is added to a small volume of melted nutrient agar to which a few drops of a healthy bacterial culture have been added. The solution is then poured evenly over a base of solid nutrient agar in a Petri dish

and allowed to solidify before incubation. As described in the preceding section, viral plaques occur at each place where a single virus initially infected one bacterium in the lawn that has grown up during incubation. If the dilution factor is too low, the plaques are plentiful and will fuse, lysing the entire lawn. This has occurred in the 10^{-3} dilution in Figure 16–16. On the other hand, if the dilution factor is increased, plaques can be counted and the density of viruses in the initial culture can be estimated:

$$\text{(plaque number/ml)} \times \text{(dilution factor)}$$

Using the results shown in Figure 16–16, it is observed that there are 23 phage plaques derived from the 0.1-ml

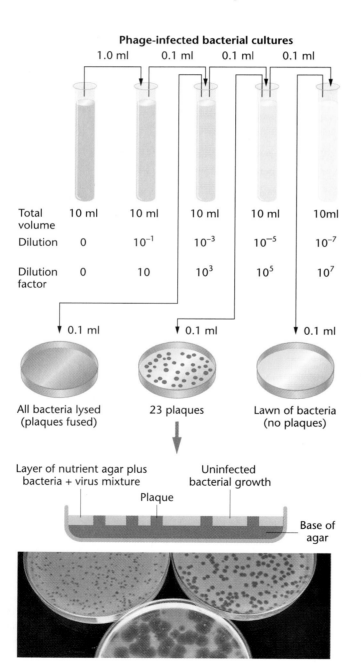

Phage-infected bacterial cultures

	1.0 ml	0.1 ml	0.1 ml	0.1 ml
Total volume	10 ml	10 ml	10 ml	10 ml	10ml
Dilution	0	10^{-1}	10^{-3}	10^{-5}	10^{-7}
Dilution factor	0	10	10^3	10^5	10^7

All bacteria lysed (plaques fused)   23 plaques   Lawn of bacteria (no plaques)

Layer of nutrient agar plus bacteria + virus mixture   Uninfected bacterial growth

Plaque

Base of agar

■ Figure 16–16
Diagrammatic illustration of the plaque assay for bacteriophage analysis. Serial dilutions of a bacterial culture infected with bacteriophages are first made. Then three of the dilutions (10^{-3}, 10^{-5}, and 10^{-7}) are analyzed using the plaque assay technique. In each case, 0.1 ml of the diluted culture is used. Each plaque represents the initial infection of one bacterial cell by one bacteriophage. In the 10^{-3} dilution, so many phages are present that all bacteria are lysed. In the 10^{-5} dilution, 23 plaques are produced. In the 10^{-7} dilution, the dilution factor is so great that no phages are present in the 0.1-ml sample, and thus no plaques form. From the 0.1-ml sample of the 10^{-5} dilution, the original bacteriophage density can be calculated as $23 \times 10 \times 10^5$ phages/ml (23×10^6 or 230×10^5), as described in the text. The photo illustrates a plaque assay involving phage T2 plaques on *E. coli*.

aliquot of the 10^{-5} dilution. Therefore, we can estimate that there are 230 phages/ml at this dilution. The initial phage density in the undiluted sample, where 23 plaques are observed from 0.1 ml, is calculated as

$$(230/\text{ml}) \times (10^5) = 230 \times 10^5/\text{ml}$$

Because this figure is derived from the 10^{-5} dilution, we can estimate that there will be only 0.23 phage/0.1 ml in the 10^{-7} dilution. As a result, when 0.1 ml from this tube is assayed, it is not unexpected that no phage particles will be present. This possibility is borne out in Figure 16–16, where an intact lawn of bacteria is depicted. The dilution factor is simply too great.

The use of the plaque assay has been invaluable in mutational and recombinational studies of bacteriophages.

Lysis and Lysogeny

The relationship between virus and bacterium does not always result in viral reproduction and lysis. As early as the 1920s, it was known that a virus can enter a bacterial cell and establish a symbiotic relationship with it. The precise molecular basis of this symbiosis is now well understood. Upon entry, the viral DNA, instead of replicating in the bacterial cytoplasm, is integrated into the bacterial chromosome, a step that characterizes the developmental stage referred to as **lysogeny**. Subsequently, each time the bacterial chromosome is replicated, the viral DNA is also replicated and passed to daughter bacterial cells following division. No new viruses are produced, and no lysis of the bacterial cell occurs. However, under certain stimuli, such as chemical or ultraviolet light treatment, the viral DNA may lose its integrated status and initiate replication, phage reproduction, and lysis of the bacterium.

Several terms are used to describe this relationship. The viral DNA that is integrated into the bacterial chromosome is called a **prophage**. Viruses that can either lyse the cell or behave as a prophage are called **temperate**. Those that can only lyse the cell are referred to as **virulent**. A bacterium harboring a prophage is said to be **lysogenic**; that is, it is capable of being lysed as a result of induced viral reproduction. The viral DNA, which can replicate either in the bacterial cytoplasm or as part of the bacterial chromosome, is classified as an **episome**.

Transduction: Virus-Mediated Bacterial DNA Transfer

In 1952, Joshua Lederberg and Norton Zinder were investigating possible recombination in the bacterium *Salmonella typhimurium*. Although they recovered prototrophs from mixed cultures of two different auxotrophic strains, subsequent investigations revealed that recombination was occurring in a manner different from that attributable to the presence of an F factor, as in *E. coli*. What they were to discover was still a third mode of bacterial recombination, one mediated by bacteriophages and now called **transduction**.

The Lederberg–Zinder Experiment

Lederberg and Zinder mixed the *Salmonella* auxotrophic strains LA-22 and LA-2 together and recovered prototroph cells when the mixture was plated on minimal medium. LA-22 was unable to synthesize the amino acids phenylalanine and tryptophan (*phe⁻ trp⁻*), and LA-2 could not synthesize the amino acids methionine and histidine (*met⁻ his⁻*). Prototrophs (*phe⁺ trp⁺ met⁺ his⁺*) were recovered at a rate of about 1 in 100,000 cells (a rate of 10^{-5}).

Although these observations at first appeared to suggest that the type of recombination involved was the kind observed earlier in *E. coli*, experiments using the Davis U-tube soon showed otherwise (Figure 16–17). When the two auxotrophic strains were separated by a sintered glass filter, thus preventing cell contact but allowing growth to occur in a common medium, a startling observation was made. When samples were removed from both sides of the filter and plated independently on minimal medium, prototrophs were recovered, but only from one side of the tube. Recall that if conjugation

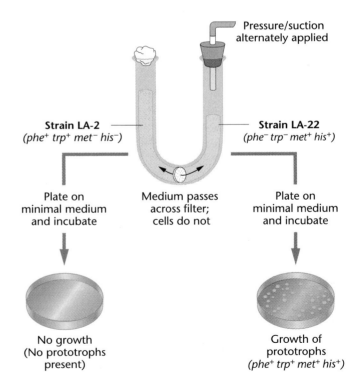

■ Figure 16–17 The Lederberg–Zinder experiment using *Salmonella*. After placing two auxotrophic strains on opposite sides of a Davis U-tube, Lederberg and Zinder recovered prototrophs from the side containing LA-22 but not from the side containing LA-2. These initial observations led to the discovery of the phenomenon called transduction.

were responsible, the conditions in the Davis U-tube would be expected to prevent recombination.

Prototrophs were recovered only when cells from the side of the tube containing LA-22 bacteria were plated. The presence of LA-2 cells on the other side of the tube was essential for recombination, since LA-2 cells were the source of the new genetic information. Because the genetic information responsible for recombination had somehow passed across the filter but in an unknown form, it was initially designated simply as a **filterable agent (FA)**.

Three subsequent observations made it clear that this recombination was quite distinct from any other form of recombination:

1. The FA would not pass across filters with a pore diameter of less than 100 nm, a size that normally allows passage of small DNA molecules.

2. Testing the FA in the presence of DNase, which will enzymatically digest DNA, showed that the FA was not destroyed. These first two observations demonstrate that FA is not naked DNA.

3. The third observation was particularly important. It was observed that FA is produced by the LA-2 cells only when they are grown in association with LA-22 cells. If LA-2 cells are grown independently and that culture medium is then added to LA-22 cells, recombination is not observed. Therefore, LA-22 cells play some role in the production of FA by LA-2 cells and do so only when sharing common growth medium. This was a key observation in determining the basis of genetic recombination.

These observations are explained by the existence of a prophage (P22) present in the LA-22 *Salmonella* cells. Rarely, P22 prophages entered the vegetative or lytic phase, reproduced, and lysed some of the LA-22 cells. This phage, being much smaller than a bacterium, was then able to cross the filter and lyse some of the LA-2 cells, because this strain is not immune to attack. In the process of lysis, the P22 phages produced in LA-2 often acquired a region of the LA-2 chromosome along with their own genetic material. If this region contained the *phe*⁺ and *trp*⁺ genes present in LA-2, and if the phages subsequently passed back across the filter and reinfected LA-22 cells, prototrophs were produced.

The Nature of Transduction

Further studies have revealed the existence of transducing phages in other species of bacteria. For example, *E. coli*, *Bacillus subtilis*, and *Pseudomonas aeruginosa* can be transduced by the phages P1, SP10, and F116, respectively. The precise mode of transfer of DNA during transduction has also been established. The process most often begins when a prophage enters the lytic cycle and progeny viruses subsequently infect and lyse other bacteria. During infection, the bacterial DNA is degraded into small fragments and the viral DNA is replicated. As packaging of the viral chromosomes in the protein head of the phage occurs, errors are sometimes made. Rarely, instead of viral DNA, segments of bacterial DNA are packaged.

Even though the initial discovery of **transduction** involved lysogenic bacteria, the same process can occur during the normal lytic cycle. Following infection, the bacterial chromosome is degraded into small pieces. If, during bacteriophage assembly, a small piece of bacterial DNA is packaged along with the viral chromosome, subsequent transduction is possible.

Sometimes, *only* bacterial DNA is packaged. Regions as large as 1 percent of the bacterial chromosome may become enclosed randomly in the viral head. Following lysis, these aberrant phages, lacking their own genetic material, are released into the culture medium. Because the ability to infect is independent of the contents of the capsid of the phage, these phages can initiate infection of other unlysed bacteria. When this occurs, bacterial rather than viral DNA is injected into the bacterium and can either remain in the cytoplasm or recombine with its homologous region of the bacterial chromosome. If the bacterial DNA remaining in the cytoplasm does not replicate, it may still remain in one of the progeny cells following each division. When this happens, only a single cell, partially diploid for the transduced genes, is produced, a phenomenon called **abortive transduction**. If the bacterial DNA recombines with its homologous region of the bacterial chromosome, the transduced genes are replicated as part of the chromosome and passed to all daughter cells. This process is called **complete transduction**. Both abortive and complete transduction are subclasses of the broader category called **generalized transduction**. As described above, transduction is characterized by the random nature of DNA fragments and genes transduced. Each fragment has a finite but small chance of being packaged in the phage head. Most cases of generalized transduction are of the abortive type; some data suggest that complete transduction occurs 10 to 20 times less frequently. This finding may be related to the fact that double-stranded DNA is involved. In comparison with single-stranded DNA, which is integrated during transformation, it may be much more difficult for double-stranded DNA to become integrated.

Mutation and Recombination in Viruses

Much of what is known concerning viral genetics has been derived from studies of bacteriophages. Phage mutations often affect the morphology of the plaques formed following lysis of the bacterial cells. For example, in 1946, Alfred Hershey observed unusual T2 plaques on plates of *E. coli* strain B. While the normal T2 plaques are small and have a clear center surrounded by a turbid, diffuse halo, the unusual plaques

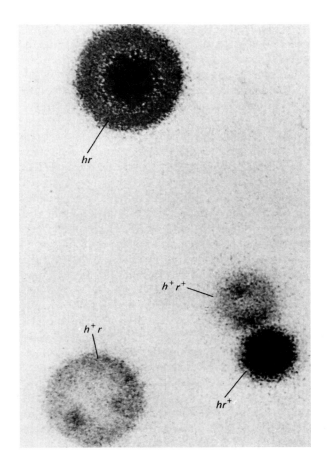

■ Figure 16–18 Actual plaque phenotypes observed following simultaneous infection of *E. coli* by two strains of phage T2, *h⁺r* and *hr⁺*. In addition to the parental genotypes, recombinant plaques *hr* and *h⁺r⁺* were also recovered.

TABLE 16.1	Some mutant types of T-even phages
Name	*Description*
minute	Small plaques
turbid	Turbid plaques on *E. coli* B
star	Irregular plaques
uv-sensitive	Alters UV sensitivity
acriflavin-resistant	Forms plaques on acriflavin agar
osmotic shock	Withstands rapid dilution into distilled water
lysozyme	Does not produce lysozyme
amber	Grows in *E. coli* K12, but not B
temperature-sensitive	Grows at 25°C, but not at 42°C

mutations have been particularly valuable in the study of essential genes in which mutations eliminate the ability of the phage to reproduce.

Following Hershey's work, several research teams demonstrated that recombination occurs between bacteriophages. This discovery of genetic exchange was made during experiments in which two mutant strains were allowed to infect the same bacterial culture simultaneously. These **mixed infection experiments** were designed so that the number of viral particles exceeded the number of bacterial cells sufficiently to ensure simultaneous infection of most cells by both viral strains.

For example, in one study using the T2/*E. coli* system, the viruses were of either the *h⁺r* or *hr⁺* genotype. If no recombination occurred, these two parental genotypes (wild-type host range restriction, rapid lysis; and extended host range, normal lysis) would be the only expected phage progeny. However, the recombinant *h⁺r⁺* and *hr* were detected in addition to the parental genotypes (Figure 16–18). The percentage of recombinant plaques divided by the total number of plaques reflects the relative distance between the genes:

$$\frac{(b^+r^+) + (br)}{\text{total plaques}} \times 100 = \text{recombinational frequency}$$

Sample data for the *h* and *r* loci are shown in Table 16.2. Similar recombinational studies have been performed with large numbers of mutant genes in a variety of bacteriophages. Data are analyzed in much the

are larger and display a sharp outer perimeter (see Figure 16–18). When the viruses are isolated from these plaques and replated on *E. coli* B cells, an identical plaque morphology is noted. Thus, the plaque phenotype is an inherited trait. Hershey named the mutant *rapid lysis (r)* because the plaques are larger, apparently resulting from a more rapid or more efficient life cycle of the phage (Figure 16–18). It is now known that wild-type phages undergo an inhibition of reproduction once a plaque achieves a particular size. The *r* mutant T2 phages are able to overcome this inhibition, producing larger plaques with a sharp perimeter.

Another bacteriophage mutation, *host range (h)*, was discovered by Luria. This mutation extends the range of bacterial hosts that the phage can infect. Although wild-type T2 phages can infect *E. coli* B, they cannot normally attach (or be adsorbed) to the surface of *E. coli* B-2. The *h* mutation, however, provides the basis for adsorption and subsequent infection.

Table 16.1 lists other representative types of mutations that have been isolated and studied in the T-even series of bacteriophages (T2, T4, T6, etc.). These mutations are important to the study of genetic phenomena in bacteriophages. Conditional, temperature-sensitive

TABLE 16.2	Results of a cross involving the *h* and *r* genes in phage T2 (*hr⁺ × h⁺r*)	
Genotype	*Plaques*	*Designation*
hr⁺	42	Parental progeny
h⁺r	34	76%
h⁺r⁺	12	Recombinants
hr	12	24%

Source: Data derived from Hershey and Rotman, 1949.

same way as they are in eukaryotic mapping experiments. Two- and three-point mapping crosses are possible, and the percentage of recombinants in the total number of phage progeny is calculated. This value is proportional to the relative distance between two genes along the chromosome.

An interesting observation in phage crosses is **negative interference**. Recall that, in eukaryotic mapping crosses, positive interference is the rule, and fewer than expected double-crossover events are observed. In phage crosses, often just the reverse occurs. In three-point analysis, a greater than expected frequency of double exchanges is observed. This negative interference is explained on the basis of the dynamics of the conditions leading to recombination within the bacterial cell.

How phage recombination occurs has been investigated. Available evidence supports a model in which recombination between phage chromosomes involves a breakage and reunion process similar to that of eukaryotic crossing over. The process is facilitated by nucleases that nick and reseal DNA strands. A fairly clear picture of the dynamics of viral recombination has emerged.

Following the early phase of infection, the chromosomes of each phage begin replication. As this stage progresses, a pool of chromosomes accumulates in the bacterial cytoplasm. If double infection by phages of two different genotypes has occurred, then the pool of chromosomes initially consists of the two parental types. Genetic exchange between these two types will occur, producing recombinant chromosomes.

In the case of the h^+r and hr^+ example just discussed, h^+r^+ and hr chromosomes are produced. Each of these recombinant chromosomes may undergo replication and is also free to undergo new exchange events with the other and with the parental types. Furthermore, recombination is not restricted to exchange between two chromosomes—three or more may be involved. As phage development progresses, chromosomes are removed randomly from the pool and packaged into the phage head, forming mature phage particles. Both parental and recombinant genotypes are produced.

Recombination events in viruses are not restricted to exchanges between genes (intergenic recombination). With powerful selection systems, it is possible to demonstrate **intragenic recombination**, in which exchanges occur within the region represented by a single gene.

Other Strategies for Viral Reproduction

In contrast to the T-even phages, other bacterial viruses have evolved many different strategies that allow them to reproduce. For example, when **bacteriophage φX** infects *E. coli*, only a circular single strand of DNA, called the plus (+) strand, is injected into the host cell. This then serves as the template for the synthesis of the complementary negative (−) strand. This results in a circular double-stranded DNA molecule called the **replicative form (RF)**. The RF is itself replicated semiconservatively, producing about 50 progeny molecules. These then serve as templates for the production of only (+) strands. As hundreds of these DNA strands are produced, they are packaged into protein capsids, synthesized under the direction of the viral genetic information.

While DNA viruses can pirate the use of the host DNA polymerase in order to replicate their nucleic acid, no comparable enzyme is present when RNA-containing viruses infect host cells. Some RNA viruses utilize an RNA-dependent RNA polymerase, which is also called **RNA replicase**. For example, when the small **bacteriophage Qβ** infects *E. coli*, the infective single-stranded RNA molecule first serves as a messenger RNA (mRNA), encoding three proteins. Two are components of the coat protein, while one is a subunit of RNA replicase. This subunit combines with three other subunits that are bacterial in origin to form a functional enzyme that is specific for the replication of the phage RNA (as compared with the bacterial RNAs that are present). First, complementary RNA virus strands (−) are created. These serve as the template for the synthesis of many (+) strands. Ultimately, the (+) strands are encapsulated by coat proteins as mature viruses are formed. A similar mode of replication occurs in the RNA-containing poliovirus, which infects mammalian cells.

Reverse Transcriptase

The final type of viral reproduction to be discussed occurs following the infection of animal cells rather than bacteria and involves an extraordinary finding made in the early 1960s. A long series of investigations by Howard Temin and David Baltimore led to the discovery of **reverse transcriptase**, an enzyme that is capable of synthesizing DNA on an RNA template. This enzyme is also called **RNA-directed DNA polymerase**. Temin proposed that RNA, which serves as the genetic material of certain animal tumor viruses, is initially transcribed into DNA, which then serves as the template for replication and transcription of RNA during viral infection. These viruses are called **oncogenic** because they induce malignant growth.

Temin's proposal concerning reverse transcriptase was largely ignored because it contradicted the previously accepted notion that genetic information flows only from DNA to RNA. Because many other RNA-containing viruses (e.g., polio virus and double- and single-stranded RNA bacteriophages), whose RNA is replicated by the enzyme RNA replicase, failed to demonstrate reverse transcriptase activity, molecular biologists questioned why the tumor viruses should require such an unusual enzyme. Working with the RNA-containing oncogenic Rous sarcoma virus (RSV), which produces cancer in chickens, Temin added the antibiotic actinomycin D to cultures of cells that had been infected. He found that the antibiotic inhibited viral multiplication. Actinomycin D is known to specifically inhibit DNA-directed RNA synthesis and not to affect RNA-directed RNA synthesis. Temin also found that inhibitors of DNA synthesis such as 5-fluorodeoxyuridine also block infection. Thus, he reasoned that DNA must serve as an essential intermediate molecule during infection.

Temin subsequently demonstrated directly that DNA is synthesized by RSV during infection. He announced these results in 1970 and found that other workers, using a mouse leukemia virus, had arrived at the same conclusions. These reports stimulated a tremendous amount of work on the enzyme and its role in tumor production. Reverse transcriptase was subsequently isolated independently by Temin and Baltimore. It has

since been found in all RNA oncogenic viruses. Because these viruses "reverse" the flow of genetic information, they are called **retroviruses**.

The enzyme uses the infecting (+) RNA strand as a template, synthesizing DNA in the 5′–3′ direction. Another enzyme then uses this DNA strand as a template and synthesizes the complementary strand, creating a double-stranded DNA molecule. This is capable of integrating into the genome of the host cell, forming what is called a **provirus**. If the proviral DNA is transcribed into RNA, (+) RNA strands are produced that may be translated into viral proteins. Packaging of other newly formed RNA molecules into the viral capsid occurs to form mature viruses. Such a host cell is said to be **transformed**. A similar life cycle occurs in all retroviruses.

Chapter Summary

1. Genetic recombination in bacteria may be accomplished as a result of three different modes: conjugation, transformation, and transduction.

2. Conjugation is initiated by a bacterium housing a plasmid called the F factor. If the F factor is in the cytoplasm (F⁺), following its replication, one copy is transferred to the recipient cell, converting it to the F⁺ status.

3. If the F factor is integrated into the donor cell chromosome (Hfr), recombination is initiated with the recipient cell, with genetic information flowing unidirectionally. Time mapping of the bacterial chromosome is based on the orientation and position of the F factor in the donor chromosome.

4. The products of a group designated as the *rec* genes have been found to be directly involved in the process of recombination between the invading DNA and the recipient bacterial chromosome.

5. Plasmids, such as the F factor, are autonomously replicating DNA molecules found in the bacterial cytoplasm. Some plasmids may contain genes such as those conferring antibiotic resistance, as well as those necessary for their transfer during conjugation.

6. The phenomenon of transformation, which does not require cell contact, involves the entry of single-stranded exogenous DNA into the host chromosome of a recipient bacterial cell. Linkage mapping of closely aligned genes may be performed using this process.

7. Bacteriophages are studied using the plaque assay. The discovery of mutations in phages, such as those causing variation in plaque morphology and the alteration of the host range of infectivity, have allowed the analysis of recombination between viruses.

8. Transduction, or virus-mediated bacterial DNA transfer, requires the formation of a symbiotic relationship between bacteria and bacteriophages that infect them. The viral genetic material may become integrated into the bacterial chromosome as a prophage and lysogenize the host cell.

9. Retroviruses are RNA oncogenic viruses that depend on reverse transcriptase activity for infection. This enzyme reverses the flow of genetic information from RNA to DNA. This newly synthesized DNA may integrate into the genome of the host chromosome and, upon transcription, become activated.

Key Terms

abortive transduction, 364
adaptation hypothesis, 349
auxotroph, 350
bacteriophage, 360
bacteriophage Qβ, 366
bacteriophage φX, 366
Col plasmid, 357
colicin, 357

competence, 358
complete transduction, 364
conjugation, 350
cotransformation, 359
D-loop, 356
dilution factor, 350
episome, 363
F⁻ cell, 351

F⁺ cell, 351
F′ cell, 355
F factor, 352
F sex pilus, 352
fertility factor, 352
filterable agent (FA), 364
fluctuation test, 349
generalized transduction, 364

INSIGHTS and SOLUTIONS

1. Time mapping was performed in a cross involving the *his, leu, mal,* and *xyl* auxotrophs. After 25 minutes, mating was interrupted with the following results in recipient cells:

90% were *xyl*⁺

80% were *mal*⁺

20% were *his*⁺

none were *leu*⁺

What are the positions of these genes relative to the origin (*O*) of the F factor and to one another?

Solution: Because the *xyl* gene was transferred most frequently, it is closest to *O* (very close). The *mal* gene is next and reasonably close to *xyl*, followed by the *his* gene. The *leu* gene is well beyond these three, since no recombinants are recovered that include it.

2. In four Hfr strains of bacteria, all derived from an original F⁺ culture grown over several months, a group of hypothetical genes was studied and shown to be transferred in the following orders:

Hfr Strain	Order of Transfer					
1	E	R	I	U	M	B
2	U	M	B	A	C	T
3	C	T	E	R	I	U
4	R	E	T	C	A	B

Assuming that *B* is the first gene along the chromosome, determine the sequence of all genes shown. One strain creates an apparent dilemma. Which one is it? Explain why the dilemma is only apparent and not real.

Solution: The solution is arrived at by overlapping the genes in each strain in the sequence in which they were transferred:

```
         2  U M B A C T
Strain:  3        C T E R I U
         1              E R I U M B
```

Starting with *B*, the sequence of genes is: *BAC-TERIUM!*

Strain 4 creates an apparent dilemma, which is resolved by realizing that the F factor is integrated in the opposite orientation; thus the genes enter in the opposite sequence, starting with gene R.

3. Three strains of bacteria, each bearing a separate mutation, *a*⁻, *b*⁻, or *c*⁻, were used as the sources of donor DNA in a transformation experiment involving recipient cells that were wild type for those genes, but which expressed the mutant *d*⁻.

(a) Based on the following data and assuming that the location of *d* precedes the *a, b,* and *c* genes, propose a linkage map for the four genes:

DNA Donor	Recipient	Transformants	Frequency of Transformants
a⁻*d*⁺	*a*⁺*d*⁻	*a*⁺*d*⁺	0.21
b⁻*d*⁺	*b*⁺*d*⁻	*b*⁺*d*⁺	0.18
c⁻*d*⁺	*c*⁺*d*⁻	*c*⁺*d*⁺	0.63

Solution: These data reflect the relative distances between each of the four genes. The *a* and *b* genes are about the same distance away from the *d* gene and are tightly linked to one another. The *c* gene is the most distant. Assuming that the *d* gene precedes the others, the map looks like this:

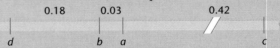

(b) If the donor DNA was wild type and recipient cells were either *a*⁻*b*⁻, *a*⁻*c*⁻, or *b*⁻*c*⁻, in which case would wild-type transformants be expected most frequently?

Solution: Because the *a* and *b* genes are closely linked, they are most likely to be cotransformed in a single event. Thus, recipient cells of *a*⁻*b*⁻ are most likely to be converted to wild type.

Problems and Discussion Questions

1. Distinguish between the three modes of recombination in bacteria.

2. With respect to F^+ and F^- bacterial matings, answer the following questions: (a) How was it established that physical contact was necessary? (b) How was it established that chromosome transfer was unidirectional? (c) What is the basis of a bacterium being F^+ or Hfr?

3. List the major differences between: (a) the $F^+ \times F^-$ and the Hfr $\times F^-$ bacterial crosses and (b) F^+, F^-, Hfr, and F' bacteria.

4. Describe the basis for chromosome mapping in the Hfr $\times F^-$ crosses.

5. When the interrupted mating technique was used with five different strains of Hfr bacteria, the following order of gene entry and recombination was observed. On the basis of these data, draw a map of the bacterial chromosome. Do the data support the concept of circularity?

Hfr Strain			Order		
1	T	C	H	R	O
2	H	R	O	M	B
3	M	O	R	H	C
4	M	B	A	K	T
5	C	T	K	A	B

6. Why are the recombinants produced from an Hfr $\times F^-$ cross, never F^+?

7. Describe the origin of F' bacteria and merozygotes.

8. Explain the observations that led Zinder and Lederberg to conclude that the prototrophs recovered in their transduction experiments were not the result of F-mediated conjugation.

9. Define plaque, lysogeny, and prophage.

10. If a single bacteriophage infects one *E. coli* cell present on a lawn of bacteria and upon lysis yields 200 viable viruses, how many phages will exist in a single plaque if only three more lytic cycles occur?

11. A culture of an auxotrophic *leu⁻* strain of bacteria was irradiated and incubated until it reached the stationary phase. A control culture, which was not irradiated, was also studied. These cultures were then serially diluted, and 0.1 ml of various dilutions was plated on minimal medium plus leucine and separately on minimal medium. The results, shown below, were used to determine the spontaneous and X-ray-induced mutation rate of *leu⁻* to *leu⁺*.

Culture Condition	Culture Medium	Dilution	Number of Colonies
Irradiated	(1) Minimal medium plus leucine	10^{-9}	24
	(2) Minimal medium	10^{-2}	12
Control	(1) Minimal medium plus leucine	10^{-9}	12
	(2) Minimal medium	10^{-1}	3

(a) Describe what is represented by each value obtained. Which values should be approximately equal? Are they?

(b) Determine the induced and spontaneous mutation rate leading to prototrophic growth (*leu⁻* to *leu⁺*).

12. Describe the role of reverse transcriptase during the infection of an animal cell by a retrovirus.

13. In a transformation experiment, donor DNA was obtained from a prototrophic bacterial strain ($a^+b^+c^+$), and the recipient was auxotrophic for the three genes ($a^-b^-c^-$). The following data were obtained:

$a^+b^-c^-$	180
$a^-b^+c^-$	150
$a^+b^+c^-$	210
$a^-b^-c^+$	179
$a^+b^-c^+$	2
$a^-b^+c^+$	1
$a^+b^+c^+$	3

What general conclusions can you draw about the linkage relationships among the three genes?

14. Two theoretical phage strains ($a^-b^-c^-$ and $a^+b^+c^+$) were used to simultaneously infect a bacterial culture. Of 10,000 plaques scored, the following data were obtained:

$a^+b^+c^+$	4100	$a^-b^+c^-$	160
$a^-b^-c^-$	3990	$a^+b^-c^+$	140
$a^+b^-c^-$	740	$a^-b^-c^+$	90
$a^-b^+c^+$	670	$a^+b^+c^-$	110

Determine the genetic map of the three genes on the viral chromosome.

15. An Hfr strain is used to map three genes in an interrupted mating experiment. The cross is $Hfr\ a^+b^+c^+rif^S \times F^-a^-b^-c^-rif^R$. (No map order is implied in the listing of the alleles.) rif = the antibiotic rifamycin. The a^+ gene is required for the biosynthesis of nutrient A, the b^+ gene for nutrient B, and c^+ for nutrient C. The minus alleles indicate mutations.

The cross is started at time 0, and at various times the mating mixture is plated on three types of plates. Each plate contains minimal medium *and* rifamycin *plus* specific supplements, which are indicated below. The results for each time point are shown as number of colonies growing on each plate.

Supplements Added	*Time of Interruption*			
to MM	5 min	10 min	15 min	20 min
Nutrients A and B	0	0	4	21
Nutrients B and C	0	5	23	40
Nutrients A and C	4	25	60	82

(a) What genotype is being selected for on each type of medium?

(b) Draw a map of the *E. coli* chromosome showing the relative positions of genes *a*, *b*, and *c* approximately to scale (it takes about 90 minutes to transfer the F factor genes in). If the data do not give an unambiguous position for a specific gene, indicate so on the map. Also indicate the position and orientation of the F factor.

Selected Readings

Adelberg, E. A. 1960. *Papers on bacterial genetics.* Boston: Little, Brown.

Birge, E. A. 1988. *Bacterial and bacteriophage genetics—An introduction.* New York: Springer-Verlag.

Broda, P. 1979. *Plasmids.* New York: W. H. Freeman.

Bukhari, A. I., Shapiro, J. A., and Adhya, S. L., eds. 1977. *DNA insertion elements, plasmids, and episomes.* Cold Spring Harbor, NY: Cold Spring Harbor Laboratory.

Cairns, J., Stent, G. S., and Watson, J. D., eds. 1966. *Phage and the origins of molecular biology.* Cold Spring Harbor, NY: Cold Spring Harbor Laboratory.

Campbell, A. M. 1976. How viruses insert their DNA into the DNA of the host cell. *Sci. Am.* (Dec.) 235:102–13.

Cavalli-Sforza, L. L., and Lederberg, J. 1956. Isolation of preadaptive mutants in bacteria by sib selection. *Genetics* 41:367–81.

Hayes, W. 1968. *The genetics of bacteria and their viruses,* 2nd ed. New York: Wiley.

Hershey, A. D., and Rotman, R. 1949. Genetic recombination between host range and plaque-type mutants of bacteriophage in single cells. *Genetics* 34:44–71.

Jacob, F., and Wollman, E. L. 1961. Viruses and genes. *Sci. Am.* (June) 204:92–106.

Lederberg, J. 1986. Forty years of genetic recombination in bacteria: A fortieth anniversary reminiscence. *Nature* 324:627–28.

Luria, S. E., and Delbruck, M. 1943. Mutations of bacteria from virus sensitivity to virus resistance. *Genetics* 28:491–511.

Lwoff, A. 1953. Lysogeny. *Bacteriol. Rev.* 17:269–337.

Morse, M. L., Lederberg, E. M., and Lederberg, J. 1956. Transduction in *Escherichia coli* K12. *Genetics* 41:141–56.

Novick, R. P. 1980. Plasmids. *Sci. Am.* (Dec.) 243:102–27.

Smith-Keary, P. F. 1989. *Molecular genetics of* Escherichia coli. New York: Guilford Press.

Stahl, F. W. 1987. Genetic recombination. *Sci. Am.* (Nov.) 256:91–101.

Stent, G. S. 1963. *Molecular biology of bacterial viruses.* New York: W. H. Freeman.

———. 1966. *Papers on bacterial viruses,* 2nd ed. Boston: Little, Brown.

Temin, H. 1972. RNA-directed DNA synthesis. *Sci. Am.* (Jan.) 24–33.

Varmis, H. 1987. Reverse transcription. *Sci. Am.* (Sept.) 257:56–64.

Visconti, N., and Delbruck, M. 1953. The mechanism of genetic recombination in phage. *Genetics* 38:5–33.

Wollman, E. L., Jacob, F., and Hayes, W. 1956. Conjugation and genetic recombination in *Escherichia coli* K12. *Cold Spring Harb. Symp. Quant. Biol.* 21:141–62.

Zinder, N. D. 1958. Transduction in bacteria. *Sci. Am.* (Nov.) 199:38–46.

———, and Lederberg, J. 1952. Genetic exchange in *Salmonella. J. Bacteriol.* 64:679–99.

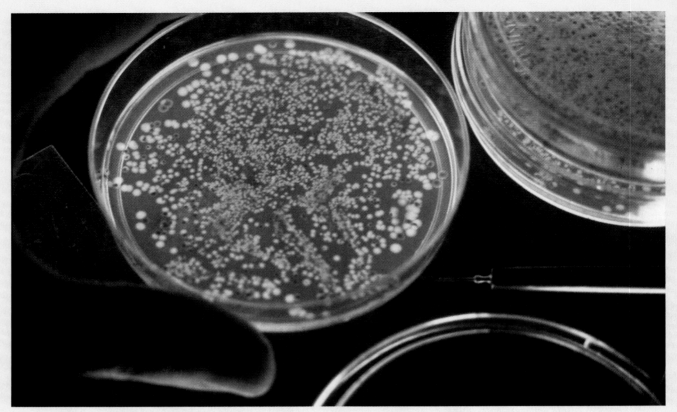

A Petri plate showing the growth of host cells after uptake of recombinant plasmids.

CHAPTER OUTLINE

CHAPTER
17

DNA Biotechnology:
Techniques and Analysis

Chapter Concepts

Recombinant DNA technology depends in part on the ability to cleave and rejoin DNA molecules at specific nucleotide sequences. Once they are created, recombinant DNA molecules can be transferred to host cells such as bacteria or yeast and amplified, isolated, and characterized. This technology has revolutionized gene mapping, disease diagnosis, the commercial production of human gene products, and the transfer of genes between species of plants and animals.

In 1982 the U.S. Food and Drug Administration approved the marketing of human insulin produced by genetically engineered bacterial cells. Previously, insulin was chemically extracted from the pancreatic glands of pigs and cows, supplied by slaughterhouses. The production of insulin by recombinant DNA technology was a milestone in the biotechnology revolution, and ensures a steady supply of a purified gene product used to treat diabetes. Today, many human proteins used for the treatment of diseases are produced by biotechnology.

The use of recombinant DNA technology makes it possible to identify and isolate a single gene from the 100,000 or so present in the human genome, and to produce large quantities of this gene in the form of cloned DNA molecules. In addition, the gene can be transferred into cells that will synthesize its gene product. **Clones** are identical organisms, cells, or molecules descended from a single ancestor. Cloning a gene involves producing many identical copies that can be used for numerous purposes, including research into the structure and organization of a gene, or the commercial production of proteins such as insulin.

In this chapter, we will review the basic methods of recombinant DNA technology that are used to isolate, replicate, and analyze genes. In the chapter that follows, we will discuss some of the applications of this technology to research, medicine, the legal system, agriculture, and industry.

Recombinant DNA Technology: An Overview

The term **recombinant DNA** refers to the creation of a new combination of DNA molecules that are not found together naturally. Although processes such as crossing over technically produce recombinant DNA, the term is generally reserved for DNA molecules produced by joining molecules derived from different biological sources.

Recombinant DNA technology uses methods derived from nucleic acid biochemistry coupled with genetic techniques originally developed for the study of bacteria and viruses. As described below, recombinant DNA technology is a powerful tool for the isolation of potentially unlimited quantities of a gene. The basic procedures involve a series of steps:

1. DNA is purified from cells or tissues.

2. DNA fragments are generated using enzymes called **restriction endonucleases** or **restriction enzymes**, which recognize and cut DNA molecules at specific nucleotide sequences.

3. The fragments produced by digestion with restriction enzymes are joined to other DNA molecules that serve as **vectors**, or carrier molecules. A vector with an inserted DNA segment is a **recombinant DNA molecule**.

4. The recombinant DNA molecule, consisting of a vector carrying an inserted DNA fragment, is trans-

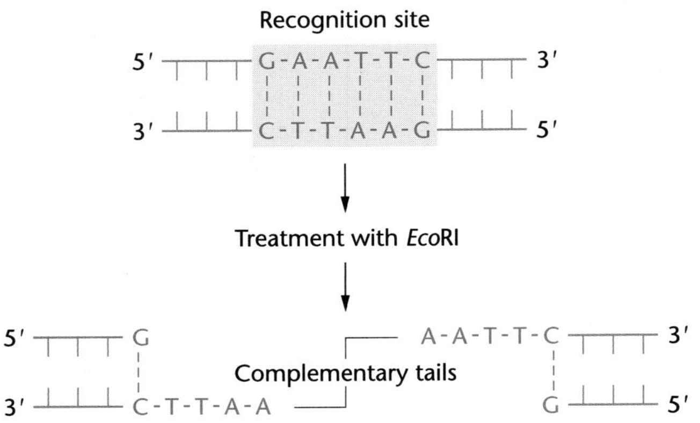

Recognition site

5' ⊤⊤⊤ G - A - T - T - C ⊤⊤⊤ 3'
        | | | | |
3' ⊥⊥⊥ C - T - T - A - A - G ⊥⊥ 5'

↓

Treatment with *Eco*RI

↓

5' ⊤⊤⊤ G   ⌐ A - A - T - T - C ⊤⊤⊤ 3'
                                                |
Complementary tails

3' ⊥⊥⊥ C - T - T - A - A ⌐   G ⊥⊥⊥ 5'

■ Figure 17–1 The restriction enzyme *Eco*RI recognizes and binds to the palindromic nucleotide sequence GAATTC. Cleavage of the DNA at this site produces complementary single-stranded tails. These single-stranded tails can anneal with single-stranded tails from other DNA fragments to form recombinant DNA molecules.

ferred to a host cell. Within the host cell, the recombinant molecule replicates, producing dozens of identical copies known as clones.

5. As host cells replicate, the recombinant DNA molecules within them are passed on to all their progeny cells, creating a population of host cells, each of which carries many copies of a cloned DNA sequence.

6. The cloned DNA can be recovered from host cells, purified, and analyzed.

7. Potentially, the cloned DNA can be transcribed, its mRNA translated, and the gene product isolated and used for research, or sold commercially.

Constructing Recombinant DNA Molecules

The development of recombinant DNA and gene cloning technology has provided scientists with methods for the isolation of large quantities of specific genes or other DNA sequences, facilitating studies of gene organization, structure, and expression.

Restriction Enzymes

The cornerstone of recombinant DNA technology is a class of enzymes called restriction endonucleases. These enzymes, isolated from bacteria, received their name because they restrict or prevent viral infection by degrading the invading viral DNA. Restriction enzymes recognize a specific nucleotide sequence and cut both strands of the DNA within that sequence. The 1978 Nobel Prize was awarded to Werner Arber, Hamilton Smith, and Daniel Nathans for their work on restriction enzymes. To date, more than 200 different

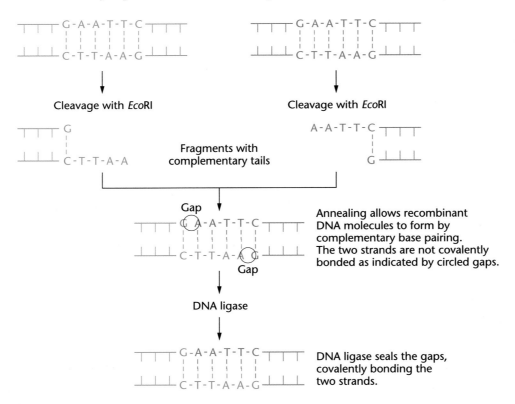

■ Figure 17–2 In forming recombinant DNA molecules, DNA from different sources is cleaved with *Eco*RI and mixed together to allow annealing into recombinant molecules. The enzyme DNA ligase is then used to chemically bond these annealed fragments into an intact recombinant DNA molecule.

restriction enzymes have been identified. Their usefulness in cloning derives from their ability to reproducibly cut DNA into fragments at specific sites.

One of the first restriction enzymes identified was isolated from *Escherichia coli* and was designated **EcoRI** (pronounced echo-r-one). Its recognition site and points of cleavage are shown in Figure 17–1. The DNA fragments produced by *Eco*RI digestion have overhanging single-stranded tails that can reanneal with complementary single-stranded tails on other DNA fragments. If they are mixed together under proper conditions, DNA fragments from two sources can form recombinant molecules by hydrogen bonding of their sticky ends. The enzyme T4 **DNA ligase** can be used to link these fragments covalently to form recombinant DNA molecules (Figure 17–2). Figure 17–3 shows some common restriction enzymes and their recognition sequences.

Vectors

Fragments of DNA produced by restriction digestion cannot enter bacterial cells; when a DNA fragment is

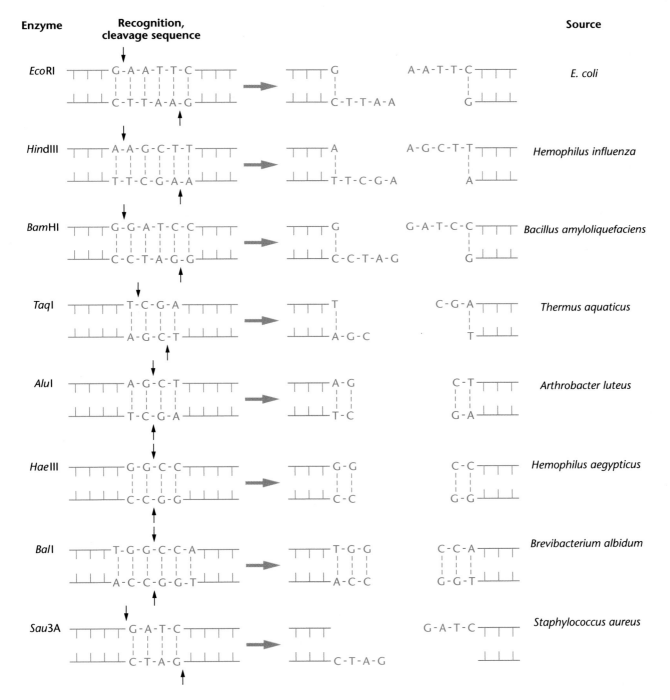

■ Figure 17–3 Some common restriction enzymes, with their recognition sequences and cutting sites, cleavage patterns, and sources.

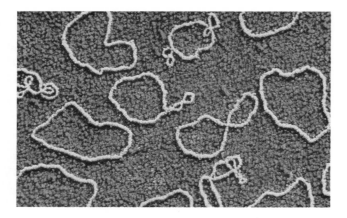

■ Figure 17–4   A color-enhanced electron micrograph of circular plasmid molecules isolated from the bacterium *E. coli.* Genetically engineered plasmids are used as vectors for cloning DNA.

joined to a vector, however, it can gain entry to a host cell where it can be replicated or cloned. Vectors are, in essence, carrier DNA molecules. A number of vectors are currently in use, including those derived from **plasmids** and **bacteriophages**.

Plasmids are naturally occurring, extrachromosomal, double-stranded DNA molecules that replicate autonomously within bacterial cells (Figure 17–4). Although only a single plasmid may enter a host cell, many plasmids can increase their copy number in the host cell so that up to 1000 copies are present. When used as a vector, such plasmids allow more copies of cloned DNA to be produced. Plasmids used as vectors have been modified in the laboratory to contain a limited number of restriction sites for the insertion of DNA fragments and marker genes that can be used to detect the presence of plasmids in host cells.

Many genetically engineered plasmid vectors are now available, and they offer a number of useful features that make it easier to identify host cells carrying a plas-

mid with an inserted DNA fragment. One such plasmid is pUC18 (Figure 17–5). As a vector, this plasmid has several useful properties:

1. The plasmid is small, allowing it to carry relatively large DNA inserts.

2. In a host cell, pUC18 can replicate to form about 500 copies per cell, producing many clones, or copies of inserted DNA fragments.

3. A large number of unique restriction sites has been engineered into pUC18, and these are conveniently clustered in one region, called a **polylinker site**.

4. The pUC18 plasmid carries a fragment of a bacterial gene called *lacZ*, and the polylinker region is inserted into this fragment. Through the action of the *lacZ* gene, bacterial host cells carrying pUC18 produce blue colonies when grown on medium containing a compound called X-gal, which is a synthetic substrate of β-galactosidase, the enzyme encoded by the *lacZ* gene. DNA segments inserted into the polylinker site disrupt and inactivate the *lacZ* gene, and a bacterial cell carrying pUC18 with an inserted DNA fragment will not metabolize the X-gal and will form white colonies, making them easy to identify (Figure 17–6). The *lacZ* gene is described in more detail in Chapter 13.

As mentioned above, bacteriophage (or phage for short) vectors are also used in recombinant DNA work. As discussed in Chapter 16, phages are viruses that infect and kill bacterial cells. One of the most widely used of these vectors is **phage lambda** (Figure 17–7). Its genes have all been identified and mapped, and the DNA sequence of the entire genome is known. The middle one-third of the lambda chromosome contains a cluster of dispensable genes, and this region can be replaced with foreign DNA without affecting the ability of the phage to infect cells and form plaques (Figure 17–8). Over 100 vectors based on phage lambda have been developed by removing various portions of the

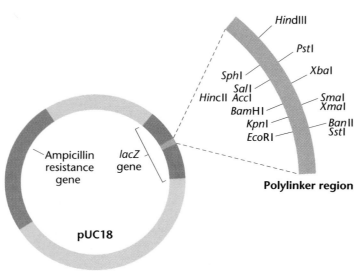

■ Figure 17–5   The plasmid pUC18 offers several advantages as a vector for cloning. Because of its small size, it can accept relatively large DNA fragments for cloning; it replicates to a high copy number, and has a large number of restriction sites within the polylinker, located within a *lacZ* gene. Bacteria carrying pUC18 produce blue colonies when grown on a medium containing X-gal. DNA inserted into the polylinker site disrupts the *lacZ* gene, resulting in white colonies, allowing direct identification of colonies carrying cloned DNA inserts.

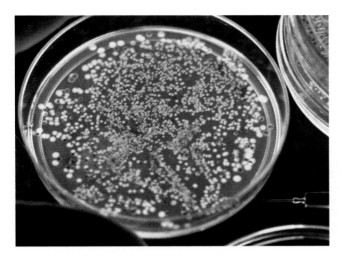

■ Figure 17–6 A Petri plate showing the growth of host cells after uptake of recombinant plasmids. The medium on the plate contains a compound called X-gal. DNA inserts into the pUC18 vector disrupt the gene responsible for the formation of blue colonies. Cells in the blue colonies do not carry any cloned DNA inserts, whereas the white colonies contain vectors carrying DNA inserts.

central gene cluster. DNA to be cloned can be inserted in place of this region. Lambda vectors containing inserted DNA can be introduced into bacterial host cells, where they reproduce to form many particles of infective phage, each of which carries a DNA insert. As they reproduce, the phages form plaques from which the cloned DNA can be recovered.

Phage vectors offer the advantage of being able to carry much longer DNA fragments than plasmids. This is an important consideration when attempting to clone entire genomes. In addition, some phage vectors will

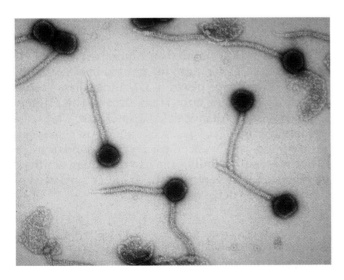

■ Figure 17–7 A colorized electron micrograph showing a cluster of the bacteriophage lambda, widely used as a vector in recombinant DNA work.

only accept inserts of a minimum size. This means that in a cloning experiment, where one is trying to isolate large inserts, no vectors will end up carrying relatively useless inserts of a few hundred or a few thousand nucleotides in length.

Cloning in *E. coli* Host Cells

As discussed earlier, biotechnology involves not only the construction of recombinant DNA molecules, but also the replication of these molecules to produce many copies, or clones. This is accomplished by transferring the recombinant molecules into host cells where replication takes place.

A variety of prokaryotic and eukaryotic cells can be used as hosts for the replication of recombinant vectors. One of the most commonly used hosts is a laboratory strain of the bacterium *E. coli* known as K12. *E. coli* strains such as K12 are genetically well characterized, and they can serve as hosts for a wide range of vectors. Several steps are required to create recombinant DNA molecules and propagate them in *E. coli* host cells (Figure 17–9):

1. The DNA to be cloned is isolated and treated with a restriction enzyme to create fragments ending in a specific sequence.

2. The fragments are ligated to plasmid molecules that have been cut with the same restriction enzyme, creating a recombinant vector molecule.

3. The bacterial cells are heat-shocked to facilitate uptake of the recombinant vector into the cell.

4. The bacterial cells are cultured to allow the resident recombinant DNA molecules to replicate and form dozens of copies.

5. The bacteria are grown on a nutrient plate where they form colonies, and are screened to identify those that have taken up recombinant plasmids. Because the cells in each colony are derived from a single ancestral cell, all cells in the colony, and the plasmids they contain, are genetically identical clones.

Similarly, phage containing foreign DNA can be grown in *E. coli,* and the resulting plaques each represent a cloned descendant of a single ancestral phage, and are therefore regarded as clones.

Cloning in Eukaryotic Host Cells

To study the expression and regulation of eukaryotic genes, it is often convenient and even necessary to use eukaryotic cells as hosts. Several cloning systems using eukaryotic vectors and host cells have been developed. One of these is based on the yeast, *Saccharomyces cerevisiae.*

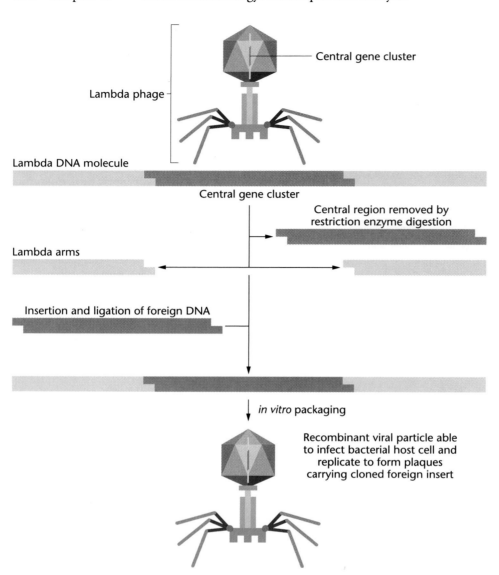

Lambda phage

Central gene cluster

Lambda DNA molecule

Central gene cluster

Central region removed by restriction enzyme digestion

Lambda arms

Insertion and ligation of foreign DNA

in vitro packaging

Recombinant viral particle able to infect bacterial host cell and replicate to form plaques carrying cloned foreign insert

■ Figure 17–8 Steps in cloning phage lambda as a vector. DNA is extracted from a preparation of lambda phage, and the central gene cluster is removed by treatment with a restriction enzyme. The DNA to be cloned is cut with the same enzyme and ligated into the arms of the lambda chromosome. The recombinant chromosome is then packaged into phage proteins to form a recombinant virus. This virus is able to infect bacterial cells and replicate its chromosome, including the DNA insert.

Although yeast is a eukaryotic organism, it can be grown and manipulated in much the same way as bacterial cells. Further, the genetics of yeast has been intensively studied, providing a large catalog of mutations and a highly developed genetic map. In addition, the entire yeast genome has recently been sequenced. Naturally occurring yeast plasmids have been used to construct a number of yeast cloning vectors. By combining bacterial plasmid sequences with those of yeast plasmids, vectors for growth in yeast host cells have been produced.

A second type of yeast vector is the **yeast artificial chromosome (YAC)**. In linear form (Figure 17–10), a YAC contains telomeres at each end, an origin of replication (called an autonomously replicating sequence or ARS), and a yeast centromere. The YAC also contains two selectable markers and a cluster of restriction sites for DNA inserts.

Segments of DNA more than 1 megabase long (1 Mb = 1 million base pairs) can be inserted into YACs. The ability to clone large pieces of DNA into these vectors has made them an important tool in the Human Genome Project (discussed in Chapter 18).

Although cloning in yeast vectors and host cells is currently the most advanced for cloning in eukaryotes, other systems, including the use of human artificial chromosomes as vectors with tissue culture cells as hosts, are being developed.

Constructing Libraries of Cloned DNA Sequences

In order to identify individual DNA sequences of interest, and ultimately to clone these sequences, it is necessary to create what is called a *library* of cloned DNA

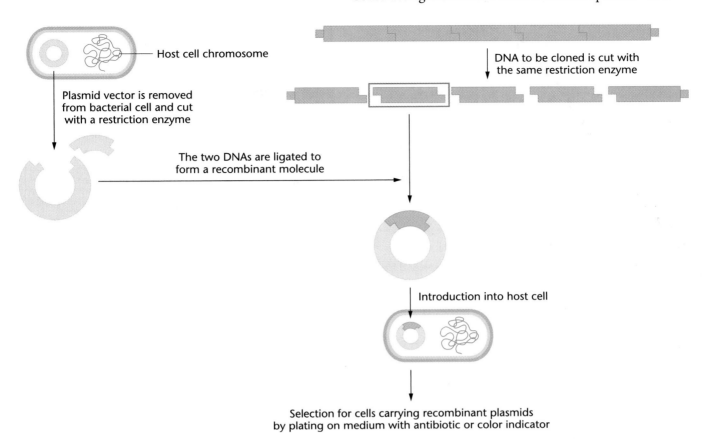

Figure 17–9 A summary of steps involved in cloning with a plasmid vector. Plasmid vectors are isolated and cut with a restriction enzyme. The DNA to be cloned is cut with the same restriction enzyme, producing a collection of fragments. These fragments are spliced into the vector and transferred to a bacterial host for replication. Bacterial cells carrying plasmids with DNA inserts can be identified by growth on selective medium and isolated. The cloned DNA can be recovered from the bacterial host for further analysis.

sequences. These libraries can represent either genomic DNA cut into fragments with restriction enzymes, or DNA copies of the mRNAs found in a particular cell or tissue type. The first type of library, called a **genomic library**, can represent an entire genome, or the sequences contained on a single chromosome. The second type of library, called a **cDNA library**, represents the set of genes actively transcribed in a particular cell type, that is, those genes that have been transcribed into mRNA in that cell type. We will now describe how these two kinds of libraries are constructed.

Genomic Libraries

Ideally, a genomic library contains at least one copy of all sequences represented in the genome. Genomic libraries are constructed by extracting DNA from cells or tissues, cutting the DNA with restriction enzymes, and ligating the fragments into vectors. Since each vector can contain only a few thousand bases of foreign DNA, selection of the proper vector to contain a genome in the smallest number of clones is an important consideration in preparing a genomic library.

The number of clones required to carry all sequences in a genome is dependent on (1) the average size of the cloned inserts carried by the vector and (2) the size of the genome to be cloned. The number of clones in a library can be calculated as:

$$N = \frac{\ln(1 - P)}{\ln(1 - f)}$$

where N is the number of required clones, P is the probability of recovering a given sequence, and f represents the fraction of the genome in each clone. Suppose that we wish to prepare a library of the human genome using a lambda phage vector. The human genome contains 3.0×10^6 kb of DNA, and the average size of cloned inserts that can be carried by the vector is 17 kb; therefore, a library of about 8.1×10^5 phages would be required to contain the library. If we had selected a plasmid vector, most of which can carry inserts of about 5 kb, several million clones would be needed to contain the library. Thus, in this example, a phage vector would be the best choice for constructing a library of the human genome.

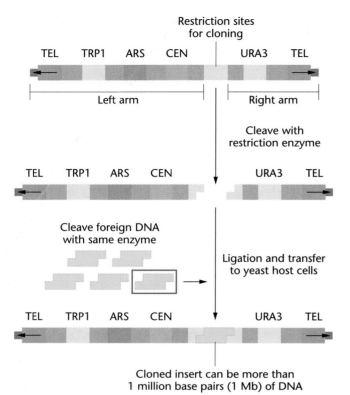

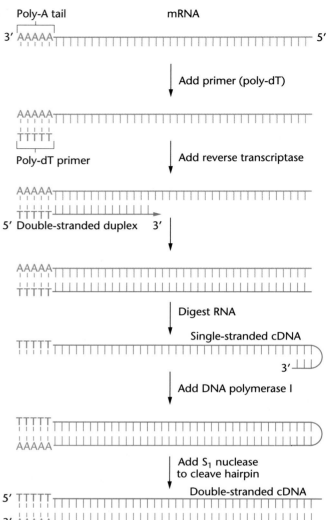

■ Figure 17–10 Cloning into a yeast artificial chromosome. The chromosome contains telomere sequences (TEL) derived from a ciliated protozoan, *Tetrahymena*, a centromere (CEN), and an origin of replication (ARS). These elements give the cloning vector the properties of a chromosome. TRP1 and URA3 are yeast genes that are selectable markers for the left and right arms of the chromosome, respectively. Near the centromere is a region containing several restriction sites. Cleavage with an enzyme in this region breaks the artificial chromosome into two arms. The DNA to be cloned is treated with the same enzyme, producing a collection of fragments. The arms and fragments can be ligated together, and the artificial chromosome can be inserted into yeast host cells. Because yeast chromosomes are large, the artificial chromosome can accept inserts up to several million base pairs (Mb = megabases).

cDNA Libraries

The construction of cDNA libraries offers several advantages. The library will contain only those genes that are expressed in a particular cell type or tissue at a specific time. In addition, the clones in a cDNA library do not contain introns, and represent only the coding regions of genes. Often, this makes it easier to derive the amino acid sequence of the gene and deduce the function of the gene product. To begin making a cDNA library from a specific cell type, the poly A-containing mRNA molecules are isolated from that cell type. This mRNA is used to synthesize **complementary DNA (cDNA)** molecules, which are subsequently cloned to create the cDNA library. After the poly A-containing

■ Figure 17–11 The production of cDNA from mRNA. Because many eukaryotic mRNAs have a polyadenylated tail of variable length (A_n) at their 3′-end, a short poly-dT oligonucleotide can be annealed to this tail to serve as a primer for the enzyme reverse transcriptase. Reverse transcriptase uses the mRNA as a template to synthesize a complementary DNA strand (cDNA), forming an mRNA/cDNA double-stranded duplex. The mRNA can be digested with alkali treatment, or by the enzyme RNAseH. The 3′ end of the cDNA often folds back to form a hairpin loop. The loop serves as a primer for DNA polymerase, which is used to synthesize the second DNA strand. The S1 nuclease is used to open the hairpin loop; the result is a double-stranded cDNA molecule that can be cloned into a suitable vector, or used as a probe for library screening.

mRNA is isolated, a poly dT primer is used to pair with the poly A residues (Figure 17–11). The poly dT primer serves as the starting point for the synthesis of a complementary DNA strand using the enzyme **reverse transcriptase**. The result is an RNA-DNA double-stranded duplex molecule. The RNA strand is removed, and the remaining single-stranded DNA is used as a template to make the DNA double-stranded, using the enzyme DNA polymerase I. The 3′ end of the single-stranded DNA often loops back upon itself and serves as a primer for the synthesis of the second strand. The result is a DNA duplex with the strands joined together at one end. This hairpin loop can be opened using the enzyme S1 nuclease, producing a double-stranded DNA molecule that can be cloned into a plasmid or phage vector.

Recovering Cloned Sequences from a Library

A genomic or a cDNA library can contain several thousand or several hundred thousand clones. To use a clone library effectively, one must have a method for identifying and selecting only the clone or clones that contain a gene of interest, and determining whether a clone contains all or only part of that gene. There are several ways to sort through a library to recover clones of interest, and the choice often depends on circumstances and available information about the gene being sought.

Probes to Identify Specific Clones

Many procedures employ a **probe** to select specific clones from a library. A probe is any piece of DNA or RNA that has been labeled and is complementary to some part of a cloned sequence present in the library. The probe is used to identify complementary nucleic acid sequences present in one or more clones. Probes can be prepared as single-stranded or double-stranded molecules, but are used in a single-stranded form in hybridization reactions (hybridization reactions were described in Chapter 9). Often, probes are radioactive polynucleotides, but other methods use probes that depend on chemical or color reactions to indicate the location of a specific clone.

Probes can be derived from a variety of sources; even related genes isolated from other species can serve as probes if enough of the DNA sequence has been conserved. For example, extrachromosomal copies of the ribosomal genes of the clawed frog *Xenopus laevis* can be isolated by centrifugation and cloned into plasmid vectors. Because ribosomal gene sequences have been highly conserved during evolution, clones carrying the human ribosomal genes can be recovered from a human genomic library by using cloned *Xenopus* ribosomal DNA fragments as probes.

Screening a Cloned Library

To screen a plasmid library, clones are grown on nutrient plates, forming hundreds or thousands of colonies (Figure 17–12). A replica of the colonies on the plate is made by gently pressing a nylon or nitrocellulose filter onto the surface of the plate, transferring bacterial cells from the colonies to the filter. The filter is transferred through solutions to lyse the bacteria, denature the double-stranded DNA, convert it to single strands, and bind these strands to the filter.

The DNA from the colonies, which is now bound to the filter, is screened by incubation with a nucleic acid probe. The probe is denatured to form single strands, and added to a solution containing the filter. If the DNA sequence of any of the cloned DNA on the filter is complementary to the probe, a DNA-DNA hybrid molecule will form between the probe and the cloned DNA (review the details of hybridization in Chapter 9). After unbound and excess probe is washed away, the filter is assayed to detect any hybridization that has taken place. In many types of assays, the filter is overlaid with a piece of photographic film. If a radioactive probe was used, decay in probe molecules bound to DNA on the filter will expose the film and produce dark spots on the developed film, representing colonies on the plate containing the cloned gene of interest (Figure 17–12). The corresponding colony can be identified and recovered from the original nutrient plate, and the cloned DNA it contains can be used in further experiments. Nonradioactive probes use a chemical reaction that emits photons of light (chemiluminescence) to expose the photographic film and reveal the location of colonies carrying the gene of interest.

To screen a phage library, a slightly different method, called **plaque hybridization**, is used. A solution of phage carrying DNA inserts is spread over a lawn of bacteria growing on a plate. The phages infect the bacterial cells on the plate, and form plaques as they replicate. Each plaque, which appears as a clear spot on the plate, represents the progeny of a single phage, and is a clone. The plaques are transferred to a nylon or nitrocellulose membrane, and the phage DNA is denatured into single strands and screened with a labeled probe. Phage plaques are much smaller than plasmid colonies, and many more plaques can be screened on a single filter, making this method more efficient for screening large genomic libraries.

Chromosome Walking

In some cases, when the approximate location of a gene is known, it is possible to clone the gene by first cloning nearby sequences. Often, these nearby sequences are identified by linkage analysis, and serve as a starting point for **chromosome walking**, which is the isolation of adjacent clones from a library. In a chromosome walk, the end piece of a cloned DNA fragment (that is, the fragment containing the linked marker) is subcloned, and used as a probe to recover overlapping clones from a library (Figure 17–13). The overlapping clones are analyzed by restriction mapping to determine the degree of overlap. A subfragment from one

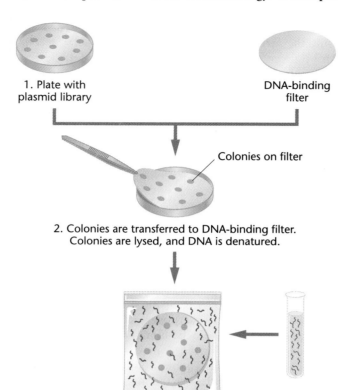

1. Plate with plasmid library

DNA-binding filter

Colonies on filter

2. Colonies are transferred to DNA-binding filter. Colonies are lysed, and DNA is denatured.

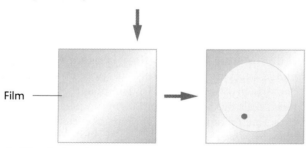

3. Filter and labeled probe in solution are transferred to heat-sealed food bag. Probe hybridizes to denatured DNA of colonies.

Film

4. Filter is rinsed to remove excess probe, and dried. Film is applied to filter for autoradiography.

5. Pick cells with plasmids that hybridized to probe.

6. Transfer cells to medium for growth, further analysis.

■ Figure 17–12 Screening a plasmid library to recover a cloned gene. The library, present on nutrient plates, is overlaid with a DNA binding filter, and colonies are transferred to the filter. Colonies on the filter are lysed, and the DNA is denatured to single strands. The filter is placed in a hybridization bag along with buffer and a labeled single-stranded DNA probe. During incubation, the probe forms a double-stranded hybrid with complementary sequences on the filter. The filter is removed from the bag and washed to remove excess probe. Hybrids are detected by placing a piece of X-ray film over the filter and exposing for a short time. The film is developed, and hybridization events are visualized as spots on the film. From the orientation of spots on the film, colonies containing the insert that hybridized to the probe can be identified. Cells are picked from this colony for growth and further analysis..

end of the overlapping clone is used as a probe to recover another set of overlapping clones, and the analysis is repeated. In this way, it is possible to "walk" along the chromosome, clone by clone. The gene in question can be identified by nucleotide sequencing of the recovered clones and searching for a potential gene-coding sequence, or **open reading frame (ORF)**. An ORF is defined as a stretch of nucleotides that begins with a start codon that is followed by amino acid-encoding codons, and ends with one or more stop codons. Although laborious and time-consuming, chromosome walking has been used to recover genes for several human genetic disorders, including those for cystic fibrosis and muscular dystrophy.

There are several limitations to chromosome walking in complex eukaryotic genomes, including the human genome. If a probe contains a repetitive sequence such as an *Alu* sequence (see Chapter 11 for a discussion of these and other repetitive sequences), it will hybridize to other clones in the genomic library containing that sequence. Most of these clones will not be adjacent to the clone from which the probe was derived, and the walk will therefore have to be terminated. In some cases, a technique called **chromosome jumping** can often be used to skip over the region containing the repetitive sequences and continue the walk.

PCR-Based DNA Cloning

Recombinant DNA techniques were developed in the early 1970s, and in subsequent years they revolutionized the way geneticists and molecular biologists conduct research and gave birth to the booming biotechnology industry. In 1986, another technique, called the **polymerase chain reaction (PCR)**, was developed that further accelerated the pace of biological research. The significance of this advance is underscored by the fact that the 1993 Nobel Prize in Chemistry was

Initial adjacent sequence (B)

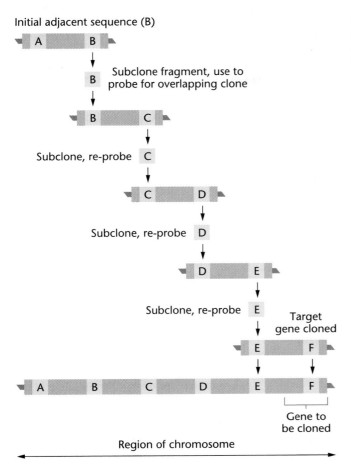

■ Figure 17–13 In chromosome walking, the approximate location of a gene to be cloned is known. A subcloned fragment of a linked adjacent sequence is used to recover overlapping clones from a genomic library. This process of subcloning and probing the genomic library is repeated to recover overlapping clones until the gene in question has been reached.

awarded to Kary Mullis for developing the PCR technique.

PCR is a rapid, cell-free method of DNA cloning that has to a certain extent replaced some of the methods of cloning that use host cells. PCR has been applied to a wide range of disciplines, including molecular biology, human genetics, evolution, development, and forensics (the gathering of evidence in criminal cases). Some of these applications will be discussed in Chapter 18.

PCR allows the direct amplification of specific target DNA sequences within a population of DNA molecules, and can be used on fragments of DNA that are initially present in infinitesimally small quantities. In order to amplify a sequence by PCR, some information about the nucleotide sequence of the target DNA is required. This information is used to synthesize two primers 15–30 nucleotides in length (nucleotide

chains in this size range are called **oligonucleotides**) that are added to a denatured DNA sample. The primers hybridize to complementary sequences that flank the sequence to be amplified. A heat-stable form of DNA polymerase is used to synthesize the complementary strands of the target DNA (Figure 17–14). There are three basic steps in the PCR reaction, and the amount of amplified DNA produced is limited theoretically only by the number of times these steps are repeated.

1. In the first step, DNA to be amplified is denatured into single strands. This DNA does not have to be purified and can come from any number of sources, including genomic DNA, forensic samples such as dried blood or semen, dried samples stored as part of medical records, single hairs, mummified remains, and fossils. The double-stranded DNA is denatured by heating (at 90-95°C) until it dissociates into single strands (usually about 4–5 minutes).

2. Primers are annealed to the single-stranded DNA. These primers are synthetic oligonucleotides that anneal to sequences flanking the segment to be amplified (Figure 17–14). To amplify a specific DNA sequence, some information about the target sequence is required to prepare the primers. Most often, two different primers are used in PCR. Each primer has a sequence that is complementary to one of the two strands of DNA being amplified. Because they anneal to opposite strands, the primers align themselves with their 3′ ends facing each other. The annealing usually takes place at a lower temperature, from about 50 to 70°C.

3. A heat-stable form of DNA polymerase (Taq polymerase, an enzyme from a bacterium that lives in hot springs) is added to the reaction mixture and DNA synthesis is carried out at temperatures between 70–75°C (Figure 17–14). The polymerase extends the primers in the 5′ to 3′ direction, making a double-stranded copy of the target DNA.

Each set of three steps—**denaturation** of the double-stranded product, **annealing** of primers, and **extension** by polymerase—is referred to as a **cycle**. Each cycle, which takes 4–5 minutes, can be repeated by carrying out each of the steps again. Twenty-five cycles of amplification result in more than a 1 million-fold increase in the amount of DNA. The process has been automated through the development of machines called *thermocyclers*, which can be programmed to carry out a predetermined number of cycles, yielding large amounts of the target DNA which can be used in other procedures such as cloning, sequencing, clinical diagnosis, and genetic screening.

DNA cloning using the PCR reaction has several advantages over host cell-based cloning. The PCR reaction is fast and can be carried out in a few hours, rather

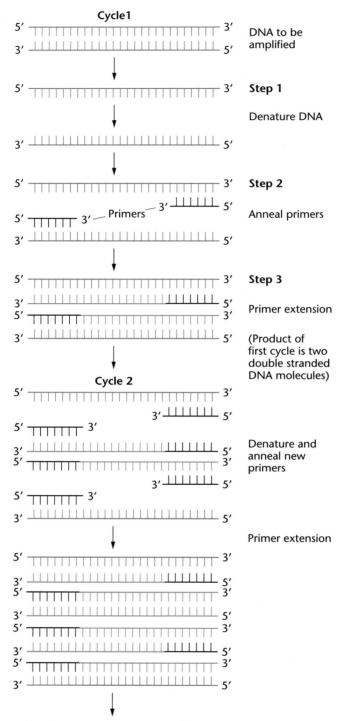

Cycles 3–25 for greater than a 10^6-fold increase in DNA

■ Figure 17–14 PCR amplification. In the polymerase chain reaction (PCR), the DNA to be amplified is denatured into single strands; each strand is then annealed to a short, complementary primer. The primers are synthetic oligonucleotides that are complementary to sequences flanking the region to be amplified. DNA polymerase and nucleotides are added to extend the primers in the 3′ direction, using the single-stranded DNA as a template. The result is a double-stranded DNA molecule with the primers incorporated into the newly synthesized strand. In a second PCR cycle, the products of the first cycle are denatured into single strands, primers are added, and they are then extended by DNA polymerase. Repeated cycles can amplify the original DNA sequence by more than a million-fold.

molecular paleontology. In addition, PCR can be used on DNA samples that have been partly degraded, contaminated with other materials, or embedded in a medium which would make DNA extraction for conventional cloning difficult or impossible.

Although PCR is a major advance, the technique has some limitations. In addition to the requirements for some information about the nucleotide sequence of the target DNA, even minor contamination of the sample with DNA from other sources can cause difficulties. Cells shed from the skin of a laboratory worker while performing PCR can contaminate samples, and this can become a critical issue when samples have been gathered from a crime scene or taken from fossils. Contamination can make it difficult to obtain accurate results. PCR must always be run with carefully designed and appropriate controls.

Analysis of Cloned Sequences

The recovery and identification of genes and other specific DNA sequences by cloning or by PCR is a powerful tool for the analysis of genomic structure and function. In fact, much of the Human Genome Project is based on such techniques. In the following sections, we will consider some of the methods used to answer questions about the organization and function of cloned sequences.

Restriction Mapping

One of the first steps in characterizing a DNA clone is the construction of a **restriction map**. A restriction map is a compilation of the number, order, and distance between restriction enzyme cutting sites along a cloned segment of DNA. Map units are expressed in base pairs (bp) or, for longer distances, kilobase pairs (kb). Restriction maps provide information about the length of a cloned insert, and the location of restriction enzyme sites within the cloned DNA. This information can be used for subcloning fragments of a gene, or for compar-

than the weeks required for host cell-based cloning. The design of PCR primers can be done using computer software, and commercial synthesis of the oligonucleotides is also fast and economical.

PCR is very sensitive, and can be used to amplify target DNA from vanishingly small DNA samples, even the DNA in a single cell. This feature of PCR is valuable in several areas, including genetic testing, forensics, and ing its internal organization with that of other cloned sequences. Restriction maps can also be used to compare a gene and its cDNA to identify exons and introns in the genomic copy of the gene.

Figure 17–15 shows the steps in constructing a restriction map from a cloned DNA segment. For this map, let us begin with a cloned DNA segment 7.0 kb in length. Three samples of the cloned DNA are digested with restriction enzymes: one is digested with *Hin*dIII, one with *Sal*I, and one with both *Hin*dIII and *Sal*I. The fragments generated by digestion with the restriction enzymes are separated by **gel electrophoresis**, and appear as a series of bands when the DNA is stained with ethidium bromide and visualized under ultraviolet light (Figure 17–16). The sizes of the resulting fragments are calculated by comparison to a set of standards run in adjacent lanes. To construct the map, the fragments generated by the restriction enzymes are analyzed.

1. When the DNA is cut with *Hin*dIII, two fragments are produced, one of 0.8 kb and one of 6.2 kb, confirming that the cloned insert is 7.0 kb in length, and indicating that there is only one cutting site for this enzyme (located 0.8 kb from one end).

2. When the DNA is cut with *Sal*I, two fragments result, one of 1.2kb and one of 5.8 kb, meaning that there is one cutting site for this enzyme, located 1.2 kb from one end of the insert.

Taken together, the results show that there is one restriction site for each enzyme, but the relationship between the two sites is unknown. From the information available, two different maps are possible. In one map (model 1), the *Hin*dIII site is located 0.8 kb from one end, and the *Sal*I site is located 1.2 kb from the same end. In the alternative map (model 2), the *Hin*dIII site is located 0.8 kb from one end, and the *Sal*I site is located 1.2 kb from the other end.

The correct model can be selected by considering the results from the sample digested with both *Hin*dIII and *Sal*I. Model 1 predicts that digestion with both enzymes will generate three fragments: 0.4, 0.8, and 6.2 kb in length; model 2 predicts that digestion with both enzymes will generate three different fragments: 0.8, 1.2, and 5.0 kb. The actual fragment pattern observed after cutting with both enzymes indicates that model 1 is correct (Figure 17–15).

Restriction maps are an important way of characterizing a cloned DNA segment, and can be constructed in the absence of any information about the coding capacity or function of the mapped DNA. In conjunction with other techniques, restriction mapping can be used to define the boundaries of a gene, to dissect the molecular organization of a gene and its flanking regions, and to locate mutational sites within genes.

Restriction maps can also be used to refine genetic maps. To a large extent, the accuracy of genetic maps depends on the frequency of recombination between genetic markers and on the number of markers used in construction of the map. In humans, for example, the genome size is large (3.2×10^9 bp), and the number of known genes is small, resulting in long distances between markers on the genetic map. Restriction enzyme cutting sites are inherited, and can be used as genetic markers, reducing the distance between markers, increasing their accuracy, and providing reference points for the correlation of genetic and physical maps.

Restriction sites have played an important role in mapping genes to specific human chromosomes and to defined regions of individual chromosomes. In addition, if a restriction site maps close to a mutant gene, it can be used as a marker in a diagnostic test. These sites, known as **restriction fragment length polymorphisms** or **RFLPs**, will be described in Chapter 18. The use of RFLPs has proven especially useful because the mutant genes underlying many human genetic diseases are poorly characterized at the molecular level, and restriction sites closely linked to mutant genes have been used to identify heterozygotes at risk for having affected children.

The DNA inserts cloned into vectors can be used in hybridization experiments to characterize the identity of specific genes, to locate coding regions or flanking regulatory regions within cloned sequences, and to study the molecular organization of genomic sequences.

Nucleic Acid Blotting

Many of the techniques described in this chapter rely on hybridization between complementary nucleic acid (DNA or RNA) molecules. We will now describe one of the most widely used methods for detecting the formation of such hybrids. This technique, called Southern blotting (after its inventor, Edward Southern), involves separating DNA fragments by gel electrophoresis, transferring them to filters, and screening the fragments with probes. The **Southern blot** procedure (Figure 17–17) has many applications.

To make a Southern blot, cloned DNA is cut into fragments with one or more restriction enzymes, and the fragments are separated by gel electrophoresis. The DNA in the gel is denatured into single-stranded fragments by treatment with an alkaline solution, and transferred to a sheet of a DNA-binding membrane (usually nitrocellulose or a nylon derivative). To transfer the fragments, the sheet of membrane is placed on top of the gel, and a buffer solution flows through the gel and the membrane by capillary action (Figure 17–17). As the buffer solution flows through both, the DNA fragments move out of the gel and become immobilized on the membrane.

The DNA fragments bound to the membrane are hybridized with a labeled single-stranded DNA probe. Only the DNA fragments bound to the membrane that are complementary to the nucleotide sequence of the

1. A population of cloned DNA fragments is prepared

Cloned 7.0 kb DNA fragments

2. DNA fragments are cut with restriction enzymes

Cut with restriction enzymes and loaded on gel for electrophoresis

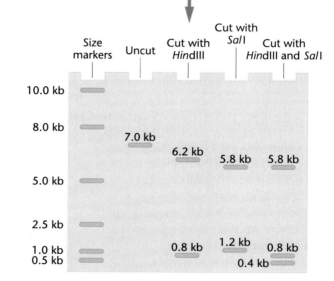

3. The restriction fragments are separated by gel electrophoresis

Uncut	*Hind*III	*Sal*I	*Hind*III + *Sal*I
7.0 kb	6.2 kb	5.8 kb	5.8 kb
	0.8 kb	1.2 kb	0.8 kb
			0.4 kb

4. Results

5. Results are used to construct models, and compared with results of double enzyme digests

Model 1

*Hind*III *Sal*I

| 0.8 | 0.4 | 5.8 |

0   0.8 1.2   7.0 kb

Predicted fragments from digestion with *Hind*III and *Sal*I: 0.4 kb, 0.8 kb, and 5.8 kb

Model 2

*Hind*III *Sal*I

| 0.8 | 5.0 | 1.2 |

0   0.8   5.8   7.0 kb

Predicted fragments from digestion with *Hind*III and *Sal*I: 0.8 kb, 1.2 kb, and 5.0 kb

6. Conclusion: model 1 is correct

Fragments generated by cutting with *Hind*III and *Sal*I are 0.4, 0.8 and 5.8 kb in length, indicating that model 1 is correct.

■ Figure 17–15 Construction of a restriction map. Samples of the 7.0-kb DNA fragments are digested with restriction enzymes: One sample is digested with *Hind*III, one with *Sal*I, and one with both *Hind*III and *Sal*I. The resulting fragments are separated by gel electrophoresis. The sizes of the separated fragments can be measured by comparison with molecular-weight standards in an adjacent lane. Cutting the DNA with *Hind*III generates two fragments: 0.8 kb and 6.2 kb. Cutting with *Sal*I produces two fragments: 1.2 kb and 5.8 kb. Models are constructed to explain the fragments generated by cutting with *Hind*III and with *Sal*I. These models can be used to predict the fragment sizes generated by digestion with both *Hind*III and *Sal*I. Model 1 predicts that the following fragments would result from cutting with both enzymes: 0.4 kb, 0.8 kb, and 5.8 kb; model 2 predicts that the fragments would be: 0.8 kb, 1.2 kb, and 5.0 kb. When the DNA is cut with both *Hind*III and *Sal*I, three fragments are produced: 0.4 kb, 0.8 kb, and 5.8 kb. Comparison of the predicted fragments with those observed on the gel indicates that model 1 is the correct restriction map.

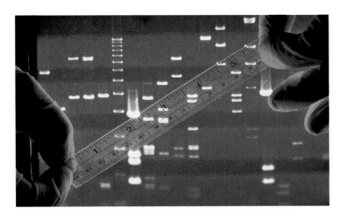

■ Figure 17–16 An agarose gel containing separated DNA fragments, stained with a dye (ethidium bromide) and visualized under ultraviolet light. Smaller fragments migrate faster and farther than do larger fragments, resulting in the distribution shown.

probe will form double-stranded hybrids. The unbound probe is washed away, and the hybridized fragments are visualized on a piece of film (Figure 17–18).

In addition to characterizing cloned DNAs, Southern blots are used for many other purposes, including the mapping of restriction sites within and near a gene, the identification of DNA fragments carrying a single gene from a mixture of many fragments, and the identification of related genes in different species. Southern blots are also used to detect rearrangements and duplications in genes associated with human genetic disorders and cancers.

A related blotting technique can be used to determine whether a cloned gene is transcriptionally active in a given cell or tissue type, by probing for the presence of RNA that is complementary to a cloned gene. This is accomplished by extracting RNA from one or several cell or tissue types. The RNA is then fractionated by gel electrophoresis and the pattern of RNA bands is transferred to a sheet of membrane as in the Southern blot. The membrane is then hybridized to a single-stranded DNA probe, derived from the cloned gene. If RNA that is complementary to the DNA probe is present, it will be detected as a band on the film. Because the original procedure using DNA bound to a filter became known as a Southern blot, this inverse procedure using RNA bound to a filter was called a **northern blot**. (Following this somewhat perverse logic, another procedure involving proteins bound to a filter is known as a **western blot**.)

Northern blots provide information about the expression of specific genes, and are used to study patterns of gene expression in embryonic and adult tissues. Northern blots can also be used to detect alternatively spliced mRNAs and multiple types of transcripts derived from a single gene. Northern blots also provide other information about transcribed mRNAs. If marker RNAs of known size are run in an adjacent lane, the size of the mRNA of a gene of interest can be calculated. In addi-

tion, the amount of transcribed RNA present in the cell or tissue being studied is related to the density of the RNA band on the film. This can be quantified by measuring the density of the band, providing a relative measurement of transcriptional activity. Thus, northern blots can be used to characterize and quantify the transcriptional activity of a given gene in different cells, tissues, or organisms.

DNA Sequencing

The ultimate characterization of a cloned DNA sequence is the determination of its nucleotide sequence. The ability to sequence cloned DNA has added immensely to our understanding of gene structure, gene function, and the mechanisms of regulation.

The most widely used method of **DNA sequencing** is based on the elongation of single-stranded DNA templates by the enzyme DNA polymerase. In this method, a series of four reactions is used. Each reaction tube contains a DNA template (the DNA being sequenced), a base-specific analog called a **dideoxynucleotide**, the normal deoxynucleotides, and DNA polymerase. As DNA synthesis takes place, the DNA polymerase occasionally inserts a dideoxynucleotide analog into a growing DNA strand instead of a deoxynucleotide. Since the analog lacks a 3′ hydroxyl group, it cannot participate in the formation of a 3′ bond, and DNA synthesis is therefore terminated. The result is a series of DNA molecules that differ in length at their 3′ ends. The fragments from each reaction, differing in length by as little as one nucleotide, are separated from each other by gel electrophoresis in four adjacent lanes. After visualization, the result is a series of bands forming a ladderlike pattern. The sequence can be read directly from bottom to top (corresponding to the 5′ to 3′ sequence of the DNA strand complementary to the template) (Figure 17–19).

For sequencing genomes, large-scale DNA sequencing uses automated DNA sequencers that employ fluorescent dyes. Four different colored dyes are used (one for each of the four nucleotides in DNA), producing a colored pattern of peaks that can be read to provide the sequence (Figure 17–20).

DNA sequencing provides information about the organization of genes and the nature and number of mutational events that alter both genes and gene products. Such information has confirmed the conclusion that genes and proteins are colinear molecules. Sequencing has also been used to study the organization of regulatory regions that flank prokaryotic and eukaryotic genes, and to infer the amino acid sequence of proteins.

Computer analysis of nucleotide sequences may be used to determine whether a cloned segment of DNA contains all or just part of a gene. This can be done by searching for exon/intron junctions, or by comparing the DNA sequence and the inferred amino acid se-

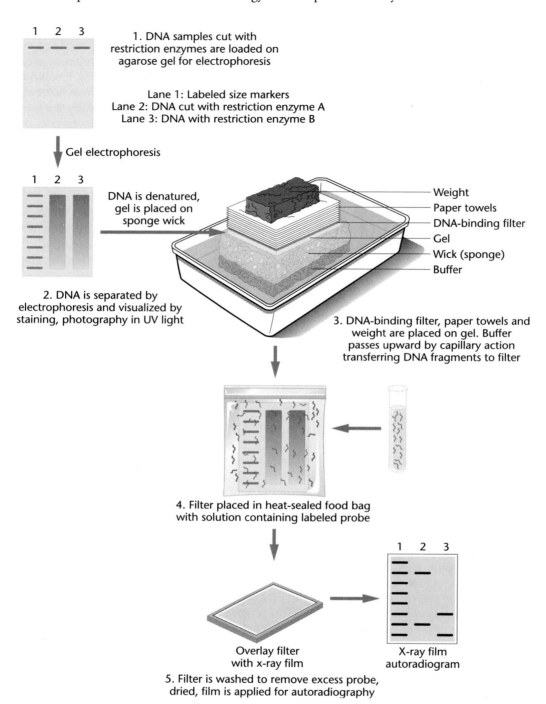

1. DNA samples cut with restriction enzymes are loaded on agarose gel for electrophoresis

Lane 1: Labeled size markers
Lane 2: DNA cut with restriction enzyme A
Lane 3: DNA with restriction enzyme B

Gel electrophoresis

DNA is denatured, gel is placed on sponge wick

Weight
Paper towels
DNA-binding filter
Gel
Wick (sponge)
Buffer

2. DNA is separated by electrophoresis and visualized by staining, photography in UV light

3. DNA-binding filter, paper towels and weight are placed on gel. Buffer passes upward by capillary action transferring DNA fragments to filter

4. Filter placed in heat-sealed food bag with solution containing labeled probe

Overlay filter with x-ray film

X-ray film autoradiogram

5. Filter is washed to remove excess probe, dried, film is applied for autoradiography

■ Figure 17–17 The Southern blotting technique. Samples of the DNA to be probed are cut with restriction enzymes and the fragments are separated by gel electrophoresis. The pattern of fragments is visualized and photographed under ultraviolet illumination by staining the gel with ethidium bromide. The gel is then placed on a sponge wick in contact with a buffer solution and covered with a DNA-binding filter. Layers of paper towels or blotting paper are placed on top of the filter and held in place with a weight. Capillary action draws the buffer through the gel, transferring the pattern of DNA fragments from the gel to the filter. The DNA fragments on the binding filter are then denatured into single strands and hybridized with a labeled DNA probe, washed, and overlaid with a piece of X-ray film for autoradiography. The hybridized fragments show up as bands on the X-ray film.

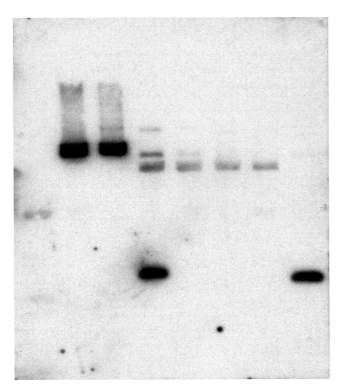

■ Figure 17–18 A Southern blot after hybridization, exposure, and development. The bands show DNA fragments that are complementary to the nucleotide sequence of the probe.

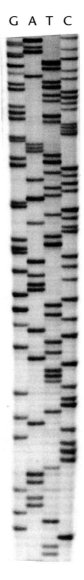

■ Figure 17–19 Photograph of a DNA sequencing gel showing the separation of bases in the four sequencing reactions (one per lane). The gel is analyzed to reveal the base sequence of the DNA fragment. To do this, the gel is read from the bottom, beginning with the lowest band in any lane, then the next lowest, then the next, etc. For example, the sequence of the DNA on this gel begins with TTCGTGAAGAA and so on.

quence with the information in various databases that contain the sequences of previously identified genes and proteins.

For example, this method was used to scan the nucleotide sequence of cloned DNA segments to find the gene for cystic fibrosis (CF), an autosomal recessive human genetic disorder which maps to a region on the long arm of chromosome 7. Since the protein product of the gene was unknown, DNA sequence analysis was used to identify a protein-coding region. The DNA sequence from this gene was used to derive a possible amino acid sequence. The amino acid sequence was used to search computer databases containing the sequences of known proteins. The results indicated that the CF protein has an amino acid sequence similar to membrane proteins that play a role in ion transport. Further work confirmed that the product of the cystic fibrosis locus is a protein inserted into the plasma membrane that regulates the transport of chloride ions across the cell membrane. In most cases of CF, the mutant gene has an altered DNA sequence that causes the production of a defective protein.

In addition to identifying DNA defects that cause mutant phenotypes, DNA sequencing is used to study the organization of genes (the number of introns and exons and their boundaries). It is also used to provide information about the nature and function of proteins encoded by genes, and to study evolutionary relationships to similar proteins from other organisms.

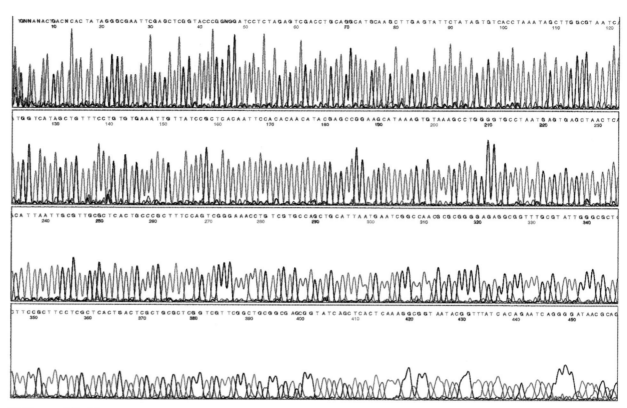

■ Figure 17–20 DNA sequencing has been automated with the use of fluorescent dyes, one for each base; the separated bases are read in order along the axis from left to right.

G E N E T I C S , T E C H N O L O G Y , A N D S O C I E T Y

DNA Fingerprints in Forensics: The Case of the Telltale Palo Verde

The use of DNA analysis in forensics is making it harder and harder to get away with many crimes. It's now a simple matter to isolate DNA from tissue left at a crime scene, a splattering of blood, some skin left under a victim's fingernails, or even the cells clinging to the base of a hair shaft. A variety of techniques can be used to determine the likelihood that the sample came from a suspect in the case. In recent years the PCR technique has been increasingly used in forensics, both because it is fast (taking only about a week) and because it requires only small amounts of DNA (about a nanogram). One variation of the PCR method that may be especially valuable is called the RAPD (pronounced "rapid") procedure, an abbreviation of random amplified polymorphic DNA. Under the right conditions, the RAPD

procedure generates a DNA profile unique to each individual, which can thus be used for personal identification.

The RAPD procedure involves the amplification of a set of DNA fragments of unknown sequence. First, an amplification primer is mixed with a sample of DNA. The primer will anneal to all sites in the DNA sample that have a matching base sequence. For primers 10 nucleotides in length, binding will occur at a few thousand sites scattered randomly throughout the genome. When primers happen to bind to DNA sites that are not too far apart (within about 2000 nucleotides), DNA amplification can occur between these sites, eventually producing millions of copies of the DNA sequence lying between the binding sites. When the products of such a reaction are separated by electrophoresis on an agarose gel, each amplified DNA segment will appear as a distinct band.

The locations of the primer binding sites in the genome are likely to differ between any two individuals in a pop-

ulation, due to minor DNA sequence differences (base changes, insertions, and deletions). Since these affect initial binding to the primers, the number and locations of the DNA bands on the gel will vary from one individual to another. The resulting DNA profile is referred to as a DNA fingerprinting. It represents a characteristic "snapshot" of the genome of an individual, allowing it to be distinguished from other individuals in the population.

When a DNA profile from tissue found at a crime scene matches that of a suspect, it does not *prove* that the tissue belongs to the suspect; instead, it *excludes* all those who have a different pattern. The strategy, therefore, is to generate five or more DNA profiles from the same sample using different amplification primers. The more profiles that match between the sample and the suspect, the more unlikely it is that the sample at the crime scene came from someone other than

(continues)

(*continued*)
the suspect. The likelihood that a complete match of all the banding patterns occurs simply by chance can be calculated, resulting in, for example, a 1-in-100,000 to a 1-in-1,000,000 probability of a "random match."

One of the most interesting uses of DNA fingerprinting in a criminal case did not involve the suspect's own DNA, but rather the DNA of plants growing at the crime scene. On the night of May 2, 1992, a Phoenix woman was strangled and her body dumped near an abandoned factory in the surrounding desert. Investigators discovered a pager near the body, making its owner a prime suspect. When questioned, this man admitted being with the woman the day of the murder, but claimed he had not been near the factory and suggested that the woman must have stolen his pager from his pickup truck. A search of the pickup provided the crucial clue, placing the suspect at the factory: In its bed were two seed pods from a palo verde tree.

The palo verde tree (*Cercidium floridum*), native to the desert of the Southwest, is a member of the bean family. In an adaptation to the hot, dry climate, it puts out very small leaves. To compensate for this, its stems become green and carry out photosynthesis. This gives the plant its name: palo verde is Spanish for "green stem." The plant may not resemble a bean in some respects, but, like a typical bean, it produces seeds in pods that drop to the ground when mature.

The homicide investigators assigned to the case wondered whether it could be proved that the seed pods found in the bed of the suspect's truck had fallen from a palo verde tree near where the body was found. If so, it would be a key piece of evidence placing the suspect at the scene. But how could this be demonstrated? The investigators turned to Dr. Timothy Helentjaris, then at the University of Arizona, in nearby Tucson. Helentjaris proceeded to determine whether the DNA profile of the seed pods from the pickup truck matched that of any of the palo verde trees near the scene. He understood that for a match between a seed pod and a tree to have significance, he would first have to show that palo verdes differ genetically from one tree to the next. Otherwise, a pod could never be unambiguously assigned to one particular tree. Fortunately, he found considerable variation in DNA profiles between the palo verde plants.

Helentjaris was then given the two seed pods from the suspect's truck along with pods collected from 12 palo verde trees from the vicinity of the factory. The investigators knew which of the 12 trees was the key one at the crime scene, but they did not tell Helentjaris. He extracted DNA from seeds from each pod and performed RAPD reactions using an amplification primer he had earlier found to reproducibly yield profiles of 10–15 distinct bands. The result of this "controlled" experiment was unmistakable: The DNA profile from one of the pods found in the truck exactly matched that of only one of the 12 trees, the one nearest where the body was found. In an important additional test, Helentjaris showed that this DNA profile differed from that of pods collected from 18 other trees located at random sites around Phoenix. Collectively, Helentjaris' analysis led to the estimate of the likelihood of a "random match" at a little less than 1 in 1,000,000.

Helentjaris' analysis was admitted as evidence in the trial, placing the suspect at the crime scene. At the completion of the five-week trial the suspect was found guilty of first-degree murder. His conviction was upheld upon appeal, and he is currently serving a life sentence without parole.

References
Ayala, F. J., and Black, B. 1993. Science and the courts. *Am. Sci.* 81:230–39.

Krawczak, M., and Schmidtke, J. 1994. *DNA fingerprinting.* Oxford: BIOS Scientific.

Marx, L. 1988. DNA fingerprinting takes the witness stand. *Science* 240:1616–18.

Chapter Summary

1. The cornerstone of recombinant DNA technology is a class of enzymes called restriction endonucleases, which cut DNA at specific recognition sites. The fragments produced are joined with DNA vectors to form recombinant DNA molecules.

2. Vectors can replicate autonomously in host cells, and facilitate the manipulation of the newly created recombinant DNA molecules. Vectors have been constructed from many sources, including bacterial plasmids and viruses.

3. Recombinant DNA molecules are transferred into a host, and cloned copies are produced during host cell replication. A variety of host cells may be used for replication, including bacteria, yeast, and mammalian cells. Cloned copies of foreign DNA sequences can be recovered, purified, and analyzed.

4. The polymerase chain reaction (PCR) is a method for amplifying a specific DNA sequence that is present in a collection of DNA sequences, such as genomic DNA. The PCR method allows cloning of DNA without the need for host cells, and is a rapid, sensitive method with wide-ranging applications.

5. Once cloned, DNA sequences can be analyzed using a variety of methods, including restriction mapping and DNA sequencing. Other methods such as blotting hybridization can be used to identify genes and flanking regulatory regions within the cloned sequences.

INSIGHTS
and
SOLUTIONS

The recognition site for the restriction enzyme *Sau*3A is GATC (Figure 17–3). The recognition site for the enzyme *Bam*HI is GGATCC, where the four internal bases are identical to the *Sau*3A site. This means that the single-stranded ends produced by the two enzymes are identical. Suppose you have a cloning vector containing a *Bam*HI site and foreign DNA that you have cut with *Sau*3A.

1. Can this DNA be ligated into the *Bam*HI site of the vector? Why?

  Solution: DNA cut with *Sau*3A can be ligated into the *Bam*HI site of the vector because the single-stranded ends generated by the two enzymes are identical.

2. Can the DNA segment cloned into this site be cut from the vector with *Sau*3A? With *Bam*HI? What potential problems do you see with the use of *Bam*HI?

Solution: The DNA can be cut from the vector with *Sau*3A because the recognition site for this enzyme is maintained. The DNA may be recovered from the vector by *Bam*HI, but there is a potential problem with the use of this enzyme. When the *Sau*3A fragment is ligated into the *Bam*HI site, a base at each end of the *Bam*HI site must be added (arrows). Even though a template is present on the opposite strand, insertion of the wrong base will occur in a fraction of cases, destroying the *Bam*HI site and making it impossible to remove the insert with *Bam*HI.

```
  GGATC ——————  Insert  —————— GATCC
  CCTAG ——————          —————— CTAGG

                    Vector
```

Key Terms

annealing, 383
bacteriophage, 376
cDNA library, 379
chromosome jumping, 382
chromosome walking, 381
clone, 373
complementary DNA (cDNA), 380
cycle, 383
denaturation, 383
dideoxynucleotide, 387
DNA ligase, 375
DNA sequencing, 387
*Eco*RI, 375

extension, 383
gel electrophoresis, 385
genomic library, 379
northern blot, 387
oligonucleotides, 383
open reading frame (ORF), 382
phage lambda, 376
plaque hybridization, 381
plasmid, 376
polylinker site, 376
polymerase chain reaction
  (PCR), 382
probe, 381

recombinant DNA, 373
recombinant DNA molecule, 373
restriction endonuclease, 373
restriction enzyme, 373
restriction fragment length polymorphism (RFLP), 385
restriction map, 384
reverse transcriptase, 380
Southern blot, 385
vector, 373
western blot, 387
yeast artificial chromosome, 378

Problems and Discussion Questions

1. In recombinant DNA studies, what is the role of each of the following: restriction endonucleases, vectors, and host cells?

2. Why is poly dT an effective primer for reverse transcriptase?

3. The human insulin gene contains a number of introns. In spite of the fact that bacterial cells will not excise introns from mRNA, how can a gene like this be cloned into a bacterial cell and produce insulin?

4. Restriction enzymes recognize palindromic sequences in intact DNA molecules and cleave the double-stranded helix at these sites. Inasmuch as the bases are internal in a DNA double helix, how is this recognition accomplished?

5. Although the potential benefits of cloning in higher plants are obvious, the development of this field has lagged behind cloning in bacteria, yeast, and mammalian cells. Can you think of any reason for this?

6. Using DNA sequencing on a cloned DNA segment, you recover the following nucleotide sequence:

CAGTATCCTAGGCAT

Does this segment contain a palindromic recognition site for a restriction enzyme? What is the double-stranded sequence of the palindrome? What enzyme would cut at this site? (Consult Figure 17–3 for a list of restriction enzyme recognition sites.)

7. Figure 17–3 lists restriction enzymes that recognize sequences of four bases and six bases. How frequently should four- and six-base recognition sequences occur in a genome? If the recognition sequence consists of eight bases, how frequently would such sequences be encountered? Under what circumstances would you select such an enzyme for use?

8. You are given a cDNA library of human genes prepared in a bacterial plasmid vector. You are also given the cloned yeast gene that encodes EF-1α, a protein that is highly conserved in protein sequence among eukaryotes. Outline how you would use these resources to identify the human cDNA clone encoding EF-1α.

9. You have recovered a cloned DNA segment of interest, and determined that the insert is 1300 bp in length. To characterize this cloned segment, you have isolated the insert and decide to construct a restriction map. Using enzyme I and enzyme II, followed by gel electrophoresis, you determine the number and size of the fragments produced by enzymes I and II alone and in combination as follows:

Enzymes	Restriction Fragment Sizes (bp)
I	350 bp, 950 bp
II	200 bp, 1100 bp
I and II	150 bp, 200 bp, 950 bp

Construct a restriction map from these data, showing the positions of the restriction sites relative to one another, and the distance between them in base pairs.

10. cDNA can be cloned into vectors to create a cDNA library. In analyzing cDNA clones, it is often difficult to find clones that are full length—that is, extend to the 5′-end of the mRNA. Why is this so?

11. Although the capture and trading of great apes has been banned in 112 countries since 1973, it is estimated that about 1000 chimpanzees are removed annually from Africa and smuggled into Europe, the United States, and Japan. This illegal trade is often disguised by private (such as zoo or circus) owners by simulating births in captivity. Until recently, genetic identity tests to uncover these illegal activities were not used because of the lack of availability of highly polymorphic markers and the difficulties of obtaining chimp blood samples. Recently a study was reported in which DNA samples were extracted from freshly plucked chimp hair roots and used as templates for the polymerase chain reaction. The primers used in these studies flank highly polymorphic sites in human DNA resulting from variable numbers of tandem nucleotide repeats. Several suspect chimp offspring and their supposed parents were tested to determine if the offspring were "legitimate" or were the product of the illegal trading and not the offspring of the putative parents. A sample of the data is shown below. Examine these data carefully and choose the best conclusion:

(a) None of the offspring are legitimate.

(b) Offspring B and C are not the products of these parents and were probably purchased on the illegal market. The data are consistent with offspring A being legitimate.

(c) Offspring A and B are products of the parents shown, but C is not and was therefore probably purchased on the illegal market.

(d) There is not enough data to draw any conclusions. Additional polymorphic sites should be examined.

(e) No conclusion can be drawn because "human" primers were used.

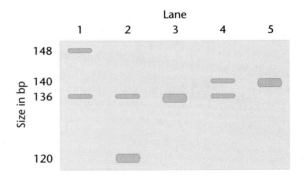

Lane 1: father chimp

Lane 2: mother chimp

Lanes 3-5: putative offspring A, B, C

12. Briefly describe the problem that a stretch of repeated sequences would cause in a chromosome walk and name the procedure that is used to overcome this problem.

Selected Readings

Alton, E. W., and Geddes, D. M. 1995. Gene therapy for cystic fibrosis: A clinical perspective. *Gene Ther.* 2:88–95.

Antonarakis, S. 1989. Diagnosis of genetic disorders at the DNA level. *N. Engl. J. Med.* 320:153–63.

Barker, D., et al. 1987. Gene for von Recklinghausen neurofibromatosis is in the pericentromeric region of chromosome 17. *Science* 236:1001–162.

Burger, S. L., and Kimmel, A. R. 1987. Guide to molecular cloning tecnhiques. *Methods in enzymology,* vol. 152. San Diego, CA: Academic Press.

Dube, I. D., and Cournoyer, D. 1995. Gene therapy: Here to stay. *Can. Med. Assoc. J.* 152:1605–13.

Eisensmith, R. C., and Woo, S. 1995. Molecular genetics of phenylketonuria: from molecular anthropology to gene therapy. *Adv. Genet.* 32:199–271.

Guyer, M. S., and Collins, F. S. 1993. The Human Genome Project and the future of medicine. *Am. J. Dis. Child.* 147:1145–52.

Knorr, D., and Sinskey, A. J. 1985. Biotechnology in food production and processing. *Science* 229:1224–29.

McKusick, V. A. 1988. The new genetics and clinical medicine. *Hosp. Pract.* 23:177–91.

Mullis, K. B. 1990. The unusual origin of the polymerase chain reaction. *Sci. Am.* (April) 262:56–65.

Old, R. W., and Primrose, S. B. 1994. *Principles of genetic manipulation: An introduction to genetic engineering.* 5th ed. Palo Alto, CA: Blackwell Scientific.

Oste, C. 1988. Polymerase chain reaction. *BioTechniques* 6:162–67.

Reiss, J., and Cooper, D. N. 1990. Application of the polymerase chain reaction to the diagnosis of human genetic disease. *Hum. Genet.* 85:1–8.

Sambrook, J., Fitch, E. F., and Maniatis, T. 1989. *Molecular cloning: A laboratory manual.* 2nd ed. Cold Spring Harbor, NY: Cold Spring Harbor Press.

Southern, E. 1975. Detection of specific sequences among DNA fragments separated by gel electrophoresis. *J. Mol. Biol.* 98: 503–507.

Thomas, T. L., and Hall, T. C. 1985. Gene transfer and expression in plants: Implications and potential. *BioEssays* 3:149–53.

Torrey, J. G. 1985. The development of plant biotechnology. *Am. Sci.* 73:354–63.

Verma, I. 1990. Gene therapy. *Sci. Am.* (Nov.) 263–68.

Welsh, J., and McClelland, M. 1990. Fingerprinting genomes using PCR with arbitrary primers. *Nucleic Acids Res.* 18:7213–18.

White, R. 1985. DNA sequence polymorphisms revitalize linkage approaches in human genetics. *Trends Genet.* 1:177–80.

Williams, J. G. K., et al. 1990. DNA polymorphisms amplified by arbitrary primers are useful as genetic markers. *Nucleic Acids Res.* 18:6531–35.

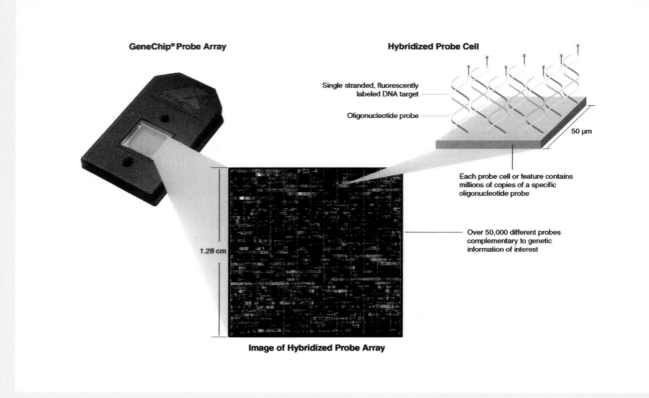

GeneChip® Probe Array

Hybridized Probe Cell

Single stranded, fluorescently labeled DNA target

Oligonucleotide probe

50 µm

Each probe cell or feature contains millions of copies of a specific oligonucleotide probe

Over 50,000 different probes complementary to genetic information of interest

1.28 cm

Image of Hybridized Probe Array

Affymetrix GeneChipR probe arrays are used to detect DNA mutations and monitor gene expression.

CHAPTER OUTLINE

18

DNA Biotechnology: Applications and Ethics

Chapter Concepts

Recombinant DNA technology has brought about significant advances in experimental genetics, in gene mapping, and in the diagnosis and treatment of disease. These techniques also form the basis for the biotechnology industry and the commercial production of human gene products for therapeutic uses, and the transfer of genes across species in agricultural plants and animals.

In 1971, a paper published by Hamilton Smith, Daniel Nathans, and Walter Arber marked the beginning of the recombinant DNA era. The paper described the isolation of an enzyme from a strain of bacteria and the use of this enzyme to cleave viral DNA. It contained the first published photograph of DNA cut with a restriction enzyme. From this modest beginning, recombinant DNA technology has revolutionized all fields of experimental biology and spread beyond the research laboratory to many other fields. In the intervening years, this technology has expanded to become part of everyday life; recombinant DNA techniques are now being used to help produce medicines, milk, and even the food at the supermarket.

In the previous chapter we discussed the methods used to create and analyze recombinant DNA molecules. This chapter describes how these tools are being used in research and commercial applications, and the ethical problems posed by the use of this technology. We will consider a cross section of applications illustrating the power of recombinant DNA technology to map and identify human genes, diagnose and treat disease, and generate new plants and animals. We will begin by considering how recombinant DNA has changed some of the basic methods of genetic mapping. Next, we will examine the impact of recombinant DNA on human genetics and medicine, from the prenatal analysis of genotypes to the treatment of genetic disorders by gene transfer, and the mapping of the human

genome. Finally, we will discuss some of the products and methods being used in the multibillion-dollar biotechnology industry.

Mapping Human Genes

An increasing number of the 50,000 to 100,000 human genes have been localized to chromosomal sites. However, in the majority of human genetic disorders, the function of the normal gene product is unknown. As a result, methods of mapping that begin with an identified gene product as a way of mapping the chromosomal locus of the gene cannot be used to map the genes responsible for many human genetic disorders.

RFLPs as Genetic Markers

Variations in nucleotide sequence occur throughout the human genome (mostly in noncoding regions) with a frequency of about 1 in every 200 nucleotides. By changing the sequence at a specific site, single nucleotide changes can create or destroy restriction enzyme sites. If a restriction site created by such variation is present on one chromosome but absent on its homolog, the two chromosomes can be distinguished from one another by their pattern of restriction fragments on a DNA blot (Figure 18–1). The region of chromosome *A* shown in the figure contains three *Bam*HI sites; its homolog, chromosome *B*, contains two such sites. When

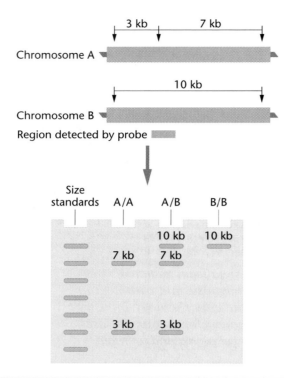

■ Figure 18–1 Restriction fragment length polymorphisms (RFLPs). The alleles on chromosome A and chromosome B represent DNA segments from homologous chromosomes. The region that hybridizes to a probe is shown below. Arrows indicate the location of restriction enzyme cutting sites that define the alleles. On chromosome A, three cutting sites generate fragments of 7 kb and 3 kb. On chromosome B, only two cutting sites are present, generating a fragment 10 kb in length. The absence of the cutting site in B could be the result of a single base mutation within the enzyme recognition/cutting site. Because these differences in restriction cutting sites are inherited in a codominant fashion, there are three possible genotypes: AA, AB, and BB. The allele combination carried by any individual can be detected by restriction digestion of genomic DNA (obtained from a blood sample or skin fibroblasts), followed by gel electrophoresis, transfer to a DNA-binding filter, and hybridization to the appropriate probe. The fragment patterns for the three possible genotypes are shown as they would appear on a Southern blot.

chromosome *A* is cut with *Bam*HI, 3-kb and 7-kb fragments are generated, whereas only a single 10-kb fragment is generated when chromosome *B* is cut. Using a probe from this chromosomal region, these fragments can be visualized on a Southern blot (Figure 18–1).

These variations in DNA fragment length generated by cutting with a restriction enzyme are called **restric-**

tion fragment length polymorphisms (RFLPs). RFLPs are quite common, and represent natural variations brought about by changes in a single nucleotide pair, or by deletions or insertions of one or more nucleotide pairs. Several thousand RFLPs have been identified in the human genome, and many of these have been assigned to individual chromosomes. These variations are inherited as codominant alleles. They can be mapped to specific regions on individual chromosomes and used as markers to follow the inheritance of genetic disorders from generation to generation in an affected family.

Linkage Analysis Using RFLPs

Finding the chromosomal locus of a genetic disorder using RFLPs involves using multigenerational families (three generations or more) in which both a chromosomally assigned RFLP marker and the genetic disorder are cosegregating. Selecting which RFLP to use in the family being studied is a matter of trial and error. The most useful RFLPs are those for which most members of the family are heterozygous, so that each member of a chromosome pair can be identified.

To map a genetic disorder, the inheritance of a given RFLP and the disorder are traced through the family (Figure 18–2). Each RFLP represents a specific human chromosome or chromosome region. If the loci of the RFLP and the disorder are near each other on the same chromosome, they will show linkage. In such studies, most RFLPs will not show linkage; these results serve to identify chromosomes that do *not* carry the disease gene. Probability analysis is used to determine whether an RFLP marker and a genetic disorder are linked. First, the probability that the observed pattern of inheritance (of the gene and the RFLP) could occur by chance alone is calculated, assuming that the RFLP marker and the gene are unlinked. These probabilities are expressed as the **logarithm of the odds** or **lod** score. A lod score of 3 means that the odds are 1000:1 in favor of linkage between the gene and the RFLP marker. A lod score of 4 means that the odds are 10,000:1 in favor of linkage. Lod scores of 3 to 4 are usually taken as evidence that the gene and the marker are linked.

In mapping human chromosomes, the unit of linkage is the **centimorgan (cM)**, named after the geneticist T. H. Morgan. One centimorgan is equal to a recombination frequency of 1 percent between two loci. The genetic distance between two genes expressed in centimorgans is not directly correlated with the physical distance expressed in nucleotides, but in human chromosomes, a distance of 1 cM corresponds roughly to 1 to 3 million nucleotides of DNA. By compiling studies from many families, the locations of many markers can be determined, and genetic maps for human chromosomes can be constructed (Figure 18–3).

Genotypes	Fragment sizes
Homozygous for chromosome A (A/A)	3 kb, 7 kb
Heterozygous (A/B)	3 kb, 7 kb, 10 kb
Homozygous for chromosome B (B/B)	10 kb

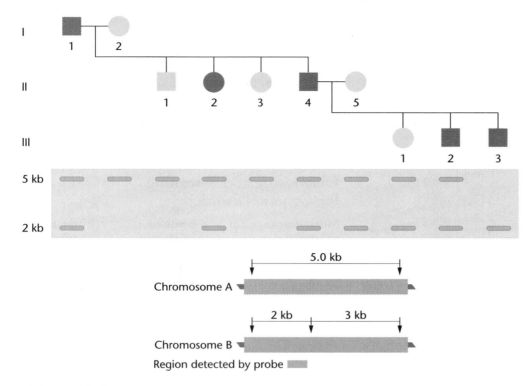

■ Figure 18–2   RFLP analysis and the inheritance of a dominant trait. The family shown in the pedigree has members affected by a dominant trait, as indicated by the filled symbols. Members of this family also carry two alleles for a restriction site: The A allele is represented as a 5.0-kb fragment, and the B allele is composed of two fragments, one of 2 kb and the other of 3 kb. The Southern blot pattern of RFLP alleles for each family member is shown below the pedigree symbol. In analyzing this pedigree, note that individual II-5 is from outside the family and is homozygous for the normal alleles, even though she carries the RFLP B allele. Her daughter, III-1, who is unaffected, probably received the A allele from her father and the B allele from her mother. Her oldest son (III-2) is affected, and probably received an A allele from his mother and the B allele from his father. The youngest son (III-3), who is affected, received a B allele from each parent. Examination of the pedigree and the results of the Southern blot indicate that the mutant allele responsible for the disorder is carried on a chromosome with the B (2.0-kb) allele. Establishing linkage to a chromosome is the first step in mapping a gene using RFLP analysis.

Positional Cloning: The Gene for Neurofibromatosis

The use of RFLP analysis in gene mapping can be illustrated by the search for the chromosomal locus of the gene for **type 1 neurofibromatosis (NF1)**. This disorder is inherited as an autosomal dominant condition, with an incidence of about 1 in 3000. NF1 is associated with a range of nervous system defects, including benign tumors and an increased incidence of learning disorders.

Mapping of the NF1 gene was accomplished in a series of steps. First, a group of laboratories used multigenerational families to compare the inheritance of NF1 with dozens of RFLP markers, with each RFLP marker representing a specific human chromosome or chromosome region. This produced an **exclusion map**, indicating which RFLPs were not linked to the disease, and therefore, which chromosomes and chromosome regions did *not* carry the NF1 locus. This work also produced some

evidence of linkage, and pointed to chromosomes 5, 10, and 17 as candidates for carrying the NF1 gene. The next stage focused on these chromosomes, and produced conclusive evidence that the disorder was closely linked to an RFLP known to reside near the centromere of chromosome 17 (Figure 18–4). Then, more than 30 RFLP markers from this region of chromosome 17 were used to analyze 13,000 individuals from NF1 families, and the gene was mapped to region 17q11.2. Finally, using a collection of genomic clones that spanned a subregion of 17q11.2, the locus for NF1 was identified in 1990 by chromosome walking and DNA sequencing. The identification of the gene was confirmed by finding mutations of the gene in individuals with NF1.

Once the gene was identified, the amino acid sequence of the gene product was reconstructed from the DNA sequence, and found to include 2485 amino acids. By searching databases containing protein sequence in-

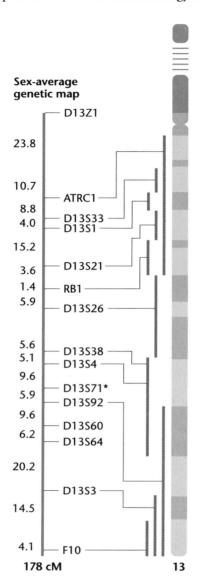

Sex-average genetic map

D13Z1
23.8
10.7
ATRC1
8.8
4.0
D13S33
D13S1
15.2
3.6
D13S21
1.4
RB1
5.9
D13S26

5.6
D13S38
5.1
D13S4
9.6
5.9
D13S71*
D13S92
9.6
D13S60
6.2
D13S64
20.2

D13S3
14.5

4.1
F10
178 cM **13**

■ FIGURE 18–3 A genetic and a physical map of human chromosome 13. The genetic map is 178 cM in length, as shown at the left. The locations of markers on the physical map are indicated by brackets adjacent to the chromosome.

formation, the NF1 gene product was found to be similar to proteins that play a role in signal transduction. Further analysis confirmed that the NF1 protein, **neurofibromin**, is involved in the transduction of intracellular signals and the regulation of a gene that controls cell division. Mutation in the NF1 gene leads to a loss of control of cell growth, causing the production of small tumors that are characteristic of this disorder.

The mapping, cloning, sequencing of this gene, and identification of the gene product took a little over three years, beginning with no direct knowledge of the nature of the gene product or the mutational events that result in the production of the NF1 phenotype. The mapping and cloning of the gene for NF1 described

above is an example of **positional cloning**. This recombinant DNA-based method is a departure from methods of mapping which work from an identified gene product to the gene locus. In positional cloning, the gene can be mapped, isolated, and cloned with no knowledge of the gene product. Using this strategy, an ever-increasing number of human genes are being mapped and isolated.

Diagnosing and Screening for Genetic Disorders

Much diagnosis for genetic diseases is performed prenatally. The most widely used methods for prenatal diagnosis of genetic disorders are **amniocentesis** and **chorionic villus sampling (CVS)**. In amniocentesis, a needle is used to withdraw amniotic fluid (Figure 18–5). The fluid and the cells it contains can be analyzed for chromosomal or single-gene disorders. In CVS, a catheter is inserted into the uterus and used to retrieve a small tissue sample of the fetal chorion. This tissue is used for cytogenetic, biochemical, and recombinant DNA-based testing.

Coupled with these methods of recovering samples, recombinant DNA technology has proven to be a highly sensitive and accurate tool for the prenatal detection of genetic disorders. The use of cloned DNA sequences allows direct examination of the genotype, expanding the range of prenatal testing. For example, disorders of β-globin (one of which will be described below), a protein component of hemoglobin, cannot be detected prenatally by testing for an abnormal gene product, because the β-globin gene is not expressed until a few days after birth.

Sickle-Cell Anemia and Prenatal Genotyping

Sickle-cell anemia is an autosomal recessive condition common in people with family origins in areas of West Africa, the Mediterranean basin, and parts of the Middle East and India (see discussion in Chapter 14). Sickle-cell anemia is caused by a single amino acid substitution in the β-globin protein. The mutation is caused by a single nucleotide change, which, coincidentally, eliminates a restriction enzyme cutting site for the restriction enzymes *Mst*II and *Cvn*I. As a result, this mutation alters the pattern of restriction fragments seen in Southern blots. These characteristic RFLP patterns can be used for prenatal diagnosis of sickle-cell anemia, and to determine the genotypes of parents and other family members who may be heterozygous carriers of this condition.

For prenatal diagnosis, fetal cells are obtained by amniocentesis or by CVS. DNA is extracted from these cells and digested with a restriction enzyme such as

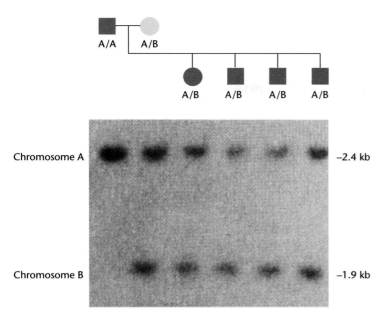

Chromosome A — -2.4 kb

Chromosome B — -1.9 kb

■ Figure 18–4 The segregation of a 2.4-kb RFLP allele with type 1 neurofibromatosis (NF1) in each of four affected offspring and their father. This RFLP is detected by probe pA10-41, which is known to be a DNA segment near the centromere of human chromosome 17. On the basis of this and results from the use of other probes, the locus for NF1 was assigned to chromosome 17.

*Mst*II. This enzyme cuts twice within the normal β-globin gene, producing two small DNA fragments. In the mutant allele, the second *Mst*II site has been destroyed by the mutation, producing one large restriction fragment (Figure 18–6). The restriction-digested DNA fragments are separated by gel electrophoresis, transferred to a nylon membrane, and visualized by Southern blot hybridization.

In the example shown in Figure 18–6, the parents (I-1 and I-2) are both heterozygous for the mutation. Digestion of DNA from each parent produces a large band (the mutant allele) and two smaller bands (the normal allele). Their first child (II-1) is homozygous normal because she has only the two smaller bands. The second child (II-2) has sickle-cell anemia; he has only one large band, and is homozygous for the mutant allele. The fetus (II-3) has a large band and two small bands and is, therefore, heterozygous for sickle-cell anemia. He or she will be unaffected, but will be a carrier (Figure 18–6).

Only about 5 to 10 percent of all nucleotide substitutions can be detected by restriction enzyme analysis.

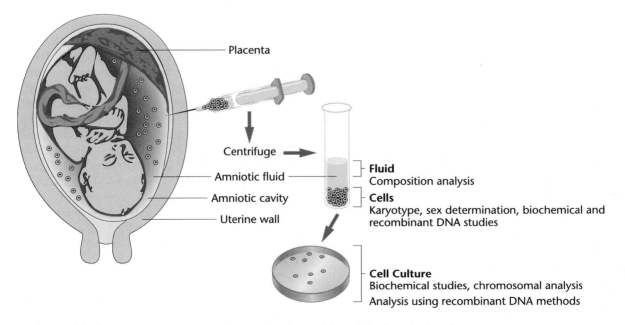

■ Figure 18–5 The technique of amniocentesis. The position of the fetus is first determined by ultrasound, and then a needle is inserted through the abdominal and uterine walls to recover fluid and fetal cells for cytogenetic and/or biochemical analysis.

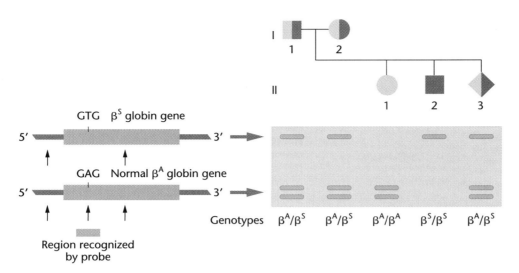

GTG β^S globin gene

GAG Normal β^A globin gene

Region recognized
by probe

Genotypes β^A/β^S   β^A/β^S   β^A/β^A   β^S/β^S   β^A/β^S

■ Figure 18–6 Southern blot diagnosis of sickle-cell anemia. Arrows represent the location of restriction enzyme cutting sites. In the mutant (β^S) globin gene, a point mutation (GAG → GTG) has destroyed a restriction enzyme cutting site, resulting in a single large fragment on a Southern blot. In the pedigree, the family has one unaffected homozygous normal daughter (II-1), an affected son (II-2), and an unaffected fetus (II-3). The genotypes of each family member can be read directly from the blot, and they are shown below the blot.

However, if a mutant gene has been well characterized and its nucleotide sequence is known, synthetic oligonucleotides can be used as probes to detect mutant alleles as described in the next section.

Allele-Specific Nucleotides and Genetic Screening

A method for distinguishing between alleles that differ by as little as a single nucleotide involves the use of synthetic probes known as **allele-specific oligonucleotides (ASO)**. In contrast to restriction enzyme analysis, the use of ASOs offers increased resolution, and wider application. Under the proper conditions, an ASO will hybridize only with its complementary sequence, and not with other sequences that might vary by as little as a single nucleotide. A method using ASOs and PCR analysis is now available to screen for sickle-cell anemia. In this method, DNA from white blood cells is extracted and denatured. A region of the β-globin gene from these cells is amplified by PCR. An aliquot of the amplified DNA is spotted onto filters, and each filter is hybridized to an ASO (Figure 18–7). After visualization, the genotype can be read directly from the filters. Using an ASO for the normal sequence [Figure 18–7(a)], the homozygous normal (AA) genotype produces a dark spot (two copies of the normal allele), and the heterozygous genotype (AS) produces a light spot (one copy of the normal allele). The homozygous recessive sickle-cell genotype will not bind the probe, and no spot will be visualized. Using a probe for the mutant allele [Figure 18–7(b)], the pattern is reversed. This rapid, inexpensive, and highly accurate technique will probably be the method of choice for diagnosis of a wide range of genetic disorders caused by point mutations.

In cases where the nucleotide sequence of the normal gene is known, and where the molecular nature of the mutant gene is known, ASOs can be synthesized directly from the wild type and mutant copies of the gene and used to screen for heterozygous carriers of the genetic disorder. For example, in cystic fibrosis, a deletion (called Δ508) represents about 70 percent of all mutant alleles. Cystic fibrosis is an autosomal recessive disorder associated with a defect in a protein called the **cystic fibrosis transmembrane conductance regulator (CFTR)**, which regulates chloride ion transport across the plasma membrane. To detect heterozygous carriers for the Δ508 mutation, allele-specific oligonucleotides are made by PCR from cloned samples of the normal allele and the mutant allele. DNA prepared from white blood cells of individuals to be tested is applied to a nylon filter and hybridized to each of the ASOs (Figure 18–8). In affected individuals, only the ASO made from the mutant allele will hybridize. In heterozygotes, both ASOs will hybridize, and in normal homozygotes, only the ASO from the normal allele will hybridize.

Because CF affects approximately 1 in 2000 individuals of northern European descent, screening for CF may be used in certain populations to detect and advise heterozygous carriers of their status for CF. However, since not all of the known alleles of this gene (over 300 mutations have been identified) can be screened, a negative result does not necessarily eliminate someone as a heterozygous carrier. In addition, it is possible that not all mutations of the CF gene have been identified. Such screening will probably be com-

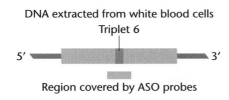

(a) Genotypes   AA   AS   SS

Normal (β^A) ASO: 5′ – CTCCTGAGGAGAAGTCTGC – 3′

(b) Genotypes   AA   AS   SS

(β^S) ASO: 5′ – CTCCTGTGGAGAAGTCTGC – 3′

■ Figure 18–7   Genotype determinations using allele-specific oligonucleotides (ASOs). In this technique, the β-globin gene is amplified by PCR using DNA extracted from blood cells. The amplified DNA is denatured and spotted onto strips of DNA-binding filters. Each strip is hybridized to a specific ASO and visualized on X-ray film after hybridization and exposure. If all three genotypes are hybridized to an ASO from the normal β-globin gene, the pattern in (a) would be observed: AA-homozygous normal hemoglobin has two copies of the normal β-globin gene and would show heavy hybridization; AS-heterozygous individuals carry one normal β-globin gene and one mutant gene, and would show weaker hybridization; SS-sickle-cell homozygotes carry no normal copy of the β-globin gene and would show no hybridization to the ASO probe for the normal β-globin gene. (b) The same genotypes hybridized to the probe for the sickle-cell β-globin gene would show the reverse pattern: no hybridization by the AA genotype, weak hybridization by the heterozygote (AS), and strong hybridization by the homozygous sickle-cell genotype (SS).

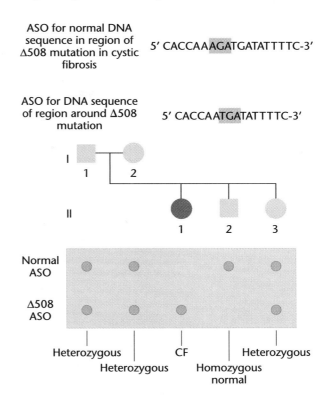

■ Figure 18–8   Screening for cystic fibrosis (CF) by allele-specific oligonucleotides (ASOs). ASOs for the region spanning the most common mutation in CF, a three-nucleotide deletion (Δ508), are prepared from normal CF genes and Δ508 CF genes. In screening, the CF gene is amplified by PCR using DNA extracted from blood samples and spotted on a DNA-binding membrane. The membrane is hybridized to a mixture of the two ASOs. The genotypes of each family member can be read directly from the filter. DNA from I-1 and I-2 hybridizes to both ASOs, indicating that they carry a normal allele and a mutant allele and are therefore heterozygous. The DNA from II-1 hybridizes only to the Δ508 ASO, indicating that she is homozygous for the mutation and has cystic fibrosis. The DNA from II-2 hybridizes only to the normal ASO, indicating that he carries two normal alleles. II-3 has two hybridization spots, and is heterozygous.

monplace once tests can cover 98 to 99 percent of all possible CF mutations.

DNA Chips and Genetic Screening

DNA probes similar to allele-specific nucleotides are being coupled with the technology of the semiconductor industry to produce **DNA chips**. The chips themselves are made of glass, and are about one-half inch square. The chip is divided into fields (smaller squares), each of which is about half the width of a human hair. Each field contains linker molecules to which are attached DNA probes about 20 nucleotides in length. Along a row of fields, the sequence of the probe differs by one nucleotide. Thus, a set of four fields (one for each nucleotide) is needed to test a given position for its nucleotide content. The current generation of chips can hold 400,000 fields, but chips with 1.6 million fields

are now planned. For testing, DNA is extracted from cells and cut with one or more restriction enzymes. The resulting fragments are tagged with a fluorescent dye, melted into single strands, and pumped into the chip. Fragments with a nucleotide sequence that exactly matches the probe sequence will bind, and those with a sequence that do not match are washed off the chip. A laser scanner reads the chip and detects the pattern of fluorescence (Figure 18–9). The fluorescent pattern is analyzed by a linked software program, and the data are presented in the form of a nucleotide sequence.

DNA chips are already being used to scan for mutations in the *p*53 gene, which is mutated in 60 percent of all cancers, and to screen for mutations in the *BRCA*1 gene, which predispose women to breast cancer. In ad-

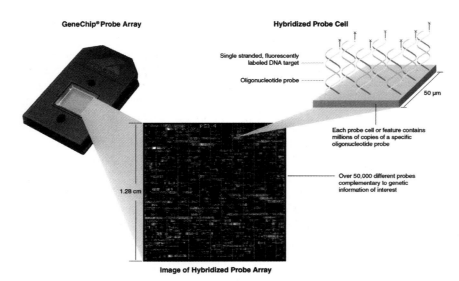

■ Figure 18–9 Affymetrix GeneChipR probe arrays are used to detect DNA mutations and monitor gene expression. Each chip holds between 50,000 to 500,000 probes. The nucleic acids to be analyzed are denatured and tagged with fluorescence which is hybridized to the probes on the chips. A laser scan detects the fluorescence.

dition to testing for mutations in single genes, DNA chips can be made to screen thousands of genes simultaneously. Chips have been programmed with the 6500 human genes that have already been sequenced, and chips with 50,000 genes are in the planning stages.

This technology will make it possible to analyze someone's DNA for dozens or hundreds of diseases, and to detect alleles of genes that predispose an individual to heart attacks, diabetes, Alzheimer disease, and other genetically defined disease subtypes.

The Ethics of Genetic Testing

In an earlier section, we described examples of two types of genetic testing, prenatal diagnosis and screening for heterozygous carriers of recessive disorders. Using current technology, genetic testing can also be used to predict risk of disease, identifying those who are presently healthy but who are at high risk of contracting a genetic disease in the future, and to establish a prognosis for those with a genetic disorder. In the next few years, DNA chips may be used to test for 50 to 100 diseases, including many that may not develop for years or decades. These advances will profoundly affect our health, reproductive patterns, and medical care. However, the use of this technology also raises legal, social, and ethical issues that have not been resolved (see the essay on genetic testing, p. 416). For example, what should people know before deciding to have a genetic test? How can we protect the information revealed by a genetic test? How can we define and prevent genetic discrimination?

Many of the potential risks and benefits of genetic testing are still unknown. Although we have the ability to test for many genetic diseases, there are almost no effective treatments to cure or improve most of these disorders. It is important to remember that with present technology, the fact that a genetic test gives a negative result does not rule out future development of the disease, nor does a positive result always mean that one will get the disease.

The development and implementation of public policy and laws about genetic testing have lagged behind the development and application of the technology for testing. The **Ethical**, **Legal**, **and Social Implications (ELSI)** Program of the Human Genome Project (described below) has set up task forces to identify issues related to genetic testing. This program is charged with developing recommendations for policy makers and lawmakers that will retain the benefits of genetic testing, but reduce or eliminate the potential harm. In addition, other groups made up of scientists, health care professionals, lawmakers, ethicists, and consumers are debating these issues and formulating policy options.

Gene Therapy

Gene products, such as insulin, have been used for decades in therapeutic treatment. Methods for the isolation and cloning of specific genes, originally developed as a research tool, are now being used to treat genetic disorders, a process known as **gene therapy**. In theory, gene therapy transfers a normal allele into a somatic cell that carries one or more mutant alleles. The delivery of these structural genes and their regulatory sequences is accomplished using a vector or gene transfer system.

Several methods can be used to transfer genes and their regulatory sequences into human cells. These include the use of viruses as vectors, the chemically assisted transfer of genes across cell membranes, and the fusion of cells with artificial vesicles containing cloned DNA sequences.

At present, the most common method of gene transfer uses a retroviral vector such as the Maloney murine leukemia virus (Figure 18–10). A cluster of three genes is removed from the virus, and the cloned human gene is inserted. After packaging into a viral protein coat, the recombinant vector can infect cells but is replication-deficient because of the missing viral genes. Once it is in the cell, the viral genome carrying the cloned human gene moves to the nucleus and integrates into a chromosome.

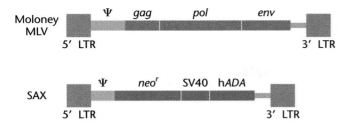

■ Figure 18–10 Retroviral vectors constructed from the Moloney murine leukemia virus (Moloney MLV). The native MLV genome contains a psi sequence required for encapsulation and genes that encode viral coat proteins (gag), an RNA-dependent DNA polymerase (pol), and surface glycoproteins (env). At each end, the genome is flanked by long terminal repeat (LTR) sequences that control transcription and integration into the host genome. The SAX vector retains the LTR and psi sequences, and it includes a bacterial neomycin resistance (neor) gene that can be used as a selective marker. As shown, the vector carries a cloned human adenosine deaminase (hADA) gene, fused to an SV40 early-region promoter/enhancer. The SAX construct is typical of retroviral vectors now being used in human gene therapy.

Several heritable disorders are currently being treated with gene therapy, including **severe combined immunodeficiency (SCID)**, **familial hypercholesterolemia**, and **cystic fibrosis**. In the next sections, we will describe how gene therapy is being used to treat a genetic disorder, and then discuss some of the ethical issues raised by the use of gene therapy.

Severe Combined Immunodeficiency (SCID): First-Generation Vectors

In severe combined immunodeficiency (SCID), affected individuals have no functional immune system and usually die from otherwise minor infections (Figure 18–11). An autosomal form of SCID is caused by a mutation in the gene that encodes the enzyme **adenosine deaminase**

■ Figure 18–11 David, a boy born with severe combined immunodeficiency (SCID), survived by being isolated in a germ-free environment. He died at the age of 12 following a bone marrow transplant undertaken to provide him with a functional immune system.

(ADA). Treatment starts with the isolation from the patient of a subpopulation of white blood cells called T cells (Figure 18–12). These cells, which are part of the immune system, are mixed with a genetically modified retrovirus carrying a normal copy of the human ADA gene (Figure 18–12). The virus infects the T cells, inserting a functional copy of the ADA gene into the cell's genome. The genetically modified T cells are grown in the laboratory to ensure that the transferred gene is expressed, and the patient is treated by injecting a billion or so of the altered T cells into the bloodstream.

Gene therapy began in 1990, with the treatment of a young girl suffering from SCID. Three years after treatment she had a new ADA gene in more than 50 percent of her T cells, and she is leading a normal life. A second child, treated a short time later, had the normal gene in only 0.1 to 1 percent of her white blood cells, a level that was not high enough to be effective. Other gene therapy trials, such as those for cystic fibrosis, have also had poor results, most of which have been traced to inefficient vectors.

The Future of Gene Therapy: New Vectors and Target-Cell Strategies

Unfortunately, as outlined above, gene therapy using modified retroviruses to treat genetic disorders has been only marginally successful. In addition, these first-generation vectors have several drawbacks that limit their widespread use. First, integration of the retroviral genome (including the cloned human gene) into the host cell genome occurs only if the host cells are replicating their DNA. Under *in vivo* conditions, there is little DNA synthesis in many highly differentiated cell types that are suitable target tissues. Second, insertion of viral genomes into the host chromosome can inactivate or mutate an indispensable host gene. Third, retroviruses have a low cloning capacity and cannot carry inserted sequences much larger than 8 kb. Many human genes, even without introns, exceed this size. Finally, there is the possibility of producing an infectious virus if a recombination event takes place between the vector and retroviral genomes already present in the host cell. For these reasons, other viral vectors and strategies for targeting cells are being developed. Some of the second-generation vectors are listed in Table 18.1. It is hoped that these vectors will overcome the problems associated with the first-generation retroviral vectors and fulfill the promise of gene therapy.

In addition to treating heritable genetic disorders, gene therapy is now being used or contemplated as a treatment for skin cancer, breast cancer, brain cancer, and AIDS. Treatment of these conditions by gene therapy has been more successful than that of metabolic disorders. In the last few years, at least a dozen biotech companies have been founded specifically to develop products for gene therapy, and a number of such therapies should reach the marketplace soon. In the twenty-

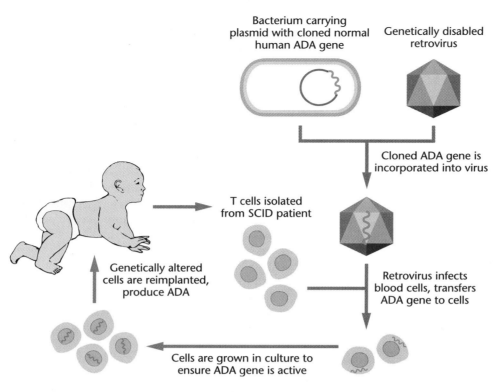

Bacterium carrying plasmid with cloned normal human ADA gene

Genetically disabled retrovirus

Cloned ADA gene is incorporated into virus

T cells isolated from SCID patient

Genetically altered cells are reimplanted, produce ADA

Retrovirus infects blood cells, transfers ADA gene to cells

Cells are grown in culture to ensure ADA gene is active

■ Figure 18–12 Gene therapy for treatment of severe combined immunodeficiency (SCID), a fatal disorder of the immune system caused by lack of the enzyme adenosine deaminase (ADA). The cloned human ADA gene is transferred into a viral vector, which is used to infect white blood cells removed from the patient. The transferred ADA gene is incorporated into a chromosome and becomes active. After growth to enhance their numbers, the cells are reimplanted in the patient, where they produce ADA, allowing the development of an immune response.

TABLE 18.1	**New viral vectors for gene therapy**		
	Adenovirus	*Adeno-Associated Virus*	*Herpes Virus*
Cell targets	Lung cells, cells of respiratory tract	Fibroblasts, T cells	Nerve cells, glial cells
Cloning capacity	7–35 kb	3 kb	Up to 150 kb
Integration into chromosomal DNA	No	Yes	No, but retained in nucleus

first century, gene therapy will be a commonplace method of treatment for a large number of disorders.

Ethical Issues and the Future of Gene Therapy

The ethical guidelines for gene therapy in its present form are well established. This experimental form of treatment is undertaken only after extensive reviews at several levels, and the trials are monitored to protect the interests of the patient. There are, however, two types of gene therapy that are not currently used because ethical issues surrounding them have not been resolved. The first is **germ-line gene therapy**, and the second is termed **enhancement gene therapy**.

All gene therapy trials currently underway or in the planning stages use only somatic cells as targets for gene transfer. In **somatic gene therapy**, only one individual is affected, and the therapy is done with permission and informed consent. One of the ethical issues surrounding germ line therapy is the issue of informed consent. In germ-line gene therapy, germ cells or mature gametes are used as targets, and individuals in future generations are affected, without their consent. Is

this ethical? Do we have the right to make this decision for future generations?

A second issue is whether gene therapy should be used to enhance human potential, rather than treat a disease. In other words, if genes thought to enhance desirable characteristics were identified and cloned, would it be permissible to use gene therapy to increase height, or to enhance athletic ability or intelligence? Presently, the consensus is that germ line and enhancement therapy are unacceptable uses for gene therapy, but the issues are still being debated, and are unresolved. The outcome of these debates may affect not only the fate of individuals, but also that of our species.

DNA Fingerprints

As discussed earlier, restriction fragment length polymorphisms (RFLPs) are common in the human genome, and can be used as genetic markers. A second type of nucleotide variation, discovered in the mid-1980s, depends on variability in the length of repetitive DNA sequence clusters. These and other polymorphisms in human DNA led to the development of methods for identifying individuals and establishing degrees of genetic relatedness between individuals. One of these techniques is called **DNA fingerprinting**.

Minisatellites and VNTRs

One of the most useful forms of restriction fragment length polymorphisms arises from variations in the number of tandemly repeated DNA sequences present between two restriction enzyme sites. These sequences are **minisatellites**, or clusters of nucleotides from 2 to 100 nucleotides in length. For example, the base sequence

GGAAGGGAAGGGAAGGGAAG

is composed of four tandem repeats of the five-nucleotide sequence GGAAG. Clusters of such sequences are widely dispersed in the human genome. Typically, each repeat contains between 14 and 100 nucleotides, and the number of repeats at each locus ranges from 2 to more than 100. These loci are known as **variable-number tandem repeats (VNTRs)**. The number of repeats at a given locus is variable, and each variation is a VNTR allele. Many loci have dozens of alleles each; as a result, heterozygosity is common.

A pattern of bands is produced when VNTR sequences are cut with restriction enzymes and visualized by Southern blotting. This pattern is one example of what is known as a **DNA fingerprint** (Figure 18–13). These patterns are the equivalent of fingerprints because the pattern of bands is always the same for a given individual, no matter what tissue is used as the source of DNA. As with real fingerprints, the pattern varies from individual to individual. In fact, there is so much variation in the band pattern from individual to individual that, theoretically, each person's pattern is unique. DNA fingerprint analysis can be performed on very small samples of material (less than 60 microliters [μl] of blood). Analysis can also be performed using samples that are quite old (VNTR analysis has been performed on Egyptian mummies over 2400 years old), thus increasing its usefulness in legal cases.

Forensic Applications

Since 1988, DNA fingerprints have been used as evidence in criminal trials in the United States. The method has also been used in a wide range of other applications, including immigration cases, disputes involving pure-bred dogs, paternity cases, and in animal conservation studies. At present, just over a dozen different VNTR probes are used in standard forensic test-

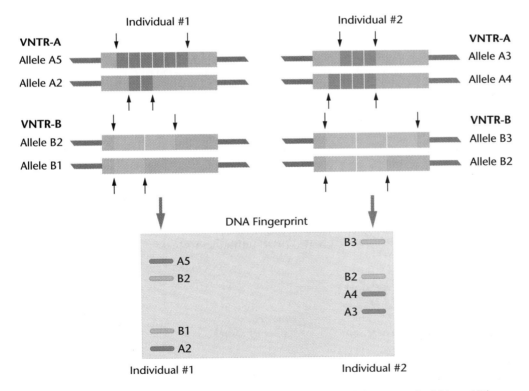

■ Figure 18–13 VNTR loci and DNA fingerprints. VNTR alleles at two loci (A and B) are shown for each individual. Arrows mark restriction cutting sites flanking the VNTRs. Restriction digestion produces a series of fragments that can be detected as bands on a Southern blot (below). Because of differences in the number of repeats at each locus, the overall pattern of bands is distinct for each individual, even though one band is shared (the band representing the B2 allele). Such a pattern is known as a DNA fingerprint.

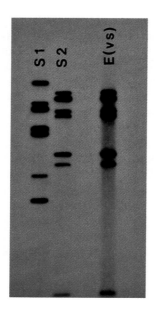

■ Figure 18–14 DNA fingerprinting in a forensic case. The DNA profile of suspect 2 (S2) matches that of the blood sample obtained as evidence, E(VS).

ing (Figure 18–14). Most of these are on different chromosomes, and each probe can be used to visualize a VNTR pattern at a particular locus. Standard forensic tests employing four to six different VNTR probes are used in developing a detailed DNA fingerprint profile.

Results of DNA fingerprints are analyzed and interpreted using statistics, probability, and population genetics. In this interpretation, the frequency of each VNTR allele in the population is calculated, and these are multiplied together to give a combined frequency, which is a calculation of how often the observed DNA fingerprint might occur in the general population. The use of a number of VNTR probes increases the accuracy of the combined frequency. For example, if the frequency of locus 1 is determined to be 1 in 333, and the frequency of locus 2 is 1 in 83, their combined frequency is equal to the product of their individual frequencies, or 1 in 28,000. While this overall frequency may not serve

as convincing evidence, if the frequency of a third locus (1 in 100) and a fourth locus (1 in 25) are included in the calculations, the combined frequency becomes quite convincing, 1 in 70 million, that is, the chance that anyone has this combination of alleles is about 1 in 70 million (about 4 individuals in the U.S. population).

Genome Projects

Over the past 90 years, geneticists have developed efficient methods for generating and mapping mutants. One drawback to these methods is that genes can be characterized only if mutants are isolated. To characterize a genome exhaustively, at least one mutation for each of the genes in the genome is required, and because of lethal alleles and other problems, this is probably an impossible goal to achieve.

Geneticists are now using recombinant DNA techniques to identify directly all the genes in an organism's genome. Instead of screening for mutants and constructing genetic maps, a genomic library is established, and overlapping clones are assembled to establish genetic and physical maps encompassing the entire genome. The final step involves sequencing the entire genome. The result is a genomic map, with all genes in the genome identified by their location and their nucleotide sequence. Analysis of derived amino acid sequences and the generation of gene-specific mutants can be used to study the function of the gene product.

Genome projects are underway for dozens of species, some of which are listed in Table 18.2. Several of these genome projects are being carried out under the auspices of the Human Genome Project (discussed below). Those model organisms that are part of the Genome Project were selected for several reasons, including genome size, the availability of detailed genetic maps, and the value of making comparisons about gene number, location, and function across species ranging from bacteria to humans.

Three basic approaches are used in genome projects: (1) the **bottom-up approach**, which starts with small, overlapping clones and assembles them into larger segments that eventually cover the entire genome; (2) the

TABLE 18.2	Some of the organisms studied by genome projects	
Organism	**Genome Size (bp)**	**Estimated Number of Genes**
Escherichia coli (bacterium)	4×10^6	4,000
Saccharomyces cerevisiae (yeast)	1.5×10^7	6,000
Arabidopsis thaliana (plant)	1×10^8	25,000
Caenorhabditis elegans (nematode)	1×10^8	13,000
Drosophila melaogaster (insect)	1.2×10^8	10,000
Mus musculus (mouse)	3×10^9	80,000
Homo sapiens (human)	3.2×10^9	80,000

top-down approach, which isolates large, randomly cloned DNA segments that are broken down into smaller units for mapping and sequence analysis; and (3) the **shotgun approach**, in which cloned fragments are selected randomly from a genomic library and sequenced. The nucleotide sequences of the fragments are then analyzed by an assembler software program, and organized into a complete genome sequence. This "shotgun" sequencing method is the latest to be developed; it greatly shortens the time required for sequencing a genome, and it bypasses the need for some intermediate stages of mapping.

A brief discussion of each of these approaches will serve as an introduction to the Human Genome Project and its goal of isolating and identifying the 50,000 to 100,000 genes in the human genome.

The Bottom-Up Method: The E. coli Project

The bottom-up method was used to sequence the *E. coli* genome. A cloned genomic library of the *E. coli* K12 strain was prepared in lambda vectors by Yuji Kohara and his colleagues. Clones from this library were extensively characterized by restriction mapping, and computer analysis of the maps was used to identify overlapping clones by the presence of shared restriction sites (Figure 18–15). The resulting contiguous segment of DNA covered by the overlapping clones is called a **contig**. Originally, 70 contigs were identified, covering 94 percent of the genome. Eventually, clones that spanned the small gaps between contigs were identified, and the contigs were arranged in a physical map that covers the entire 4700 kb of DNA in the *E. coli* genome.

The second stage of the project involved sequencing the clones in the contig library. The nucleotide sequencing of two different strains of *E. coli* was completed in early 1997. The genome contains 4,638,858 nucleotides, and approximately 4300 genes. In spite of decades of work on *E. coli*, almost 2500 of its genes were discovered for the first time during the sequencing project. All of these new genes have unknown functions, providing a wealth of new research opportunities for those studying this bacterium.

The Top-Down Approach: The Drosophila Genome

For the *Drosophila* genome project, a cloned genomic library has been constructed using **yeast artificial chromosome (YAC)** vectors (see Chapter 17). The chromosomal origin of the cloned insert DNA in each YAC has been determined by *in situ* hybridization to the banded polytene chromosomes of larval salivary glands (Figure 18–16). The average YAC covers six to eight chromosome bands; contigs were constructed by identifying YACs that have two or more chromosome bands in common.

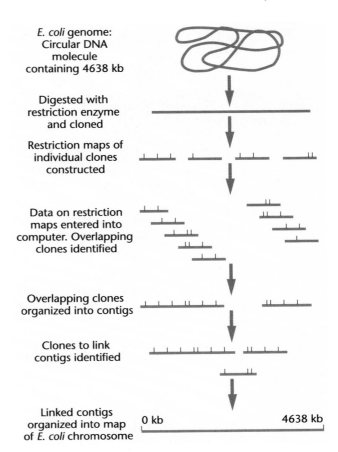

■ Figure 18–15 Bottom-up strategy for the *E. coli* genome project. The 4700-kb *E. coli* chromosome is digested with restriction enzymes into small fragments that are cloned into a vector. Individual clones are characterized by restriction mapping. Information about these restriction maps is analyzed by computer to identify overlapping clones. These overlapping clones are organized into contiguous segments of DNA called contigs. Next, more clones are screened to identify those spanning contigs. These are used to reduce gaps, bringing the number of contigs from 70 to 7 and finally to 1, which corresponds to the entire *E. coli* chromosome.

To study individual genes or regions, YACs are broken down by restriction digestion and subcloned into other vectors. Identifying genes by DNA sequencing in *Drosophila* is more difficult than in *E. coli*, in part because the relatively large size of the genome (see Table 18.2 to compare the genome sizes) makes sequencing a less efficient method of mapping. To make the task easier, each *Drosophila* gene has been defined as any stretch of unique DNA sequence (each gene should occur only once in a haploid genome). Therefore, any short DNA segment within a gene can be used as a marker, called a **sequence tagged site (STS)**. The location of genes can therefore be identified without knowing the nature of the gene, the gene product, or its phenotype, and these regions can be sequenced before the noncoding regions.

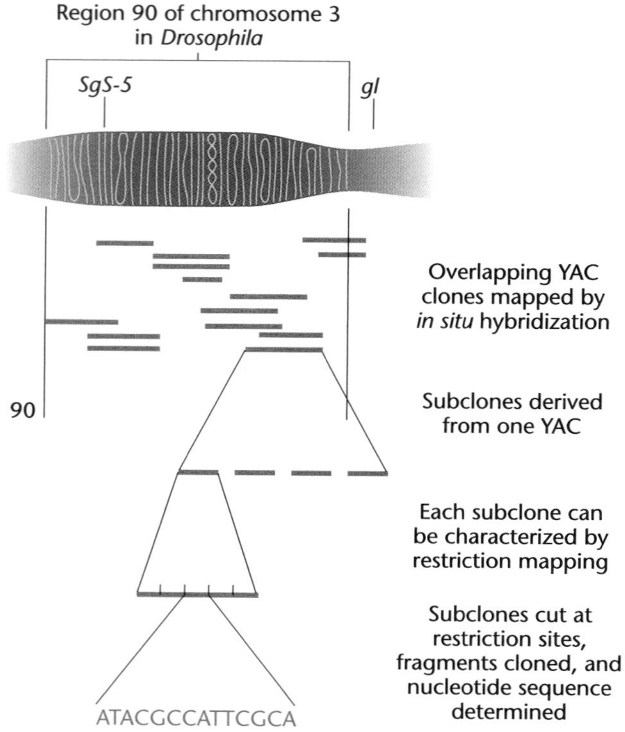

■ Figure 18–16 Top-down strategy for the *Drosophila* genome project. A genomic library is constructed using large (>200 kb) fragments in YACs. The physical locations of YACs are mapped to polytene chromosomes by *in situ* hybridization. Shown is region 90, located on chromosome 3. This region contains about 890 kb of DNA and approximately 34 genes. Approximately 14 percent of the region is transcribed as RNA. Each YAC covers several polytene bands and presumably includes several genes. Individual YACs are digested with restriction enzymes and the individual fragments are subcloned. The subclones are characterized by restriction mapping, and restriction fragments from these subclones are used for nucleotide sequence analysis.

The Shotgun Method: Prokaryotic Genomes

The genomes of two species of bacteria, *Hemophilus influenzae* and *Mycoplasma genitalium,* were both completely sequenced in less than a year using the shotgun method. In conventional genome sequencing, a genomic library is organized into a series of contigs. Each of these is then broken down into smaller clones, which are sequenced and put into the correct order. In sequencing the *M. genitalium* genome, a genomic library was prepared from restriction digests of DNA. From this library, randomly selected fragments were sequenced, and the nucleotide sequence of 8472 such fragments was analyzed by an assembler software program, and organized into a complete genome sequence.

The shotgun method, also used on the *H. influenzae* genome, greatly shortens the time required for sequencing a genome, and bypasses the need to prepare physical maps (see below). Investigators are now working to determine whether the shotgun method can be adapted

to sequencing parts of the human genome, reducing the time and cost for this part of the project.

The Human Genome Project

The **Human Genome Project** is an international, coordinated effort to determine the sequence of the 3.2 billion nucleotide pairs in the haploid human genome. In the United States, the project began in 1990 as a joint program of the National Institutes of Health and the U.S. Department of Energy. Other countries, notably France, Britain, and Japan, began similar projects, which are now coordinated by an international organization, the Human Genome Organization (HUGO).

The task of sequencing the human genome is proceeding in a series of well-defined stages. The first stage involved the construction of high-resolution genetic maps for each human chromosome (Figure 18–17). These maps were constructed using identified genes, RFLPs, STSs, and other markers. Genetic maps for each chromosome, incorporating about 15,000 markers, were completed in 1995. These maps, with an average distance of 2 million base pairs of DNA between markers, are being used to sort out and organize the contigs generated by physical mapping.

The second level of the Human Genome Project is the construction of physical maps, using either the top-down approach beginning with chromosomes isolated in somatic cell hybrids, or the bottom-up approach of assembling contigs from DNA segments randomly cloned in cosmids or YACs. The goal is the creation of a physical map of the entire genome consisting of 30,000 STSs spaced at intervals of about 100 kb.

An STS-based physical map covering 95 percent of the genome, and encompassing over 23,000 STS sites, has been recently assembled. The STSs in this map are each composed of 200 to 500 nucleotides, and are stored in databases as a nucleotide sequence. The sequences can be downloaded, and for use as a probe, the sequence of any STS can be generated by PCR. Sequence tagged sites serve as landmarks for researchers using a variety of vectors, including YACs, phages, or plasmids, allowing information from many sources to be integrated into maps with higher and higher resolution.

The ultimate goal of the project is to sequence the 3.2 billion nucleotides in the human genome by 2005. This task may require the development of new technology for sequencing DNA, as well as new technology for information storage, analysis, and retrieval. This last stage of the project has started, with the establishment of six centers in the United States to improve technology related to sequencing, and to begin large-scale sequencing of the genome. Regions on eight chromosomes have been selected for sequencing, covering about 3 percent of the genome. It is expected that these regions will be completely sequenced by the end of 1999. A similar project, covering other regions of the genome, is underway in Europe.

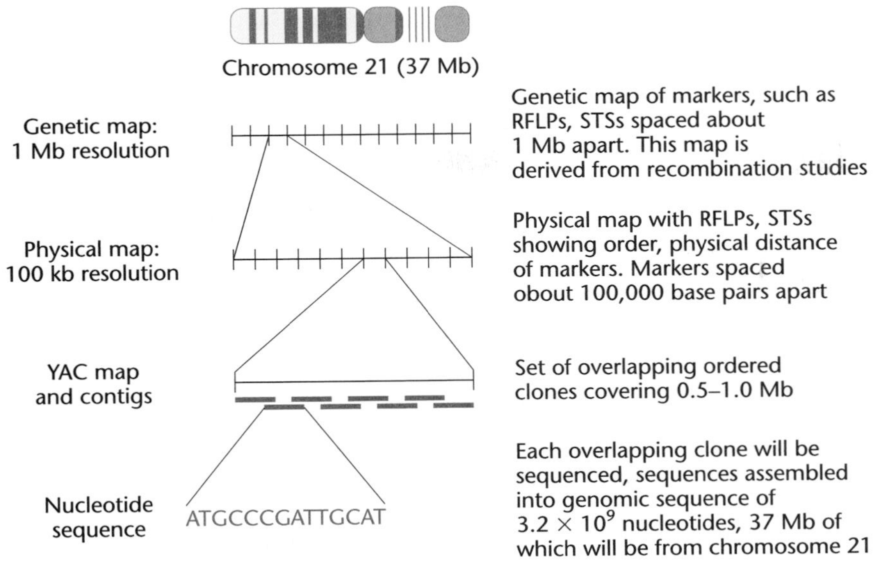

Chromosome 21 (37 Mb)

Genetic map:
1 Mb resolution

Genetic map of markers, such as RFLPs, STSs spaced about 1 Mb apart. This map is derived from recombination studies

Physical map:
100 kb resolution

Physical map with RFLPs, STSs showing order, physical distance of markers. Markers spaced obout 100,000 base pairs apart

YAC map and contigs

Set of overlapping ordered clones covering 0.5–1.0 Mb

Nucleotide sequence

ATGCCCGATTGCAT

Each overlapping clone will be sequenced, sequences assembled into genomic sequence of 3.2×10^9 nucleotides, 37 Mb of which will be from chromosome 21

■ Figure 18–17   An overview of the strategy used in the Human Genome Project. The first goal, achieved in 1995, was to have a genetic map of each chromosome, with markers spaced at distances of about 1 Mb (1 million base pairs of DNA). This work was accomplished by finding markers such as RFLPs and STSs and assigning them to chromosomes. Once assigned to chromosomes, the markers' inheritance was observed in heterozygous families to establish the order and distance between them (a genetic map). In the second stage, the goal is to prepare a physical map of each chromosome (our example uses chromosome 21, the smallest chromosome), containing the location of markers spaced about 100,000 base pairs apart. The third stage involves the construction of a set of overlapping clones, in yeast artificial chromosomes (YACs) or other vectors that cover the length of the chromosome. The last stage will be the sequencing of the entire genome. Sequencing on selected parts of the genome has started, and about 3 percent of the genome is expected to be sequenced in the next few years.

Recently, J. Craig Ventner, the scientist who developed the shotgun method of genome sequencing has formed a partnership with a company that manufactures automated DNA sequencing machines. This partnership proposes to use the shotgun method to sequence the human genome in 3 years at a cost of $200–300 million (compared with about $3 billion for the NIH-sponsored project). The partnership plans to recover its costs by patenting genes it discovers during sequencing and by subscriptions to sequence data bases.

The Ethical, Legal, and Social Implications of the Human Genome Project

At the time the Human Genome Project was being established, scientists raised concerns about how genome information would be used, and how the interests of both individuals and society could be protected. To address these concerns, the Ethical, Legal, and Social Implications (ELSI) Program was established as part of the Human Genome Project. The ELSI program began by considering a number of issues, including the impact of genetic information on individuals, the privacy and confidentiality of genetic information, the impact of genome informa-

tion and technology on medical practice, genetic counseling, and reproductive decision making. Through research grants, workshops, and public forums, ELSI has started to formulate policy options addressing these issues.

Currently, ELSI is focusing on four areas: (1) privacy and fairness in the use and interpretation of genetic information, (2) the transfer of genetic knowledge from the research laboratory to clinical practice, (3) issues of informed consent for participants in genetic research, and (4) public and professional education.

The ELSI program is working to ensure that as the Human Genome Project moves from generating information about the genetic basis of disease to the long-range goals of improved treatment, prevention, and cures, that ethical issues of concern are identified and policies and laws are instituted to deal with them.

After the Genome Projects: What Next?

Several genome projects have already reached their goals. The genomes of two bacteria, *Hemophilus influenzae* and *Mycoplasma genitalium,* were both completely sequenced in 1995. The nucleotide sequence of the yeast genome was reported in 1996, the *E. coli* genome was sequenced in 1997, and the sequence for the *C. elegans* genome should be available before the end of 1998. Work on sequencing the human genome is proceeding, and by 1999, regions on several chromosomes, covering some 3 percent of the genome, should be completely sequenced. As data from these projects become available, it is apparent they are rich sources of information about genome organization, function, and evolution.

From a biological perspective, the *M. genitalium* genome is of interest because it is one of the smallest genomes of any free-living organisms, containing only 482 genes. Ninety of these genes encode proteins used in translation, and about 30 encode products for DNA replication. Because of the small size of the genome, these numbers may represent close to the minimum number of genes necessary to complete such tasks. Somewhat surprisingly, 140 of the 482 genes (about 30 percent of the genome) encode membrane-inserted proteins, emphasizing the importance of interaction between the cell and its environment for survival. Altogether, 260 of the 482 genes in this organism (54%) function in three processes: translation, DNA replication, and signal transduction. This leaves less than half the genome for all other functions. Further work on this organism and those with similar-sized genomes may help define the minimum number of genes necessary for life.

Although the flow of information on the human genome is still in its early stages, several insights into human biology and genetic disorders have already been gained. Previously unknown mechanisms of mutation, including the expansion of trinucleotide repeats, have provided insights into the nature of several neurodegenerative diseases including Huntington disease, myotonic dystrophy, and spinal/bulbar muscular atrophy. Another new mutational mechanism, an increase in gene dosage caused by duplication of small regions of chromosomes, has been identified and associated with specific disorders such as Charcot–Marie–Tooth syndrome. In addition, it is now known that different kinds of mutations in a single gene can give rise to different genetic disorders (familial medullary thyroid carcinoma, multiple endocrine neoplasias 2A/2B, and Hirschprung disease). In addition, mutations in members of a gene family can give rise to diseases with related phenotypes (Crouzon syndrome, Apert syndrome, and achondroplasia).

Biotechnology

Although recombinant DNA techniques were originally developed to facilitate basic research on gene organization and regulation of expression, scientists have not been blind to the commercial possibilities of this technology. As a result, researchers have participated in the formation of biotechnology companies using recombinant DNA technology to develop products including hormones, clotting factors, herbicide-resistant plants, enzymes for food production, and vaccines. In the last decade, the biotechnology industry has grown into a multibillion-dollar segment of the economy.

Insulin Production

The first human gene product manufactured using recombinant DNA and licensed for therapeutic use was human insulin, which became available in 1982. Insulin is a protein hormone that regulates sugar metabolism, and an inability to produce insulin results in diabetes, a disease that in its more severe form affects more than 2 million individuals in the United States.

A look at the method originally used for producing insulin by recombinant DNA methods is instructive, as it shows both the promises and the difficulties of this technology. The functional insulin protein is made of two polypeptide chains, A and B. The A polypeptide of insulin has 21 amino acids, and the B subunit has 30. Synthetic genes for the A and B subunits were constructed by oligonucleotide synthesis (63 nucleotides for the A polypeptide and 90 nucleotides for the B polypeptide). Each synthetic oligonucleotide was inserted into a vector at a position adjacent to a gene encoding the bacterial form of the enzyme, β-galactosidase. When transferred to a bacterial host, the β-galactosidase gene and the synthetic oligonucleotide were transcribed and translated as a unit. The product is a **fusion polypeptide**, consisting of the amino acid sequence for β-galactosidase attached to the amino acid sequence for one of the insulin subunits (Figure 18–18). The fusion proteins were purified from bacterial extracts and treated with cyanogen bromide, to cleave the fusion protein from the β-galactosidase.

Each insulin subunit was produced separately by this method. When mixed together, the two subunits spontaneously unite, forming an intact, active insulin molecule. The purified insulin is then packaged for use by those with diabetes, who must take insulin injections one or more times daily.

A number of genetically engineered proteins for therapeutic use have been produced by similar methods or are in clinical trials (Table 18.3). In most cases, these proteins are produced by cloning a human gene into a plasmid and inserting the recombinant vector into a bacterial host. After ensuring that the transferred gene is expressed, large quantities of the transformed bacteria are produced, and the human protein is recovered and purified.

Pharmaceutical Products in Animal Hosts Bacterial hosts were used to produce the first generation of recombinant proteins, even though there are some disadvantages in using prokaryotic hosts to synthesize eukaryotic proteins. For example, bacterial cells are unable to process and modify eukaryotic proteins, and they cannot add sugars and phosphate groups that are

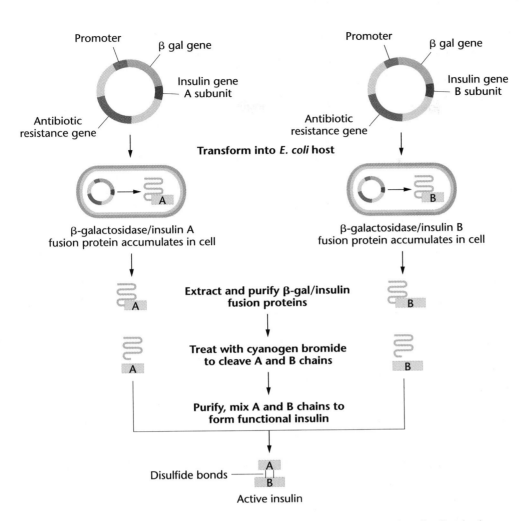

■ Figure 18–18   The method used to synthesize recombinant human insulin. Synthetic oligonucleotides encoding the insulin A and B chains were inserted at the tail end of a cloned *E. coli* β-galactosidase (β-gal) gene. These recombinant plasmids were transferred to *E. coli* hosts, where the β-gal/insulin fusion protein was synthesized and accumulated in the host cells. Fusion proteins were extracted from the host cells and purified. Insulin chains were released from the β-galactosidase by treatment with cyanogen bromide. The insulin subunits were purified and mixed to produce a functional insulin molecule.

TABLE 18.3	Genetically engineered pharmaceutical products available or in clinical testing
Gene Product	*Condition Being Treated*
Atrial natriuretic factor	Heart failure, hypertension
Epidermal growth factor	Burns, skin transplants
Erythropoietin	Anemia
Factor VIII	Hemophilia
Gamma interferon	Cancer
Granulocyte colony–stimulating factor	Cancer
Hepatitis B vaccine	Hepatitis
Human growth hormone	Dwarfism
Insulin	Diabetes
Interleukin–2	Cancer
Superoxide dismutase	Transplants
Tissue plasminogen activator	Heart attacks

often needed for full biological activity. In addition, eukaryotic proteins produced in prokaryotic cells often do not fold into the proper three-dimensional configuration, and as a result, are inactive. To overcome these difficulties and to increase yields, second-generation methods use eukaryotic hosts. Rather than being produced by host cells grown in tissue culture, human proteins such as alpha-1-antitrypsin are being produced in the milk of livestock, as described below.

A deficiency of the enzyme alpha-1-antitrypsin is associated with the heritable form of emphysema, a progressive and fatal respiratory disorder common among those of European ancestry. To produce alpha-1-antitrypsin by genetic engineering, the human gene was cloned into a vector at a site adjacent to a promoter sequence from sheep that regulates expression of milk-associated proteins. The gene adjacent to this promoter is expressed only in mammary tissue. This fusion gene was then microinjected into sheep zygotes fertilized *in vitro* (Figure 18–19), which in turn were implanted into foster mothers. The resulting **transgenic** sheep developed normally and, after mating, produced milk that contained high concentrations of functional human alpha-1-antitrypsin. This human protein is present in concentrations up to 35 g per liter of milk. It is easy to envision that a small herd of lactating sheep could easily provide an adequate supply of this protein, and that herds of other transgenic animals, acting as biofactories, might become part of the pharmaceutical industry.

Herbicide-Resistant Crop Plants

In agriculture, vectors have been used to transfer traits for herbicide resistance to crop plants. The herbicide **glyphosate** is widely used for weed control, but it cannot be used on fields containing crop plants because it kills

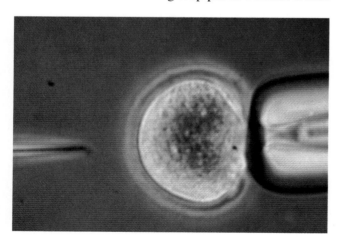

■ Figure 18–19 Injection of cloned DNA. A micropipette is used to transfer cloned genes into the nucleus of a mammalian zygote. The injected zygote will then be transferred to the uterus of a surrogate mother for development.

them along with the weeds. Weed growth is a serious agricultural problem, and damage from weeds reduces the yield on many crops by more than 10 percent. As a herbicide, glyphosate is effective at very low concentrations, is not toxic to humans, and is rapidly degraded by soil microorganisms. Glyphosate works by inhibiting the action of a chloroplast enzyme called EPSP synthase, which is active in amino acid synthesis. Deprived of these vital amino acids, plants wither and die.

Glyphosate resistance results from increasing the synthesis of EPSP synthase. This was accomplished by cloning the EPSP synthase gene into a vector at a site adjacent to a plant virus promoter. The recombinant vector was transferred into the bacterium *Agrobacterium tumifaciens* (Figure 18–20). Plasmid-carrying bacteria were in turn used to infect cells in discs cut from plant leaves. Calluses formed from these discs were then selected for their ability to grow on glyphosate. Transgenic plants generated from glyphosate-resistant calluses were grown and sprayed with glyphosate at concentrations four times higher than needed to kill wild-type plants. The transgenic plants overproducing the EPSP synthase grew and developed, while the control plants withered and died. Glyphosate-resistant corn and soybeans developed in this way are now available for planting. Use of these crops should reduce costs to the farmer for herbicides and lower the concentration of herbicides in the water supply.

Similar methods have been used to transfer resistance to virus infection, insects, and drought to crop plants. Other work has been directed at improving the nutritional value of crops such as soybeans and corn. Many of these projects are in the developmental stage, and these transgenic products will reach the marketplace over the next few years.

Transgenic Plants and Vaccines

One of the most beneficial applications of recombinant DNA technology may be in the production of vaccines. Vaccines stimulate the immune system to produce antibodies against a disease-causing organism and thereby confer immunity against the disease (see Chapter 19). Two types of vaccines are commonly used: **inactivated vaccines**, prepared from killed samples of the infectious virus or bacteria, and **attenuated vaccines**, which are live viruses or bacteria that can no longer reproduce and cause disease when present in the body.

Using recombinant DNA technology, a new type of vaccine called a **subunit vaccine** is being produced. These vaccines consist of one or more surface proteins of the virus or bacterium. The protein acts as an antigen to stimulate the immune system to make antibodies against the virus or bacterium. One of the first licensed subunit vaccines is for a surface protein of hepatitis B, a virus that causes liver damage and cancer. The gene for this hepatitis B protein was cloned into a yeast-expression vector and produced using yeast as a host. It is ex-

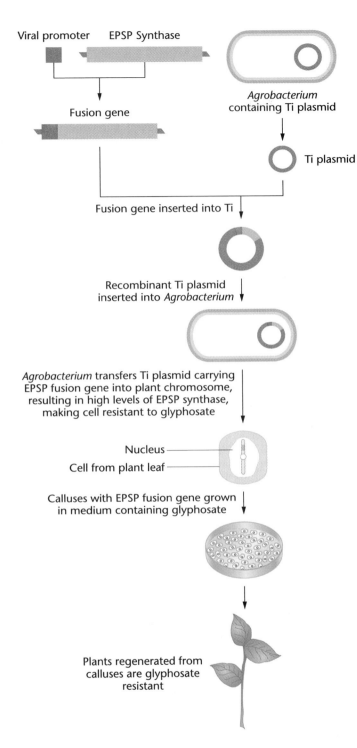

Figure 18–20 Gene transfer of glyphosate resistance. The EPSP gene is fused to a promoter from cauliflower mosaic virus. This chimeric or fusion gene is then transferred to a Ti plasmid vector, and the recombinant vector is inserted into an *Agrobacterium* host. *Agrobacterium* infection of cultured plant cells transfers the EPSP fusion gene into a plant cell chromosome. Cells that acquire the gene are able to synthesize large quantities of EPSP synthase, making them resistant to the herbicide glyphosate. Resistant cells are selected by growth in herbicide-containing medium. Plants regenerated from these cells are herbicide resistant.

tracted and purified from the host cells and then packaged for use as a vaccine.

Recombinant DNA technology is currently being used to produce the hepatitis B surface protein in edible plants. The idea is to have plant leaves or fruits serve as the source of an oral vaccine. Plant-produced vaccines offer several advantages. They would be inexpensive, as industrial investment for production and extensive purification would be unnecessary. In addition, such vaccines would not require injection and would be useful in developing countries, where medical services and delivery of health care are not highly developed.

As a model system, the antigenic subunit of hepatitis B vaccine has been transferred to tobacco plants and expressed in the leaves (Figure 18–21). For use as a source of vaccine, the gene would be inserted into food plants such as grains or vegetables.

Recent clinical tests have used genetically engineered potatoes that carry recombinant bacterial antigens. Human volunteers ate small quantities of these potatoes (50–100 g) and developed antibodies to the antigen, establishing the validity of this approach. Similar tests are bing planned using genetically engineered bananas. If further tests are successful, genetically engineered edible plants will soon be used to vaccinate people against many diseases.

Figure 18–21 Tobacco leaves expressing the hepatitis B antigen. Transgenic tobacco plants carrying an antigenic subunit of hepatitis B virus were generated. Leaves from transgenic plants were treated with antibodies against hepatitis B antigen, showing that the plants produced the antigen. The central leaf in the photograph is a leaf from a normal tobacco plant, and is unstained.

As discussed in Chapter 1, plants and animals were domesticated some 8000 to 10,000 years ago, and by selective breeding we have been modifying these organisms ever since, producing the diversity of domesticated plants and animals present today. Recombinant DNA technology has changed the rate at which new plants and animals can be developed, and by transfer of genes between species, has altered the types of changes that can be made. Unlike selective breeding programs, biotechnology has generated concerns about the release of genetically modified organisms into the environment, and about the safety of eating such products. If biotechnology is to achieve a new "green revolution," these concerns need to be addressed through prudent research and education.

Chapter Summary

1. Recombinant DNA technology offers a new approach to genetic analysis. Instead of relying on isolation and mapping of mutant genes, it is now possible to manipulate large cloned segments of the genome to make genetic and physical maps using molecular markers rather than phenotypes visible at the level of the organism.

2. Genome projects are completed or underway for several organisms, including *E. coli*, yeast, *Drosophila*, the mouse, and humans. The goals of these genome projects are to identify and map all genes in the genome and to determine the complete nucleotide sequence of the genome.

3. Cloned DNA is being used in a wide variety of applications, including gene mapping and the identification and isolation of genes responsible for genetic disorders. The method known as positional cloning, based on recombinant DNA methodology, allows the mapping and identification of a gene without any knowledge of the nature or function of the gene product.

4. Recombinant DNA techniques are being used in the prenatal diagnosis of human genetic disorders. This method allows direct examination of the genotype, whereas previous methods relied on gene expression and the identification of the gene product. These methods can also be used to identify carriers of genetic disorders, and they are the basis of proposals to screen the population for a number of genetic disorders, including sickle-cell anemia and cystic fibrosis.

5. The availability of cloned human genes has led to their use in replacement of mutant genes in somatic tissues. This somatic gene therapy is carried out by transferring a cloned normal copy of a gene into a vector and using the vector as a means of transferring the gene to a target tissue that takes up and expresses the cloned copy of the gene, altering the mutant phenotype. The development of new and more effective vector systems probably means that gene therapy will become a standard method for the treatment of genetic disorders in the near future.

6. The use of recombinant DNA techniques in detecting allelic variants of variable tandem nucleotide repeats (DNA fingerprints) has found applications in forensics, paternity testing, and a wide range of fields including archeology, conservation biology, and public health.

7. The biotechnology industry is using recombinant DNA methods to produce human gene products in a variety of hosts, ranging from bacteria to farm animals. In addition, gene transfer techniques are being used to improve crop plants by transfer of herbicide resistance. In the near future, it may be possible to offer vaccination against infectious disease through food plants, making resistance to infectious agents almost universal.

Key Terms

adenosine deaminase (ADA), 405
allele-specific oligonucleotide (ASO), 402
amniocentesis, 400
attenuated vaccine, 414
bottom-up approach, 408
centimorgan (cM), 398
chorionic villus sampling (CVS), 400
cystic fibrosis, 404
cystic fibrosis transmembrane conductance regulator (CFTR), 402
DNA chips, 403
DNA fingerprinting, 406

enhancement gene therapy, 406
exclusion map, 399
familial hypercholesterolemia, 404
fusion polypeptide, 412
gene therapy, 404
germ-line gene therapy, 406
glyphosate, 414
Human Genome Project, 410
inactivated vaccine, 414
logarithm of the odds (lod) score, 398
minisatellite, 407
neurofibromin, 400
positional cloning, 400

restriction fragment length polymorphism (RFLP), 398
sequence tagged site (STS), 409
severe combined immunodeficiency (SCID), 404
shotgun approach, 409
somatic gene therapy, 406
subunit vaccine, 414
top-down approach, 409
transgenic, 414
variable-number tandem repeat (VNTR), 407
yeast artificial chromosome, (YAC), 409

GENETICS, TECHNOLOGY, AND SOCIETY

Beyond Dolly: The Cloning of Humans

On February 23, 1997, the world was stunned by the announcement in the journal *Nature* of the birth of a sheep named Dolly that was cloned from the cells of an adult sheep. This feat was accomplished by a group of embryologists headed by Ian Wilmut and Keith Campbell at the Roslin Institute in Scotland. Cloning sheep and other animals was not an end unto itself, they said, but rather a means of replicating animals that are genetically engineered to produce pharmaceutical products such as blood clotting factors or insulin. However, the reaction to the announcement had little to do with Wilmut and Campbell's intentions; rather, was based on fears of the cloning of humans, which has been variously termed "immoral," "repugnant," "morally despicable," and "ethically wrong." Within days, bills were introduced into the U.S. Congress to prohibit human cloning, and worldwide bans were called for.

The method Wilmut and Campbell used to create Dolly, a procedure called nuclear transfer, was first suggested by embryologist Hans Spemann in 1938. The idea is simply to replace the nucleus of a newly fertilized egg (or an unfertilized egg) with the nucleus of an adult cell, creating a "reconstructed" zygote. In theory, the genetic information of this donor nucleus is comparable to that it replaces. Therefore, it should be able to direct embryonic development. A "reconstructed" zygote should therefore develop into an exact genetic copy of the adult organism from which the donor cell nucleus was isolated.

Finding out whether this would work turned out not to be simple. It took many years for lab procedures to be perfected for collecting and manipulating fragile eggs, removing their genetic material, transferring in donor nuclei, and implanting the "reconstructed" cell into the uterus of a "surrogate mother." One by one, these technical obstacles were overcome. However, cloning by nuclear transfer never seemed to work when adult cells were used. Instead, the technique seemed to work only when embryonic cells were nuclear donors.

The possible explanation for the failure of adult cells to work was based on decades of research with many animal species. As cells differentiate during embryonic development, they begin to take on specialized functions related to chemical changes in the nucleus. In addition to "household genes" that are needed by all cells, a subset of genes corresponding to each function, perhaps 10 percent of the total number of genes, are switched on. The remainder are switched off, perhaps irreversibly. Unless this genetic program can be reversed, the nucleus of a differentiated cell will thus be incapable of directing development from the very beginning, when *all* the genes must be potentially active. Wilmut's group shocked not just the public but also their fellow biologists, by showing that the programming of the genome during development is not necessarily permanent, and that under just the right conditions, the molecular switches can be reset, making the nucleus embryonic once again.

The procedure used by Wilmut and his colleagues is a variation on a method proposed by Spemann. They isolated eggs (actually secondary oocytes) from a Poll Dorset ewe (the egg donor) and removed their nuclei with a fine pipette. Next they inserted an udder cell from a six-year-old Finn Dorset ewe (the nucleus donor) under the protective covering of the enucleated eggs. Applying an electrical pulse caused the membranes of the two cells to fuse, releasing the udder cell's nucleus into the cytoplasm of the enucleated egg. This electrical pulse also stimulated the "reconstructed" egg to begin to divide. After one week, enough time for three cell divisions, the 8-cell embryos were implanted into the wombs of "surrogate mother" Scottish blackface sheep. All told, 277 nuclear transfers were done using udder cells, resulting in 29 developing embryos that were implanted into 13 "surrogate mother" ewes. One pregnancy resulted, which culminated on July 5, 1996, in the birth of lamb 6LL3, renamed Dolly.

The reason why these researchers succeeded where so many before had failed probably had to do with the way they treated the udder cells before the transfer. These cells had been maintained in tissue culture growing in a chemically defined medium supplemented with fetal calf serum at 10 percent. Before fusion with the egg, the serum was reduced to 0.5 percent. Such serum starvation caused the udder cells to exit the cell cycle and enter the G0 stage, which is a resting stage during which cells remain metabolically active but no longer divide. Whatever chemical changes occur in the nucleus of G0 cells apparently make their genes amenable to reprogramming. The protein factors present in the cytoplasm of the egg cell thus are able to reset the molecular switches in the udder cell nucleus, making it capable of directing embryonic development.

The complete reprogramming of the udder cell nucleus takes time. In the case of sheep, however, gene transcription does not begin until after at least three cell divisions. There is thus a "window of opportunity" for reprogramming of the transferred nucleus lasting about a week; if all the switches are reset by the time transcription must start, the transferred nucleus can direct the rest of development, just like a genuine zygote nucleus.

Of course, the general public isn't particularly interested in the technical details of Dolly's creation. What they want to know is, if cloning from adult cells work in sheep, will it also work in humans? And more to the point, should the cloning of humans be allowed? The answer to the first question is simple: There is reason to believe that if cloning is possible in sheep, it can also be accomplished in humans. Most of the necessary technical procedures are already being used during *in vitro* fertilization (IVF). Whether or not somebody attempts it is the difficult question.

Some objections to human cloning have been based on safety concerns. Would a cloned human be healthy? Or would there be a higher risk of genetic disease and cancer due to the accumulation of mutations in the nucleus of the donor cell? If the donor cell is from an adult, would the clone show accelerated aging, possibly due to shortened telomeres at the ends of the donor cell's chromosomes? If the donor nucleus is not completely reprogrammed,

(continues)

(continued) would that lead eventually to abnormal development, either in the embryo or even during childhood?

The recent history of reproductive technology suggests that after some initial delays, research into human cloning will very likely move forward. Previous advances likewise caught us unprepared and were at first opposed and even outlawed. Examples include artificial insemination, IVF, and surrogate motherhood. As time passed and we became more accustomed to these technologies and familiar with their benefits, acceptance grew. But is cloning just the next step in reproductive biology, or is it something altogether new? Is the creation of a genetic copy of a person, regardless of the justification, simply wrong? Will the benefits of this technology ever outweigh the risks? This is what we must decide, and soon.

References

Kolata, G. 1998. *Clone: The road to Dolly and the path ahead.* New York: William Morrow.

Marshall, E. 1997. Mammalian cloning debate heats up. *Science* 275:1733.

Silver, L. M. 1997. *Remaking Eden: Cloning and beyond in a brave new world.* New York: Avon Books.

Wilmut, I., et al. 1997. Viable offspring derived from fatal and adult mammalian cells. *Nature.*

INSIGHTS *and* SOLUTIONS

1. DNA fingerprints have been used to test forensic specimens, to identify criminal suspects, and to settle immigration cases and paternity disputes. Probes for DNA fingerprinting can be derived from a single locus or multiple loci. Two multiple-loci probes have been widely employed in both criminal and civil cases and derive from minisatellite loci on chromosome 1 (1cen-q24) and chromosome 7 (7q31.3). These probes, which are widely used because they produce a highly individual fingerprint, have determined paternity in thousands of cases over the last few years.

(a) Shown below are the results of DNA fingerprinting of a mother (M), putative father (F), and child (C) using these probes.

alleged father. Based on this fingerprint, can you conclude that the man tested is the father of the child?

Solution: All bands present in the child can be assigned as coming from either the mother or the father. In other words, all the bands in the child's DNA fingerprint that are not maternal are present in the father. Because the father and the child share 11 bands in common and the child has no unassigned bands, paternity can be assigned with confidence. In fact, the chance that the man tested is not the father is on the order of 10^{-13}.

(b) In a second case, the fingerprints of the mother, the alleged father, and the child are shown below.

As shown, the child has 6 maternal bands, 11 paternal bands, and 5 bands shared between the mother and the

Legend: Maternal band / Paternal band / Band from either or both parents / Unassigned band / Band common to both parents / M Mother / C Child / F Alleged father

(continues)

(continued)

In this case, the child has 8 maternal bands, 15 paternal bands, 6 bands that are common to both the mother and the alleged father, and 1 band that is not present in either the mother or the alleged father. What are the possible explanations for the presence of this band? Based on your analysis of the band pattern, which explanation is most likely?

> **Solution:** In this case, one band in the child cannot be assigned to either parent. Two possible explanations for this finding are that the child is mutant for one band, or that the man tested is not the father. In estimating probability of paternity, the mean number of resolved bands (n) is determined, and the mean probability (x) that a band in individual A matches that in a second unrelated individual, B, is calculated. In this case, because the child and the father share 15 bands in common, the probability that the man tested is not the father is very low (probably 10^{-7} or lower). As a result, the most likely explanation is that the child is a mutant for a single band. In fact, in 1419 cases of genuine paternity resolved by the minisatellite probes on chromosomes 1 and 7, single-mutant bands in the children were recorded in 399 cases, accounting for 28 percent of all cases.

2. Infection by HIV-1 (human immunodeficiency virus) is responsible for destruction of cells in the immune system and results in the symptoms of AIDS (acquired immunodeficiency syndrome). HIV infects and kills cells of the immune system that carry a cell-surface receptor known as CD4. An HIV surface protein known as gp120 binds to the CD4 receptor and allows the virus entry into the cell. The gene that encodes the CD4 protein has been cloned. How can this clone be used along with recombinant DNA techniques to combat HIV infection?

> **Solution:** Several methods utilizing the CD4 gene are being explored to combat HIV infection. First, because infection depends on an interaction between the viral gp120 protein and the CD4 protein, the cloned CD4 gene has been modified to produce a soluble form of the protein (sCD4). The idea is that HIV can be prevented from infecting cells if the gp120 protein of the virus is bound up with the soluble form of the CD4 protein, and thus made unable to bind to CD4 proteins on the surface of immune system cells. Studies in cell culture systems indicate that the presence of sCD4 effectively prevents HIV infection of tissue culture cells. However, studies in HIV-positive humans have been somewhat disappointing, mainly because the strains of HIV used in the laboratory are different from those found in infected individuals.
>
> HIV-infected cells carry the viral gp120 protein on their surface. To kill such cells, a second approach involves fusing the CD4 gene with those that encode bacterial toxins. The resulting fusion protein contains the CD4 regions that bind to gp120 and the toxin regions that kill the infected cell. In tissue culture experiments, cells infected with HIV are killed by the fusion protein, whereas noninfected cells survive. It is hoped that the targeted delivery of drugs and toxins can be used in therapeutic applications to treat HIV infection.

Problems and Discussion Questions

1. In attempting to vaccinate against diseases by eating antigens, the antigen (such as the cholera toxin) must be presented to the cells of the small intestine. What are some potential problems in this method? Why don't absorbed food molecules stimulate the immune system and make you allergic to the food you eat?

2. Outline the steps involved in transferring glyphosate resistance to a crop plant. Do you envision that this trait could escape from the crop plant and make weeds glyphosate resistant? Why or why not?

3. What lessons from the *E. coli* genome project can be applied to the Human Genome Project?

4. Gene therapy for human genetic disorders involves transferring a copy of the normal human gene into a vector and using the vector to transfer the cloned human gene into target tissues. Presumably, the gene enters the target tissue, becomes active, and the gene product relieves the symptoms. Although gene therapy is now being used to treat several disorders, there are some unresolved problems with this method.

   (a) Why are disorders such as muscular dystrophy difficult to treat by gene therapy?

   (b) What are the potential problems in using retroviruses as vectors?

   (c) In the long run, should gene therapy involve germ tissue instead of somatic tissue?

   (d) What are some of the potential ethical problems associated with this approach?

5. In producing physical maps of markers and cloned sequences, what advantage does *Drosophila* offer that other organisms including humans do not?

6. Outline the steps in identifying a gene by positional cloning. What steps may cause difficulty in this process? Once a region on a chromosome has been identified as containing a given gene, what kinds of mutations would speed the process of identifying the locus?

7. The phenotype of many behavior traits such as bipolar disorder or schizophrenia may be controlled by several genes, each at a different locus. Can positional cloning be used to map and isolate such genes? What if a trait is controlled by six genes, each contributing equally in an additive way to the phenotype? Can positional cloning be used in this case? Why or why not?

8. Suppose that you develop a screening method for cystic fibrosis that allows you to identify the predominant mutation (Δ508) and the next six most prevalent mutations. What do you need to consider before using this method in screening the population at large for this disorder?

9. The DNA sequence surrounding the site of the sickle-cell mutation in β-globin is shown for normal and mutant genes:

5′ G A C T C C T G A G G A G A A G T 3′

3′ C T G A G G A C T C C T C T T C A 5′

Normal DNA

5′ G A C T C C T G T G G A G A A G T- 3′

3′ C T G A G G A C A C C T C T T C A- 5′

Sickle-cell DNA

Each type of DNA is denatured into single strands and applied to a filter. The paper containing the two spots is hybridized to an ASO of the following sequence: 5′-GACTCCTGAGGAGAAGT-3′. Which (if either) spot will hybridize to this probe? Why?

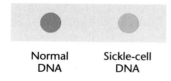

Normal DNA   Sickle-cell DNA

10. Dominant mutations can be categorized according to whether they increase or decrease overall activity of a gene or gene product. Although a "loss of function" mutation (that is, a mutation that inactivates the gene product) is usually recessive, for some genes one dose of the gene product is not sufficient to produce a normal phenotype. In this case, a loss-of-function mutation in the gene will be dominant and the gene is said to be *haploinsufficient*. A second category of dominant mutations are "gain of function" mutations that result in an increased activity or expression of the gene or gene product. The phenotype of such a mutation results from too much gene product. The gene therapy technique currently used in clinical trials involves the "addition" to somatic cells of a normal copy of a gene. In other words, a normal copy of the gene is inserted into the genome of the mutant somatic cell, but the mutated copy of the gene is not removed or replaced. Will this strategy work for either of these two types of dominant mutations?

11. One form of hemophilia, an X-linked disorder of blood clotting, is caused by mutation in clotting factor VIII. Many single-nucleotide mutations of this gene have been described, making detection of mutant genes by Southern blots inefficient. There is, however, an RFLP for the enzyme HindIII contained in an intron of the factor VIII gene that can often be used in screening, as shown below.

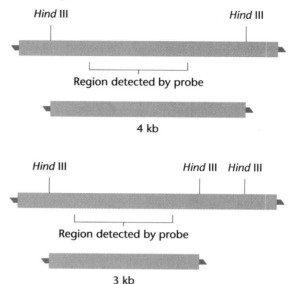

A female whose brother has hemophilia is at 50 percent risk of being a carrier of this disorder. To test her status, DNA is obtained from white blood cells from family members, cut with HindIII, and the fragments probed and visualized by Southern blotting. The results are shown below.

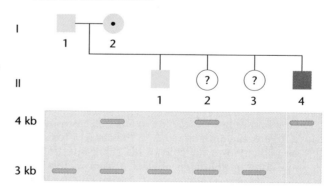

Determine whether any of the females in generation II are carriers for hemophilia.

12. The human insulin gene contains a number of introns. Despite the fact that bacterial cells will not excise introns from mRNA, how can a gene like this be cloned into a bacterial cell and produce insulin?

13. In mice transfected with the rabbit β-globin gene, the rabbit gene is active in a number of tissues including spleen, brain, and kidney. In addition, some mice suffer from thalassemia, caused by an imbalance in the coordinate production of α- and β-globins. What problems associated with gene therapy are illustrated by these findings?

Selected Readings

Amos, B., and Pemberton, J. 1993. DNA fingerprinting in non-human populations. *Curr. Opin. Genet. Dev.* 2:857–60.

Barker, D., et al. 1987. Gene for von Recklinghausen neurofibromatosis is in the pericentromeric region of chromosome 17. *Science* 236:1001–02.

Cawthon, R. M., et al. 1990. A major segment of the neurofibromatosis type 1 gene: cDNA sequence, genomic structure, and point mutations. *Cell* 62:193–201.

Chang, J. C., and Kan, Y. W. 1981. Antenatal diagnosis of sickle-cell anemia by direct analysis of the sickle mutation. *Lancet* 2:1127–29.

Collins, F. S. 1995. Ahead of schedule and under budget: The Genome Project passes its fifth birthday. *Proc. Natl. Acad. Sci. USA* 92:10821–23.

Crystal, R. G. 1995. Transfer of genes to humans: early lessons and obstacles to success. *Science* 270:404–10.

Danna, K., and Nathans, D. 1971. Specific cleavage of simian virus 40 DNA by restriction endonuclease of *Hemophilus influenzae. Proc. Natl. Acad. Sci. USA* 68:2913–17.

Davidson, B., et al. 1993. A model system for *in vivo* gene transfer into the central nervous system using an adenoviral vector. *Nature Genet.* 3:219–23.

Dietrich, W. F., et al. 1995. Mapping the mouse genome: current status and future prospects. *Proc. Natl. Acad. Sci. USA* 92:10849–53.

Dujon, B. 1996. The yeast genome project: What did we learn? *Trends Genet.* 12:263–70.

Fleischmann, R. D., et al. 1995. Whole genome random sequencing and assembly of *Haemophilus influenzae* Rd. *Science* 269:496–512.

Flotte, T. 1993. Prospects for virus-based gene therapy for cystic fibrosis. *J. Bioenerg. Biomembr.* 25:37–42.

Fraser, C. M., et al. 1995. The minimal gene complement of *Mycoplasma genitalium. Science* 270:397–403.

Goodman, H. M., Ecker, J. R., and Dean, C. 1995. The genome of *Arabadopsis thaliana. Proc. Natl. Acad. Sci. USA* 92:10831–35.

Guyer, M. and Collins, F. S. 1995. How is the Human Genome Project doing, and what have we learned so far? *Proc. Natl. Acad. Sci. USA* 92:10841–48.

Hartl, D., and Lozovskaya, E. 1992. The *Drosophila* genome project: Current status of the physical map. *Comp. Biochem. Physiol. [B]* 103:1–8.

Hudson, T. J., et al. 1995. An STS-based map of the human genome. *Science* 270:1945–54.

Inbal, A., et al. 1993. Identification of three candidate mutations causing type IIA von Willebrand disease using a rapid, nonradioactive, allele-specific hybridization method. *Blood* 82:830–36.

Krumlauff, R., Jeanpierre, M., and Young, B. D. 1982. Construction and characterization of genomic libraries from specific human chromosomes. *Proc. Natl. Acad. Sci. USA* 79:2971–75.

Lewontin, R., and Hartl, D. 1991. Population genetics in forensic DNA typing. *Science* 254:1745–50.

Maniatis, T., Fritsch, E. F., and Sambrook, J. 1988. *Molecular cloning: A laboratory manual,* 2nd ed. Cold Spring Harbor, NY: Cold Spring Harbor Laboratory Press.

Marshall, E. 1995. Gene therapy's growing pains. *Science* 269:1050–55.

———, and Pennisi, E. 1996. NIH launches the final push to sequence the genome. *Science* 272:188–89.

———, and Pennisi, E. 1998. Hubris and the human genome. *Science* 280:994–995.

Mason, H., Lam, D., and Arntzen, C. 1992. Expression of hepatitis B surface antigen in transgenic plants. *Proc. Natl. Acad. Sci. USA* 89:11745–49.

Moldoveanu, Z., et al. 1993. Oral immunization with influenza virus in biodegradable microspheres. *J. Infect. Dis.* 167:84–90.

Mullis, K. B. 1990. The unusual origin of the polymerase chain reaction. *Sci. Am.* (Apr.) 262:56–65.

Nahreini, P., et al. 1993. Versatile adeno-associated virus 2-based vectors for constructing recombinant virions. *Gene* 124:257–62.

Reiss, J., and Cooper, D. N. 1990. Application of the polymerase chain reaction to the diagnosis of human genetic disease. *Hum. Genet.* 85:1–8.

Saiki, R. K., et al. 1986. Analysis of enzymatically amplified beta-globin and HLA-DQ alpha DNA with allele-specific oligonucleotide probes. *Nature* 324:163–66.

Tacket, C. O., Mason, H. S., Losonsky, G., Clements, J. D., Levine, M. M., and Arntzen, C. J. 1998. Immunogenicity in humans of a recombinant bacterial antigen delivered in a transgenic potato. *Nature Medicine* 4:607–609.

Travis, J. 1998. Another human genome project. *Science News* 153:334–335.

Villa-Komaroff, L., et al. 1978. A bacterial clone synthesizing proinsulin. *Proc. Natl. Acad. Sci. USA* 75:3727–31.

Waterson, R., and Sulston, J. 1995. The genome of *Caenorhabditis elegans. Proc. Natl. Acad. Sci.* 92:10836–40.

Williams, J. G. K., et al. 1990. DNA polymorphisms amplified by arbitrary primers are useful as genetic markers. *Nucleic Acids Res.* 18:6531–35.

Wu, C.-L., and Melton, D. W. 1993. Production of a model for Lesch-Nyhan syndrome in hypoxanthine phosphoribosyltransferase-deficient mice. *Nature Genet.* 3:235–39.

Zuo, J., et al. 1993. Construction of cosmid contigs and high-resolution restriction mapping of the Huntington disease region of human chromosome 4. *Hum. Mol. Genet.* 2:889–99.

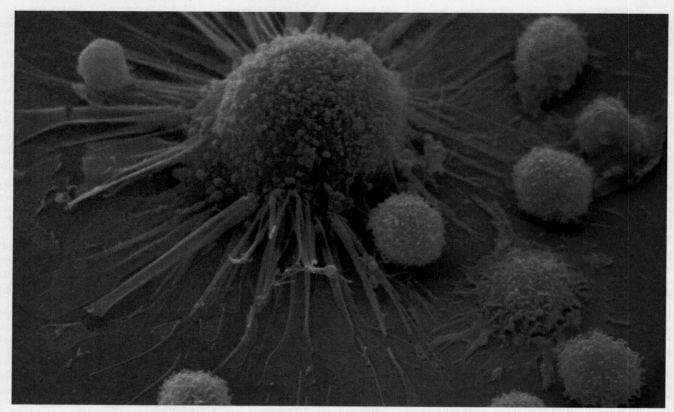

A large cancer cell, with cellular processes extended, has been detected and surrounded by an array of killer cells.

CHAPTER OUTLINE

CHAPTER
19

Genetics of Cancer and Immunology

Chapter Concepts

Cancer is now recognized as a genetic disorder at the cellular level that involves the mutation of a small number of genes. Many of these genes normally act to suppress or stimulate progress through the cell cycle, and loss or inactivation of these genes causes uncontrolled cell division and tumor formation. The immune system of vertebrates is a genetically controlled mechanism that defends the body against disease-causing organisms. Extensive genetic recombination in somatic cells reshuffles a limited number of genes into new combinations that produce literally millions of different antibodies.

Although often viewed as a single disease, cancer is actually a complex group of diseases that affect a wide range of cells and tissues. A genetic link to cancer was first proposed early in the twentieth century, and this idea has served as one of the foundations of cancer research. Mutations that alter the genome or gene expression are now regarded as a common feature of all cancers. In some cases, such mutations are part of the germ line and are inherited. More often, mutations arise in somatic cells and are not passed on to future generations through the germ cells. Sometimes, the inherited mutation must be accompanied by a somatic mutation at the homologous locus, creating homozygosity. Whichever the case, *cancer is now considered a genetic disorder at the cellular level*.

Familial forms of cancer have been known for over 200 years. In many of these cases, no well-defined pattern of inheritance can be established. In a small number of cases, a Mendelian pattern of dominant or recessive inheritance can be established, indicating the hereditary nature of the cancer.

Because mutation is the underlying cause of cancer, there will always be a baseline rate of cancer, because there is a background rate of spontaneous mutation. Over and above this baseline rate, environmental agents that promote mutation also play a role in the development of cancer. Environmental factors such as ionizing radiation, chemicals, and viruses are cancer-causing agents, and almost all of these act by generating mutations. Given that mutations play a role in cancer, this chapter will explore answers to a series of questions about how mutations initiate the changes that convert normal cells into malignant tumors, which mutant genes are most likely to result in cancer, and how many mutations are required to cause cancer.

Genes and Cancer

Genetic studies of several different cancers have identified a small number of genes that must be mutated in order to bring about the development of cancer or maintain the growth of malignant cells. It is clear that the two main properties of cancer, uncontrolled cell division and the ability to spread or metastasize, are the result of genetic alterations. These alterations can involve large-scale mutational events such as chromosome loss, rearrangement, or the insertion of foreign (often viral) DNA sequences into loci on human chromosomes. Smaller-scale alterations, such as changes in nucleotide sequence or more subtle modifications that alter only the amount of a gene product that is present or the time over which the gene product is active, may also be involved. Two interrelated questions arise from considering cancer as a genetic disorder at the cellular level: (1) Are there mutant alleles that predispose an organism or specific cell types to cancer, and (2) if so, how many mutational events are necessary to cause cancer?

Genes That Predispose to Cancer

Mutations in single genes do indeed predispose cells to becoming malignant. Many studies have documented families with high frequencies of certain types of cancers, such as breast, colon, or kidney cancers. In most cases, however, it is difficult to identify a clear, simple pattern of inheritance. One example of such a predisposition is the inheritance of **retinoblastoma (RB)**, a cancer of the retinal cells of the eye. Retinoblastoma occurs with a frequency ranging from 1 in 14,000 to 1 in 20,000, and most often it appears between the ages of 1 and 3 years. Two forms of retinoblastoma are known. One is a predisposition to retinoblastoma (about 40 percent of all cases) inherited as an autosomal dominant trait, though the disorder itself is recessive, as will be explained below. Because the trait is dominant, those who inherit the mutant RB allele are predisposed to develop eye tumors, and in fact, 90 percent of these individuals will develop retinal tumors, usually in both eyes. In addition, those family members who inherit the mutant allele are predisposed to develop other forms of cancer, such as osteosarcoma, a bone cancer, even if they do not develop retinoblastoma.

A second form of retinoblastoma is also known, amounting to 60 percent of all cases. This form is not familial, and tumors develop spontaneously. The sporadic form is characterized by the formation of tumors in only one eye, and onset occurs at a much later age than in the familial form.

Other types of dominant predispositions to cancer are also known (Table 19.1). These include **Wilms tumor (WT)**, a cancer of the kidney inherited as an autosomal dominant condition, and **Li–Fraumeni syndrome**, a rare autosomal dominant condition that predisposes to a number of different cancers, including breast cancer.

How Many Mutations Are Needed to Cause Cancer?

The study of genes that predispose an individual to cancer has allowed scientists to estimate the number and sequence of mutational events necessary to trigger different forms of cancer. By studying the two different types of retinoblastoma, Alfred Knudson and his colleagues developed a model that requires the presence of two mutant copies of the RB gene in the same retinal cell for tumor formation to occur (*i.e.*, the disorder is expressed as a recessive trait). In the familial predisposition, one mutant RB allele is inherited and is carried by all cells of the body, including the retinal cells (Figure 19–1). If the normal allele of the RB gene becomes mutated in such a retinal cell, formation of retinal tumors will result. Individuals carrying an inherited single mutant of the RB gene are predisposed to develop RB because only one additional mutational event is re-

TABLE 19.1	Dominantly inherited predispositions to tumors
Tumor Predisposition Gene	**Chromosome**
Early onset familial breast cancer	17q
Familial adenomatous polyposis	5q
Familial melanoma	9p
Gorlin syndrome	9q
Hereditary nonpolyposis colon cancer	2p
Li–Fraumeni syndrome	17p
Multiple endocrine neoplasia, type 1	11q
Multiple endocrine neoplasia, type 2	22q
Neurofibromatous, type 1	17q
Neurofibromatous, type 2	22q
Retinoblastoma	13q
von Hippel–Lindau syndrome	3p
Wilms tumor	11p

quired to cause tumor formation. This does not happen in all cases, since 10 percent of those who inherit a mutant RB gene do not develop retinal cancer. This model explains how retinoblastoma can be inherited as a recessive trait (both alleles of the RB gene must be inactivated by mutation to express the mutant phenotype), and how carrying one mutant allele acts as a dominant trait in predisposing an individual to cancer (only one additional mutation is needed).

In sporadic cases, both copies of the RB gene must undergo mutation in the same retinal cell in order for tumor formation to occur. As might be expected, these events are less frequent, and occur at a later age, so sporadic forms of retinoblastoma are more likely to occur in a single eye, and at a later age than with familial forms.

Similar studies on inherited predispositions to other cancers have led to the conclusion that the number of mutations necessary for the development of cancer ranges from as few as 2 to perhaps as many as 15-20 (Table 19.2).

Tumor Suppressor Genes

In general, mitosis can be regulated in two ways: (1) by genes that normally function to suppress cell division, and (2) by genes that normally function to promote cell division. The first group, called **tumor suppressor genes**, inactivates or represses passage through the cell cycle and the resulting cell division. These genes and/or their gene products must be absent or inactive for cell division to take place. If these genes become permanently inactivated or deleted through mutation, control over cell division is lost, and the mutant cell begins to proliferate in an uncontrolled fashion.

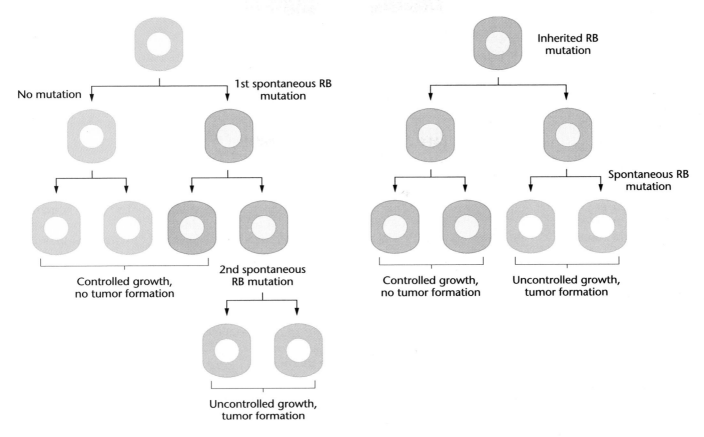

■ Figure 19–1   In spontaneous cases of retinoblastoma (left), two mutations in the retinoblastoma gene are acquired in a single cell, causing uncontrolled cell growth and division, resulting in tumor formation. In familial cases of retinoblastoma (at right), one mutation is inherited and present in all cells. A second mutation at the retinoblastoma locus in any retinal cell will result in uncontrolled cell growth and tumor formation.

TABLE 19.2	Number of mutations associated with some cancers	
Cancer	*Chromosome Sites*	*Minimum Number of Mutations Required*
Retinoblastoma	13q	2
Wilms tumor	11p	2
Colon cancer	5q, 12p, 17p, 18q	5–7
Small-cell lung cancer	3p, 11p, 13q, 17p	10–15

Genes belonging to the second class, called **proto-oncogenes**, normally function to promote cell division. These genes can be "on" or "off," and when they are "on" they promote cell division. To halt cell division, these genes and/or their gene products must be inactivated. If these genes become permanently switched on, then uncontrolled cell division occurs, leading to tumor formation. The mutant forms of proto-oncogenes are known as **oncogenes**.

We shall consider some examples of how mutations in tumor suppressor genes can lead to a loss of control over cell division and the development of cancer. Following that discussion, we will consider proto-oncogenes and oncogenes.

Retinoblastoma (RB)

As described previously, RB is a tumor of the retinal layer of the eye. The RB gene is located on chromosome 13 and encodes a protein (pRb) 928 amino acids long, which is confined to the nucleus. The pRb gene product is present in all cell and tissue types examined to date and is found in both resting cells and cells active in the cell cycle. Because the gene is expressed at all times, regulation occurs by reversible chemical modifications to the RB gene product and not by transcriptional control. The activity of the pRb protein is regulated by alternately adding or removing phosphate groups. The pRb is phosphorylated in S phase and at the G2/M junction of the cell cycle, but is not phosphorylated in the G0 and G1 stages of the cycle.

The observation that pRb has phosphate groups added and removed in an alternating fashion, and that this pattern of modification occurs in synchrony with phases of the cell cycle, suggests that pRb may be part of a cell cycle control point in G1. In cells at G0 or G1, when pRb is not phosphorylated, the protein is active and stops passage through the cell cycle. Just after the cell enters S phase, pRb is chemically modified by the additon of phosphate groups and becomes inactive. When pRb activity is suppressed, cells pass through S phase and the subsequent events leading to mitosis. Direct evidence for the role of pRb in controlling passage through the cell cycle (at the G1/S control point) has come from experiments in which normal copies of the RB gene have been introduced into cultured tumor cells. Osteosarcoma cells carry two mutated RB genes and do not make pRb. When a normal RB gene is transferred into cultured osteosarcoma cells, pRb is synthesized and cell division stops.

Breast Cancer Genes

A small proportion of all cases of breast cancer, especially those in younger women, is related to the inheritance of dominant genes that confer a predisposition to the disease. Mutations in *BRCA1*, a gene that maps to the long arm of chromosome 17, are associated with a predisposition to breast cancer, and this predisposition is inherited as an autosomal dominant trait. About 90 percent of women with a mutant *BRCA1* gene will develop breast cancer; these women also have an increased risk of ovarian cancer.

A second breast cancer gene, *BRCA2*, which maps to the long arm of chromosome 13, is also inherited as an autosomal dominant predisposition to breast cancer, but is not associated with an increased risk of ovarian cancer. Together, these genes account for a large majority of all inherited cases of breast cancer, but have little or no role in sporadic cases of breast and ovarian cancer, which make up 80 percent of all cases.

Both the *BRCA1* and the *BRCA2* genes are expressed in rapidly dividing cells, with expression at its highest at the G1/S boundary in the cell cycle. This common pattern of expression, the similar phenotype of the mutant alleles (breast cancer), and experimental evidence from knockout mice indicates that both genes may work in the same pathway. Both the BRCA1 and BRCA2 proteins bind to another protein, RAD51, which is involved in the repair of breaks in the double-stranded DNA, which may explain their normal function as tumor suppressor genes.

The role of BRCA1 and BRCA2 gene products in DNA repair may also explain why they are not associated with sporadic cases of breast cancer. In sporadic cases, at least three mutations would have to accumulate in a breast cell for it to become cancerous: Both copies of *BRCA1* or *BRCA2* would need to be mutant, and at least one other mutation in a cell cycle regulatory gene would have to occur. On the other hand, those who inherit a mutant copy of *BRCA1* or *BRCA2* already have one mutation, and need only two more to trigger breast cancer.

The p53 Gene and the Cell Cycle

The *p53* gene encodes a nuclear protein that acts as a transcription factor. Normally, *p53* is a tumor suppressor gene that controls passage of the cell from G1 into S phase of the cell cycle. Mutations of *p53* are found in a wide range of cancers, including breast, lung, bladder, and colon cancers. It is estimated that over half of all cancers are associated with mutations in the *p53* gene, suggesting that it controls a key event in cell proliferation and is not involved in a cell- or tissue-specific form of regulation. Inherited mutations in *p53* are also associated with the Li–Fraumeni syndrome, an autosomal dominant trait associated with a predisposition to develop cancers in a number of tissues at a high frequency.

Normal cells contain low levels of p53 protein, but levels rise dramatically after irradiation of cells by ultraviolet (UV) light. Irradiated cells temporarily arrest in G1 to allow repair of DNA damaged by UV light. Cells that have no functional p53 protein are unable to arrest in G1 following irradiation, and they move immediately from G1 into S. These cells do not repair DNA damage caused by UV; as a result, they have a high rate of mutation. These studies have led to the conclusion that *p53* controls passage through the cell cycle to ensure that DNA damage is repaired before the cell enters S phase. In this role, *p53* is often referred to as a guardian of the genome.

The central role of the *p53* gene in controlling the cell cycle emphasizes the relationship between cancer and the cell cycle outlined in the chapter introduction. It also highlights the relationship between genes that regulate cell growth and cancer.

Oncogenes

Proto-oncogenes normally function to turn on cell growth and division. In mutant form, as **oncogenes**, they induce or maintain uncontrolled cellular proliferation associated with cancer. The existence of specific genes associated with the transformation of normal cells into cancerous cells was inferred by Peyton Rous in 1910–1911. Using cells from a tumor of chickens known as a **sarcoma**, Rous demonstrated that injection of cell-free extracts from these tumors could induce sarcoma formation in healthy chickens. He postulated the existence of a "filterable agent" that was responsible for transmitting the disease.

Decades later, other investigators showed that his filterable agent is a virus, now known as the **Rous sarcoma virus (RSV)**. The RNA genome of these viruses is first transcribed by the enzyme **reverse transcriptase**, into a single-stranded DNA molecule. The single-stranded DNA is converted to double-stranded DNA and then integrated into the genome of the infected cell, forming a **provirus**. At a later time, the viral DNA can be transcribed to produce RNA, which is packaged into viral proteins to form new virus particles. Because RSV and related viruses "reverse" the normal flow of genetic information, they are called **retroviruses**.

In the case of RSV, the tumor-forming ability results from a single gene (called the *src* gene) acquired from the host genome. This gene, responsible for the induction of sarcoma formation in chickens, is designated as an oncogene, and retroviruses that carry oncogenes are known as **acute transforming viruses**. Other retroviruses that can cause tumor formation do not carry oncogenes, but are able to induce the activity of cellular genes that brings about tumor formation. They are designated as **nonacute** or **nondefective** viruses.

Origins of Oncogenes

Oncogenes carried by acute transforming viruses are acquired from the host's genome during the process of infection, when a portion of the viral genome and host genome are exchanged (Figure 19–2). The oncogenes (*onc*) carried by retroviruses are called *v-onc*, and the normal cellular version of the gene is called a *c-onc* or proto-oncogene. Retroviruses carrying a *v-onc* gene are able to infect and transform a specific type of host cell. For RSV, the oncogene captured from the chicken genome is called *v-src*, and it confers the ability to transform chicken cells after infection. The normal, cellular version of the same gene, found in the chicken genome, is called *c-src*. More than 20 oncogenes have

been identified by their presence in retroviral genomes, and over 50 oncogenes have been identified overall. Some of these are listed in Table 19.3.

Proto-Oncogenes and Oncogenes

At least three mechanisms can explain the conversion of proto-oncogenes into oncogenes. These mechanisms include **point mutations**, **translocations**, and **overexpression**. Although some of these are mediated by viruses, others are generated solely by intracellular events in the absence of retroviruses (Table 19.4).

The *ras* gene family, which encodes a protein of 189 amino acids involved in the transduction of signals across the plasma membrane, illustrates how a single nucleotide change can convert a *c-onc* sequence into the oncogenic version of the gene. The normal *ras* protein functions as a molecular switch, alternating between an "on" and an "off" state (Figure 19–3). Oncogenic mutations of the *ras* gene encode a protein that becomes stuck in the "on" position, stimulating cell growth. In some tumors, this is the result of a somatic mutation, and is not mediated by a retrovirus. Comparison of the amino acid sequence of ras proteins from a number of different human carcinomas reveals that all mutant *ras* gene products have a single amino acid substitution at position 12 or 61 that is caused by a single nucleotide change in the *ras* gene (Figure 19–4).

At least three separate mechanisms of proto-oncogene activation are associated with overexpression. First, the *c-onc* may acquire a new promoter, causing an increase in the level of transcript production or activating a silent locus. This is the case in avian leukosis, where strong viral promoters become integrated upstream from a proto-oncogene, causing an increase in mRNA production and the amount of the gene product. A second mechanism of overexpression involves acquisition of new upstream regulatory sequences including enhancers. The third mechanism involves amplification of the proto-oncogene. In

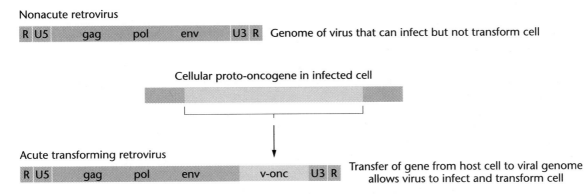

■ Figure 19–2   A transforming retrovirus has acquired a copy of a gene from the host genome, converting it from a *c-onc* or proto-oncogene into an oncogene that confers on the virus the ability to transform a specific type of host cell into a cancerous cell.

TABLE 19.3	Oncogenes		
c-onc	*Origin*	*Species*	*Human Chromosome Location*
	A. Cellular Oncogenes with Viral Equivalents		
src	Rous sarcoma virus	Chicken	20
fos	FBJ osteosarcoma virus	Mouse	2
sis	Simian sarcoma virus	Monkey	22
fes	ST feline virus	Cat	15
abf	Abelson murine leukemia virus	Mouse	9
erbB	Avian erythrobeastosis virus	Chicken	7
Ha-ras-I	Harvey murine sarcoma virus	Mouse	11
myc	Avian MC29 myelocytomatosis virus	Chicken	8
	B. Oncogenes Without Viral Equivalents		
N-ras	Neuroblastoma, leukemia	Human	1
N-myc	Neuroblastoma	Human	12
neu	Neuroglioblastoma	Human, rat	17
man	Mammary carcinoma	Human, mouse	2

TABLE 19.4	Conversion of proto-oncogenes to oncogenes
Mechanism	*Example*
Point mutation	*ras*
Translocation	*abl*
Overexpression of gene product	
New promotor by viral insertion	*mos, myh*
New enhancer by viral insertion	*myc*
Amplification of proto-oncogene	*myc*

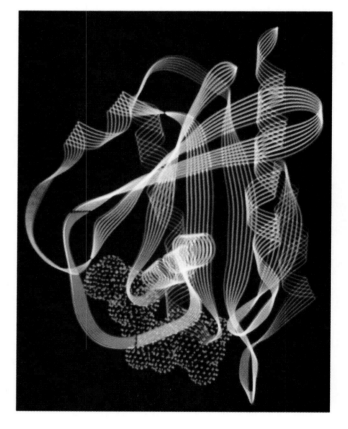

human tumors, members of the *myc* family of oncogenes are frequently amplified up to several hundred times.

The protein products of proto-oncogenes are found in, and are associated with, the plasma membrane, cytoplasm, and nucleus (Table 19.5). Although their specific functions vary widely, all products characterized to date alter gene expression in a direct or an indirect manner.

A Genetic Model for Colon Cancer

Even in the small number of cases where it has been studied in detail, it is clear that cancer is a multistep process resulting from a number of specific genetic alterations. Although the studies of tumors such as retinoblastoma and Wilms tumor have established that a small number of steps is required to transform a normal cell into a malignant one, they are of limited value in dissecting the order of the mutational events leading to tumor formation. For these questions, the study of colon cancer offers several advantages. First, malignant tumors of the colon and rectum develop from preexisting benign tumors, and tumors at all stages of development are available for study. Moreover, two forms of

■ Figure 19–3 A three-dimensional computer-generated image of ras proteins in two different conformations. Normal ras proteins act as molecular switches controlling cell growth and differentiation. The switch is "on" when GTP binds to the protein, and "off" when the GTP is hydrolyzed to GDP. Switching the protein between states alters the conformation of the protein in the two regions shown in blue and orange. Oncogenic mutations of *ras* make proteins stuck in the "on" state, continuously signaling for cell growth.

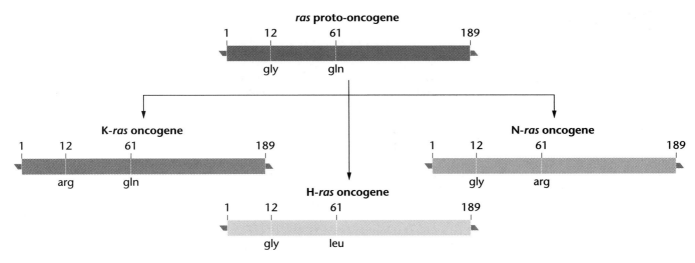

■ Figure 19–4   The *ras* proto-oncogene encodes a protein of 189 amino acids. In the normal protein, glycine is encoded at position 12, and glutamine is encoded at position 61. Analysis of *ras* oncogene proteins from several tumors shows a single amino acid substitution at one of these positions. A single amino acid substitution resulting from a single base change can convert a proto-oncogene into a tumor-promoting oncogene.

TABLE 19.5	Cellular location of *c-onc* and *v-onc* proteins	
Gene	*Location of c-onc Protein*	*Location of v-onc Protein*
src	Membranes	Membranes
ras	Membranes	Membranes
myc	Nucleus	Nucleus
fps	Cytoplasm	Cytoplasm and membranes
abl	Nucleus	Cytoplasm
erbB	Plasma membrane	Plasma membrane and Golgi

genetic predisposition to colon cancer are known: an autosomal dominant trait, known as **familial adenomatous polyposis (FAP)**, and a genetically complex trait known as **hereditary nonpolyposis colorectal cancer (HNPCC)**. In addition, there are spontaneous forms of colon cancer. With this array of causes, it is possible to study the interaction of genetic and environmental factors in the genesis of tumors.

The Steps in Colon Cancer

Through an analysis of mutations in tumors at various stages, the number and nature of the genetic steps involved in changing normal intestinal epithelial cells into tumor cells has been defined. This has led to the development of a genetic model for colon cancer. The model is shown in Figure 19–5. The first feature of this model to be noted is that multiple mutations are required. Mutations in five to seven specific genes are needed to bring about malignant growth. If fewer changes are present, benign growths or intermediate stages of tumor formation result. Second, based on the

analysis of many tumors, the order of mutations usually follows a sequence indicated in the figure, and suggests that both the accumulation of mutations and the order in which they occur are important in the development of colon cancer.

Mutation in the *APC* gene on the long arm of chromosome 5 gene converts normal epithelium into benign tumors called adenomas. An individual with FAP inherits a single *APC* mutation, and develops hundreds or thousands of benign adenomas in the colon and rectum. In spontaneous cases, the evidence suggests that the initial mutational event occurs in a single cell and that the resulting adenoma consists of a clone of cells, all carrying the mutation. Mutation in the second allele is not necessary for proliferation and adenoma formation, suggesting that the *APC* gene can be classified as a tumor suppressor gene. If other mutations accumulate in adenomas as shown in Figure 19–5, they will form cancerous tumors. The relative order of subsequent mutations is shown in Figure 19–5. A mutation in one copy of the K-*ras* oncogene is enough to cause progression from an early adenoma to an intermediate ade-

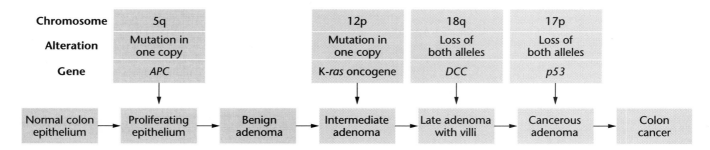

■ Figure 19–5 A model for the stepwise production of colon cancer. The first step is the loss or inactivation of one allele of the APC gene on chromosome 5. In familial cases, one mutation of APC is inherited, producing the dominantly inherited syndrome of familial adenomatous polyposis (FAP), associated with the formation of thousands of benign adenomas. Subsequent mutations involving genes on chromosomes 12, 18, and 17 in cells of the benign adenomas can lead to a malignant transformation resulting in colon cancer. A mutation in one allele of the K-*ras* gene causes progression of adenomas to the next stage. Both copies of the genes on chromosome 17 and 18 must be lost for further progression to colon cancer. Chromosome 18 contains several genes that may be involved in colon cancer. The figure shows one of these, the *DCC* gene.

noma, a benign tumor with a number of fingerlike villous outgrowths. The other genes shown in the figure are all tumor suppressor genes, and *both* alleles of these genes must be mutated for tumor progression. The 18q21 region contains a number of tumor suppressor genes involved in colon cancer, including *DCC*, *DPC4*, and *JV18-1*. Mutation in at least one of these genes is required for the formation of late adenomas. Finally, mutations involving the loss or inactivation of the *p53* gene (on chromosome 17p) causes the transition to a cancerous cell. Mutations in the *p53* gene are pivotal to the development of a number of cancers, including lung, brain, and breast cancers. Recall that *p53* is a tumor suppressor gene that normally functions to regulate the passage of cells from the late G1 to the S phase. The *p53* gene product has DNA-binding properties, and mutations in *p53* may alter DNA binding to confer a new function on the gene product.

Mutations and Colon Cancer

The process of metastasis occurs after the formation of cancerous cells and involves an unknown number of mutational steps, although work on other tumors suggests that one or two mutations may be sufficient to allow tumor cells to detach and spread to secondary sites.

The genetic model for colon cancer involves mutations in proto-oncogenes, tumor suppressor genes, and disruption of the cell cycle at a specific transition point. In cases where there is a familial predisposition to colon cancer, the first mutation is inherited, and the rest occur as a result of the action of environmental factors including amounts of dietary fiber and fat. This multistep model appears to have applications to other forms of cancer, but many questions remain unanswered, including the normal functions of some of the genes involved and the molecular mechanisms by which the

mutated genes and/or their gene products bring about tumor formation.

Cancer and Genomic Instability

Most cancers require mutations in two or more genes that accumulate over a period of time. If, as in colon cancer or retinoblastoma, one of these mutations is inherited, fewer mutations are needed to initiate tumor formation. The result is a genetic predisposition to cancer. But how do the multiple mutations necessary for cancer formation accumulate? Recent work on another inherited form of colon cancer has provided evidence that mutations in DNA repair genes may be important in destabilizing the genome, allowing mutations to accumulate.

Hereditary nonpolyposis colorectal cancer (HNPCC) has been mapped by linkage analysis to loci at 2p16 and 3p21. This form of cancer may be one of the most common human hereditary disorders, affecting about 1 in 200 individuals. Mutations at these two loci generate a cascade of mutations in short, tandemly repeated microsatellite sequences located throughout the genome.

The gene on chromosome 2 associated with HNPCC is a DNA repair gene called *hMSH2*. Inactivation of this gene causes a rapid accumulation of mutations and also causes the development of colorectal cancer. A second DNA repair gene associated with HNPCC has been mapped to chromosome 3. Mutations in this gene, called *hMLH1*, are also associated with the development of HNPCC. At least two other DNA repair genes related to *hMLH1* and *hMLH2* have been identified, and remain to be mapped. Mutations in any of these repair genes promote genome-wide genetic instability, accelerating the rate at which mutations accumulate, with colon cancer as one of the outcomes. It may turn out that mutations in any one of these genes is enough to cause HNPCC.

The discovery of these DNA repair genes and their association with cancer will make it possible to screen individuals with a family history of colon cancer to identify those who have inherited a mutant allele of *hMLH1* or *hMLH2* and who are thus at high risk for colon cancer.

Gatekeeper Genes and Caretaker Genes in Cancer

The two pathways to colon cancer in FAP and HNPCC offer an insight into the nature of genes that control cancer predisposition. In FAP, an inherited or acquired mutation in the *APC* gene causes the formation of thousands of adenomas. These adenomas progress slowly to a malignant condition by the accumulation of mutations in other genes (K-*ras*, *p53*, etc). Because there are thousands of such adenomas, there is a high risk that at least one will progress to form a malignant tumor. In HNPCC, adenomas form at a slow rate during the aging process. However, mutations accumulate at a magnitude two to three times faster than in normal cells, and the accumulation of mutations in genes such as K-*ras* and *p53* leads to malignancy.

These differences have led to the idea that mutations in two different types of genes cause predisposition to cancer: **gatekeeper genes** and **caretaker genes**. The *FAP* gene is a gatekeeper, inhibiting cell growth or promoting cell death. In general, tumor suppressor genes are gatekeepers. In different cell types, only one or a few genes serve as gatekeepers, and if both copies of the gatekeeper are mutated, a specific cancer, such as retinoblastoma or colon cancer, develops. Individuals predisposed to a specific form of cancer inherit one mutant copy of such a gatekeeper, and need only one additional mutation to initiate tumor formation. In spontaneous cases, both copies of the gatekeeper must be mutated for cancer to develop.

Caretaker genes are those that maintain the integrity of the genome, e.g., as DNA repair genes. These genes normally function to repair damage to DNA caused by environmental agents such as ultraviolet light, or mismatches that occur during DNA replication. Mutation of a caretaker gene does not promote tumor formation directly, but leads to genetic instability that increases the mutation rate of all genes, including gatekeeper genes. If a gatekeeper gene is mutated and begins tumor formation, the process is accelerated by the high rate of mutation in the cell.

Genomic Changes and Cancer

Alterations in chromosome structure and/or number are associated with many forms of cancer. In most cases, the relationship between changes in chromosome number or structure and the development of cancer is not clear. For example, individuals with Down syndrome carry an extra copy of chromosome 21. This quantitative alteration in genetic content is also associated with a 20-fold increased risk of leukemia as compared to the general population. In a limited number of cases in which specific alterations in chromosome structure are associated with cancer, more direct information is available to indicate how structural rearrangements are associated with the development and/or maintenance of the cancerous state.

Chromosome Rearrangements and Leukemias

In selected cases where specific chromosome rearrangements are associated with cancer, the relationship between the chromosomal aberration and the development and/or maintenance of the cancerous state is known.

The connection between chromosome aberrations and cancer is most clearly seen in leukemias, where the presence of specific chromosome changes is well defined and often diagnostic (Table 19.6). One of the best-studied examples of the association between a chromosome rearrangement and the development of cancer is the translocation between chromosomes 9 and 22 that is associated with **chronic myelogenous leukemia (CML)** (Figure 19–6). Originally, this translocation was described as an abnormal chromosome 21 and called the **Philadelphia chromosome**. Later, it was shown by Janet Rowley that the Philadelphia chromosome results from the exchange of genetic material between chromosomes 9 and 22. This translocation is never seen in a normal cell and is found only in the white blood cells involved in CML. These observations indicate that the translocation is a primary and causal event in the generation of CML, and that this form of cancer may originate from a single cell bearing this translocated chromosome.

TABLE 19.6	Specific chromosome aberrations and cancer	
Cancer	***Chromosome Alteration***
Chronic myelogenous leukemia	t(9;22)
Acute promyelocytic leukemia	t(15;17)
Acute lymphocytic leukemia	t(4;11)
Acute myelogenous leukemia	t(8;21)
Prostate cancer	del(10q)
Synovial cancer	t(X;18)
Testicular cancer	i(12p)
Retinoblastoma	del(13q)
Wilms tumor	del(11p)

t = translation, del = deletion, i = inversion

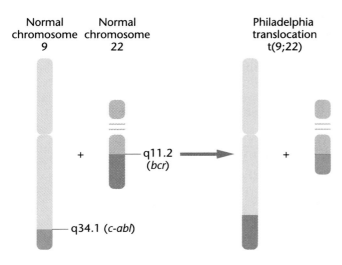

Normal chromosome 9   Normal chromosome 22   Philadelphia translocation t(9;22)

q11.2 (*bcr*)

q34.1 (*c-abl*)

■ Figure 19–6   A reciprocal translocation involving the long arms of chromosomes 9 and 22 results in the production of a characteristic chromosome called the Philadelphia chromosome, which is associated with cases of chronic myelogenousleukemia.

Translocations and Hybrid Genes

By examining a large number of cases involving the Philadelphia chromosome, it was possible to establish the exact location of the break points on chromosomes 9 and 22 (Figure 19–6). Genetic mapping studies using recombinant DNA techniques established that the proto-oncogene *c-abl* maps to the breakpoint region on chromosome 9, and the gene *bcr* maps near the break-

point on chromosome 22. In the translocation event, most or all of the *c-abl* gene is moved to a region within the *bcr* gene. This generates a hybrid *bcr/c-abl* gene that is transcriptionally active and produces a hybrid 200-kDa protein product (Figure 19–7) that has been implicated in the generation of CML.

Other genes at translocation breakpoints associated with other forms of leukemia have also been isolated and characterized. In these cases, and in CML, the translocation results in the formation of a hybrid gene that is transcribed to form a gene product that causes the cell to undergo a malignant transformation even though a second and presumably normal copy of the gene is present and active. Identification and characterization of these gene products is important so that therapeutic strategies can be developed to inactivate the abnormal gene product, or provide enough normal gene product to prevent tumor formation.

The Immune System

Genomic alterations are one of the hallmarks of cancer. These changes include chromosomal translocations, aneuploidy resulting from mitotic recombination or aberrant segregation, or from mutations in specific genes that promote genomic instability in DNA sequences throughout the genome. Genomic alterations are also hallmarks of *normal* events in the maturation and function of the immune system. These alterations include rounds of recombination-associated deletion of DNA sequences in so-

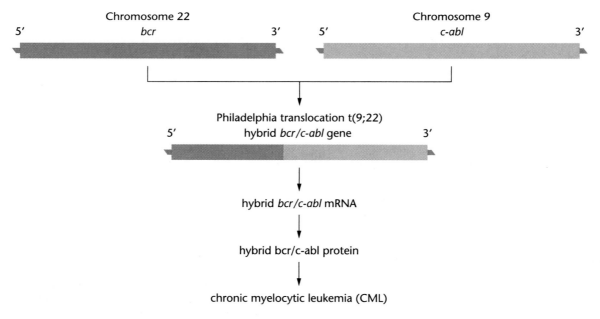

Chromosome 22   5′   *bcr*   3′   5′   Chromosome 9   *c-abl*   3′

Philadelphia translocation t(9;22)
5′   hybrid *bcr/c-abl* gene   3′

hybrid *bcr/c-abl* mRNA

hybrid bcr/c-abl protein

chronic myelocytic leukemia (CML)

■ Figure 19–7   The t(9:22) translocation associated with chronic myelogenous leukemia (CML) results in the fusion of the *c-abl* oncogene on chromosome 9 with the *bcr* gene on chromosome 22. The normal c-abl protein is a kinase and the normal bcr protein activates a phosphorylation reaction. The fusion protein is a powerful hybrid that allows cells to escape control of the cell cycle, resulting in leukemia.

matic cells, and a lack of precision in cutting and joining segments during excision.

Instead of producing a pathological condition, genomic alterations associated with the immune system allow the body to produce several billion combinations of DNA sequences encoding antibodies and cell surface receptors. From a set of less than 300 genes, recombination events, deletions, and mismatching of rejoined segments within immature cells of the immune system create a vast immune potential.

The immune system is an effective barrier against the successful invasion of potentially harmful foreign substances. Invading viruses, bacteria, fungi, and parasites are recognized as nonself and subsequently sequestered, inactivated, and destroyed by the immune system. This immune response usually involves antibodies specific to foreign substances. Agents that elicit antibody production are termed **antigens**. **Antibodies** are proteins produced and secreted by specific cells of the immune system. Many different molecules can act as antigens, including proteins, polysaccharides and, rarely, other molecules such as nucleic acids. Usually, a distinctive structural feature of an antigen, called an **epitope**, stimulates antibody production. Invading antigens may be free molecules, or they may be part of the surface of a cell, microorganism, or virus. Whatever the case, organisms with an immune system have the capacity to make antibodies against any antigen they encounter.

Genetic Diversity in the Immune System

Among vertebrates, each individual can produce millions of different types of antibodies, each responding to a different antigen. In the following sections we will examine the molecular basis for diversity in antibodies, the genetic basis of organ and tissue transplantation, and disorders of the immune system, including AIDS.

Antibodies

In humans, antibodies are produced by plasma cells, a type of lymphocyte (white blood cell). Five classes of antibodies or **immunoglobulins (Ig)** are recognized: **IgG, IgA, IgM, IgD, and IgE** (Table 19.7). The first class, IgG, represents about 80 percent of the antibodies found in the blood and is the most extensively characterized class of antibodies. IgG is associated with immunological memory. The IgA class of immunoglobulins (found in breast milk) can be secreted across plasma membranes and is associated with immunological resistance to infections of the respiratory and digestive tracts. IgM antibodies are usually the first secreted by a plasma cell in response to an antigen, and they are associated with the early stages of the immune response. At this time, little is known about the role of the IgD class of immunoglobulins. They are associated with the surface of B cells, and may regulate their action. IgE

TABLE 19.7	Categories and components of immunoglobins			
Ig Class	**Light Chain**	**Heavy Chain**	**Tetramers**	
IgA		α	$\kappa_2\alpha_2$	$\lambda_2\alpha_2$
IgD		δ	$\kappa_2\delta_2$	$\lambda_2\delta_2$
IgE		ε	$\kappa_2\varepsilon_2$	$\lambda_2\varepsilon_2$
IgG	κ or λ	γ	$\kappa_2\gamma_2$	$\lambda_2\gamma_2$
IgM		μ	$\kappa_2\mu_2$	$\lambda_2\mu_2$

antibodies are involved with fighting parasitic infections. IgE antibodies are also associated with allergic responses.

A typical IgG molecule (Figure 19–8) consists of two different polypeptide chains, each present in two copies. The larger or **heavy chain (H)** contains approximately 440 amino acids. The sequence of the first 110 amino acids at the N-terminus differs among different heavy chains, and is known as the **variable region (V_H)**. The remaining C-terminal amino acids are the same in all H chains, and make up the **constant region (C_H)**. In humans, the H chains are encoded by genes on the long arm of chromosome 14.

The **light chain (L)** is composed of 220 amino acids, with the first 110 amino acids making up the variable region (V_L) (Figure 19–8). The rest of the amino acids at the C-terminus make up the constant region (C_L) of the light chain. There are two types of L chains: **kappa chains**, encoded by genes on human chromosome 2, and **lambda chains**, encoded by genes on the long arm of chromosome 22.

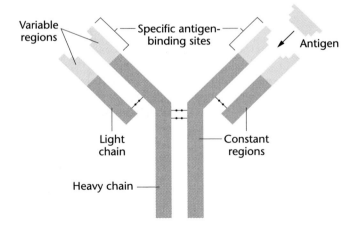

■ Figure 19–8 A typical antibody (immunoglobulin) molecule. The molecule is Y-shaped, and contains four polypeptide chains. The longer arms are heavy chains, and the shorter arms are light chains. Each chain contains a constant region and a variable region. The variable regions form an antibody combining site that interacts with a specific antigen, similar to a lock-and-key mechanism. The chains are joined together by disulfide bonds indicated as ←→ .

Two heavy and two light chains make up a functional IgG molecule, and these are held together by disulfide bonds. The variable regions of the heavy and light chains form the **antibody combining site**. The combining site of each antibody has a unique conformation that allows it to bind to a specific antigen, like a key in a lock.

Theories of Antibody Formation

One of the dominant features of the immune system is the generation of new cells that contain different combinations of antibodies. Because there are billions of such combinations, it is impossible that each combination is coded directly within the genome. There simply is not enough DNA in the human genome to encode tens or hundreds of millions of antibodies. One of the long-standing questions in immunogenetics is how this vast array of molecular variability in antibodies and antigen receptors is encoded in the genome.

The diversity of antibodies is the result of genetic recombination in the three clusters of antibody genes: the H-chain genes on chromosome 14, the kappa L genes on chromosome 2, and the lambda L genes on chromosome 22. At each locus, there is a limited number of genes present, encoding various portions of the H and L chains. As antibody-forming B cells mature, DNA recombination rearranges these genes so that each mature B lymphocyte comes to encode, synthesize, and secrete only one specific type of antibody. Each mature B cell can make one type of light chain (kappa or lambda) and one type of heavy chain. When an antigen is present, it stimulates the B cell, which encodes an antibody against that antigen, to divide and differentiate. As a result, populations of differentiated plasma cells are produced, all of which synthesize one type of antibody that interacts with the antigen.

Evidence for genetic recombination in antibody genes was first provided by Susumu Tonegawa and his colleagues in the mid-1970s. By comparing the size of restriction fragments, these researchers demonstrated that DNA segments coding for parts of the light-chain gene are far away from each other in embryonic cells, but are adjacent to one another in antibody-producing cells. In their experiments, they isolated and characterized a cloned DNA fragment specifying a lambda V chain derived from embryonic cells. The organization of this cloned fragment was compared with the equivalent region isolated from an antibody-producing cell. In the embryo DNA, the variable region of an L-chain gene was separated by 4.5 kb of DNA from the region encoding another part of the L chain (Figure 19–9). However, in the DNA isolated from the antibody-producing cell, these regions were joined together to form a single transcription unit encoding a specific L chain. This observation strongly supports the hypothesis that the diverse array of immunoglobulin genes found in an-

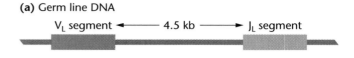

(a) Germ line DNA

V$_L$ segment ◄——— 4.5 kb ———► J$_L$ segment

(b) DNA from antibody-producing cell

V$_L$ segment   J$_L$ segment

■ Figure 19–9   (a) In DNA isolated and cloned from embryonic DNA, the V region of an L gene was located 4.5 kb away from the DNA encoding the J region. (b) In DNA isolated from an antibody-producing cell, the V and J regions were contiguous, forming a single transcription unit. This evidence provided support for the hypothesis that recombination is involved in forming antibody genes in mature B cells.

tibody-producing cells are formed through a process of somatic recombination.

Organization of the Immunoglobulin Genes

The work of Tonegawa and his colleagues inspired a large-scale research effort to study the organization and rearrangement of the immunoglobulin genes in mice and humans. The coding sequences for kappa and lambda genes have now been isolated from germ line and B-cell DNA in a variety of higher organisms. Comparative studies indicate that the basic organizational plan and mechanism of formation is similar in most mammals. In humans, the kappa L genes (Figure 19–10) consist of several elements: **L-V (leader-variable)** regions, **J (joining)** regions, and a **C (constant)** region. There are 70 to 300 V-L segments in each light-chain gene, each with a different nucleotide sequence. The J region contains six different segments, and there is one C segment.

During maturation in a B cell, one of the 300 L-V regions is randomly joined by a recombination event to one of the six J genes to form a functional light chain gene (Figure 19–10). The remaining gene segments are excised and destroyed. This form of recombination is different from the reciprocal recombination occurring in meiosis. This newly created L-V-J region becomes an exon in a gene that includes the C region. This light-chain gene is transcribed and translated to form kappa L chains that become part of an antibody molecule. This rearranged gene is stable, and is passed on to all progeny of the B cell.

Diversity in kappa L-chain production is the result of combining any of the 300 V genes with any of the six J genes. The joining of one of 300 V genes with any of the six J segments generates about 1800 different kappa genes. The organization of the lambda L-chain genes is somewhat different from that of the kappa genes, but

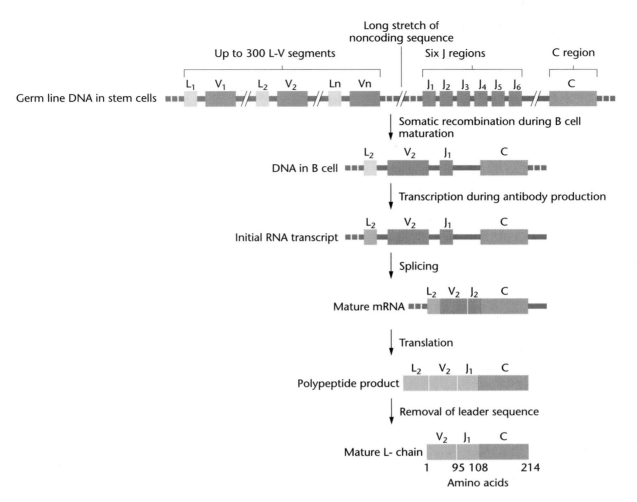

Figure 19–10 Formation of the DNA segments encoding a human kappa light chain and the subsequent transcription, mRNA splicing, and translation leading to the final polypeptide chain. In germ line DNA, up to 300 different L-V (Leader-Variable) segments are present. These are separated from the J regions by a long noncoding sequence. The J regions are separated from a single C gene by an intervening sequence (intron) that must be spliced out of the initial mRNA transcript. Following translation, the amino acid sequence derived from the leader RNA is cleaved off as the mature polypeptide chain passes across the cell membrane.

recombination events also generate about 1800 different lambda genes.

The heavy-chain genes in humans extend over a large region of DNA and include four types of segments: V (variable), D (diversity), J (joining), and C (constant) regions. There are approximately 300 different V genes, 10 to 50 different D genes, and four different J genes. In addition, there are five different C genes, one for each class of immunoglobulins (Figure 19–11). During B-cell maturation, recombination randomly joins a V region with one of the D sequences and one of the J sequences.

The V-D-J composite assembled by recombination lies adjacent to the C regions of the five classes of heavy chains. A V-D-J segment can be joined to any one of the C segments to form a gene for an IgM, IgD, IgG, IgE, or IgA antibody, respectively. This joining can occur in several ways. One is by transcription of a long pre-mRNA molecule that begins at the 5′-end of the V region and terminates at the 3′-end of the C segment. Splicing of this pre-mRNA yields mRNA molecules that can encode the same V-D-J sequence in combination with any of the five different C sequences. An alternative method involves a second round of recombination and excision in the H gene that places one of the C segments adjacent to the joined V-D-J segment and eliminates the other C regions (Figure 19–12).

The random recombination of H-gene components generates a large number of H-chain proteins. If we assume there are 25 D regions, then combining these with any of the 300 V regions and 4 J regions, 30,000 possible H chains can be generated. The potential for overall antibody diversity can be estimated by calculating the combination of all heavy-chain genes (30,000) with all light-chain genes (3600), resulting in 108 million possi-

Germ-line DNA (300 L-V segments, 10–50 D segments, 4 J segments, 5 C segments)

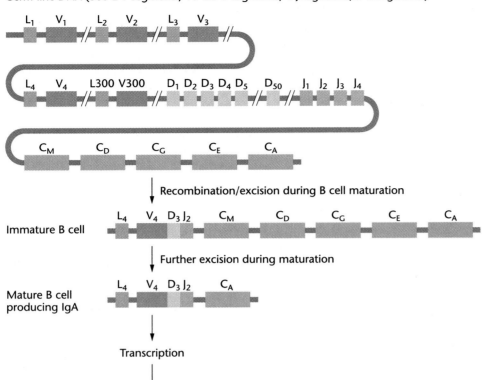

Figure 19–11 The variable region of the H chain is assembled by joining three different DNA segments together: L-V, D, and J. In embryonic DNA, these segments occur in clusters separated by long intervals of other DNA sequences adjacent to the C region. In a maturing B cell, a random combination of one L-V region with one of the 20 D regions and one of the four J regions produces an H-chain gene that is transcribed and translated in the B cell. In other B cells, different combinations of H-chain segments are joined together, producing a large number of different H chains.

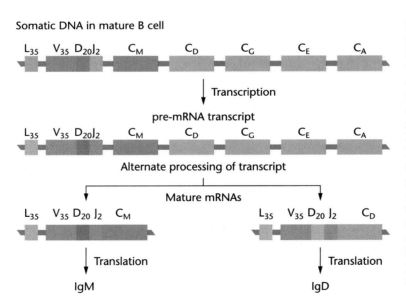

Figure 19–12 An alternative method of producing heavy chains with the same variable regions but different constant regions involves alternative splicing of heavy-chain transcripts during B-cell maturation. In the somatic cell DNA, the rearranged (L-V)-D-J segment is adjacent to all five heavy-chain constant regions. Transcription and alternate processing of the transcripts lead to the production of IgM and IgD containing different constant regions but the same variable region. This phenomenon is called class switching.

ble antibody genes (30,000 × 3600) from a few hundred coding sequences.

The HLA System

Organ transplants and skin grafts between unrelated individuals are usually rejected within a few weeks. Second attempts are rejected in a matter of days. The speed of the second rejection indicates that the immune sys-

tem is involved in accepting or rejecting grafts and transplants. The interaction of genetically encoded cell-surface antigens of the donor with the immune system of the recipient determines whether grafts will be accepted or rejected. In laboratory mice, these antigens, known as **histocompatibility antigens**, are encoded by 20 to 40 genes, many of which have a large number of alleles, producing an enormous number of genotypic combinations. In humans, a group of closely linked genes on chromosome 6, known as the **HLA (human leukocyte antigen) complex**, plays a critical role in histocompatible transplants.

HLA Genes

The HLA complex in humans consists of four closely linked major genes: HLA-A, HLA-B, HLA-C, and HLA-D (Figure 19–13). HLA-D is subdivided into the HLA-DR, HLA-DQ, and HLA-DP genes.

A large number of alleles has been identified in each of the HLA genes; it is one of the most highly polymorphic gene systems in the human genome. There are at least 23 alleles in HLA-A, 47 in HLA-B, 8 in HLA-C, 14 alleles in HLA-DR, 3 in HLA-DQ, and 6 in HLA-DP, making literally millions of genotypic combinations possible. Each of these alleles encodes a distinct antigen identified by a letter and a number. For example, A6 is allele number 6 at the HLA-A locus, B2 is allele 2 at the HLA-B locus, etc.

The genes of the HLA system are closely linked and inherited in a codominant fashion. This linkage means that recombination in this region is a rare event, and that the allelic combination on a single chromosome tends to be inherited as a unit. The array of HLA alleles on a given copy of chromosome 6 is known as a **haplotype**. Since humans have two copies of chromosome 6, we each have two HLA haplotypes (Figure 19–14). The codominant pattern of inheritance means that each haplotype is fully and completely expressed.

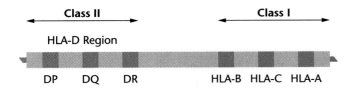

■ Figure 19–13 The organization of the HLA region on human chromosome 6.

The inheritance of HLA haplotypes is shown in Figure 19–14. Because of the large number of alleles that are possible, it is rare that anyone will be homozygous at any of these loci. In the example shown, the unrelated parents have completely different haplotypes, and each child receives one haplotype from each parent, carried by the copy of chromosome 6, which is incorporated into the parental sperm or egg. The result is four new haplotype combinations possible in the offspring. Siblings, therefore, have a one-in-four chance of sharing the same combination of haplotypes.

Organ and Tissue Transplantation

Successful transplantation of organs and tissues depends to a large extent on matching HLA haplotypes between donor and recipient. Because of the number of HLA alleles, the best chance for a match is between siblings or close relatives. Identical twins always have a perfect match. Parents have only one haplotype in common with a child, whereas siblings have a one-in-four chance of a perfect match. Thus, the order of preference for organ and tissue donors among relatives is: identical twin > sibling > parent > unrelated donor. Among unrelated donors and recipients, the chances for a successful match are between 1 in 100,000 and 1 in 200,000. Because HLA allele frequency differs widely between racial and ethnic groups (for example, B27 is found in 8 percent of American whites, but only 4 percent of American blacks), matches between groups is often difficult.

When HLA types are matched, the survival of transplanted organs is improved dramatically. Figure 19–15 shows the four-year survival rates for matched and unmatched kidney transplants. In HLA-matched transplants, over 90 percent of the transplanted kidneys survived the four-year period, but in unmatched transplants, less than 50 percent of the kidneys were functional after four years. The major causes of rejection are mismatching the HLA and/or ABO alleles. Other

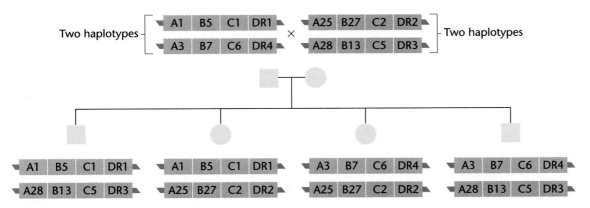

■ Figure 19–14 Transmission of HLA haplotypes. The cluster of HLA alleles on a single chromosome 6 is called a haplotype, each containing four major loci, encoding a different cell-surface antigen. The arrangement of HLA haplotypes and their distribution to the offspring result in new combinations of haplotypes in each generation.

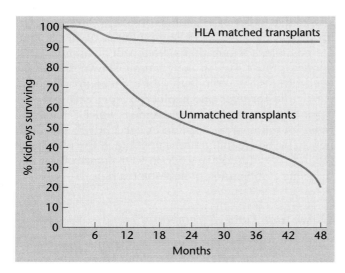

■ Figure 19–15 The survival of kidney transplants when they are HLA matched surpasses that of unmatched transplants.

causes are more subtle and often difficult to define. For example, if the recipient has had blood transfusions, memory cells against HLA antigens might be present. If these antibodies act against HLA antigens on the grafted tissue, rejection is more likely.

At present, drugs are used to improve the survival of transplants even when HLA matching is somewhat imperfect. The most widely used drug is **cyclosporine**, first isolated from a soil fungus. This drug selectively inactivates the T-cell subpopulation that is most active in tissue rejection, while not impairing other cells of the immune system. Other drugs to suppress the immune rejection of transplants are now under development, and if successful they may reduce the need for precise HLA matching for transplants, making it possible for more individuals to receive needed organs without prolonged waits for a proper donor.

Disorders of the Immune System

Genetically determined immunodeficiency disorders are caused by mutations that inactivate or destroy some component of the immune system. Most often, these disorders cause deficiencies in the production and functioning of one or more of the cell types in the immune system (Table 19.8). Other disorders of the immune system are due to factors that include infections, cancer chemotherapy, and developmental errors.

The Genetics of Immunodeficiency

A genetic disorder in which B cells are missing is **X-linked agammaglobulinemia**, or Bruton disease. Affected individuals are almost always male, with an age of onset somewhere between 6 and 12 months. Both B cells and plasma cells are completely absent or immature B cells are present, but they never become functional. The result is a complete lack of circulating antibodies and an inability to make antibodies. Affected individuals are highly susceptible to infections from microorganisms such as bacteria and fungi. However, these same individuals have normal levels of T cells and retain the part of the immune system that confers immunity to most viral infections. Currently, the most effective method of treatment is a bone marrow transplant to provide a stem-cell population that can give rise to functional B cells.

A genetically controlled form of **T-cell immunodeficiency** has been described, although the number of reported cases is very small. Affected individuals have a deficiency of the enzyme nucleoside phosphorylase, part of a biochemical pathway that salvages nucleotides. The number of B and T cells in these individuals is normal at birth, but thereafter a gradual decrease occurs in the number of T cells. The number of B cells is not affected, and antibody-mediated immunity is maintained. With a T-cell deficiency, there is an increased frequency of viral infections and a high risk of certain forms of cancer. It has been suggested that the decline in T cells is caused by the buildup of purines as a result of the enzyme deficiency and that the resulting toxicity kills T cells but does not affect B cells. The structural gene for nucleoside phosphorylase maps to the long arm of human chromosome 14, and the condition is inherited as an autosomal dominant trait.

In **severe combined immunodeficiency syndrome (SCID)**, T- and B-cell populations are absent or nonfunctional, and affected individuals have neither antibody-mediated nor cell-mediated immunity. As a result, they are susceptible to recurring and severe bacterial, viral, and fungal infections. There are autosomal and X-linked forms of SCID, indicating that this disorder is genetically heterogenous. In the X-linked form (XSCID), there are persistent infections beginning about 6 months after birth, with no T cells present.

TABLE 19.8	Cell types affected in immune disorders	
Immune Disorder	**Affected Cell Type**
X-linked agammaglobulinemia	B cells missing
Nucleoside phosphorylase deficiency	T cells reduced after birth
Severe combined immunodeficiency	T and/or B cells missing or nonfunctional
DiGeorge syndrome	T cells missing
Acquired immunodeficiency syndrome	T cells decline after infection

There are normal or even elevated levels of B cells present, but there is no antibody production by these cells. The longest-surviving individual with this form of SCID was a boy named David, who was born in Texas and isolated from the outside world by being placed in germ-free isolation (see Figure 18–11). David died at age 12 of complications following a bone marrow transplant.

In cases of autosomally inherited SCID associated with a deficiency of the enzyme adenosine deaminase (ADA), there are no T cells present, and B cells (if present) are not functional. Again, the result is a complete lack of cell-mediated and antibody-mediated immunity, and severe, recurring infections lead to death, usually by the age of 7 months. Children affected with ADA-deficient SCID are currently receiving gene therapy to provide them with a normal copy of the gene. This treatment represents the first use of gene therapy in humans, a technique that will undoubtedly become widely used as a treatment for certain genetic disorders in the future. The recombinant DNA techniques used in gene therapy were discussed in Chapter 18.

Acquired Immunodeficiency Syndrome (AIDS)

AIDS is a collection of disorders that develop as a result of infection with a retrovirus known as the **human immunodeficiency virus (HIV)**. The HIV virus consists of a protein coat, enclosing an RNA molecule that serves as the genetic material, and an enzyme, reverse transcriptase. The entire viral particle is enclosed in a lipid coat derived from the plasma membrane of a T cell (Figure 19–16). The virus selectively infects a subset of the T lymphocytes, known as T4 cells. Once inside the cell, the RNA is transcribed into a DNA molecule by reverse transcriptase, and the viral DNA is inserted into a human chromosome, where it can remain for months or years.

At a later time, when the infected T cell is called upon to participate in an immune response, the cellular and viral DNAs are transcribed. The viral RNA transcript is translated into viral proteins, and new viral particles are formed (Figure 19–17). These bud off the surface of the

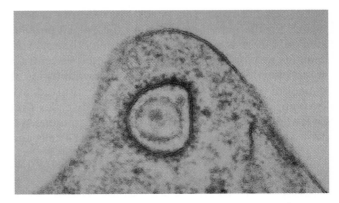

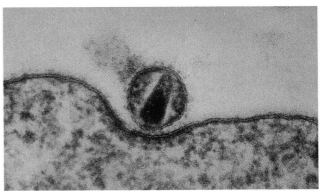

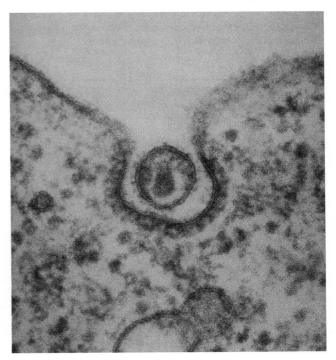

■ Figure 19–17 Release of new HIV particles from an infected cell. (a) The virus approaches the cell surface. (b) The virus buds off from the cell surface. (c) The virus, free from the cell surface, can now infect another T cell. Each round of replication releases several dozen new virus particles, gradually decreasing the number of T4 cells and reducing the body's ability to mount an immune response.

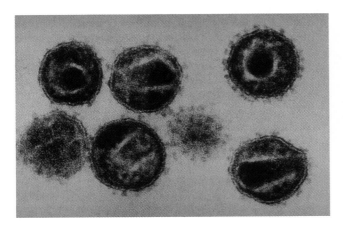

■ Figure 19–16 Transmission electron micrograph of the human immunodeficiency virus (HIV).

T cell, rupturing and killing the cell and setting off a new round of T-cell infection. Gradually, over the course of HIV infection, there is a decrease in the number of helper T4 cells. These cells act as the master "on" switch for the immune system. As the T4 cell population falls, there is a decrease in the ability to mount an immune response. The results are increased susceptibility to infection and increased risk of certain forms of cancer. Eventually, the outcome is premature death brought about by any of a number of diseases that overwhelm the body and its compromised immune system. The relationship between the loss of T4 cells and the progression of HIV infection is strong and can be used to monitor the status of infected individuals (Figure 19–18).

HIV is transmitted through body fluids from infected individuals to noninfected individuals; these fluids include blood, semen, vaginal secretions, and breast milk. The virus is not viable for more than 1 or 2 hours outside the body and cannot be transmitted by food, water, or casual contact. At this time, treatment options are rather limited, although a combination of drugs is effective at slowing or stopping reproduction of HIV in infected individuals.

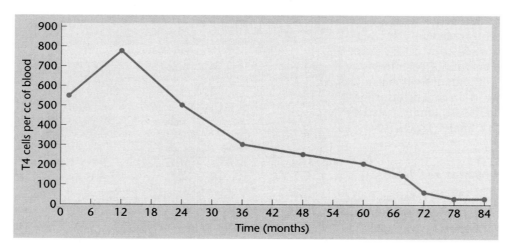

■ Figure 19–18   T4-cell levels during the course of an HIV infection. Following HIV infection, the number of T4 cells rises, then begins a slow decline. When levels fall below 200, the clinical symptoms of AIDS usually appear.

Chapter Summary

1. Cancer can be defined as a genetic disorder at the cellular level. Although some forms of cancer show familial patterns of inheritance, few show clear evidence of Mendelian inheritance.

2. Although Mendelian patterns of cancer development are not clear-cut, there are genes that predispose to cancer. Study of these genes, such as the gene for retinoblastoma, has provided insight into the number and sequence of mutations that result in the development of tumors.

3. Tumor suppressor genes normally act to suppress cell division. When these genes are mutated or have altered expression, control over cell division is lost. Molecular analysis of retinoblastoma indicates that this gene works to control gene expression.

4. Oncogenes are genes that normally function to maintain cell division, and these genes must be mutated or inactivated to halt mitosis. If these genes escape control and become permanently switched on, cell division occurs in an uncontrolled fashion.

5. The cells of most tumors have visible chromosomal alterations, and the study of these aberrations has provided some insight into the steps involved in the development of cancer. The relationship between cancer and chromosome alterations has been well studied in leukemias, in which the formation of hybrid genes or substitution of regulatory sequences is associated with the transformation of normal cells into malignant tumors.

6. One of the most complex families of genes specifying proteins are those that participate in the immune response. Human antibodies or immunoglobulins are characterized by amino acid sequence diversity. DNA recombination during maturation of B cells produces the genetic variation required to match millions of different antigens with corresponding antibodies.

7. A class of cell-surface antigens is produced by a complex of genes on human chromosome 6 called the HLA complex. With a large number of alleles, the combination carried by an individual as haplotypes constitutes a genetic signature for each individual. Certain HLA alleles are associated with specific diseases and can be used to diagnose and predict risk factors for these diseases.

8. Mutations that disrupt the immune system work by altering the development and/or function of the compo-

nent cells that participate in the immune response. These mutations can affect one cell type, a subsystem, or the entire immune system. In addition to genetic forms of disruption, the immune system can be inactivated by external factors, most notably by infection with the human immunodeficiency virus (HIV). This virus selectively infects and destroys the T cells that act as the "on" switch for the immune system, producing a gradual loss in the ability of an infected individual to mount an immune response. As a result, infectious diseases that would ordinarily be suppressed by the immune system can pose a serious risk to the lives of HIV-infected individuals.

GENETICS, TECHNOLOGY, AND SOCIETY

The Double-Edged Sword of Genetic Testing: The Case of Breast Cancer

As a result of the Human Genome Project, the sequence of all 75,000 or so genes residing on each set of 23 chromosomes will be known, among which will be thousands that potentially contribute to disease. This is expected to revolutionize medicine, as it will open the door to treating genetic diseases at their source by replacing or even correcting defective genes. However, the lag between the identification of a gene and the development of a new treatment is likely to be quite lengthy, measured in decades rather than years. In the meantime, it is already becoming possible to test people to see whether they carry mutant alleles that impart a high risk of developing disease. In some cases, those with a genetic predisposition to disease then have the opportunity to take preventive measures to lower their risk. Although genetic testing has the potential to save or extend lives, at the same time it may raise difficult questions for which the answers are not always clear. To illustrate how genetic testing may create troubling choices, consider recent advances in the genetics of breast cancer.

Breast cancer is the most common type of cancer among women and the second leading cause of cancer deaths (after lung cancer). Each year, more than 180,000 new cases are diagnosed and about 43,000 women die of breast cancer. The likelihood of developing breast cancer is low before age 35, but the risk rises after that. If a woman lives a long life, her risk of breast cancer is about 12 percent.

For many years it has been known that 5 to 10 percent of cases of breast cancer are familial, meaning that they are clustered in families in which generation after generation of women are stricken. The familial form of breast cancer also tends to develop earlier than spontaneous occurrences, often in a woman's 30s or 40s, and is more often bilateral (involving both breasts) and multifocal (involving several separate tumors).

The clustering of breast cancer in certain families suggested that genetic mutations are involved in the development of these cases of breast cancer. The study of such families subsequently led to the identification of two genes that, when mutated, impart an increased susceptibility to breast cancer. These genes, named *BRCA1* and *BRCA2*, were isolated in 1994 and 1995, respectively. Since then several other genes have been found that affect the risk of breast cancer to a lesser degree.

Mutations in *BRCA1* and *BRCA2* result in a greatly increased likelihood of developing breast cancer over the lifetime risk in the general population. The exact risk is somewhere between 56 and 85 percent, depending on the specific mutation and the extent of the family history of breast cancer. Mutations in *BRCA1* and *BRCA2* are responsible for about two-thirds of all familial breast cancer. In addition, *BRCA1* mutations also result in a 10-fold increased risk of ovarian cancer.

The role of *BRCA1* and *BRCA2* in the origin of cancer is indirect. The protein products of both normal genes appear to be involved in the repair of damaged DNA. They thus help prevent the accumulation of mutations in somatic cells, including those that cause the cell to lose control over the cell cycle and become cancerous.

BRCA1 also behaves as a tumor suppressor gene. Why mutations in these two genes increase the likelihood of specifically cancer in the breast (and in the case of *BRCA1* the ovary) is not yet clearly understood, but it may be related to the physiological responses of these tissues to hormones.

The identification of breast cancer susceptibility genes was hailed as a breakthrough in the understanding of this disease at the molecular level, an understanding that would pave the way for new therapies and better preventive measures. But unraveling the exact chain of events from a mutated *BRCA* gene to runaway cell division will take many years. In the short term, women in high-risk families now have the option of being tested to see whether they carry a mutant allele of *BRCA1* or *BRCA2*. If so, they face a lifetime risk of breast cancer as high as 85 percent; if not, they still have a 12 percent risk.

There are some very good reasons why a woman in a high-risk family would want to be tested. Foremost among them is that a negative test result gives tremendous peace of mind, allowing a woman to live her life out from under a cloud of fear. She will also be secure in the knowledge that she will not pass along the mutant allele to her children. It's the consequence of a positive test result that makes the decision whether to be tested so difficult. For women with *BRCA1* or *BRCA2* mutations, there are two alternative courses of action, neither very appealing. The conservative approach is close medical surveillance, including frequent breast self-exams, clinical exams, and mammographies, with the goal of identifying a tumor at its earliest stages, when it

(continues)

(continued)

is most treatable. The more radical strategy is the surgical removal of the breasts (bilateral prophylactic mastectomy) and, in the case of a mutation in *BRCA1*, possibly the ovaries. Unfortunately, there currently seems to be no middle ground.

Recent studies have suggested that women in high-risk families who undergo a bilateral prophylactic mastectomy reduce their breast cancer risk by about 90 percent. Even so, it is unclear whether the personal and psychological costs of such extreme surgery are outweighed by the increased life expectancy. There are other fears, as well. A positive test result, some believe, may lead to discrimination in employment and to higher insurance premiums or loss of coverage altogether.

Issues raised in genetic testing for breast cancer susceptibility are likely to be played out again and again with other genes predisposing one to other cancers. Already, genes that increase the risk of cancers of the prostate, colon, skin, and lung are in hand, are more will soon follow. In addition, several genes have been identified that increase the risk of other late-onset disorders, including Alzheimer disease, Parkinson disease, and heart disease. This list is sure to grow with each passing year.

We are now on the verge of having the ability to predict diseases based on genotype, many years before they might occur. From our experience with testing for breast cancer susceptibility, we must learn how to extend the use of this new technology by gaining the wisdom to know when to test and when not to test.

References

Kahn, P. 1996. Coming to grips with genes and risk. *Science* 274:496–98.

Marx, J. 1997. Possible function found for breast cancer genes. *Science* 276:531–32.

Miki, Y., et al. 1994. A strong candidate for the breast and ovarian cancer susceptibility gene *BRCA1*. *Science* 266:66–71.

Schrag, D., Kuntz, K. M., Garber, J.E., and Weeks, J.C. 1997. Decision analysis—Effects of prophylactic mastectomy and oophorectomy on life expectancy among women with *BRCA1* or *BRCA2* mutations. *New Engl. J. Med.* 336:1465–71.

Wopster, R., et al. 1995. Identification of the breast cancer susceptibility gene *BRCA2*. *Nature* 378:789–92.

Key Terms

adenoma, 430

AIDS, 439

antibodies, 433

antibody combining site, 434

antibody-mediated immunity, 433

antigen receptor, 433

antigens, 433

B cells, 438

chronic myelogenous leukemia (CML), 431

constant region, 433

familial adenomatous polyposis (FAP), 439

haplotype, 437

heavy chain (H), 433

hereditary nonpolyposis colorectal cancer (HNPCC), 429

histocompatibility antigens, 436

immunoglobulins: IgG, IgA, IgM, IgD, IgE, 433

J genes, 434

kappa chains, 433

lambda chains, 433

light chain, 433

nonacute or nondefective virus, 427

oncogene, 425

Philadelphia chromosome, 431

point mutation, 427

proto-oncogene, 425

retinoblastoma (RB), 424

retrovirus, 427

reverse transcriptase, 427

Rous sarcoma virus (RSV), 427

sarcoma, 426

SCID, 438

T cells, 438

T-cell immunodeficiency, 438

translocation, 427

tumor suppressor, 424

variable region, 433

V genes, 434

V(D)J recombination, 434

X-linked agammaglobulinemia, 438

INSIGHTS
and
SOLUTIONS

1. In disorders such as retinoblastoma, a mutation in one allele of the retinoblastoma (RB) gene can be inherited from the germ line, causing an autosomal dominant predisposition to the development of eye tumors. In order to develop tumors, a somatic mutation in the second copy of the RB gene is necessary. In sporadic cases, two independent mutational events involving both RB alleles is necessary for tumor formation. Given that the first mutation can be inherited, what are the ways in which a second mutational event can occur?

Solution: In considering how the second mutation can arise, several levels of mutational events need to be considered, including changes in nucleotide sequence as well as events that involve whole chromosomes or chromosome parts. Retinoblastoma results when both copies of the RB locus have been lost or inactivated. With this in mind, perhaps the best way to proceed is to prepare a list of the phenomena that can result in a mutational loss or inactivation of a gene.

The first and most obvious way for the second RB mutation to occur is by a nucleotide alteration that converts the remaining normal RB allele to a mutant allele. This alteration may occur through a base substitution or by a frameshift mutation caused by the insertion or deletion of one or more nucleotides. A second method involves loss of the chromosome carrying the normal allele. Because the retinal cells are active in cell division, this event would take place during mitosis, resulting in chromosome 13 monosomy, with the mutant copy of the gene left as the only RB allele. This mechanism does not necessarily have to involve loss of the entire chromosome; deletion of the long arm (RB is on 13q) or an interstitial deletion involving the RB locus and some surrounding material would have the same result.

As an alternative, a chromosome aberration involving loss of the normal copy of the RB gene might be followed by a duplication of the chromosome carrying the mutant allele. This would restore two copies of chromosome 13 to the cell, but no normal allele of the RB gene would be present. Lastly, a recombination event followed by chromosome segregation could produce a homozygous combination of mutant RB alleles.

2. An antibody molecule contains two identical H chains and two identical L chains. This results in antibody specificity, with the antibody binding to a specific antigen. Recall that there are five classes of H genes and two classes of L genes. Because of the high degree of variability in the genes that encode the H and L chains, it is possible (even likely) that an antibody-producing cell contains two different alleles of the H gene and two different alleles of the L gene. Yet the antibodies produced by the plasma cell contain only a single type of H chain and a single type of L chain. How can you account for this, based on the number of classes of H and L genes and the possibility of heterozygosity?

Solution: Although an antibody-producing cell contains different classes of H and L genes, and although it is entirely possible and even likely that a given antibody-producing cell will contain different alleles for the H gene and for the L gene, a phenomenon known as allelic exclusion allows expression of only one H allele and one L allele in any given plasma cell at a given time. That is not to say that the same antibody is produced over the life span of the antibody-producing cell, however. At first, many plasma cells produce antibodies of the IgM class, and at later times they may produce antibodies of a different class (e.g., IgG). Even in switching between different H genes, allelic exclusion is maintained, with only one allele of an H gene (or L gene) being expressed at a given time.

Problems and Discussion Questions

1. As a genetic counselor, you are asked to assess the risk for a couple who plan to have children, but where there is a family history of retinoblastoma. In this case, both the husband and wife are phenotypically normal, but the husband has a sister with bilateral familial retinoblastoma. What is the probability that this couple will have a child with retinoblastoma? Are there any tests you can think of that you could recommend to help in this assessment?

2. What is the difference between saying that cancer is inherited and saying that predisposition to cancer is inherited?

3. Define tumor suppressor genes. Why are most tumor suppressor genes expected to be recessive?

4. Distinguish between oncogenes and proto-oncogenes. In what ways can proto-oncogenes be converted to oncogenes?

5. How do translocations such as the Philadelphia chromosome lead to oncogenesis?

6. The compound benzo[a]pyrene is found in cigarette smoke. This compound chemically modifies guanine bases in DNA. Such abnormal bases are typically removed by an enzyme that hydrolyzes the base leaving an apurinic site. If such a site is left unrepaired, an adenine is preferentially inserted across from the apurinic site. In a study of lung cancer patients, tumor cells from 15 of 25 patients had a G-to-T transversion in a gene called *p53*, which has a known role in cancer formation. You are asked to testify as an expert witness in the following court case. The widow of a man who died of lung cancer is suing a tobacco company for selling tobacco products that killed her husband (who was a lifelong smoker). What do you tell the jury? (Science only, no personal expositions on lawyers or the legal system.) [Reference: *Nature* 350:377–78 (1991)]

7. What do the following symbols represent in immunoglobulin structure: VL, CH, IgG, J, and D?

8. Figure 19–11 shows two unique gene sequences formed by recombination. How many other unique sequences can be formed?

9. If germ line DNA contains 10 V, 30 D, 50 J, and 3 C genes, how many unique DNA sequences can be formed by recombination?

10. If there are 5 V, 10 D, and 20 J genes available to form a heavy-chain gene, and 10 V and 100 J genes available to form a light chain, how many unique antibodies can be formed?

11. Distinguish between an antigen and an antibody.

12. How many polypeptide chains are present in an IgG antibody molecule? How many different polypeptide chains are represented in this molecule? How many gene segments from a germ line cell were combined to produce the polypeptide chains in this molecule?

13. What is an HLA haplotype? How many of these haplotypes do you carry? How are these haplotypes inherited?

14. Why is the idea that every cell carries a complete set of genetic information not true for lymphocytes?

Selected Readings

Aaltonen, L. A., et al. 1993. Clues to the pathogenesis of familial colorectal cancer. *Science* 260:812–16.

Ames, B., Magraw, R., and Gold, L. 1990. Ranking possible cancer hazards. *Science* 236:71–80.

———, Profet, M., and Gold, L. 1990. Dietary pesticides (99.99% all natural). *Proc. Natl. Acad. Sci. USA* 87:7777–81.

Benedict, W., et al. 1990. Role of the retinoblastoma gene in the initiation and progression of a human cancer. *J. Clin. Invest.* 85:988–93.

Birrer, M., and Minna, J. 1989. Genetic changes in the pathogenesis of lung cancer. *Annu. Rev. Med.* 40:305–17.

Croce, C., and Klein, G. 1985. Chromosome translocations and human cancer. *Sci. Am.* (Mar.) 252:54–60.

Felman, M., and Eisenbach, L. 1988. What makes a tumor cell metastatic? *Sci. Am.* (Nov.) 259:60–85.

Kinzler, K.W., and Vogelstein, B. 1996. Lessons from hereditary colorectal cancer. *Cell* 87:156–70.

Kinzler, K.W. ,and Vogelstein, B. 1997. Gatekeepers and caretakers. *Nature* 386:761–63

Lasko, D., Cavanee, W., and Nordenskjold, M. 1991. Loss of constitutional heterozygosity in human cancer. *Annu. Rev. Genet.* 25:281–314.

Lengauer, C., Kinzler, K.W., and Vogelstein, B. 1997. Genetic instability in colorectal cancers. *Nature* 386:623–27.

Murray, A., and Kirschner, M. 1991. What controls the cell cycle. *Sci. Am.* (Mar.) 264:56–63.

Nurse, P. 1997. Checkpoint pathways come of age. *Cell* 91:865–67.

Papadopoulos, N., et al. 1994. Mutation of a mutL homolog in hereditary colon cancer. *Science* 263:625–29.

Rustgi, A., and Podolsky, D. 1992. The molecular basis of colon cancer. *Annu. Rev. Med.* 43:61–68.

Solomon, E., et al. 1987. Chromosome 5 allele loss in human colorectal carcinomas. *Nature* 328:616–29.

Stanbridge, E. 1992. Functional evidence for human tumor suppressor genes: Chromosome and molecular genetic studies. *Cancer Surv.* 12:43–57.

Sugimura, T. 1990. Cancer prevention: Underlying principles. *Basic Life Sci.* 52:225–32.

Thibodeau, S. N., Bren, G., and Schaid, D. 1993. Microsatellite instability in cancer of the proximal colon. *Science* 260:816–19.

Thompson, M. A., and Ramsey, R. G. 1995. Myb: An old oncoprotein with new roles. *Bioessays* 17:341–50.

Weinberg, R. 1985. The action of oncogenes in the cytoplasm and nucleus. *Science* 203:770–76.

White, R. 1992. Inherited cancer genes. *Curr. Opin. Genet. Dev.* 2:53–57.

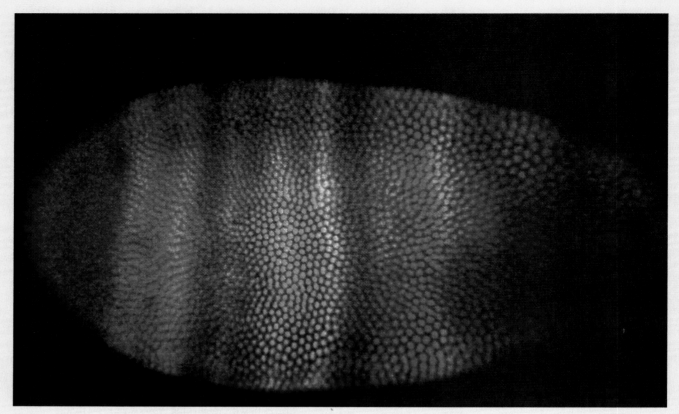

Regional differentiation of various genes during early differentiation of the Drosophila *embryo.*

CHAPTER OUTLINE

CHAPTER
20

Genetics of Development and Behavior

Chapter Concepts

The genetic basis of development is differential gene action. Gene action varies over time in the same tissue and between different tissues present at the same developmental stage. Differential gene expression is required to bring about and maintain adult structures. Behavior in many organisms is controlled by both single genes and polygenic systems. At the biochemical level, these genes work by controlling cascades of metabolic reactions.

In multicellular plants and animals, a fertilized egg, without further external stimulus, begins a cycle of developmental events that ultimately give rise to an adult member of the species from which the egg and sperm were derived. Thousands, millions, or even billions of cells are organized into a cohesive unit that we perceive as a living organism. The series of events whereby organisms attain their final adult form is studied by developmental biologists. This is perhaps one of the most intriguing areas in biology, because an understanding of developmental processes requires knowledge of many biological disciplines.

In the past hundred years, investigations in embryology, genetics, biochemistry, molecular biology, cell physiology, and biophysics have contributed to the study of development. Largely, the findings have pointed out the tremendous complexity of developmental processes. Unfortunately, this description of what actually happens does not answer the "why" and "how" of development. Nevertheless, further hypotheses and experimentation will be based on this knowledge and will lead us closer to a comprehensive understanding of development.

Geneticists and neurobiologists interested in the study of behavior have increasingly focused their attention on the analysis of gene action as a means of explaining how organisms react to stimuli and/or their environment. In humans, it is hoped that this approach will allow an objective evaluation of the genetic elements that contribute to intelligence and behavior. The

use of genetics to study behavior minimizes the effect of environment on a behavioral response and allows a specific pattern of action to be dissected into its components. In turn, many behavioral phenotypes are actually the result of developmental events that alter or restrict the function of the nervous system or its interactions with other body systems, emphasizing the relationship between cellular structure and function. In this chapter, we will first focus on the role of genetic information during development. After establishing how the genome controls the cellular phenotype, we will turn our attention to how genetics is being used to probe the relationship between the genotype and behavior.

Basic Concepts of Developmental Genetics

Several processes govern the formation of embryonic and adult structures from the fertilized egg: determination, differentiation, and cell–cell interaction. **Determination** can be defined as one or more regulatory events that establish a specific pattern of gene activity and developmental fate for a given cell. Through determination, an undifferentiated embryonic cell is programmed to develop into a muscle cell, a liver cell, an epidermal cell, or any of the other cell types of the adult. **Differentiation** is the process of genetic and morphological change by which cells attain their adult form and function. It is the mechanism by which the determined state becomes expressed. **Cell–cell interaction** represents a form of intercellular communication and can be involved in both determination and differentiation.

The current thrust in developmental genetics is to provide a molecular explanation of the events in embryogenesis and morphogenesis. This includes the description of temporal changes in gene expression and cellular differentiation to establish a causal relationship between the presence or absence of molecules, receptors, transcriptional events, and cell and tissue interactions. It also includes the observable morphological events that accompany the process of development. To explain the sequence of events that take place in a developing embryo, geneticists rely on certain heuristic concepts. These include (1) the theory of variable gene activity and (2) the principles of differential transcription and the stability of the differentiated state. These concepts have been derived from the study of a wide range of organisms used as model systems, and form the theoretical framework used to explain how the developmental potential present in a single cell becomes elaborated into a recognizable individual with a certain behavioral repertoire we identify as an adult organism.

The Variable Gene Activity Theory

From a genetic perspective, development may be described as the attainment of a differentiated state. An erythrocyte that is active in hemoglobin synthesis is differentiated, but a cell in a blastula-stage embryo is undifferentiated. In order to accomplish this specialization, certain genes are actively transcribed but many other genes are not.

The concept of differential transcription has led to **the variable gene activity hypothesis** of differentiation. The theory holds that the adult form assumed by any specific cell type is determined by those genes that are actively transcribed. Its underlying assumption is that each cell carries an intact diploid genome in the nucleus and only certain gene products characteristic of the cell type are produced. The rest of the genome is actively shut down and not transcribed. Given current knowledge of molecular biology, the variable gene activity theory is readily acceptable. Every cell is a biochemical entity composed of macromolecules that are informational, structural, catalytic, or metabolic. These molecules constitute the cell. Within the cell, the presence or absence of any given type of molecule is directly influenced by gene activity or inactivity. Therefore, a cell is what its active genes direct it to be.

The variable gene activity theory is a useful model for experimental design. The first part of this chapter examines the validity of this theory and its premises and provides examples that offer experimental support. It is important to remember, however, that this type of approach initially tells us what happens. We are only beginning to learn why certain genes are active and others inactive, and why patterns of gene activity differ among different cells.

Genomic Equivalence

The variable gene activity hypothesis is based on the premise that somatic cells in multicellular organisms all contain a complete set of genetic information. Analysis has shown that in the cells of most species the quantity of nuclear DNA is equal to the diploid ($2n$) content. But is it possible to show that differentiated cells are qualitatively equivalent? In other words, is a complete set of genes present in each cell?

One approach to this question would be to show that a differentiated cell is **totipotent**—that is, capable of giving rise to a complete organism under the proper conditions. In the early part of this century, Hans Spemann demonstrated that a nucleus derived from one cell in the 16-cell stage of a newt embryo is capable of supporting development to the adult stage. By constricting a newly fertilized egg prior to the first cell division, Spemann partitioned the zygote nucleus into one-half of the cell. The half containing the nucleus proceeded to undergo division and produce 16 cells. At this stage of development, the constriction was loosened, and one of the 16 nuclei was allowed to pass back into the nonnucleated cytoplasm on the other side. Both halves subsequently produced complete but separate embryos.

The results of Spemann's experiment showed that in newts, at the 16-cell stage, each nucleus is equivalent to the one contained in the zygote. In the 1950s and 1960s, more sophisticated experiments were performed by Robert Briggs and Thomas King using the grass frog *Rana pipiens,* and by John Gurdon using the African frog *Xenopus laevis.* These investigators inactivated or removed the nucleus from an unfertilized egg. Then they transplanted nuclei isolated from cells at various stages of development into the enucleated oocytes. Nuclei derived from the blastula stage were capable of supporting the development of complete and normal adults. In *Rana* (Figure 20–1), transplanted nuclei derived from later stages such as the gastrula and neurula usually allowed only partial development.

In *Xenopus,* however, Gurdon's experiments showed that epithelial gut nuclei of tadpoles will direct the development of an adult frog when transplanted into enucleated oocytes. To test whether the nucleus of a differentiated *adult* somatic cell can direct the development of a normal embryo, Gurdon used nuclei from adult frog skin cells in serial transplant experiments (Figure 20–2). In these experiments, a donor nucleus was transplanted into an enucleated egg, and the cell was allowed to develop to the blastula stage. The blastula cells

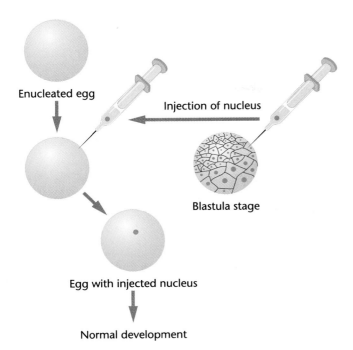

■ Figure 20–1 The process of nuclear transplantation in the frog. The unfertilized egg is activated by needle puncture, and the nucleus is removed with a micropipette. A nucleus from a blastula stage embryo is injected.

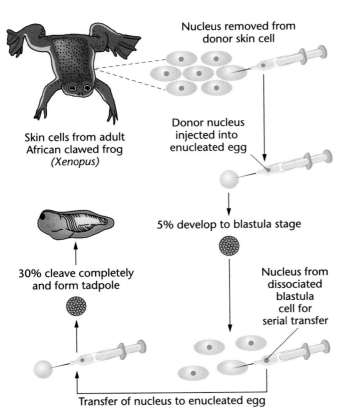

■ Figure 20–2 Serial transplantation experiments using adult frog skin cell nuclei. Serial transplantation dramatically increases the percentage (from 5 to 30 percent) of recipient eggs that undergo complete cleavage.

were then dissociated and a nucleus removed from one of them. This nucleus was transplanted into still another enucleated egg, and development was allowed to occur. Such serial transfers were repeated a number of times. Subsequently, the blastula was not dissociated, but instead was allowed to continue development.

After several rounds of nuclear transfer, Gurdon found that 30 percent of the transplants developed into advanced tadpoles. Because nuclei from specialized adult epidermal cells can direct the synthesis of gene products and promote the organization of cells and tissues into a tadpole, we can be confident that such cells contain a complete copy of the genome. Under the proper circumstances, it appears that genes that are not expressed in specialized cells can be reactivated.

These and similar studies using other organisms argue convincingly that differentiated adult cells have not lost any of the genetic information present in the zygote. Instead, some genes in any given cell type are shut off or repressed, but can be reprogrammed to direct normal development.

Genetics of Embryonic Development in *Drosophila*

Why a certain cell is fated to turn on or off specific genes during development is a central question in developmental biology. As alluded to earlier, there is no simple answer to this question. However, information derived from the study of many different organisms provides a starting point. We shall examine one of these examples, embryonic development in *Drosophila*, which reflects the direction and scope of our knowledge in this area. We will begin with an overview of development, and then describe how the analysis of mutations has provided insight into the number and action of genes that regulate the processes of determination and differentiation.

Overview of Drosophila *Development*

Beginning with the fertilized egg, *Drosophila* passes through a pre-adult period of development with several distinct phases: the embryo, three larval stages, and the pupal stage. The adult fly emerges from the pupal case 10 to 14 days after fertilization.

The adult body of *Drosophila* is composed of head, thoracic, and abdominal segments. Each segment is formed from the progeny of cells set aside during embryogenesis and incorporated into discrete structures called **imaginal discs** (Figure 20–3). These discs, formed from hollow sacs of cells, are determined to form specific parts of the adult body. There are 12 bilaterally paired discs and one genital disc. For example, there are eye-antennal discs, leg discs, wing discs, and so forth. Other cells in the embryo form the internal and

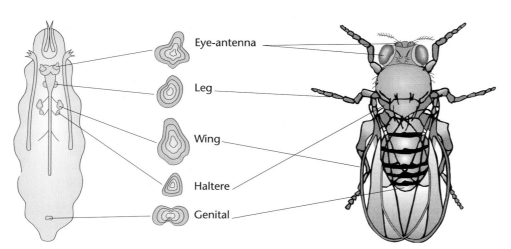

external structures of the larva. During metamorphosis, most of the larval body parts histolyze, or break down, and the imaginal discs undergo mitosis and differentiate to form the head, thorax, abdomen, and appendages of the adult fly.

Externally, the *Drosophila* egg has a number of structures that delineate the anterior, posterior, dorsal, and ventral regions (Figure 20–4). The anterior end of the egg contains the micropyle, a specialized conical structure for the entrance of sperm into the egg, while the posterior end is rounded and marked by a series of aeropyles (openings that allow gas exchange during development). The dorsal side of the egg is flattened and contains the chorionic appendages, while the ventral side is curved.

■ Figure 20–4 Scanning electron micrograph of the anterior pole of a newly oviposited *Drosophila* egg.

Internally, the egg cytoplasm is organized into a series of maternally derived molecular gradients. These gradients play a key role in establishing the developmental fates of nuclei that migrate into specific regions of the embryo.

Immediately after fertilization, the zygote nucleus undergoes a series of divisions. After nine rounds of division, most of the approximately 512 nuclei migrate to the egg's outer surface or cortex, where further divisions take place. The nuclei then become enclosed in membranes, forming a single layer of cells over the embryo surface (Figure 20–5). This is the **blastoderm** stage of embryonic development. As nuclei migrate into different regions of the egg's cytoplasm, maternally derived transcripts and gene products localized in these regions initiate a program of gene expression in the nuclei. These transcriptional programs result in the formation of the anterior–posterior axis and the dorsal–ventral axis of the embryo. The formation of these axes requires the action of maternal gene products and zygotic genes, whose discovery is discussed below. Table 20.1 lists some of the maternal genes that are required for formation of the anterior–posterior and dorsal–ventral embryonic axes.

The body axis of the larva, pupa, and adult forms along the anterior–posterior axis. As soon as the blastoderm forms, cells that will form adult structures are set aside. Two-dimensional maps of the adult structures that will be formed by regions of the embryo are known as **fate maps** (Figure 20–6). The existence of a fate map implies that the developmental outcome for a given nucleus is governed by the location to which it migrates during embryonic development, and that positional information (where a given nucleus ends up along the anterior–posterior axis and dorsal–ventral axis) directs the nucleus to initiate a specific developmental program.

At a stage after blastoderm formation, the embryo becomes organized into a series of segments (Figure 20–7). Within each segment, cells first become determined to form either an anterior or a posterior portion

(a)

Diploid zygote nucleus produced by
fusion of parental gamete nuclei

(b)

Nine rounds of nuclear divisions
produce multinucleated syncytium

(c)

Nuclei migrate to outer surface

(d)

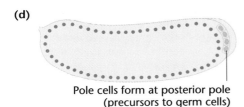

Pole cells form at posterior pole
(precursors to germ cells)
Approximately four further
divisions take place at surface

(e)   pole cells

Nuclei become enclosed in
membranes, forming a single layer of
cells over embryo surface

■ **Figure 20–5 (left)** Early stages of embryonic development in *Drosophila*. (a) Fertilized egg with zygotic nucleus, about 30 minutes after fertilization. (b) Nuclear divisions occur about every 10 minutes, producing a multinucleate cell. (c) After approximately nine divisions (512 nuclei), the nuclei migrate to the outer surface or cortex of the egg. (d) At the surface, four additional rounds of nuclear division occur. A small cluster of cells, the pole cells, form at the posterior pole about 2-1/2 hours after fertilization. These cells will form the germ cells of the adult. (e) About 3 hours after fertilization, the nuclei become enclosed in membranes, forming a single layer of cells over the embryo surface. This stage is known as the blastoderm.

(called a **compartment**) of the segment. In the following stages of development, further restrictions of developmental options occur so that cells in each compartment are eventually restricted to forming a single structure in the adult body.

Genetic Analysis of Embryogenesis

Although development in *Drosophila* has been examined anatomically (as described above) and biochemically, genetic analysis has provided the most information about the events in embryogenesis.

Genes that control embryonic development fall into two classes: maternal-effect genes and zygotic genes. **Maternal-effect genes** are those whose gene products (mRNA and proteins) are deposited in the developing egg. These products may be distributed in a gradient, or concentrated in specific regions of the egg. The phenotype of flies with mutations in maternal-effect genes is expected to be female sterility, since none of the embryos of females homozygous for a recessive mutation would receive wild-type gene products from their mother, and would not develop normally. In

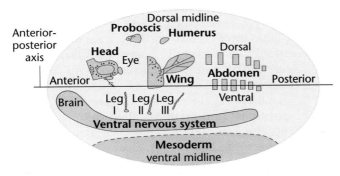

■ Figure 20–6 Diagram showing the origin of some *Drosophila* structures from the embryonic blastoderm.

TABLE 20.1	Maternally transcribed and zygotically transcribed genes that control the anterior–posterior axis of the *Drosophila* embryo		
	Anterior	*Posterior*	*Terminal*
Maternal	bicoid	staufen	trunk
	exuperantia	oskar	fs(1)Nasrat
	swallow	vasa	fs(1)polehole
		valois	torso-like
		tudor	l(1)polehole
		mago nashi	
		nanos	
		pumilio	
Zygotic	hunchback	knirps	tailless
		giant	huckebein

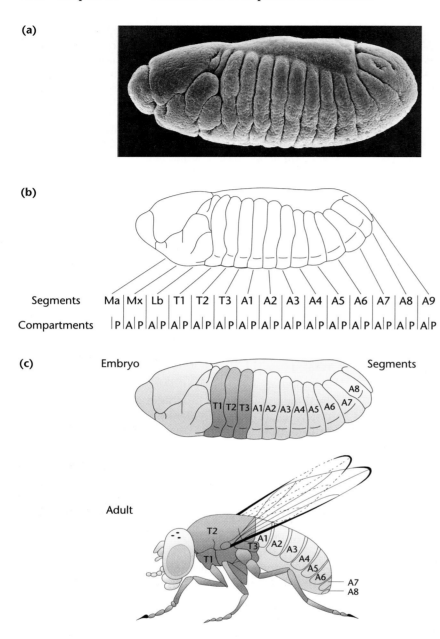

(a)

(b)

Segments	Ma	Mx	Lb	T1	T2	T3	A1	A2	A3	A4	A5	A6	A7	A8	A9
Compartments	P A	P A	P A	P A	P A	P A	P A	P A	P A	P A	P A	P A	P A	P A	P A	P

(c) Embryo Segments

Adult

■ Figure 20–7

Segmentation in *Drosophila*. (a) Scanning electron micrograph of *Drosophila* embryo at about 10 hours after fertilization. By this stage, the segmentation pattern of the body is clearly established. (b) The segments, compartments, and parasegments of the *Drosophila* embryo. Ma, Mx, and Lb represent the segments that will form head structures. T1–T3 are thoracic segments, and A1–A9 are abdominal segments. Each segment is divided into anterior (A) and posterior (P) compartments. Parasegments represent an early pattern specification that is later refined into the segmental plan of the body. Note that all of the parasegments are shifted forward by one compartment to form the segments. (c) The segmented embryo and the adult structures that will form each segment.

Drosophila, these maternal-effect genes encode transcription factors, receptors, and proteins that regulate translation. During embryonic development, these gene products work to activate or repress the expression of zygotic genes in a temporal and spatial sequence.

Zygotic genes encode the proteins produced by the developing embryo itself, and are therefore transcribed after fertilization. The phenotype of mutations in this class show embryonic lethality: In a cross between two flies that are heterozygous for a recessive zygotic mutation, one-fourth of the embryos (the homozygotes) will fail to develop. In *Drosophila*, many of the zygotic genes are transcribed in patterns that depend on the distribution of maternal-effect proteins.

To identify as many zygotic genes as possible, Christiane Nusslein-Volhard and Eric Weischaus systematically screened for mutations that affect embryonic development. The F2 offspring of mutagenized flies were examined for recessive embryonic lethal mutations with defects in external structures. Thousands of dead embryos were screened, and the mutants recovered fell into three classes, which were named gap, pairrule, and segment polarity. In a paper published in 1980, Nusslein-Volhard and Weischaus proposed a model in which embryonic development is initiated by gradients of maternal-effect gene products (Figure 20–8).

The positional information laid down by molecular gradients along the anterior–posterior axis of the embryo is interpreted by two gene sets: (1) zygotic **segmentation genes**, which divide the embryo into a series of stripes or segments and define the number, size, and polarity of each segment; and (2) **selector genes**, which specify the identity or fate of each segment but do not affect the number or size of the segments (Figure 20–9).

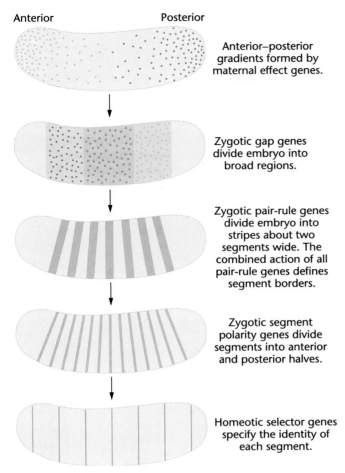

Anterior   **Posterior**

Anterior–posterior gradients formed by maternal effect genes.

Zygotic gap genes divide embryo into broad regions.

Zygotic pair-rule genes divide embryo into stripes about two segments wide. The combined action of all pair-rule genes defines segment borders.

Zygotic segment polarity genes divide segments into anterior and posterior halves.

Homeotic selector genes specify the identity of each segment.

■ Figure 20–8 Overview of progressive restriction of cell fate during development in *Drosophila*. The process begins with the anterior–posterior axis laid down in oogenesis and activated immediately after fertilization. This axis determines cells to form either anterior or posterior structures. Gene products from this gradient activate transcription of the gap genes that divide the embryo into a limited number of broad regions. Gap proteins are transcription factors that activate pair-rule genes whose products divide the embryo into regions about two segments wide. The combined action of all gap genes defines segment borders. The pair-rule genes in turn activate the segment polarity genes that divide the segments into anterior and posterior compartments. The collective action of the maternal genes that form the anterior–posterior axis and the segmentation genes define fields of action for the homeotic selector genes that specify the identity of each segment.

In a second screening, Weischaus and Trudi Schupbach screened thousands of flies for maternal-effect mutations that affected the external structures of the embryo. Based on these exhaustive screens for mutants, it is estimated that there are about 40 maternal effect genes and 50 to 60 zygotic genes. This means that normal embryonic development is controlled by a relatively small number of genes (about 100), a surprisingly low number given the complexity of the structures formed from the fertilized egg. For their

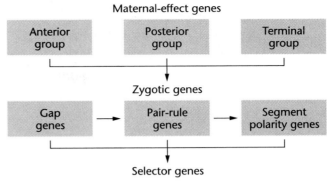

■ Figure 20–9 The hierarchy of genes involved in embryogenesis in *Drosophila*. Gene products of the maternal genes regulate the expression of the first zygotic (segmentation) genes, which in turn control expression of the selector genes.

work on genetic control of development in *Drosophila*, Nusslein-Volhard and Weischaus, along with E.B. Lewis, were awarded the 1995 Nobel Prize for Physiology or Medicine.

Maternal-Effect Genes and the Basic Body Plan in *Drosophila*

Maternal genes are those transcribed by the maternal genotype during oogenesis and transported as mRNA or as translated protein into the oocyte for storage. These transcripts and gene products are utilized during early stages of development, and regulate expression of zygotic genes.

Formation of the Anterior–Posterior Axis

The maternal-effect genes that control the anterior–posterior axis are grouped into three classes: anterior, posterior, and terminal. The following discussion will emphasize the formation of the anterior–posterior axis as an example of the interaction of maternal gene products and the zygotic genes.

Of the genes required to form the anterior portion of the embryo, the most important is *bicoid* (Table 20.1). Embryos of *bicoid* mutants lack head and thoracic structures and also have abnormalities of the first four segments of the abdomen. During oogenesis and the first hours of embryonic development, *bicoid* mRNA is localized to a small region at the anterior pole of the egg [Figure 20–10(a)]. The actions of two genes, *swallow* and *exuperantia*, are involved in localizing the *bicoid* mRNA by docking it to anterior cytoskeletal components. In mutants of these genes, the *bicoid* mRNA is distributed to more posterior areas of the egg, and the mutant embryos form enlarged anterior structures.

During early cleavages, *bicoid* mRNA is translated, and the bicoid protein becomes distributed in a gradi-

(a) Localization of *bicoid* mRNA in oocyte

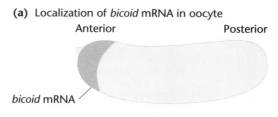

(b) Localization of bicoid protein

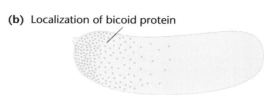

■ Figure 20–10   Formation of the anterior–posterior axis of the *Drosophila* egg depends on the action of three independent systems. The anterior portion of the axis is formed by a gradient of *bicoid* mRNA (a) that is translated into a transcription factor which distributes itself in a gradient at the anterior end of the embryo (b) .

ent along the anterior–posterior axis, with very low levels in the posterior region of the egg [Figure 20–10(b)]. This gradient plays a critical role in determining the developmental pattern of the egg. The bicoid protein is a transcription factor that contains a **homeodomain** (see Chapter 13 for a review of these domains). The bicoid protein binds to the upstream regulatory region of *hunchback*, one of the first zygotic genes to be activated in development, and which is expressed in the anterior half of the embryo. The upstream promoter region of *hunchback* contains five sites, each with the consensus sequence 5'-TCTAATCCC-3', which bind to the bicoid protein. The presence of the consensus binding sequence causes the gene to respond to binding of the *bicoid* transcription factor, and the number of sites occupied by the bicoid protein determines the degree of response. In regions of the gradient where there is a higher concentration of bicoid protein, there will be a higher response by the *hunchback* gene, leading to the formation of a gradient of hunchback protein along the anterior–posterior axis (Figure 20–11).

This interaction between the bicoid protein and the promoter region of *hunchback* was demonstrated by

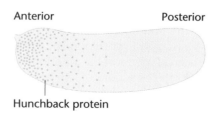

■ Figure 20–11   The gradient of bicoid protein triggers expression of the *hunchback* gene, forming a gradient of hunchback protein along with the anterior–posterior axis.

fusing a reporter gene encoding the enzyme chloramphenicol acetyltransferase (CAT) to promoter sites of the *hunchback* gene (Figure 20–12). When these constructs were injected into *bicoid* mutant embryos, no CAT was produced. When injected into wild-type embryos, the CAT constructs were expressed. Full expression of the CAT gene required the presence of three of the five promoter sites. These experiments demonstrated that the bicoid protein determines the spatial pattern of expression of the *hunchback* gene.

The gradient of bicoid protein causes transcription of *hunchback* in the anterior half of the embryo, but not in the posterior half. The hunchback protein then stimulates transcription of genes that form the head and thorax, and represses activity of genes that form the abdomen.

The Posterior and Terminal Gene Sets

The posterior of the embryo is formed through the action of eight genes (Table 20.1). Mutants of these genes all have a similar phenotype: The posterior portion of

(a)

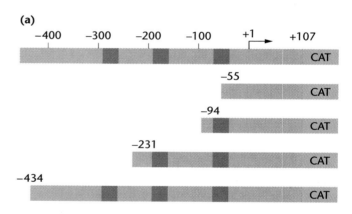

(b)

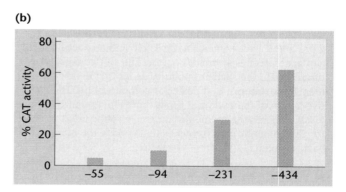

■ Figure 20–12   (a) Fusion genes between a reporter gene for chloroamphenicol acetyltransferase (CAT) with various promoter regions upstream from the *hunchback* gene. Each construct was tested by injection into *bicoid* embryos and wild-type embryos. When injected into *bicoid* embryos, the fusion gene was not transcribed, and no CAT protein was produced. When injected into wild-type embryos (b), full expression of the CAT protein required the presence of three of the five *hunchback* promoter sites.

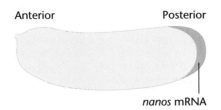

Anterior   Posterior

nanos mRNA

■ Figure 20–13 The posterior end of the anterior–posterior axis is formed by the action of the *nanos* gene. The mRNA from *nanos* transcribed during oogenesis is stored at the posterior pole of the oocyte. After fertilizztion, the mRNA is distributed throughout the posterior region of the embryo by action the of *pumplio* gene. The nanos protein acts to suppress the action of the hunchback protein, ensuring the formation of abdominal structures in the posterior region.

the body is dwarfish, and abdominal segments are lacking. The names of these mutants reflect this phenotype: *oskar* (the dwarf in the Günter Grass novel, *The Tin Drum*) and *nanos,* derived from the Spanish word for dwarf. The nanos gene is transcribed by the maternal genome during oogenesis, and its mRNA is stored at the posterior pole of the egg (Figure 20–13). The nanos protein is produced soon after fertilization, and it controls formation of posterior structures in the developing embryo, but it does so in a way different from the bicoid protein. In contrast to the action of anterior genes, the posterior genes work by a negative effect on a maternal gene transcript. The nanos protein acts to prevent the translation of any *hunchback* mRNA that might be present in the posterior region of the embryo, ensuring the expression of abdomen-forming genes (Figure 20–14). The *hunchback* gene, therefore, is regulated by maternal-effect gene products from both the anterior and posterior parts of the developing embryo.

The other genes involved in posterior formation ensure that the nanos mRNA is packaged and stored at the posterior pole (*oskar, staufen, valois, vasa*) and is distributed properly (*pumilio*).

The terminal genes represent a third set of genes controlling the formation of the anterior–posterior axis (Table 20.1). Mutations in these genes result in the loss of the unsegmented structures that develop from the

anterior and posterior poles of the embryo. Cloning of two of the terminal genes, *torso* and *fs(1)polehole,* suggests that this group contains genes that encode transmembrane signal receptors associated with the reception and transduction of an extracellular signal.

Together, these three groups of maternal genes establish the anterior–posterior axis of the embryo as a spatial pattern of gene products. The anterior of the embryo is organized by a gradient of the bicoid protein that acts to activate anterior-specific genes and to repress posterior-specific genes. The posterior of the embryo is formed through the action of the nanos protein, which, after migrating from the posterior pole, suppresses the formation of the hunchback protein, relieving the *hunchback*-mediated inhibition of abdominal-specific gene expression. The critical gene in formation of the terminal region of the embryo is the *torso* gene product, which allows expression of the genes that form the unsegmented terminal structures. Two of these systems function by sequestering a specific mRNA at the poles of the egg (*bicoid* at the anterior, *nanos* at the posterior) and by the formation of a gradient of protein. The third system works via a signal transduction system coupled to the anterior–posterior axis system during oogenesis.

The dorsal–ventral axis of the embryo is specified by a family of maternal genes that depends on the formation of a single protein gradient (dorsal) along the dorsal–ventral axis.

Genes forming the anterior–posterior axis regulate transcriptional activities of cells along the anterior–posterior and dorsal–ventral axes and illustrate the critical role of cytoplasmic gradients in determining the fates of cells in different embryonic regions. The next section describes how components of the anterior–posterior axis controls transcription of gene sets that regulate further development of the embryo.

Zygotic Genes and Segment Formation

The zygotic genes are activated or repressed in a positional gradient by the maternal-effect gene products, and they divide the embryo into a series of segments

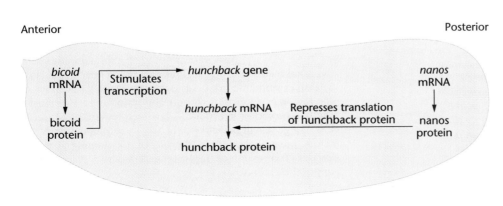

Anterior   Posterior

bicoid mRNA → bicoid protein — Stimulates transcription → *hunchback* gene → *hunchback* mRNA → hunchback protein ← Represses translation of hunchback protein — *nanos* mRNA → nanos protein

■ Figure 20–14 Coordinated action of bicoid and nanos proteins produces the anterior–posterior axis of the early embryo. Throughout the anterior end, bicoid protein activates the *hunchback* gene. The hunchback protein activates transcription of head and thorax genes. In the posterior region, the nanos protein inhibits production of hunchback protein, allowing the expression of abdominal genes.

along the anterior–posterior axis. These segmentation genes are transcribed in the developing embryo, and mutations of these genes are seen as embryonic lethal phenotypes. Action of these genes brings about determination of cell fate.

The segmentation genes are expressed soon after the bicoid gradient has been established. Over 20 segmentation loci have been identified (Table 20.2). They are classified on the basis of their mutant phenotypes: (1) the gap genes (*Krüppel, knirps*) delete a group of adjacent segments; (2) the pair-rule genes (*even-skipped, fushi-tarazu*) affect every other segment and eliminate a specific part of each affected segment; and (3) the segment polarity genes (*hedgehog, gooseberry*) cause defects in homologous portions of each segment.

Gap Genes

Transcription of the **gap genes** is activated or inactivated by gene products previously expressed along the anterior–posterior axis from *bicoid* and the other genes of the maternal gradient systems. When mutated, these genes produce large gaps in the segmentation pattern of the embryo. *Hunchback* mutants lose head and thorax structures, *Krüppel* mutants lose thoracic and abdominal structures, and mutations in *knirps* result in the loss of most abdominal structures. Transcription of gap genes divides the embryo into a series of broad regions (roughly, the head, thorax, and abdomen) within which different combinations of gene activity will eventually specify both the type of segment that will form and the proper order of segments in the body of the larva, pupa, and adult. To date, gap genes that have been cloned all encode transcription factors with zinc-finger DNA-binding motifs. Expression of the gap genes correlates roughly with their mutant phenotypes (*hunchback* at the anterior, *Krüppel* in the middle, and *knirps* at the posterior). Gap genes control transcription of pair-rule genes.

Pair-Rule Genes

The **pair-rule genes** divide the broad regions established by the gap genes into regions about one segment wide. Mutations in pair-rule genes eliminate segment-size regions at every other segment. The pair-rule genes are expressed in narrow bands or stripes of nuclei that extend circumferentially around the embryo. Expression of this gene set first establishes the boundaries of segments, then establishes the developmental fate of the cells within each segment by controlling the segment polarity genes. At least eight pair-rule genes act to divide the embryo into a series of stripes. However, the boundaries of these stripes overlap, meaning that cells in the stripes express different combinations of pair-rule genes in an overlapping fashion (Figure 20–15).

Many pair-rule genes encode transcription factors containing helix-turn-helix homeodomains. Transcription of the pair-rule genes is mediated by the action of gap gene products, but the pattern is resolved into highly delineated stripes by the interaction among the gene products of the pair-rule genes themselves (Figure 20–16).

Segment Polarity Genes

Expression of the **segment polarity genes** is controlled by transcription factors encoded by the pair-rule genes. Each segment polarity gene becomes active in a single

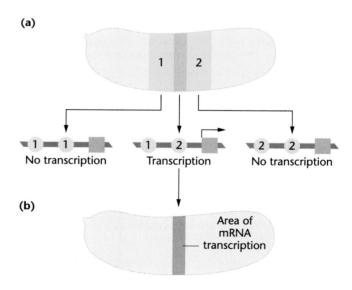

(a)

1   2

1 — 1 No transcription

1 — 2 Transcription

2 — 2 No transcription

(b)

Area of mRNA transcription

■ Figure 20–15 New patterns of gene expression can be generated by overlapping regions containing gene products. (a) Transcription factors 1 and 2 are present in an overlapping region of expression. If both transcription factors must bind to a promoter in order to trigger expression of a target gene, the gene will be active only in cells containing both factors (most likely, only in the zone of overlapping expression). (b) Expression of the target gene in the restricted region of the embryo.

TABLE 20.2	Segmentation genes in *Drosophila*	
Gap Genes	**Pair-Rule Genes**	**Segment Polarity Genes**
Krüppel	hairy	engrailed
knirps	even-skipped	wingless
hunchback	runt	cubitis interruptus[D]
giant	fushi-tarazu	hedgehog
tailless	odd-paired	fused
huckebein	odd-skipped	armadillo
	sloppy-paired	patched
		gooseberry
		paired
		naked
		disheveled

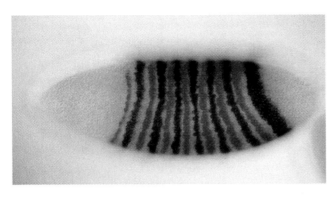

■ Figure 20–16 Stripe patterns of gene expression in a *Drosophila* embryo. This embryo is stained to show expression of the pair-rule genes *even-skipped* and *fushi-tarazu*.

band of cells within each segment that extends around the embryo. This concerted action divides the embryo into 14 segments and controls the cellular identity within each segment. Some of the segment polarity genes, including *engrailed*, encode transcription factors. Rather than activating transcription, however, the engrailed protein competitively inhibits activation by other homeodomain proteins, resulting in the establishment of a segment border.

To summarize, the genes that control development in *Drosophila* act in a temporally and spatially ordered cascade, beginning with the genes that establish the anterior–posterior and dorsal–ventral axes of the egg and early embryo. The gradients of mRNAs and proteins along the anterior–posterior axis activate the gap genes, which subdivide the embryo into broad bands. The gap genes in turn activate the pair-rule genes, which divide the embryo into segments. Finally, the segment polarity genes divide each segment into anterior and posterior regions arranged linearly along the anterior–posterior axis. This progressive restriction of the

developmental potential of the cells in the *Drosophila* embryo during the first one-third of embryogenesis. The restriction involves transcriptional control of gene sets and is accomplished by regulating the activity of proteins that bind to promoter and enhancer regions of genes. Most of these proteins act as transcription factors.

Selector Genes

As segment boundaries are being established by the action of the segmentation genes, the selector genes are activated. Expression of these genes determines the structures to be formed by each segment, including the antennae, mouth parts, legs, wings, thorax, and abdomen. Mutants of these genes are known as **homeotic mutants**. For example, the wild-type allele of the *Antennapedia (Antp)* gene is required to specify structures in the second thoracic segment (which carries a leg). Dominant gain of function *Antp* mutations lead to expression of the gene in the head and thorax, and the antennae of the fly are transformed into legs (Figure 20–17).

The *Drosophila* homeotic selector genes are found in two clusters on chromosome 3. The *Antp* complex contains five genes (Table 20.3) required for specification of structures in the head and first two thoracic segments. The **bithorax (BX-C) complex** contains three

TABLE 20.3	Homeotic selector genes of *Drosophila*
Antennapedia Complex	***Bithorax Complex***
labial	*Ultrabithorax*
Antennapedia	*abdominal A*
Sex-comb reduced	*Abdominal B*
Deformed	
proboscipedia	

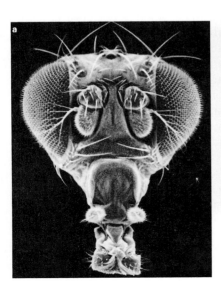

■ Figure 20–17 *Antennapedia (Antp)* mutation in *Drosophila*. (a) Head from wild-type *Drosophila* showing the structure of the antenna and other head parts. (b) Head from an *Antp* mutant, showing the replacement of normal antenna structures with a leg. This is caused by activation of the *Antp* gene in the head region.

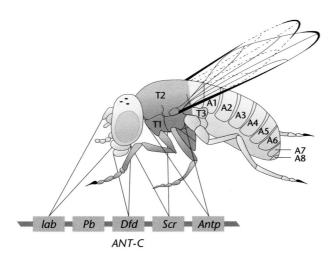

■ Figure 20–18 Genes of the *Antennapedia* complex (ANT-C) and the adult structures they specify. The *labial (lab)* and *Deformed (Dfd)* genes control the formation of head segments. The *Sex comb reduced (Scr)* and *Antennapedia (Antp)* genes specify the identity of the first two thoracic segments. The remaining gene in the complex, *proboscipedia,* may not act during embryogenesis, but may be required to maintain the differentiated state in adults. In mutants, the labial palps are transformed into legs.

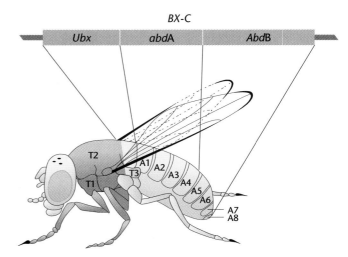

■ Figure 20–19 Genes of the *bithorax* complex *(BX-C)* and the adult structures they specify. *Ultrabithorax (Ubx)* controls structures in the posterior compartment of T2 and structures in T3. The two other genes, *abdominal A (abdA)* and *Abdominal B (AbdB),* specify the segmental identities of the eight abdominal segments (A1–A8) .

genes required for specification of structures formed by the posterior portion of the second thoracic segment, the entire third thoracic segment, and the abdominal segments (Figures 20–18 and 20–19).

Activation of the selector genes is under the control of the gap genes and the pair-rule genes. Although selector gene expression is confined to certain domains in the embryo, the timing and patterns of expression are rather complex and involve interactions between segmentation and selector genes as well as interactions among the various selector genes. For example, the *Antennapedia* gene has two promoters (P1 and P2) and can be transcribed to produce two different pre-mRNAs (Figure 20–20). Transcription from P1 is stimulated by Krüppel protein (a gap-gene product) and repressed by Ultrabithorax protein (a selector-gene product). Transcription from P2 is activated by hunchback and fushi-tarazu proteins (maternal-effect and gap-gene products) and inhibited by oskar protein (a maternal-effect gene product). Paradoxically, the protein produced from these two pre-mRNAs is identical.

Regulation of expression of selector genes is also accomplished by alternative splicing of pre-mRNAs to yield different transcripts (Figure 20–21). The *Ultrabithorax* transcripts produced during early embryogenesis can be processed at varying 5′ sites in exon 1 and will include two "micro-exons," producing a number of related proteins. Transcripts from *Ultrabithorax* later in development (during formation of the nervous system) include one or neither of the micro-exons.

A number of genes that control the expression of selector genes have been identified, including *extra sex combs (esc), Polycomb (Pc), super sex combs (sxc),* and *trithorax (trx).* In the second chromosomal mutant *extra sex combs (esc),* some of the head and all of the thoracic and abdominal segments develop as posterior abdominal segments (Figure 20–22), indicating that this gene normally controls the expression of *BX-C* genes in all body segments. The mutation does not affect either the number or the polarity of segments. It does affect their developmental fate, indicating that the *esc*+ gene product that is synthesized and stored in the egg by the maternal genome may be required for correct interpretation of the information gradient in the egg cortex.

■ Figure 20–20 Transcription and processing of mRNA from the *Antp* gene. Transcription from P1 or P2 in the *Antp* gene results in two different pre-mRNAs.

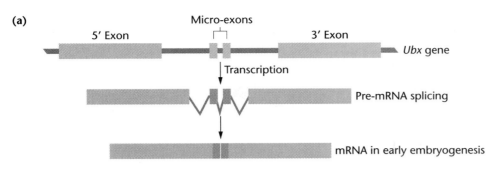

(a)

Micro-exons

5' Exon   3' Exon

Ubx gene

↓ Transcription

Pre-mRNA splicing

↓

mRNA in early embryogenesis

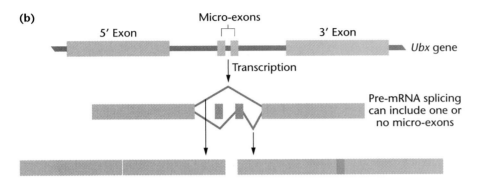

(b)

Micro-exons

5' Exon   3' Exon

Ubx gene

↓ Transcription

Pre-mRNA splicing
can include one or
no micro-exons

↓   ↓

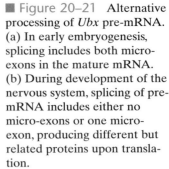

■ Figure 20–21 Alternative processing of *Ubx* pre-mRNA. (a) In early embryogenesis, splicing includes both micro-exons in the mature mRNA. (b) During development of the nervous system, splicing of pre-mRNA includes either no micro-exons or one micro-exon, producing different but related proteins upon translation.

E. B. Lewis has proposed that genes in the bithorax complex, and perhaps other genes involved in segmentation, arose from a common ancestral gene by tandem duplication and subsequent divergence of structure and function. In fact, each of the homeotic selector genes listed in Table 20.3 encodes a transcription factor that includes a DNA-binding domain encoded by a 180-bp sequence known as a **homeobox**. The homeobox encodes a 60-amino acid sequence known as a **homeodomain**. Similar sequences have been found in the genomes of other eukaryotes with segmented body plans, including *Xenopus*, chicken, mice, and humans. In mammals, the complexes of selector genes are called *Hox* gene clusters (Figure 20–23). Homeodomains from all organisms examined to date are very similar in amino acid sequence and encode a protein associated with the transcriptional regulation of a specific gene set. This suggests that the metameric or segmented body plan may have evolved only once.

Genetic Analysis of Behavior

Behavior is defined generally as reaction to stimuli or environment. In broad terms, every action, reaction,

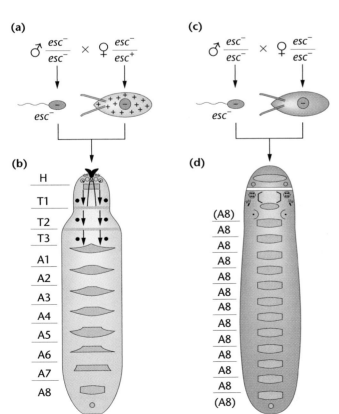

(a)

♂ $\frac{esc^-}{esc^-}$ × ♀ $\frac{esc^-}{esc^+}$

esc⁻

(c)

♂ $\frac{esc^-}{esc^-}$ × ♀ $\frac{esc^-}{esc^-}$

esc⁻

(b)

H
T1
T2
T3
A1
A2
A3
A4
A5
A6
A7
A8

(d)

(A8)
A8
A8
A8
A8
A8
A8
A8
A8
A8
A8
(A8)

■ Figure 20–22 (left) Action of *the extra sex combs* mutation in *Drosophila*. (a) Heterozygous females form wild-type *esc* gene product and store it in the oocyte. (b) Fertilization of esc⁻ egg formed at meiosis by the heterozygous female by esc⁻ sperm still produces wild-type larva with normal segmentation pattern (maternal rescue). (c) Homozygous esc⁻ females produce defective eggs, which, when fertilized by esc⁻ sperm, (d) produce a larva in which most of the segments of the head, thorax, and abdomen are transformed into the eighth abdominal segment. Maternal rescue demonstrates that the esc⁻ gene product is produced by the maternal genome and is stored in the oocyte for use in the embryo (H, head; T1–3, thoracic segments; A1–8, abdominal segments). Borderlines between the head and thorax and between the thorax and abdomen are marked with arrows.

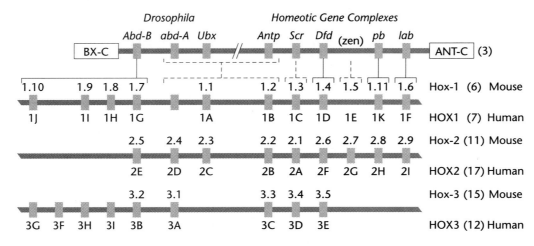

Figure 20–23 The *Hox* genes of mammals and their organizational alignment with the *BX-C* and *ANT-C* complexes of *Drosophila*.

and response represents a type of behavior. Animals run, remain still, or counterattack in the presence of a predator; birds build complex and distinctive nests; fruit flies execute intricate courtship rituals; plants bend toward light; and humans reflexively avoid painful stimuli as well as "behave" in a variety of ways as guided by their intellect, emotions, and culture.

Since about 1950, studies of the genetic component of behavioral patterns have intensified, and support for the importance of genetics in understanding behavior has increased. The prevailing view is that all behavior patterns are influenced both genetically and environmentally. The genotype provides the physical basis that is essential to execute the behavior and further determines the limitations of environmental influences. Behavior genetics has blossomed into a distinct specialty within the larger field of genetics as more behaviors have been found to have genetic components.

Genetics of Geotaxis in *Drosophila*

Several approaches have been used to study the genetics of behavior. In one approach, behavior is compared in two closely related strains. Studies on alcohol preference in mice, and on open field behavior in different strains of mice, are examples of this approach. A second approach uses selection for behavior traits from a heterozygous population. This approach—selection for modified behavior from a genetically heterogeneous population and subsequent inbreeding—leads to the production of strains with significant differences in behavior. Genetic analysis of these strains can identify genes that control these behavior differences. This method has been used in studying a form of behavior in *Drosophila*.

Taxis is the movement toward or away from an external stimulus. The response may be positive or nega-

tive, and the sources of stimulation may include chemicals **(chemotaxis)**, gravity **(geotaxis)**, light **(phototaxis)**, and so on.

To investigate geotaxis in *Drosophila*, Jerry Hirsch and his colleagues designed a mass screening device that tests about 200 flies per trial, as shown in Figure 20–24. In this test, flies are added to a vertical maze. Flies that turn up at each intersection will arrive at the top of the maze; those that always turn down will arrive at the bottom; and those making both "up" and

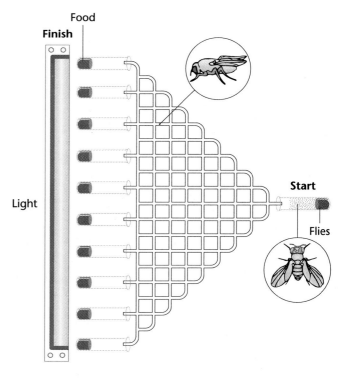

Figure 20–24 Schematic drawing of a maze used to study geotaxis in *Drosophila*.

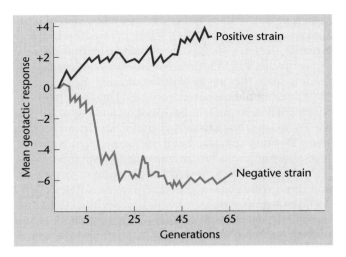

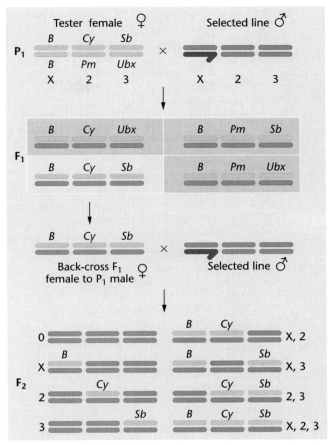

■ Figure 20–25 Selection for positive and negative geotaxis in *Drosophila* over many generations.

■ Figure 20–26 In this mating scheme in *Drosophila*, the effect of genes located on specific chromosomes that contribute to geotaxis can be assessed. The progeny produced by backcrossing the F1 female contain all combinations of chromosomes. Examination of the phenotypes of these flies makes it possible to determine which chromosomes from the selected strain are present. Subsequent testing for geotaxis is then performed *(B, Bar eyes; Cy, Curly wing; Sb, Stubble bristles)*. The designations alongside each genotype (X, 2, 3, etc.) indicate which chromosomes are heterozygous.

"down" decisions will end up somewhere in between. Flies can be selected for both positive and negative geotropism, establishing the genetic influence on this behavioral response.

As shown in Figure 20–25, mean scores may vary from +4.0 to –6.0, corresponding to the number of T-junctions the fly encounters. These data show that selection for negative geotropism is stronger than selection for positive geotropism. The two lines have now undergone selection for almost 30 years, encompassing over 500 generations and the testing of more than 80,000 flies. Throughout the experiment, clear-cut but fluctuating differences have been observed. These results indicate that geotropism in *Drosophila* is controlled in a polygenic fashion.

Hirsch and his colleagues have analyzed the relative contribution of loci on different chromosomes to geotaxis. Loci on chromosomes 2 and 3 and the X were identified in an ingenious way. Crosses were performed to produce flies that were either heterozygous or homozygous for a given chromosome. Figure 20–26 shows how this is accomplished. From a selected line, a male is crossed to a "tester" female, whose chromosomes are each marked with a dominant mutation. Each marked chromosome carries an inversion to suppress the recovery of any crossover products. One of the F1 females heterozygous for each chromosome is backcrossed to a male from the original line. The resulting female offspring contain all combinations of chromosomes. The dominant mutations make it possible to recognize which chromosomes from the selected lines are present in homozygous or heterozygous configurations.

Combinations O and X (Figure 20–26) are homozygous and heterozygous, respectively, for the X chromosome. Similar combinations exist for chromosome 2 (0 and 2) and chromosome 3 (0 and 3). By testing these flies in the maze, it is possible to assess the behavioral influence of genes on any given chromosome.

For negatively geotaxic flies (the ones that go up in the maze), genes on the second chromosome make the largest contribution to the phenotype, followed by loci on chromosome 3 and the X (a gradient of chromosomes 2 > 3 > X). For the positive line (flies that go down in the maze), the reverse arrangement (a gradient of chromosomes X > 3 > 2) is true. The overall results indicate that geotaxis is under polygenic control, and that the loci controlling this trait are distributed on all three major chromosomes of *Drosophila*.

Further genetic testing in which each chromosome from a selected line has been isolated in homozygous form in an unselected background has been used to estimate the number of genes controlling the geotaxic response in *Drosophila*. This work indicates that a small number of genes, perhaps two to four loci, is responsible for this behavior.

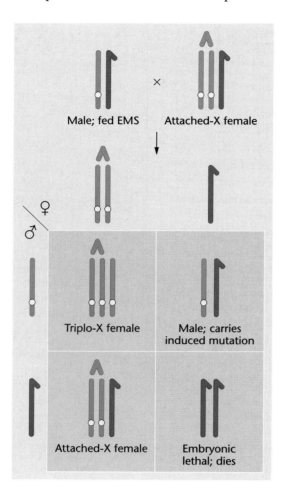

Figure 20–27 Genetic cross in *Drosophila* that facilitates the recovery of X-linked induced mutations. The female parent contains two X chromosomes that are attached, in addition to a Y chromosome. In a cross between this female and a normal male that has been fed the mutagen ethylmethanesulfonate, all surviving males receive their X chromosomes from their father and express all mutations induced on that chromosome.

Male; fed EMS

Attached-X female

♀

♂

Triplo-X female

Male; carries induced mutation

Attached-X female

Embryonic lethal; dies

Genetic Dissection of Behavior in *Drosophila*

In 1967, Seymour Benzer and his colleagues initiated a study of behavior genetics in *Drosophila*. Benzer's approach uses mutant alleles of behavior to "dissect" a complex biological phenomenon into its simpler components. This approach has several goals:

1. Isolate mutants that disrupt normal behavior.
2. Identify the mutant genes by chromosome localization and mapping.
3. Determine the level and structural component at which the gene expression influences the behavioral response.
4. Establish a causal link between the mutant allele and its behavioral phenotype.

All four steps can be illustrated in a discussion of phototaxis, one of the first behaviors studied by Benzer. Normal flies are positively phototactic; that is, they move toward a light source. Mutations were induced by feeding male flies sugar water containing **EMS** (ethylmethanesulfonate, a potent mutagen) and mating them to attached-X virgin females. As shown in Figure 20–27, the F1 males receive their X chromosome from their fathers. Because they are hemizygous, induced X-linked recessive mutations are phenotypically expressed.

The F1 males were tested for phototaxic responses, and those with abnormal behavior were isolated. Benzer found runner mutants, which move quickly to and from light; negative phototactic mutants, which move away from light; and nonphototactic mutants, which show no preference for light or darkness. The genetic basis of these behavioral changes was confirmed by mating these F1 males to attached-X virgin females. Male progeny of this cross also showed the abnormal phototactic responses, indicating that these behavioral alterations are the result of X-linked recessive mutations.

Nonphototactic mutants walk normally in the dark, but show no phototactic response to light. Benzer and Yoshiki Hotta tested electrical activity at the surface of mutant eyes in response to a flash of light. The pattern of electrical activity was recorded as an **electroretinogram**. Various types of abnormal responses were detected in different nonphototactic mutant strains. None of the mutants had a normal pattern of electrical activity. When these mutations were mapped, they were not all allelic; instead, they were shown to occupy several loci on the X chromosome. This analysis indicates that several gene products contribute to the phototactic response.

Mosaics

Where, within the fly, must gene expression occur to produce a normal electroretinogram? In an ingenious approach to answer this question, Benzer turned to the use of **mosaics**. In mosaic flies, some tissues are mutant and others are wild type. If it can be ascertained which part must be mutant in order to yield the abnormal behavior, the **primary focus** of the genetic alteration can be determined.

To produce mosaic flies, Benzer used a *Drosophila* strain that carries one of its X chromosomes in an unstable ring shape. When the ring-X is present in a zygote undergoing cell division, it is frequently lost by nondisjunction. If the zygote is female and has two X chromosomes (one normal and one ring-X), loss of the ring-X at the first mitotic division will result in two cells—one with a single X (normal X) and one with two X chromosomes (one normal and one ring-X). The former cell goes on to produce male tissue (XO) and expresses all alleles on the remaining X, whereas the latter produces female tissue and does not express heterozygous recessive X-linked genes. This is illustrated in Figure 20–28. Loss of the ring-X results in mosaic

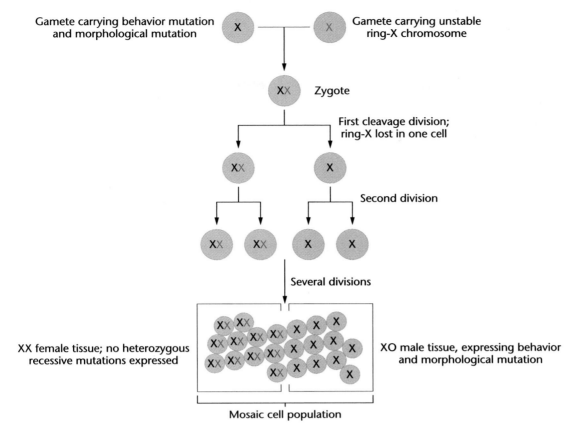

Gamete carrying behavior mutation and morphological mutation

Gamete carrying unstable ring-X chromosome

XX   Zygote

First cleavage division; ring-X lost in one cell

XX   X

Second division

XX   XX   X   X

Several divisions

XX female tissue; no heterozygous recessive mutations expressed

XO male tissue, expressing behavior and morphological mutation

Mosaic cell population

■ Figure 20–28   Production of a mosaic fruit fly as a result of fertilization by a gamete carrying an unstable ring-X chromosome (shown in color). If this chromosome is lost in one of the two cells following the first mitotic division, the body of the fly will consist of one part which is male (XO) and the other part which is female (XX). The male side will express all mutations contained on the X chromosome.

flies with male and female parts that differ in the expression of X-linked recessive genes.

When and where the ring-X is lost in development determines the pattern of mosaicism. The loss usually occurs early in development, before the cells migrate to the embryo surface. As shown in Figure 20–29, different types of mosaics are produced, depending on the orientation of the spindle when loss of the ring-X takes place. If the normal X chromosome carries the behavior mutation and an obvious mutant allele (*yellow,* for example), the pattern of mosaicism will be easy to distinguish. By examining the distribution of body color, flies with a combination of normal and mutant structures can be identified, such as a fly with a mutant head on a wild-type body, or a normal head on a mutant body, or a fly with one normal and one mutant eye on a normal or mutant body, and so on.

When nonphototactic mosaics were studied, the focus of the genetic defect was found to be in the eye itself. In mosaics in which every part of the fly except the eye was normal, abnormal behavior was still detected. When one eye was mutant and the other normal, the fly exhibited an unusual behavior. Instead of crawling straight up toward light as the normal fly does, the mosaic fly with one mutant eye crawled upward to light in a spiral pattern. In the dark, this fly moved in a straight line.

Using a combination of genetics, physiology, and biochemistry, Benzer and other workers in the field have identified and analyzed a large number of genes affecting behavior in *Drosophila.* As shown in Table 20.4, mutants that affect locomotion, response to stress, circadian rhythm, sexual behavior, visual behavior, and even learning have been isolated.

Many mutations have been analyzed with the mosaic technique to localize the focus of gene expression. While it was easy to predict that the focus of the nonphototactic mutant would be in the eye, other mutants are not so predictable. For example, the focus of mutants affecting circadian rhythms has been located in the head, presumably in the brain. A mutant with a wings-up phenotype might have a defect in the wings, muscle attachments in the thorax, abnormal muscle formation, or a neuromuscular defect. Analysis of mosaics established that the defect is in flight muscles of the thorax. Cytological studies have confirmed this finding, showing a complete lack of myofibrils in these muscles.

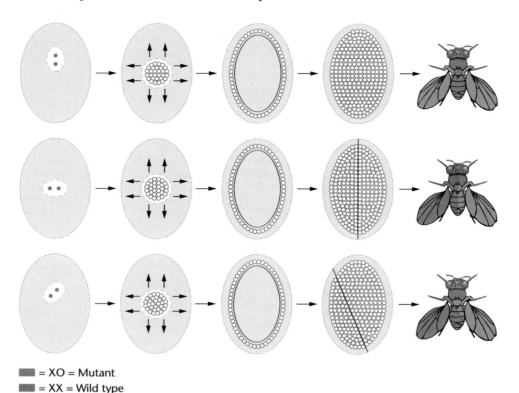

= XO = Mutant
= XX = Wild type

■ Figure 20–29 The effect of spindle orientation on the production of mosaic flies as shown in Figure 20–28.

The mosaic technique has also been used to determine which regions of the brain are associated with sex-specific aspects of courtship and mating behavior. Jeffrey Hall and his associates have shown that mosaics with male cells in the most dorsal region of the brain, the protocerebrum, exhibit the initial stages of male courtship toward females. Later stages of male sexual behavior, including wing vibrations and attempted copulations, require male cells in the thoracic ganglion. Similar studies of female-specific sexual behavior have shown that the ability of a mosaic to induce courtship by a male depends on female cells in the posterior thorax or abdominal region. A region of the brain within the protocerebrum must be female for receptivity to copulation. Anatomical studies have confirmed that there are fine structural differences between the brains of male and female *Drosophila*, indicating that some forms of behavior may be dependent on the development and maturation of specific parts of the nervous system.

Neurogenetics

Analysis of behavior mutants in *Drosophila* has led to an understanding of fundamental mechanisms in the animal nervous system. Nerve impulses are generated at one end of a neuron and move along the cell to the opposite end (Figure 20–30). During this process, sodium and potassium ions move across the plasma membrane. The movement of these electrically charged ions can be monitored by measuring changes in the electrical potential of the neuron's membrane. To screen for genes that control the generation and transmission of nerve impulses, Barry Ganetzky and his colleagues screened behavioral mutants of *Drosophila* to identify those with electrophysiological abnormalities in the generation and propagation of nerve impulses. Two general classes of such mutants have been isolated: those with defects in the movement of sodium, and those with defects in potassium transport (Table 20.5).

One mutant identified in this screening procedure is a temperature-sensitive allele called *paralytic*. Flies that are homozygous for this mutant allele become paralyzed when exposed to temperatures at or above 29°C but they recover rapidly when the temperature is lowered to 25°C. Mosaic studies revealed that both the brain and thoracic ganglia are the focus of this abnormal behavior. Electrophysiological studies showed that mutant flies have defective sodium transport associated with the conduction of nerve impulses. Subsequently, Ganetzsky and his colleagues mapped, isolated, and cloned the paralytic gene. This locus encodes a protein, called the sodium channel, that controls the movement of sodium across the membrane of nerve cells.

A second mutant, called *Shaker*, originally isolated over 40 years ago as a behavioral mutant, encodes a potassium channel gene that has also been cloned and characterized. Because the mechanism of nerve im-

TABLE 20.4	Some behavioral mutants of *Drosophila*	
Class	**Name**	**Characteristics**
Locomotor	sluggish	Moves slowly
	hyperkinetic	High consumption of O_2; shaking of legs; early death
	wings up	Wings perpendicular to body
	flightless	Does not fly well, although wings are well developed
	uncoordinated	Lacks coordinated movements
	nonclimbing	Fails to climb
Response to stress	easily shocked	Mechanical shock induces "coma"
	stoned	Stagger induced by mechanical shock
	shaker	Vibrates all legs while etherized
	freaked out	Grotesque, random gyrations under the influence of ether
	paralyzed	Collapses above a critical temperature
	parched	Dies quickly in low humidity conditions
	tko	Epileptic-like response
	comatose	Paralyzed by cool temperature
	out-cold	Similar to comatose
Circadian rhythm	period^o	Eclosion at any time; locomotor activity spread randomly over the day
	period^s	19-hour cycle rather than 24-hour cycle
	period^l	28-hour cycle rather than 24-hour cycle
Sexual	savior-faire	Males unsuccessful in courtship
	fruitless	Males pursue each other
	stuck	Male is often unable to withdraw after copulation
	coitus interruptus	Males disengage in about half the normal time
Visual	nonphototactic	Blind
	negatively phototactic	Moves away from light
Learning	dunce	Fails to learn conditioned response

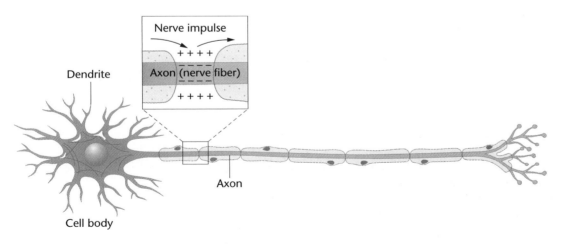

■ Figure 20–30 A nerve cell (neuron) has a cell body and extensions called dendrites that carry impulses toward the cell body and one or more axons that carry impulses away from the cell body. Electrodes placed on either side of the plasma membrane can record the electrical potential across the membrane. At rest, there is more sodium outside the cell, and more potassium inside the cell. When a nerve impulse is generated, sodium moves into the cell, and potassium moves out. As the impulse moves away, ions are pumped across the membrane to restore the electrical potential.

TABLE 20.5	Behavioral mutants of *Drosophila* affecting nerve impulse transmission		
Mutation	*Map Location*	*Ion Channel Affected*	*Phenotype*
nap[ts]	2–56.2	sodium	Adults, larvae paralyzed at 37.5°C. Reversible at 25°C.
para[ts]	1–53.9	sodium	Adults paralyzed at 29°C, larvae at 37°C. Reversible at 25°C.
tip-E	3–13.5	sodium	Adults, larvae paralyzed at 39–40°C. Reversible at 25°C.
sei[ts]	2–106	sodium	Adults paralyzed at 38°C, larvae unaffected. Adults recover at 25°C.
Sh	1–57.7	potassium	Aberrant leg shaking in adults exposed to ether.
eag	1–50.0	potassium	Aberrant leg shaking in adults exposed to ether.
Hk	1–30.0	potassium	Ether-induced leg shaking.
sio	3–85.0	potassium	At 22°C, adults are weak fliers; at 38°C, adults are weak, uncoordinated.

pulse conduction has been highly conserved during animal evolution, the cloned *Drosophila* genes were used as probes to isolate the equivalent human ion channel genes. The cloned human genes are being used to provide new insights into the molecular basis of neuronal activity. One of the human ion-channel genes first identified in *Drosophila* is defective in a heritable form of cardiac arrhythmia. Identification and cloning of the human gene now makes it possible to screen for family members at risk for this potentially fatal condition.

Human Behavior Genetics

Genetic control of behavior in humans has proven more difficult to characterize than that of other organisms. Not only are humans unavailable as experimental subjects in genetic investigations, but the types of responses considered to be the most interesting forms of behavior, including some aspects of intelligence, language, personality, and emotion, are difficult to study. Two problems arise in studying such behaviors. First, all are difficult to define objectively and to measure quantitatively. Second, they are affected by environmental factors. In each case, the environment is extremely important in shaping, limiting, or facilitating the final phenotype for each trait.

Historically, the study of human behavior genetics has been hampered by other factors. First, many studies of human behavior were performed by psychologists without adequate input from geneticists. Second, traits involving intelligence, personality, and emotion have the greatest social and political significance. As such, these traits are more likely to be the subject of sensationalism when reported to the lay public. Because the study of these traits comes closest to infringing upon individual liberties such as the right to privacy, they are the basis of the most controversial investigations.

In lamenting the gulf between psychology and genetics in explaining human behavior genetics, C. C. Darlington in 1963 wrote, "Human behavior has thus become a happy hunting ground for literary amateurs. And the reason is that psychology and genetics, whose business it is to explain behavior, have failed to face the task together." Since 1963, some progress has been made in bridging this gap, but the genetics of human behavior remains a controversial area.

Single-Gene Disorders

Many genetic disorders in humans result in a behavioral abnormality. One of the most prominent examples is **Huntington disease (HD)**. Inherited as an autosomal dominant neurodegenerative disorder, HD affects the nervous system, including the brain. Symptoms usually appear in the fifth decade of life as a gradual loss of motor function and coordination. Structural degeneration of the nervous system is progressive, and personality changes accompany this degeneration. Most victims die within 10 to 15 years after onset of the disease. Because HD usually appears after a family has been started, all children of an affected person must live with the knowledge that they face a 50 percent probability of developing the disorder (affected individuals are usually heterozygotes).

The HD gene has been identified and cloned. It encodes a large protein (348 kDa) called **huntingtin**, which is unrelated to any known gene product. The mutant form of the gene is associated with the presence of extra CAG trinucleotide repeats near the 5′-end (see Chapter 15). Normal individuals have 11 to 24 repeats, but those affected by HD carry 42 to 86 CAG repeats. The CAG repeats add glutamine residues to the protein encoded by the gene. The mutant proteins accumulate in the nuclei of affected brain cells and form spherical deposits that may inhibit normal function, leading to cell death. However, there seems to be no correlation between the number of repeats carried by affected individuals and the severity of the behavioral and psychiatric symptoms. The expansion of a trinucleotide repeat in HD is similar to expansions of other repeats in several disorders that affect the brain and nervous system, indicating that such "stutter" mutations may be a common form of mutation in neurobehavioral traits.

Monoamine oxidases are enzymes that degrade chemical signals (called neurotransmitters) in the nervous system. Recently, studies have established a link between mutation in monoamine oxidase A (MAOA), an

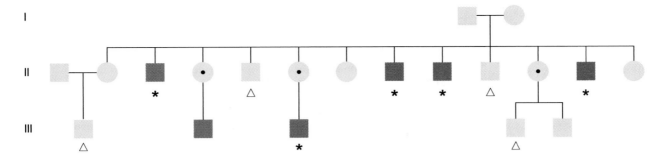

■ Figure 20–31 Partial pedigree of a family in which mental retardation and aggressive behavior segregates with a mutation in the monoamine oxidase A gene (MAOA). Affected males are indicated by filled symbols. Those marked with an asterisk were analyzed and found to be hemizygous for a mutant allele of MAOA. Those marked with a triangle have the normal allele. Female heterozygotes have one normal and one mutant allele.

X-linked gene, and a syndrome that involves mild mental retardation and control of aggressive behavior. A pedigree of one affected family is shown in Figure 20–31. Six males are affected, and all showed a characteristic pattern of aggressive behavior and lack of impulse control. Using a variety of RFLP markers, the locus for this behavior was mapped to the short arm of the X chromosome, near the MAOA locus. Analysis of the structural gene for MAOA in five of the six affected males showed that they carry a single nucleotide substitution that introduces a stop codon causing premature termination of translation, resulting in a lack of MAOA activity. This mutation is carried by heterozygous females, but is not found in normal males. Other studies show that mutations and polymorphisms in the MAOA gene are not common, strengthening the link between a mutation in this gene and an abnormal form of behavior.

Other metabolic disorders are also known to affect mental function. These include **Lesch–Nyhan syndrome**, an X-linked recessive disorder. Onset is within the first year, and the disease is most often fatal early in childhood. The disorder involves a defect in purine biosynthesis. Affected individuals lack an enzyme, hypoxanthine-guanine phosphoribosyltransferase (HGPRT), and accumulate high levels of uric acid. Symptoms include mental and physical retardation, and affected individuals demonstrate uncontrolled self-mutilation. **Tay–Sachs disease**, an autosomal recessive disorder that maps to chromosome 15, is characterized by severe mental retardation. The disease is apparent soon after birth and is fatal. The autosomal recessive disease **phenylketonuria**, unless detected and treated early, also results in mental retardation. All of these disorders alter the normal biochemistry of the affected individuals and are inherited in a Mendelian fashion.

Multifactorial Traits

Other aspects of human behavior, notably **schizophrenia** and **bipolar disorder**, have been the subject of extensive investigations. Twins studies and studies on adopted and natural siblings have supported the idea that bipolar disease has a genetic component. Mapping studies originally identified loci on the X chromosome and chromosome 11 as being associated with bipolar disorder, but these studies were later invalidated. Several laboratories are using RFLP markers (RFLP markers are discussed in Chapters 17 and 18) to screen large pedigrees in which manic-depression is segregating in an attempt to find loci for this disorder, which affects about 1 percent of the U.S. population. Linkage between susceptibility to bipolar disorder and one or more loci on chromosome 18 have recently been reported. Work is now progressing to identify the genes at these loci, and their role in this complex disorder.

Schizophrenia is a mental disorder characterized by withdrawn, bizarre, and sometimes delusional behavior. Affected individuals are unable to lead organized lives and are periodically disabled by the condition. It is clearly a familial disorder, with relatives of schizophrenics having a much higher incidence of this disorder than the general population. Furthermore, the closer the relationship to the index case or proband, the greater is the probability of the disorder occurring.

The concordance of schizophrenia in monozygotic and dizygotic twins has been the subject of many studies. In almost every investigation, concordance has been higher in monozygotic twins than in dizygotic twins reared together. Although these results suggest that a genetic component exists, they do not reveal the precise genetic basis of schizophrenia. Simple monohybrid and dihybrid inheritance as well as multiple-gene control have been proposed for schizophrenia. It seems unlikely that only one or two loci are involved, nor is it likely that the control is strictly quantitative, as in polygenic inheritance. As with bipolar illness, a large-scale collaborative effort is using DNA markers to identify genomic regions that may contain loci controlling this behavioral disorder. Once they are identified, these regions will be studied in detail to identify, isolate, and clone genes for schizophrenia.

GENETICS, TECHNOLOGY, AND SOCIETY

Coming to Terms with the Heritability of IQ

Few topics in human heredity are as contentious as the genetics of intelligence. The debate concerning "genetic" vs. "environmental" influences over intelligence can have far-reaching societal ramifications, and may even affect the formulation of educational policy. What does genetic analysis tell us about the hereditary basis of intelligence?

To answer this question, the first challenge is to define and assess intelligence. From a genetic standpoint, is it a single phenotypic entity, and if so, can it be measured? Obviously, intelligence is not a physical characteristic like height or weight, which are easy to see and easy to measure. However, performance on IQ tests has been taken as a measure of intelligence since Alfred Binet's original test was introduced in the United States in 1916. Despite many objections, this practice is unlikely to change. The debate about the nature of intelligence, whether or not it is an innate, mental process that can be reliably assessed, is certain to continue for some time.

What about the genetic basis for IQ scores? For a complex trait such as intelligence, there is no readily discernible chain of cause-and-effect events leading from genotype to phenotype, as there is for, say, albinism or hemophilia. Genetic analysis of intelligence therefore requires the assumption that it is a quantitative trait, whereby the final phenotype is influenced by the additive contributions of many genes, as well as a variety of environmental factors. Statistical techniques can then be employed that were first developed for the analysis of other quantitative traits in plants and animals. The most frequently used statistic is heritability.

Heritability is the fraction of the total variance in the phenotypic expression of a trait within a specific population that is due to genetic factors. The heritability of many traits in plants and animals can be estimated from the results of breeding experiments. Several methods to estimate heritabilities have been used in humans.

One method involves examining sets of identical twins and fraternal twins.

Identical (monozygotic) twins develop from a single zygote and thus are genetically identical, whereas fraternal (dizygotic) twins are no more alike genetically than any other siblings, sharing, on average, 50 percent of their genes. If the trait in question has a genetic basis, then, given common environments, identical twins should resemble each other more closely than fraternal twins do. A still better way is to study identical twins who have been separated very early in life and reared apart. In theory, this allows the effects of a common genetic inheritance to be distinguished from the effects of different environments.

Reflecting the uncertainties of such approaches, these and similar methods yield a range of heritability values, from a low of around 0.30 to a high of around 0.80. A commonly accepted heritability value for IQ score is 0.60. So let's consider what a heritability of 0.60 for IQ score really means, and, perhaps more important, what it does *not* mean. Strictly speaking, a heritability of 0.60 for IQ score means that *within the population under study*, 60 percent of the variance in IQ scores can be attributed to different genotypes within the population.

It should be understood, first of all, that heritability is a statistic pertaining to variation within a population and has no meaning when applied to an individual. It is therefore not correct to say that 60 percent of a person's IQ is determined by genes and the other 40 percent by environment.

Second, heritability estimates pertain to variation only *within* the specific population that was measured and not to variation *between* populations. A heritability value of 0.60 *does not* mean that 60 percent of the difference in IQ scores between two separate populations (or groups) is due to different genes.

Third, the heritability of a trait is not a permanent characteristic of a population. In fact, a heritability value, whether for IQ score or anything else, pertains to that particular population *at a specific time*, under the conditions then prevailing. Heritability values may change as environmental factors change, and may thus vary from one year to the next. Perhaps most important, "heritable" does not therefore equal "inevitable." A high heritability

value, such as 0.60, does not mean that the trait cannot be modified substantially by environmental factors. The genotype may determine an upper and lower limit within a broad range of possible phenotypes, but exactly what phenotype occurs in an individual depends on interactions with the environment.

Attempts to elucidate the genetic basis of human intelligence will remain controversial, and for good reason. The results of such studies, especially when misinterpreted, can be taken as a biological justification for social policies that may be viewed as discriminatory. For example, some people prefer to believe that the problems of the disadvantaged in our society are not the result of social forces beyond their control, but are based on their genes. Recently, heritability estimates of IQ scores have been used to bolster claims that intelligence is largely genetic and not significantly influenced by the environment in which a child finds himself or herself. On the basis of such faulty reasoning, it has been concluded that efforts to boost IQ through better educational and social interventions are doomed to failure.

It should be clear that the heritability of intelligence, even an estimate as high 0.60, is an inadequate foundation on which to base such claims and upon which to construct social policy. This is not to say that intelligence, however it is defined and measured, is not influenced to a substantial degree by the genes each of us has inherited, or that differences in IQ score will not be found in populations that are studied. However, we cannot allow the heritability of IQ score to serve as the rationale for making policies that discriminate against anyone in our society.

References

Herrnstein, R. J., and Murray, C. 1994. *The bell curve: Intelligence and class structure in American life.* New York: The Free Press.

Jacoby, R., and Glauberman, N. (eds.). 1995. *The bell curve debate.* New York: Times Books.

Terman, L. M., and Merrill, M. A. 1973. *Stanford-Binet intelligence scale: 1972 norms edition.* Boston: Houghton Mifflin.

INSIGHTS
and
SOLUTIONS

1. The timing of differential gene action during development is a key factor in the normal developmental program. If a given gene has been cloned, the time of action and range of cell types in which the gene is active can be determined using a number of molecular techniques. However, when a cloned gene or its transcripts are not available, genetic methods can be used to establish a comparative order of gene action among two or more genes. By extending this analysis to include several genes, a relative order of gene action can be established. This order can serve as a starting point for investigations at the molecular level by providing a developmental time scale for gene action. The following example shows how such a relative time scale for the action of two genes can be constructed.

In *Drosophila*, the autosomal recessive gene *lozenge-clawless (lzcl)* produces multiple abnormalities of the female genitals, eyes, and tarsal regions of the legs. Homozygous females have abnormal genitals, no sperm storage organs, no ovarian glands, and are sterile. Homozygous mutant males, on the other hand, have normal genitals and are fully fertile. A second autosomal recessive gene, *transformer (tra),* converts XX females into phenotypic males (recall that in *Drosophila,* XX flies are female, XY flies are male). XX flies that are homozygous for both *lzcl* and *tra* are phenotypically male with normal genitals. What do these results say about the normal sequence of action of these two genes? Which one acts first during development?

Solution: The flies in question are genetically female, since they have two X chromosomes. Females that are homozygous for the *lzcl* gene are expected to have abnormal genitalia. In this case, the action of the *tra* gene in changing the female phenotype into the male phenotype (with the development of male genitalia) must take place before the action of the *lzcl* gene, since the XX flies have a normal male phenotype.

2. Bipolar illness is an affective disorder associated with recurring mood changes. It is estimated that one in four individuals will suffer from some form of affective disorder at least once in his or her lifetime. Genetic studies indicate that bipolar illness is familial, and that single genes may play a major role in controlling this behavioral disorder. In 1987 two separate studies using RFLP analysis and other genetic markers reported linkage between bipolar illness and markers on the X chromosome and to the short arm of chromosome 11. At the time, these reports were hailed as landmark discoveries, opening the way to the isolation and characterization of genes that control specific forms of behavior, and to the development of therapeutic strategies based on knowledge of the nature of the gene product and its action. However, further work on the same populations (reported in 1989 and 1990) demonstrated that the original results were invalid, and it was concluded that no linkage to markers on the X chromosome and to markers on chromosome 11 could be established. These findings do not exclude the role of major genes on the X chromosome and autosomes in bipolar illness, but they do exclude linkage to the markers used in the original reports. This setback not only caused embarrassment and confusion but also forced reexamination of the validity of the methods used in mapping human genes, and the analysis of data from linkage studies involving complex behavioral traits. Several factors have been proposed to explain the flawed conclusions reported in the original studies. What do you suppose some of these factors to be?

Solution: While some of the criticisms were directed at the choice of markers, most of the factors at work in this situation appear to be related to the phenotype of bipolar illness. At least three confounding elements have been identified. One is age of onset. There is a positive correlation between age and the appearance of bipolar illness. Therefore, at the time of a pedigree study, younger individuals who will be affected later in life may not show any signs of bipolar illness. Another confusing factor is that in the populations studied, there may in fact be more than one major X-linked gene and more than one major autosomal gene that trigger bipolar illness. A third factor relates to the diagnosis of bipolar illness itself. The phenotype is complex and not as easily quantified as height or weight. In addition, mood swings are a universal part of everyday life, and it is not always easy to distinguish transient mood alterations and the role of environmental factors from affective disorders having a biological and/or genetic basis. These factors point up the difficulty in researching the genetic basis of complex behavioral traits. Individually or in combinations, the factors described above might skew the results enough so that guidelines for proof that are adequate for other traits are not stringent enough for behavioral traits with complex phenotypes and complex underlying causes.

Chapter Summary

1. The variable gene activity model assumes that all somatic cells in an organism contain equivalent but differentially expressed genetic information. Nuclear transplantation studies have demonstrated that somatic cells undergoing differentiation retain an intact genome, which can be reprogrammed to direct the development of the entire organism.

2. In *Drosophila*, genetic and molecular studies have confirmed that the egg contains information that specifies the body plan of the larva and adult, and that interaction of embryonic nuclei with the maternal cytoplasm initiates a transcriptional program characteristic of a specific developmental pathway.

3. Behavioral genetics has emerged as an important specialty within the field of genetics. Both genotype and environment have been found to have an impact in determining an organism's behavioral response.

4. By isolating mutations that cause deviations from normal behavior, the role of a corresponding wild-type allele in the respective response is established. Numerous examples have been examined covering a variety of behaviors in *Drosophila*.

5. Aspects of human behavior are difficult to study because the individual's environment makes an important contribution toward trait development. In humans, studies using twins have shown that while a family may have a predisposition to manic-depression or schizophrenia, the expression of this disorder may be modified by the environment.

Key Terms

behavior, 459
bipolar disorder, 467
bithorax complex (BX-C), 457
cell–cell interaction, 448
chemotaxis, 460
compartment, 451
determination, 447
differentiation, 447
electroretinogram, 462
emotion, 466
EMS, 462
fate map, 450
gap gene, 456

geotaxis, 460
homeobox, 459
homeodomain, 454
homeotic mutant, 457
Hox gene clusters, 459
huntingtin, 466
Huntington disease (HD), 466
imaginal disc, 449
Lesch–Nyhan syndrome, 467
monoamine oxidases, 466
mosaics, 462
pair-rule genes, 456
phenylketonuria, 467

phototaxis, 460
primary focus, 462
schizophrenia, 467
segment polarity, 456
segmentation gene, 452
selector gene, 452
taxis, 460
Tay–Sachs disease, 467
totipotent, 448
variable gene activity
   hypothesis, 448

Problems and Discussion Questions

1. Carefully distinguish between the terms *differentiation* and *determination*. Which phenomenon occurs earlier in development?

2. Nuclei from almost any species can be injected into frog oocytes. Studies have shown that these nuclei remain active in transcription and translation. How can such an experimental system be useful in developmental genetic studies?

3. The concept of epigenesis indicates that an organism develops by forming cells that acquire new structures and functions, which become greater in number and complexity as development proceeds. This theory is in contrast to the preformationist doctrine that miniature adult entities are contained in the egg and must merely unfold and grow to give rise to a mature organism. Why is the epigenetic theory held as correct today?

4. Both the *ftz* gene and the *engrailed* gene encode homeobox transcription factors and are capable of eliciting the expression of other genes. Both genes work at about the same time during development and in the same region to specify cell fate in body segments. The question is: Does *ftz* regulate the expression of *engrailed,* or does *engrailed* regulate *ftz*? Or are they both regulated by another gene? To answer these questions, mutant analysis is performed. In *ftz2* embryos (*ftz/ftz*), engrailed protein is absent; in *engrailed2* embryos (*eng/eng*), *ftz* expression is normal. What does this tell you about the regulation of these two genes? Does the *engrailed* gene regulate *ftz*? Does the *ftz* gene regulate *engrailed*?

5. What are the advantages of using *Drosophila* for studying behavior genetics?

6. Assume that you discovered a fruit fly that walked with a limp and was continually off balance as it moved. Describe how you would determine if this behavior was due to an injury (induced in the environment) or was an

inherited trait. Assuming that it is inherited, what are the various possibilities for the focus of gene expression that causes the imbalance? Describe how you would locate the focus experimentally if it were X-linked.

7. In humans, the chemical phenylthiocarbamide (PTC) is either tasted or not. When the offspring of various combinations of taster and nontaster parents are examined, the following data are obtained:

Parents	*Offspring*
Both tasters	All tasters
Both tasters	1/2 tasters 1/2 nontasters
Both tasters	3/4 tasters 1/4 nontasters
One taster/One nontaster	All tasters
One taster/One nontaster	1/2 tasters 1/2 nontasters
Both nontasters	All nontasters

Based on these data, how is PTC tasting behavior inherited?

8. J. P. Scott and J. L. Fuller studied 50 traits in five pure breeds of dogs. Almost all traits varied significantly in the five breeds, but very few bred true in crosses. What can you conclude with respect to the genetic control of these behavioral traits?

Selected Readings

Bastock, M. 1956. A gene which changes a behavior pattern. *Evolution* 10:421–29.

Beachy, P., Helfand, S., and Hogness, D. 1985. Segmental distribution of bithorax complex proteins during *Drosophila* development. *Nature* 313:545–51.

Beerman, W., and Clever, U. 1964. Chromosome puffs. *Sci. Am.* (April) 210:50–58.

Benzer, S. 1973. Genetic dissection of behavior. *Sci. Am.* (Dec.) 229:24–37.

Berg, H. 1988. A physicist looks at bacterial chemotaxis. *Cold Spring Harbor Symp. Quant. Biol.* 53(1):1–9.

Brenner, S. 1974. The genetics of *Caenorhabditis elegans*. *Genetics* 77:71–94.

Briggs, R., and King, T. 1952. Transplantation of living nuclei from blastula cells into enucleated frog eggs. *Proc. Natl. Acad. Sci. USA* 38:455–63.

DeRobertis, E., Oliver, G., and Wright, C. 1990. Homeobox genes and the vertebrate body plan. *Sci. Am.* (July) 262:46–52.

Devor, E., and Cloninger, C. 1989. The genetics of alcoholism. *Annu. Rev. Genet.* 23:19–36.

Duboule, D., and Morata, G. 1994. Colinearity and functional hierarchy among genes of the homeotic complexes. *Trends Genet.* 10:358–64.

Farmer, A. E., Williams, J., and Jones, I. 1994. Phenotype definitions of psychotic illness for molecular genetic research. *Am. J. Med. Genet.* 54:365–71.

Gurdon, J. 1968. Transplanted nuclei and cell differentiation. *Sci. Am.* (Dec.) 219:24–35.

Herrnstein, R. J., and Murray, C. 1994. *The bell curve: Intelligence and class structure in American life.* New York: The Free Press.

Jacoby, R., and Glaubermann, N. 1995. *The bell curve debate.* New York: Times Books.

Lawrence, P. A., and Morata, G. 1994. Homeobox genes: Their function in *Drosophila* segmentation and pattern formation. *Cell* 78:181–89.

Manseau, L., and Schupbach, T. 1989. The egg came first, of course! Anterior-posterior pattern formation in *Drosophila* embryogenesis and oogenesis. *Trends Genet.* 5:400–5.

McGinnis, W., and Krumlauf, R. 1992. Homeobox genes and axial patterning. *Cell* 68:283–302.

Schultz, C., and Tantz, D. 1995. Zygotic caudal regulation by hunchback and its role in abdominal segment formation of the *Drosophila* embryo. *Development* 121:1023–28.

Slack, J., and Tannahill, D. 1992. Mechanism of anteroposterior axis specification in vertebrates. Lessons from the amphibians. *Development* 114:285–302.

Tantz, D., and Sommer, R. J. 1995. Evolution of segmentation genes in insects. *Trends Genet.* 11:23–27.

Wilson, D. S., and Desplan, C. 1995. Homeodomain proteins. Cooperating to be different. *Curr. Biol.* 5:32–34.

Lady-bird beetles capping a daisy in the Chiricahua Mountains in Arizona.

CHAPTER OUTLINE

CHAPTER
21

Population Genetics

Chapter Concepts

Individuals can carry only two different alleles of a given gene. A group of individuals can carry a larger number of different alleles, giving rise to a reservoir of genetic diversity. The diversity contained in the population can be measured by the Hardy–Weinberg law. Mutation is the ultimate source of genetic variation, and other factors such as drift, migration, and selection can alter the amount of genetic variation in populations.

When Charles Darwin published *The Origin of Species* in 1859 following work co-authored with Alfred Russel Wallace on the likely mechanism of natural selection, the foundation for the modern interpretation of evolution was established. Although organisms are capable of reproducing in an exponential fashion, Wallace and Darwin observed that this growth potential of species is not realized. Instead, population numbers remain relatively constant in nature. Both Wallace and Darwin deduced that some form of competitive struggle for survival must therefore occur. Darwin observed that there is a natural variation between individuals within a species; on this observation, he based his theory of natural selection: "... that any being, if it vary however slightly in any manner profitable to itself ... will have a better chance of surviving." Darwin included in his concept of survival "not only the life of the individual, but success in leaving progeny."

While Gregor Mendel was familiar with Darwin's work, Darwin was unaware of any underlying mechanism to account for the morphological variation he had observed. However, with the development of the concept of genes and alleles, the genetic basis of inherited variation was established.

As others pursued the study of evolution, it became apparent that the population rather than the individual was the unit of study in this process. In order to study the role of genetics in the process of evolution, therefore, it was necessary to consider allele frequencies in populations rather than offspring from individual matings. Thus arose the discipline of **population genetics**.

Early in the twentieth century, a number of workers, including Gudny Yule, William Castle, Godfrey Hardy, and Wilhelm Weinberg, formulated the basic principles of this field. In the early years of population genetics, the emphasis was on theory and the development of mathematical models to describe the genetic structure of populations. Workers in this field include Sewall Wright, Ronald Fisher, and J. B. S. Haldane. Following their work, experimentalists and field workers have used biochemical and molecular techniques to measure variation at the protein and DNA levels directly, in order to test these theories and models. Allele frequencies and the forces that alter these frequencies, such as mutation, migration, selection, and random genetic drift, have been and are being examined. In this chapter we shall consider some general aspects of population genetics and also discuss other areas of genetics relating to evolution.

Populations and Gene Pools

Members of a species are often distributed over a wide geographic range. A **population** is a local group belonging to a single species, within which mating is actually or potentially occurring. The set of genetic information carried by all interbreeding members of a population is called the **gene pool**. For a given locus, this pool includes all the alleles of that gene that are present in the population. In population genetics the focus is on groups rather than on individuals, and on the measurement of allele and genotype frequencies in succeeding generations rather than on the distribution of genotypes resulting from a single mating. The term **allele frequency** will be used throughout the chapter and will represent the frequency of alleles, in contrast to genotype frequencies. Gametes produced by one generation form the zygotes of the next generation. This new generation has a reconstituted gene pool that may differ from that of the preceding generation.

Populations are dynamic; they may grow and expand or diminish and contract through changes in birth or death rates, by migration, or by merging with other populations. This has important consequences and, over time, can lead to changes in the genetic structure of the population.

Calculating Allele Frequencies

One approach used to study a population's genetic structure is to measure the frequency of a given allele. This is possible once the mode of inheritance and the number of different alleles of this gene present in the population have been established. Allele frequencies cannot always be determined directly because in many cases only phenotypes, and not genotypes, can be observed. If, however, alleles expressed in a codominant fashion are considered, there is a direct relationship between phenotypes and genotypes, with each phenotype having a unique genotype. Such is the case with the autosomally inherited **MN blood group** in humans.

In this case, the gene L on chromosome 2 has two alleles, L^M and L^N (often referred to as M and N, respectively).* Each allele controls the production of a distinct antigen on the surface of red blood cells. Thus, the genotype of any individual in a population may be type M ($L^M L^M$), N ($L^N L^N$), or MN ($L^M L^N$).

The genotypes, phenotypes, and immunological reactions of the MN blood group are shown in Table 21.1. Because they are codominant (and genotypes may be inferred from phenotypes), the frequency of the M and N alleles in a population can be determined simply by counting the number of individuals with each phenotype. As an example, consider a population of 100 individuals of which 36 are type M, 48 are type MN, and 16 are type N. The 36 type M individuals represent 72 M alleles, and the 48 MN heterozygotes represent an additional 48 M alleles. This means there are 72 + 48 or 120 M alleles in the population, of a total of 200 alleles at this locus (100 individuals, each with two alleles). Therefore, the frequency of the M allele in this population is 0.6 (120/200 = 0.6 = 60%). The frequency of the

*L stands for Karl Landsteiner, the geneticist for whom the locus is named.

TABLE 21.1	**MN blood groups**		
		Reaction with Antibodies	
Genotype	**Blood Type**	**Anti-M**	**Anti-N**
$L^M L^M$	M	+	–
$L^M L^N$	MN	+	+
$L^N L^N$	N	–	+

N allele can be estimated in a similar fashion (80/200 = 0.4 = 40%) and is 0.4.

Table 21.2 illustrates two methods for computing the frequency of M and N alleles in a hypothetical population of 100 individuals, and Table 21.3 lists the frequencies of M and N alleles actually measured in several human populations.

The Hardy–Weinberg Law

In the case of the MN blood group, M and N are codominant alleles. If, on the other hand, one allele were recessive, the heterozygotes would be phenotypically identical to the homozygous dominant individuals, and the frequency of the alleles could not have been determined directly. However, a mathematical model developed independently by the British mathematician Godfrey H. Hardy and the German physician Wilhelm Weinberg can be used to calculate allele frequencies in this case. The **Hardy–Weinberg law (HWL)** is one of the fundamental concepts in population genetics. Under a set of assumptions, the HWL predicts that genotype and allele frequencies will remain constant from generation to generation. The HWL has three important properties:

1. Allele frequencies predict genotype frequencies.

2. At equilibrium, allele and genotype frequencies do not change from generation to generation.

3. Equilibrium is reached in one generation of random mating.

Assumptions for the Hardy–Weinberg Law

In the Hardy–Weinberg law, the following conditions are presumed:

1. The population is infinitely large, which in practical terms means that the population is large enough that sampling errors and random effects are negligible.

2. Mating within the population occurs at random.

3. There is no selective advantage for any genotype; that is, all genotypes produced by random mating are equally viable and fertile.

4. There is an absence of other factors, including mutation, migration, and random genetic drift.

TABLE 21.2   Methods of determining allele frequencies for codominant alleles

A. Counting Alleles

Genotype/Phenotype	MM	MN	NN	Total
Number of individuals	36	48	16	100
Number of M alleles	72	48	0	20
Number of N alleles	0	48	32	80
Total number of alleles	72	96	32	200

Frequency of M in population: $\dfrac{120}{200} = 0.6 = 60\%$

Frequency of N in population: $\dfrac{80}{200} = 0.4 = 40\%$

B. From Genotypes

Genotype/Phenotype	MM	MN	NN	Total
Number of individuals	36	48	16	100
Genotype frequency	36/100 = 0.36	48/100 = 0.48	16/100 = 0.16	1.00

Frequency of M in population: 0.36 + (1/2)0.48 = 0.36 + 0.24 = 0.60 = 60%

Frequency of N in population: 0.16 + (1/2)0.48 = 0.16 + 0.24 = 0.40 = 40%

TABLE 21.3   Frequencies of M and N alleles in various populations

Population	Genotype Frequency (%)			Allele Frequency	
	MM	MN	NN	M	N
Eskimos (Greenland)	83.48	15.64	0.88	0.913	0.087
U.S. Indians	60.00	35.12	4.88	0.776	0.224
U.S. whites	29.16	49.38	21.26	0.540	0.460
U.S. blacks	28.42	49.64	21.94	0.532	0.468
Ainus (Japan)	17.86	50.20	31.94	0.430	0.570
Aborigines (Australia)	3.00	29.60	67.40	0.178	0.822

In such an ideal population, suppose that a locus has two alleles, A and a. The frequency of the allele A in both eggs and sperm is represented by p, and the frequency of a in gametes is represented by q. Because the sum of p and q represents 100 percent of the alleles for that gene in the population, $p + q = 1$. A diagram can be used to represent the random combinaton of gametes containing these alleles and the resulting genotypes (Figure 21–1).

In the random combination of gametes in the population, the probability that sperm and egg both contain the A allele is $p \times p = p^2$. Similarly, the chance that gametes will carry unlike alleles is $(p \times q) + (p \times q) = 2pq$, and the chance that a homozygous recessive individual will result is $q \times q = q^2$. Note that these terms also describe one of the characteristics of the HWL: Allele frequencies determine genotype frequencies. In other words, while the value p^2 is the probability that both gametes in a fertilization event will carry the A allele, it is also a measure of the frequency of the AA homozygous genotype in the ensuing generation. In a similar way, $2pq$ describes the frequency of Aa heterozygotes, and

q^2 is a measure of the frequency of homozygous recessive (aa) zygotes. Thus, the distribution of homozygous and heterozygous genotypes in the next generation can be expressed as

$$p^2 + 2pq + q^2 = 1$$

■ Figure 21–1   Gametes represent withdrawals from the gene pool to form the genotypes of the next generation. In this example, males and females have the same frequency (p) of the dominant allele A, and the same frequency (q) for the recessive allele a. After mating, the three genotypes, AA, Aa, and aa, have the frequencies of p^2, $2pq$, and q^2, respectively.

Next, consider a population in which 70 percent of the alleles for a given gene are *A,* and 30 percent are *a.* Thus, $p = 0.7$ and $q = 0.3$, and $p(0.7) + q(0.3) = 1$. The distribution of genotypes produced by random mating is shown in Figure 21–2. In the new generation, 49 percent (p^2) of the individuals will be homozygous dominant, 42 percent ($2pq$) will be heterozygous, and 9 percent (q^2) will be homozygous recessive. The frequency of the *A* allele in the new generation can be calculated as

$$p^2 + \tfrac{1}{2}(2pq)$$
$$0.49 + \tfrac{1}{2}(0.42)$$
$$0.49 + 0.21 = 0.70$$

For *a,* the frequency is

$$q^2 + \tfrac{1}{2}(2pq)$$
$$0.09 + \tfrac{1}{2}(0.42)$$
$$0.09 + 0.21 = 0.30$$

Since $p + q = 1$, the value for *a* could have been calculated as

$$q = 1 - p$$
$$= 1 - 0.70 = 0.30$$

The frequencies of *A* and *a* in the new generation are the same as in the previous generation.

If Hardy–Weinberg conditions are assumed, allele frequencies and the frequencies of other genotypes can be calculated from knowing only the frequency of one genotype (such as the homozygous recessive). In human genetics, this relationship is used to calculate the frequency of heterozygotes carrying a recessive allele for a genetic disorder (see below).

A population in which the frequency of a given allele remains constant from generation to generation is said to be in a state of **genetic equilibrium**. In this case, the frequencies of *A* and *a* remain constant.

Several important points are relevant to the preceding example. First, while the hypothetical alleles we

have considered were at equilibrium, not all alleles in a population are. This is particularly true when the assumptions made in the Hardy–Weinberg law do not hold. Second, the examples illustrate why dominant traits do not tend to increase in frequency as new generations are produced. Finally, the examples demonstrate that genetic equilibrium and genetic variability can be maintained in a population. Once they are established in a population, allelic frequencies remain unchanged during equilibrium—a factor that is important to the evolutionary process.

Testing for Equilibrium

The Hardy–Weinberg Law can also be used to determine whether genotypes in a given population are in equilibrium. In a natural population, any of the assumptions of the HWL (size, random mating, no selection) may not be valid. To do this, one must be able to identify heterozygotes phenotypically. If this is possible, then it must be ascertained whether the existing population fits a $p^2 + 2pq + q^2 = 1$ relationship. If not, some factor (presumably selection, mutation, or migration) is causing allele frequencies to shift with each successive generation.

MN blood group distribution of allele frequencies in Australian aborigines is a good example of this application of the Hardy–Weinberg law. From the values of allele frequencies in Table 21.3 (0.178 for *M* and 0.822 for *N*), the expected frequencies of blood types M, MN, and N can be calculated to determine if the population is in equilibrium:

Expected frequency
of type M $\quad = p^2$
$\quad = (0.178)^2 = 0.032$
$\quad = 3.2\%$

Expected frequency
of type MN $\quad = 2pq$
$\quad = 2(0.178)(0.822) = 0.292$
$\quad = 29.2\%$

Expected frequency
of type N $\quad = q^2$
$\quad = (0.822)^2 = 0.676$
$\quad = 67.6\%$

These **expected frequencies** are nearly identical to the **observed frequencies** shown in Table 21.3, confirming that the population is in equilibrium. If there were a question as to whether the observed frequencies varied significantly from the expected frequencies, a chi-square analysis could be performed (see Chapter 3).

On the other hand, if the Hardy–Weinberg test demonstrates that the population is not in equilibrium, one or more of these necessary conditions are not being met. Using the values listed for *M* and *N* frequencies in Table 21.3, we can construct a hypothetical mixed population of 500 Australian aborigines and 500 Native Americans. The frequencies for *M* and *N* for this hypo-

	Sperm	
	$A(p = 0.7)$	$a(q = 0.3)$
$A(p = 0.7)$	*AA* p^2 (0.49)	*Aa* pq (0.21)
Eggs		
$a(q = 0.3)$	*Aa* pq (0.21)	*aa* q^2 (0.09)

■ Figure 21–2 In this population, the frequency of the *A* allele is $p = 0.7$, and the frequency of the *a* allele is $q = 0.3$. Using the formula $p^2 + 2pq + q^2$, the frequencies of the genotypes in the next generation can be calculated as $AA = 0.49$, $Aa = 0.42$ and $aa = 0.09$. The frequencies of *A* and *a* remain constant from generation to generation.

TABLE 21.4	Frequencies of *M* and *N* alleles and genotypes in natural and artificial populations				
	Genotype Frequencies			*Allele Frequencies*	
Population	**MM**	**MN**	**NN**	**p**	**q**
Australian aborigines	0.03	0.296	0.674	0.178	0.822
American Indians	0.60	0.351	0.049	0.776	0.224
Mixed population (500 + 500):					
Observed	0.315	0.324	0.361	0.477	0.523
Expected	0.228	0.498	0.274		

Figure 21–3 If mating in a population with differences in allele frequencies follows the expectations of the Hardy–Weinberg law, an equilibrium will be reached in one generation. Compare the genotype frequencies after one generation of mating with those in Table 21.4 for a population in equilibrium.

thetical population are shown in Table 21.4. In such a mixed population, the frequency of *M* (*p*) would be

$$0.315 + \tfrac{1}{2}(0.324) = 0.477$$

and that of *N* (*q*) would be

$$0.361 + \tfrac{1}{2}(0.324) = 0.523$$

If these allelic frequencies were derived by random mating, then the proportions of phenotypes we should expect are MM (p^2), 22.8 percent; MN ($2pq$), 49.8 percent; and NN (q^2), 27.4 percent. Because the expected genotype frequencies do not fit those observed (and a chi-square test would confirm this), we would conclude that the population is in a state of nonequilibrium brought about, obviously, by a lack of random mating in this hypothetical population.

What would happen, however, if such a mixed population were to be established on an island with no previous inhabitants and mating did take place at random? How long would it take for equilibrium to be established? Applying the Hardy–Weinberg equation, we can predict that after one generation the observed and expected genotype frequencies would converge, illustrating the third important characteristic of the HWL. Using the *p* and *q* values just calculated, and remembering that the allele frequencies represent those of the total gene pool of the first-generation mixed population, the frequencies after one generation of random mating are shown in Figure 21–3. We can see that after one generation, the observed genotype frequencies of 22.8 percent for MM (p^2); 49.8 percent for MN ($2pq$); and 27.4 percent for NN (q^2) are those expected for an equilibrium population. The allele frequencies of $M = 0.477$ and $N = 0.523$ are also those expected for an equilibrium population. However, to demonstrate HWL equilibrium rigorously, we would need to calculate the frequency of matings.

Extensions of the Hardy–Weinberg Law

In considering genotype and allele frequencies for autosomal loci using the Hardy–Weinberg equation,

we assumed that the frequency of *A* is the same in both sperm and eggs. What about X-linked genes? In species that have two X chromosomes, such as humans and *Drosophila,* females carry two copies of all genes on the X chromosome, whereas males carry only one copy of all X-linked genes. Genes on the X chromosome are therefore distributed unequally in the population, and when there are equal numbers of males and females, the females carry two-thirds of all X-linked genes, and males carry one-third of the total number. Can the Hardy–Weinberg law be applied to calculate allelic frequencies and genotypic frequencies in such circumstances?

X-Linked Genes

It is easy to determine frequencies for X-linked genes in males. Because they have only one copy of all genes on this chromosome, the phenotype reveals both dominant and recessive alleles. Therefore, in males, the frequency of an X-linked allele is the same as the phenotypic frequency. In Western Europe, a form of X-linked color blindness occurs with a frequency of 8 percent in males ($q = 0.08$). Because females have two doses of all genes on the X chromosome, color-blind females exhibit the phenotypic frequencies expected for complete dominance, and the genotype and allele frequencies can be calculated by using the standard Hardy–Weinberg equation. For example, as color blindness in males has a frequency of 0.08, at equilibrium, the expected frequency in females is q^2, or 0.0064. This means that 800 out of 10,000 males surveyed would be expected to be color-blind, but only 64 of 10,000 females would show this trait. Expected values of the frequency of X-linked traits in males and females are compared in Table 21.5.

If the frequency of an X-linked allele differs between males and females, then the population is not in equilibrium. In contrast to alleles of autosomal loci, equilibrium in each group will not be reached in a single generation, but will be approached over a series of succeeding generations. Because males inherit their X

TABLE 21.5	Expected relative frequency of X-linked traits in males and females
Frequency of Males with Trait	**Expected Frequency in Females**
90/100	81/100
50/100	25/100
10/100	1/100
1/100	1/10,000
1/1000	1/1,000,000
1/10,000	1/100,000,000

chromosome maternally, the allelic frequency in females will determine the frequency in males of the next generation. Daughters will inherit both a maternal and a paternal X, and their allelic frequency is the average of that found in the parents. Even though the population's overall allele frequency remains constant, there is an oscillation in frequencies for the two sexes in each generation, with the differences being halved in each succeeding generation until an equilibrium is reached. This concept is illustrated in Figure 21–4 for the simple case where an allele has an initial frequency of 1.0 in females and 0.0 in males.

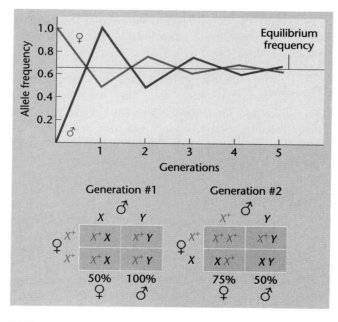

■ Figure 21–4 Approach to equilibrium for an X-linked trait with an initial frequency of 1.0. The fluctuations in frequency in the first two generations are shown. In the parents of generation 1, the X-linked trait (X^+) is present in 100 percent of the females and 0 percent of the males. In the progeny, the trait is present in 50 percent of all female X chromosomes and in 100 percent of all male X chromosomes. Thus, as the graph shows, the frequency in females drops from 1.0 to 0.5 in a generation, while the frequency in males rises from 0.0 to 1.0. In generation 2, the frequency in females is 0.75, and in males 0.5. This oscillation continues until eventually an equilibrium is reached.

Multiple Alleles

In addition to autosomal recessive and sex-linked genes, it is common to find several alleles of a single locus in a population. The ABO blood group in humans is such an example. The locus *I* (isoagglutinin) has three alleles (I^A, I^B, and I^O), yielding six possible genotypic combinations (I^AI^A, I^BI^B, I^OI^O, I^AI^B, I^AI^O, I^BI^O). Recall that in this case *A* and *B* are codominant alleles, and both of these are dominant to *O*. The result is that homozygous *AA* and heterozygous *AO* individuals are phenotypically identical, as are *BB* and *BO* individuals, so we can distinguish only four phenotypic combinations.

By adding another variable to the Hardy–Weinberg equation, we can calculate both genotype and allele frequencies for the situation involving three alleles. In an equilibrium population, the frequency of the three alleles can be described by

$$p(A) + q(B) + r(O) = 1$$

and the distribution of genotypes will be given by

$$(p + q + r)^2$$

In our hypothetical population, the genotypes *AA*, *AB*, *AO*, *BB*, *BO*, and *OO* will be found in the ratio

$$p^2(AA) + 2pq(AB) + 2pr(AO) + q^2(BB) + 2qr(BO) + r^2(OO) = 1$$

If we know the frequencies of blood types for a population, we can then estimate the frequencies for the three alleles of the ABO system. For example, in one population sampled, the following blood types were observed: A = 0.53, B = 0.13, AB = 0.08, and O = 0.26. Because the *O* allele is recessive, the frequency of type O blood in the population is equal to the frequency of the recessive genotype, r^2. Thus,

$$r^2 = 0.26$$
$$r = \sqrt{0.26}$$
$$= 0.51$$

By using the value estimated for *r*, we can estimate the allele frequencies for the A (*p*) and B (*q*) alleles. The *A* allele is present in two genotypes, *AA* and *AO*. The frequency of the *AA* genotype is represented by p^2, and the *AO* genotype by $2pr$. Therefore,

$$\text{Frequency of } A + O = p^2 + 2pr + r^2$$

where *A* and *O* represent the phenotypic frequencies. This can be rearranged to give

$$p = \sqrt{\text{Frequency of } A + O} - r$$
$$= \sqrt{0.53 + 0.26} - 0.51$$
$$= 0.89 - 0.51 = 0.38$$

Having estimated the frequencies for A (*p*) and O (*r*), the frequency for the B allele can be estimated from the following relationship:

$$p + q + r = 1$$
$$q = 1 - (p + r)$$
$$= 1 - (0.38 + 0.51)$$
$$= 1 - 0.89$$
$$= 0.11$$

The phenotypic frequencies and genotypic frequencies for this population are summarized in Table 21.6.

Using the Hardy–Weinberg Law: Calculating Heterozygote Frequency

One of the practical applications of the Hardy–Weinberg law is the estimation of heterozygote frequency in a population. The frequency of a recessive phenotype usually can be determined by counting such individuals in a sample of the population. This information and the Hardy–Weinberg equation can be used to calculate the allele and genotype frequencies.

Albinism, an autosomal recessive trait, has an incidence of about 1/10,000 (0.0001) in some populations. Albinos are easily distinguished from the population at large by a lack of pigment in skin, hair, and iris. Because this is a recessive trait, albino individuals must be homozygous. Their frequency in a population is represented by q^2, provided that mating has been random and all Hardy–Weinberg conditions have been met in the previous generation.

The frequency of the recessive allele therefore is

$$\sqrt{q^2} = \sqrt{0.0001}$$
$$q = 0.01 \quad \text{or} \quad 1/100$$

Since $p + q = 1$, the frequency of p is

$$p = 1 - q$$
$$= 1 - 0.01$$
$$= 0.99 \quad \text{or} \quad 99/100$$

In the Hardy–Weinberg equation, the frequency of heterozygotes is given as $2pq$. Therefore, we have

Heterozygote frequency $= 2pg$

$$= 2[(0.99)(0.01)]$$
$$= 0.02 \text{ or } 2\% \text{ or } 1/50$$

Thus, heterozygotes for albinism are rather common in the population (2 percent), even though the incidence of homozygous recessives is only 1/10,000.

In general, the frequencies of all three genotypes can be estimated once the frequency of either allele is known and Hardy–Weinberg conditions are assumed. The relationship between genotype frequency and allele frequency is shown in Figure 21–5. It is important to note how fast heterozygotes increase in a population as the values of p and q move away from zero. This observation confirms our conclusion that when a recessive trait like albinism is rare, the majority of those carrying the allele are heterozygotes. In cases where the frequencies of p and q are between 0.33 and 0.67, heterozygotes actually constitute the major class in the population.

Factors That Alter Allele Frequencies in Populations

The Hardy–Weinberg law establishes a set of ideal conditions that allows us to estimate allele frequencies and genotype frequencies in populations in which initial assumptions about random mating, absence of selection and mutation, and equal viability and fecundity hold true. Obviously, it is difficult to find natural populations in which all these conditions are met. In nature, populations are dynamic, and changes in size and structure are part of their life cycles. This situation allows us to use the Hardy–Weinberg law to investigate exceptions to the initial assumptions. In this section we will discuss factors that prevent populations from reaching equilibrium, or drive populations toward a different equilibrium, and the relative contribution of these factors to evolutionary change.

Mutation

Within a population, the gene pool is reshuffled each generation to produce new combinations in the genotypes of the offspring. Because the number of possible genotypic combinations is so large, the members of a

TABLE 21.6	Calculating genotypic frequencies for multiple alleles where the frequency of A (p) is 0.38, of B (q) is 0.11, and of O (r) is 0.51			
Genotype	*Genotypic Frequency*	*Phenotype*	*Phenotypic Frequency*
AA *AO*	$p^2 = (0.38)^2 = 0.14$ $2pr = 2(0.3 \times 0.51) = 0.39$	*A*	0.53
BB *BO*	$q^2 = (0.11)^2 = 0.01$ $2qr = 2(0.11 \times 0.51) = 0.11$	*B*	0.12
AB *OO*	$2pq = 2(0.38 \times 0.11) = 0.084$ $r^2 = (0.51)^2 = 0.26$	*AB* *O*	0.08 0.26

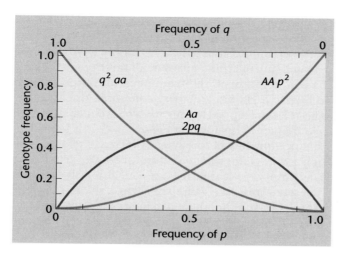

Figure 21–5 Relationship between genotype frequency and allele frequencies derived from the Hardy–Weinberg equation.

population alive at any given time represent only a fraction of all possible genotypes. Although an enormous genetic reserve is present in the gene pool, new genotypic combinations are produced by Mendelian assortment and recombination, but these processes do not produce any new alleles. **Mutation** alone acts to create new alleles, but in the absence of other forces, mutation has a negligible effect on allele frequencies.

For our purposes, we need only consider that mutational events occur at random—that is, without regard for any possible benefit or disadvantage to the organism. Here we will discuss only the generation of mutant alleles; in a later section we will consider the spread and distribution of new alleles through the population.

To know whether mutation is a significant force in changing allele frequencies, we must measure the rate at which mutations are produced. As most mutations are recessive, it is difficult to observe mutation rates directly in diploid organisms. Indirect methods using probability and statistics or large-scale screening programs must be employed. For certain dominant mutations, however, a direct method of measurement can be used. To ensure accuracy, several conditions must be met:

1. The trait must produce a distinctive phenotype that can be distinguished from similar traits produced by recessive alleles.

2. The trait must be fully expressed or completely penetrant, so that mutant individuals can be identified.

3. An identical phenotype must never be produced by nongenetic agents such as drugs or chemicals.

Mutation rates can be expressed as the number of new mutant alleles per given number of gametes. Suppose for a given gene that undergoes mutation to a dominant allele, 2 of 100,000 births exhibit a mutant phenotype. In these two cases, the parents are phenotypically normal. Because the zygotes that produced

these births each carry two copies of the gene, we have actually surveyed 200,000 copies of the gene (or 200,000 gametes). If we assume that the affected births are each heterozygous, we have uncovered 2 mutant alleles out of 200,000. Thus the mutation rate is 2/200,000 or 1/100,000, which in scientific notation would be written as 1×10^{-5}.

In humans, a dominant form of dwarfism known as **achondroplasia** fulfills the requirements outlined above for the measurement of mutation rates. Individuals with this skeletal disorder have an enlarged skull, short arms and legs, and can be diagnosed by radiologic examination at birth. In a survey of almost 250,000 births, the mutation rate (μ) for achondroplasia has been calculated as

$$1.4 \times 10^{-5} \pm 0.5 \times 10^{-5}$$

Knowing the rate of mutation, we can then estimate the change in frequency for each generation. If initially only the normal d allele exists (i.e., all individuals are homozygous dd), the frequency of d (q_0) is 1.0 and the frequency of D (dwarf) is $p_0 = 0$. If the rate of mutation from d to D is μ, then in the next generation,

$$\text{Frequency of } D = p_1 = q_0\mu$$

The new frequency for d, which will be lowered by the rate of mutation, can be expressed as

$$\text{Frequency of } d = q_0\mu$$

Although the rate of mutation is constant from generation to generation, the rate of change in the mutant allele is initially high, but declines to zero at mutational equilibrium.

As a more general example, let us assume that the mutation rate for genes in the human genome is 1.0×10^{-5}. With such a low mutation rate, changes in allele frequency brought about by mutation alone are very small. Figure 21–6 shows that if a population begins with only one allele (A) at a locus ($p = 1$), and a mutation rate of 1.0×10^{-5} for A to a, it would require about 70,000 generations to reduce the frequency of A to 0.5.

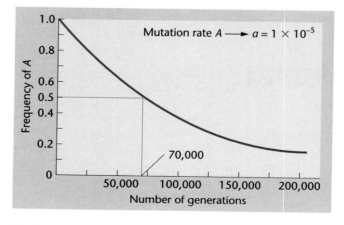

Figure 21–6 Rate of replacement of an allele by mutation alone, assuming an average mutation rate of 1.0×10^{-5}.

Even if the rate of mutation increased through exposure to higher levels of radioactivity or chemical mutagens, the impact of mutation on allele frequencies would be extremely weak. This example once again emphasizes that mutation is a major force in generating genetic variability, but by itself plays a relatively insignificant role in changing allele frequencies.

Migration

Frequently, a species of plant or animal becomes divided into subpopulations that to some extent are separated geographically. Differences in mutation rate and selective pressures can establish different allele frequencies in the subpopulations. **Migration** occurs when individuals move between these populations. Consider a single pair of alleles, A (p) and a (q). The change in the frequency of A in one generation can be expressed as

$$\Delta p = m(p_m - p)$$

where

p = frequency of A in existing population

p_m = frequency of A in immigrants

Δp = change in one generation

m = coefficient of migration (proportion of migrant genes entering the population per generation)

For example, assume that $p = 0.4$ and $p_m = 0.6$, and that 10 percent of the parents giving rise to the next generation are immigrants ($m = 0.1$). Then the change in the frequency of A in one generation is

$$\begin{aligned}\Delta p &= m(p_m - p)\\ &= 0.1(0.6 - 0.4)\\ &= 0.1(0.2)\\ &= 0.02\end{aligned}$$

In the next generation, the frequency of A (p_1) will increase as follows:

$$\begin{aligned}p_1 &= p + \Delta p\\ &= 0.40 + 0.02\\ &= 0.42\end{aligned}$$

If either m is large or p is very different from p_m, then a rather large change in the frequency of A will occur in a single generation. All other factors being equal (including migration), an equilibrium will be attained when $p = p_m$. These calculations reveal that the change in allele frequency attributable to migration is proportional to the differences in allele frequency between the donor and recipient populations and to the rate of migration. As the migration coefficient can have a wide range of values, the effect of migration can substantially alter allele frequencies in populations (Figure 21–7).

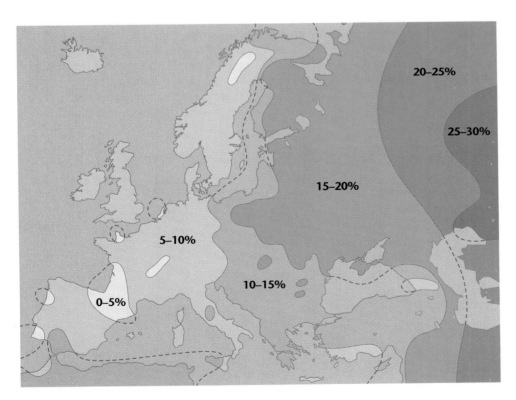

20–25%

25–30%

15–20%

5–10%

10–15%

0–5%

■ Figure 21–7 The frequency of the *B* allele of the *ABO* locus is present in a gradient from east to west. This allele has its highest frequency in Central Asia, and declines to its lowest point in northeastern Spain. This gradient parallels the waves of Mongol migration into Europe following the fall of the Roman Empire, and is a genuine relic of human history.

Although migration can be somewhat difficult to quantify, it can often be estimated.

Migration can also be regarded as the flow of genes between two populations that were once but are no longer isolated geographically. For example, most blacks in the United States are descended from ancestors in West Africa. In Africa, the frequency of the Duffy blood group allele Fy^0 is almost 100 percent. In Europe, the source of most U.S. whites, the frequency is close to 0 percent, and almost all individuals are Fy^a or Fy^b. By measuring the frequency of Fy^a or Fy^b among U.S. blacks, we can estimate the amount of gene flow into the black population. Figure 21–8 shows the frequency of the Fy^a allele among blacks in three African nations and in several regions of the United States, along with the frequency in a white population. Using the average frequency and assuming an equal rate of gene flow in each generation, we can estimate the migration of the Fy^a allele at about 5 percent per generation.

Natural Selection

Mutation and migration introduce new alleles into populations. **Natural selection**, on the other hand, is the principal force that shifts allele frequencies within large populations and is one of the most important factors in evolutionary change.

Wallace and Darwin's main contribution to the study of evolution was their recognition that natural selection and reproductive isolation are the mechanisms that lead to the divergence and eventual separation of populations into distinct species. In any population at a given time, there are individuals with different genotypes. Because of these inherent genetic differences, some of the individuals in this population will be better adapted to the environment than will others, leading to the differential survival and reproduction of some genotypes over others. Natural selection is thus the consequence of the differential reproduction of genotypes. Therefore, allelic frequencies will change over time. Natural selection, then, is a major force in changing allelic frequencies and therefore represents a departure from the Hardy–Weinberg assumption that all genotypes have equal viability and fertility.

Because selection is a consequence of the organism's genotypic/phenotypic combination, polygenic or quantitative traits (those that are controlled by a number of genes and that might be susceptible to environmental influences) also respond to selection. Such quantitative traits, including adult body height and weight in humans, often demonstrate a continuously varying distribution resembling a bell-shaped curve. Selection for such traits can be classified as (1) directional, (2) stabilizing, or (3) disruptive.

In **directional selection**, which is important to plant and animal breeders, desirable traits, often representing phenotypic extremes, are selected. In fact, if the trait is polygenic, the most extreme phenotypes that the genotype can express will appear in the population only after prolonged selection. An example of directional selection is the long-running experiment at the State Agricultural Laboratory in Illinois to select for high and low oil content in corn kernels. Beginning in 1896, a population of 163 corn ears was surveyed, and the 24 ears that were highest in oil content were used as the parents of the next generation. In each succeeding generation, the ears with the highest oil content were used for breeding. In this example of upward selection, the oil content was raised after 50 generations from about 4 percent to just over 16 percent, with no sign that a plateau has been reached (Figure 21–9). This experiment also selected 12 ears with the lowest oil content, and downward selection from this

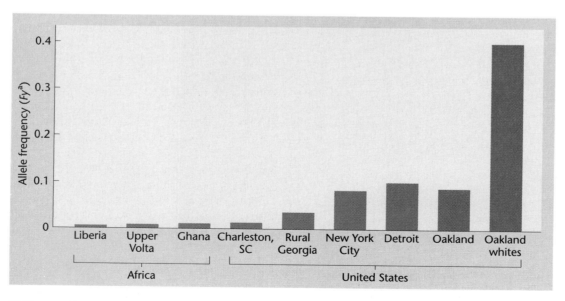

■ Figure 21–8   Frequencies of the Fy^a allele in African and U.S. black populations.

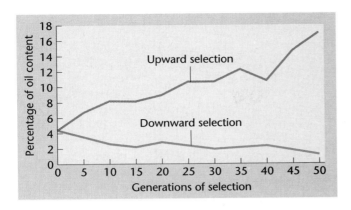

■ Figure 21–9 Long-term selection to alter the oil content of corn kernels.

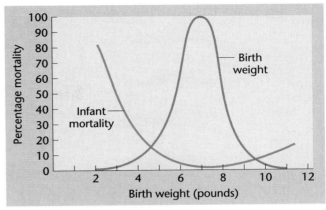

■ Figure 21–10 Relationship between birth weight and mortality in humans.

parental group has lowered the oil content from 4 percent to less than 1 percent over the period of 50 generations (Figure 21–9). In this case, the high and low lines have diverged to the point where the distribution of oil content in each line does not overlap with that of the other line.

In upward selection, alleles for high oil content increase in frequency, and replace alleles for low oil content. At some point, all individuals in the population will have the genotype for the highest oil content, and all will express the most extreme phenotype for high oil content. In downward selection, the opposite situation will occur, eventually producing a population with a genotype and phenotype for the lowest oil content.

When the lines fail to respond to further selection, there will be no remaining genetic variance for oil content, and each line will have a homozygous genotype for oil content.

In nature, directional selection can occur when one of the phenotypic extremes becomes selected for or against, usually as a result of changes in the environment.

Stabilizing selection, on the other hand, tends to favor intermediate types, with both extreme phenotypes being selected against. One of the clearest demonstrations of stabilizing selection is the data of Mary Karn and Sheldon Penrose on human birth weight and survival. Figure 21–10 shows the distribution of birth weight for 13,730 children born over an 11-year period and the percentage of mortality at 4 weeks of age. Infant mortality increases dramatically on either side of the optimal birth weight of 7.5 pounds. At the genetic level, stabilizing selection acts to keep a population well adapted to its environment. In this situation, individuals who are closer to the average for a given trait will have higher fitness (a concept discussed below).

Disruptive selection is selection against intermediates and for both phenotypic extremes. It may be viewed as the opposite of stabilizing selection, because the intermediate types are selected against. In one set of experiments, John Thoday applied disruptive selection to progeny of *Drosophila* strains with high and low bristle

number. Matings were carried out using females from one strain and males from the other. Progeny were selected for high and low bristle number, matings were carried out for a number of generations, and the two lines diverged rapidly (Figure 21–11). In natural populations, such a situation might exist for a population in a heterogeneous environment.

The types and effects of selection are summarized in Figure 21–12.

Fitness and Selection

When a particular genotype/phenotype combination confers an advantage to organisms in competition with others harboring an alternative combination, selection occurs. The relative strength of selection varies with the amount of advantage provided. The probability that a particular phenotype will survive and leave offspring is a measure of its fitness. Thus, fitness refers to total reproductive potential or efficiency. The concept of fitness is usually expressed in relative terms by comparing a particular genotypic/phenotypic combination with one regarded as optimal. Fitness is a relative concept because as environmental conditions change, so does the advantage conferred by a particular genotype.

Mathematically, the difference between the fitness of a given genotype and another, reference genotype is called the **selection coefficient (s)**. For a phenotype conferred by the genotype *aa*, when 99 of every 100 organisms reproduce successfully, $s = 0.01$. If the *aa* genotype is a homozygous lethal and *AA* and *Aa* have equal fitness, then $s = 1.0$, and the *a* allele is propagated only in the heterozygous carrier state. Starting with any original frequencies of p_0 and q_0 in a population, when $s = 1.0$, the effect of selection on any successive generation can be calculated using the formula

$$q_n = \frac{q_0}{1 + nq_0}$$

where *n* is the number of generations elapsing since p_0 and q_0.

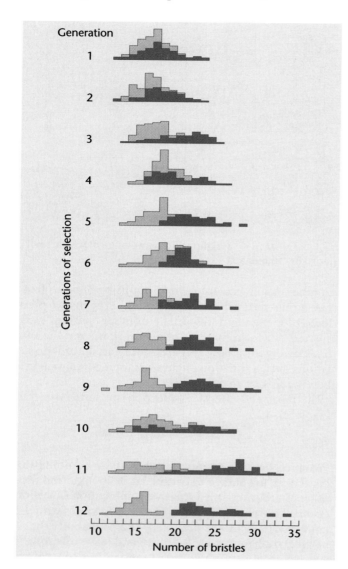

Generation

Generations of selection

Number of bristles

■ Figure 21–11 The effect of disruptive selection on bristle number in *Drosophila*. By selecting lines with the highest and lowest bristle number, the population showed a nonoverlapping divergence in only 12 generations.

Let us consider another example. Figure 21–13 demonstrates the change in allele frequencies when A $(p_0) = a$ $(q_0) = 0.5$. Initially, because of the high percentage of aa genotypes, the frequency of the a allele is reduced rapidly. The frequency of a is halved (0.25) in only two generations. By the sixth generation, the frequency of a is again reduced twofold (0.12). By now, however, the majority of a alleles are carried by heterozygotes. Because selection operates on phenotypes and not on components of the genotype, the heterozygotes are not selected against. Subsequent reductions in frequency occur very slowly in successive generations and depend on the rate at which heterozygotes are removed from the population. For this reason, it is difficult to eliminate a recessive allele from the population by selection as long as heterozygotes continue to mate.

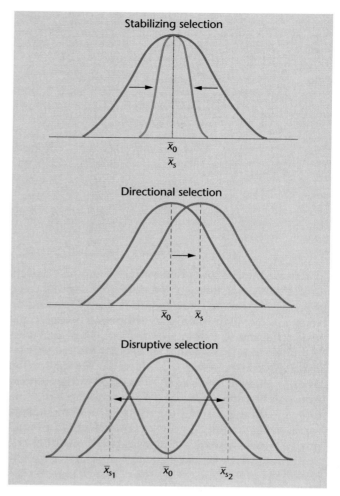

Stabilizing selection

$\bar{x}_0$
$\bar{x}_s$

Directional selection

$\bar{x}_0$ $\bar{x}_s$

Disruptive selection

$\bar{x}_{s_1}$ $\bar{x}_0$ $\bar{x}_{s_2}$

■ Figure 21–12 Comparison of the impact of stabilizing, directional, and disruptive selection. In each case, $\bar{x}_0$ is the mean of an original population, and $\bar{x}_s$ is the mean of the population following selection.

When s is less than 1.0, it is possible to calculate the effects of selection on each successive generation with any s, p_0, and q_0 values using the formula

$$q_1 = \frac{q_0 - sq_0^2}{1 - sq_0^2}$$

Figure 21–14 demonstrates a variety of initial conditions and the change of the frequency of q (a) through a large number of generations. If q_0 is initially 0.5 and $s = 0.5$, five generations must elapse before q is halved. Then, another nine generations must elapse to cut q in half again. Figure 21–14 shows the relative trends as q_0 and s change.

Selection in Natural Populations

There has been much study of selection on both laboratory and natural populations. A classic example of selection in natural populations is that of the peppered

Generation	p	q	p^2	$2pq$	q^2
0	0.50	0.50	0.25	0.50	0.25
1	0.67	0.33	0.44	0.44	0.12
2	0.75	0.25	0.56	0.38	0.06
3	0.80	0.20	0.64	0.32	0.04
4	0.83	0.17	0.69	0.28	0.03
5	0.86	0.14	0.73	0.25	0.02
6	0.88	0.12	0.77	0.21	0.01
10	0.91	0.09	0.84	0.15	0.01
20	0.95	0.05	0.91	0.09	<0.01
40	0.98	0.02	0.95	0.05	<0.01
70	0.99	0.01	0.98	0.02	<0.01
100	0.99	0.01	0.98	0.02	<0.01

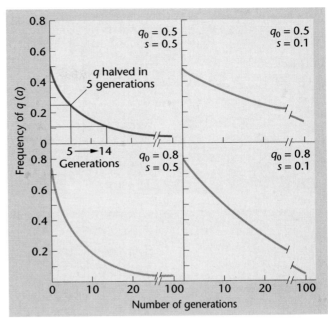

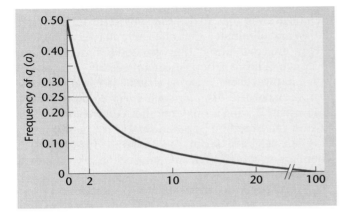

■ Figure 21–13   Changes in allele frequency under selection where s = 1.0. The frequency of *a* is halved in two generations, and halved again by the sixth generation. Subsequent reductions occur slowly, because the majority of *a* alleles are carried by heterozygotes.

■ Figure 21–14   Changes in allele frequencies under different selection coefficients and different initial gene frequencies. When $q_0 = 0.5$ and $s = 0.5$, the frequency of q is halved in five generations. When $q_0 = 0.8$ and $s = 0.5$, the frequency of q is halved in five generations and halved again in an additional five generations.

moth, *Biston betularia*, in England. Before 1850, 99 percent of the moth population was light-colored, allowing the nocturnal moths to rest undetected during the day on lichen-covered trees. As toxic gases produced by industry killed the lichens growing on trees and buildings, and soot deposits darkened the landscape, the light-colored moths became easy targets for their predators. The rare dark-colored (melanic) moths suddenly gained a great selective advantage because of their natural camouflage. A rapid shift in frequency of this phenotype occurred, probably in 50 generations or less.

Laboratory experiments have demonstrated that the dark form is due to a single dominant allele, *C*. Figure 21–15 illustrates the protective advantage from predators provided by the dark phenotype.

Following the enactment of laws in the mid-1960s to restrict environmental pollution, the frequency of non-melanic forms of the moth has started to increase in many areas of England, underscoring the close relationship between the environment and selection. In one region south of Liverpool, the frequency of the dark-colored form decreased from over 90 percent to just over 50 percent in the period from 1959 to 1985, showing that genotype frequencies can respond quickly to a shift in the environment.

Genetic Drift

In laboratory crosses, one condition essential to realizing theoretical genetic ratios (e.g., 3:1, 1:2:1, 9:3:3:1) is a fairly large sample size. This sample size is also important to the study of population genetics as allele and genotype frequencies are examined or predicted. For example, if a population consists of 1000 randomly mating heterozygotes (*Aa*), the next generation will consist of approximately 25 percent *AA*, 50 percent *Aa*, and 25 percent *aa* genotypes. Provided that the initial and subsequent populations are large, there will be at most only minor deviations from this mathematical ratio, and the frequency of *A* and *a* will remain about equal (0.50).

However, if a population is formed from only one set of heterozygous parents and they produce only two offspring, allele frequency can change drastically. In this case, genotypes and allele frequencies in the offspring can be predicted, as shown in Table 21.7. As shown, in 10 of 16 times such a cross is made, the new allele frequencies would be altered. In 2 of 16 times, either the *A* or *a* allele would be eliminated in a single generation.

Using only one set of heterozygous parents that produce two offspring is an extreme example, but has been

■ Figure 21–15 The speckled and melanic forms of the peppered moth, *Biston,* seen on a normal, lichen-covered tree trunk (left) and on a darkened, soot-covered trunk (right).

TABLE 21.7	All possible pairs of offspring produced by two heterozygous parents (*Aa* × *Aa*)		

Possible Genotypes of Two Offspring	Probability	Allele Frequency	
		A	*a*
AA and *AA*	(1/4)(1/4) = 1/16	1.00	0.00
aa and *aa*	(1/4)(1/4) = 1/16	0.00	1.00
AA and *aa*	2(1/4)(1/4) = 2/16	0.50	0.50
Aa and *Aa*	(2/4)(2/4) = 4/16	0.50	0.50
AA and *Aa*	2(2/4)(1/4) = 4/16	0.75	0.25
aa and *Aa*	2(1/4)(2/4) = 4/16	0.25	0.75

chosen to illustrate the point that large interbreeding populations are essential to the Hardy–Weinberg equilibrium. In small populations, significant random fluctuations in allelic frequencies are possible by chance deviation. The degree of fluctuation will increase as the population size decreases. Such changes illustrate the concept of **genetic drift**. In the extreme case, genetic drift may lead to the chance fixation of one allele to the exclusion of another allele.

To study genetic drift in laboratory populations of *Drosophila melanogaster,* Warwick Kerr and Sewall Wright set up over 100 lines, with four males and four females as the parents for each line. Within each line, the frequency of the sex-linked bristle mutant *forked* (*f*) and its wild-type allele (*f⁺*) was 0.5. In each generation, four males and four females were chosen at random to be parents of the next generation. After 16 generations, fixation had occurred in 70 lines—29 in which only the *forked* allele was present and 41 in which the wild-type allele had become fixed. The rest of the lines were still segregating the two alleles or had been lost. If fixation had occurred randomly, then an equal number of lines should have become fixed for each allele. In fact, the experimental results do not differ statistically from the expected ratio of 35:35, demonstrating that alleles can spread through a population and eliminate other alleles by chance alone.

How are small populations created in nature? In one instance, a large group of organisms may be split by some natural event, creating a small, isolated subpopulation. A natural disaster such as an epidemic might also occur, leaving a small number of survivors to constitute the breeding population. Third, a small group might emigrate from the larger population, as founders, to a new environment, such as a volcanically created island.

Allelic frequencies in certain human isolates best support the role of drift as an evolutionary force in natural populations. The Pingelap atoll in the western Pacific Ocean (lat. 6° N, long. 160° E) has in the past been devastated by typhoons and famine, and in about 1780 there were only about 9 surviving males. Today there are fewer than 2000 inhabitants, and their ancestry can be traced to the typhoon survivors. From 4 to 10 percent of the current population is blind from infancy. These people are affected by an autosomal recessive disorder, **achromatopsia**, which causes ocular disturbances, a form of color blindness, and cataract formation. This disorder is extremely rare in the human population as a whole. However, the mutant allele is present at a relatively high frequency in the Pingelap population. From genealogical reconstruction, it was found that one of the original survivors (about 30 individuals) was a chief who was heterozygous for the condition. If we assume that he was the only carrier in the founding population, the initial gene frequency was 1/60, or 0.016. On average, about 7 percent of the current population is affected (a homozygous recessive genotype), and the frequency of the allele in the present population is calculated as 0.26. The story of the inhabitants of Pingelap has been told by Oliver Sacks in his recent book, *The Island of the Colorblind.*

Another example of genetic drift involves the Dunkers, a small, isolated, religious community who emigrated from the German Rhineland to Pennsylvania. Because their religious beliefs do not permit marriage with outsiders, the population has grown only through marriage within the group. When the frequencies of the ABO and MN blood group alleles are compared, significant differences are found among the Dunkers, the German population, and the U.S. population. The frequency of blood group A in the Dunkers is about 60 percent, in contrast to 45 percent in the U.S. and German populations. The I^B allele is nearly absent in the Dunkers. Type M blood is found in about 45 percent of the Dunkers, compared with about 30 percent in the U.S. and German populations. As there is no evidence for a selective advantage of these alleles, it is apparent that their observed frequencies are the result of chance events in a relatively small, isolated population.

Inbreeding

One of the assumptions in the Hardy–Weinberg law is random mating in a large population. In the previous section we considered how allele frequencies in small populations can be altered by drift, leading to a reduction in genetic variability, and a decrease in heterozygosity. The same effect can be produced by inbreeding and nonrandom mating. In a small population, potential mates are more likely to be related to each other than in large populations. In large populations distributed over a wide geographic range, individuals tend to mate with those nearby, rather than with those living at a great distance. If the mobility of individuals is restricted, a pattern of nonrandom mating can cause genetic drift, and subdivide the population into a number of smaller, interbreeding subpopulations, differing from each other in the frequency of some alleles, and leading to the chance elimination of alleles.

Nonrandom Mating

Nonrandom mating occurs in many species, including humans. One form of nonrandom mating is called **assortative**. In humans, pair bonds may be established by religious practices, physical characteristics, professional interests, and so forth on a nonrandom basis. In nature, phenotypic similarity plays the same role.

Another form of nonrandom mating is **inbreeding**, where mating occurs between relatives. One of the consequences of inbreeding is an increase in the chance that an individual will be homozygous for a recessive deleterious allele. For a given allele, inbreeding increases the proportion of homozygotes in the population. Over time, with complete inbreeding, only homozygotes will remain in the population. To illustrate this concept, we shall consider the most extreme form of inbreeding, **self-fertilization**.

Figure 21–16 shows the results of four generations of self-fertilization starting with a single individual that is heterozygous for one pair of alleles. By the fourth generation, only about 6 percent of the individuals are still

	P_1	Self-fertilization Aa	
	AA	*Aa*	*aa*
F_1	0.250	0.500	0.250
F_2	0.375	0.250	0.375
F_3	0.437	0.125	0.437
F_4	0.468	0.063	0.468
F_n	$\dfrac{1-\frac{1}{2}n}{2}$	$\dfrac{1}{2^n}$	$\dfrac{1-\frac{1}{2}n}{2}$

■ Figure 21–16 Reduction in heterozygote frequency brought about by self-fertilization. After *n* generations, the frequencies of the genotypes can be calculated according to the formulas in the bottom row.

heterozygous, and 94 percent of the population is homozygous. Note, however, that the frequencies of *A* and *a* still remain at 50 percent.

In humans, inbreeding (called **consanguineous marriages**) is related to population size, mobility, and social customs governing marriages among relatives. To determine the probability that two alleles at the same locus in an individual are derived from a common ancestor, Sewall Wright devised the **coefficient of inbreeding**. Expressed as *F*, the coefficient can be defined as the probability that two alleles of a given gene in an individual are derived from a common allele in an ancestor. For example, an F2 generation produced by brother–sister matings will have an *F* value of 1/4. If *F* = 1, all genotypes are homozygous and both alleles are derived from the same ancestor. If *F* = 0, no alleles present are derived from a common ancestor. Shown in Figure 21–17 are pedigrees of first- and second-cousin marriages. A lethal allele, *a*, is assumed to be present in one heterozygous parent. For each relative, the probability of the presence of this lethal allele is shown. The pedigrees then culminate in the final probability, *F*, that the lethal allele will become homozygous in the offspring of first or second cousins. Because every population carries as its genetic burden a number of heterozygous recessive alleles that are lethal or deleterious in homozygotes, mortality will rise in matings between relatives.

Genetic Effects of Inbreeding

From an evolutionary standpoint, if a population is split into smaller subpopulations, the effect of inbreeding will become evident. Inbreeding results in the production of individuals who are homozygous for recessive alleles that were previously concealed in heterozygotes. Because many recessive alleles are deleterious when

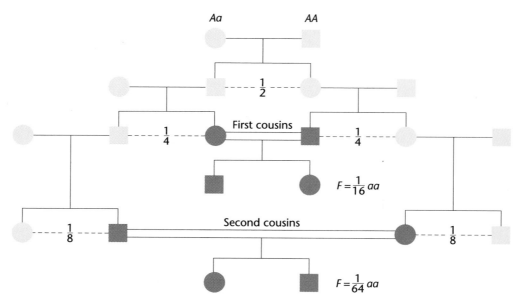

■ Figure 21–17 Changes in the coefficient of inbreeding (F) brought about by consanguineous matings. Numbers indicate the probability of carrying the recessive allele a.

they are homozygous, inbred populations usually have a lowered fitness. **Inbreeding depression** is a measure of the loss of fitness caused by inbreeding. In domesticated plants and animals, inbreeding and selection have been used for thousands of years, and these organisms already have a high degree of homozygosity at many loci. Further inbreeding will usually produce only a small loss of fitness. However, inbreeding among individuals from large, randomly mating populations can produce high levels of inbreeding depression. This effect can be seen by examining the mortality rates in offspring of inbred animals in zoo populations (Table 21.8). To counteract this effect of inbreeding, many zoos are using DNA fingerprinting to estimate the relatedness of animals used in breeding programs, and choosing the most unrelated animals as parents.

As natural populations of endangered species become smaller, concerns about inbreeding are a factor in designing programs to restore these species. For example, the black rhinoceros population in Africa declined from 65,000 in 1970 to less than 4500 in 1986. These remaining rhinos are distributed in small, isolated populations subject to inbreeding depression. Early in this century, taxonomists used small morphological differences to divide black rhinos into several subspecies. DNA analysis of the mitochondrial genome was used to determine whether the subspecies should be managed separately to maintain genetic diversity, or whether all remaining black rhinos should be regarded as a single population for breeding purposes. The results indicate that there is little divergence in the mitochondrial genomes of these subspecies, and that the

TABLE 21.8	Mortality in offspring of inbred zoo animals				
Species	*N*		*Noninbred*	*Inbred*	*Inbreeding Coefficient*
Zebra	32	Lived:	20	3	0.250
		Died	7	2	
Eld's deer	24	Lived:	13	0	0.250
		Died:	4	7	
Giraffe	19	Lived:	11	2	0.250
		Died:	3	3	
Oryx	42	Lived:	35	0	0.250
		Died:	2	5	
Dorcas gazelle	92	Lived:	36	17	0.269
		Died:	14	25	

G E N E T I C S ,   T E C H N O L O G Y ,   A N D   S O C I E T Y

What Can We Learn from the Failure of the Eugenics Movement?

The eugenics movement had its origins in the ideas of the English scientist Francis Galton, who became convinced from his study of the appearance of geniuses within families (including his own) that intelligence is inherited. Galton concluded in his 1869 book *Hereditary Genius* that it would be "quite practicable to produce a highly-gifted race of men by judicious marriages during several consecutive generations." The term eugenics, coined by Galton in 1883, refers to the improvement of the human species by such selective mating. Once Mendel's principles were rediscovered in 1900, the *eugenics* movement flourished.

The eugenicists believed that a wide range of human attributes were inherited as Mendelian traits, including many aspects of behavior, intelligence, and moral character. Their overriding concern was that the presumed genetically "feebleminded" and immoral in the population were reproducing faster than the genetically superior, and that this differential birthrate would result in the progressive deterioration of the intellectual capacity and moral fiber of the human race. Several remedies were proposed. *Positive eugenics* called for the encouragement of especially "fit" parents to have more children. More central to the goals of the eugenicists, however, was the *negative eugenics* approach aimed at discouraging the reproduction of the genetically inferior or, better yet, eliminating it altogether.

In the United States, the eugenics movement was never highly organized, but for a time it enjoyed wide popular support and had a significant impact on public policy. Partially at the urging of prominent eugenicists, 30 states passed laws compelling the sterilization of criminals, epileptics, and inmates in mental institutions; most states enacted laws invalidating marriages between the "feebleminded" and others considered eugenically unfit. The crowning legislative achievement of the eugenics movement, however, was the passage of the Immigration Restriction Act of 1924, which severely limited the entry of immigrants from eastern and southern Europe due to their perceived mental inferiority.

Throughout the first two decades of the century, most geneticists passively accepted the views of eugenicists, but by the 1930s critics recognized that the goals of the eugenics movement were determined more by racism, class prejudice, and anti-immigrant sentiment than by sound genetics. Increasingly, prominent geneticists began to speak out against the eugenics movement, among them William Castle, Thomas Hunt Morgan, and Hermann Muller. When the horrific extremes to which the Nazis took eugenics became known, a strong reaction developed that all but ended the eugenics movement.

Paradoxically, the eugenics movement arose at the same time basic Mendelian principles were being developed, principles that eventually undermined the theoretical foundation of eugenics. Today, any student who has completed an introductory course in genetics should be able to identify several fundamental mistakes the eugenicists made:

1. They assumed that complex human traits such as intelligence and personality were strictly inherited, completely disregarding any environmental contribution to the phenotype. Their reasoning was that because certain traits ran in families, they must be genetically determined.

2. They assumed that these complex traits were determined by single genes with dominant and recessive alleles. This belief persisted despite research showing that multiple gene pairs contribute to many phenotypes.

3. They assumed that a single ideal genotype existed in humans. Presumably, such a genotype would be highly homozygous in order to be sustained. This precept runs counter to current evidence suggesting that a high level of homozygosity is often deleterious, supporting the superiority of the heterozygote.

4. They assumed that the frequency of recessively inherited defects in the population could be significantly lowered by preventing homozygotes from reproducing. In fact, for recessive traits that are relatively rare, most of the recessive alleles in the population are "carried" by asymptomatic heterozygotes who are spared from such selection. Negative eugenic practices, no matter how harsh, are relatively ineffective at eliminating such traits.

5. They assumed that those deemed genetically unfit in the population were out-reproducing those thought to be genetically fit. This is the exact reverse of the Darwinian concept of fitness, which equates reproductive success with fitness. (Galton should have understood this, being Darwin's first cousin!)

More than seven decades have passed since the eugenics movement was in full bloom. We now have a much more sophisticated understanding of genetics, as well as a greater awareness of its potential misuses. But the application of current genetic technologies makes possible a "new eugenics" of a scope and power that Francis Galton could not have imagined. In particular, prenatal genetic screening and *in vitro* fertilization enable the selection of children according to their genotype, a power that will dramatically increase as more and more genes are associated with inherited diseases and perhaps even behaviors.

As we move into this new genetic age, we must not forget the mistakes made by the early eugenicists. We must remember that phenotype is a complex interaction between the genotype and the environment, and not lapse into a new hereditarianism that treats a person as only a collection of genes. We must keep in mind that many genes may contribute to a particular phenotype, whether a disease or a behavior, and that the alleles of these genes may interact in unpredictable ways. We must not fall prey to the assumption that there is an ideal *(continues)*

(continued)
genotype. The success of all populations in nature is correlated with genetic diversity. And most of all, we must not use genetic information to advance ideological goals. We may find that there is a fine line between the legitimate uses of genetic technologies, such as having healthy children, and other eugenic practices. It will be up to us to decide exactly where the line falls.

References
Allen, G. E., Jacoby, R., and Glauberman, N. (eds.). 1995. Eugenics and American social history, 1880-1950. *Genome* 31:885-889.

Hartl, D. L. 1988. A prime of population genetics. 2nd ed. Sunderland, MA: Sinauer.
Kevles, D. J. 1985. In the name of eugenics: Genetics and the uses of human heredity. Berkeley: Univ. of California Press.

rhinos can be grouped into a single population for breeding programs.

In humans, the effects of inbreeding are an increase in the risks for spontaneous abortions, neonatal deaths, and congenital deformities and recessive genetic disorders. Although it is less common than in the past, inbreeding occurs in many regions of the world where social custom favors marriage between a man and his first cousin (his father's brother's daughter). Over many generations, inbreeding should eventually reduce the frequency of deleterious recessive alleles (if homozygotes die before reproducing). However, some studies indicate that such families tend to have more children to compensate for those lost to such disorders. On average, two-thirds of these are heterozygotes, carrying the deleterious allele.

Inbreeding is not always harmful, however, and it has been used in breeding programs for domesticated plants and animals. When an inbreeding program is initiated, homozygosity increases, and some breeding stocks become fixed for favorable alleles and others for unfavorable alleles. By selecting the more viable and vigorous plants or animals, the proportion of individuals carrying desirable traits can be increased.

If members of two favorable inbred lines are mated, hybrid offspring are often more vigorous in desirable traits than either of the parental lines. This phenomenon has been called **hybrid vigor**. Such an approach was used in the breeding programs established for corn, and crop yields increased tremendously. Unfortunately, the hybrid vigor extends only through the first generation. Many hybrid lines are sterile, and those that are fertile

show subsequent declines in yield. Consequently, the hybrids must be regenerated each time by crossing the inbred parental lines.

Hybrid vigor has been explained in two ways. The first hypothesis, the **dominance hypothesis**, incorporates the obvious reversal of inbreeding depression, which inevitably must occur in outcrossing. Consider a cross between two strains of corn with the following genotypes:

$$\text{Strain A} \qquad\qquad \text{Strain B}$$
$$aaBBCCddee \quad \times \quad AAbbccDDEE$$
$$\downarrow$$
$$AaBbCcDdEe$$

The F1 hybrids are heterozygotes at all loci shown. Deleterious recessive alleles present in the homozygous form in the parents will be masked by the more favorable dominant alleles in the hybrids. Such masking is thought to cause hybrid vigor.

The second theory, **overdominance**, holds that in many cases the heterozygote is superior to either homozygote. This result may be related to the fact that in the heterozygote, there may be two forms of a gene product present, providing a form of biochemical diversity. Thus, the cumulative effect of heterozygosity at many loci accounts for the hybrid vigor. Most likely, such vigor results from a combination of phenomena explained by both hypotheses.

Chapter Summary

1. The idea that the population rather than the individual is the effective unit in the process of evolution has led to the study of allele frequencies and to the emergence of the discipline of population genetics.

2. The Hardy–Weinberg law established the conditions for genetic equilibrium within a population. These conditions include a large, randomly breeding population, and the absence of selection, mutation, migration, and drift. If any of these conditions is not met, allele or genotype frequencies will change from generation to generation.

3. The Hardy–Weinberg formula is also useful in demonstrating whether a population is in equilibrium and in calculating the degree of heterozygosity within a population for any particular locus.

4. The process of evolution requires changes in allele frequencies. The mathematical derivation of the Hardy–Weinberg equilibrium allows assessment of the forces of evolution: selection, mutation, migration, genetic drift, and inbreeding.

5. Mutation and migration introduce new alleles into a population, but the retention or loss of these alleles depends on the degree of fitness conferred and the action of selection.

6. Natural selection is the most powerful force altering a population's genetic constitution. In small populations, however, inbreeding and genetic drift can be important factors.

Key Terms

achondroplasia, 480
achromatopsia, 486
allele frequency, 474
assortative mating, 487
coefficient of inbreeding, 487
consanguineous marriages, 487
directional selection, 482
disruptive selection, 483
dominance hypothesis, 490
expected frequencies, 476

fitness, 483
gene pool, 473
genetic drift, 486
genetic equilibrium, 476
Hardy–Weinberg law (HWL), 474
hybrid vigor, 490
inbreeding, 487
inbreeding depression, 488
migration, 481
MN blood group, 474

mutation, 480
natural selection, 482
observed frequencies, 476
overdominance, 490
population, 473
population genetics, 473
selection coefficient, 483
self-fertilization, 487
stabilizing selection, 483

INSIGHTS
and
SOLUTIONS

1. Color blindness is caused by a recessive allele on the X chromosome. In a survey, 16 women out of 20,000 were found to be color-blind. What is the expected frequency of color-blind males in this population?

Solution Remember that for X-linked traits, females exhibit the phenotypic frequencies expected for complete dominance. This means that the frequency of color blindness in women ($16/20,000 = 0.0008$) is equal to q^2. If $0.0008 = q^2$, then $q = 0.0282$, the value for color blindness in males. The expected frequency in males would be 2.82 percent or about 1 in 34.

2. *Eugenics* is the term employed for the selective breeding of humans to bring about improvements in the species. As a eugenic measure, it has been suggested (sometimes by force of law) that individuals suffering from serious genetic disorders should be prevented from reproducing (by sterilization, if necessary) in order to reduce the frequency of the disorder in future generations. Suppose that such a trait were present in the population at a frequency of 1 in 40,000, and that affected individuals did not reproduce. In 100 generations, or about 2500 years, what would be the frequency of the condition? What are your assumptions about this condition? Are the eugenic measures effective in this case?

Solution: Let us assume that the trait is recessive, and that homozygous recessive individuals are prevented from reproducing. In this case, selection against the homozygous recessive phenotype is equal to 1. Using formula (21.19), we can calculate the frequency after 100 generations as follows:

$$q_n = \frac{q_0}{1 - nq_0}$$

$$= \frac{0.000025}{1 + 100(0.000025)}$$

$$= \frac{0.000025}{1.0025}$$

where

$$q_n = \frac{1}{41,666}$$

$$s = 1$$
$$n = 100$$
$$q_0 = 0.000025$$

The frequency of the condition has been reduced from 1 in 40,000 to 1 in 41,666 in 100 generations, indicating that this eugenic measure is ineffective.

Problems and Discussion Questions

1. The ability to taste the compound PTC is controlled by a dominant allele T, while individuals who are homozygous for the recessive allele t are unable to taste this compound. In a genetics class of 125 students, 88 were able to taste PTC, 37 could not. Calculate the frequency of the T and t alleles in this population, and the frequency of the genotypes.

2. Calculate the frequencies of the AA, Aa, and aa genotypes after one generation if the initial population consists of 0.2 AA, 0.6 Aa, and 0.2 aa genotypes. What frequencies will occur after a second generation?

3. Consider rare disorders in a population caused by an autosomal recessive mutation. From the following frequencies of the disorder in the population, calculate the percentage of heterozygous carriers:
   (a) 0.0064
   (b) 0.000081
   (c) 0.09
   (d) 0.01
   (e) 0.10

4. What must be assumed in order to validate the answers in Problem 3?

5. In a population in which only the total number of individuals with the dominant phenotype is known, how can you calculate the percentage of carriers and homozygous recessives?

6. For the following two sets of data, determine whether they represent populations that are in equilibrium. Use chi-square analysis if necessary.
   (a) MN blood groups: MM, 60%; MN, 35.1%; NN, 4.9%
   (b) Sickle-cell hemoglobin: AA, 75.6%; AS, 24.2%; SS, 0.2%

7. If 4 percent of the population in equilibrium expresses a recessive trait, what is the probability that the offspring of two individuals who do not express the trait will express it?

8. Consider the allele frequencies $p = 0.7$ and $q = 0.3$, where selection occurs against the homozygous recessive genotype. What will be the allele frequencies after one generation if
   (a) $s = 1.0$?
   (b) $s = 0.5$?
   (c) $s = 0.1$?
   (d) $s = 0.01$?

9. If initial allele frequencies are $p = 0.5$ and $q = 0.5$, and $s = 1.0$, what will be the frequencies after 1, 5, 10, 25, 100, and 1000 generations?

10. Determine the frequency of allele A after one generation under the following conditions of migration:
    (a) $p = 0.6$; $p_m = 0.1$; $m = 0.2$
    (b) $p = 0.2$; $p_m = 0.7$; $m = 0.3$
    (c) $p = 0.1$; $p_m = 0.2$; $m = 0.1$

11. Assume that a recessive autosomal disorder occurs in one of 10,000 individuals (0.0001) in the general population. Assume that in this population about 2 percent (0.02) of the individuals are carriers for the disorder. Estimate the probability of this disorder occurring in the offspring of a marriage between first cousins and of one between second cousins. Compare these probabilities to those for the population at large.

12. What is the basis of inbreeding depression?

13. Does inbreeding cause an increase in frequency of recessive alleles within a population?

14. Describe how inbreeding may be used in the domestication of plants and animals. Discuss the theories underlying these techniques.

15. Evaluate the following statement: Inbreeding increases the frequency of recessive alleles in a population.

16. In a breeding program to improve crop plants, which of the following mating systems should be employed to produce a homozygous line in the shortest possible time?
    (a) Self-fertilization
    (b) Brother–sister matings
    (c) First-cousin matings
    (d) Random matings
    Illustrate your choice with pedigree diagrams.

17. If the recessive trait albinism (a) is present in 1/10,000 individuals, calculate the frequency of
    (a) The recessive mutant allele
    (b) The normal dominant allele
    (c) Heterozygotes in the population
    (d) Matings between heterozygotes

18. One of the first Mendelian traits identified in humans was a dominant condition known as *brachydactyly*. This gene causes an abnormal shortening of the fingers and/or toes. At the time, it was thought by some that this dominant trait would spread until 75 percent of the population would be affected (since the phenotypic ratio of dominant to recessive is 3:1). Show why this line of reasoning is incorrect.

19. Achondroplasia is a dominant trait that causes a characteristic form of dwarfism. In a survey of 50,000 births, five infants with achondroplasia were identified. Three of the affected infants had affected parents, while the rest had normal parents. Calculate the mutation rate for achondroplasia and express the rate as the number of mutant genes per given number of gametes.

20. A prospective groom, who is normal, has a sister with cystic fibrosis, an autosomal recessive disease state. Both of his parents are normal. He plans to marry a woman who has no history of cystic fibrosis (CF) in her family. What is the probability that they will produce a CF child? They are both Caucasian, and the overall frequency of CF in the Caucasian population is 1/2500—that is, one affected child per 2500. (Assume that the population meets the Hardy–Weinberg conditions.)

Selected Readings

Amos, W., and Harwood, J. 1998. Factors affecting levels of genetic diversity in natural populations. *Philos. Trans. R. Soc. London B Biol. Sci.* 353:177–186.

Ballou, J., and Ralls, K. 1982. Inbreeding and juvenile mortality in small populations of ungulates: A detailed analysis. *Biol. Conserv.* 24:239–72.

Barton, N. H., and Turelli, M. 1989. Evolutionary quantitative genetics: How little do we know? *Annu. Rev. Genet.* 23: 337–70.

Bodmer, W. F., and Cavalli-Sforza, L. L. 1976. Genetics, evolution and man. New York: W. H. Freeman.

Charlesworth, B. 1998. Measures of divergence between populations and the effect of forces that reduce variablitiy. *Mol. Biol. Evol.* 15: 538–543.

Cook, L. M., Askew, R. R., and Bishop, J. A. 1970. Increasing frequency of the typical form of the peppered moth in Manchester. *Nature* 227:1155.

Crow, J. F. 1986. *Basic concepts in population, quantitative and evolutionary genetics.* New York: W. H. Freeman.

———, and Kimura, M. 1970. *An introduction to population genetic theory.* New York: Harper & Row.

David, J. R., and Capy, P. 1988. Genetic variation of *Drosophila melanogaster* natural populations. *Trends Genet.* 4:106–11.

Dudley, J. W. 1977. 76 generations of selection for oil and protein percentage in maize. In E. Pollack, O. Kempthorne, and T. B. Bailey, Eds. *Proceedings of the International Conference on Quantitative Genetics,* pp. 459–73. Ames: Iowa State University Press.

Edwards, J. H. 1989. Familarity, recessivity and germline mosaicism. *Ann. Hum. Genet.* 53:33–47.

Frankham, R. 1995 Conservation genetics. *Annu. Rev. Genet.* 29: 305–327.

Freire-Maia, N. 1990. Five landmarks in inbreeding studies. *Am. J. Med. Genet.* 35:118–20.

Gayle, J. S. 1990. *Theoretical population genetics.* Boston: Unwin-Hyman.

Hartl, D. L. 1988. *A primer of population genetics,* 2nd ed. Sunderland, MA: Sinauer Associates.

Jones, J. S. 1981. How different are human races? *Nature* 293:188–90.

Karn, M. N., and Penrose, L. S. 1951. Birth weight and gestation time in relation to maternal age, parity and infant survival. *Ann. Eugen.* 16:147–64.

Kerr, W. E., and Wright, S. 1954. Experimental studies of the distribution of gene frequencies in very small populations of *Drosophila melanogaster.* I. *Forked. Evolution* 8:172–77.

Kettlewell, H. B. D. 1961. The phenomenon of industrial melanism in Lepidoptera. *Annu. Rev. Entomol.* 6:245–62.

———. 1973. The evolution of melanism: The study of a recurring necessity, with special reference to industrial melanism in the Lepidoptera. London: Oxford University Press.

Kevles, D. J. 1985. *In the name of eugenics: Genetics and the uses of human heredity.* Berkeley: University of California Press.

Khoury, M. J., and Flanders, W. D. 1989. On the measurement of susceptibility to genetic factors. *Genet. Epidemiol.* 6: 699–701.

Li, C. C. 1977. *First course in population genetics.* Pacific Groves, CA: Boxwood Press.

Mettler, L. E., Gregg, T., and Schaffer, H. E. 1988. *Population genetics and evolution,* 2nd ed. Englewood Cliffs, NJ: Prentice-Hall.

Reed, T. E. 1969. Caucasian genes in American Negroes. *Science* 165:762–68.

Renfew, C. 1994. World linguistic diversity. *Sci. Am.* January 270:116–23.

Roberts, D. F. 1988. Migration and genetic change. *Hum. Biol.* 60:521–39.

Spiess, E. B. 1989. *Genes in populations,* 2nd ed. New York: Wiley.

Wallace, B. 1989. One selectionist's perspective. *Quart. Rev. Biol.* 64:127–45.

Woodworth, C. M., Leng, E. R., and Jugenheimer, R. W. 1952. Fifty generations of selection for protein and oil in corn. *Agron. J.* 44:60–66.

Zhu, J., Nestor, K. E., and Moritsu, Y. 1996. Relationship between band sharing levels of DNA fingerprints and inbreeding coefficients and estimation of true inbreeding in turkey lines. *Poult. Sci.* 75:25–28.

The monkey flower Mimulus lewisii.

CHAPTER OUTLINE

CHAPTER
22

Genetics and Evolution

Chapter Concepts

Evolution is brought about by natural selection acting on the gene pool of a population. Selection leads to changes in genetic diversity and the ability to undergo evolutionary divergence. The process of speciation divides one gene pool into two or more reproductively separate gene pools. This division may be accompanied by changes in morphology, physiology, and adaptation to the environment.

In Chapter 21 we described populations in terms of allele frequencies and genetic equilibria, and outlined the forces that alter the genetic makeup of populations. Mutation, migration, selection, and drift individually and collectively alter allele frequencies and bring about evolutionary divergence and species formation. This process depends not only on genetic divergence but also on environmental or ecological diversity. If a population is spread over a geographic range encompassing a number of subenvironments or **niches**, populations occupying these niches will adapt and become genetically differentiated. Over time, this process may lead to a change in allele frequencies and, eventually, the formation of a phenotypic variant of the species. Because both the formation and the maintenance of these variants depend ultimately on the interaction of the genotype and environment, these variant populations are dynamic entities that may remain in existence, become extinct, merge with the parental population, or continue to diverge until they become reproductively isolated and form new species.

A **species** is defined as one or more groups of interbreeding or potentially interbreeding organisms that are reproductively isolated in nature from all other organisms. In sexually reproducing organisms, the process of *speciation* divides a single gene pool into two or more reproductively isolated subunits (separate gene pools). Changes in morphology, physiology, and adaptation to an ecological niche may also occur, but are not necessary components of the speciation event. Specia-

tion can take place gradually over a long time period or within a few generations. Throughout this process, the degree of genetic diversity is the key factor. Individuals of a species are members of a common gene pool that distinguishes them from members of other species.

In this chapter we will focus on three topics: first, the role of natural selection in the process of species formation; second, analysis of genetic variation and its role in evolution; and third, an example of evolution in *Drosophila*. In addition, we will consider molecular aspects of human evolution.

Wallace, Darwin, and the Origin of Species

At the age of 22, Charles Darwin accepted a position as a naturalist aboard the *Beagle,* a British ship that sailed around the world between 1831 and 1836 on a surveying expedition. During the voyage, Darwin collected plants, animals, and fossils, and he kept a detailed journal about the geology and geographic distribution of the organisms he saw and collected. When he returned home, he began to sort through his journals and collected specimens, with the intention of investigating the origin of species. Some 22 years later, in 1859, he published his book, *The Origin of Species by Means of Natural Selection,* which outlined a mechanism for evolution, and which provided a detailed and meticulous examination of the evidence for how new species arose.

While Darwin labored over the problem of evolution, Alfred Russel Wallace, a self-taught naturalist, spent the years between 1848 and 1852 exploring the Amazon Basin, collecting animals and plants in an attempt to explain how new species arise. He then spent seven years in the islands of Indonesia and Malaysia and formulated a theory on the role of natural selection in species formation, which he sent to Darwin. Their work was read at a meeting of the Linnean Society in London in July 1858. The work of Wallace motivated Darwin to finish his book, published in the following year.

For both Wallace and Darwin, the idea that natural selection is the driving force of evolution was developed by careful observation of plants and animals in their environment, coupled with readings in geology and an essay by Thomas Malthus on the factors that limit human population growth. Wallace and Darwin extracted from this essay the idea that when resources are limited, the organisms that survive and reproduce, on average, are better adapted to the environment.

The Wallace–Darwin theory of natural selection as a force in evolution can be summarized in a series of statements:

1. Minor variations in phenotype exist among individuals of a species. These variations can include differences in size, agility, coloration, or ability to obtain food. Most of these variations, although small and seemingly insignificant, are heritable, and are passed on to offspring.

2. Organisms tend to reproduce in an exponential fashion. More offspring are produced than can survive. This causes members of a species to engage in a struggle for survival, competing with other members of the species for scarce resources.

3. In the struggle for survival, some individuals will be more successful than others, allowing them to survive and reproduce.

4. Those individuals that survive to reproduce because of some favorable phenotypic variation leave behind more offspring than those with less favorable variations, eventually eliminating some variants from the species.

5. Over time, these heritable variations in phenotype can transform a species into a new species, similar to but distinct from the species it has replaced or from which it has diverged.

Although Wallace and Darwin proposed that natural selection can explain how evolution occurs, they could not explain the origin of the variations that provided the raw material for evolution, nor how such variation is maintained in a population or a species. In the twentieth century, as the principles of Mendelian genetics were applied to populations, the source of variation (mutation) and the maintenance of variation (allele frequencies) were both explained, and evolution was seen as changes in the genetic structure of populations

over time. This union of population genetics and natural selection generated a new view of the evolutionary process, called *neo-Darwinism*.

Models of Speciation

The process of splitting a genetically homogeneous population into two or more populations that undergo genetic differentiation and eventual reproductive isolation is called speciation. According to Ernst Mayr, species originate predominantly in two ways. In the first, often called **phyletic evolution** or **anagenesis**, species A becomes transformed into species B over a long period of time. In **cladogenesis**, the second way that species originate, one species gives rise to two or more species (Figure 22–1). This second process can occur over a long period of time or, rarely, in a generation or two.

Allopatric Speciation

The most developed model of speciation is **geographic** or **allopatric speciation**, first proposed by Moritz Wagner in 1868. According to Wagner, geographic features such as lakes, rivers, or mountains act as barriers to gene flow between populations, and physical isolation is the first step in the process of evolution. In a second step, the isolated populations undergo independent genetic changes and diverge to produce two distinct species.

Twentieth-century workers such as Mayr and Theodosius Dobzhansky have refined and updated this model but have retained the general features of Wagner's hypothesis. In the refined model, the first step requires that populations become separated; that is, gene flow (interbreeding) must be interrupted. The absence or interruption of gene flow is a prerequisite for the development of genetic differences generated by adaptation to local conditions. Genetic diversity arising by mutation, or by random genetic drift, will be reflected

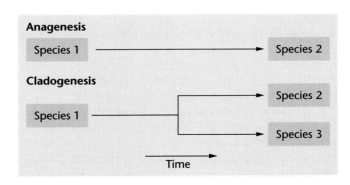

■ Figure 22–1 In anagenesis or phyletic evolution, one species transformed over time into another species. At all times, only one species exists. In cladogenesis, one species splits into two or more species.

in the presence of new alleles, changes in allele frequency, or the presence of new chromosomal arrangements. Eventually, a point will be reached when the populations have enough genetic differences that they can be identified as distinct races or semispecies. This process may continue uninterrupted until two or more species are present.

If at any time during the process of genetic divergence, the conditions that prevent gene flow between the populations are removed, two outcomes are possible: (1) The two populations may fuse into a single gene pool, because hybridization does not necessarily reduce fertility or viability; or (2) the gene pools of the populations may have diverged to the point where biological isolating mechanisms may have arisen, the two populations cannot interbreed, and two species are present.

Sympatric Speciation

A second model of speciation, called **sympatric speciation**, involves the formation of species from populations that live in the same geographic range and do not become geographically isolated. In this model, the first step is the formation of a population with a variant phenotype (with a degree of genetic divergence), and a shift to a new subenvironment or niche. Whether these populations can diverge to the point of becoming separate species depends on several factors, including how quickly reproductive isolation can be achieved.

There is agreement that the development of variant phenotypes and shifts to new niches occurs frequently within a species range, but there is some disagreement on how often this results in sympatric speciation. The development of over 200 species of cichlid fish in Lake Victoria in the Rift Valley of East Africa is often cited as an example of sympatric speciation (Figure 22–2). The lake is less than a million years old, but is home to over 200 species of cichlids (Figure 22–3). These species have some morphological diversity, but are highly specialized for different niches. Members of some species eat algae floating on the water surface, others are bottom feeders, insect feeders, mollusc eaters, and predators on other fish species. The possibility that these species evolved from a single ancestral species is supported by an examination of protein variation and nucleotide variation (both of which are discussed below) in 14 species from Lake Victoria, and in species from nearby lakes. The lack of biochemichal variation between species in Lake Victoria, and the high degree of variation between species in Lake Victoria and those in nearby lakes, supports the idea that the species in Lake Victoria have a single ancestral species.

■ Figure 22–2 Lake Victoria in the Rift Valley of East Africa is home to more than 200 species of cichlids.

■ Figure 22–3 Cichlids occupy a diverse array of niches, and each species is specialized for a distinct food source.

Statispatric Speciation

The third model of speciation that we will discuss, **statispatric speciation**, was originally derived to account for the evolution of flightless grasshoppers in Australia. In this model, a chromosomal aberration such as a translocation arises by chance in a small population. If heterozygotes for the translocation have a slightly reduced fitness, perhaps caused by abnormal meiosis, selection will favor either homokaryotype (two copies of the translocation or two normal chromosomes) by divergent selection. A translocation homokaryotype, with a chromosome number different from the original, may arise, spread, and partially displace the ancestral population. Semispecies would then be reproductively isolated because hybrids would have an unbalanced chromosome complement and a lowered fitness. In this model, reproductive isolation precedes the development of genetic variability. In addition to grasshoppers, the statispatric model has been applied to explain the origin of closely related species of naked mole rats, the *Spalax ehrenbergi* complex (Figure 22–4), which differ in chromosome number and do not interbreed.

Isolating Mechanisms

By whatever model speciation occurs, the final step involves the development of mechanisms that prevent breeding between individuals of different species. Physiological, behavioral, or mechanical barriers are effective at preventing interbreeding. The biological and behavioral properties of organisms that prevent or reduce interbreeding are called **reproductive isolating mechanisms**. These mechanisms are classified in Table 22.1. For example, genetic divergence may have reached the stage where the viability or fertility of hybrids is reduced. Hybrid zygotes may be formed, but all or most may be inviable. Alternatively, the hybrids may be viable but be sterile or have reduced fertility. In an-

TABLE 22.1	**Reproductive isolating mechanisms**

Prezygotic mechanisms (prevent fertilization and zygote formation)

1. Geographic or ecological: The populations live in the same regions but occupy different habitats.
2. Seasonal or temporal: The populations live in the same regions but are sexually mature at different times.
3. Behavioral (only in animals): The populations are isolated by different and incompatible behavior before mating.
4. Mechanical: Cross-fertilization is prevented or restricted by differences in reproductive structures (genitalia in animals, flowers in plants).
5. Physiological: Gametes fail to survive in alien reproductive tracts.

Postzygotic mechanisms (fertilization takes place and hybrid zygotes are formed, but these are nonviable or give rise to weak or sterile hybrids)

1. Hybrid nonviability or weakness.
2. Developmental hybrid sterility: Hybrids are sterile because gonads develop abnormally or meiosis breaks down before completion.
3. Segregational hybrid sterility: Hybrids are sterile because of abnormal segregation to the gametes of whole chromosomes, chromosome segments, or combinations of genes.
4. F_2 breakdown: F_1 hybrids are normal, vigorous, and fertile, but F_2 contains many weak or sterile individuals.

Source: From G. Ledyard Stebbins. *Processes of organic evolution,* 3rd edition, copyright 1977, p. 143. Reprinted by permission of Prentice-Hall, Englewood Cliffs, N.J.

other possibility, the hybrids themselves may be fertile, but their progeny may have lowered viability or fertility. These **postzygotic isolating mechanisms** act at or beyond the level of the zygote and are generated by genetic divergence. Postzygotic isolating mechanisms waste gametes and zygotes and lower the reproductive fitness of hybrid survivors. Selection will therefore favor the spread of alleles that reduce the formation of hybrids, leading to the development of **prezygotic isolating mechanisms**. These mechanisms prevent interbreeding and the formation of hybrid zygotes and offspring. In animal evolution, the most effective of these is behavioral isolation, involving courtship.

The Rate of Speciation

At the heart of the allopatric model of speciation is the concept of **phyletic gradualism**. According to this concept, speciation is a microevolutionary event resulting from the accumulation of many small gene differences

■ Figure 22–4   Members of a colony of naked mole rats from Kenya.

over a long period of time under the influence of natural selection. Recently, however, more **stochastic** or **catastrophic models** of speciation have been proposed. These models emphasize the role of evolutionary events that occur suddenly and intermittently and are therefore referred to as **quantum speciation**. We will briefly discuss one model derived from the fossil record, and an example that depends on chromosome rearrangement as an isolating mechanism. Finally, we will consider a mechanism that has long been known to exist in angiosperm plants—speciation through polyploidy.

Drawing on evidence from the fossil record, Niles Eldredge and Stephen Jay Gould have proposed a mechanism of species formation known as **punctuated equilibrium**. According to this model, evolutionary changes are not gradual and continuous, but occur intermittently and rapidly as events that punctuate or interrupt long periods of evolutionary stasis during which little or no change occurs. In paleontological terms, *rapidly* means within thousands or even hundreds of thousands of years. Results of such evolutionary alterations are thought to cause abrupt changes in the fossil record, as relatively few fossils of a given type will be formed during such changes, compared to the large numbers of fossils formed during long periods (extending into millions of years) of evolutionary equilibrium.

The evolutionary spurts that punctuate periods of stasis are thought to be related to catastrophic events that produce mass extinctions, or events that take place in small populations located at the fringes of the species range. In these locales, environmental stresses disrupt genetic stability, causing the emergence of new species. Proponents of this hypothesis point out that such changes leading to speciation take place through natural selection of individual variations and do not invoke any unusual genetic mechanisms.

Evidence for punctuated evolution caused by the appearance of rare, beneficial mutations has recently been gathered in laboratory studies using the bacterium, *Escherichia coli*. In a study that measured changes in cell size over 3000 generations of bacteria in a constant environment, researchers found that periods of stasis were interrupted by periods of rapid change. Cultures of bacteria, each founded from a single cell, were grown in a continuous culture, and samples were frozen at regular intervals. Measurements of cell size were done every 100 generations; size remained constant for long periods. However, several major increases in cell size occurred over the course of 3000 generations. It is thought that these changes in cell size were the result of direct selection (or a side effect) for a rare, beneficial mutation that caused an increase in cell size. This mutation swept through the population, producing a change in cell size in 100 generations or less. The patterns of change observed in these laboratory cultures of asexually reproducing bacteria are similar to those seen in the fossil record of some eukaryotes. Whether these patterns have similar underlying causes,

and the role, if any, of punctuated evolution in eukaryotic speciation has yet to be determined.

A second model of quantum speciation, proposed by Hampton Carson, is called the **founder-flush theory**, and is derived from his studies on Hawaiian *Drosophila*. According to this model, populations and even species can be started by a single individual. The founder principle is not new, as this idea was advanced much earlier by Ernst Mayr. In Carson's proposal, however, gradualism is not a necessary component of speciation; more important, reproductive isolation precedes adaptation rather than being a consequence of genetic diversification. In other words, Carson has changed the order of steps in the classical model of speciation.

According to the founder-flush theory, a single fertilized female or a mating pair can colonize an isolated territory previously unoccupied by members of this species. If conditions are favorable, the population founded by this individual will undergo a **flush**, or rapid expansion. After several generations, it is likely that the population growth will outstrip the environment's carrying capacity, causing a **population crash**. The crash causes the death or dispersal of almost the entire population. A few survivors may rebuild the population, which eventually undergoes several flush–crash cycles before coming to equilibrium with the environment. This cycle is diagrammed in Figure 22–5. Through the genetic revolution it has undergone, the colony will, in all probability, have acquired adaptations that will make it unable to interbreed with the parental population.

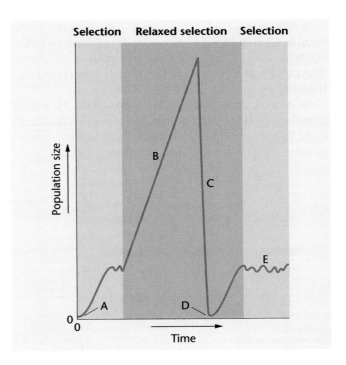

■ Figure 22–5 The flush–crash cycle. Small populations descended from a single founder (A) undergo a population flush (B), followed by a crash (C). A single survivor in the form of a fertilized female (D) builds a new population (E).

According to Carson's theory, genetic changes are brought about in two ways. First, the founding of the population by a very few individuals can establish allele frequencies very different from those in the ancestral population. Second and most important, selection is relaxed during a population flush, because resources are in excess. If descendants of the founder can invade a new niche, they expand their numbers in a flush. Carson proposes that normally, some blocks of genes on chromosomes are tightly linked or "closed" to recombination because of some selective advantage conferred by this configuration. New genotypes produced by recombination within these closed areas have reduced fitness. In fact, the advantage conferred by balanced polymorphisms for inversion heterozygotes may derive from the protection of such closed gene complexes (Figure 22–6).

During the flush period, genotypes produced by recombination in closed regions of the genome may survive because selection is relaxed at such times. These new genotypic combinations may be incompatible with the normal closed system of the species, but they survive because there is reduced competition for resources. After a crash, the reshuffled genome of survivors is acted upon by selection to produce a new combination of open and closed gene groups adapted to the environment. These periods of genomic reorganization may affect the timing and/or order of gene expression, bringing about developmental changes that accelerate the pace of evolution.

Several such cycles can produce enough genetic differences so that crosses between the ancestral population and the colony populations produce hybrids with lowered fitness commonly displayed by interspecific hybrids. Carson views cycles of disorganization and reorganization of the genomes as the essence of speciation rather than as the gradual genetic divergence of isolated populations over long periods of time.

The evolution of *Drosophila* species in the Hawaiian Islands appears to have followed such a founder-flush cycle. Geologic evidence indicates that the northwest-ernmost islands are the oldest, and that the southeast-

ern island of Hawaii is the youngest at about 700,000 years old. The relationships among species of *Drosophila* can be traced by mapping the location and frequency of inversions in the banded polytene chromosomes of larval salivary glands. One group, the *planitibia* complex, has three species with the same basic set of chromosome inversions: *D. planitibia, D. heteroneura,* and *D. silvestris*. The *D. planitibia* is found on the older island of Maui, and the other two are found on the younger island of Hawaii. It is postulated that a fertilized female belonging to an ancestral stock on Maui (chromosomally related to the present-day *D. planitibia*) crossed the Alenuihaha Channel to Hawaii. Subsequent flush–crash cycles led to the reconstruction of the colony's genotype, giving rise to the two species found on Hawaii—*D. heteroneura* and *D. silvestris* (Figure 22–7). An alternative hypothesis proposes that one of the species on Hawaii could have arisen as the result of a second colonization from Maui. Similar evidence indicates that two other groups on Maui have given rise to a total of five species on Hawaii.

In laboratory tests of this theory, Jeffrey Powell has found that 15 generations after several flush–crash cycles, laboratory strains of *Drosophila* species showed significant behavioral (prezygotic) isolation from other strains. Subsequent testing several months later showed that these differences were stable. These experimental results are important in establishing that the first stages of speciation can occur rapidly under certain circumstances.

The final example of quantum speciation that we will discuss involves polyploidy in plants. Formation of animal species by polyploidy is rare, but polyploidy is an

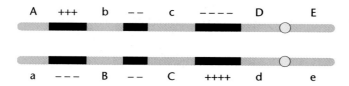

■ Figure 22–6 Model for open and closed regions of chromosomes. Products of crossing over anywhere within the open system (tan) result in offspring with high fitness, whereas crossovers in the closed regions (black) produce zygotes with low fitness. The letters represent genes or polygenes, and the pluses and minuses represent internally balanced gene complexes. Recombination within these blocks produces unfit gene combinations.

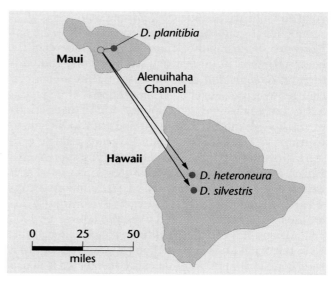

■ Figure 22–7 Proposed pathway of colonization of Hawaii by members of the *D. planitibia* species complex. Open circles represent a population ancestral to the three present-day species.

important factor in the evolution of plants. It is estimated that one-half of all flowering plants have evolved by this mechanism. One form of polyploidy is **allopolyploidy** (see Chapter 8), produced by doubling the chromosome number in an interspecific hybrid.

If two species of related plants have the genetic constitution *SS* and *TT* (where *S* and *T* represent the haploid set of chromosomes in each species), then the F1 hybrid would have the chromosome constitution *ST*. Normally such a plant would be sterile because there are few or no homologous chromosome pairs, and aberrations would arise during meiosis. However, if the hybrid undergoes a spontaneous doubling of chromosome number, a tetraploid *SSTT* plant would be produced. This chromosomal aberration might occur during mitosis in somatic tissue, giving rise to a partially tetraploid plant that would produce some tetraploid flowers. Alternatively, aberrant meiotic events may produce *ST* gametes, which when fertilized would yield *SSTT* zygotes. The *SSTT* plants would be fertile because they would possess homologous chromosomes producing viable *ST* gametes. This new, true-breeding tetraploid would have a combination of characters derived from the parental species, and would be reproductively isolated from them because F1 hybrids would be triploids and consequently sterile.

The tobacco plant *Nicotiana tabacum* ($2n = 48$) is the result of the doubling of the chromosome number in the hybrid between *N. otophora* ($2n = 24$) and *N. silvestris* ($2n = 24$) (Figure 22–8).

■ Figure 22–8 The cultivated tobacco plant, *Nicotiana tabacum,* is the result of hybridization between two other species.

Measuring Genetic Variation

If a population of organisms constituting a species is well suited to inhabit and reproduce successfully in its environment, how much genetic variation does it possess? We might assume that the members of a well-adapted population are genetically homogeneous because the most favorable allele at each locus has become fixed. Certainly, an examination of most populations of plants and animals reveals them to be phenotypically similar. However, current evidence supports the opinion that a high degree of heterozygosity is maintained within the gene pool of a diploid population. This built-in diversity is concealed, so to speak, because it is not necessarily apparent in the phenotype. Such diversity may better adapt a population to the inevitable changes in the environment and may promote the survival of the species. The detection of this concealed genetic variation is not a simple task. Nevertheless, such investigation has been successful using several techniques, discussed in the following sections.

Inbreeding Depression

A genetic approach to measuring allelic diversity can be used to determine what fraction of the alleles carried by individuals of a population will be deleterious. Deleterious alleles, such as lethals, can be detected by monitoring the effects (death or disease) of increases in homozygosity that result from inbreeding. Because inbreeding involves mating between relatives, mates have genotypic similarities that extend across all segregating loci. As a result, inbreeding causes recessive alleles to become homozygous at a high frequency in the offspring. If these alleles are deleterious, inbreeding depression is evident in successive generations (review inbreeding depression in Chapter 21).

In using inbreeding depression to study genetic diversity, Theodosius Dobzhansky and his colleagues succeeded in making groups of alleles and sometimes entire chromosomes of *Drosophila* species completely homozygous. When this is accomplished, viability is depressed. As shown in Table 22.2, such studies reveal that lethal, semilethal, subvital, and male and female sterility alleles are prominent in natural populations of different species. In *Drosophila pseudoobscura,* almost 100 percent of the copies of chromosomes 2 and 3 taken from a sample of the population cause a depression in viability when made homozygous. In *Drosophila willistoni,* almost 70 percent of chromosomes 2 and 3 contain male sterile alleles.

Although these studies reveal that organisms carry deleterious genes (mostly in the heterozygous condition), we discuss them here as examples of genetic variation maintained in natural populations. If detrimental

TABLE 22.2	Percentage of homozygous chromosomes tested that reveal recessive mutations of viability and fertility

Drosophila Species	Chromosome	Lethals and Sublethals	Subvitals	Sterility	Female Sterility	Male Sterility
D. pseudoobscura	2	33.0	62.6	<0.1	10.6	8.3
(Mather, California)	3	25.0	58.7	<0.1	13.6	10.5
	4	22.7	51.8	<0.1	4.3	11.8
D. persimilis	2	25.5	49.8	0.2	18.3	13.2
(Mather, California)	3	22.7	61.7	2.1	14.3	15.7
	4	28.1	70.7	0.3	18.3	8.4
D. prosultans	2	32.6	33.4	<0.1	9.2	11.0
(Brazil)	3	9.5	14.5	3.0	6.6	4.2
D. willistoni	2	38.8	57.5	<0.1	40.5	64.8
(Venezuela, British Guiana, and Trinidad)	3	34.7	47.1	1.0	40.5	66.7

Note: Lethal alleles cause premature death, often before sexual maturity and reproduction; subvital mutations kill less than 50% of carriers before maturation; supervital mutations increase viability above the wild-type level.
Source: From Spiess, 1977, p.288.

alleles are maintained in a heterozygous state, certainly alleles that are effectively neutral or potentially advantageous must also be retained in the heterozygous condition. As recessive alleles, they represent concealed genetic variation within the "normal" heterozygous genotypes characterizing the population.

Protein Polymorphisms

Gel electrophoresis is a simple technique that can be used to separate protein molecules on the basis of differences in size and electrical charge. If a nucleotide variation in a structural gene results in the substitution of a charged amino acid such as glutamic acid for an uncharged amino acid such as glycine, the net electrical charge on the protein will be altered. This difference in charge can be detected as a change in the rate at which proteins migrate through an electrical field in electrophoresis. In the mid-1960s, John Hubby and Richard Lewontin used gel electrophoresis to measure protein variation in natural populations of *Drosophila*. Since

then, genetic variation has been studied using gel electrophoresis in a wide range of organisms (Table 22.3).

The electrophoretic forms of a protein produced by different alleles are designated as **allozymes**. As shown in Table 22.3, a surprisingly large percentage of loci examined from diverse species produce distinct allozymes. Of the populations shown in Table 22.3, approximately 30 loci per species were examined, and about 30 percent of the loci were polymorphic. An approximate average of 10 percent allozyme heterozygosity per diploid genome was revealed.

These values apply only to genetic variation detectable by altered protein migration in an electric field. Electrophoresis probably detects only about 30 percent of the variation that is due to amino acid substitutions, because many substitutions do not change the net electric charge on the molecule. Richard Lewontin has estimated that about two-thirds of all loci are polymorphic in a population, and that in any individual within that group, about one-third of the loci exhibit genetic variation in the form of heterozygosity.

TABLE 22.3	Heterozygosity at the molecular level			
Species	Number of Populations Studied	Number of Loci Examined	Loci per Population	Hetero-zygosity per Locus
Homo sapiens (humans)	1	71	28	6.7
Mus musculus (house mouse)	4	41	29	9.1
Drosophila pseudoobscura (fruit fly)	10	24	43	12.8
Limulus polyphemus (horseshoe crab)	4	25	25	6.1

Source: From Lewontin, 1974, p. 117.

There is some controversy over the significance of genetic variation as detected by electrophoresis. Some argue that allozymes are biochemically equivalent and therefore do not play any role in evolution. We shall address this argument later in the chapter.

Variations in Nucleotide Sequence

The most direct way to estimate genetic variation is to compare the nucleotide sequence of genes carried by individuals in a population. With the development of techniques for cloning and sequencing DNA, nucleotide sequence variations have been cataloged for an increasing number of gene systems. Using restriction endonucleases to detect polymorphisms, Alec Jeffrey surveyed 60 unrelated individuals to estimate the total number of DNA sequence variants in humans. His results show that within the genes of the beta-globin cluster, 1 in 100 base pairs shows polymorphic variation. If this region is representative of the genome, this indicates that at least 3×10^7 nucleotide variants per genome are possible.

In another study, Martin Kreitman studied the alcohol dehydrogenase locus (*Adh*) in *Drosophila melanogaster* (Figure 22–9). This locus encodes two allozymic variants: the *Adh-f* and the *Adh-s* alleles. These differ by a single amino acid (Thr versus Lys at codon 192). To determine whether the amount of genetic variation detectable at the protein level (one amino acid difference) corresponds to the variation at the nucleotide level, Kreitman cloned and sequenced *Adh* loci from five natural populations of *Drosophila*.

The 11 cloned loci contained 43 nucleotide variations from the consensus *Adh* sequence of 2721 base pairs. These variations are distributed throughout the *Adh* gene: 14 in the exon-coding regions, 18 in the introns, and 11 in the nontranslated and flanking regions. Of the 14 in the coding regions, only one leads to an amino acid replacement, accounting for the two observed electrophoretic variants. The other 13 nucleotide substitutions do not lead to amino acid replacements. Are the differences in the number of allozyme and nucleotide variants the result of natural selection? If so, of what significance is this? These issues will be discussed below.

Studies of other organisms including the rat and the mouse have produced similar estimates of nucleotide diversity, indicating that there is an enormous reservoir of genetic variability within a population and that, at the level of DNA, most and perhaps all genes exhibit diversity from individual to individual.

Evolution and Genetic Variation: A Dilemma

A population of interbreeding organisms is more or less adapted to its surrounding environment. Phenotypes that enhance adaptation to a given environment are controlled by the genotypes carried by the individuals constituting the group. Ideally, each individual should possess a genotype and phenotype best suited to the immediate environment. However, the preceding discussion on genetic variability in natural populations suggests that there is wide variation in genotypes within populations.

Evolution depends on the presence of genetic variation within a population of organisms and the action of natural selection on that variation. Although genetic variation can be measured in a variety of ways, there is some controversy about whether it is important to maintain a large amount of variation in a population. This controversy centers particularly on the variation detected at the molecular level. On the one hand, natural selection should favor homozygotes carrying the most favorable alleles at each locus, so that organisms can be best adapted to their environment. Homozygosity, therefore, should be the rule rather than the exception. If this is true, why is heterozygosity so prevalent?

On the other hand, alleles that at one point in time are detrimental to individuals may be of great value to the population in future generations, as shown by the melanic forms of the peppered moth, discussed in Chapter 21. Under changing environmental conditions, previously insignificant or even detrimental alleles may become essential to the maintenance of fitness. Thus, selection should favor the maintenance of a high degree of heterozygosity in populations, so that members of the population have the genetic diversity to respond to changes in environmental conditions. In addition, heterozygotes are often superior in many ways to homozygotes (see the discussion of hybrid vigor in Chapter 21).

The apparent dilemma is that organisms should have low levels of heterozygosity to be well adapted to their environment, but should have a high level of heterozygosity in order to respond to changes in the environment. This problem has generated several points of view about the significance of variation. Two of these will be discussed below.

Neutralists and Variation

One point of view argues that the existence of a high degree of genetic variation does not prove that it is important in evolutionary change. The **neutralist hypothesis**, originally proposed by James Crow and Motoo

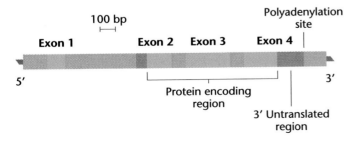

■ Figure 22–9 The *Adh* gene of *Drosophila melanogaster*.

Kimura, argues that mutations leading to amino acid substitutions are rarely favorable. They are sometimes detrimental but most often neutral or genetically equivalent to the allele that is replaced, and they are therefore unimportant from an evolutionary perspective. Polymorphisms that are favorable or detrimental are either preserved or removed from the population, respectively, by natural selection. However, neutral genetic changes will not be affected by selection and will instead accumulate randomly in the population. Their frequency will be determined by the rate of mutation and the principles of random genetic drift.

Selectionists and Variation

Opposed to the neutralist theory are the **selectionists**. These geneticists point out examples where enzyme or protein polymorphism is associated with adaptation to certain environmental conditions. The well-known advantage of sickle-cell anemia heterozygotes to infection by malarial parasites is such an example.

Selectionists also stress that enzyme polymorphisms may often appear to offer no advantage, but exist with such a frequency that they are impossible to explain as a random occurrence. Thus, even though no currently available analytical technique can detect any physiological difference, some slight advantage associated with any given amino acid substitution may exist. There may in fact be an advantage to having two forms of a given protein, allowing optimum performance under a wider range of cellular conditions.

Even though this controversy seems highly theoretical and somewhat esoteric, it is important that we not lose sight of several factors. First, the neutralists do not discount natural selection as a guiding force in evolution. Rather, they suggest that some features of organisms' genotypes are nonadaptive, and may have been fixed by genetic drift. On the other hand, the selectionists certainly do not deny that genetic drift is an important factor in establishing differences in allele frequencies.

Finally, it should be pointed out that the two theories are not mutually exclusive. It is difficult to argue against the notion that some genetic variation must be neutral. The difference between the two theories is in the degree of neutrality that exists. Although current data are insufficient to resolve the problem, one important point has emerged from the arguments. In considering natural selection, there are clearly two levels that must be examined: the phenotypic level, including all morphological and physiological characteristics imparted by the genotype; and the molecular level, represented by the precise nucleotide and amino acid sequence of DNA and proteins. There is no question about selection acting at the phenotypic level, but at the genotypic level, does it act on individual loci, or on the entire genome?

Formation of Species

Although the allopatric model of speciation is well developed, it has been more difficult to observe directly the processes involved. Diversification of isolated populations occurs gradually over thousands or hundreds of thousands of years. In addition, the geographic changes that paralleled this divergence may be complex or unknown. In most cases, therefore, the formation of species is a historical event, and biologists studying this process must rely on the present-day distribution of races, subspecies, and sibling species to reconstruct stages in the evolutionary process.

To study speciation, biologists must find examples in nature where all or most of the stages of race formation and speciation can be documented. The intensive studies carried out on natural populations of *Drosophila* provide a good example of the stages involved in **allopatric speciation**.

To illustrate the first step, namely, the formation of populations with substantial genetic differences (race formation), we shall consider studies on *Drosophila pseudoobscura* conducted by Theodosius Dobzhansky and his colleagues. This species is found over a wide range of environmental habitats, including the western and southwestern United States. Although the flies throughout this range are morphologically similar, Dobzhansky discovered that populations from different locations vary in the arrangement of genes on chromosome 3. He found several different inversions in this chromosome that can be detected by loop formations in larval polytene chromosomes. Each inversion sequence was named after the locale in which it was first discovered (e.g., AR 5 Arrowhead, British Columbia; CH 5 Chiricahua Mountains, etc.). The inversion sequences were compared with one standard sequence, designated ST.

Figure 22–10 shows a comparison of three arrangements found in populations at three different elevations in the Sierra Nevada chain in California. The ST arrangement is most common at low elevations; at 8000 feet, AR is the most common and ST least common. In these populations, the frequency of the CH arrangement gradually increases with elevation. The gradual change in inversion frequencies is probably the result of natural selection and parallels the gradual environmental changes occurring at ascending elevations. As the populations along this gradient show a continuous and gradual change in inversion frequencies, it is difficult to classify a fly as a member of one racial group.

Dobzhansky also found that if populations were collected at a single site throughout the year, inversion frequencies also changed. Through the seasons, cyclic variation in chromosome arrangements occurred, as shown in Figure 22–11. Such variation was consistently observed over a period of several years. The frequency

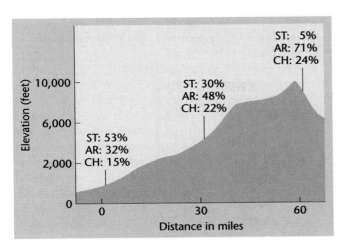

■ Figure 22–10   Inversions in chromosome 3 of *D. pseudoobscura* at different elevations in the Sierra Nevada range near Yosemite National Park.

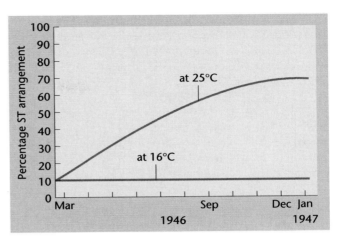

■ Figure 22–12   Increase in the ST arrangement of *D. pseudoobscura* in population cages under laboratory conditions. .

of ST always declined during the spring, with a concomitant increase in CH during the same period.

To test the hypothesis that this cyclic change is a response to natural selection, Dobzhansky devised the following laboratory experiment. He constructed a large population cage from which samples of *D. pseudoobscura* could be periodically removed and studied. He began with a population of a known inversion frequency, 88 percent CH and 12 percent ST. He reared it at 25°C and sampled it over a one-year period. As shown in Figure 22–12, the frequency of ST increased gradually until it was present at a level of 70 percent. At this point, an equilibrium between ST and CH was reached. When the same experiment was performed at 16°C, no change in inversion frequency occurred. It can be concluded that the equilibrium reached at 25°C was in response to the elevated temperature, the only variable in the experiment.

The results are evidence that a balance in the frequency of the two inversions and their respective gene arrangements in a population is superior to either inversion by itself. The equilibrium attained presumably represents the greatest degree of fitness in the population under controlled laboratory conditions. This interpretation of the experiment suggests that natural selection is the driving force toward equilibrium.

In a more extensive study, Dobzhansky and his colleagues sampled *D. pseudoobscura* populations over a broader geographic range. Twenty-two different chromosome arrangements were found in populations from 12 locations. In Figure 22–13, the frequencies of five of these inversions are shown according to geographic location. The differences are largely quantitative, with most populations differing only in relative frequencies of inversions.

Because these locations represent varied environments and because inversions preserve different gene combinations, we can conclude that numerous races of *D. pseudoobscura* have been formed.

For speciation to occur, the development of races must be followed by a second step, reproductive isolation. We might then ask whether the evolution of *D. pseudoobscura* has gone beyond the formation of races. Dobzhansky investigated this question by examining the chromosome structure of other, closely related species called **sibling species**. Sibling species are reproductively isolated from one another, but remain very similar morphologically.

Drosophila persimilis and *D. pseudoobscura* each have five pairs of chromosomes and carry inversions within chromosome 3. When Dobzhansky examined the chromosome 3 inversions carefully, one arrangement (ST) was found in both species. Using this common inversion as a starting point, a phylogenetic sequence for all arrangements found in both species was constructed. Only one hypothetical arrangement is necessary to complete the continuity of the tree. It appears that an ancestral population with the ST arrangement gave rise to many different inversions. Some were

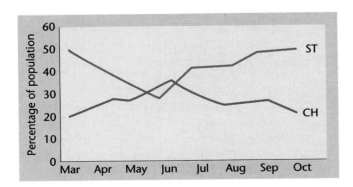

■ Figure 22–11   Changes in the ST and CH arrangements in *D. pseudoobscura* throughout the year.

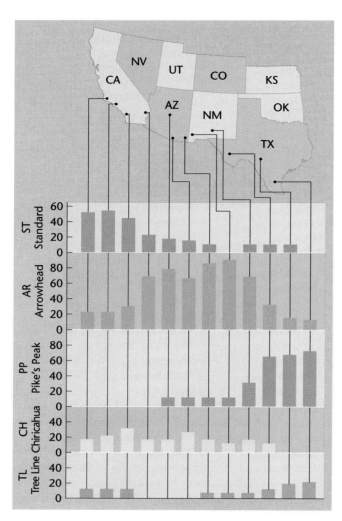

■ Figure 22–13 Frequencies of five chromosomal inversions in *D. pseudoobscura* in different geographic regions.

incorporated into members of the *D. pseudoobscura* species, and others gave rise to the *D. persimilis* species (Figure 22–14).

Today, even when the geographic distributions of these sibling species overlap (Figure 22–15), several isolating mechanisms keep them from interbreeding. The species are isolated by prezygotic mechanisms such as habitat selection, with *D. persimilis* preferring high elevations and cooler temperatures. Differences in courtship rituals allow females to distinguish between males of the two species and to choose only males of their own species for mating. Even the time of day when courtship and mating occur is different in the species, with *D. persimilis* tending to court in the morning and *D. pseudoobscura* more active in the evening.

Postzygotic mechanisms also maintain reproductive isolation in these species. When cross-fertilized in the laboratory, the species produce sterile F1 hybrid males, with male sterility being associated with interactions between the X chromosome and chromosome 2. Backcrosses between F1 females and parental males exhibit hybrid breakdown through lowered viability of the offspring.

Using Molecular Techniques to Study Evolution

Differences between two species can be measured in several ways. Taxonomists use several ways of evaluating differences, and determining whether two organisms being compared are phenotypic variants of the same species, or members of different species. In some cases, morphological differences are used to define and classify organisms; in addition, cytogenetics, geographical distribution, and behavior are used as measures of differences.

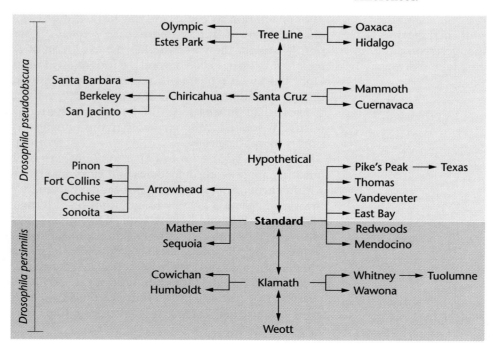

■ Figure 22–14 Inversion phylogeny for arrangements of chromosome 3 in *D. pseudoobscura and D. persimilis.* The ST arrangement is shared by both species.

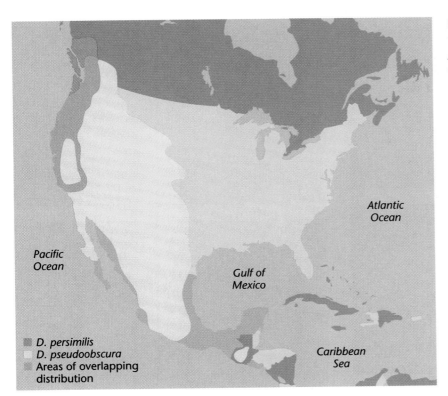

■ Figure 22–15 Although the geographic ranges of *D. persimilis* and *D. pseudoobscura* overlap, there is no interbreeding between the species.

Molecular biology and recombinant DNA technology are now being used to measure evolutionary relationships. These methods offer an increase in resolving power, and can be used to measure the degree of evolutionary divergence and to extrapolate the time scale over which genetic differences developed.

Measuring the Genetic Distance Between Species

The evidence from studies on *D. persimilis and D. pseudoobscura* indicates that chromosome rearrangements help maintain genetic differences between species, but they cannot provide information about the degree of genetic difference between these species. In other cases, cytogenetic studies of closely related species are uninformative, and genetic analysis has been used to measure the genetic differences between species.

Drosophila heteroneura and *D. silvestris,* found only on the island of Hawaii, are estimated to have diverged from a common ancestral species only about 300,000 years ago, but it is difficult to demonstrate significant differences between these species in chromosomal inversion patterns or protein polymorphisms. When DNA hybridization studies are carried out on these species, the sequence diversity between the two is only about 0.55 percent (Figure 22–16). Thus, nucleotide sequence diversity may precede the development of protein or chromosomal polymorphisms.

The fact that *D. heteroneura* and *D. silvestris* share identical chromosome arrangements, that they have almost no detectable protein differences, are 99 percent homologous in DNA sequence, and yet are classified as separate species may seem paradoxical. However, the

two species are clearly separated from each other by different and incompatible courtship and mating behaviors (a prezygotic isolating mechanism), by morphology, and by pigmentation of the body and wings (Figure 22–17). Genetic evidence suggests that these differences are controlled by a relatively small number of genes. Only about 15 to 19 major loci may be re-

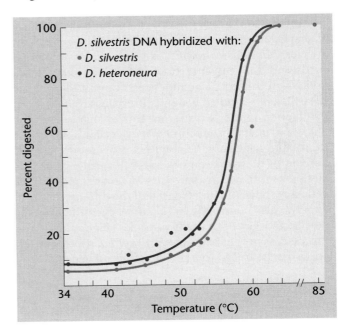

■ Figure 22–16 Nucleotide sequence diversity in the *D. planitibia* species complex. The degree of shift to the left by the heterologous hybrids is an indication of the degree of nucleotide sequence divergence.

(a)

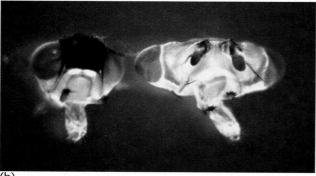

(b)

■ Figure 22–17   (a) Differences in pigmentation patterns in *D. sylvestris* (left) and *D. heteroneura* (right). (b) Head morphology in *D. sylvestris* (left) and *D. heteroneura* (right).

(a)

(b)

■ Figure 22–18   The flowers of two closely related species of monkey flowers. (a) *Mimulus cardinalis.* (b) *Mimulus lewisii.*

sponsible for the morphological differences between these species, demonstrating that the process of speciation need involve only a small number of genes.

More recent studies using two closely related species of monkey flowers, a plant that grows in the Rocky Mountains and areas west of the mountains, confirm that species can be separated by only a few genes. One species, *Mimulus cardinalis,* is fertilized by hummingbirds and does not interbreed with *Mimulus lewisii,* which is fertilized by bumblebees. Toby Bradshaw and his colleagues studied genetic differences related to reproduction in the two species: flower shape, size, and color, and nectar production. For each trait, a difference in a single gene or gene complex provided at least 25 percent of the variation observed. *M. cardinalis* makes 80 times more nectar than does *M. lewisii,* and a single gene is responsible for at least half the difference. A single gene also controls a large part of the differences in flower color between the two species (Figure 22–18). In this case, as in the Hawaiian *Drosophila,* species differences can be traced to a small number of genes.

Protein Evolution

The degree of evolutionary relatedness between two species can be measured by comparing the amino acid sequences of proteins found in both species.

Cytochrome c is a respiratory pigment found in the mitochondria of eukaryotes. The amino acid sequence of cytochrome c has changed very slowly during evolution. The amino acid sequence in humans and chimpanzees is identical; between humans and rhesus monkeys only one amino acid is different. This is remarkable considering that the fossil record indicates that the lines leading to humans and monkeys diverged from a common ancestral species approximately 20 million years ago.

Table 22.4 shows the number of amino acid differences in cytochrome c among a variety of organisms. Distantly related organisms such as humans and yeasts have 38 amino acid differences (out of 104 amino acids), but more closely related species have few, if any, differences.

The Molecular Clock

In addition to comparing amino acid differences, it is possible to assess evolutionary relationships by measuring the minimum number of nucleotide changes that have occurred during the evolution of a protein. This requires sequencing the gene encoding a protein in a number of different species, and then comparing the degree of nucleotide differences. This method is more complex than measuring amino acid differences. More than one nucleotide change may be required to change a given amino acid. When the nucleotide changes necessary for all amino acid differences observed in a protein are totaled, the **minimal mutational distance** between any two species is established. Table 22.4

	a. Number of Amino Acid Differences	b. Minimal Mutational Distance
Organism		
Human	0	0
Chimpanzee	0	0
Rhesus monkey	1	1
Rabbit	9	12
Pig	10	13
Dog	10	13
Horse	12	17
Penguin	11	18
Moth	24	36
Yeast	38	56

TABLE 22.4 A comparison of (a) the number of amino acid differences and (b) the minimal mutational distance in cytochrome c

Source: From W. M. Fitch and E. Margoliash, Construction of phylogenetic trees, *Science* 155:279–84, 20 January 1967. Copyright 1967 by the American Association for the Advancement of Science.

shows such an analysis of the genes coding for cytochrome c in 10 organisms. As expected, these values are larger than the corresponding number of amino acids separating humans from the other nine organisms.

Data on amino acid substitutions and mutational distance can be combined with paleontology to construct

a **molecular clock**. Information on amino acid or DNA differences measures the number of mutational events that have accumulated since any two organisms shared a common ancestor. The fossil record provides information about the time that has elapsed since the two organisms shared a common ancestor. Assuming that the rate of amino acid replacements occurred at a regular rate proportional to absolute time, differences in amino acid content can be used as a molecular clock, measuring the time since the two species diverged.

Phylogenetic Trees

On the basis of information provided by a molecular clock, it is possible to construct **divergence dendrograms** or **phylogenetic trees** based on the analysis of amino acid sequences of a single protein from diverse organisms. The underlying assumption in this analysis is that all present-day sequences from different species represent gene products that diverged from common ancestral sequences at various points in evolutionary time. By determining minimal mutational distances among all species under analysis, and taking into account which amino acids have changed, the most likely relationships among the species can be determined. This analysis can also establish the point in these relationships at which a now-extinct ancestral sequence must have existed in order to lead to evolutionary divergence.

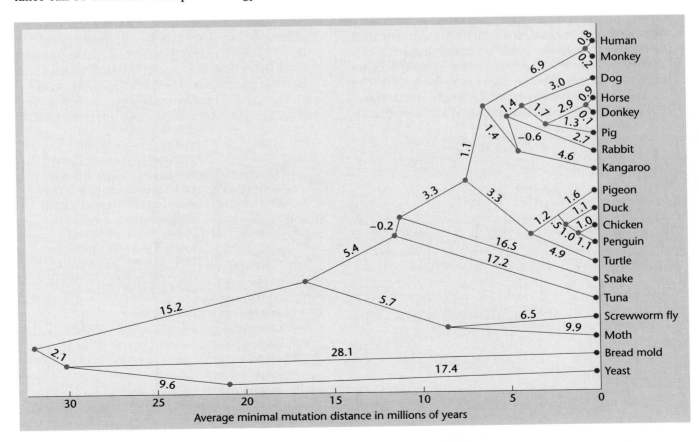

■ Figure 22–19 Phylogenetic sequence constructed by comparison of homologies in cytochrome c amino acid sequences.

This information is usually summarized in the form of a phylogenetic tree (see Figure 22–19 on page 509). This tree (based on the data of Table 22.4) is plotted so that the ordinate represents proportional amounts of evolutionary distance. If a constant substitution rate is assumed, the ordinate represents a relative estimate of geological time.

Phylogenetic trees agree remarkably well with trees constructed using more conventional approaches such as morphological and paleontological evidence. Once a number of proteins from a variety of organisms are sequenced and analyzed simultaneously, even more accurate trees can be constructed.

When closely related species are to be examined in this way, proteins that have evolved more rapidly than cytochrome c are more useful. The 115-amino-acid protein carbonic anhydrase has been used to analyze more accurately the relationship between humans and several other primates (Figure 22–20).

Molecular Studies on Human Evolution

The hominoid primates include chimpanzees, gorillas, orangutans, gibbons, and humans. Studies using protein differences have failed to resolve the taxonomic relationships among the chimpanzee, gorilla, and humans, because they are so closely related. Using hybridization of DNA sequences from a large number of individuals and calibrations derived from the fossil record, Charles Sibley and Jon Ahlquist have clarified the evolutionary branching pattern in the hominoids and have estimated the times at which divergences occurred (Figure 22–21). According to their model, humans and chimpanzees are more closely related than either is to the gorilla. The measure of relatedness is called the $\Delta T_{50}H$, and is related to the nucleotide matching of hybrid DNA molecules

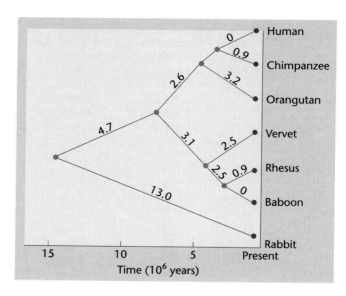

■ Figure 22–20 Phylogenetic sequence of carbonic anhydrase amino acid substitutions.

formed when single-stranded DNA from the two species is mixed. Between humans and chimpanzees, the $\Delta T_{50}H$ is 1.8, somewhat less than the value of 2.3 for the distance between the gorilla line and the chimpanzee/human line. Using the fossil record, this places the separation of the line leading to the gorilla at 8 million to 10 million years (MY) ago. The lines leading to the chimpanzees diverged about 6.3 to 7.7 MY ago, and the chimpanzee and pygmy chimpanzee diverged about 2.4 to 3.0 MY ago.

The close evolutionary relationship between the two species of chimpanzees and the human species revealed by DNA hybridization poses an interesting taxonomic problem. In other areas of taxonomy, differences of less than 2 in the $\Delta T_{50}H$ for two species are usually accompanied by classification in the same genus and family.

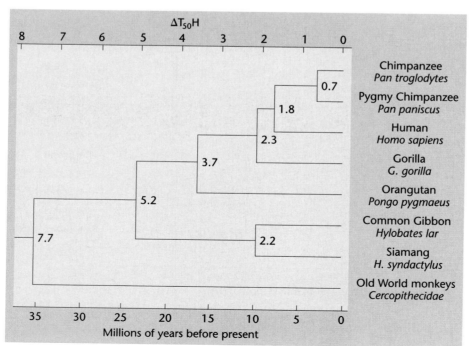

■ Figure 22–21 Phylogenetic sequence of the hominoid primates and the Old World monkeys as estimated by DNA hybridization. The numbers at the branch points are $\Delta T_{50}H$ measurements. The evolutionary branch points are dated by reference to the fossil record and nucleotide divergence.

As it stands, current taxonomy places humans not only in a different genus, but also in a different family than gorillas and chimpanzees (humans in Hominidae, gorillas and chimps in Pongidae). The problems and challenges posed by this anthropomorphic inconsistency and the evolutionary relationship between humans and chimpanzees are explored in the book, *The Third Chimpanzee,* by Jared Diamond.

The question of where the modern human species (*Homo sapiens*) originated has generated several theories. One model holds that there was no single site of origin, and that a gradual transition to *H. sapiens* occurred simultaneously in geographically dispersed but genetically linked populations of a precursor species, *Homo erectus.* Another model suggests that *H. sapiens* was present in southern Africa more than 100,000 years ago, and was found in Asia at least 50,000 years ago, and that *H. sapiens* replaced Neandertals in Europe about 30,000 to 40,000 years ago. This model suggests a single point of origin for *H. sapiens,* in Africa, with subsequent migrations to other regions, where *H. erectus* was displaced.

There is little dispute over the evidence that earlier ancestors of the human species arose in Africa, and that an ancestral species, *H. erectus,* spread out of Africa to populate Europe and Asia beginning about 2 million years ago. What is in dispute is how and where *H. sapiens* originated.

Over the past two decades, molecular techniques, including restriction fragment length polymorphism (RFLP) analysis, have been used to provide evidence for the origin and spread of the human species across the globe. In Chapter 18 we discussed the use of RFLPs to map genes to specific chromosomes and/or chromosomal regions. The RFLPs can be generated by single nucleotide base changes in a DNA sequence that is recognized by a restriction enzyme. These polymorphisms are inherited in a codominant Mendelian fashion. Also, RFLP analysis has been used to construct both a molecular clock and phyletic trees in order to reconstruct events in human evolution.

Phyletic trees constructed from RFLPs in mitochondrial genomic DNA suggest that the modern human lineage arose in Africa about 200,000 years ago. According to the results of these studies, the modern human species, *H. sapiens,* arose in Africa. From here, modern *H. sapiens* spread throughout the Old World, replacing the several human lineages descended from *H. erectus.*

The opposing hypothesis agrees with the idea that humans originated in Africa and spread throughout Europe and Asia as *H. erectus.* However, this hypothesis holds that the transitions in the fossil record in Asia and the Middle East from *H. erectus* to *H. sapiens* support the idea that modern humans arose in multiple regions of the Old World as part of an interbreeding network of lineages descended from *H. erectus.* This debate is often contentious and bitterly argued. Proponents of the multiregional model have called into question the accuracy of the molecular clock as measured by nucleotide substitutions in mitochondria, as well as the methods used to construct the mitochondrial tree.

Data gathered from the larger and more complex nuclear genome are now being used to resolve the debate. Recent studies have used RFLP sites on the Y chromosome and in the region surrounding the CD4 locus on chromosome 12. Two polymorphisms at the CD4 locus were studied in more than 1600 individuals from 42 geographically diverse populations. Results support the out-of-Africa model for the origin of *H. sapiens* and place the migration from Africa at about 100,000 to 300,000 years ago, dates that are much more recent than suggested by the multiregional model. Results from studies on the Y chromosome give similar results. Thus, studies on the nuclear genome support the results from studies on the mitochondrial genome, and both suggest a recent, common origin in Africa for all human populations.

Chapter Summary

1. After extensive observation of natural populations, Alfred Russel Wallace and Charles Darwin formulated a theory on the role of natural selection in the formation of species. The genetic basis of evolution and the role of natural selection in changing allele frequencies were discovered in the twentieth century.

2. Several models of speciation have been proposed, differing in the order and rate at which evolutionary processes take place.

3. The amount of genetic variation present in a population can be studied by genetic analysis of inbreeding depression, protein polymorphisms revealed by gel electrophoresis, or by DNA restriction mapping and DNA sequencing.

4. There are two main views about whether the nucleotide and amino acid changes present in a population are adaptive. Some (the neutralists) believe that the majority of these changes are genetically equivalent, while others (the selectionists) believe that all variations have a selective value.

5. Speciation involves the partitioning of a population's gene pool into two or more gene pools that become reproductively isolated. The stages of species formation have been carefully studied in *Drosophila.*

6. Two approaches have been especially fruitful in studying evolution at the level of the genome: (1) comparison of amino acid substitutions in proteins common to a number of organisms, and (2) comparison of complementary nucleotide sequences present in the DNA of related organisms. These studies help measure evolutionary relatedness and allow the amount of genetic variation present in different individuals to be measured.

Key Terms

allopatric speciation, 496
allopolyploidy, 501
allozymes, 502
cladogenesis, 496
cytochrome c, 508
divergence dendrogram, 509
founder-flush theory, 499
geographic (allopatric)
    speciation, 496
minimal mutational distance, 508
molecular clock, 509

neutralist hypothesis, 503
niches, 495
phyletic evolution (anagenesis), 496
phyletic gradualism, 498
phylogenetic trees, 509
population crash, 499
postzygotic isolating
    mechanisms, 498
prezygotic isolating
    mechanisms, 498
punctuated equilibrium, 499

quantum speciation, 499
reproductive isolating
    mechanisms, 498
selectionists, 504
sibling species, 505
species, 495
statispatric speciation, 498
stochastic (catastrophic) models of
    speciation, 499
sympatric speciation, 497

INSIGHTS and SOLUTIONS

1. Sequence analysis of DNA can be accomplished by a number of techniques. Protein sequencing, on the other hand, is made more complex by the fact that 20 different subunits need to be unambiguously identified and enumerated, rather than the four nucleotides of DNA. Because of their unique properties, the N-terminal and C-terminal amino acids in a protein are easy to identify, but the array in between offer a difficult challenge, since many proteins contain hundreds of amino acids. How is it that protein sequencing is accomplished?

Solution: The strategy for protein sequencing is the same as for DNA sequencing: divide and conquer. To accomplish this, specific enzymes are used that reproducibly cleave proteins between certain amino acids. The use of different enzymes produces overlapping fragments. Each fragment is isolated and its amino acid sequence is determined by chemical means. Sequences from overlapping fragments are then assembled to give sequence for the entire protein.

2. A single plant twice the size of others in the same population suddenly appears. Normally, plants of this species reproduce by self-fertilization and by cross-fertilization. Is this new giant plant simply a variant, or could it be a new species? How would you determine this?

Solution: One of the most widespread mechanisms of speciation in higher plants is polyploidy, the multiplication of entire chromosome sets. The result of polyploidy is usually a larger plant with larger flowers and seeds. There are two ways of testing this new variant to determine whether it is a new species. First, the giant plant should be crossed with a normal-sized plant to see whether it produces viable, fertile offspring. If it does not, the two different types of plants would appear to be reproductively isolated. Second, the giant plant should be cytogenetically screened to examine its chromosome complement. If it has twice the number of its normal-sized neighbors, it is a tetraploid that may have arisen spontaneously. If the chromosome number differs by a factor of 2, and the new plant is reproductively isolated from its normal-sized neighbors, it is a new species.

Problems and Discussion Questions

1. Discuss the rationale behind the statement that inversions in chromosome 3 of *Drosophila pseudoobscura* represent genetic variation.

2. Describe how populations with substantial genetic differences can form. What is the role of natural selection?

3. Contrast the classical and balance hypotheses.

4. What types of nucleotide substitutions will not be detected by electrophoretic studies of a gene's protein product?

5. In a sequencing experiment (using the numbers 1 through 6 to represent amino acids), the following two sets of peptide fragments were obtained in independent experiments with different proteolytic enzymes:

Proteins	Fragments			
Enzyme I	624	246	35	135
Enzyme II	136	356	524	24

Determine the sequence of fragments and amino acids in the protein.

6. Shown below are two homologous lengths of the alpha and beta chains of human hemoglobin. Consult the genetic code dictionary (Figure 12–6) and determine how many amino acid substitutions may have occurred as a result of a single nucleotide substitution. For any that cannot occur as the result of a single change, determine the minimal mutational distance.

Alpha:	Ala	Val	Ala	His	Val	Asp	Asp	Met	Pro
Beta:	Gly	Leu	Ala	His	Leu	Asp	Asn	Leu	Lys

7. Determine the minimal mutational distances between the following amino acid sequences of cytochrome c from various organisms. Compare the distance between humans and each organism.

Human:	Lys	Glu	Glu	Arg	Ala	Asp
Horse:	Lys	Thr	Glu	Arg	Glu	Asp
Pig:	Lys	Gly	Glu	Arg	Glu	Asp
Dog:	Thr	Gly	Glu	Arg	Glu	Asp
Chicken:	Lys	Ser	Glu	Arg	Val	Asp
Bullfrog:	Lys	Gly	Glu	Arg	Glu	Asp
Fungus:	Ala	Lys	Asp	Arg	Asn	Asp

8. The genetic difference between two species of *Drosophila, D. heteroneura* and *D. sylvestris,* as measured by nucleotide diversity, is about 1.8 percent. The difference between chimpanzees *(Pan troglodytes)* and humans *(Homo sapiens)* is about the same, yet the latter species are classified in different genera. In your opinion, is this valid? If so, why; if not, why not?

9. As an extension of the previous question, consider the following. In sorting out the complex taxonomic relationships among birds, species with $\Delta T_{50}H$ values of 4.0 are placed in the same genus, even by traditional taxonomy based on morphology. Using the data in Figure 22–17, construct a phylogeny that obeys this rule, using any of the appropriate genus names *(Pongo, Pan, Homo),* or constructing new ones.

10. The use of nucleotide sequence data to measure genetic variability is complicated by the fact that the genes of higher eukaryotes are complex in organization and contain 5' and 3' flanking regions as well as introns. Slightom and colleagues have compared the nucleotide sequence of two cloned alleles of the gamma-globin gene from a single individual and found a variation of 1 percent. Those differences include 13 substitutions of one nucleotide for another, and three short DNA segments that have been inserted in one allele or deleted in the other. None of the changes takes place in the exons (coding regions) of the gene. Why do you think this is so, and should it change the concept of genetic variation?

11. Discuss the arguments supporting the neutralist hypothesis. What counterarguments are proposed by the selectionists?

12. Of what value to our understanding of genetic variation and evolution is the debate concerning the neutralist hypothesis?

Selected Readings

Anderson W., et al. 1975. Genetics of natural populations: XLII. Three decades of genetic change in *Drosophila pseudoobscura. Evolution* 29:24–36.

Armour, J. A., et al. 1996. Minisatellite diversity supports a recent African origin for modern humans. *Nat. Genet.* 13:154–60.

Ashian, R. E., and Carter, N. D. 1976. Biochemical genetics of carbonic anhydrase. *In Advances in human genetics,* eds. H. Harris and K. Hirschhorn, pp. 1–56. New York: Plenum Press.

Avise, J. C. 1990. Flocks of African fishes. *Nature* 347: 512–13.

Ayala, F. J., and Escalante, A. A. 1996. The evolution of Human Populations: a molecular perspective. *Mol. Pylogenet. Evol.* 5: 188–201.

———. 1984. Molecular polymorphism: How much is there, and why is there so much? *Dev. Genet.* 4:379–91.

Barton, N. H., and Hewitt, G. M. 1989. Adaptation, speciation and hybrid zones. *Nature* 341:497–503.

Bowcock, A. M., et al. 1994. High resolution of human evolutionary trees with polymorphic microsatellites. *Nature* 368:455–57.

Bult, C., et al. 1996. Complete genome sequence of the methanogenic Archeon, *Methanococcus jannaschii. Science* 273:1058–73.

Carson, H. 1970. Chromosome tracers of the origin of species. *Science* 168:1414–18.

———. 1975. The genetics of speciation at the diploid level. *Am. Natural.* 109:83–92.

Coyne, J. A. 1992. Genetics and speciation. *Nature* 355:511–15.

Dayhoff, M. O. 1969. Computer analysis of protein evolution. *Sci. Am.* (July) 221:86–95.

Diamond, J. 1992. *The third chimpanzee: The evolution and future of the human animal.* New York: HarperCollins.

Dobzhansky, T. 1947. Adaptive changes induced by natural selection in wild populations of *Drosophila. Evolution* 1:1–16.

———. 1948. Genetics of natural populations, XVI. Altitudinal and seasonal changes produced by natural selection in certain populations of *Drosophila pseudoobscura* and *Drosophila persimilis. Genetics* 33:158–76.

———. 1955. *Genetics of the evolutionary process.* New York: Columbia University Press.

———, et al. 1966. Genetics of natural populations: XXXVIII. Continuity and change in populations of *Drosophila pseudoobscura* in western United States. *Evolution* 20:418–27.

Eldredge, N. 1985. *Time frames: The evolution of punctuated equilibria.* Princeton, NJ: Princeton University Press.

Elena, S. F., Cooper, V. S., and Lenski, R. E. 1996. Punctuated evolution caused by selection of rare beneficial mutations. *Science* 272:1802–4.

Fitch, W. M. 1973. Aspects of molecular evolution. *Annu. Rev. Genet.* 7:343–80.

———, and Margoliash, E. 1967. Construction of phylogenetic trees. *Science* 155:279–84.

——— 1970. The usefulness of amino acid and nucleotide sequences in evolutionary studies. *Evol. Biol.* 4:67–109.

Gillespie, J. H. 1992. *The causes of molecular evolution.* New York: Oxford University Press.

Gould, S. J. 1982. Darwinism and the expansion of evolutionary theory. *Science* 216:380–87.

Hunt, J., et al. 1981. Evolution distance in Hawaiian *Drosophila. J. Mol. Evol.* 17:361–67.

Kimura, M. 1979a. Model of effectively neutral mutations in which selective constraint is incorporated. *Proc. Natl. Acad. Sci. USA* 76:3440–44.

———. 1979b. The neutral theory of molecular evolution. *Sci. Am.* (Nov.) 241:98–126.

———. 1989. The neutral theory of molecular evolution and the world view of the neutralists. *Genome* 31:24–31.

King, M. C., and Wilson, A. C. 1975. Evolution at two levels: Molecular similarities and biological differences between humans and chimpanzees. *Science* 188:107–16.

Kreitman, M. 1983. Nucleotide polymorphism at the alcohol dehydrogenase locus of *Drosophila melanogaster. Nature* 304:412–17.

Lande, R. 1989. Fisherian and Wrightian theories of speciation. *Genome* 31:221–27.

Lewin, R. 1993. *Human evolution,* 3rd ed. Cambridge, MA: Blackwell Scientific.

Lewontin, R. C., and Hubby, J. L. 1966. A molecular approach to the study of genic heterozygosity in natural populations: II. Amount of variation and degree of heterozygosity in natural populations of *Drosophila pseudoobscura. Genetics* 54:595–609.

Mayr, E. 1963. *Animal species and evolution.* Cambridge, MA: Harvard University Press.

Meyer, A., et al. 1990. Monophyletic origin of Lake Victoria cichlid fishes suggested by mitochondrial DNA sequences. *Nature* 347:550–53.

Paabo, S. 1993. Ancient DNA. *Sci. Am.* (Nov.) 269:86–92.

Pagel, M. D., and Harvey, P. H. 1989. Comparative methods for examining adaptation depend on evolutionary models. *Folia Primatol.* 53:203–20.

Powell, J. 1978. The founder-flush speciation theory: An experimental approach. *Evolution* 32:465–74.

Ridley, M. 1993. *Evolution.* Cambridge: Blackwell Scientific.

Sibley, C., and Ahlquist, J. 1984. The phylogeny of the hominoid primates, as indicated by DNA-DNA hybridization. *J. Mol. Evol.* 20:2–15.

Sibley, C. G., Comstock, J. A., and Ahlquist, J. E. 1990. DNA evidence of hominoid phylogeny: A re-analysis of the data. *J. Mol. Evol.* 30:202–236.

Smithies, O., et al. 1981. Co-evolution and control of globin genes. In *Levels of genetic control in development,* eds. S. Subtelny and U. Abbot, pp. 185–200. New York: Alan R. Liss.

Stebbins, G. L. 1977. *Processes of organic evolution,* 3rd ed. Englewood Cliffs, NJ: Prentice-Hall.

Stoneking, M. 1995. Ancient DNA: How do you know when you have it and what can you do with it? *Am. J. Hum. Genet.* 57:1259–62.

Templeton, A. R. 1985. Phylogeny of the hominoid primates: A statistical analysis of the DNA-RNA hybridization data. *Mol. Biol. Evol.* 2:420–33.

———. 1994. The role of molecular genetics in speciation studies. *EXS* 69:455–57.

Thorne, A. G., and Wolpoff, M. H. 1992. The multiregional evolution of humans. *Sci. Am.* (Apr.) 266:76–83.

Tishkoff, S. A., et al. 1996. Global patterns of linkage disequilibrium at the CD4 locus and modern human origins. *Science* 271:1380–87.

Val, F. C. 1977. Genetic analysis of the morphological differences between two interfertile species of Hawaiian *Drosophila. Evolution* 31:611–29.

White, M. J. D. 1977. *Modes of speciation.* New York: W. H. Freeman.

Wilson, A. C., and Cann, R. L. 1992. The recent African genesis of humans. *Sci. Am.* (Apr.) 266:68–73.

Yunis, J. J., and Prakash, O. 1982. The origin of man: A chromosomal pictorial legacy. *Science* 215:1525–30.

Appendix
Answers to Selected Problems

Glossary

Credits

Index

Appendix

Answers to Selected Problems

CHAPTER 1

2. *Pangenesis* refers to a theory that various parts of the body contain "humors" that bear the hereditary traits and gather in the reproductive organs. *Epigenesis* refers to the theory that organisms are derived from the assembly and reorganization of substances in the egg that eventually lead to the development of the adult. *Preformationism* is a 17th century theory that states that the sex cells (eggs of sperm) contain miniature adults, called homunculi, which grow in size to become the adult.

4. The theory of natural selection proposed that more offspring are produced than can survive, and that in the competition for survival, those with favorable variations survive. Over many generations, this will produce a change in the genetic make-up of populations if the favorable variations are inherited. Darwin did not understand the nature of heredity and variation.

CHAPTER 2

10. Compared with mitosis, meiosis provides for a reduction in chromosome number and provides an opportunity for exchange of genetic material from homologous chromosomes. In mitosis there is no change in chromosome number or kind in the two daughter cells, whereas in meiosis numerous potentially different haploid (n) cells are produced. During oogenesis, only one of the four meiotic products is functional; however, four of the four meiotic products of spermatogenesis are potentially functional.

12. Notice that for a cell with 4 chromosomes, there are two tetrads each comprised of a homologous pair of chromosomes.

   (a) 8 tetrads
   (b) 8 dyads
   (c) 8 monads migrating to *each* pole

14. In meiosis various chromosomal arrangements are possible because of random alignment of homologous chromosomes at metaphase I. In addition, crossing over, which introduced additional variation, is virtually absent in mitotic processes.

16. There would be 16 combinations possible.

20. p53 represents a 53 kilodalton protein which acts as a tumor suppressor by regulating the transition between G1 and S.

CHAPTER 3

2. (a) Parents must both be heterozygous (*Aa*).
   (b) To start out, the normal male could have either the *AA* or *Aa* genotype. The female must be *aa*. Since all the children are normal, one would consider the male to be *AA* instead of *Aa*. However, the male *could* be *Aa*. Under that circumstance, the likelihood of having six children, all normal is 1/64.

4. *Pisum sativum* is easy to cultivate. It is naturally self-fertilizing, but it can be crossbred. It has several visible features (e.g., tall or short, red flowers or white flowers that are consistent under a variety of environmental conditions yet contrast due to genetic circumstances. Seeds could be obtained from local merchants.

6. F₂:

9/16	*W_G_*	round seeds, yellow cotyledons
3/16	*W_gg*	round seeds, green cotyledons
3/16	*wwG_*	wrinkled seeds, yellow cotyledons
1/16	*wwgg*	wrinkled seeds, green cotyledons

8. In Problem 7, **(c)** is a test cross.

10. 1. Factors occur in pairs. Notice *A* and *a*. 2. Some genes are dominant to their alleles. Notice *A* and *a*. 3. Alleles segregate from each other during gamete formation. 4. One gene pair separates independently from other gene pairs. Different gene pairs on the same homologous pair of chromosomes (if far apart) or on nonhomologous chromosomes will separate independently from each other during meiosis.

14. (a) 4: *AB, Ab, aB, ab*
   (b) 2: *AB, aB*
   (c) 8: *ABC, ABc, AbC, Abc, aBC, aBc, abC, abc*
   (d) 2: *ABc, aBc*
   (e) 4: *ABc, Abc, aBc, abc*
   (f) $2^5 = 32$

16. P₁: *WWgg × wwGG*
   F₁: *WwGg* cross to *wwgg*
   The offspring will occur in a typical 1:1:1:1 as
      1/4 *WwGg* (round, yellow)
      1/4 *Wwgg* (round, green)
      1/4 *wwGg* (wrinkled, yellow)
      1/4 *wwgg* (wrinkled, green)

18. $\chi^2 = 0.5$, for 3 degrees of freedom (because there are now four classes in the χ^2 test). The observed and expected values do not deviate significantly. For **(b)** the χ^2 value is 0.35 and for 1 degree of freedom, the P value is greater than 0.50 and less than 0.90. We fail to reject the null hypothesis and are confident that the observed values do not differ significantly from the expected values. **(c)** For the yellow:green portion the χ^2 value is 0.01 and for 1 degree of freedom, the P value is greater than 0.90. We fail to reject the null hypothesis and are confident that the observed values do not differ significantly from the expected values.

24. (a) Since in the F₁ only round, axial, violet, and full phenotypes were expressed, they must each be dominant.
   (b) Round, axial, violet, and full would be the most frequent phenotypes: 3/4 × 3/4 × 3/4 × 3/4
   (c) Wrinkled, terminal, white, and constricted would be the least frequent phenotypes: 1/4 × 1/4 × 1/4 × 1/4
   (d) 3/4 × 1/4 × 3/4 × 1/4
   (e) There would be 16 different phenotypes in the test cross offspring just as there are 16 different phenotypes in the F₂ generation.

CHAPTER 4

2. *Incomplete dominance* can be viewed more as a quantitative phenomenon where the heterozygote is intermediate (approximately) between the limits set by the homozygotes. *Codominance* can be viewed in a more qualitative manner where both of the alleles in the heterozygote are expressed. For example in the AB blood group, both the I^A and I^B genes are expressed.

4. In *discontinuous variation* each phenotypic class is clearly discernible while in *continuous variation* various genes interact to give a "blending" phenotype. *Epistasis* is a phenomenon involving gene interaction where one gene or gene pair "masks" or inhibits expression of a nonallelic gene or gene pair(s) and as such is most likely grouped in the category of discontinuous inheritance.

6. **(a)** $RRpp$ $\times$ $rrPP$ $\longrightarrow$ $RrPp$
   (b) $RRPP$ $\times$ $rrpp$ $\longrightarrow$ $RrPp$
   (c) $RrPp$ $\times$ $RRpp$ $\longrightarrow$ $RRPp$
   $RRpp$
   $RrPp$
   $Rrpp$
   (d) $RrPp$ $\times$ $rrpp$ $\longrightarrow$ $rrPp$
   $rrpp$
   $RrPp$
   $Rrpp$

8. **(a)** $AaBbCc$ $\longrightarrow$ gray (C allows pigment)
   (b) A_B_Cc $\longrightarrow$ gray (C allows pigment)
   (c) 16/32 albino;
   9/32 gray;
   3/32 yellow;
   3/32 black;
   1/32 cream

10. **(a)** 1/4
    (b) 1/2
    (c) 1/4
    (d) zero

12. RR = red, Rr = red in females,
    Rr = mahogany in males,
    rr = mahogany.
    P_1: female: RR (red) $\times$ male: rr (mahogany)
    F_1: 1/2 females (red), 1/2 males (mahogany)
    F_2: 1/4 RR; 2/4 Rr, 1/4 rr

	1/2 females	*1/2 males*
1/4 RR	1/8 red	1/8 red
2/4 Rr	2/8 red	2/8 mahogany
1/4 rr	1/8 mahogany	1/8 mahogany

14. **(a)** P_1: $X^{sd}X^{sd}$ $\times$ X^+/Y $\longrightarrow$
    F_1: 1/2 X^+X^{sd} (female, normal)
    1/2 X^{sd}/Y (male, scalloped)
    F_2: 1/4 X^+X^{sd} (female, normal)
    1/4 $X^{sd}X^{sd}$ (female, scalloped)
    1/4 X^+/Y (male, normal)
    1/4 X^{sd}/Y (male, scalloped)
    (b) P_1: X^+/X^+ $\times$ X^+/Y $\longrightarrow$
    F_1: 1/2 X^+X^{sd} (female, normal)
    1/2 X^+/Y (male, normal)
    F_2: 1/4 X^+X^+ (female, normal)
    1/4 X^+X^{sd} (female, normal)
    1/4 X^+/Y (male, normal)
    1/4 X^{sd}/Y (male, scalloped)

16. It is extremely important that one account for both the mutant genes and each of their wild-type alleles.

(a) P_1: X^vX^v; +/+ $\times$ X^+/Y; b^r/b^r $\longrightarrow$
F_1: 1/2 X^+X^v; +/b^r (female, normal)
1/2 X^v/Y; +/b^r (male, vermilion)
F_2: 3/16 = females, normal
1/16 = females, brown eyes
3/16 = females, vermilion eyes
1/16 = females, white eyes
3/16 = males, normal
1/16 = males, brown eyes
3/16 = males, vermilion eyes
1/16 = males, white eyes

(b) P_1: X^+X^+; b^r/b^r $\times$ X^v/Y; +/+ $\longrightarrow$
F_1: 1/2 X^+X^v; +/b^r (female, normal)
1/2 X^+/Y; +/b^r (male, normal)
F_2: 6/16 = females, normal
2/16 = females, brown eyes
3/16 = males, normal
1/16 = males, brown eyes
3/16 = males, vermilion eyes
1/16 = males, white eyes

(c) P_1: X^v/X^v; b^r/b^r $\times$ X^+/Y; +/+ $\longrightarrow$
F_1: 1/2 X^+X^v; +/b^r (female, normal)
1/2 X^v/Y; +/b^r (male, vermilion)
F_2: 3/16 = females, normal
1/16 = females, brown eyes
3/16 = females, vermilion eyes
1/16 = females, white eyes
3/16 = males, normal
1/16 = males, brown eyes
3/16 = males, vermilion eyes
1/16 = males, white eyes

18. **(a)** Because the denominator in the ratios is 64 one would begin to consider that there are three independently assorting gene pairs. Because there are only two characteristics (eye color and croaking), however, one might consider two gene pairs interacting for one trait.
    (b) Croaking is due to one gene pair while eye color is due to two gene pairs.
    (c) Croaking: $R\_$ = rib-it; rr = knee-deep
    $A\_B\_$ = blue-eyed
    A_bb = purple
    $aaB\_$ and $aabb$ = green

20. $AabbRr$

22. **(a) (i)** $BbYy$ = green parents
    (ii, iii)
    $B\_Y\_$ = green progeny (9/16)
    B_yy = blue progeny (3/16)
    $bbY\_$ = yellow progeny (3/16)
    $bbyy$ = albino progeny (1/16)
    (b) $BByy \times bbYY$ or $BBYY \times bbyy$

24. Cross 1: complementing (nonallelic)
    Cross 2: noncomplementing (allelic)
    $r2 \times r3$: complementing (nonallelic)

CHAPTER 5

4. **(a)** It *is possible* that two parents of moderate height can produce offspring that are much taller or shorter than either parent as illustrated below:
   $rrSsTtuu$ $\times$ $RrSsTtUu$
   (moderate) (moderate)
   (b) If the individual with a minimum height, $rrssttuu$, is married to an individual of intermediate height

RrSsTtUu, the offspring can be no taller than the height of the tallest parent.

6. **(a)** Because the extreme phenotypes (6 cm and 30 cm) each represent 1/64 of the total, it is likely that there are three gene pairs in this cross. The genotypes of the parents would be combinations of alleles which would produce a 6 cm (*aabbcc*) tail and a 30 cm (*AABBCC*) tail while the 18 cm offspring would have a genotype of *AaBbCc*.

   (b) A mating of an *AaBbCc* (for example) pig with the 6-cm *aabbcc* pig would result in a 1:3:3:1 ratio. However had a different 18 cm-tailed pig been selected a different ratio would occur:

$$AABbcc \times aabbcc$$

Gametes (18 cm tail)	Gamete (6 cm tail)	Offspring
ABc	*abc*	*AaBbcc* (14 cm)
Abc		*Aabbcc* (10 cm)

8. *Monozygotic twins* are derived from a single fertilized egg and are thus genetically identical to each other. They provide a method for determining the influence of genetics and environment on certain trains. *Dizygotic twins* arise from two eggs fertilized by two sperm cells. They have the same genetic relationship as siblings. The higher concordance value for monozygotic twins as compared to the value for dizygotic twins indicates a significant genetic component for a given trait.

10. **(a)** mean of 140 cm

    (b) variance = 374.18

    (c) The *standard deviation* is the square root of the variance or 19.34

    (d) The *standard error of the mean* is about 0.70.

12. The formula for estimating heritability is

$$H^2 = V_G/V_P$$

V_P is the combination of genetic and environmental variance. Because the two parental strains are inbred, they are assumed to be homozygous and the variance of 4.2 and 3.8 considered to be the result of environmental influences. The average of these two values is 4.0. The F_1 is also genetically homogeneous and gives us an additional estimation of the environmental factors.

   By averaging with the parents we obtain

$$(4.0 + 5.6)/2 = 4.8$$

The phenotypic variance in the F_2 is the sum of the genetic (V_G) and environmental (V_E) components. We have estimated the environmental input as 4.8, so 10.3 minus 4.8, gives us an estimate of (V_G) which is 5.5. Heritability then becomes 5.5/10.3 or 0.53. This value, when viewed in percentage form indicates that about 53% of the variation in plant height is due to genetic influences.

14. $h^2 = (7.5 - 8.5)/(6.0 - 8.5) = 0.4$

   Selection will have little relative influence on olfactory learning in Drosophilia.

CHAPTER 6

2. $1/X^2$

4. dp——cl————ap
       3 mu   39mu

6. The original parental arrangement in the test cross was *PZ/pz* ×*pz/pz*. Adding the crossover percentages together (6.9 + 7.1) gives 14%, which would be the map distance between the two genes.

8. **(a)** $P_1: sc\ s\ v\ /\ sc\ s\ v\ \ \times\ \ +++/Y$
   $F_1: +++/\ sc\ s\ v\ \ \times\ \ sc\ s\ v/Y$

   (b) $\dfrac{sc\quad v\quad s}{+\quad +\quad +}$

   $sc - v = 33\%$ (map units)
   $v - s = 10\%$ (map units)

   (c) The coefficient of coincidence = .727

10. 10 map units.

12. One would use the typical test cross arrangement with the *curled* gene so the arrangement would be + *cu*/ +*cu*.

14. The map for parts **(a)** and **(b)** is the following:

d	*b*	*pr*	*vg*	*c*	*adp*
31	48	54	67	75	83

   Map Units

   The expected map units between *d* and *c* would be 44, *d* and *vg* would be 36, and *d* and *adp* 52; however, because there is a theoretical maximum of 50 map units possible between two loci in any one cross, that distance would be below the 52 determined by simple subtraction.

16. **(a)** The *short* gene is on chromosome 2 with the *black* gene.

   (b) The parental cross is now the following:
   Females: $\dfrac{b\ sh\ p}{+\ +\ +}$ × Males: $\dfrac{b\ sh\ p}{b\ sh\ p}$
   The map distance between the two genes is 15.

CHAPTER 7

2. All of the offspring must have the phenotype of the mother's genotype, which is dextral.

4. **(a)** green
   (b) white
   (c) variegated (patches of white and green)
   (d) green

6. The segregational mode is dependent on nuclear genes while that of the neutral type is dependent on cytoplasmic influences, namely mitochondria. If the two are crossed as stated in the problem, then one would expect, in the diploid zygote, the *segregational* allele to be "covered" by normal alleles from the neutral strain. On the other hand, as the nuclear genes are again "exposed" in the haploid state of the ascospores, one would expect a 1:1 ratio of normals to petites. the petite phenotype is caused by the nuclear, *segregational* gene.

8. **(a)** It is likely that mitochondria sand chloroplasts evolved from bacteria in a symbiotic relationship; therefore it is not surprising that certain antibiotics which influence bacteria will also influence all mitochondria and chloroplasts.

   (b) The *mt*⁺ strain is the donor of the cpDNA since the inheritance of resistance or sensitivity is dependent on the status of the *mt*⁺ gene.

10. It appears as if some factor normally provided by the *gs*⁺ allele is necessary for normal development and/or functioning of the female offspring's gonads. Without this product, the daughters are sterile, thus the term "grandchildless."

12. Developmental phenomena which occur early are more likely to be under maternal influence than those occurring late. Anterior/posterior and dorsal/ventral orientations are among the earliest to be established and in organisms where their study is experimentally and/or genetically approachable, they often show considerable maternal influence. Thus, patterns resulting from ex-

tranuclear inheritance differ from typical Mendelian (biparental) schemes in that the maternal genotype is often expressed preferentially in offspring. In reciprocal crosses, the maternal contribution may be stronger than the paternal. In Mendelian inheritance, it is expected that both parents (except for X-linked genes, etc.) contribute equally to the characteristics of the offspring.

CHAPTER 8

2. Klinefelters syndrome (XXY)   = 1
Turners syndrome (XO)   = 0
47,XYY   = 0
47,XXX   = 2
48,XXXX   = 3

8. As stated in the text, at least 20 percent of all conceptions are terminated in natural abortion. Of these, thirty percent show some chromosomal anomaly. Of the chromosomal anomalies that occur, approximately ninety percent are eliminated by spontaneous abortion. Aneuploidy contributes to the majority of spontaneous abortions. Trisomy for every human chromosome has been observed, however, monosomy, the reciprocal meiotic event of trisomy, is rare. This observation probably results from gametic or early embryonic inviability.

10. Temperature shock applied during meiosis or colchicine applied during mitosis may lead to chromosome doubling. Colchicine interferes with spindle fiber formation.

12. If the diploid chromosome number is 18, $2n = 18$, then in the somatic nuclei of
haploid individuals $n = 9$,
triploid individuals $(3n) = 27$,
tetraploid individuals $(4n) = 36$.
A trisomic has one extra chromosome, therefore it is $2n + 1 = 19$, and a monosomic is $2n - 1 = 17$.

14. Basically the synaptic configurations produced by chromosomes bearing a deletion or duplication (on one homologue) are very similar. Inversion loops, formed from the combination of an "outside" loop synapsing with an "inside" loop, are illustrated in the text.

16. Duplicated genes, or the original genes themselves, would be able to undergo mutational "experimentation" without necessarily threatening the survival of the organism.

18. The primrose, *Primula kewensis,* with its 36 chromosomes, is likely to have formed from the hybridization and subsequent chromosome doubling of a cross between the two other species, each with 18 chromosomes.

20. Dosage compensation and the formation of Barr bodies occur only when there are two or more X chromosomes. Males normally have only one X chromosome; therefore such mosaicism cannot occur. Females normally have two X chromosomes. There are cases of male calico cats which are XXY.

22. Since there is a region of synapsis close to the SRY-containing section on the Y chromosome, crossing over in the region would generate XY translocations which would lead to the condition described.

24. In mammals, the scheme of sec determination is dependent on the presence of a piece of the Y chromosome. If present a male is produced. In *Bonellia viridis,* the female proboscis produces some substance which triggers a morphological, physiological and behavioral developmental pattern which produces males.

To elucidate the mechanism, one could attempt to isolate and characterize the active substance by testing different chemical fractions of the proboscis. Secondly, mutant analysis usually provides critical approaches into developmental processes. Depending on characteristics of the organism, one could attempt to isolate mutants that lead to changes in male or female development. Third, by using micro-tissue transplantations, one could attempt to determine which "centers" of the embryo respond to the chemical cues of the female.

CHAPTER 9

4. Nucleic acids contain large amounts of phosphorus and no sulfur, whereas proteins contain sulfur and no phosphorus. therefore the radioisotopes ^{32}P and ^{35}S will selectively label nucleic acids and proteins, respectively. The Hershey and Chase experiment demonstrated that most of the ^{32}P-labeled material (DNA) was injected while the phage ghosts (protein coats) remained outside the bacterium. Therefore the nucleic acid must be the genetic material.

6. Some viruses contain a genetic material composed of RNA. The tobacco mosaic virus is composed of an RNA core and a protein coat. "Crosses" can be made in which the protein coat and RNA of TMV are interchanged with another strain (Holmes ribgrass). The source of the RNA determines the type of lesion; thus, RNA is the genetic material in these viruses. Retroviruses contain RNA as the genetic material and use an enzyme known as *reverse transcriptase* to produce DNA which can be integrated into the host chromosome.

8. Guanine: 2-amino-6-oxypurine
Cytosine: 2-oxy-4-aminopyrimidine
Thymine: 2,4-dioxy-5-methylpyrimidine
Uracil: 2,4-dioxypyrimidine

10. In addition to creative "genius" and perseverance, model building skills, and the conviction that the structure would turn out to be "simple" and have a natural beauty in its simplicity, Watson and Crick employed the X-ray diffraction information of Franklin and Wilkins, along with the base ratio information of Chargaff.

12. Three main differences between RNA and DNA are the following:
(1) Uracil in RNA replaces thymine in DNA.
(2) Ribose in RNA replaces deoxyribose in DNA.
(3) RNA often occurs as both single- and double-stranded forms, whereas DNA most often occurs in a double-stranded form.

14. In order for hydrogen bonds to form between complementary base pairs, complementary strands of nucleic acids must be in proximity. for a given concentration of nucleic acids, the more copies of a given type of nucleic acid that are present, the higher the likelihood that complementary strands will be close to each other. Conversely, unique sequences of nucleic acids have a lower probability of interacting because there are fewer of them; thus, the likelihood of forming hydrogen bonds (hybridizing) is less.

18. Since cytosine pairs with guanine and uracil pairs with adenine, the result would be a base substitution of G:C to A:T.

CHAPTER 10

2. After one round or replication in the ^{14}N medium the conservative scheme can be ruled out. After one round of replication in ^{14}N under a dispersive model, the DNA is of intermediate density, just as it is in the semiconservative model. However, in the next round of replication in ^{14}N medium, the density of the DNA is between the intermediate and "light" densities.

4. The *in vitro* replication requires a DNA template, a divalent cation (Mg^{2+}), and all four of the deoxyribonucleoside triphosphates: dATP, dCTP, dTTP, and dGTP. The lowercase "d" refers to the deoxyribose sugar.

6. φX174 is a well-studies single-stranded virus (phage) which can be easily isolated. It has a relatively small DNA genome (5500 nucleotides) which , if mutated, usually alters its reproductive cycle.

12. Eukaryotic DNA is replicated in a manner which is very similar to that of *E. coli:* bidirectional, continuous on one strand and discontinuous on the other. The requirements of synthesis (four deoxyribonucleoside triphosphates, divalent cation, template, and primer) are the same, Okazaki fragments of eukaryotes are about one-tenth the size of those in bacteria, and there are multiple initiation sites for replication in eukaryotes in contrast to the single replication origin in prokaryotes.

14. Gene conversion is considered a result of heteroduplex formation which is accompanied by mismatched bases. When these mismatches are corrected, the "conversion" occurs.

16. (a) In *E. coli,* 100kb are added to each growing chain per minute. Therefore the chain should be about 4,000,000bp.

  (b) Given $(4 \times 10^{6}\text{bp}) \times 0.34\text{nm/bp} =$ $1.36 \times 10^{6}\text{nm}$ or 1.3mm

CHAPTER 11

4. *Heterochromatin* stains deeply and remains condensed when other parts of chromosomes, euchromatin, are otherwise pale and decondensed. Heterochromatic regions replicate late in S phase and are relatively inactive in a genetic sense because there are few genes present or if they are present, they are repressed. Telomeres and the areas adjacent to centromeres are composed of heterochromatin. Examples include regions at the same sites on homologous chromosomes that are genetically "inert" such as telomers and centromeric regions; certain chromosomal regions have the potential to become heterochromatic such as the X chromosome in female mammals (Barr body), and one haploid set of chromosomes in the mealy bug.

6. The first part of this problem is to convert all of the given values to cubic angstroms (Å) remembering that 1μm = 1000 Å. Using the formula πr^2 for the area of a circle and $4/3\ \pi r^3$ for the volume of a sphere, the following calculations apply:

  Volume of DNA: $3.14 \times 10\ \text{Å} \times 10\ \text{Å} \times (50 \times 10^4\ \text{Å})$
  $= 1.57 \times 10^8\ \text{Å}^3$
  Volume of capsid: $4/3\ (3.14 \times 400\ \text{Å} \times 400\ \text{Å} \times 400\ \text{Å})$
  $= 2.67 \times 10^8\ \text{Å}^3$

Because the capsid head has a greater volume than the volume of DNA, the DNA will fit into the capsid.

8. While the β-globin gene family is relatively large (60kb) sequence and restriction analyses show that it is com-

posed of six genes, one is a pseudogene and therefore does not produce a product.

The five functional genes each contain two similarly-sized introns which when included with non-coding flanking regions (5′ and 3′), and spacer DNA between genes, accounts for the 95% mentioned in the question.

CHAPTER 12

2. (a)

GGG	=	$3/4 \times 3/4 \times 3/4$	=	27/64
GGC	=	$3/4 \times 3/4 \times 1/4$	=	9/64
GCG	=	$3/4 \times 1/4 \times 3/4$	=	9/64
CGG	=	$1/4 \times 3/4 \times 3/4$	=	9/64
CCG	=	$1/4 \times 1/4 \times 3/4$	=	3/64
CGC	=	$1/4 \times 3/4 \times 1/4$	=	3/64
GCC	=	$3/4 \times 1/4 \times 1/4$	=	3/64
CCC	=	$1/4 \times 1/4 \times 1/4$	=	1/64

  (b) Glycine: GGG and one G_2C
  Alanine: one G_2C and one C_2G
  Arganine: one G_2C and one C_2G
  Proline: one C_2G and CCC

  (c) Glycine: GGG, GGC
  Alanine: CGG, GCC, CGC, GCG
  Arganine: GCG, GCC, CGC, CGG
  Proline: CCC, CCG

4. Coupling information with that of the AAG copolymer, GAA must also code for glu, and AAG must code for lys.

6. Apply the most conservative pathway of change.

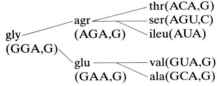

8. mRNA, charged tRNA, large and small ribosomal subunits, elongation and perhaps initiation factors, peptidyl transferase, GTP, Mg^{2+}, nascent proteins, possibly GTP-dependent release factors.

10. With an adaptor molecule, specific hydrogen bonding could occur between nucleic acids, and specific covalent bonding could occur between an amino acid and a nucleic acid tRNA.

12. Approximately 20

14. The amino acid is not involved in recognition of the anticodon.

16. (a) #1: *nonsense mutation*
  #2: *frameshift mutation*
  #3: *missense mutation*

  (b) #1: mutation in third position to A or G
  #2: removal of a G in the UGG triplet (trp)
  #3: change from U to C

  (c) Promoter or enhancer, although many posttranscriptional alterations are possible.

CHAPTER 13

2. Under *negative* control the regulatory molecule interferes with transcription, whereas under *positive* control the regulatory molecule stimulates transcription. Negative control is seen in the *lactose* and *tryptophan* systems as well as a portion of the *arabinose* regulation. Catabolite repression and a portion of the *arabinose* regulatory systems are examples of positive control.

4. $I^+O^+Z^+$ = **Inducible**
$I^-O^+Z^+$ = **Constitutive**
$I^+O^cZ^+$ = **Constitutive**
$I^-O^+Z^+ / F'I^+$ = **Inducible**
$I^+O^cZ^+ / F'O^+$ = **Constitutive**
$I^sO^+Z^+$ = **Repressed**
$I^sO^+Z^+ / F'I^+$ = **Repressed**

6. The *c* gene codes for the **structural gene.** The *a* locus is the **operator** and the *d* locus is the **repressor** gene.

10. *Promoters* are conserved DNA sequences that influence transcription from the "upstream" side (5′) of mRNA coding genes. They are usually fixed in position and within 100 base pairs of the initiation site for mRNA synthesis. Examples of such promoters are the following: TATA, CAAT, and GC boxes.

   Enhancers are *cis*-acting sequences of DNA which stimulate the transcription from most, if not all, promoters. They are somewhat different from promoters in that the position of the enhancer need not be fixed; it may be upstream, downstream, or within the gene being regulated. The orientation may be inverted without the gene being regulated. The orientation may be inverted without significantly influencing its action. Enhancers can work on different genes; that is, they are not gene-specific.

16. Your response should deal with the following issues:
   —differences in basic chromosome structure
   —differences in gene structure
   —"C" value paradox and its implications
   —cell structure (nucleus in eukaryotes)
   —levels of potential regulation
      —transcriptional
      —mRNA processing
      —transport
      —selection for processing and translation
      —mRNA stability
      —posttranslational processing
   —genomic aspects (amplification, etc.)
   —biological context in terms of multicellular interactions *versus* single cell survival.

CHAPTER 14

2. (a)

$$\underset{\text{(white)}}{X} \xrightarrow{\quad aa \quad \nmid \quad} \underset{\text{(white)}}{Y} \xrightarrow{\quad bb \quad \nmid \quad} \text{Purple pigment}$$

(b)

$$\underset{\text{(white)}}{X} \xrightarrow{\quad aa \quad \nmid \quad} \underset{\text{(pink)}}{Y} \xrightarrow{\quad bb \quad \nmid \quad} \text{Purple pigment}$$

4.

Precursor $\xrightarrow{\;\; thi\text{-}2 \;\;\nmid\;}$ pyrimidine $\searrow thi\text{-}3$

$\qquad\qquad\qquad\qquad\qquad\qquad \longrightarrow$ thiamine

Precursor $\xrightarrow{\;\; thi\text{-}1 \;\;\nmid\;}$ thiazole $\nearrow$

6. Sickle-cell anemia is termed a *molecular* disease because it is well understood at the molecular level; at the level of a base change in DNA which leads to an amino acid change in the b chain of hemoglobin. It is a *genetic* disease in that it is inherited from one generation to the next. It is not contagious as might be the case of a disease caused by a microorganism. Diseases caused by microorganisms may not necessarily follow family blood lines whereas genetic diseases do.

8. *Primary:* the sequence of amino acids. *Secondary:* α-helix and β-pleated-sheet structures. *Tertiary:* folding that occurs as a result of interactions of the amino acid side chains with each other. *Quaternary:* the association of two or more polypeptide chains. Called *oligomeric,* such a protein is made up of more than one *promoter.*

10. Enzymes function to regulate catabolic and anabolic activities of cells. They influence (lower) the *energy of activation* thus allowing chemical reactions to occur under conditions which are compatible with living systems. Enzymes possess *active sites* and/or other domains that are sensitive to the environment. The active site is considered to be a crevice, or pit, which binds reactants, thus enhancing their interaction. The other domains mentioned above may influence the conformation and therefore function of the active site.

CHAPTER 15

2. A coordinated output of each gene product is required for life. Deviations from the norm, caused by mutation, are likely to be disruptive because of the complex and interactive environment in which each gene product must function. However, on occasion a beneficial variation occurs.

4. Let *II* indicate a mutagenized second chromosome
   Females *Cy L/Pm* × males *II/II*
   F_1 males *Cy L/II* × *Cy L/Pm*
      (individual crosses)
   F_2 females *Cy L/II* × *Cy L/II*
      (individual crosses)
   F_3 genotypes:
      Cy L/Cy L   (lethal)
      Cy L/II (Curly, Lobe)
      II/II   (dies if recessive lethal)
   The *II/II* class will be present in crosses where no recessive lethal was introduced. If a recessive lethal had been introduced, only Curly/Lobe flies would be seen in the cultures. In addition, recessive morphological mutations will be expressed in the *II/II* offspring.

6. Frameshift mutations are likely to change more than one amino acid in a protein product because as the reading frame is shifted, new codons are generated. In addition, there is the possibility that a nonsense triplet could be introduced, thus causing premature chain termination. If a single pyrimidine or purine has been substituted, then only one amino acid is influenced.

8. Because mammography involves the use of X-rays and X-rays are known to be mutagenic, it has been suggested that frequent mammograms may do harm.

10. $2(5 \times 10^{-5})(1 \times 10^5) = 10$
   Assuming 4.3×10^9 individuals, there would be 4.3×10^{10} new mutations in the current populace.

12. Your study should include examination of the following short-term aspects: immediate assessment of radiation amounts distributed in a matrix of the bomb sites as well as a control area not receiving bomb-induced radiation, radiation exposure as measured by radiation sickness and evidence of radiation poisoning from tissue samples, abortion rates, birthing rates, and chromosomal studies. Long-term assessment should include: sex-linked distortion (males being more influenced by X-linked recessive lethals than females), chromosomal studies, birth and abortion rates, cancer frequency and type and, genetic disorders. In each case data should be compared to the control site to see if changes are bomb-related. In addition, to attempt to determine cause-effect, it is often helpful to show a dose response. Thus, by comparing the location of individuals at the time exposure to the matrix of radiation amounts, one may be able to determine whether those most exposed to radiation suffer the most physiologically and genetically. If a positive correlation is observed, then statistically significant conclusions may be possible.

CHAPTER 16

2. **(a)** The requirement for physical contact between bacterial cells during conjugation was established by placing a filter in a U-tube such that the medium can be exchanged but the bacteria cannot come in contact. Under this condition, conjugation does not occur.

   (b) By treating cells with streptomycin, an antibiotic, it was shown that recombination would not occur if one of the two bacterial strains was inactivated. However, if the other was similarly treated, recombination would occur. Thus, directionality was suggested, with one strain being a donor strain and the other being the recipient.

   (c) An F^+ bacterium contains a circular, double-stranded, structurally independent, DNA molecule which can direct recombination. In Hfr cells the F factor is integrated into the bacterial chromosome.

4. Mapping the chromosome in an Hfr $\times$ F^- cross takes advantage of the oriented transfer of the bacterial chromosome through the conjugation tube. for each F type, the point of insertion and the direction of transfer are fixed, therefore breaking the conjugation tube at different times produces partial diploids with corresponding portions of the donor chromosome being transferred. The length of the chromosome being transferred is contingent on the duration of conjugation, thus mapping of genes is based on time.

6. Because the F factor is the last element to be transferred and the conjugation tube is fragile, the likelihood for complete transfer is low.

8. F-mediated conjugation requires contact; without that contact, such conjugation cannot occur. The treatment with DNase showed that the filterable agent was not naked DNA.

10. Starting with a single bacteriophage, one lytic cycle produces 200 progeny phage, three more lytic cycles would produce $(200)^4$ or 1,600,000,000 phage.

12. Reverse transcriptase is required to make a DNA molecule from the RNA genome (template).

14. The *c* gene must be in the middle.

   a to c $= (740 + 670 + 90 + 110)/10,000$
   $= 16.1$ map units
   c to b $= (160 + 140 + 90 + 110)/10,000$
   $= 5$ map units

   The *expected* frequency of double crossovers is .000805, which when multiplied by 10,000 gives approximately 80. The *observed* number of double crossovers is $90 + 110$ or

200. Since many more double crossovers are observed than expected, negative interference is occurring.

CHAPTER 17

2. *Reverse transcriptase* is often used to promote the formation of cDNA (complementary DNA) from a mRNA molecule. Eukaryotic mRNAs typically have a 3' polyA tail. The poly dT segment provides a double-stranded section which serves to prime the production of the complementary strand.

4. The question of protein/DNA recognition and interaction is a difficult one to answer. Much research has been done to attempt to understand the nature of the specificity of such interactions. In general it is believed that the protein interacts with the major groove of the DNA helix. This information comes from the structure of the few proteins which have been sufficiently well studied to suggest that the DNA major groove and "fingers" or extensions of the protein form the basis of interaction.

6. The segment contains the palindromic sequence CCTAGG which is recognized by the restriction enzyme *Bam*HI.

8. A filter is used to bind the DNA from the colonies and a labeled probe is used to detect, through hybridization, the DNA of interest.

10. There may be several factors contributing to the lack of representation of the 5' end of the mRNA. One has to do with the possibility that the reverse transcriptase may not completely synthesize the DNA from the RNA template. The other reason may be that the 3' end of the copied DNA tends to fold back on itself thus providing a primer for the DNA polymerase. Additional preparation of the cDNA requires some digestion at the folded region. Since this folded region corresponds to the 5' end of the mRNA, some of the message is often lost.

CHAPTER 18

4. (a,b) (1) Integration into the host must be cell specific so as to not damage non-target cells. (2) Retroviral integration into host cell genomes only occurs if the host cell is replicating. (3) Insertion of the viral genome might influence non-target but essential genes. (4) Retroviral genomes have a low cloning capacity and cannot carry large inserted sequences as are many human genes. (5) There is a possibility that recombination with host viruses will produce an infectious virus that may do harm.

6. Positional cloning is a technique whereby the linkage group of a genetic disorder is determined by association of certain RFLPs (as markers). A more precise location is obtained by association (linkage studies in kindreds) with additional RFLP markers. It is unlikely that this technique will be successfully applied to genetically complex traits in the near future.

8. Even though you have developed a method for screening seven of the mutations described, it is possible that negative results can occur even though the person carries the gene for CF.

10. In the case of haplo-insufficient mutation, gene therapy holds promise; however in "gain-of-function" mutations, in all probability the mutant gene's activity or product must be compromised.

12. One method is to use the amino acid sequence of the protein to produce the gene synthetically. Alternatively, since the introns are spliced out of the hnRNA in the production of mRNA, if mRNA can be obtained, it can be used to make DNA through the use of reverse transcriptase.

CHAPTER 19

2. To say that a particular trait is inherited conveys the assumption that when a particular genetic circumstance is present, it will be revealed in the phenotype. For instance, albinism is inherited in such a way that individuals who are homozygous recessive, express albinism. When one discusses an inherited predisposition, one usually refers to situations where a particular phenotype is expressed in families in some consistent pattern, however, the phenotype may not always be expressed or may manifest itself in different ways. In retinoblastoma, the gene is inherited as an autosomal dominant and those that inherit the mutant RB allele are predisposed to develop eye tumors. However, approximately 10 percent of the people known to inherit the gene don't actually express it, and in some cases expression involves only one eye rather than two.

4. Oncogenes induce or maintain uncontrolled cellular proliferation associated with cancer. They are mutant forms of proto-oncogenes that normally function to regulate cell division.

6. Any agent that causes damage to DNA is a potential carcinogen since cell cycle control is achieved by gene (DNA) products, known as proteins. Since cigarette smoke is known to contain an agent that changes DNA, in this case transversions, numerous modified gene products (including cell cycle controlling proteins) are likely to be produced. The fact that many cancer patients have such transversions in *p53* strongly suggests that cancer is caused by agents in cigarette smoke.

8. There would be a possibility of 18 more combinations, 27 total possible.

10. 10^6

12. There are four polypeptide chains in each IgG antibody molecule. There are three different types of polypeptide chains represented in the IgG molecule: κ, λ, γ. Each chain is encoded by at least three gene segments (C, V, and J) while each heavy chain is encoded by at least four gene segments (C, V, D, and J).

CHAPTER 20

2. The fact that nuclei from almost any source remain transcriptionally and translationally active substantiates the fact that the genetic code and the ancillary processes of transcription are compatible throughout the animal kingdom. Because the egg represents an isolated, "closed" system which can be mechanically, environmentally, and to some extent biochemically manipulated, various conditions may be developed which allow one to study facets of gene regulation. Combinations of injected nuclei may reveal nuclear-nuclear interactions that could not normally be studied by other methods.

4. The *ftz* gene product regulates, either directly or indirectly, *eng*.

6. One of the easiest ways to determine whether a genetic basis exists for a given abnormality is to cross the abnor-mal fly to a normal fly. If the trait is determined by a dominant gene, the trait should appear in the offspring, probably half of them if the gene was in the heterozygous state. If the gene is recessive and homozygous, then one may not see expression in the offspring of the first cross. However, if one crosses the F_1 one should see the trait appearing in approximately 1/4 of the offspring. Modifications of these patterns would be expected if the mode if inheritance is X-linked or shows other modifications of typical Mendelian ratios.

One might hypothesize that the focus is in the nervous or muscular system. Mapping the primary focus of the gene could be accomplished with patience and the use of the unstable ring-X chromosome to generate gynandromorphs. Given that the gene is X-linked, one would use classical recombination methods to place a recessive X-linked marker, such as *singed bristles*, on the X chromosome with the gene causing the limp. This would help one identify the male/female boundaries. One would then cross homozygous (for the trait and markers) females to ring-X males, or the reciprocal, then examine the offspring for synandromorphs (and singed mosaics).

If one obtained a pool of gynandromorphs, one could then assess the phenotype (limp or normal) with respect to exposure of the recessive gene in the male tissue. Correlating such would allow one to provide an educated guess as to the primary focus of the gene causing the limp.

8. Usually, if traits fail to breed true in animals that are *considered to be* genetically homozygous (the pure breeds of dogs), it is likely that the trait is determined by conditioning or complex genetic factors that each have minor influences. Even though the dogs are "pure breeds" it is likely that alleles are segregating and combining in various ways to produce the variations noted. There may also be different interactions with environmental stimuli from such subtle genetic variations.

CHAPTER 21

2. AA = .25 or 25%, Aa = .5 or 50%, aa = .25 or 25%. The initial population was not in equilibrium; however, after one generation of mating under the Hardy-Weinberg assumptions, the population is in equilibrium.

4. In order for the Hardy-Weinberg equations to apply, the population must be in equilibrium.

6. **(a)** MM = .6014 or 60.14%, MN = .3482 or 34.82%, NN = .0504 or 5.04%. The population is in equilibrium.

(b) AA = .7691 or 76.91%, AS = .2157 or 21.57%, SS = .0151 or 1.51%. χ^2 = 1.47, the frequencies of AA, AS, SS sampled a population that is in equilibrium.

8. **(a)** q_1 = .23, p_1 = .77
(b) q_1 = .267, p_1 = .733
(c) q_1 = .293, p_1 = .707
(d) q_1 = .299, p_1 = .701

10. **(a)** $p_1 = 0.6 + 0.2(0.1 - 0.6) = 0.5$
(b) $p_1 = 0.2 + 0.3(0.7 - 0.2) = 0.35$
(c) $p_1 = 0.1 + 0.1(0.2 - 0.1) = 0.11$

12. *Inbreeding depression* refers to the reduction in fitness observed in populations which are inbred.

14. Inbreeding schemes will often be used to render strains homozygous so that such recessive genes can be expressed.

16. *Self-fertilize* or brother-sister matings

18. The frequency of a gene is determined by a number of factors including the fitness it confers, mutation rate, and input from migration. There is no tendency for a gene to reach any artificial frequency such as 0.5.

20. The overall probability of the couple producing a CF child is $98/2500 \times 2/3 \times 1/4$.

CHAPTER 22

4. Because of degeneracy in the code, there are some nucleotide substitutions, especially in the third base, that do not change amino acids. In addition, if there is not change in the overall charge of the protein, it is likely that electrophoresis will not separate the variants. If a positively charged amino acid is replaced by an amino acid of like charge, then the overall charge on the protein is unchanged. The same may be said for other, negatively charged and neutral amino acid substitutions.

6. All of the amino acid substitutions (Ala $\longrightarrow$ Gly, Val $\longrightarrow$ Leu, Asp $\longrightarrow$ Asn, Met $\longrightarrow$ Leu) require only one nucleotide change. the last change from Pro (CC–) $\longrightarrow$ Lys (AAA,G) requires two changes (the minimal mutational distance).

8. The classification of organisms into different species is based on evidence (morphological, genetic, ecological, etc.) that they are reproductively isolated. Classifications above the species level (genus, family, etc.) are not based on such empirical data. DNA sequence divergence is not always directly proportional to morphological, behavioral, or ecological divergence. As more information is gained on the meaning of DNA sequence differences (ΔTm) in comparison to morphological factors, many physiogenetic relationships will be reconsidered.

10. There are many sections of DNA in a eukaryotic genome that are not reflected in a protein product. Indeed, there are many sections of DNA that are not even transcribed and/or have no apparent physiological role.

12. It is likely that some genes (like histories) will not tolerate nucleotide substitutions to a significant degree and the neutral mutation theory will not hold. However, there are other genes which produce quite variable products and provide support for the neutral mutation theory. It is controversy which stimulates a desire to seek answers.

Glossary

A-DNA An alternate form of the right-handed double-helical structure of DNA in which the helix is more tightly coiled, with 11 base pairs per full turn of the helix. In this form, the bases in the helix are displaced laterally and tilted in relation to the longitudinal axis. It is not yet clear whether this form has biological significance.

abortive transduction An event in which transducing DNA fails to be incorporated into the recipient chromosome (See *transduction*.)

acentric chromosome Chromosome or chromosome fragment with no centromere.

acquired immunodeficiency syndrome (AIDS) An infectious disease caused by a retrovirus designated as human immunodeficiency virus (HIV). The disease is characterized by a gradual depletion of T lymphocytes, recurring fever, weight loss, multiple opportunistic infections, and rare forms of pneumonia and cancer associated with collapse of the immune system.

acrocentric chromosome Chromosome with the centromere located very close to one end. Human chromosomes 13, 14, 15, 21, and 22 are acrocentric.

active immunity Immunity gained by direct exposure to antigens followed by antibody production.

active site That portion of a protein, usually an enzyme, whose structural integrity is required for function (e.g., the substrate binding site of an enzyme).

adaptation A heritable component of the phenotype which confers an advantage in survival and reproductive success. The process by which organisms adapt to the current environmental conditions.

additive genes See *polygenic inheritance*.

albinism A condition caused by the lack of melanin production in the iris, hair, and skin. In humans, most often inherited as an autosomal recessive.

aleurone layer In seeds, the outer layer of the endosperm.

alkaptonuria An autosomal recessive condition in humans caused by the lack of an enzyme, homogenistic acid oxidase. Urine of homozygous individuals turns dark upon standing due to oxidation of excreted homogenistic acid. The cartilage of homozygous adults blackens from deposition of a pigment derived from homogenistic acids. Such individuals often develop arthritic conditions.

allele One of the possible mutational states of a gene, distinguished from the other alleles by phenotypic effects.

allele frequency Measurement of the proportion of individuals in a population carrying a particular allele.

allelic exclusion In plasma cell heterozygous for an immunoglobulin gene, the selective action of only one allele.

allopatric speciation Process of speciation associated with geographic isolation.

allopolyploid Polyploid condition formed by the union of two or more distinct chromosome sets with a subsequent doubling of chromosome number.

allosteric effect Conformational change in the active site of a protein brought about by interaction with an effector molecule.

allotetraploid Diploid for two genomes derived from different species.

Allozyme An allelic form of a protein derived from different species.

alpha fetoprotein (AFP) A 70-kd glycoprotein synthesized in embryonic development by the yolk sac. High levels of this protein in the amniotic fluid are associated with neural tube defects such as spina bifida. Lower than normal levels may be associated with Down syndrome.

Alu sequence An interspersed DNA sequence of approximately 300 bp found in the genome of primates that is cleaved by the restriction enzyme *Alu* I. *Alu* sequences are composed of a head to tail dimer, with the first monomer approximately 140 bp and the second approximately 170 bp. In humans, they are dispersed throughout the genome and are present in 300,000 to 600,000 copies, constituting some 3 to 6 percent of the genome. See *SINEs*.

amber codon The codon UAG, which does not code for an amino acid but for chain termination.

Ames test An assay developed by Bruce Ames to detect mutagenic and carcinogenic compounds using reversion to histidine independence in the bacterium *Salmonella typhimurium*.

amino acid Any of the subunit building blocks that are covalently linked to form proteins.

aminoacyl tRNA Covalently linked combination of an amino acid and a rTNA molecule.

amniocentesis A procedure used to test for fetal defects in which fluid and fetal cells are withdrawn from the amniotic layer surrounding the fetus.

amphidiploid See *allotetraploid*.

anabolism The metabolic synthesis of complex molecules from less complex precursors.

analogue A chemical compound structurally similar to another, but differing by a single functional group (e.g., 5-bromodeoxyuridine is an analogue of thymidine).

anaphase Stage of cell division in which chromosomes begin moving to opposite poles of the cell.

aneuploidy A condition in which the chromosome number is not an exact multiple of the haploid set.

angstrom Unit of length equal to 10^{-10} meter. Abbreviated Å.

antibody Protein (immunoglobulin) produced in response to an antigenic stimulus with the capacity to bind specifically to the antigen.

anticodon The nucleotide triplet in a tRNA molecule which is complementary to, and binds to, the codon triplet in a mRNA molecule.

antigen A molecule, often a cell surface protein, that is capable of eliciting the formation of antibodies.

antiparallel Describing molecules in parallel alignment, but running in opposite directions. Most commonly used to describe the opposite orientations of the two strands of a DNA molecule.

apoenzyme The protein portion of an enzyme that requires a cofactor or prosthetic group to be functional.

ascospore A meiotic spore produced in certain fungi.

ascus In fungi, the sac enclosing the four or eight ascospores.

asexual reproduction Production of offspring in the absence of any sexual process.

assortative mating Nonrandom mating between males and females of a species. Selection of mates with the same genotype is positive; selection of mates with opposite genotypes is negative.

ATP Adenosine triphosphate.

attached-X chromosome Two conjoined X chromosomes that share a single centromere.

attenuator A nucleotide sequence between the promoter and the structural gene of some operons that can act to regulate the transit of RNA polymerase and thus control transcription of the structural gene.

autogamy A process of self-fertilization resulting in homozygosis.

autoimmune disease The production of antibodies that results from an immune response to one's own molecules, cells, or tissues. Such a response results from the inability of the immune system to distinguish self from nonself. Diseases such as arthritis, scleroderma, systemic lupus erythematosus, and perhaps diabetes are considered to be autoimmune diseases.

autopolyploidy Polyploid condition resulting from the replication of one diploid set of chromosomes.

autoradiography Production of a photographic image by radioactive decay. Used to localize radioactively labeled compounds within cells and tissues.

autosomes Chromosomes other than the sex chromosomes. In humans, there are 22 pairs of autosomes.

auxotroph A mutant microorganism or cell line which requires a substance for growth that can be synthesized by wild-type strains.

B-DNA See *double helix*.

back-cross A cross involving an F_1 heterozygote and one of the P_1 parents (or an organism with a genotype identical to one of the parents).

bacteriophage A virus that infects bacteria (synonym is *phage*).

bacteriostatic A compound that inhibits the growth of bacteria, but does not kill them.

balanced lethals Recessive, nonallelic lethal genes, each carried on different homologous chromosomes. When organisms carrying balanced lethal genes are interbred, only organisms with genotypes identical to the parents (heterozygotes) survive.

balanced polymorphism Genetic polymorphism maintained in a population by natural selection.

Barr body Densely staining nuclear mass seen in the somatic nuclei of mammalian females. Discovered by Murray Barr, this body is thought to represent an inactivated X chromosome.

base analogue See *analogue*.

base substitution A single base change in a DNA molecule that produces a mutation. There are two types of substitutions: *transitions*, in which a purine is substituted for a purine or a pyrimidine for a pyrimidine; and *transversions*, in which a purine is substituted for a pyrimidine, or vice versa.

biotechnology Commercial and/or industrial processes that utilize biological organisms or products.

bivalents Synapsed homologous chromosomes in the first prophase of meiosis.

BrdU (5-bromodeoxyuridine) A mutagenically active analogue of thymidine in which the methyl group at the 5′ position in thymine is replaced by bromine.

buoyant density A property of particles (and molecules) that depends upon their actual density, as determined by partial specific volume and degree of hydration. Provides the basis for density gradient separation of molecules or particles.

CAAT box A highly conserved DNA sequence found about 75 base pairs 5′ (upstream) to the initiation site of transcription in eukaryotic genes.

canonical sequence See *consensus sequence*.

cAMP Cyclic adenosine monophosphate. An important regulatory molecule in both prokaryotic and eukaryotic organisms.

CAP Catabolite activator protein. A protein that binds cAMP and regulates the activation of inducible operons.

carcinogen A physical or chemical agent that causes cancer.

carrier An individual heterozygous for a recessive trait.

cassette model First proposed to explain mating type interconversion in yeast, this model proposes that both genes for mating types, *a* and *alpha* are present as silent or unexpressed genes in transposable DNA segments (cassettes) that are activated (played) by transposition to the mating type locus.

catabolism A metabolic reaction in which complex molecules are broken down into simpler forms, often accompanied by the release of energy.

catabolite activator protein See *CAP*.

catabolite repression The selective inactivation of an operon by a metabolic product of the enzymes encoded by the operon.

cdc mutation A class of mutations in yeast that affect the timing and progression through the cell type.

cDNA DNA synthesized from an RNA template by the enzyme reverse transcriptase.

cell cycle Sum of the phases of growth of an individual cell type; divided into G1 (gap 1), S (DNA synthesis), G2 (gap 2), and M (mitosis).

cell-free extract A preparation of the soluble fraction of cells, made by lysing cells and removing the particulate matter, such as nuclei, membranes, and organelles. Often used to carry out the synthesis of proteins by the addition of specific, exogenous mRNA molecules.

CEN In yeast, fragments of chromosomal DNA, about 120 bp in length, that when inserted into plasmids confer the ability to segregate during mitosis. These segments contain at least three types of sequence elements associated with centromere function.

centimeter A unit of length equal to 10^{-2} meter. Abbreviated cm.

centimorgan A unit of distance between genes on chromosomes. One centimorgan represents a value of 1 percent crossing over between two genes.

central dogma The concept that information flow progresses from DNA to RNA to proteins. Although exceptions are known, this idea is central to an understanding of gene function.

centric fusion   See *Robertsonian translocation.*

centriole   A cytoplasmic organelle composed of nine groups of microtubules, generally arranged in triplets. Centrioles function in the generation of cilia and flagella and serve as foci for the spindles in cell division.

centromere   Specialized region of a chromosome to which the spindle fibers attach during cell division. Location of the centromere determines the shape of the chromosome during the anaphase portion of cell division. Also known as the primary construction.

centrosome   Region of the cytoplasm containing the centriole.

character   An observable phenotypic attribute of an organism.

charon phage   A group of genetically modified lambda phage designed to be used as vectors for cloning foreign DNA. Named after the ferryman in Greek mythology who carried the souls of the dead across the River Styx.

chemotaxis   Negative or positive response to a chemical gradient.

chiasma (pl., **chiasmata**)   The crossed strands of nonsister chromatids seen in diplotene of the first meiotic division. Regarded as the cytological evidence for exchange of chromosomal material, or crossing over.

chi-square (χ^2) analysis   Statistical test to determine if an observed set of data fits a theoretical expectation.

chloroplast   A cytoplasmic self-replicating organelle containing chlorophyll. The site of photosynthesis.

chorionic villus sampling (CVS)   A technique of prenatal diagnosis that intravaginally retrieves fetal cells from the chorion and uses them to detect cytogenetic and biochemical defects in the embryo.

chromatid   One of the longitudinal subunits of a replicated chromosome, joined to its sister chromatid at the centromere.

chromatin   Term used to describe the complex of DNA, RNA, histones, and nonhistone proteins that make up chromosomes.

chromatography   Technique for the separation of a mixture of solubilized molecules by their differential migration over a substrate.

chromocenter   An aggregation of centromeres and heterochromatic elements of polytene chromosomes.

chromomere   A coiled, beadlike region of a chromosome most easily visible during cell division. The aligned chromomeres of polytene chromosomes are responsible for their distinctive banding pattern.

chromosomal aberration   Any change resulting in the duplication, deletion, or rearrangement of chromosomal material.

chromosomal mutation   See *chromosomal aberration.*

chromosomal polymorphism   Alternate structures or arrangements of a chromosome that are carried by members of a population.

chromosome   In prokaryotes, an intact DNA molecule containing the genome; in eukaryotes, a DNA molecule complexed with RNA and proteins to form a threadlike structure containing genetic information arranged in a linear sequence.

chromosome banding   Technique for the differential staining of mitotic or meiotic chromosomes to produce a characteristic banding pattern or selective staining of certain chromosomal regions such as centromeres, the nucleolus

organizer regions, and GC- or AT-rich regions. Not to be confused with the banding pattern present in unstained polytene chromosomes, which is produced by the alignment of chromomeres.

chromosome map   A diagram showing the location of genes on chromosomes.

chromosome puff   A localized uncoiling and swelling in a polytene chromosome, usually regarded as a sign of active transcription.

***cis* configuration**   The arrangement of two mutant sites within a gene on the same homologue, such as

$$\frac{a^1 \; a^2}{+\; -}$$

Contrasts with a *trans* arrangement, where the mutant alleles are located on opposite homologues.

***cis* dominance**   The ability of a gene to affect the expression of other genes adjacent to it on the chromosome.

***cis-trans* test**   A genetic test to determine whether two mutations are located within the same cistron.

cistron   That portion of a DNA molecule that codes for a single polypeptide chain; defined by a genetic test as a region within which two mutations cannot complement each other.

cline   A gradient of genotype or phenotype distributed over a geographic range.

clonal selection   Theory of the immune system that proposes that antibody diversity precedes exposure to the antigen, and that the antigen functions to select the cells containing its specific antibody to undergo proliferation.

clone   Genetically identical cells or organisms all derived from a single ancestor by asexual or parasexual methods. For example, a DNA segment that has been enzymatically inserted into a plasmid or chromosome of a phage or a bacterium and replicated to form many copies.

cloned library   A collection of cloned DNA molecules representing all or part of an individual's genome.

code   See *genetic code.*

codominance   Condition in which the phenotypic effects of a gene's alleles are fully and simultaneously expressed in the heterozygote.

codon   A triplet of bases in a DNA or RNA molecule which specifies or encodes the information for a single amino acid.

coefficient of coincidence   A ratio of the observed number of double-crossovers divided by the expected number of such crossovers.

coefficient of selection   A measurement of the reproductive disadvantage of a given genotype in a population. If for genotype *aa*, only 99 of 100 individuals reproduce, then the selection coefficient (*s*) is 0.1.

colchicine   An alkaloid compound that inhibits spindle formation during cell division. Used in the preparation of karyotypes to collect a large population of cells inhibited at the metaphase stage of mitosis.

colicin   A bacteriocidal protein produced by certain strains of E. coli and other closely related bacterial species.

colinearity   The linear relationship between the nucleotide sequence in a gene (or the RNA transcribed from it) and the order of amino acids in the polypeptide chain specified by the gene.

competence   In bacteria, the transient state or condition during which the cell can bind and internalize exogenous DNA molecules, making transformation possible.

complementarity Chemical affinity between nitrogenous bases as a result of hydrogen bonding. Responsible for the base pairing between the strands of the DNA double helix.

complementation test A genetic test to determine whether two mutations occur within the same gene. If two mutations are introduced into a cell simultaneously and produce a wild-type phenotype (i.e., they complement each other), they are often nonallelic. If a mutant phenotype is produced, the mutations are noncomplementing and are often allelic.

complete linkage A condition in which two genes are located so close to each other that no recombination occurs between them.

complexity The total number of nucleotides or nucleotide pairs in a population of nucleic acid molecules as determined by reassociation kinetics.

complex locus A gene within which a set of functionally related pseudoalleles can be identified by recombinational analysis (e.g., the *bithorax* locus in *Drosophila*).

concatemer A chain or linear series of subunits linked together. The process of forming a concatemer is called concatenation (e.g., multiple units of a phage genome produced during replication).

concordance Pairs or groups of individuals identical in their phenotype. In twin studies, a condition in which both twins exhibit or fail to exhibit a trait under investigation.

conditional mutation A mutation that expresses a wild-type phenotype under certain (permissive) conditions and a mutant phenotype under other (restrictive) conditions.

conjugation Temporary fusion of two single-celled organisms for the sexual exchange of genetic material.

consanguine Related by a common ancestor within the previous few generations.

consensus sequence A nucleotide sequence most often found in a defined segment of DNA.

cosmid A vector designed to allow cloning of large segments of foreign DNA. Cosmids are hybrids composed of the cos sites of lambda inserted into a plasmid. In cloning, the recombinant DNA molecules are packaged into phage protein coats, and after infection of bacterial cells, the recombinant molecule replicates and can be maintained as a plasmid.

coupling conformation See *cis configuration*.

covalent bond A nonionic chemical bond formed by the sharing of electrons.

cri-du-chat syndrome A clinical syndrome in humans produced by a deletion of a portion of the short arm of chromosome 5. Afflicted infants have a distinctive cry which sounds like that of a cat.

crossing over The exchange of chromosomal material (pairs of chromosomal arms) between homologous chromosomes by breakage and reunion. The exchange of material between nonsister chromatid during meiosis is the basis of genetic recombination.

cross-reacting material (CRM) Nonfunctional form of an enzyme, produced by a mutant gene, which is recognized by antibodies made against the normal enzyme.

C-terminal amino acid The terminal amino acid in a peptide chain which carries a free carboxyl group.

cytogenetics A branch of biology in which the techniques of both cytology and genetics are used to study heredity.

cytokinesis The diversion or separation of the cytoplasm during mitosis or meiosis.

cytological map A diagram showing the location of genes at particular chromosomal sites.

cytoplasmic inheritance Non-Mendelian form of inheritance involving genetic information transmitted by self-replicating cytoplasmic organelles such as mitochondria, chloroplasts, etc.

dalton A unit of mass equal to that of the hydrogen atom, which is 1.67×10^{-24} gram. A unit used in designating molecular weight.

Darwinian fitness See *fitness*.

deficiency (deletion) A chromosomal mutation involving the loss or deletion of chromosomal material.

degenerate code Term used to describe the genetic code, in which a given amino acid may be represented by more than one codon.

deletion See *deficiency*.

deme A local interbreeding population.

denatured DNA DNA molecules that have been separated into single strands.

de novo Newly arising; synthesized from less complex precursors rather than having been produced by modification of an existing molecule.

density gradient centrifugation A method of separating macromolecular mixtures by the use of centrifugal force and solvents of varying density. In sedimentation velocity centrifugation, macromolecules are separated by the velocity of sedimentation through a preformed gradient such as sucrose. In density gradient equilibrium centrifugation, macromolecules in a cesium salt solution are centrifuged until the cesium solution establishes a gradient under the influence of the centrifugal field, and the macromolecules sediment until the density of the solvent equals their own.

deoxyribonuclease A class of enzymes that breaks down DNA into oligonucleotide fragments by introducing single-stranded breaks into the double helix.

dermatoglyphics The study of the surface ridges of the skin, especially of the hands and feet.

deoxyribonucleic acid (DNA) A macromolecule usually consisting of antiparallel polynucleotide chains held together by hydrogen bonds, in which the sugar residues are deoxyribose. The primary carrier of genetic information.

determination A regulatory event that establishes a specific pattern of gene activity and developmental fate for a given cell.

diakinesis The final stage of meiotic prophase I in which the chromosomes become tightly coiled and compacted and move toward the periphery of the nucleus.

dicentric chromosome A chromosome having two centromeres.

differentiation The process of complex changes by which cells and tissues attain their adult structure and functional capacity.

dihybrid cross A genetic cross involving two characters in which the parents possess different forms of each character (e.g., tall, round × short, wrinkled peas).

diploid A condition in which each chromosome exists in pairs; having two of each chromosome.

diplotene A stage of meiotic prophase 1 immediately after pachytene. In diplotene, one pair of sister chromatids begins separating from the other, and chiasmata become visible. These overlaps move laterally toward the ends of the chromatids (terminalization).

directional selection A selective force that changes the frequency of an allele in a given direction, either toward fixation or toward elimination.

discontinuous variation Phenotypic data that fall into two or more distinct classes that do not overlap.

discordance In twin studies, a situation where one twin shows a trait but the other does not.

disjunction The separation of chromosomes at the anaphase stage of cell division.

disruptive selection Simultaneous selection for phenotypic extremes in a population, usually resulting in the production of two discontinuous strains.

dizygotic twins Twins produced from separate fertilization events; two ova fertilized independently. Also known as fraternal twins.

DNA See *deoxyribonucleic acid.*

DNA footprinting See *footprinting.*

DNA gyrase One of the DNA topoisomerases that functions during DNA replication to reduce molecular tension caused by supercoiling. DNA gyrase produces, then seals, double-stranded breaks.

DNA ligase An enzyme that forms a covalent bond between the 5′ end of one polynucleotide chain and the 3′ end of another polynucleotide chain. Also called polynucleotide-joining enzyme.

DNA polymerase An enzyme that catalyzes the synthesis of DNA from deoxyribonucleotides and a template DNA molecule.

DNase Deoxyribonuclease, a class of enzymes that degrades or breaks down DNA into fragments or constitutive nucleotides.

dominance The expression of a trait in the heterozygous condition.

dosage compensation A genetic mechanism that regulates the levels of gene products at certain autosomal loci; this results in homozygous dominants and heterozygotes having the same amount of a gene product. In mammals, random inactivation of one X chromosome in females leads to equal levels of X chromosome-coded gene products in males and females.

double-crossover Two separate events of chromosome breakage and exchange occurring within the same tetrad.

double helix The model for DNA structure proposed by James Watson and Francis Crick, involving two antiparallel, hydrogen-bonded polynucleotide chains wound into a right-handed helical configuration, with 10 base pairs per full turn of the double helix. Often called B-DNA.

duplication A chromosomal aberration in which a segment of the chromosome is repeated.

dyad The products of tetrad separation or disjunction at the first meiotic prophase. Consists of two sister chromatids joined at the centromere.

effector molecule Small, biologically active molecule that acts to regulate the activity of a protein by binding to a specific receptor site on the protein.

electrophoresis A technique used to separate a mixture of molecules by their differential migration through a stationary phase in an electrical field.

endogenote The segment of the chromosome in a partially diploid bacterial cell (merozygote) which is homologous to the chromosome transmitted by the donor cell.

endomitosis Chromosomal replication that is not accompanied by either nuclear or cytoplasmic division.

endonuclease An enzyme that hydrolyzes internal phosphodiester bonds in a polynucleotide chain or nucleic acid molecule.

endoplasmic reticulum A membranous organelle system in the cytoplasm of eukaryotic cells. The outer surface of the membranes may be ribosome-studded (rough ER) or not (smooth ER).

endopolyploidy The increase in chromosome sets that results from endomitotic replication within somatic nuclei.

endosymbiont theory The proposal that self-replicating cellular organelles such as mitochondria and chloroplasts were originally free-living organisms that entered into a symbiotic relationship with nucleated cells.

enhancer Originally, a 72-bp sequence in the genome of the virus, SV40, that increases the transcriptional activity of nearby structural genes. Similar sequences that enhance transcription have been identified in the genomes of eukaryotic cells. Enhancers can act over a distance of thousands of base pairs and can be located 5′ or 3′ to the gene they affect, and thus are different from promoters.

enhanson The DNA sequence that represents the core sequence of an enhancer.

environment The complex of geographic, climatic, and biotic factors within which an organism lives.

enzyme A protein or complex of proteins that catalyzes a specific biochemical reaction.

epigenesis The idea that an organism develops by the appearance and growth of new structures. Opposed to preformationism, which holds that development is the growth of structures already present in the egg.

episome A circular genetic element in bacterial cells that can replicate independently of the bacterial chromosome or integrate and replicate as part of the chromosome.

epitope That portion of a macromolecule or cell that acts to elicit an antibody response; an antigenic determinant. A complex molecule or cell can contain several such sites.

epistasis Nonreciprocal interaction between genes such that one gene interferes with or prevents the expression of another gene. For example, in *Drosophila*, the recessive gene *eyeless*, when homozygous, prevents the expression of eye color genes.

equatorial plate See *metaphase plate.*

euchromatin Chromatin or chromosomal regions that are lightly staining and are relatively uncoiled during the interphase portion of the cell cycle. The region of the chromosomes thought to contain most of the structural genes.

eukaryotes Those organisms having true nuclei and membranous organelles and whose cells demonstrate mitosis and meiosis.

euploid Polyploid with a chromosome number that is an exact multiple of a basic chromosome set.

evolution The origin of plants and animals from preexisting types. Descent with modifications.

excision repair Repair of DNA lesions by removal of a polynucleotide segment and its replacement with a newly synthesized, corrected segment.

exogenote In merozygotes, the segment of the bacterial chromosome contributed by the donor cell.

exon The DNA segment(s) of a gene that are transcribed and translated into protein.

exonuclease An enzyme that breaks down nucleic acid molecules by breaking the phosphodiester bonds at the 3′ or 5′ terminal nucleotides.

expression vector Plasmids or phage carrying promoter regions designed to cause expression of cloned DNA sequences.

expressivity The degree or range in which a phenotype for a given trait is expressed.

extranuclear inheritance Transmission of traits by genetic information contained in cytoplasmic organelles such as mitochondria and chloroplasts.

F⁺ cell A bacterial cell having a fertility (F) factor. Acts as a donor in bacterial conjugation.

F⁻ cell A bacterial cell that does not contain a fertility (F) factor. Acts as a recipient in bacterial conjugation.

F factor An episome in bacterial cells that confers the ability to act as a donor in conjugation.

F′ factor A fertility (F) factor that contains a portion of the bacterial chromosome.

F₁ generation First filial generation; the progeny resulting from the first cross in a series.

F₂ generation Second filial generation; the progeny resulting from a cross of the F₁ generation.

F pilus See *pilus.*

facultative heterochromatin Chromatin that may alternate in form between euchromatic and heterochromatic. The Y chromosome of many species contains facultative heterochromatin.

familial trait A trait transmitted through and expressed by members of a family.

fate map A diagram or "map" of an embryo showing the location of cells whose developmental fate is known.

fertility (F) factor See *F factor.*

filial generations See *F₁, F₂ generations.*

fingerprint The pattern of ridges and whorls on the tip of a finger. The pattern obtained by enzymatically cleaving a protein or nucleic acid and subjecting the digest to two-dimensional chromatography or electrophoresis.

fitness A measure of the relative survival and reproductive success of a given individual or genotype.

fixation In population genetics, a condition in which all members of a population are homozygous for a given allele.

fixity of species The idea that members of a species can give rise only to other members of the species, thus implying that all species are independently created.

fluctuation test A statistical test developed by Salvadore Luria and Max Delbruck to determine whether bacterial mutations arise spontaneously or are produced in response to selective agents.

fMet See *formylmethionine.*

footprinting A technique for identifying a DNA sequence that binds to a particular protein, based on the idea that the phosphodiester bonds in the region covered by the protein are protected from digestion by deoxyribonucleases.

formylmethionine (fMet) A molecule derived from the amino acid methionine by attachment of a formyl group to its terminal amino group. This is the first amino acid inserted in all bacterial polypeptides. Also known as *N*-formyl methionine.

founder effect A form of genetic drift. The establishment of a population by a small number of individuals whose genotypes carry only a fraction of the different kinds of alleles in the parental population.

fragile site A heritable gap or nonstaining region of a chromosome that can be induced to generate chromosome breaks.

frameshift mutation A mutational event leading to the insertion of one or more base pairs in a gene, shifting the codon reading frame in all codons following the mutational site.

fraternal twins See *dizygotic twins.*

gamete A specialized reproductive cell with a haploid number of chromosomes.

gene The fundamental physical unit of heredity whose existence can be confirmed by allelic variants and which occupies a specific chromosomal locus. A DNA sequence coding for a single polypeptide.

gene amplification The process by which gene sequences are selected for differential replication either extrachromosomally or intrachromosomally.

gene conversion A directional process that changes one allele into another specific allele. In fungi, this process involves meiosis and a recombinational step, but in other organisms, such as trypanosomes, meiosis may not be required.

gene duplication An event in replication leading to the production of a tandem repeat of a gene sequence.

gene flow The gradual exchange of genes between two populations, brought about by the dispersal of gametes or the migration of individuals.

gene frequency The percentage of alleles of a given type in a population.

gene interaction Production of novel phenotypes by the interaction of alleles of different genes.

gene mutation See *point mutation.*

gene pool The total of all genes possessed by reproductive members of a population.

generalized transduction The transduction of any gene in the bacterial genome by a phage.

genetic burden Average number of recessive lethal genes carried in the heterozygous condition by an individual in a population. Also called genetic load.

genetic code The nucleotide triplets that code for the 20 amino acids or for chain initiation or termination.

genetic counseling Analysis of risk for genetic defects in a family and the presentation of options available to avoid or ameliorate possible risks.

genetic drift Random variation in gene frequency from generation to generation. Most often observed in small populations.

genetic engineering The technique of altering the genetic constitution of cells or individuals by the selective removal, insertion, or modification of individual genes or gene sets.

genetic equilibrium Maintenance of allele frequencies at the same value in successive generations. A condition in which allele frequencies are neither increasing nor decreasing.

genetic fine structure Intragenic recombinational analysis that provides mapping information at the level of individual nucleotides.

genetic load See *genetic burden.*

genetic polymorphism The stable coexistence of two or more discontinuous genotypes in a population. When the frequencies of two alleles are carried to an equilibrium, the condition is called balanced polymorphism.

genetics The branch of biology that deals with heredity and the expression of inherited traits.

genome The array of genes carried by an individual.

genotype The specific allelic or genetic constitution of an organism; often, the allelic composition of one or a limited number of genes under investigation.

Goldberg-Hogness box A short nucleotide sequence 20 to 30 bp 5′ to the initiation site of eukaryotic genes to which RNA polymerase II binds.; The consensus sequence is TATAAAA. Also known as TATA box.

graft versus host disease (GVHD) In transplants, reaction by immunologically competent cells of the donor against the antigens present on the cells of the host. In human bone marrow transplants, often a fatal condition.

gynandromorph An individual composed of cells with both male and female genotypes.

gyrase One of a class of enzymes known as topoisomerases. Gyrase converts closed circular DNA to a negatively supercoiled form prior to replication, transcription, or recombination.

H substance The carbohydrate group present on the surface of red blood cells. When unmodified, it results in blood type O; when modified by the addition of monosaccharides, it results in type A, B, and AB.

haploid A cell or organism having a single set of unpaired chromosomes. The gametic chromosome number.

haplotype The set of alleles from closely linked loci carried by an individual and usually inherited as a unit.

Hardy-Weinberg law The principle that both gene and genotype frequencies will remain in equilibrium in an infinitely large population in the absence of mutation, migration, selection, and nonrandom mating.

heat shock A transient response following exposure of cells or organisms to elevated temperatures. The response involves activation of a small number of loci, inactivation of previously active loci, and selective translation of heat shock mRNA. Appears to be a nearly universal phenomenon observed in organisms ranging from bacteria to humans.

helicase An enzyme that participates in DNA replication by unwinding the double helix near the replication fork.

hemizygous Conditions where a gene is present in a single dose. Usually applied to genes on the X chromosome in heterogametic males.

hemoglobin (Hb) An iron-containing, conjugated respiratory protein occurring chiefly in the red blood cells of vertebrates.

hemophilia A sex-linked trait in humans associated with defective blood-clotting mechanisms.

heredity Transmission of traits from one generation to another.

heritability A measure of the degree to which observed phenotypic differences for a trait are genetic.

heterochromatin The heavily staining, late replicating regions of chromosomes that are prematurely condensed in interphase.

heteroduplex A double-stranded nucleic acid molecule in which each polynucleotide chain has a different origin. These structures may be produced as intermediates in a recombinational event, or by the *in vitro* reannealing of single-stranded, complementary molecules.

heterogametic sex The sex that produces gametes containing unlike sex chromosomes.

heterogenote A bacterial merozygote in which the donor (exogenote) chromosome segment carries different alleles than does the chromosome of the recipient (endogenote). A heterozygous merozygote.

heterogeneous nuclear RNA (hnRNA) The collection of RNA transcripts in the nucleus, representing precursors and processing intermediates to rRNA, mRNA, and tRNA. Also represents RNA transcripts that will not be transported to the cytoplasm, such as snRNA (small nuclear RNAs).

heterokaryon A somatic cell containing nuclei from two different sources.

heterosis The superiority of a heterozygote over either homozygote for a given trait.

heterozygote An individual with different alleles at one or more loci. Such individuals will produce unlike gametes and therefore will not breed true.

Hfr A strain of bacteria exhibiting a high frequency of recombination. These strains have a chromosomally integrated F factor that is able to mobilize and transfer all or part of the chromosome to a recipient F⁻ cell.

histocompatibility antigens See *HLA.*

histones Proteins complexed with DNA in the nucleus. They are rich in the basic amino acids arginine and lysine and function in the coiling of DNA to form nucleosomes.

HLA Cell surface proteins, produced by histocompatibility loci, which are involved in the acceptance or rejection of tissue and organ grafts and transplants.

hnRNA See *heterogeneous nuclear RNA.*

holandric A trait transmitted from males to males. In humans, genes on the Y chromosome are holandric.

homeotic mutation A mutation that causes a tissue normally determined to form a specific organ or body part to alter its differentiation and form another structure. Alternately spelled: homoeotic.

homogametic sex The sex that produces gametes that do not differ with respect to sex chromosome content: in mammals, the female is homogametic.

homogeneously staining regions (hsr) Segments of mammalian chromosomes that stain lightly with Giemsa following exposure of cells to a selective agent. These regions arise in conjunction with gene amplification and are regarded as the structural locus for the amplified gene.

homogenote A bacterial merozygote in which the donor (exogenote) chromosome carries the same alleles as the chromosome of the recipient (endogenote). A homozygous merozygote.

homologous chromosomes Chromosomes that synapse or pair during meiosis. Chromosomes that are identical with respect to their genetic loci and centromere placement.

homozygote An individual with identical alleles at one or more loci. Such individuals will produce identical gametes and will therefore breed true.

homunculus The miniature individual imagined by preformationists to be contained within the sperm or egg.

human immunodeficiency virus (HIV) A human retrovirus associated with the onset and progression of acquired immunodeficiency syndrome (AIDS).

hybrid An individual produced by crossing two parents of different genotypes.

hybridoma A somatic cell hybrid produced by the fusion of an antibody-producing cell and a cancer cell, specifically, a myeloma. The cancer cell contributes the ability to divide indefinitely, and the antibody cell confers the ability to synthesize large amounts of a single antibody.

hybrid vigor See *heterosis*.

hydrogen bond An electrostatic attraction between a hydrogen atom bonded to a strongly electronegative atom such as oxygen or nitrogen and another atom that is electronegative or contains an unshared electron pair.

identical twins See *monozygotic twins*.

Ig See *immunoglobulin*.

imaginal disk Discrete groups of cells set aside during embryogenesis in holometabolous insects which are determined to form the external body parts of the adult.

immunoglobulin The class of serum proteins having the properties of antibodies.

inborn error of metabolism A biochemical disorder that is genetically controlled; usually an enzyme defect that produces a clinical syndrome.

inbreeding Mating between closely related organisms.

inbreeding depression A loss or reduction in fitness that usually accompanies inbreeding of organisms.

incomplete dominance Expression of heterozygous phenotype which is distinct from, and often intermediate to, that of either parent.

incomplete linkage Occasional separation of two genes on the same chromosome by a recombinational event.

independent assortment The independent behavior of each pair of homologous chromosomes during their segregation in meiosis I. The random distribution of genes on different chromosomes into gametes.

inducer An effector molecule that activates transcription.

inducible enzyme system An enzyme system under the control of a regulatory molecule, or inducer, which acts to block a repressor and allow transcription.

initiation codon The triplet of nucleotides (AUG) in an mRNA molecule that codes for the insertion of the amino acid methionine as the first amino acid in a polypeptide chain.

insertion sequence See *IS element*.

in situ hybridization A technique for the cytological localization of DNA sequences complementary to a given nucleic acid or polynucleotide.

intercalating agent A compound that inserts between bases in a DNA molecule, disrupting the alignment and pairing of bases in the complementary strands (e.g., acridine dyes).

interference A measure of the degree to which one cross-over affects the incidence of another crossover in an adjacent region of the same chromatid. Positive interference decreases the chances of another crossover; negative interference increases the probability of a second crossover event.

interferon One of a family of proteins that act to inhibit viral replication in higher organisms. Some interferons may have anticancer properties.

interphase That portion of the cell cycle between divisions.

intervening sequence See *intron*.

intron A portion of DNA between coding regions in a gene which is transcribed, but which does not appear in the mRNA product.

inversion A chromosomal aberration in which the order of a chromosomal segment has been reversed.

inversion loop The chromosomal configuration resulting from the synapsis of homologous chromosomes, one of which carries an inversion.

in vitro Literally, in glass; outside the living organism; occurring in an artificial environment.

in vivo Literally, in the living; occurring within the living body of an organism.

IS element A mobile DNA segment that is transposable to any of a number of sites in the genome.

isoagglutinogen An antigenic factor or substance present on the surface of cells that is capable of inducing the formation of an antibody.

isochromosome An aberrant chromosome with two identical arms and homologous loci.

isolating mechanism Any barrier to the exchange of genes between different populations of a group of organisms. In general, isolation can be classified as spatial, environmental, or reproductive.

isotopes Forms of a chemical element that have the same number of protons and electrons, but differ in the number of neurons contained in the atomic nucleus. Unstable isotopes undergo a transition to a more stable form with the release of radioactivity.

isozyme Any of two or more distinct forms of an enzyme that have identical or nearly identical chemical properties, but differ in some property such as net electrical charge, pH optima, number and type of subunits, or substrate concentration.

kappa particles DNA-containing cytoplasmic particles found in certain strains of *Paramecium auelia*. When these self-reproducing particles are transferred into the growth medium, they release a toxin, paramecin, which kills other sensitive strains. A nuclear gene, *K,* is responsible for the maintenance of kappa particles in the cytoplasm.

karyokinesis The process of nuclear division.

karyotype The chromosome complement of a cell or an individual. Often used to refer to the arrangement of metaphase chromosomes in a sequence according to length and position of the centromere.

kilobase A unit of length consisting of 1000 nucleotides. Abbreviated kb.

kinetochore A fibrous structure with a size of about 400 mn, located within the centromere. It appears to be the location to which microtubules attach.

Klenow fragment   A part of bacterial DNA polymerase that lacks exonuclease activity, but retains polymerase activity. It is produced by enzymatic digestion of the intact enzyme.

Klinefelter syndrome   A genetic disease in human males caused by the presence of an extra X chromosome. Klinefelter males are XXY instead of XY. This syndrome is associated with enlarged breasts, small testes, sterility, and occasionally, mental retardation.

lagging strand   In DNA replication, the strand synthesized in a discontinuous fashion. 5′ to 3′ away from the replication fork. Each short piece of DNA synthesized in this fashion is called an Okazaki fragment.

lampbrush chromosomes   Meiotic chromosomes characterized by extended lateral loops, which reach maximum extension during diplotene. Although most intensively studied in amphibians, these structures occur in meiotic cells of organisms ranging from insects through humans.

leader sequence   That portion of an mRNA molecule from the 5′ end to the beginning codon; may contain regulatory or ribosome binding sites.

leading strand   During DNA replication, the strand synthesized continuously 5′ to 3′ toward the replication fork.

lectins   Carbohydrate-binding proteins found in the seeds of leguminous plants such as soybeans. Although their physiological role in plants is unclear, they are useful as probes for cell surface carbohydrates and glycoproteins in animal cells. Proteins with similar properties have also been isolated from organisms such as snails and eels.

leptotene   The initial stage of meiotic prophase I, during which the chromosomes become visible and are often arranged in a bouquet configuration, with one or both ends of the chromosomes gathered at one spot on the inner nuclear membrane.

lethal gene   A gene whose expression results in death.

leucine zipper   A structural motif in a DNA binding protein that is characterized by a stretch of leucine residues spaced at every seventh amino acid residue, with adjacent regions of positively charged amino acids. Leucine zippers on two polypeptides may interact to form a dimer that binds to DNA.

LINEs   Long interspersed repetitive sequences found in the genomes of higher organisms, such as the 6-kb KpnI sequences found in primate genomes.

linkage   Condition in which two or more nonallelic genes tend to be inherited together. Linked genes have their loci along the same chromosome, do not assort independently, but can be separated by crossing over.

linkage group   A group of genes that have their loci on the same chromosome.

linking number   The number of times that two strands of a closed, circular DNA duplex cross over each other.

locus   The site or place on a chromosome where a particular gene is located.

long period interspersion   Pattern of genome organization in which long stretches of single copy DNA are interspersed with long segments of repetitive DNA. This pattern of genome organization is found in *Drosophila* and the honeybee.

long terminal repeat (LTR)   Sequence of several hundred base pairs found at the ends of retroviral DNAs.

Lutheran blood group   One of a number of blood group systems inherited independently of the ABO, MN, and Rh systems. Alleles of this group determine the presence or absence of antigens on the surface of red blood cells. Gene is on human chromosome 19.

Lyon hypothesis   The idea proposed by Mary Lyon that random inactivation of one X chromosome in the somatic cells of mammalian females is responsible for dosage compensation and mosaicism.

lysis   The disintegration of a cell brought about by the rupture of its membrane.

lysogenic bacterium   A bacterial cell carrying a temperate bacteriophage integrated into its chromosome.

lysogeny   The process by which the DNA of an infecting phage becomes repressed and integrated into the chromosome of the bacterial cell it infects.

lytic phase   The condition in which a temperate bacteriophage loses its integrated status in the host chromosome (becomes induced), replicates, and lyses the bacterial cell.

major histocompatibility loci   See *MHC*.

map unit   A measure of the genetic distance between two genes, corresponding to a recombination frequency of 1 percent. See *centimorgan*.

maternal effect   Phenotypic effects on the offspring produced by the maternal genome. Factors transmitted through the egg cytoplasm which produce a phenotypic effect in the progeny.

maternal influence   See *maternal effect*.

maternal inheritance   The transmission of traits via cytoplasmic genetic factors such as mitochondria or chloroplasts.

mean   The arithmetic average.

median   The value in a group of numbers below and above which there is an equal number of data points or measurements.

meiosis   The process in gametogenesis or sporogenesis during which one replication of the chromosomes is followed by two nuclear divisions to produce four haploid cells.

melting profile   See T_m.

merozygote   A partially diploid bacterial cell containing, in addition to its own chromosome, a chromosome fragment introduced into the cell by transformation, transduction, or conjugation.

messenger RN Asee mRNA.

metabolism   The sum of chemical changes in living organisms by which energy is generated and used.

metacentric chromosome   A chromosome with a centrally located centromere, producing chromosome arms of equal lengths.

metafemale   In *Drosophila*, a poorly developed female of low viability in which the ratio of X chromosomes to sets of autosomes exceeds 1.0. Previously called a superfemale.

metamale   In *Drosophila*, a poorly developed male of low viability in which the ratio of X chromosomes to sets of autosomes is less than 0.5. Previously called a supermale.

metaphase   The stage of cell division in which the condensed chromosomes lie in a central plane between the two poles of the cell, and in which the chromosomes become attached to the spindle fibers.

metaphase plate   The arrangement of mitotic or meiotic chromosomes at the equator of the cell during metaphase.

MHC   Major histocompatibility loci. In humans, the HLA complex, and in mice, the H2 complex.

micrometer   A unit of length equal to 1×10^{-6} meter. Previously called a micron. Abbreviated μm.

micron   See *micrometer.*

migration coefficient   An expression of the proportion of migrant genes entering the population per generation.

millimeter   A unit of length equal to 1×10^{-3} meter. Abbreviated mm.

minimal medium   A medium containing only those nutrients that will support the growth and reproduction of wild-type strains of an organism.

missense mutation   A mutation that alters a codon to that of another amino acid, causing an altered translation product to be made.

mitochondrion   Found in the cells of eukaryotes, a cytoplasmic, self-reproducing organelle that is the site of ATP synthesis.

mitogen   A substance that stimulates mitosis in nondividing cells; e.g., phytohemagglutinin.

mitosis   A form of cell division resulting in the production of two cells, each with the same chromosome and genetic complement as the parent cell.

mode   In a set of data, the value occurring in the greatest frequency.

monohybrid cross   A genetic cross between two individuals involving only one character (e.g., $AA \times aa$).

monosomic   An aneuploid condition in which one member of a chromosome pair is missing; having a chromosome number of $2n - 1$.

monozygotic twins   Twins produced from a single fertilization event; the first division of the zygote produces two cells, each of which develops into an embryo. Also known as identical twins.

mRNA   An RNA molecule transcribed from DNA and translated into the amino acid sequence of a polypeptide.

mtDNA   Mitochondrial DNA.

multiple alleles   Three or more alleles of the same gene.

multiple-factor inheritance   See *polygenic inheritance.*

multiple infection   Simultaneous infection of a bacterial cell by more than one bacteriophage, often of different genotypes.

mu phage   A phage group in which the genetic material behaves like an insertion sequence, capable of insertion, excision, transposition, inactivation of host genes, and induction of chromosomal rearrangements.

mutagen   Any agent that causes an increase in the rate of mutation.

mutant   A cell or organism carrying an altered or mutant gene.

mutation   The process which produces an alternation in DNA or chromosome structure; the source of most alleles.

mutation rate   The frequency with which mutations take place at a given locus or in a population.

muton   The smallest unit of mutation in a gene, corresponding to a single base change.

nanometer   A unit of length equal to 1×10^{-9} meter. Abbreviated nm.

natural selection   Differential reproduction of some members of a species resulting from variable fitness conferred by genotype differences.

nearest-neighbor analysis   A molecular technique used to determine the frequency with which nucleotides are adjacent to each other in polynucleotide chains.

neutral mutation   A mutation with no immediate adaptive significance or phenotypic effect.

noncrossover gamete   A gamete which contains no chromosomes that have undergone genetic recombination.

nondisjunction   An accident of cell division in which the homologous chromosomes (in meiosis) or the sister chromatids (in mitosis) fail to separate and migrate to opposite poles; responsible for defects such as monosomy and trisomy.

nonsense codon   The nucleotide triplet in an mRNA molecule that signals the termination of translation. Three such codons are known: UGA, UAG, and UAA.

nonsense mutation   A mutation that alters a codon to one which encodes no amino acid, i.e., VAG (amber codon), UAA (ochre codon), or UGA (opal codon). Leads to premature termination during the translation of mRNA.

NOR   See *nucleolar organizer region.*

normal distribution   A probability function that approximates the distribution of random variables. The normal curve, also known as a Gaussian or bell-shaped curve, is the graphic display of the normal distribution.

np   See *nucleotide pair.*

N-terminal amino acid   The terminal acid in a peptide chain that carries a free amino group.

nu body   See *nucleosome.*

nuclease   An enzyme that breaks bonds in nucleic acid molecules.

nucleoid   The DNA-containing region within the cytoplasm in prokaryotic cells.

nucleolar organizer region (NOR)   A chromosomal region containing the genes for rRNA; most often found in physical association with the nucleolus.

nucleolus   A nuclear organelle that is the site of ribosome biosynthesis; usually associated with or formed in association with the NOR.

nucleoside   A purine or pyrimidine base covalently linked to a ribose or deoxyribose sugar molecule.

nucleosome   A complex of four histone molecules, each present in duplicate, wrapped by two turns of a DNA molecule. One of the basic units of eukaryotic chromosome structure. Also known as a nu body.

nucleotide   A nucleoside covalently linked to a phosphate group. Nucleotides are the basic building blocks of nucleic acids. The nucleotides commonly found in DNA are deoxyadenylic acid, deoxycytidylic acid, deoxyguanylic acid, and uridylic acid.

nucleotide pair   The pair of nucleotides (A and T, or G and C) in opposite strands of the DNA molecule that are hydrogen-bonded to each other.

nucleus   The membrane-bounded cytoplasmic organelle of eukaryotic cells that contains the chromosomes and nucleolus.

nullisomic   Describes an individual with a chromosomal aberration in which both members of a chromosome pair are missing.

ochre codon   A codon that does not code for the insertion of an amino acid into a polypeptide chain, but signals chain termination. The ochre codon is UAA.

Okazaki fragment   The small, discontinuous strands of DNA produced during DNA synthesis.

oncogene   A gene whose activity promotes uncontrolled proliferation in eukaryotic cells.

operator region   A region of a DNA molecule that interacts with a specific repressor protein to control the expression of an adjacent gene or gene set.

operon   A genetic unit that consists of one or more structural genes (that code for polypeptides) and an adjacent operator gene that controls the transcriptional activity of the structural gene or genes.

orphon   Single copies of tandemly repeated genes found dispersed in the genome. For example, histone genes are members of a multigene family present in several hundred copies clustered in a tandem array. A single copy of a histone gene found elsewhere in the genome is said to have lost its family and is regarded as an orphon.

overlapping code   A genetic code first proposed by George Gamow in which any given nucleotide is shared by two adjacent codons.

pachytene   The stage in meiotic prophase I when the synapsed homologous chromosomes split longitudinally (except at the centromere), producing a group of four chromatids called a tetrad.

palindrome   A word, number, verse, or sentence that reads the same backward or forward (e.g., able was I ere I saw elba). In nucleic acids, a sequence in which the base pairs read the same on complementary strands ($5' \longrightarrow 3'$). For example: $5'$GAATTC$3'$; $3'$GTTAAG$5'$. These often occur as sites for restriction endonuclease recognition and cutting.

pangenesis   A discarded theory of development that postulated the existence of pangenes, small particles from all parts of the body that concentrate in the gametes, passing traits from generation to generation, blending the traits of the parents in the offspring.

paracentric inversion   A chromosomal inversion that does not include the centromere.

parasexual   Condition describing recombination of genes from different individuals which does not involve meiosis, gamete formation, or zygote production. The formation of somatic cell hybrids is an example.

parental gamete   See *noncrossover gamete*.

parthenogenesis   Development of an egg without fertilization.

partial diploids   See *merozygote*.

partial dominance   See *incomplete dominance*.

passive immunity   A form of immunity produced by receipt of antibodies synthesized by another individual.

patroclinous inheritance   A form of genetic transmission in which the offspring have the phenotype of the father.

pedigree   In human genetics, a diagram showing the ancestral relationships and transmission of genetic traits over several generations in a family.

penetrance   The frequency (expressed as a percentage) with which individuals of a given genotype manifest at least some degree of a specific mutant phenotype associated with a trait.

peptide bond   The covalent bond between the amino group of one amino acid and the carboxyl group of another amino acid.

pericentric inversion   A chromosomal inversion that involves both arms of the chromosome and thus involves the centromere.

permissive condition   Environmental conditions under which a conditional mutation (such as temperature-sensitive mutant) expresses the wild-type phenotype.

phage   See *bacteriophage*.

phenocopy   An environmentally induced phenotype (nonheritable) which closely resembles the phenotype produced by a known gene.

phenotype   The observable properties of an organism that are genetically controlled.

phenylketonuria (PKU)   A hereditary condition in humans associated with the inability to metabolize the amino acid phenylalanine. The most common form is caused by the lack of the liver enzyme phenylalanine hydroxylase.

phosphodiester bond   In nucleic acids, the covalent bond between a phosphate group and adjacent nucleotides, extending from the $5'$ carbon of one pentose (ribose or deoxyribose) to the $3'$ carbon of the pentose in the neighboring nucleotide. Phosphodiester bonds form the backbone of nucleic acid molecules.

photoreactivation repair   Light-induced repair of damage caused by exposure to ultraviolet light. Associated with an intracellular enzyme system.

phyletic evolution   The gradual transformation of one species into another over time; vertical evolution.

pilus   A filamentlike projection from the surface of a bacterial cell. Often associated with cells possessing F factors.

plaque   A clear area on an otherwise opaque bacterial lawn caused by the growth and reproduction of phages.

plasmid   An extrachromosomal, circular DNA molecule (often carrying genetic information) that replicates independently of the host chromosome.

platysome   Term originally used in electron and X-ray diffraction studies to describe the flattened appearance of the DNA in the nucleosome core.

pleiotropy   Condition in which a single mutation simultaneously affects several characters.

ploidy   Term referring to the basic chromosome set or to multiples of that set.

point mutation   A mutation that can be mapped to a single locus. At the molecular level, a mutation that results in the substitution of one nucleotide for another.

polar body   A cell produced at either the first or second meiotic division in females which contains almost no cytoplasm as a result of an unequal cytokinesis.

polycistronic mRNAA messenger   RNA molecule that encodes the amino acid sequence of two or more polypeptide chains in adjacent structural genes.

polygenic inheritance   The transmission of a phenotypic trait whose expression depends on the additive effect of a number of genes.

polymerase chain reaction (PCR)   A method for amplifying DNA segments that uses cycles of denaturation, annealing to primers, and DNA polymerase-directed DNA synthesis.

polymerases   The enzymes that catalyze the formation of DNA and RNA from deoxynucleotides and ribonucleotides, respectively.

polymorphism The existence of two or more discontinuous, segregating phenotypes in a population.

polypeptide A molecule made up of amino acids joined by covalent peptide bonds. This term is used to denote the amino acid chain before it assumes its functional three-dimensional configuration.

polyploid A cell or individual having more than two sets of chromosomes.

polyribosome See *polysome*.

polysome A structure composed to two or more ribosomes associated with mRNA, engaged in translation. Formerly called polyribosome.

polytene chromosome A chromosome that has undergone several rounds of DNA replication without separation of the replicated chromosomes, forming a giant, thick chromosome with aligned chromosomes producing a characteristic banding pattern.

population A local group of individuals belonging to the same species, which are actually or potentially interbreeding.

position effect Change in expression of a gene associated with a change in the gene's location within the genome.

postzygotic isolation mechanism Factors that prevent or reduce inbreeding by acting after fertilization to produce nonviable, sterile hybrids or hybrids of lowered fitness.

preadaptive mutation A mutational event which later becomes of adaptive significance.

prezygotic isolation mechanism Factors that reduce inbreeding by preventing courtship, mating, or fertilization.

Pribnow box A 6-bp sequence 5′ to the beginning of transcription in prokaryotic genes, to which the sigma subunit of RNA polymerase binds. The consensus sequence for this box is TATAAT.

primary protein structure Refers to the sequence of amino acids in a polypeptide chain.

primary sex ratio Ratio of males to females at fertilization.

primer In nucleic acids, a short length of RNA or single-stranded DNA which is necessary for the functioning of polymerases.

prion An infectious pathogenic agent devoid of nucleic acid and composed mainly of a protein, PrP, with a molecular weight of 27,000 to 30,000 daltons. Prions are known to cause scrapie, a degenerative neurological disease in sheep, and are thought to cause similar diseases in humans, such as kuru and Creuzfeldt-Jakob disease.

probability Ratio of the frequency of a given event to the frequency of all possible events.

proband See *propositus*.

product law The law that holds that the probability of two independent events occurring simultaneously is the product of their independent probabilities.

progeny The offspring produced from a mating.

prokaryotes Organisms lacking nuclear membranes, meiosis, and mitosis. Bacteria and blue-green algae are examples of prokaryotic organisms.

promoter Region having a regulatory function and to which RNA polymerase binds prior to the initiation of transcription.

prophage A phage genome integrated into a bacterial chromosome. Bacterial cells carrying prophage are said to be lysogenic.

propositus (female, proposita) An individual in whom a genetically determined trait of interest is first detected. Also known as a proband.

protein A molecule composed of one or more polypeptides, each composed of amino acids covalently linked together.

proto-oncogene A cellular gene that normally functions to control cellular proliferation. Proto-oncogenes can be converted to oncogenes by alterations in structure or expression.

protoplast A bacterial or plant cells with the cell wall removed. Sometimes called a spheroplast.

prototroph A strain (usually microorganisms) that is capable of growth on a defined, minimal medium. Wild-type strains are usually regarded as prototrophs.

pseudoalleles Genes that behave as alleles to one another by complementation, but that can be separated from one another by recombination.

pseudodominance The expression of a recessive allele on one homologue caused by the deletion of the dominant allele on the other homologue.

pseudogene A nonfunctional gene with sequence homology to a known structural gene present elsewhere in the genome. They differ from their functional relatives by insertions or deletions and by the presence of flanking direct repeat sequences of 10 to 20 nucleotides.

puff See *chromosome puff*.

quantitative inheritance See *polygenic inheritance*.

quantum speciation Formation of a new species within a single or a few generations by a combination of selection and drift.

quaternary protein structure Types and modes of interaction between two or more polypeptide chains within a protein molecule.

R point The point (also known as the restriction point) during the G1 stage of the cell cycle when either a commitment is made to DNA synthesis and another cell cycle, or the cell withdraws from the cycle and becomes quiescent.

race A phenotypically or geographically distinct subgroup within a species.

rad A unit of absorbed dose of radiation with an energy equal to 100 ergs per gram of irradiated tissue.

radioactive isotope One of the forms of an element, differing in atomic weight and possessing an unstable nucleus that emits ionizing radiation.

random mating Mating between individuals without regard to genotype.

reading frame Linear sequence of codons (groups of three nucleotides) in a nucleic acid.

reannealing Formation of double-stranded DNA molecules from dissociated single strands.

recessive Term describing an allele that is not expressed in the heterozygous condition.

reciprocal cross A paired cross in which the genotype of the female in the first cross is present as the genotype of the male in the second cross, and vice versa.

reciprocal translocation A chromosomal aberration in which nonhomologous chromosomes exchange parts.

recombinant DNA A DNA molecule formed by the joining of two heterologous molecules. Usually applied to DNA molecules produced by *in vitro* ligation of DNA from two different organisms.

recombinant gamete   A gamete containing a new combination of genes produced by crossing over during meiosis.

recombination   The process that leads to the formation of new gene combinations on chromosomes.

recon   A term coined by Seymour Benzer to denote the smallest genetic units between which recombination can occur.

redundant genes   Gene sequences present in more than one copy per haploid genome (e.g., ribosomal genes).

regulatory site   A DNA sequence that is involved in the control of expression of other genes, usually involving an interaction with another molecule.

rem   Radiation equivalent in man; the dosage of radiation that will cause the same biological effect as one roentgen of X-rays.

renaturation   The process by which a denatured protein or nucleic acid returns to its normal three-dimensional structure.

repetitive   DNA sequencesDNA sequences present in many copies in the haploid genome.

replicating form (RF)   Double-stranded nucleic molecules present as an intermediate during the reproduction of certain viruses.

replication   The process of DNA synthesis.

replication fork   The Y-shaped region of a chromosome associated with the site of replication.

replicon   A chromosomal region or free genetic element containing the DNA sequences necessary for the initiation of DNA replication.

replisome   The term used to describe the complex of proteins, including DNA polymerase, that assembles at the bacterial replication form to synthesize DNA.

repressible enzyme system   An enzyme or group of enzymes whose synthesis is regulated by the intracellular concentration of certain metabolites.

repressor   A protein that binds to a regulatory sequence adjacent to a gene and blocks transcription of the gene.

reproductive isolation   Absence of interbreeding between populations, subspecies, or species. Reproductive isolation can be brought about by extrinsic factors, such as behavior, and intrinsic barriers, such as hybrid inviability.

resistance transfer factor (RTF)   A component of R plasmids that confers the ability for cell-to-cell transfer of the R plasmid by conjugation.

resolution   In an optical system, the shortest distance between two points or lines at which they can be perceived to be two points or lines.

restriction endonuclease   Nuclease that recognizes specific nucleotide sequences in a DNA molecule, and cleaves or nicks the DNA at that site. Derived from a variety of microorganisms, those enzymes that cleave both strands of the DNA are used in the construction of recombinant DNA molecules.

restrict condition   Environmental conditions under which a conditional mutation (such as temperature-sensitive mutant) expresses the mutant phenotype.

restrict transduction   See *specialized transduction*.

retrovirus   Viruses with RNA as genetic material that utilize the enzyme reverse transcriptase during their life cycle.

reverse transcriptase   A polymerase that uses RNA as a template to transcribe a single-stranded DNA molecule as a product.

reversion   A mutation that restores the wild-type phenotype.

R factor (R plasmid)   Bacterial plasmids that carry antibiotic resistance genes. Most R plasmids have two components: an r-determinant, which carries the antibiotic resistance genes, and the resistance transfer factor (RTF).

Rh factor   An antigenic system first described in the rhesus monkey. Recessive *r/r* individuals produce no Rh antigens and are Rh negative, while *R/R* and *R/r* individuals have Rh antigens on the surface of their red blood cells and are classified as Rh positive.

ribonucleic acid   A nucleic acid characterized by the sugar ribose and the pyrimidine uracil, usually a single-stranded polynucleotide. Several forms are recognized, including ribosomal RNA, messenger RNA, transfer RNA, and heterogeneous nuclear RNA.

ribosomal RNA   See *rRNA*.

ribosome   A ribonucleoprotein organelle consisting of two subunits, each containing RNA and protein. Ribosomes are the site of translation of mRNA codons into the amino acid sequence of a polypeptide chain.

RNA   See *ribonucleic acid*.

RNA polymerase   An enzyme that catalyzes the formation of an RNA polynucleotide strand using the base sequence of a DNA molecule as a template.

RNase   A class of enzymes that hydrolyze RNA molecules.

Robertsonian translocation   A form of chromosomal aberration that involves the fusion of long arms of acrocentric chromosomes at the centromere.

roentgen   A unit of measure of the amount of radiation corresponding to the generation of 2.083×10^9 ion pairs in one cubic centimeter of air at 0°C at an atmospheric pressure of 760 mm of mercury. Abbreviated R.

rolling circle model   A model of DNA replication in which the growing point of replication fork rolls around a circular template strand; in each pass around the circle, the newly synthesized strand displaces the strand from the previous replication, producing a series of contiguous copies of the template strand.

rRNA   The RNA molecules that are the structural components of the ribosomal subunits. In prokaryotes, these are the 16S, 23S, and 5S molecules; and in eukaryotes, they are the 18S, 28S, and 5S molecules.

RTF   See *resistance transfer factor*.

S$_1$ nuclease   A deoxyribonuclease that cuts and degrades single-stranded molecules of DNA.

satellite DNA   DNA that forms a minor band when genomic DNA is centrifuged in a cesium salt gradient. This DNA usually consists of short sequences repeated many times in the genome.

SCE   See *sister chromatid exchange*.

secondary protein structure   The alpha helical or pleated-sheet form of a protein molecule brought about by the formation of hydrogen bonds between amino acids.

secondary sex ratio   The ratio of males to females at birth.

secretor   An individual having soluble forms of the blood group antigens A and/or B present in saliva and other body fluids. This condition is caused by a dominant, autosomal gene unlinked to the *ABO* locus (*I* locus).

sedimentation coefficient   See *Svedberg coefficient unit*.

segregation   The separation of homologous chromosomes into different gametes during meiosis.

selection The force that brings about changes in the frequency of alleles and genotypes in populations through differential reproduction.

selection coefficient(s) A quantitative measure of the relative fitness of one genotype compared with another.

selfing In plant genetics, the fertilization of ovules of a plant by pollen produced by the same plant. Reproduction by self-fertilization.

semiconservative replication A model of DNA replication in which a double-stranded molecule replicates in such a way that the daughter molecules are composed of one parental (old) and one newly synthesized strand.

semisterility A condition in which a proportion of all zygotes are inviable.

sex chromosome A chromosome, such as the X or Y in humans, which is involved in sex determination.

sexduction Transmission of chromosomal genes from a donor bacterium to a recipient cell by the F factor.

sex-influenced inheritance Phenotypic expression that is conditioned by the sex of the individual. A heterozygote may express one phenotype in one sex and the alternate phenotype in the other sex.

sex-limited inheritance A trait that is expressed in only one sex even though the trait may not be X-linked.

sex linkage The pattern of inheritance resulting from genes located on the X chromosome.

sex ratio See *primary* and *secondary sex ratio.*

sexual reproduction Reproduction through the fusion of gametes, which are the haploid products of meiosis.

Shine-Dalgarno sequence The nucleotides AGGAGG present in the leader sequence of prokaryotic genes that serves as a ribosome binding site. The 16S RNBA of the small ribosomal subunit contains a complementary sequence to which the mRNA binds.

short period interspersion Pattern of genome organization in which stretches of single copy DNA (about 1000 bp) are interspersed with short segments of repetitive DNA (300 bp). This pattern is found in *Xenopus,* humans, and the majority of organisms examined to date.

shotgun experiment The cloning of random fragments of genomic DNA into a vehicle such as a plasmid or phage, usually to produce a bank or library of clones from which clones of specific interest will be selected.

sickle-cell anemia A genetic disease in humans caused by an autosomal recessive gene, usually fatal in the homozygous condition. Caused by an alteration in the amino acid sequence of the beta chain of globin.

sickle-cell trait The phenotype exhibited by individuals heterozygous for the sickle-cell gene.

sigma factor A polypeptide subunit of the RNA polymerase which recognizes the binding site for the initiation of transcription.

SINEs Short interspersed repetitive sequences found in the genomes of higher organisms, such as the 300-bp *Alu* sequence.

sister chromatid exchange (SCE) A crossing over event which can occur in meiotic and mitotic cells; involves the reciprocal exchanges of chromosomal material between sister chromatids (joined by a common centromere). Such exchanges can be detected cytologically after BrdU incorporation into the replicating chromosomes.

site-directed mutagenesis A process that uses a synthetic oligonucleotide containing a mutant base or sequence as a primer for inducing a mutation at a specific site in a cloned gene.

small nuclear RNA (snRNA) Species of RNA molecules ranging in size from 90 to 400 nucleotides. The abundant snRNAs are present in 1×10^4 to 1×10^6 copies per cell. snRNAs are associated with proteins and form RNP particles known as snRNPs or snurps. Six uridine-rich snRNAs known as U1–U6 are located in the nucleoplasm, and the complete nucleotide sequence of these is known. snRNAs have been implicated in the processing of pre-mRNA and may have a range of cleavage and ligation functions.

snurps See *small nuclear RNA (snRNA).*

solenoid structure A level of eukaryotic chromosome structure generated by the supercoiling of nucleosomes.

somatic cell genetics The use of cultured somatic cells to investigate genetic phenomena by parasexual techniques.

somatic cells All cells other than the germ cells or gametes in an organism.

somatic mutation A mutational event occurring in a somatic cell. In other words, such mutations are not heritable.

somatic pairing The pairing of homologous chromosomes in somatic cells.

SOS response The induction of enzymes to repair damaged DNA in *E. coli.* The response involves activation of an enzyme that cleaves a repressor, activating a series of genes involved in DNA repair.

spacer DNADNA sequences found between genes, usually repetitive DNA segments.

special creation An idea that each species originated through a special act of creation by a divine force.

specialized transduction Genetic transfer of only specific host genes by transducing phages.

speciation The process by which new species of plants and animals arise.

species A group of actually or potentially interbreeding individuals that is reproductively isolated from other such groups.

spheroplast See *protoplast.*

spindle fibers Cytoplasmic fibrils formed during cell division which are involved with the separation of chromatids at anaphase and their movement toward opposite poles in the cell.

spontaneous generation The origin of living systems from nonliving matter.

spontaneous mutation A mutation that is not induced by a mutagenic agent.

spore A unicellular body or cell encased in a protective coat that is produced by some bacteria, plants, and invertebrates; is capable of survival in unfavorable environmental conditions; and can give rise to a new individual upon germination. In plants, spores are the haploid products of mitosis.

stabilizing selection Preferential reproduction of those individuals having genotypes close to the mean for the population. A selective elimination of genotypes at both extremes.

standard deviation A quantitative measure of the amount of variation in a sample of measurements from a population.

standard error A quantitative measure of the amount of variation in a sample of measurements from a population.

standard error   A quantitative measure of the amount of variation in a sample of measurements from a population.

sterility   The condition of being unable to reproduce; free from contaminating microorganisms.

strain   A group with common ancestry which has physiological or morphological characteristics of interest for genetic study or domestication.

structural gene   A gene that encodes the amino acid sequence of a polypeptide chain.

sublethal gene   A mutation causing lowered viability, with death before maturity in less than 50 percent of the individuals carrying the gene.

submetacentric chromosome   A chromosome with the centromere placed so that one arm of the chromosome is slightly longer than the other.

subspecies   A morphologically or geographically distinct interbreeding population of a species.

sum law   The law that holds that the probability of one or the other of two mutually exclusive events occurring is the sum of their individual probabilities.

supercoiled DNA   A form of DNA structure in which the helix is coiled upon itself. Such structures can exist in stable forms only when the ends of the DNA are not free, as in a covalently closed circular DNA model.

superfemale   See *metafemale.*

supermale   See *metamale.*

suppressor mutation   A mutation that acts to restore (completely or partially) the function lost by a previous mutation at another site.

Svedberg coefficient unit   A unit of measure for the rate at which particles (molecules) sediment in a centrifugal field. This unit is a function of several physico-chemical properties, including size and shape. A sedimentation value of 1×10^{-13} sec is defined as one Svedberg coefficient (S) unit.

symbiont   An organism coexisting in a mutually beneficial relationship with another organism.

sympatric speciation   Process of speciation involving populations that inhabit, at least in part, the same geographic range.

synapsis   The pairing of homologous chromosomes at meiosis.

synaptonemal complex (SC)   An organelle consisting of a tripartite nucleoprotein ribbon that forms between the paired homologous chromosomes in the pachytene stage of the first meiotic division.

syndrome   A group of signs or symptoms that occur together and characterize a disease or abnormality.

synkaryon   The nucleus of a zygote that results from the fusion of two gametic nuclei. Also used in somatic cell genetics to describe the product of nuclear fusion.

syntenic test   In somatic cell genetics, a method for determining whether or not two genes are on the same chromosome.

T_m The temperature at which a population of double-stranded nucleic acid molecules is half-dissociated into single strands. This is taken to be the melting temperature for that species of nucleic acid.

target theory   In radiation biology, a theory which states that damage and death from radiation is caused by the inactivation of specific targets within the organism.

TATA box   See *Goldberg-Hogness box.*

tautomeric shift   A reversible isomerization in a molecule brought about by a shift in the localization of a hydrogen atom. In nucleic acids, tautomeric shifts in the bases of nucleotides can cause changes in other bases at replication and are a source of mutations.

telocentric chromosome   A chromosome in which the centromere is located at the end of the chromosome.

telomere   The terminal chromomere of a chromosome.

telophase   The stage of cell division in which the daughter chromosomes reach the opposite poles of the cell and reform nuclei. Telophase ends with the completion of cytokinesis.

temperate phage   A bacteriophage that can become a prophage and confer lysogeny upon the host bacterial cell.

temperature-sensitive mutation   A conditional mutation that produces a mutant phenotype at one temperature range and a wild-type phenotype at another temperature range.

template   The single-stranded DNA or RNA molecule that specifies the nucleotide sequence of a strand synthesized by a polymerase molecule.

teratocarcinoma   Embryonal tumors that arise in the yolk sac or gonads and are able to undergo differentiation into a wide variety of cell types. These tumors are used to investigate the regulatory mechanism underlying development.

terminalization   The movement of chiasmata toward the ends of chromosomes during the diplotene stage of the first meiotic division.

tertiary protein structure   The three-dimensional structure of a polypeptide chain brought about by folding upon itself.

test cross   A cross between an individual whose genotype at one or more loci may be unknown and an individual who is homozygous recessive for the genes in question.

tetrad   The four chromatids that make up paired homologues in the prophase of the first meiotic division. The four haploid cells produced by a single meiotic division.

tetrad analysis   Method for the analysis of gene linkage and recombination using the four haploid cells produced in a single meiotic division.

tetranucleotide theory   An early theory of DNA structure which proposed that the molecule was composed of repeating units, each consisting of the four nucleotides adenosine, thymidine, cytosine, and guanine.

tetraparental mouse   A mouse produced from an embryo that was derived by the fusion of two separate blastulas.

thymine dimer   A pair of adjacent thymine bases in a single polynucleotide strand between which chemical bonds have formed. This lesion, usually the result of damage caused by exposure to ultraviolet light, inhibits DNA replication unless repaired by the appropriate enzyme system.

topoisomerase   A class of enzymes that convert DNA from one topological form to another. During DNA replication, these enzymes facilitate the unwinding of the double-helical structure of DNA.

totipotent   The ability of a cell or embryo part to give rise to all adult structures. This capacity is usually progressively restricted during development.

trailer sequence   A transcribed but nontranslated region of a gene or its mRNA that follows the termination signal.

trait   Any detectable phenotypic variation of a particular inherited character.

trans configuration   The arrangement of two mutant sites on opposite homologues, such as

$$\frac{a^1 \ +}{+ \ \ a^2}$$

Contrasts with a *cis* arrangement, where they are located on the same homologue.

transcription   Transfer of genetic information from DNA by the synthesis of an RNA molecule copied from a DNA template.

transdetermination   Change in developmental fate of a cell or group of cells.

transduction   Virally mediated gene transfer from one bacterium to another, or the transfer of eukaryotic genes mediated by retroviruses.

transfer RNA   See *tRNA*.

transformation   Heritable change in a cell or an organism brought about by exogenous DNA.

transgenic organism   An organism whose genome has been modified by the introduction of external DNA sequences into the germ line.

transition   A mutational event in which one purine is replaced by another, or one pyrimidine is replaced by another.

translation   The derivation of the amino acid sequence of a polypeptide from the base sequence of an mRNA molecule in association with a ribosome.

translocation   A chromosomal mutation associated with the transfer of a chromosomal segment from one chromosome to another. Also used to denote the movement of mRNA through the ribosome during translation.

transposable element   A defined length of DNA that translocates to other sites in the genome, essentially independent of sequence homology. Usually such elements are flanked by short, inverted repeats of 20 to 40 base pairs at each end. Insertion into a structural gene can produce a mutant phenotype. Insertion and excision of transposable elements depends on two enzymes, transposase and resolvase. Such elements have been identified in both prokaryotes and eukaryotes.

transversion   A mutational event in which a purine is replaced by a pyrimidine, or a pyrimidine is replaced by a purine.

triploidy   The condition in which a cell or organism possesses three haploid sets of chromosomes.

trisomy   The condition in which a cell or organism possesses two copies of each chromosome, except for one, which is present in three copies. The general form for trisomy is therefore $2n + 1$.

tritium (^{3}H)   A radioactive isotope of hydrogen, with a half-life of 12.46 years.

tRNA   Transfer RNA; a small ribonucleic acid molecule that contains a three-base segment (anticodon) that recognizes a codon in mRNA, a binding site for a specific amino acid, and recognition sites for interaction with the ribosome and the enzyme that links it to its specific amino acid.

Turner syndrome   A genetic condition in human females caused by a 45,X genotype (XO). Such individuals are phenotypically female but are sterile because of undeveloped ovaries.

unequal crossing over   A crossover between two improperly aligned homologues, producing one homologue with three copies of a region and the other with one copy of that region.

unique DNA   DNA sequences that are present only once per genome. Single copy DNA.

unwinding proteins   Nuclear proteins that act during DNA replication to destabilize and unwind the DNA helix ahead of the replicating fork.

up promoter   A promoter sequence, often mutant, that increases the rate of transcription initiation. Also known as strong promoter.

variable region   Portion of an immunoglobulin molecule that exhibits many amino acid sequence differences between antibodies of differing specificities.

variance   A statistical measure of the variation of values from a central value, calculated as the square of the standard deviation.

variegation   Patches of differing phenotypes, such as color, in a tissue.

vector   In recombinant DNA, an agent such as a phage or plasmid into which a foreign DNA segment will be inserted.

viability   The measure of the number of individuals in a given phenotypic class that survive, relative to another class (usually wild type).

virulent phage   A bacteriophage that infects and lyses the host bacterial cell.

W, Z chromosomes   Sex chromosomes in species where the female is the heterogametic sex (WZ).

wild type   The most commonly observed phenotype or genotype, designated as the norm or standard.

wobble hypothesis   An idea proposed by Francis Crick which states that the third base in an anticodon can align in several ways to allow it to recognize more than one base in the codons of mRNA.

writhing number   The number of times that the axis of a DNA duplex crosses itself by supercoiling.

X inactivation   In mammalian females, the random cessation of transcriptional activity of one X chromosome. This event, which occurs early in development, is a mechanism of dosage compensation. Molecular basis of inactivation is unknown, but loci on the tip of the short arm of the X can escape inactivation. (See *Barr body; Lyon hypothesis.*)

X linkage   See *sex linkage*.

x-ray crystallography   A technique to determine the three-dimensional structure of molecules through diffraction patterns produced by X-ray scattering by crystals of the molecule under study.

Y chromosome   Sex chromosome in species where the male is heterogametic (XY).

Y linkage Mode of inheritance shown by genes located on the Y chromosome.

Z-DNA An alternate structure of DNA in which the two anti-parallel polynucleotide chains form a left-handed double helix. Z-DNA has been shown to be present along with B-DNA in chromosomes and may have a role in regulation of gene expression.

zein Principal storage protein of corn endosperm, consisting of two major proteins, with molecular weights of 19,000 and 21,000 daltons.

zinc finger A DNA binding domain of a protein that has a characteristic pattern of cysteine and histidine residues that complex with zinc ions, throwing intermediate amino acid residues into a series of loops or fingers.

zygote The diploid cell produced by the fusion of haploid gametic nuclei.

zygotene A stage of meiotic prophase I in which the homologous chromosomes synapse and pair along their entire length, forming bivalents. The synaptonemal complex forms at this stage.

Credits

Chapter 1: An Introduction to Genetics

CO.1 Biophoto Associates/SS/Photo Researchers, Inc.
F1.1 The Metropolitan Museum of Art, Gift of John D. Rockefeller, Jr., 1932. (32.143.3) Photograph © 1983 The Metropolitan Museum of Art.
F1.2 Hartsoeker, N. Essay de dioptrique, Paris, 1694, p. 230.
F1.3 Archive/Photo Researchers, Inc.
F1.5 M. Wurtz/Biozentrum, University of Basel/ Science Photo Library/Photo Researchers, Inc.
F1.6 K. G. Murti/Visuals Unlimited
F1.7 Dan McCoy/Rainbow
F1.8 Edward S. Ross/California Academy of Sciences
F1.9 Grant Heilman/Grant Heilman Photography, Inc.
F1.10 Renee Lynn/Photo Researchers, Inc.

Chapter 2: Cell Division and Chromosomes

CO.2 Prof. P. Motta/Dept of Anatomy/Univ. La Sapienza, Rome/Science Photo Library/Photo Researchers, Inc.
F2.2 CNRI/Science Photo Library/Photo Researchers, Inc.
F2.4a Cytographics/Visuals Unlimited
F2.4b L. Lisco/D.W. Fawcett/Visuals Unlimited
F2.8 Dr. Andrew S. Bajer, University of Oregon
F2.8a Dr. Andrew S. Bajer, University of Oregon
F2.8b Dr. Andrew S. Bajer, University of Oregon
F2.8c Dr. Andrew S. Bajer, University of Oregon
F2.8d Dr. Andrew S. Bajer, University of Oregon
F2.8e Dr. Andrew S. Bajer, University of Oregon
F2.8f Dr. Andrew S. Bajer, University of Oregon
F2.8g Dr. Andrew S. Bajer, University of Oregon
F2.11a Biological Photo Service
F2.11b Clare A. Hasenkampf/Biological Photo Service
F2.11c Biological Photo Service
F2.11d Biological Photo Service
F2.11e Biological Photo Service
F2.11f Biological Photo Service
F2.11g Biological Photo Service
F2.11h Biological Photo Service
F2.14a Biophoto Associates/Photo Researchers, Inc.
F2.14b Walter F. Engler
F2.14c Biophoto Associates/Science Source/Photo Researchers, Inc.
F2.14e Dr. Don Fawcett/Photo Researchers, Inc.
F2.15 Andrew Syred/Science Photo Library Photo Researchers, Inc.
F2.16 Science Source/Photo Researchers, Inc.
F2.17a Omikron/Photo Researchers, Inc.
F2.18a Omikron/Science Source Photo Researchers, Inc.
F2.18b Nicole Angelier
F2.18c Nicole Angelier

Chapter 3: Mendelian Genetics

CO.3 Courtesy of Allan Gotthelf
F3.1a Alan & Linda Detrick/Photo Researchers, Inc.
F3.1a Adam Hart-Davis/Science Photo Library Photo Researchers, Inc
F3.1b Larry Lefever/Grant Heilman Photography, Inc.

Chapter 4: Modification of Medelian Ratios

CO.4 Tom Cerniglio/Oak Ridge National Laboratory
F4.1a John D. Cunningham/Visuals Unlimited
F4.3a Photo courtesy of Stanton K. Short (The Jackson Laboratory, Bar Harbor, ME)
F4.3b Tom Cerniglio/Oak Ridge National Laboratory
F4.9a Carolina Biological Supply Company/Phototake NYC
F4.9b Carolina Biological Supply Company/Phototake NYC
F4.11 Mary Teresa Giancoli
F4.12 Hans Reinhard/Bruce Coleman, Inc.
F4.13 Debra P. Hershkowitz Bruce Coleman, Inc.
UP4.1 Dr. Ralph Somes
UP4.2 Dr. Ralph Somes
UP4.3 J. James Bitgood/University of Wisconsin, Animal Sciences Dept.
UP4.4 J. James Bitgood/University of Wisconsin, Animal Sciences Dept.

Chapter 5: Quantitative Genetics

CO.5 Wolfgang Kaehler/Wolfgang Kaehler Photography
F5.1 Wolfgang Kaehler/Wolfgang Kaehler Photography
F5.2b Norm Thomas/Photo Researchers, Inc.

Chapter 6: Linkage and Chromosome Mapping

CO.6 B. John Cabisco/Visuals Unlimited
F6.7 Biological Photo Service
F6.16 Science Source/Photo Researchers, Inc.
F6.18 Dr. Sheldon Wolff & Judy Bodycote/Laboratory of Radiology and Environmental Health, University of California, San Francisco.

Chapter 7: Extranuclear Inheritance

CO.7 Dr. Don Fawcett/Kahri/Dawid/Science Source/Photo Researchers, Inc.
F7.2 George I. Bernard/Animals Animals/Earth Scenes
F7.3 Courtesy of Dr. Ed Coe, USDA-ARS, Curtis Hall, University of Missouri
F7.4 Raymond B. Otero/Visuals Unlimited
F7.5 Dr. Douglas C. Wallace/Emory University School of Medicine

F7.5in   CNRI/Phototake NYC
F7.6   John D. Cunningham/Visuals Unlimited

Chapter 8: Chromosome Variation and Sex Determination

CO.8   Evelin Shrock, Stan du Manoir and Tom Reid, NIH
F8.1a   Catherine B. Palmer. Dept. of Medical Genetics/Indiana University-Indianapolis
F8.1b   Catherine B. Palmer. Dept. of Medical Genetics/Indiana University-Indianapolis
F8.2a   M. Abbey/Photo Researchers, Inc.
F8.2b   M. Abbey/Photo Researchers, Inc.
F8.4   Hans Reinhard/Okapia/Photo Researchers, Inc.
F8.8a   Irene Uchida/McMaster University
F8.8b   Irene Uchida/McMaster University
F8.9b   Jane Grushow/Grant Heilman Photography, Inc.
F8.10a   Catherine B. Palmer, Ph. D.
F8.10b   William McCoy/Rainbow
F8.12   David D. Weaver, M.D., Indiana University
F8.15   Ken Wagner/Phototake NYC
F8.16   Pfizer, Inc./Phototake NYC
F8.19a   Mary Lilly/Carnegie Institute of Washington
F8.19b   Mary Lilly/Carnegie Institute of Washington
F8.19c   Mary Lilly/Carnegie Institute of Washington
F8.23   From Yunis and Chandler, 1979.
F8.24   Science VU/Visuals Unlimited

Chapter 9: DNA–The Physical Basis of Life

CO.9   Richard Megna/Fundamental Photographs
F9.2a   Bruce Iverson
F9.2b   Bruce Iverson
F9.4   Lee D. Simon/Science Source/Photo Researchers, Inc.
F9.11   Science Source/Photo Researchers, Inc.
F9.14   Ken Edward/Science Source/Photo Researchers, Inc.
F9.16   Oncor, Inc.

Chapter 10: DNA–Replication and Synthesis

CO.10   Omikron/Science Source/Photo Researchers, Inc.
F10.4a   Walter H. Hodge/Peter Arnold, Inc.
F10.4b   "Molecular Genetics," Pt. 1 pp. 74075, J.H. Taylor (ed). Reprinted by Permission of Academic Press (1963).
F10.4c   "Molecular Genetics," Pt. 1 pp. 74075, J.H. Taylor (ed). Reprinted by Permission of Academic Press (1963).
F10.5   Sundin and Varshavsky, *Cell* 25:659 (1981). Courtesy of A. Varshavsky.
F10.6   Dr. Gopal Murti/Science Photo Library/Photo Researchers, Inc.
F10.15   Courtesy of Dr. Ross Inman and Maria Schnos, Institiute for Molecular Virology, University of Wisconsin-Madison.

Chapter 11: Organization of DNA in Chromosomes and Genes

CO.11   Science VU/BMRL/Visuals Unlimited

F11.1a   Dr. M. Wurtz/Biozentrum, University of Basel/Science Photo Library/Photo Researchers, Inc.
F11.1a   Institut Pasteur/CNRI/Phototake NYC
F11.1b   Hans Ris, from Westmoreland, 1969.
F11.2   Science Source/PR/Photo Researchers, Inc.
F11.3   Dr. Gopal Murti/Science Photo Library/Photo Researchers, Inc.
F11.4   Dr. Don Fawcett/Kahri/Dawid/Science Source/Photo Researchers, Inc
F11.5   Dr. Richard D. Kolodnar/Dana-Farber Cancer Institute
F11.6a   From Olins and Olins, 1978, Fig. 1 and 4
F11.6b   From Olins and Olins, 1978, Fig. 1 and 4
F11.8   Chen and Ruddle, 1971, p. 54
F11.9   Dept. of Clinical Cytogenetics, Addenbrookes Hospital, Cambridge/Science Photo Library/Photo Researchers, Inc.

Chapter 12: Storage and Expression of Genetic Information

CO.12   Prof. Oscar L. Miller/Science Photo Library/Photo Researchers, Inc.
F12.9   Courtesy of Bert W. O'Malley
F12.18a   Omikron/Science Source/Photo Researchers, Inc.
F12.18b   E.V. Kiseleva

Chapter 13: Regulation of Gene Expression

CO.13   Yale University and the Howard Hughes Medical Institute, Dr. Paul Sigler
F13.4   Science Lewis, et al. 271 pg. 1247–1254/Johnson Research Foundation
F13.15   Dr. Paul Sigler

Chapter 14: Proteins: The End Product of Genes

CO.14   J. Gross/Science Photo Library/Photo Researchers, Inc.
F14.4a   Dennis Kunkel/Phototake NYC
F14.4b   Francis Leroy/Biocosmos/Science Photo Library/Photo Researchers, Inc.

Chapter 15: DNA–Mutation, Repair, and Transposable Elements

CO.15   Francis Leroy/Biocosmos/Science Photo Library/Photo Researchers, Inc.
F15.2b   Sue Ford/Science Photo Library/Photo Researchers, Inc.
F15.3a   Mary Evans Picture Library/Photo Researchers, Inc.
F15.12a   W. Clark Lambert, M.D., Ph. D./University of Medicine & Dentistry of New Jersey
F15.12b   W. Clark Lambert, M.D., Ph. D./University of Medicine & Dentistry of New Jersey

Chapter 16: Genetics of Bacteria and Bacteriophages

CO.16   Dr. L. Caro/Science Photo Library/Photo Researchers, Inc.
F16.2   Michael G. Gabridge/Visuals Unlimited
F16.5   Dennis Kunkel/Phototake NYC

F16.15b Dr. M. Wurtz/Biozentrum, University of Basel/Science Photo Library/Photo Researchers, Inc.
F16.16 Bruce Iverson
F16.18 From Hershey & Chase, 1951

Chapter 17: DNA Biotechnology: Techniques and Analysis

CO.17 Michael Gabridge/Visuals Unlimited
F17.4 K.G. Murti/Visuals Unlimited
F17.6 Michael Gabridge/Visuals Unlimited
F17.7 Dr. M. Wurtz/Biozentrum, University of Basel/Science Photo Library/Photo Researchers, Inc.
F17.16 CMSP/NIH/Custom Medical Stock Photo, Inc.
F17.18 Dr. Suzanne McCutcheon
F17.19 Dr. Suzanne McCutcheon

Chapter 18: DNA Biotechnology: Applications and Ethics

CO.18 Affymetrix, Inc.
F18.4b From Barker, D., et al. 1987. "Gene for von Recklinghausen Neurofibromatosis is in the Pericentromeric Region of Chromosome 17", *Science* 236: 1100–1102, Fig 1. © 1987 by the American Association for the Advancement of Science.
F18.9 Affymetrix, Inc.
F18.11 Baylor College of Medicine/Peter Arnold, Inc.
F18.14 Leonard Lessin/Peter Arnold, Inc.
F18.19 Hank Morgan/Photo Researchers, Inc.
F18.21 Charles J. Arntzen, President & CEO, Boyce Thompson Institute, Ithaca, NY

Chapter 19: Genetics of Cancer and Immunology

CO.19 Photo Lennart Nilsson/Albert Bonniers Forlag AB, copyright Boehringer Ingelheim International GmbH/The Body Victorious, Dell Publishing Company
F19.3 Sung-Hou Kim, University of California, Department of Chemistry and Lawrence Berkeley National Laboratory, Berkeley, California.
F19.16 Hans Gelderblom/Visuals Unlimited

F19.17a Hans Gelderblom/Visuals Unlimited
F19.17b Hans Gelderblom/Visuals Unlimited
F19.17c Hans Gelderblom/Visuals Unlimited

Chapter 20: Genetics of Development and Behavior

CO.20 Jim Langeland, Steve Paddock and Sean Carroll/University of Wisconsin/Dept. Molecular Biology
F20.4 Dr. William S. Klug, The College of New Jersey
F20.7a F.R. Turner, Department of Biology, Indiana University.
F20.16 Steve Small, Dept. of Biology, Indiana University.
F20.17a From T. Kaufmann, et al. 1990. *Advanced Genetics* 27:309–362. Department of Biology, Indiana University.
F20.17b From T. Kaufmann, et al. 1990. *Advanced Genetics* 27:309–362. Department of Biology, Indiana University.

Chapter 21: Population Genetics

CO.21 Edward S. Ross/California Academy of Sciences
F21.15a Breck P. Kent
F21.15b Breck P. Kent

Chapter 22: Genetics and Evolution

CO.22 Courtesy of Toby Bradshaw and Doug W. Schemske, University of Washington, Photo by Jordan Rehm.
F22.3a Dr. Paul V. Loiselle
F22.03b Dr. Paul V. Loiselle
F22.4 Roland Seitre/Peter Arnold, Inc.
F22.8 C.B. & D.W. Frith/Bruce Coleman, Inc.
F22.17a Kenneth Kaneshiro/University of Hawaii
F22.17b Kenneth Kaneshiro/University of Hawaii
F22.18a Courtesy of Toby Bradshaw and Doug W. Schemske, University of Washington, Photo by Jordan Rehm.
F22.18b Courtesy of Toby Bradshaw and Doug W. Schemske, University of Washington, Photo by Jordan Rehm.

Cover Gregor Zlokarnik, Ph.D.

Index